SECOND EDITION

The
STRUCTURE
of
BIOLOGICAL
MEMBRANES

SECOND EDITION

The
STRUCTURE
of
BIOLOGICAL
MEMBRANES

Edited by
Philip L. Yeagle

CRC PRESS

Boca Raton   London   New York   Washington, D.C.

Library of Congress Cataloging-in-Publication Data

The Structure of biological membranes / edited by Philip L. Yeagle.—2nd ed.
        p. cm.
    Includes bibliographical references and index.
    ISBN 0-8493-1403-8 (alk. paper)
    1. Membranes (Biology)   2. Membrane proteins. 3. Membrane lipids. I. Yeagle, Philip.

QH601.S777 2004
571.6'4—dc22   2004043576

Visit the CRC Press Web site at www.crcpress.com

© 2005 by CRC Press LLC

No claim to original U.S. Government works
International Standard Book Number 0-8493-1403-8
Library of Congress Card Number 2004043576
Printed in the United States of America 1 2 3 4 5 6 7 8 9 0
Printed on acid-free paper

Preface

The modern study of the membranes of cells from a biochemical and structural point of view is only about four decades old. The first phase was the age of lipids, commencing with definitive evidence for the lipid bilayer as the fundamental structural element of all biological membranes. In the decades of the 1970s and 1980s, an explosion of data burst forth on the fascinating properties of lipid bilayers, along with enormous speculation on the potential roles in biology for the variety of physical behaviors that had been observed. Much was discovered about this two-dimensional, highly anisotropic world that resembles no other biological structure.

Partly because of the lack of adequate analogy, the lipid bilayer was, and is today, widely misunderstood, which forms part of the rationale for the preparation of this book. Some of what was learned in that first age of lipid studies is not part of the current view on membrane structure. We would do well to review such studies, to rediscover the unique properties of the lipid bilayer that must inform our contemporary interpretation of membrane structure and function. For that reason, much fundamental lipid lore is included in this volume, most of it updated, but all of it critical to an adequate understanding of this most fundamental of biological structures.

This understanding begins with lipid structure (Chapter 1), which determines the chemistry and the thermodynamics of the collision of lipid properties with water properties. Biological lipids, because they are not soluble in water, form separate phases with behaviors directly relevant to biological function and interesting phases that are not yet related to biological function (Chapter 2, Chapter 3, and Chapter 5). Understanding this phase behavior would obviate the too common treatment of membranes as analogous to soluble components in an experimental system and help the student to appreciate that the biological membrane is one of the largest biological structures that is held together by *non*covalent forces. The multiple phase behavior of lipid bilayers is also critical to a correct understanding of "rafts," as described in Chapter 9. Chapter 4 takes us inside the lipid bilayer to explore the highly anisotropic structure and dynamics of this system. We come away with a view that the interior of a membrane is quite unlike liquid hydrocarbon, to which it has been compared, and that recourse to such inappropriate models will lead to incorrect conclusions. Chapter 6 explores the interaction between membranes at their surfaces, an interaction critical to a variety of cellular phenomena, including nonlamellar lipid phases (Chapter 5), lipid fusion (Chapter 8), and viral fusion (Chapter 16). Cholesterol has long fascinated those who study biological membranes, initially because of the interesting physical chemical phenomena exhibited by cholesterol in a bilayer, and subsequently because of the varied roles cholesterol plays in membrane function. Chapter 7 attempts to synthesize all the kinds of data on cholesterol generated over the past several decades into a unified view of several distinct functions of this molecule. One of the important functions exhibited by cell membranes is membrane fusion. Membrane lipids clearly have a role to play in this phenomenon, and much has been learned from studies on lipid fusion systems, as described in Chapter 8. This section of the book ends with a look at a topic of intense interest at the time this book is being prepared: lipid rafts (Chapter 9). Understanding this model for nonuniform lipid behavior in the plasma membrane requires virtually all the material in the previous chapters and thus is an appropriate way to close Part I.

Part II takes us into the other major component of biological membranes: membrane proteins. The study of membrane proteins has lagged seriously behind both the study of the lipid components of membranes and the study of soluble protein structure and function. This is partly a result of the historical difficulty researchers have experienced in obtaining crystals of membrane proteins suitable for X-ray diffraction experiments. As a result, membrane protein studies have in the past focused more on function than on structure. Thus, much of this section is devoted to function, in particular transport (Chapter 10 through Chapter 13). Because membranes of cells form barriers to transport, separating organelles from the cytoplasm and the cytoplasm from the exterior of the cell, movement of specific materials across membranes requires specific transport systems. Much progress has been made, which has been enhanced recently by the solution of a limited number of structures of membrane transport proteins. Chapter 14 shows an example of membrane proteins for which three-dimensional structural information is now available. Although it is not possible in this book to review exhaustively all the structural information on membrane proteins that has been developed, mostly in just the past few years, this chapter will illustrate, with one protein, how the structural information can be obtained and how it can be used to understand function.

In general, membranes are the location for linking of three functions: lipid–lipid, lipid–protein, and protein–protein interactions. These three sets of interactions are not independent; perturbation in one will affect one or both of the others. Therefore the study of lipid–protein interactions has attracted great interest over the years. Chapter 15 brings together the protein and lipid components to see how one affects the other. Finally, in the chapter on viral fusion, the influence of proteins on the mixing of membrane components, initially lipid, can lead to the fusion of two membranes into one.

This series of topics offers the student of membrane structure and function in-depth insight into a number of the topics necessary to appreciate fully biological membranes in today's research climate. With the development of new structural information on membrane proteins, we can expect a new burst of research activity and understanding in the first part of this new millennium.

I thank all the authors for their erudite contributions. I also thank readers of the previous edition of this book for their helpful comments. Appreciation is due to the editorial staff at CRC Press for their assistance and patience in this effort.

Philip Yeagle
University of Connecticut

Editor

Philip Yeagle, a National Merit finalist, graduated from St. Olaf College (*magna cum laude* with honors in chemistry) in 1971, having spent 1970 at University of Cambridge. He obtained his Ph.D. at Duke University in 1974, studying enzyme structure and function, under the support of an NDEA predoctoral fellowship. As a postdoctoral fellow he switched fields to the study of membrane structure and dynamics at the University of Virginia, joining one of the leading centers of membrane research of that time, with the support of an NIH postdoctoral fellowship, and where he was one of the early investigators to find the advantages of ^{31}P NMR for biological membrane studies.

He subsequently moved to the University at Buffalo to take an assistant professor position in 1978, initially supported by an NIH Research Career Development Award, and defined new roles for cholesterol in biological membranes. He was promoted to associate professor in 1983 and introduced magic angle spinning NMR experiments to the study of membranes starting in 1985. In 1985, he was a visiting scientist at the CSIRO, New South Wales, Australia. In 1988, he initiated the FASEB Summer Research Conference on membrane structure, which has continued in various forms to the present. In 1990, he was awarded the NYS UUP Excellence Award. In 1991, he rose to the rank of professor and studied mechanisms of membrane fusion. In 1993, Dr. Yeagle was a visiting professor, Department of Biochemistry, and associate member, Senior Common Room, St. Hughes College, University of Oxford, England, which appointment was repeated in 2003. In 1994, he began developing new approaches to the problem of membrane protein structure. In 1997, he moved to the University of Connecticut to take the position of head of the Department of Molecular and Cell Biology. The membrane protein structure work focused on rhodopsin and, together with Professor Arlene Albert, produced the first experimental three-dimensional structure for an activated G-protein coupled receptor in 2002. He was elected a member of council of the national Biophysical Society and chair of the subgroup on Membrane Structure and Assembly (which he helped form), is executive editor of *Biochimica et Biophysica Acta Biomembranes* and a member of the editorial board of the *Journal of Biological Chemistry,* and has published 140 original papers and reviews and six books.

Contributors

Joseph Bentz
Drexel University
Philadelphia, Pennsylvania

Kathleen Boesze-Battaglia
University of Pennsylvania
Philadelphia, Pennsylvania

Richard J. Cherry
University of Essex
Colchester, England, United Kingdom

Michael Edidin
Johns Hopkins University
Baltimore, Maryland

Richard M. Epand
McMaster University
Hamilton, Ontario, Canada

Klaus Gawrisch
NIAAA, NIH
Rockville, Maryland

Sol M. Gruner
Princeton University
Princeton, New Jersey

Helmut Hauser
Eidgenössische Technische Hochschule Zürich
Zürich, Switzerland

Ching-hsien Huang
University of Virginia
Charlottesville, Virginia

Michael L. Jennings
University of Arkansas
Little Rock, Arkansas

Ruthven N.A.H. Lewis
University of Alberta
Edmonton, Alberta, Canada

Ronald N. McElhaney
University of Alberta
Edmonton, Alberta, Canada

Aditya Mittal
Drexel University
Bethesda, Maryland

Shinpei Ohki
State University of New York at Buffalo
Buffalo, New York

V. Adrian Parsegian
National Institutes of Health
Bethesda, Maryland

Guy Poupart
Eidgenössische Technische Hochschule Zürich
Zürich, Switzerland

Richard P. Rand
Brock University
St. Catherines, Ontario, Canada

David P. Siegel
Givaudan Inc.
Cincinnati, Ohio

James L. Slater
National Institutes of Health
Bethesda, Maryland

Robert A. Spangler
State University of New York at Buffalo
Buffalo, New York

Philip L. Yeagle
University of Connecticut
Storrs, Connecticut

Contents

1 Lipid Structure

Helmut Hauser and Guy Poupart

CONTENTS

1.1 LIPID CLASSIFICATION

Using the broadest possible definition, a lipid (Greek *lipos = fat*) may be defined as a compound of low or intermediate molecular weight (5000), a substantial proportion of which is made up of hydrocarbons. Included are diverse compounds such as fatty acids; soaps; detergents; steroids; mono-, di-, and triacylglycerols; and more complex compounds such as phospholipids, sphingolipids, glycolipids, and lipopolysaccharides. Because this volume addresses the structure and function

0-8493-1403-8/05/$0.00+$1.50
© 2005 by CRC Press LLC

TABLE 1.1
Classification of Lipids

Nonhydrolyzable (Nonsaponifiable) Lipids

Hydrocarbons
    Simple alkanes
    Terpenes (isoprenoid compounds)
Substituted hydrocarbons
    Long-chain alcohols
    Long-chain fatty acids
    Detergents
    Steroids
    Vitamins
Simple Esters
    Acylglycerols
    Cholesteryl esters
    Waxes
Complex Lipids
    Glycerophospholipids
    Sphingolipids
Glycolipids
    Glycoglycerolipids
    Glycosphingolipids
    Cerebrosides
    Gangliosides
    Lipopolysaccharides

---

of biological membranes, we are primarily concerned with the more complex lipids that are constituents of biological membranes.

Membrane lipids consist of a wide and still expanding range of amphipathic (amphiphilic) compounds containing a nonpolar (hydrocarbon) and a polar region. It is clear that such diverse molecules as lipids will differ widely in their physicochemical behavior. Nevertheless, solubility is a property that is used as a unifying criterion. Lipids are soluble in organic solvents such as alkanes, benzene, ether, chlorinated alkanes (e.g., chloroform, tetrachlorcarbon), methanol, and mixtures of these solvents. For instance, mixtures of chloroform and methanol are used as universal lipid solvents. Lipids with long hydrocarbon chains are practically insoluble in water. Depending on their chemical structure, they are either completely immiscible in water, such as hydrocarbons, or they interact with water, forming colloidal dispersions (Small, 1986).

The classification of lipids is arbitrary. In many texts, lipids are ranked in order of increasing complexity. Sometimes, however, practical criteria are used. As mentioned above, lipids differ widely in their physicochemical properties, e.g., they vary considerably in their interaction with water and in their spreading properties at the air–water interface. Hence, lipid classification may be based on hydration, swelling of lipid in the presence of water, or spreading of lipid at the air–water interface (Small, 1986). In Table 1.1, lipid classes are listed in order of increasing complexity.

1.1.1 NONHYDROLYZABLE (NONSAPONIFIABLE) LIPIDS

1.1.1.1 Hydrocarbons

1.1.1.1.1 Simple Alkanes

Pure hydrocarbons — saturated or unsaturated, aliphatic or aromatic — are usually classified as lipids. They are rare in the animal kingdom and, with few exceptions, are not particularly important. They are, however, major constituents of petroleum deposits.

1.1.1.1.2 Terpenes (Isoprenoid Compounds)

Many plant odors are due to volatile C_{10} and C_{15} compounds termed *terpenes*. Isolation of these compounds from various parts of plants by steam distillation or ether extraction yields the so-called essential oils. They can be regarded as derivatives of isoprene, 2-methyl-1,3-butadiene (C_5H_8, see Figure 1.1A). Essential oils are obtained from cloves, roses, lavender, citronella, eucalyptus, peppermint, camphor, sandalwood, cedar, and turpentine. They are widely used in perfumery, as food flavorings, medicines, and solvents (Roberts and Caserio, 1979). The essential oils, such as these C_{10} and C_{15} compounds, may be regarded as members of a much larger class of substances with carbon skeletons that are multiples of the isoprene (C_5H_8)-unit.

These compounds are categorized as isoprenoid compounds and are widespread in both plants and animals. They comprise open-chain (acyclic) and cyclic compounds, and, in addition to the C_{10} compounds that are customarily designated terpenes, they comprise C_{15} (sesquiterpenes), C_{20} (diterpenes), C_{30} compounds (triterpenes), and so on (Figure 1.1). Important examples of isoprenoid compounds are β-carotene, vitamin A, and squalene. Carotenoids in general are tetraterpenes (C_{40} compounds) and are widespread as yellow or red pigments of plants. β-carotene (Figure 1.1A) is a precursor of vitamin A. It is oxidized in the liver at the central double-bond to yield vitamin A, which is a diterpene alcohol. Squalene is a triterpene ($C_{30}H_{50}$) and an example of an important isoprenoid compound of animal origin. It occurs in fish liver oil and is a precursor of the steroids lanosterol and cholesterol (see Figure 1.6). Also, terpene hydrocarbons and oxygenated terpenes have been isolated from insects and have hormonal and pheromonal activity.

1.1.1.2 Substituted Hydrocarbons

Substitution in position 1 of hydrocarbons with electronegative atoms or groups leads to amphipathic molecules, such as long-chain alcohols, fatty acids, and detergents. *Long-chain* is defined as a hydrocarbon chain with 12 or more carbon atoms.

1.1.1.2.1 Long-Chain Alcohols

Long-chain alcohols, e.g., octadecanol or stearyl alcohol (Figure 1.2) are amphipathic and surface active. Monolayers of long-chain alcohols spread at the air–water interface prevent water evaporation and have been widely used as antievaporants in water reservoirs. Furthermore, they are used in water-in-oil emulsions and in cosmetics. Long-chain alcohols are important constituents of complex lipids, in which they are covalently linked to glycerol via ether bonds. Such ether linkages are found in plasmalogens and in membrane lipids of bacteria (see below). These long-chain, ether-linked alcohols can be branched and of the isoprenoid-type, like the phytanyl residue (Figure 1.1B). A great profusion of oxygenated isoprenoid compounds exist: of importance are alcohols and aldehydes, which occur in plant oils and flower essences. Two important diterpene alcohols are vitamin A, mentioned previously, and phytol (Figure 1.1B). The latter is a constituent of chlorophyll, the major pigment of chloroplasts and of central importance in photosynthesis. In chlorophylls, phytol is esterified to the propanoic acid side chain of the Mg^{2+}-porphyrin ring. The phytyl-group is also a side chain of vitamin K (Figure 1.7). Isoprenoid alcohols and their cyclopentane phytanyl derivatives with an OH-group on one end (monopolar) or both ends (bipolar) of the branched hydrocarbon chains are found in the plasma membranes of archaebacteria (Figure 1.1B) (Luzzati et al., 1987; Blöcher et al., 1985). The bipolar C_{40} hydrocarbon chain that occurs in the diglyceryl tetraether compounds of archaebacteria can contain 0-4 cyclopentane rings.

1.1.1.2.2 Long-Chain Fatty Acids

Long-chain fatty acids are constituents of all esterified and complex lipids. They are covalently linked via ester or amide bonds, and in this form they are major constituents of membrane lipids. Free fatty acids rarely make up more than a few percent of membrane lipids. Long-chain fatty acids are widely used in industry as precursors of emulsifiers, food additives, and detergents. In eukaryotes, the fatty acids are straight with an even number of carbon atoms, typically 14 to 24

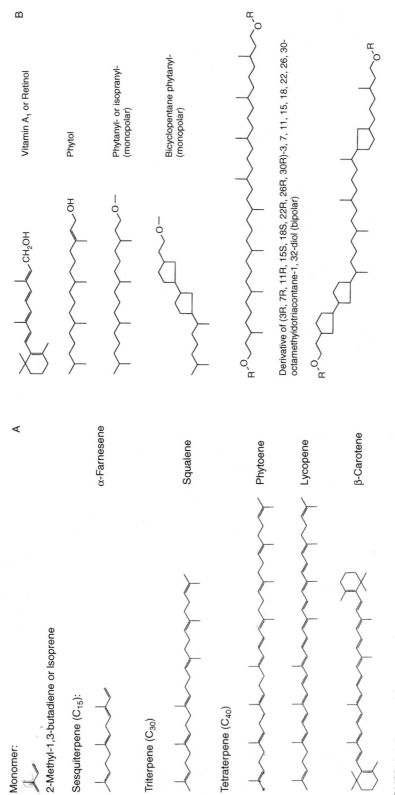

FIGURE 1.1 Isoprenoid compounds.

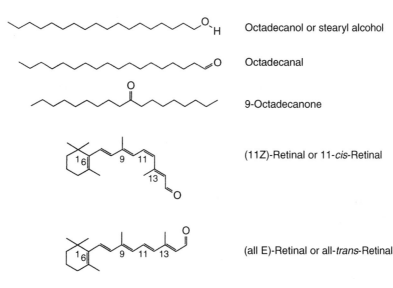

Octadecanol or stearyl alcohol

Octadecanal

9-Octadecanone

(11Z)-Retinal or 11-*cis*-Retinal

(all E)-Retinal or all-*trans*-Retinal

FIGURE 1.2 Long-chain alcohols, aldehydes, and ketones. The IUPAC-IUB Commission on Biochemical Nomenclature (CBN) recommends the use of the E/Z system to designate the configuration of double bonds. Although the system has advantages, it has not received widespread use in lipid nomenclature and has not replaced the old *cis–trans* convention. Z (for *zusammen*) corresponds to *cis,* and E (for *entgegen*) to *trans*.

and with zero to six double bonds (Table 1.2 and Figure 1.3). The double bonds of mono- and polyunsaturated fatty acyl chains usually have the *cis*-configuration, unless stated otherwise (cf. Table 1.2). It should be stressed that polyunsaturated chains are not normally conjugated. Unusual fatty acids with branched chains, cyclopropane, cyclohexyl-, and β-OH groups are found in bacterial membrane lipids (Figure 1.4) (Gennis, 1989; Jain, 1988).

1.1.1.2.3 Detergents

Substitution of hydrocarbons in the 1-position with polar groups, such as carboxylates, sulfonates, phosphates, amines, alkyl-amines, sugar residues, etc., generates the important lipid class of *detergents*. They are widely used in everyday life and also in biology. The sodium and potassium salts of fatty acids are called *soaps* and represent perhaps the oldest detergent. They are not only used in cosmetics but also as emulsifiers and lubricants. Soaps of alkaline earths (e.g., calcium distearate) are water-insoluble and used as insoluble lubricants and constituents of water-in-oil emulsions. Some of the most commonly used detergents in biology are shown in Figure 1.5. Sodium dodecylsulfate (SDS) is universally used to solubilize proteins and to determine their approximate molecular weight by electrophoresis on polyacrylamide gels. It is also a major constituent of nearly all shampoos. Detergents in biology are used to solubilize macromolecular assemblies, including proteins, DNA and RNA, lipid aggregates such as bilayers, and complexes of these different classes of compounds, e.g., biological membranes. The products of solubilization are usually small, mixed micelles containing the macromolecule(s), e.g., protein, and the detergent. In this micellar form, the macromolecule can be purified by subjecting the micellar dispersion to various separation processes, e.g., chromatography, electrophoresis, sedimentation, etc. In general, nonionic and zwitterionic detergents are milder and have proved more successful than charged ones in solubilizing and purifying membrane proteins. The purified membrane protein may then be reconstituted to a simple and well-defined membrane system using established methods of reconstitution (Lodish and Rothman, 1979).

1.1.1.2.4 Steroids

The fundamental ring system common to all steroids is the cyclopentanophenanthrene or, more precisely, its perhydrogenated form (Figure 1.6). Most steroids are alcohols, and accordingly named

TABLE 1.2
Names of Straight Fatty Acids

Carbon Atoms	Systematic Designation	Trivial Name	Abbreviation(s)[a]	Melting Point, °C
Saturated				
12	Dodecanoic acid	Lauric acid	12:0	44.2
14	Tetradecanoic acid	Myristic acid	14:0	53.9
16	Hexadecanoic acid	Palmitic acid	16:0	63.1
18	Octadecanoic acid	Stearic acid	18:0	69.6
20	Eicosanoic acid	Arachidic acid	20:0	76.5
22	Docosanoic acid	Behenic acid	22:0	79.9
24	Tetracosanoic acid	Lignoceric acid	24:0	86.0
Monoenoic				
16	9-Hexadecenoic acid	Palmitoleic acid	$16:1^9$ or 16:1 (n-7)	–0.5
18	9-Octadecenoic acid	Oleic acid	$18:1^9$ or 18:1 (n-9)	13.4
18	(11E)-Octadecenoic acid or 11-trans-Octadecenoic acid	Vaccenic acid	$18:1^{11}$t or 18:1 (n-7)t	44.0
18	(9E)-Octadecenoic acid or 9-trans-Octadecenoic acid	Elaidic acid	$18:1^9$t or 18:1 (n-9)t	46.5
24	15-Tetracosenoic acid	Nervonic acid	$24:1^{15}$ or 24:1 (n-9)	42.5–43
Dienoic				
18	9,12-Octadecadienoic acid	Linoleic acid	$18:2^{9,12}$ or 18:2 (n-6)	–5
Trienoic				
18	9,12,15-Octadecatrienoic acid	α-Linolenic acid	$18:3^{9,12,15}$ or 18:3 (n-3)	–11
18	6,9,12-Octadecatrienoic acid	γ-Linolenic acid	$18:3^{6,9,12}$ or 18:3 (n-6)	
Tetraenoic				
20	5,8,11,14-Eicosatetraenoic acid	Arachidonic acid	$20:4^{4,7,10,13,16,19}$ or 20:4 (n-6)	–49.5
Hexaenoic				
22	4,7,10,13,16,19-Docosahaenoic acid	Cervonic acid	$22:6^{4,7,10,13,16,19}$ or 22:6 (n-3)	—

[a] Key to abbreviations: The first number is n = number of C-atoms; the number after the colon gives the number of double bonds; the position of the double bond(s) is indicated by the superscript(s) to the symbol . In this shorthand notation, the position of the double bond is determined by counting from the carboxyl group (=C1). The configuration of double bonds is defined by both the E-Z and *cis-trans* conventions (see Figure 1.2). Unless indicated by the letter *t* (for *trans* configuration), the *cis* configuration is implied. In a related system widely used for fatty acids of animal lipids, the position of the double bond is determined by (n-x), where x is the position of the first double bond encountered when counting from the terminal methyl group of the fatty acid.

as *sterols*. As is evident from their structures, most have the same ring skeleton but differ considerably in their side chain and peripheral structural features, in stereochemistry, and in the number of double bonds in the rings.

Steroids are widely distributed in both plants and animals and comprise a diversity of compounds, such as cholesterol, bile acids, vitamin D (Figure 1.7), sex and corticoid hormones, and saponins. Many of these compounds are of vital importance in physiology. Others are valuable medicinals, such as cardiac glycosides, hormones, and steroidal antibiotics. The most common sterol is cholesterol (cholest-5-en-3β-ol), which is an unsaturated alcohol (for the stereochemistry of sterols, see Section 1.2). Cholesterol is found in mammalian plasma membranes to about 30% of the total lipid mass, and also in lysosomes, endosomes, and Golgi. The sterols found in higher plants are β-sitosterol and stigmasterol. These plant sterols (*phytosterols*) frequently have an additional side chain at position C-24 and/or an additional double bond at position C-22 (Figure 1.6). Ergosterol is often found in eukaryotic microorganisms, e.g., yeast.

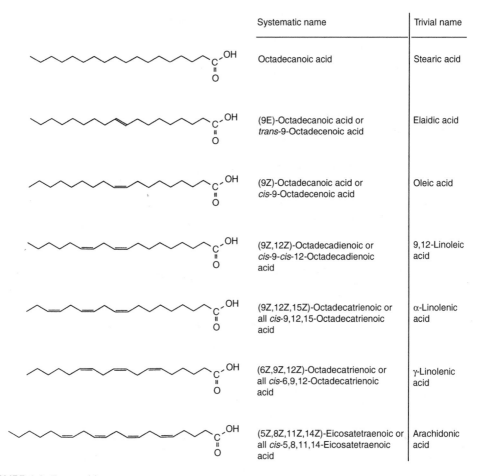

Systematic name	Trivial name
Octadecanoic acid	Stearic acid
(9E)-Octadecanoic acid or *trans*-9-Octadecenoic acid	Elaidic acid
(9Z)-Octadecanoic acid or *cis*-9-Octadecenoic acid	Oleic acid
(9Z,12Z)-Octadecadienoic or *cis*-9-*cis*-12-Octadecadienoic acid	9,12-Linoleic acid
(9Z,12Z,15Z)-Octadecatrienoic or all *cis*-9,12,15-Octadecatrienoic acid	α-Linolenic acid
(6Z,9Z,12Z)-Octadecatrienoic or all *cis*-6,9,12-Octadecatrienoic acid	γ-Linolenic acid
(5Z,8Z,11Z,14Z)-Eicosatetraenoic or all *cis*-5,8,11,14-Eicosatetraenoic acid	Arachidonic acid

FIGURE 1.3 Fatty acids.

In the last two decades, a new class of pentacyclic sterollike compounds termed *hopanoids* (Figure 1.6) was discovered. Hopanoids are very widespread in bacteria but are also found in some plants. Hopanoids were also detected in sediments and crude oils, where they account for more than 5% of the soluble organic matter. This extrapolates to 10^{11} to 10^{12} tons of hopanoids, suggesting that this class of compounds represents the most abundant one on our planet (Ourisson, 1987; Prince, 1987). Bacteriohopanetetrol is the most abundant microbial hopanoid.

Other compounds are diplopterol and tetrahymenol, which are found in *Tetrahymena* species. These compounds occur in bacterial membranes, and their function has been proposed to be the stabilization of the membrane structure, rather similar to that of membrane sterols in eukaryotes.

Both sterols and hopanoids are rigid and rather flat amphipathic molecules. The amphiphilic groups are, however, on opposite sides of the molecules: it is the 3-hydroxyl group in sterols and the polyol side chain in hopanoids. The biosynthesis of sterols and hopanes shares a common pathway from acetate to mevalonate and squalene, but differs thereafter. Direct cyclization leads to hopane; lanosterol (Figure 1.6) is produced as an intermediate in the sterol pathway.

1.1.1.2.5 Vitamins

A number of vitamins are fat-soluble and have solubility properties characteristic of lipids. Based on this, they are frequently classified as lipids. Fat-soluble vitamins contain alicyclic or aromatic rings, which are usually oxygenated and attached to an isoprenoid hydrocarbon chain or isoprenoid alcohol. Aforementioned examples are retinol (vitamin A_1) and ergocalciferol (vitamin D_2). Other

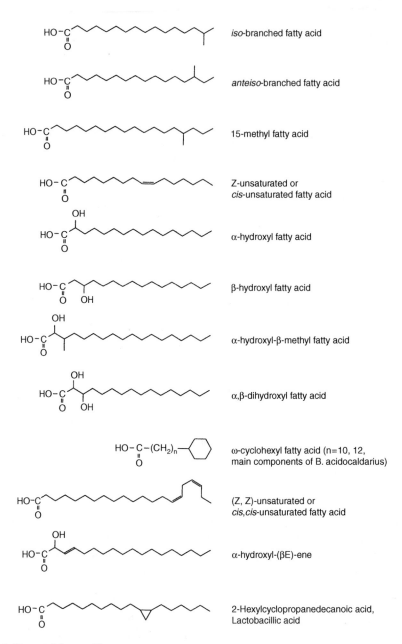

FIGURE 1.4 Bacterial fatty acids.

examples of fat-soluble vitamins are α-tocopherol (vitamin E) and vitamins K_1 and K_2 (Figure 1.7). Both vitamins K_1 and K_2 are substituted naphthoquinones with isoprenoid side chains differing in length. They are abundant in most higher plants.

1.1.2 SIMPLE ESTERS

The lipids discussed under this heading are saponifiable, that is, they are hydrolyzed by heating with alkali to yield soaps.

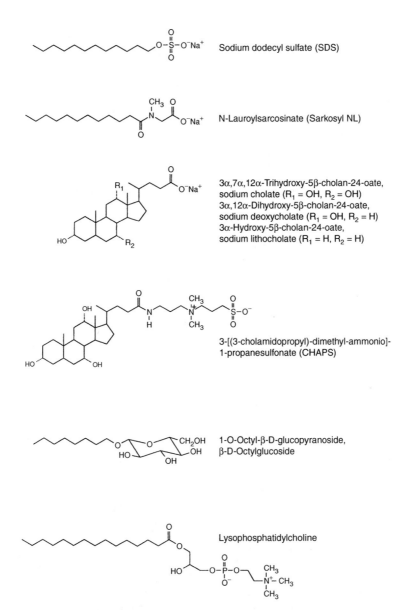

Sodium dodecyl sulfate (SDS)

N-Lauroylsarcosinate (Sarkosyl NL)

3α,7α,12α-Trihydroxy-5β-cholan-24-oate,
sodium cholate (R$_1$ = OH, R$_2$ = OH)
3α,12α-Dihydroxy-5β-cholan-24-oate,
sodium deoxycholate (R$_1$ = OH, R$_2$ = H)
3α-Hydroxy-5β-cholan-24-oate,
sodium lithocholate (R$_1$ = H, R$_2$ = H)

3-[(3-cholamidopropyl)-dimethyl-ammonio]-
1-propanesulfonate (CHAPS)

1-O-Octyl-β-D-glucopyranoside,
β-D-Octylglucoside

Lysophosphatidylcholine

FIGURE 1.5 Detergents. Lysophospholipids are a class of detergents that is biologically important. A representative of this class is lysophosphatidylcholine, which has one fatty acyl chain per polar group. The generic term *lysophospholipid* indicates the detergentlike properties and the lytic activity of this class of lipds.

1.1.2.1 Acylglycerols

A number of complex lipids have, as a central part of their structure, the glycerol group. Glycerol (1,2,3-propanetriol) has three reactive OH-groups and forms esters (ethers) with fatty acids (alcohols). Depending on the number of fatty acids (alcohols) reacting with glycerol, one obtains monoacyl-(monoalkyl-), diacyl- (dialkyl-), and triacyl- (trialkyl-) glycerols (Figure 1.8). Triacylglycerols, frequently also called triglycerides,* are the major components of dietary fats. They are

---

* This term should be avoided because it is misleading.

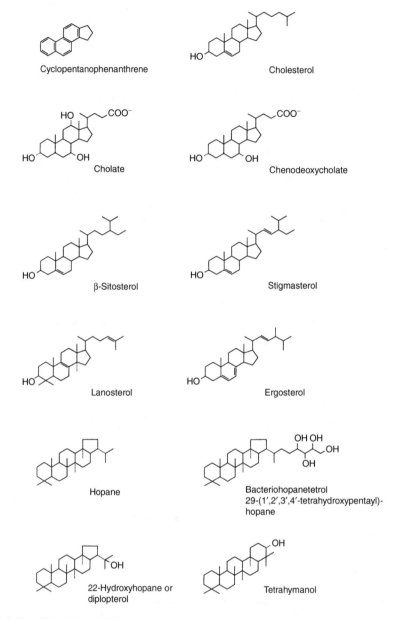

FIGURE 1.6 Steroids and hopanoids.

also the major storage lipids for both plants and higher animals. Acylglycerols, particularly mono-acyl- and diacylglycerols, are also minor components of biological membranes. Diacylglycerol serves an important function as a second messenger in signal transduction (Bell, 1986). Many biologically active substances, such as hormones and neurotransmitters, react with receptors at the cell membrane and elicit an intracellular response. The mechanism of signal transduction has been shown to involve phophatidylinositols and their products of lipolysis (see below).

1.1.2.2 Cholesteryl Esters

Cholesterol can be esterified with long-chain fatty acids to form cholesteryl esters (Figure 1.8), which are much more hydrophobic than the parent compounds. Cholesteryl esters are constituents

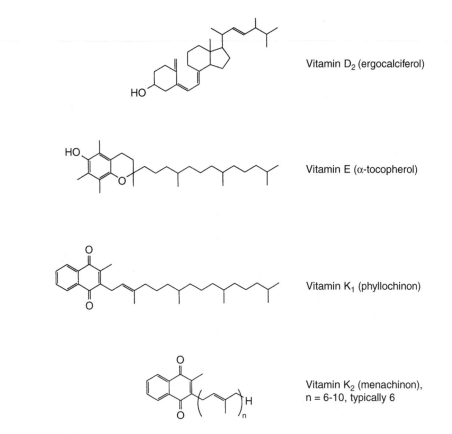

FIGURE 1.7 Vitamins.

of serum lipoproteins. For example, low density lipoproteins (LDL) are rich in cholesteryl esters. Cholesteryl esters are stored in organs such as the adrenal cortex and the corpus luteum of the ovary, where they are precursors in the synthesis of steroid hormones. They accumulate in certain disorders (Small, 1977), such as cholesteryl ester storage disease (Fredrickson and Ferrans, 1978), atherosclerosis (Small and Shipley, 1974), familial hypercholesterolemia (Fredrickson et al., 1978), and Tangier disease (Herbert et al., 1978). They are also minor components of cell membranes.

1.1.2.3 Waxes

Waxes are esters of long-chain saturated and unsaturated fatty acids having from 14 to 36 carbon atoms, with long-chain alcohols having 16 to 34 carbon atoms (Figure 1.8).

1.1.3 COMPLEX LIPIDS

1.1.3.1 Glycerophospholipids

The bulk of the membrane lipids belongs to the classes of complex lipids and glycolipids. The most commonly found membrane lipids are the glycerophospholipids, which are summarized in Figure 1.9. Glycerophospholipids are derivatives of *sn*-glycero-3-phosphoric acid (Figure 1.15), with usually two fatty acids esterified in the *sn* 1 and 2 positions of the glycerol moiety. The compound thus formed is 1,2-diacyl-*sn*-glycero-3-phosphoric acid, using stereospecific numbering (*sn*; for details, see Section 1.2). The trivial name is 3-*sn*-phosphatidic acid.

In virtually all naturally occurring glycerophospholipids, the polar group is attached to the *sn*-3-position of glycerol. One exception to this rule is the glycerol-based lipids of archaebacteria. As

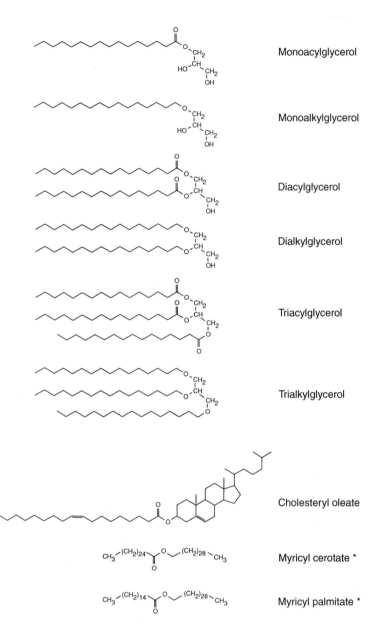

FIGURE 1.8 Acyl- and alkylglycerols, cholesteryl esters, and waxes. Cerotic acid and myricyl alcohol are trivial names for hexacosanoic acid and 1-triacontanol, respectively.

discussed below, the stereospecific numbering is restricted to glycerol-containing lipids and has certain disadvantages. A more systematic or formal designation of 3-*sn*-phosphatidic acid is 2,3-diacyl-D-glycero-1-phosphoric acid. The advantages of this nomenclature are discussed in Section 1.2. Diacylglycerophospholipids may be regarded as phosphodiester derivatives of 1,2-diacyl-*sn*-glycero-3-phosphoric acid. Even in the formal IUPAC/IUB nomenclature of glycerophospholipids (see IUPAC-IUB Commission on Biochemical Nomenclature (CBN), 1977), the alcohols esterified to diacylglycerophosphoric acid are designated by trivial names, such as choline, ethanolamine, serine, glycerol, glycerolphosphate, myoinositol, etc.

The generic term *phosphatidyl* stands for the 1,2-diacyl-*sn*-glycero-3-phosphomoiety. The diacylglycerophospholipids shown in Figure 1.9 are the predominant lipids in most eukaryotic and

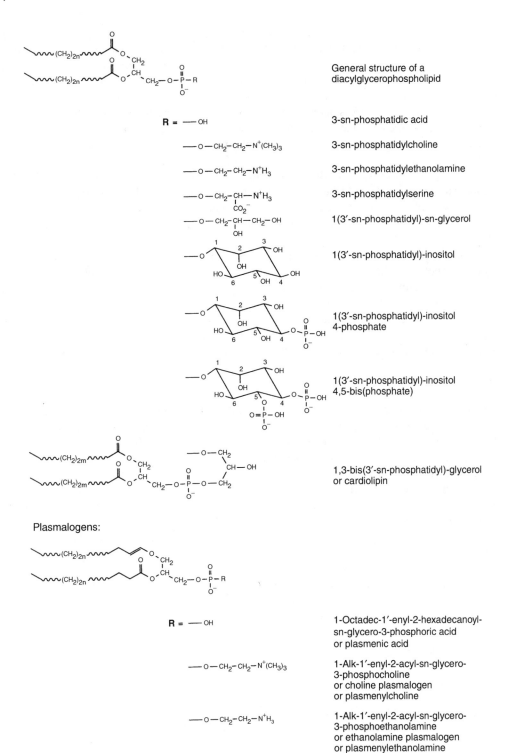

R = ── OH   3-sn-phosphatidic acid

── O ─ CH₂─ CH₂─ N⁺(CH₃)₃   3-sn-phosphatidylcholine

── O ─ CH₂─ CH₂─ N⁺H₃   3-sn-phosphatidylethanolamine

── O ─ CH₂─ CH ─ N⁺H₃   3-sn-phosphatidylserine
               |
               CO₂⁻

── O ─ CH₂─ CH ─ CH₂─ OH   1(3'-sn-phosphatidyl)-sn-glycerol
               |
               OH

── O   1(3'-sn-phosphatidyl)-inositol

── O   1(3'-sn-phosphatidyl)-inositol
                                  4-phosphate

── O   1(3'-sn-phosphatidyl)-inositol
                                  4,5-bis(phosphate)

1,3-bis(3'-sn-phosphatidyl)-glycerol
or cardiolipin

Plasmalogens:

R = ── OH   1-Octadec-1'-enyl-2-hexadecanoyl-
                                   sn-glycero-3-phosphoric acid
                                   or plasmenic acid

── O ─ CH₂─ CH₂─ N⁺(CH₃)₃   1-Alk-1'-enyl-2-acyl-sn-glycero-
                                   3-phosphocholine
                                   or choline plasmalogen
                                   or plasmenylcholine

── O ─ CH₂─ CH₂─ N⁺H₃   1-Alk-1'-enyl-2-acyl-sn-glycero-
                                   3-phosphoethanolamine
                                   or ethanolamine plasmalogen
                                   or plasmenylethanolamine

FIGURE 1.9 Glycerophospholipids. The generic term *phosphatidyl-* stands for 1,2-diacyl-*sn*-glycero-3-phos-pho-. Thus, the systematic name of 1(3'-sn-phosphatidyl)-inositol is 1-(1',2'-diacyl-sn-glycero-3-phospho)-L-myo-inositol. The aliphatic chains are saturated or unsaturated.

prokaryotic membranes, excluding archaebacteria. Phosphatidic acid with a primary phosphate group has detergentlike properties under certain conditions (Hauser and Gains, 1982) and is only a minor component of membranes. It is, however, a key intermediate in the biosynthesis of glycerophospholipids. 1,2-Diacyl-*sn*-glycero-3-phosphocholine is a major constituent of animal cell membranes, and 1,2-diacyl-*sn*-glycero-3-phosphoethanolamine is often a major constituent of bacterial membranes. 1-(3'-*sn*-phosphatidyl) inositol (Figure 1.9) constitutes only about 2% to 8% of the phospholipids in eukaryotic membranes. A small fraction of phosphatidylinositol is phosphorylated at the 4-position of myoinositol (1-(3-*sn*-phosphatidyl) inositol-4-phosphate) or at both the 4- and 5-positions (1-(3'-*sn*-phosphatidyl) inositol-4,5-bisphosphate). The latter compound, which makes up 1% to 10% of the total phosphatidylinositol lipids, is a key compound in phosphatidylinositol turnover and in the transduction of hormonal messages: upon the formation of a hormone-receptor complex at the membrane level, 1-(3-*sn*-phosphatidyl) inositol-4,5-bisphosphate is hydrolyzed to diacylglycerol and inosit-1,4,5-trisphosphate, which both serve as second messengers in the signal transduction.

1,3-Bis(3'-*sn*-phosphatidyl)-glycerol (diphosphatidyl glycerol or cardiolipin, Figure 1.9) is a major phospholipid of inner mitochondrial, chloroplast, and some bacterial membranes. *Plasmalogen* is a generic term for glycerophospholipids in which the glycerol moiety bears a 1-alkenyl ether group. Ethanolamine plasmalogens or plasmenylethanolamines (Figure 1.9) are an important component of myelin and the cardiac sarcoplasmic reticulum (Gross, 1985). The generic term for phospholipids with only one hydrocarbon chain is *lysophospholipid* (Figure 1.5). The term *lyso* was coined originally to point out the detergentlike properties and hemolytic activity of this lipid class.

1.1.3.2 Sphingolipids

The fundamental structure common to sphingolipids is sphinganine or dihydrosphingosine. Its unsaturated analog is (4E)-sphingenine (*trans*-4-sphingenine) or sphingosine (Figure 1.10). The simplest derivative of sphingosine is psychosine: a monosaccharide is bonded glycosidically to the hydroxyl of position 1 of sphingosine. An example of a psychosine is 1-*O*-(-D-galactopyranosyl)-sphingosine (Figure 1.10). Acylation of the NH$_2$-group in the 2-position of sphingosine yields another fundamental structure, *N*-acyl-sphingosine, the generic term for which is *ceramide*. Although free ceramides are rare in biological membranes, they form the basic structure of complex lipids such as sphingomyelins, cerebrosides, and gangliosides.

All three classes of lipids are sphingolipids; nevertheless, it is customary to class cerebrosides and gangliosides as glycolipids. In sphingomyelins, the terminal hydroxyl group of ceramide is esterified, usually to phosphocholine or alternatively to phosphoethanolamine. Similar to glycerophospholipids, sphingomyelins contain a phosphodiester group; because of this and their close resemblance to glycerophospholipids regarding the general physicochemical properties, sphingomyelins (Figure 1.10) are frequently grouped together with glycerophospholipids.

1.1.4 GLYCOLIPIDS

1.1.4.1 Glycoglycerolipids

Glycoglycerolipids are glycerol-based lipids in which the *sn*-3 position of glycerol is linked glycosidically to a carbohydrate, e.g., galactose. They are major constituents of the thylakoid membrane of chloroplasts. As such, they make up more than 5% of the polar plant lipids, representing one of the most abundant lipids in nature (Figure 1.11A).

Glycoglycerolipids are rare in the animal kingdom. Glycoglycerolipids containing a sulfogroup in the carbohydrate moiety are termed *sulfolipids*. An example is given in Figure 1.11B, which is a sulfonated derivative of α-D-quinovosyldiacylglycerol. Note that α-D-quinovose is 6-deoxy-α-D-glucose and that the sulfolipid has a carbon–sulfur bond. Glycoglycerolipids are also found in

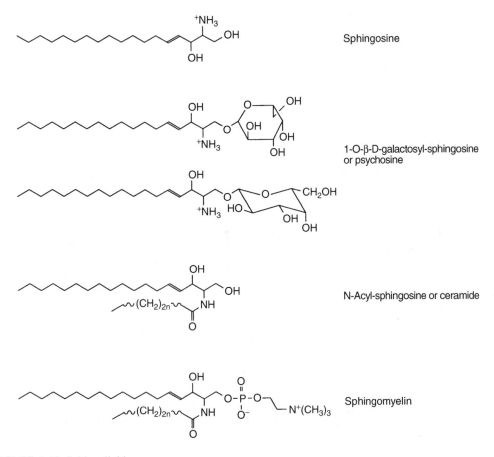

Sphingosine

1-O-β-D-galactosyl-sphingosine or psychosine

N-Acyl-sphingosine or ceramide

Sphingomyelin

FIGURE 1.10 Sphingolipids.

substantial quantity in blue-green algae and bacteria. Gram-positive bacteria contain glycoglycer-olipids with a variety of sugars. Archaebacteria also contain glycerol-based lipids (Figure 1.11C through Figure 1.11G). Besides glycerol, they may contain more complex polyols, such as tetritol, inositol, or even nonitol (carditol), as backbones.

Another remarkable feature are the hydrocarbon chains, which comprise a variety of polyiso-prenoid structures, e.g., phytanyl, biphytanyl, and cyclopentane biphytanyl. Biphytanyl chains with 40 carbon atoms having OH-functional groups on both ends are linked to polyols on both ends and are termed bipolar (Luzzati et al., 1987). Also shown in Figure 1.11 are the types of linkages between the polyol backbones and the polyisoprenoid chains that are found in archaebacteria. The bulk (nearly 90%) of the lipids consist of the bipolar tetraether lipids which, due to their complexity, are usually designated in a generic way. They are named glycerol-dialkyl-glycerol-tetraether and glycerol-dialkyl-nonitol-tetraether; in the latter case nonitol replaces on the glycerol backbones. As shown in Figure 1.11C, one biphytanyl (C_{40}-residue) is linked through two ether bonds to the 2- and 3-positions of two different *sn*-glycerol groups. The same is true for the second biphytanyl residue, resulting in a macrocyclic tetraether structure (Figure 1.11C). This structure consists of a 72-membered ring with 18 stereocenters (Luzzati et al., 1987). Each biphytanyl residue can contain 1–4 cyclopentane rings. Other types of linkages are also included in Figure 1.11. For instance, a single biphytanyl chain is linked through ether bonds to the 2- and 3-position of a single *sn*-glycerol group, forming a macrocyclic diether which consists of a 36-membered ring (Figure 1.11D). Other types of ether lipids occurring in archaebacteria are shown in Figure 1.11E through Figure 1.11G.

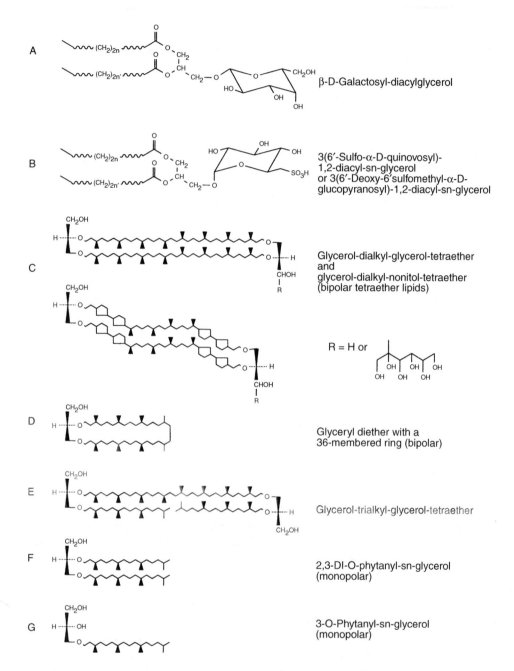

FIGURE 1.11 Glycoglycerolipids consist of diacylglycerol to which a mono- or disaccharide is linked glycosidically in the *sn*-3 position. In addition to these relatively simple structures, this class comprises more complex, glycerol-based lipids found in archaebacteria. For practical reasons, the complex macro-cyclic diether and tetraether lips of archaebacteria are designed using an abbreviated schematic nomenclature (examples are structures C, D, and E). For simple archaebacterial lipids, such as lipids F and G, a more systematic nomenclature is used. Phytanyl (C_{20}) residues with one functional OH-group are termed *monopolar*; biphytanyl (C_{40}) residues with OH-groups on both ends of the chain are called *bipolar*.

1.1.4.2 Glycosphingolipids

1.1.4.2.1 Cerebrosides

The fundamental structure of cerebrosides is ceramide. Monoglycosyl- and oligoglycosylceramides, in which a mono- or polysaccharide is bonded glycosidically to the terminal OH-group of ceramide, are defined as *cerebrosides*. Monogalactosylceramide is the largest single component of the myelin sheath of nerves (Figure 1.12). More complex cerebrosides are named as oligoglycosylceramides, using the appropriate trivial name of the mono- or oligosaccharide residues. It is understood that the sugar moiety is linked glycosidically to the C-1 hydroxyl group of ceramide. An example is lactosylceramide. For glycosphingolipids carrying two or more sugar residues, the shorthand system listed in Table 1.3 is used. Cerebrosides carrying a sulfuric ester (sulfate) group, formerly called sulfatides, also occur in the myelin sheath of nerves. These compounds are preferably named as sulfates of the parent glycosphingolipid (Figure 1.12).

Glycosphingolipids of the globo and lacto series usually lack sialic acid. They are neutral and may be regarded as complex cerebrosides with four and more carbohydrate residues. In the globo series, the sequence of carbohydrates is GalNAc-Gal-Gal-Glc-Cer; in the lacto series, it is Gal-GlcNAc-Gal-Glc-Cer (for the abbreviations see Table 1.3).

1.1.4.2.2 Gangliosides

Sialoglycosphingolipids, or gangliosides, are a class of complex glycosphingolipids that contain a polysaccharide chain of four sugars glycosidically linked to the C-1-hydroxyl group of ceramide. They are anionic lipids because the polysaccharide chain contains one or more molecules of the negatively charged sialic acid (*N*-acetylneuraminic acid, NeuAc) (Figure 1.12). The sialic acid is bound glycosidically to one of the sugar residues of the oligosaccharide chain. A shorthand notation for gangliosides has been introduced (Table 1.3). The pictorial representation of gangliosides (e.g., of the structure of G_{M1} shown in Figure 1.12) has the advantage of perspicuity compared to the shorthand notation given in Table 1.3.

Glycosphingolipids are present on the outer surface of mammalian plasma membranes, usually as minor components. In epithelial brush border membranes, they are, however, a major constituent. Glycosphingolipids have antigenic properties and act as receptors for antibodies, lectins, and certain toxins. Glycosphingolipids carrying blood group antigens are constituents of the erythrocyte membrane and exposed on the external surface of erythrocytes. The antigenic properties have been identified with oligosaccharides. Glycosphingolipids are believed to originate in the membranous system of the endoplasmic reticulum, from which they are exported to the Golgi apparatus. Secretory vesicles are assumed to bud off the Golgi cisternae and eventually fuse with the cell membrane. By this process, glycosphingolipids arrive at the site of destination and are incorporated in the correct orientation. Glycosphingolipids are therefore also found in the membranes of the previously mentioned cell organelles.

1.1.4.3 Lipopolysaccharides

Lipid X, lipid Y, and lipid A (Figure 1.13) are constituent molecules (Raetz, 1984) of lipopolysaccharides, which occur in the outer leaflet of the cell envelope of Gram-negative bacteria. Lipopolysaccharides (Figure 1.14) are complex glycolipids that are responsible for the antigenicity and pathogenicity of *Escherichia coli* and other Gram-negative bacteria. They display a wide variety of biological activities: among others, they mediate the interaction of the microorganism with the host cell, they activate the immune system of mammals, and they have endotoxic activity. A synonym for lipopolysaccharide is *endotoxin* (Rietschel et al., 1982).

These activities are responsible for the great interest that lipopolysaccharides have aroused in the last two decades. Lipid A provides the hydrophobic portion with which lipopolysaccharides are anchored in the cell envelope of Gram-negative bacteria. The chemical structure of lipid X and lipid

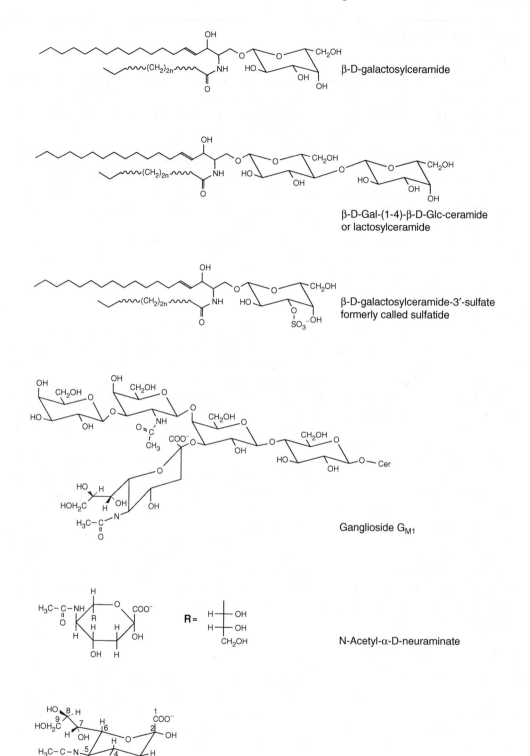

FIGURE 1.12 Glycosphingolipids (cerebrosides and gangliosides). The key to the abbreviations used is given in Table 1.3.

TABLE 1.3
Shorthand Notation of Glycosphingolipids

Structure	Series	Abbreviation
Gal(β1-4)Glc(β1-1)Cer	Lactosylceramide	
Gal(β1-3)GalNAc(β1-4)Gal(β1-4)Glc(β1-1)Cer	Ganglio series	G_{M1}
$\quad\quad\quad\quad\quad$ |		
$\quad\quad\quad$ NeuAc(α2-3)		
Fuc(α1-2 or 3)Gal(β1-3)GalNAc(β1-4)Gal(β1-4)Glc(β1-1)Cer	Ganglio series	Fuc-G_{M2}
$\quad\quad\quad\quad\quad\quad\quad\quad\quad\quad$ |		
$\quad\quad\quad\quad\quad\quad\quad$ NeuAc(α2-3)		
GalNAc(β1-4)Gal(β1-4)Glc(β1-1)Cer	Ganglio series	G_{M2}
$\quad\quad$ |		
$\quad$ NeuAc(α2-3)		
NeuAc(α2-3)Gal(β1-4)Glc(β1-1)Cer	Ganglio series	G_{M3}
Gal(β1-3)GalNAc(β1-4)Gal(β1-4)Glc(β1-1)Cer	Ganglio series	G_{D1a}
105 $\quad$ | $\quad\quad\quad\quad\quad\quad$ |		
$\quad\quad$ NeuAc(α2-3) $\quad\quad\quad$ NeuAc(α2-3)		
Gal(β1-3)GalNAc(β1-4)Gal(β1-4)Glc(β1-1)Cer	Ganglio series	G_{Db1}
107 $\quad\quad\quad\quad\quad\quad\quad\quad\quad$ |		
$\quad\quad\quad\quad\quad$ NeuAc(α2-8) NeuAc(α2-3)		
NeuAc(α2-3) Gal(β1-3)GalNAc(β1-4)Gal(β1-4)Glc(β1-1)Cer	Ganglio series	G_{T1b}
109 $\quad\quad\quad\quad\quad\quad\quad\quad\quad\quad\quad\quad$ |		
$\quad\quad\quad\quad\quad\quad\quad$ NeuAc(α2-8) NeuAc(α2-3)		
GalNAc(β1-3)Gal(α1-4)Gal(β1-4)Glc(β1-1)Cer	Globo series	
GalNAc(α1-3)GalNAc(β1-3)Gal(α1-4)Gal(β1-4)Glc(β1-1)Cer	Globo series	
Gal(β1-3)GlcNAc(β1-3)Gal(β1-4)Glc(β1-1)Cer	Lacto series	
Gal(β1-3)Gal(β1-4)GlcNAc(β1-3)Gal(β1-4)Glc(β1-1)Cer	Lacto series	

Y was recently elucidated; both lipids may be regarded as constituent molecules of lipid A. Lipid X is 2,3-bis((R)-3-hydroxymyristoyl) α-D-glucosamine-1-phosphate, and lipid Y is a triacyl-D-glucosamine-1-phosphate derivative, in which the hydroxyl of N-(R)-3′-hydroxytetradecanoyl is esterified with hexadecanoic acid (Figure 1.13). The chemical structure of lipid A varies widely from species to species. The structure of lipid A of *Escherichia coli* has been studied extensively, and its proposed structure is shown in Figure 1.13 (Rietschel et al., 1984). At the central part of the molecule are two 2-deoxy-2-amino-D-glucopyranosyl residues, which are β1′-6-interlinked. This disaccharide carries two primary phosphate groups, one in position 4′ of the nonreducing glucosamine residue (GlcN II) and another one esterified in position 1 of the reducing glucosamine (GlcN I). Another phosphate may be added to the α-linked phosphate in a nonstoichiometric way, forming a pyrophosphate. For instance, in lipid A of *Salmonella minnesota,* the pyrophosphate is replaced by ethanolamine. Lipid A has a characteristic set of long-chain fatty acids that are bound to both amino and hydroxyl groups of the two glucosamine residues. The hydroxyl and amino group of GlcN I are acylated by (R)-3-hydroxytetradecanoic acid, the hydroxyl group in position 3′ by (R)-3-tetradecanoyloxytetradecanoic acid, and the amino group of GlcN II by (R)-3-dodecanoyloxytetradecanoic acid. Lipid A of different organisms varies in the number of chain length of the fatty acids (Figure 1.13).

Of various lipopolysaccharides studied to date, those of *Salmonella* have probably been investigated most thoroughly (Lüderitz et al., 1982) (Figure 1.14E and Figure 1.14F). As evident from this figure, lipopolysaccharides consist of a hydrophobic portion, termed lipid A, and a long, hydrophilic polysaccharide chain. The latter is covalently linked to lipid A and can be subdivided into the core and the O-specific polysaccharide chain. As shown in the schematic diagram of Figure 1.14E, one to three molecules of 3-deoxy-D-manno-2-octulosonic acid, abbreviated KDO (see

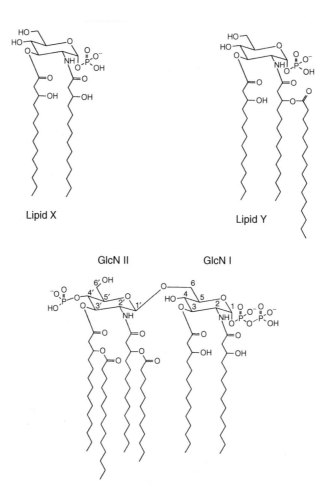

FIGURE 1.13 Lipopolysaccharides. The proposed chemical structure of lipid A of *Escherichia coli* is presented, together with the chemical structure of lipid X and lipid Y, which are constituent molecules of lipd A. GlcN = glucosamine. The dotted line in the lipid A structure indicates nonstoichiometric substitution. (From Raetz CRH; and Rietschel et al., both in *Handbook of Endotoxins*, Vol. 1. Elsevier, Amsterdam, 1984.)

Figure 1.14A through Figure 1.14C), bridge the hydrophobic lipid A and the hydrophilic polysaccharide chain. The hydroxyl in position 6′ of the disaccharide backbone of lipid A (see GlcN II of Figure 1.13) is covalently linked to KDO (see Figure 1.14E). The three regions of lipopolysaccharides differ not only in their chemical structures, but also in their functional properties (Lüderitz et al., 1982). The viability of the bacterial cell is associated with a minimal core structure, the antigenic properties are determined by the O-specific polysaccharide chain, and the endotoxic principle is lipid A. Lipopolysaccharides have been reported to form smectic (lamellar) phases both in the dry and the hydrated state, and this is probably also true for their aqueous dispersions. The dimensions of different regions of *Salmonella* lipopolysaccharides, as derived from X-ray diffraction (Labischinski et al., 1985), are shown schematically in Figure 1.14F.

1.2 THE STEREOCHEMISTRY OF LIPIDS

1.2.1 GLYCEROL-BASED LIPIDS

The following discussion deals with the stereochemistry of membrane lipids. Of particular interest here are the glycerophospholipids; related to them are the acylglycerols, sphingolipids, and gly-

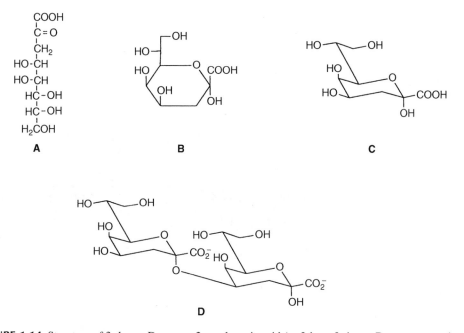

FIGURE 1.14 Structure of 3-deoxy-D-manno-2-octulosonic acid (or 2-keto-3-deoxy-D-manno-octonic acid), abbreviated KDO. In (A) the Fischer projection of the acyclic form is shown, in (B) the Haworth projection, and in (C) the chair form. (D) The chemical structure of the 2-4 linked KDO disaccharide as isolated from lipopolysaccharide of *Salmonella godesberg*. This disaccharide represents a common constituent of bacterial lipopolysaccharides. (Adapted from Brade H et al.: *Zbl. Bakt. Hyg. A* 1988, 268: 151.) (E) schematic diagram of the structure of *Salmonella* lipopolysaccharides. The number of nonhydroxylated and hydroxylated fatty acids shown in this drawing is arbitrary. All carbohydrates present in the O-specific chain and in the core are in the pyranose form. The nature of the sugar residues in the O-specific and the core polysaccharide chain, together with the glycosidic linkages between these residues, between the core acid KDO, and between the KDO-disaccharide and lipid A are all givien in shorthand notation beneath the schematic diagram of the lipopolysaccharide structre. Note that the fatty acids contain 12 or more C-atoms and may be b-hydroxylated. Abbreviations: Abe = abequosin; Glc = glucose; GlcNAc = N= acetylglucosamine; Gal = galactose; Man = mannose; Rha = rhamnose; Hep = L-glycero-α-D-mannoheptose equivalent to -L-D-Hep or -Hep. (F) Proposed structural dimensions of *Salmonella* lipopolysaccharide derived from X-ray diffraction (Labischinski et al., 1985). LPS = lipopolysaccharide; HR = hydrophilic region of lipid A, which is the phosphorylated glucosamine disaccharide; LR = lipophilic region consisting of fatty acyl chains of lipd A; KDO = 3-deoxy-D-manno-2-octulosonic acid; PS = polysaccharide chain of LPS. (From Lüderitz O et al.: in *Current Topics in Membranes and Transport,* Vol. 17, Academic Press, New York, 1982). *Continued.*

colipids. The stereospecific numbering (*sn*) of glycerol derivatives introduced by Hirschmann (1960) has proved useful in the description of steric and chemical relationships of acylglycerols. It is widely accepted and used not only for acylglycerols but also for complex glycerolipids and glycerophospholipids. In the convention of stereospecific numbering, the C-atom that appears on top in the Fischer projection, showing a vertical carbon chain with the hydroxyl group at carbon atom 2 pointing to the left, is designated C-1. To indicate such numbering and in order to distinguish it from conventional numbering conveying no steric information, the prefix *sn* is used. For instance, *sn*-glycerol-3-phosphate is the parent structure common to all naturally occurring glycerophospho-lipids. Its configuration is shown in Figure 1.15. According to the standard rules of nomenclature, this compound is named D-glycerol-1-phosphate. Its enantiomer is *sn*-glycerol-1-phosphate or L-glycerol-1-phosphate (Figure 1.15).

It has been pointed out that stereospecific numbering, though useful in some cases, is limited. For instance, it cannot be extended to sphingolipids (Hauser et al., 1981) and hence is unsuitable

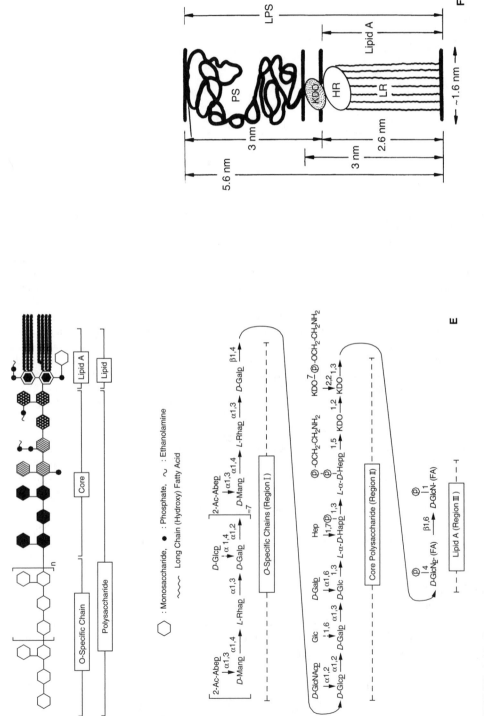

FIGURE 1.14 *Continued.*

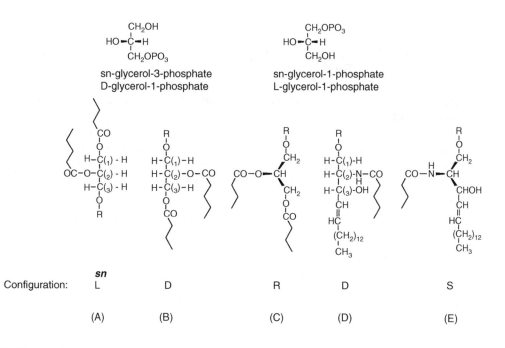

FIGURE 1.15 Stereochemistry of natural glycerophospholipids, glycoglycerolipids, and sphingolipids. Different conventions (*sn*, D/L, R/S) are used to describe the stereochemistry of the asymmetric carbon atom 2 of glycerolipids and sphingolipids. Molecules (A) to (C) represent naturally occurring glycerophospholipids; the configuration of the asymmetric C-atom 2 of glycerol is defined by the *sn* (A), the D/L (A and B), and the R/S system (C). Molecules (D) and (E) represent a naturally occurring sphingolipid and asymmetric carbon atom 2, which is defined by the D/L (D) and the R/S system (E). The use of the D/L system reveals the configurational analogy of the two classes of compounds. (A) 1,2-diacyl-sn-glycero-3-; (B) 2,3-diacyl-D-glycero-1-; (C) (2R)-2,3-diacylglycero-1-; (D) N-acyl-2D-amino-(4E)-octadecene-1,3D-diol; (E) N-acyl-(2S, 3R, 4E)-2-amino-4-octadecene-1,3-diol.

for the discussion of structural and configurational relationships between glycerophospholipids and sphingolipids. Using the standard rules of nomenclature in the discussion of lipid structure has been proposed (Sundaralingam, 1972; Hauser et al., 1981). This has the advantage that configurational analogies are disclosed. For instance, the configurational analogy between glycerophospholipids and sphingolipids is evident from Figure 1.15. The asymmetric C(2)-atom of both classes of compounds has the D-configuration. A uniform atom numbering and nomenclature is therefore a prerequisite if conformations of the two classes of compounds are discussed and compared. Using the atom numbering as shown in Figure 1.15B, the configuration of asymmetric C-atoms may be defined by either the D/L or the R/S system. The universal character of the R/S system speaks for itself. Its only disadvantage is that it might obscure configurational analogies. This is true for the configuration of the C(2)-atom of glycerophospho- and sphingolipids. As is evident from Figure 1.15, they have the R- and S-configuration, respectively (cf. Figure 1.15C and Figure 1.15E). In this case, the D/L system is preferred because it brings to the fore the configurational analogy (cf. Figure 1.15B and Figure 1.15D). In structural work on lipids, therefore, adherence to the standard rules of nomenclature and avoidance of the stereospecific numbering of glycerol is recommended.

Asymmetric C-atoms are preferably defined by the R/S system; if need arises, use of the D/L system is made. The C(2)-atom of the glycerol of monoglycosyl- and diglycosyl diacylglycerols (e.g., 3(β-D-galactosyl)-1,2-diacyl-*sn*-glycerol) has the same configuration as in glycerophospholipids. In contrast, in acylglycerols of archaebacteria the configuration of the C(2)-atom is reversed: the glycosidic bond is at the *sn*-1-position and the glycerol backbone becomes *sn*-2,3-dialkyl rather than 1,2-dialkyl as in glycerophospholipids (cf. Figure 1.11).

1.2.2 SPHINGOLIPIDS

The stereochemistry of sphingosine and its relation to glycerophospholipids have been discussed previously (cf. Figure 1.15). For reasons mentioned, it is desirable to use the D/L or the R/S system. The systematic name is then preceded by configurational prefix(es) specifying the asymmetric center(s). For instance, sphinganine is D-erythro-2-amino-1,3-octadecanediol or, alternatively, (2S, 3R)-2-amino-1,3-octadecanediol. The configurational prefix *erythro* indicates that both the amino and hydroxyl group are on the same side (Figure 1.15). Sphingosine or (4E)-sphingenine (Figure 1.10) is therefore 2D-amino-trans-4-octadecene-1,3-D-diol. Phytosphingosine with three asymmetric centers is 4D-hydroxysphinganine or (2S, 3S, 4R)-2-amino-1,3,4-octadecanetriol.

1.2.3 STEROIDS

The accepted numbering of the steroid nucleus and the side chain is shown in Figure 1.16. The alicyclic 4-ring system of steroids consists of three fused six-membered rings and one five-membered ring. Cholestane may be regarded as the fundamental steroid structure (Figure 1.16). In cholestane rings A and B, B and C and C and D are fused in the *trans*-configuration, that is, the H or CH$_3$-substituents attached to the two bridging carbon atoms of the rings project on opposite

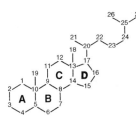

Numbering of the cholesterol ring system (cholestane)

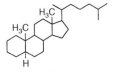

5-α-Cholestane: structure and chairform

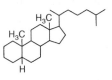

Coprostane or 5-β-cholestane: structure and chairform

Chairform of cholesterol

Chairform of cholate

FIGURE 1.16 Stereochemistry of steroids. Cholestane may be regarded as the fundamental steroid structure. Note that in the chair form of cholestane and coprostane, the H-atom at carbon atom 5 is omitted.

sides of the ring system. The fact that the ring fusions are all *trans* in cholesterol and related compounds is responsible for the compact and relatively flat arrangement of the ring system of cholesterol, shown in Figure 1.16. The stereochemistry of each atom or group is defined with respect to the plane of the four-ring system:

- The orientation of an atom or group above the ring plane is denoted β and indicated by a solid line.
- That below the ring plane is denoted α and given by a dashed line.

The methyl groups at the junction of rings A and B (C19) and C and D (C18) are termed *angular*. These two methyls, the OH-group, and the side chain of cholesterol are all on the same side of the ring (β-orientation). However, in some sterols the 3-OH group is α-oriented. Note that the configuration of the asymmetric carbon atom in position 20 is R. Cholesterol contains a double bond between C5 and C6, which distorts both the A-ring and the B-ring from the unstrained chair conformation. It was found that oxidation of 5β-cholestane yields cholanic acid, which is also obtained by dehydration of cholic acid at 300°C, followed by hydrogenation. This is chemical evidence that the stereochemistry of bile acids, such as cholic acid and related compounds, is different from that of cholesterol. In bile acids, rings A and B are fused in the *cis*-configuration (Figure 1.16). The same kind of ring fusion is present in coprostane and -coprostanol, a compound produced from cholesterol by bacterial action and found in large amounts in feces. The *cis*-fusion of rings A and B alters the overall shape of the steroid nucleus from a relatively flat to a distinctly bent arrangement (Figure 1.16).

Coprostane and 5-α-cholestane are stereochemically related like *cis*- and *trans*-decalin. For comparison, the chairform of cholate is included in Figure 1.16. All three OH-groups, as well as the side chain with the carboxyl group, lie on the α-side of the ring plane, giving rise to a hydrophilic surface. Bile salts may therefore be regarded as amphipathic molecules having a hydrophilic and a hydrophobic surface. The α-axial orientation of the mobile side chain at C17 of bile salts was inferred from monolayer experiments. In the protonated form of bile acids, the side chain is believed to attain an almost coplanar orientation with the ring plane.

1.3 THE LIPID COMPOSITION OF BIOLOGICAL MEMBRANES

The major components of biological membranes are proteins and lipids. Carbohydrates, which account for about 10% of the weight of plasma membranes, are invariably covalently bound to either protein or lipid. The relative amounts of protein and lipids vary widely, as evident from Table 1.4. In addition to natural variations, there are variations in the protein/lipid ratio, depending on the isolation procedure. There are a number of proteins which are only weakly bound to the membrane surface by mainly electrostatic forces and which are readily liberated from the membrane structure in the course of the isolation of the membrane. Changes in ionic strength, pH, or the buffer composition (e.g., the addition or removal of a chelator such as EDTA) suffice to remove these proteins from the membrane. As a result, the protein/lipid weight ratio of the membrane will decrease.

The enormous diversity of lipids is a remarkable feature of biological membranes. The reason for this diversity is still a matter of debate, despite our increasing awareness of the various roles lipids play in membranes. Table 1.4 and Table 1.5 summarize the lipid composition and the fatty acid composition, respectively, of mammalian plasma and subcellular membranes. As mentioned previously, the major phospholipids are phosphatidylcholines and phosphatidylethanolamines, with the negatively charged phospholipids contributing about 10% to 20% (Table 1.4). Table 1.5 shows that the acyl chains have an even number of C-atoms, ranging from C_{14} to C_{22}, with a predominance of acyl chains with 16 and 18 C-atoms. Nearly all double bonds have the Z- or *cis*-configuration. The degree of unsaturation can vary considerably. Most common, however, are acyl chains with

TABLE 1.4
Lipid Composition of Plasma and Subcellular Membranes

	Percentage of Phospholipids*								Phospholipids (μg/mg protein)	Cholesterol (μg/mg protein)
	PC	PE	PS	PI	PA	CL	LGP**	SM		
Rectal gland plasma membrane	50.4	35.5	8.4	<1	—	—	—	5.7	389	n.d.
Brush border membrane	33.3	35.6	7.4	8.2	1.2	n.d.	4.1	10.3	190	50
Cholinergic receptor membranes	37	40.5	17		<1	—	<1	<1	330(480)***	135(190)***
Plasma membrane	39	23	9	8	1	1	2	16	672	128
Mitochondria	40	35	1	5	—	18	1	1	175	3
Microsomes	58	22	2	10	1	1	11	1	374	14
Lysosomes	40	14	2	5	1	1	7	20	156	38
Nuclear membrane	55	13	3	10	2	4	3	3	500	
Golgi membrane	50	20	6	12	<1	1	3	8	825	38
Sarcoplasmic reticulum	72.7	13.5	1.8	8.7	<1	<1	—	1	603	78
										12

Note: n.d.: not determined. More data on the lipid compositions of enterocyte membranes can be found in the review by Pind and Kuksis (1986).

* Abbreviations of phospholipids: PC, phosphatidylcholine; PE, phosphatidylethanolamine; PS, phosphatidylserine; PI, phosphatidylinositol; PA, phosphatidic acid; CL, cardiolipin; LGP, lysoglycerophospholipid.

** Values for LGP in excess of a few percent should be viewed with caution. High contents of lysoglycerophospholipids are probably the result of phospholipid degradation during preparation.

*** Values for two different preparations by the same authors.

Source: From dog fish (Perrone et al., 1975), rabbit small intestine (Hauser et al. 1980a), *Torpedo marmorata* (calculated from Popot et al., 1978), rat liver (Daum, 1985), and rabbit (Meissner and Fleischer, 1971).

TABLE 1.5
Fatty Acid Composition of Phosphatidylcholine of Some Membranes From Rat Liver

	Fatty Acids as Weight%										
	14:0	16:0	16:1	18:0	18:1	18:2	18:3	20:0	20:3	20:4	22:6
Rat liver	0.5	29.7	1.0	16.8	10.4	16.8			1.5	18.3	3.4
Mitochondrial (outer)	0.4	27.0	4.1	21.0	13.5	13.5			1.1	15.7	3.5
Mitochondrial (inner)	0.3	27.1	3.6	18.0	16.2	15.8			1.0	18.5	3.8
Plasma membrane	0.9	36.9		31.2	6.4	12.9	tr	tr		11.1	
Microsomes											
Smooth endoplasmic reticulum	0.4	28.6	3.1	26.5	10.6	14.9			1.4	14.0	0.7
Rough endoplasmic reticulum	0.5	22.7	3.6	22.0	11.1	16.1			1.8	19.7	2.9
Golgi	0.9	34.7		22.5	8.7	18.1	tr	tr		14.5	

Note: tr = trace.

Source: The data were taken from White (1973). The fatty acid composition varies depending on the nature of the phospholipid. The reader interested in the fatty acid composition of different membrane phospholipids is referred to the review paper by White (1973).

one to four double bonds. Phospholipids usually have one unsaturated acyl chain in the *sn*-2-position. Any single membrane may contain more than 100 individual lipid species. Whether or not the lipid heterogeneity is an absolute requirement so that the membrane can fulfill its functional role is still unclear. Lipids actively participate in a number of specialized functions:

1. Membrane lipids are supposed to form liquid crystalline bilayers in which the proteins are embedded and can function adequately. Some lipids may be required for morphological reasons. They may stabilize particular regions of the bilayer, e.g., bilayers of high curvature or bilayers involved in the formation of a tight junction with another membrane.
2. Specific lipids may accumulate in the boundary layer of enzymes and may be required to optimize enzymatic activities.
3. Lipids may participate in metabolism and represent intermediate compounds. Other lipids may be involved in regulatory functions. A notable example is phosphatidylinositol and its phosphorylated analogs.
4. There is an ever-increasing body of evidence that lipids are involved in a number of specialized functions. Glycolipids have been implicated in such important functions such as cell recognition, cell differentiation, and cell growth. Isoprenoid lipids play specialized roles as vitamins; some quinones, such as ubiquinones and menaquinones, are involved in electron transport and energy transduction.

1.4 X-RAY CRYSTALLOGRAPHY OF LIPIDS

The crystal structures described here are of interest because they represent minimum free energy conformations that may be relevant to the structures of less ordered lipid aggregates, such as the liquid crystalline state. The crystal structures serve as a valuable reference in the discussion of motionally averaged structures and conformations, as, for instance, those encountered in liquid crystalline phases. They have proved most useful in the interpretation of lower-resolution studies, such as X-ray and neutron powder diffraction, magnetic resonance (NMR, ESR), IR, Raman, and fluorescence spectroscopic studies. In addition to the chemical structure of lipids discussed in Section 1.2, the knowledge of the three-dimensional structure of lipids at or near atomic resolution (~0.1 nm) is required for a complete understanding of their functional role, whether biological or technological. It is already obvious that the knowledge of the crystal structure of lipids has greatly advanced our understanding of lipid conformation, dynamics, and interactions in natural environments, e.g., in biological membranes and serum lipoproteins. The primary technique for deriving the three-dimensional structure in the solid state is X-ray crystallography. Provided that suitable single crystals are available, it allows the determination of the atoms in space with high precision, yielding bond lengths, angles, and torsion angles. Furthermore, details of the molecular packing arrangement and intermolecular distances are obtained.

At the early stages, lipid crystallography was hampered by the difficulties in synthesizing pure lipids and isolating them from natural sources. Related to this problem was the difficulty of growing single crystals suitable for X-ray crystallographic analysis. Lipids have a tendency to form very thin, platelike crystals with limited growth in the direction of the lipid long molecular axis. It was not until about 1960 that the related problems of lipid purity and crystal growth were overcome, at least partially, and a number of crystal structures of simple lipids, such as alkanes, alcohols, fatty acids, and soaps, were solved. The reader interested in these lipid structures is referred to an excellent review by G. Shipley (Shipley, 1986). It was in 1974 that the first crystal structure of a phospholipid, phosphatidylethanolamine, was reported (Hitchcock et al., 1974) and in 1977 the first and, so far, only crystal structure of a glycolipid, a cerebroside (Pascher and Sundell, 1977). The reason for the scarcity of single-crystal structures of complex lipids is primarily due to difficulties in producing suitable crystals.

Single-crystal structures of membrane lipids, particularly the more complex ones, are briefly discussed in this section. The emphasis will be on general principles governing the molecular packing arrangement and the most stable conformation (of minimum free energy) of lipids in bilayers. It should be understood that the crystallographic data vary widely in precision, due mainly to differences in crystal quality.

1.4.1 CRYSTAL STRUCTURES OF SIMPLE LIPIDS

From early X-ray crystallographic studies on simple lipids, including hydrocarbons, alcohols, fatty acids, and soaps, some key features emerged concerning the hydrocarbon chain conformation and packing. The conformation of the –C-C- bonds of the hydrocarbon (polymethylene) chain is, with very few exceptions, all-antiperiplanar.* The C-C bond lengths vary between 0.149 nm (1.49 Å) and 0.152 nm (1.52 Å) and the C-C-C bond angles between 114° and 116° (Shipley, 1986). A general principle evolving from these early studies is that lipid molecules pack in a layer arrangement. This has been confirmed by more recent single-crystal studies on complex lipids. The predominant packing arrangement is the bilayer; see, for instance, the single-crystal structures of glycerophospholipids. Less frequently, other layer arrangements are observed, e.g., lipid molecules may pack as a single layer, or *monolayer*.

The fundamental principle governing the molecular packing is maximization of the interaction energy between various portions of the amphipathic lipid molecule. This results in a packing arrangement in which the hydrocarbon chains form layers bordered by polar group layers. The long hydrocarbon chains are packed in a regular and periodic structure within the real unit cell. The repeating unit of the whole structure is defined by the unit cell, whereas the internal periodic structure of the hydrocarbon chain packing is defined by the subcell. The three axes of the subcell are a_s, b_s, c_s: the translation relating equivalent positions within a single hydrocarbon chain is given by c_s, the lateral translations relating adjacent hydrocarbon chains by a_s and b_s. Simple subcells with triclinic (T), orthorhombic (O), monoclinic (M), and hexagonal (H) symmetry are found in the crystal structure of simple lipids. It was recognized that the C-C-C zigzag plane of neighboring chains are in either a mutually parallel or perpendicular orientation, indicated by the symbols | | and ⊥, respectively. For instance, the symbols O and $O_⊥$ indicate orthorhombic subcells with the plane of the hydrocarbon chains being in parallel and perpendicular orientation, respectively. More recently, it has been shown that more complicated subcells are required to describe the mode of hydrocarbon chain packing in complex membrane lipids, notably phospho- and sphingolipids. Because these chain packing modes contain features of several simple subcells, they are termed *hybrid subcells*. A valuable review addressing simple and hybrid subcells was published by Abrahamsson et al. (1978).

Simple lipids, such as alcohols, fatty acids, methyl and ethyl esters of fatty acids, and soaps, crystallize predominantly as bilayers (Shipley, 1986). They exhibit polymorphism, which is based primarily on different modes of hydrocarbon chain packing, i.e., different subcells. Other parameters in which these crystal structures differ are the tilt angle** of the hydrocarbon chains and the hydrogen bonding within the polar group layer.

Of all the lipids, fatty acids have been studied most extensively by X-ray crystallography. A characteristic feature of long-chain fatty acids is a complex polymorphism. In most crystals of fatty

---

* The terminology of Klyne and Prelog (1960) is used here to define torsion angle ranges (cf. Table 1.6). Unfortunately, the nomenclature used in the literature is not uniform. Frequently, the shorter term *trans* is used instead of *antiperiplanar*. This may be confusing because *trans* is also used as a configurational term for defining the geometric isomerism of double bonds and ring systems (Hauser et al., 1981). We recommend the Klyne and Prelog terminology for defining torsion angle ranges and replacing the old *cis–trans* system with the (E,Z) system for defining the stereochemistry of double bonds and ring systems.

** The angle of tilt of the hydrocarbon chains is defined either with respect to the basal or polar group plane; alternatively, it is the angle between the hydrocarbon chain axis and the layer normal (Hauser et al., 1981).

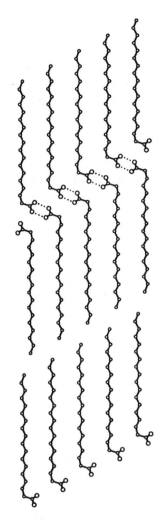

FIGURE 1.17 Molecular packing in the single-crystal structure of *n*-octadecanoic (stearic) acid viewed along the b-axis. The dotted lines indicate hydrogen bonds that link fatty acid molecules of opposing bilayers. In this way, fatty acid dimmers are formed. (Adapted from Goto M and Asada E: *Bull. Chem. Soc. Jpn.* 1978, 51: 2456.)

acids, hydrogen-bonded dimers occur as a structural element. As a representative structure of a fatty acid, the crystal structure of *n*-octadecanoic (stearic) acid is shown in Figure 1.17. This fatty acid crystallizes in a monoclinic unit cell, space group $P2_1/a$, containing four molecules in the unit cell (Goto and Asada, 1978). The hydrocarbon chain packing mode is defined by the orthorhombic $O_\perp$ subcell; note that the axes of all hydrocarbon chains are aligned in parallel. Two fatty acid molecules form a centrosymmetrical dimer, held together by two hydrogen bonds between the oxygen atoms of opposing carboxyl groups. The hydrocarbon chains do not have a complete all-antiperiplanar conformation; the $C(2)-C(3)$ bond is in the synclinal conformation.

1.4.2 CRYSTAL STRUCTURES OF CHOLESTEROL

The crystal structures of cholesterol and cholesterol monohydrate are of considerable interest. Crystalline cholesterol monohydrate is the main constituent of gallstones. It crystallizes in large single crystals of ~1 mm in length inside the gall bladder and gall bladder duct. Microcrystals of cholesterol monohydrate also occur in atherosclerotic lesions. The precision of the single-crystal structure determination of cholesterol and its derivatives varies widely, not only from structure to structure but even within the same structure. Within the dimension of one cholesterol molecule, the precision is best for the atoms of the fused ring (tetracyclic) system, which represents a rigidly bonded framework: for room-temperature studies, bond length and angles of the ring system are determined with standard deviations, typically $2 \cdot 10^{-3}$ nm and 1.5°. For low-temperature studies at −150°C, the estimated standard deviations for bond lengths and angles are $4 \cdot 10^{-4}$ nm and 0.2°; see, for instance, the single-crystal structure of cholesteryl acetate determined at −150°C, in which all the hydrogen atoms could be located experimentally (Sawzik and Craven, 1979).

In contrast, the cholesteryl side chain attached to C(17) (cf. Figure 1.16) has considerable thermal motion at room temperature, including large-amplitude torsional vibrations, librations, and translations. For instance, the isopropyl group in single crystals of cholesteryl myristate is highly disordered and virtually in the liquid state (Craven and DeTitta, 1976). The apparent root mean square amplitudes of its thermal vibrations range from 0.08 to 0.13 nm at room temperature. In comparison, the carbon atoms of the C(17) cholesteryl side chain are clearly resolved in the single-crystal structure of cholesteryl acetate at −150°C, with root mean square amplitudes of thermal vibrations ranging between 0.012 and 0.025 nm (Sawzik and Craven, 1979). Both cholesterol and its monohydrate crystallize in a rare triclinic space group P1, with the lowest possible symmetry. In both crystals, the unit cell contains eight molecules of cholesterol which are not related by true crystallographic symmetry operations (Shieh et al., 1977; Craven, 1976, 1979, 1986).

Figure 1.18A shows the single-crystal structure of cholesterol monohydrate. The crystal consists of cholesterol bilayers of thickness D = 3.39 nm. The ring systems and the molecular long axes of all cholesterol molecules are nearly parallel. The molecular long axis has a tilt angle of about 17° with respect to the (001) plane, which is parallel to the plane containing the hydrogen-bonded oxygen atoms. The eight molecules in the unit cell are related by the following sublattice translations, which are as rigorous as those of a true crystal lattice:

- The molecules A and B, and also the molecules E and F, are related by translations b/2.
- The molecules C and D, as well as the molecules G and H, are related by translations a/2.
- The subcell repeats are then (a/2,b) and (a,b/2) for the top and bottom half of the cholesterol bilayer, respectively (Figure 1.18A).

The crystal structures of cholesterol monohydrate and cholesterol are quite different regarding the molecular packing and the nature of pseudosymmetry relating the eight molecules in the unit cell (for details see Shieh et al., 1977). The cholesteryl side chain at C(17) of cholesterol monohydrate is disordered due to thermal motion. The root mean square amplitudes of vibration range from 0.025 to 0.04 nm. As evident from Figure 1.18B, there are marked differences in conformation in this side chain between the four pairs of molecules. The hydrogen bonding within the plane containing the OH-groups of cholesterol and the water molecules is shown in Figure 1.18C. The C(3) hydroxyl groups and water molecules are hydrogen-bonded to form a pleated sheet. Each oxygen atom forms three hydrogen bonds, with O-O distances ranging from 0.268 to 0.301 nm. The average distance of 0.286 nm and the high values for the thermal vibration of the oxygen atoms suggest that the hydrogen-bonding network is weak and that the water molecules have an important space-filling role in stabilizing the crystal structure.

1.4.3 CRYSTAL STRUCTURES OF ACYLGLYCEROLS

Three single-crystal structures of monoacylglycerols have been reported by Larsson (Larsson, 1964a, 1964b, 1966). In all three crystal structures, the molecules pack in bilayers. As a representative structure, the crystal structure of L-1-(11-bromo-undecanoyl)glycerol (Larsson, 1966) is shown in Figure 1.19. This compound crystallizes in the monoclinic space group $P2_1$, with four molecules in the unit cell. The molecules pack in a typical bilayer, with the hydrocarbon chains

FIGURE 1.18 (A) Molecular packing in the single-crystal structure of cholesterol monohydrate viewed along an axis that is close to the a-axis and produces a "best plane" projection. Superposing the four molecules at the right (H, D, F, E) onto the left gives the complete structure. Molecules in each pair (A, B) and (E, F) are related by a translation b/2, molecules in pairs (C, D) and (G, H) by a/2. The hydrogen-bonded oxygen atoms are near the plane z = c/2. Full circles are oxygen atoms of water molecules. (B) Conformation of the cholesteryl side chain attached to C(17) in single crystals of cholesterol monohydrate. The cholesterol ring systems (incompletely shown) are in the same orientation. (C) Hydrogen bonding (viewed along the c-axis) in the ab plane in single crystals of cholesterol monohydrate. Full circles represent the oxygen atoms of cholesterol molecules; open circles are those of water molecules. (Adapted from Craven BM: *Nature* 1976, 260:727 and Craven BM: *Acta Crystallogr. B* 1979, 35:1123.)

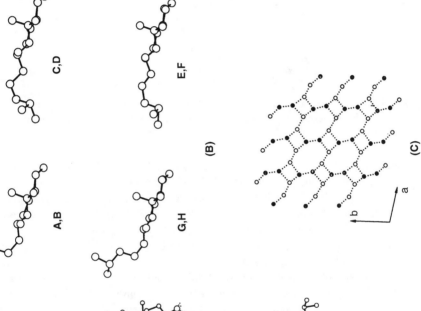

(B)

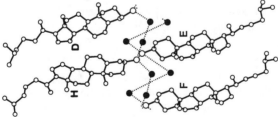

(C)

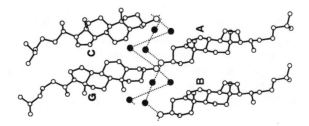

(A)

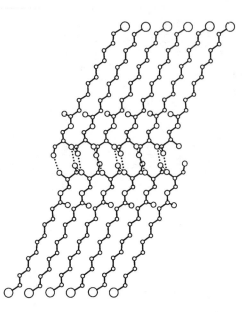

FIGURE 1.19 Molecular packing in the single-crystal structure of L-1-(11-bromoundecanoyl) glycerol viewed along the b-axis. The dotted lines represent hydrogen bonds. (Adapted from Larsson K: *Acta Crystallogr.* 1966, 21:267.)

tilted at an angle of 26.7° with respect to the bilayer normal. The hydrocarbon chains are packed according to the common orthorhombic subcell $O_\perp$. In the polar group region, the molecules are linked through hydrogen bonds. Both free hydroxyl groups are involved in hydrogen bonding. However, only one free electron pair of each hydroxyl is hydrogen-bonded, so there are two hydrogen bonds per oxygen atom (Figure 1.19). In this way, each molecule is linked to adjacent molecules within one bilayer and with one molecule in the opposite bilayer forming an infinite, two-dimensional network of hydrogen bonds.

The crystal structure of the diacylglycerol, 2,3-dilauroyl-D-glycerol, is shown in Figure 1.20 (Pascher et al., 1981a). It crystallizes in the monoclinic space group $P2_1$, with two molecules per unit cell. The molecules are packed in a typical bilayer structure. The hydrocarbon chains are parallel to the arc plane but are tilted by an angle of 26.5° with respect to the bilayer normal. The tilt arises because the molecular area in the layer plane is S = 0.414 nm² and exceeds the sum of the cross-sectional area of the two hydrocarbon chains nΣ = 0.37 nm² (see discussion below). The hydrocarbon chains pack according to the $O_\perp$ subcell, i.e., the C-C-C zigzag planes of the two hydrocarbon chains of each molecule are in a mutually perpendicular orientation. Adjacent molecules are linked by hydrogen bonds of length 0.28 nm between the free hydroxyl group O(11) and the carbonyl oxygen O(32) of the nearest neighbor in the b-direction. The atom numbering and notation for torsion angles of glycerophospholipids, according to Hauser et al. (1981), are illustrated in Figure 1.22. The definition of torsion angles and the notation for torsion angle ranges are given in Table 1.6. In contrast to the monoacylglycerol structure, there are no hydrogen bonds between adjacent bilayers.

A key feature of this crystal structure, which warrants a more detailed discussion, is the chain stacking, i.e., the parallel alignment of the two hydrocarbon chains. The diacylglycerol molecule may be regarded as a constituent of the more complex lipids, such as glycerophospholipids and glycoglycerolipids. The glycerol C(2)-C(3) bond is the central link between the two fatty acyl chains esterified to the glycerol oxygen atoms O(21) and O(31). The conformation of this bond, defined by torsion angles Θ_3 and Θ_4 (cf. Table 1.6), is therefore of interest and should be compared with that of the same bond in glycerophospholipids (cf. Section 1.4.4). For 2,3-dilauroyl-D-glycerol,

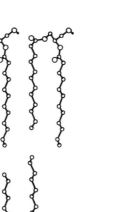
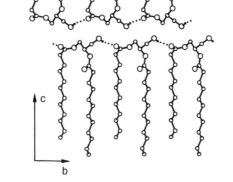

FIGURE 1.20 Molecular packing in the single-crystal structure of 2,3-dilauroyl-D-glycerol projected onto the bc-plane. The intermolecular hydrogen bonds between the hydroxyl group O(11) and the carbonyl oxygen atom O(32) are indicated by dotted lines. (Adapted from Pascher I et al.: *J. Mol. Biol.* 1981a, 153:791.)

these torsion angles are $\Theta_3 = 173°$ and $\Theta_4 = 54°$ and consistent with the predominant conformation of this bond found in glycerophospholipids. By comparison with the single-crystal structure of glycerophospholipids, it is seen that there is a preference for Θ_3/Θ_4 = antiperiplanar/synclinal (cf. Section 1.4.4). With Θ_4 = O(21)-C(2)-C(3)-O(31)±synclinal, the two ester oxygens O(21) and O(31) are synclinal substituents on glycerol C(2) and C(3), respectively. Apparently, the two acyl or alkyl chains attached to oxygen atoms in this kind of conformation can readily be aligned in parallel, i.e., give rise to parallel chain stacking. In 2,3-dilauroyl-D-glycerol, parallel chain stacking is achieved as follows: the β-fatty acyl chain (bonded to C(2)) extends at a right angle from the glycerol group. The γ-fatty acyl chain continues the C-C-C zigzag of the glycerol group, but at carbon atom C(31), the γ-fatty acyl chain makes a 72° bend and runs parallel to the -chain. Such a bend is accomplished by torsion angles γ_1 = C(2)-C(3)-O(31)-C(31) = 91° and γ_3 = O(31)-C(31)-C(32)-C(33) = 65°, both deviating markedly from the antiperiplanar zigzag conformation.

Perhaps the most striking feature of the crystal structure of 2,3-dilauroyl-D-glycerol is the almost coplanar orientation of the glycerol group, i.e., the glycerol group is oriented parallel to the plane of the bilayer (Figure 1.20). In this respect, the diacylglycerol molecule is quite different from glycerophospholipids; in the latter class of compounds, the orientation of the glycerol group is predominantly perpendicular to the bilayer plane (or parallel with respect to the bilayer normal).

The crystal structure of the triacylglycerol, (tridecanoyl)glycerol, is depicted in Figure 1.21. This triacylglycerol crystallizes in a triclinic space group P1, with two molecules in the unit cell (Jensen and Mabis, 1966). The (tridecanoyl)glycerol molecule has a turning fork conformation: the fatty acyl chains ester-linked to the C(1) and C(2) atoms of glycerol form an extended, almost

TABLE 1.6
Notation for Torsion Angles

Θ_1	O(11)-C(1)-C(2)-C(3)	β_1	C(1)-C(2)-O(21)-C(21)
Θ_2	O(11)-C(1)-C(2)-O(21)	β_2	C(2)-O(21)-C(21)-C(22)
Θ_3	C(1)-C(2)-C(3)-O(31)	β_3	O(21)-C(21)-C(22)-C(23)
Θ_4	O(21)-C(2)-C(3)-O(31)	β_4	C(21)-C(22)-C(23)-C(24)
		β_5	C(22)-C(23)-C(24)-C(25)
α_1	C(2)-C(1)-O(11)-P	γ_1	C(2)-C(3)-O-(31)-C(31)
α_2	C(1)-O(11)-P-O(12)	γ_2	C(3)-O(31)-C(31)-C(32)
α_3	O(11)-P-O(12)-C(11)	γ_3	O(31)-C(31)-C(32)-C(33)
α_4	P-O(12)-C(11)-C(12)	γ_4	C(31)-C(32)-C(33)-C(34)
α_5	O(12)-C(11)-C(12)-N	γ_5	C(32)-C(33)-C(34)-C(35)
α_6	C(11)-C(12)-N-C(13)		

For atom numbering, see Figure 1.22.

Torsion Angle Range	Notation
$0 \pm 30°$	Synperiplanar (sp)
$+60 \pm 30°$	+Synclinal (sc) or +gauche
$+120 \pm 30°$	+Anticlinal (ac)
$180 \pm 30°$	Antiperiplanar
$+240 \pm 30°$ or $-120 \pm 30°$	−Anticlinal
$+300 \pm 30°$ or $-60 \pm 30°$	−Synclinal

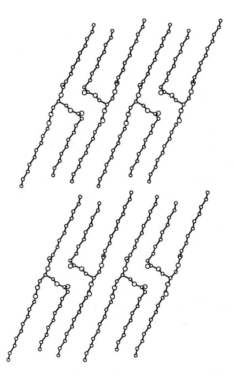

FIGURE 1.21 Molecular packing in the single-crystal structure of triclinic (tridecanoyl) glycerol projected onto the bc-plane. (Adapted from Jensen LH and Mabis AJ: *Acta Crystallogr.* 1966, 21:770.)

straight chain, whereas the fatty acyl chain linked to glycerol C(3) is oriented at right angles to the straight chain and folds over at the carboxyl-C-atom in such a way that it aligns parallel to the fatty acyl chain at glycerol C(1). As shown in Figure 1.21, two symmetry-related molecules form a pair, and these pairs are packed together laterally to produce a single-layer arrangement. The resulting layered structure consists of a hydrophobic region of hydrocarbon chains and a more polar region containing the glycerol and the ester groups.

1.4.4 CRYSTAL STRUCTURES OF GLYCEROPHOSPHOLIPIDS

Glycerophospholipids are main constituents of cell membranes and play a key role in both the structure and function of these membranes. They are also major components of plasma lipoproteins. As mentioned before, it was not until 1974 that the first single-crystal structure of a glycerophospholipid, 2,3-dilauroyl-DL-phosphatidylethanolamine, was solved by Hitchcock et al. (1974). The first few single-crystal structures of glycerophospholipids have been discussed in some detail in reviews by Hauser et al. (1981) and Shipley (1986). In discussing the single-crystal structures of glycerophospholipids, the contribution of Pascher and Sundell and their coworkers of the University of Gothenburg is outstanding: they take the credit for all but one of the single-crystal structures of glycerophospholipids published to date. Crystal data of glycerophospholipids are summarized in Table 1.7, including the crystallographic system, space group, number Z of molecules per unit cell, the unit cell parameters, the molecular area S, the hydrocarbon chain cross-sectional area Σ, the angle of tilt ϕ, and the hydrocarbon chain packing mode.

1.4.4.1 Molecular Packing

Most glycerophospholipids pack tail-to-tail, forming a typical bilayer structure (Figure 1.22 through Figure 1.24). The single crystal consists of stacks of bilayers. For instance, 2,3-dilauroyl-DL-phosphatidylethanolamine forms a bilayer structure (Figure 1.22) of thickness of 4.77 nm with the hydrocarbon chains precisely parallel to the bilayer normal (or perpendicular to the plane of the bilayer). The angle of tilt of the hydrocarbon chain axis with respect to the bilayer normal is given by $\phi = \cos^{-1} n\Sigma/S$ (see legends of Table 1.7), i.e., $\cos \phi$ is given by the ratio of the cross-sectional areas $n\Sigma/S$. As can be seen from Table 1.7, this relation is strictly obeyed for all single-crystal structures listed. Accordingly, for 1 n/S > 0.5, ϕ varies between 0 and 60°. In the case of the crystal structure of phosphatidylethanolamine, $n\Sigma$ equals precisely the molecular area S, hence no chain tilt is observed ($\phi = 0°$; Table 1.7). In contrast, an extreme chain tilt of 57.5° is present in the crystal structure of 3-palmitoyl-DL-glycero-1-phosphoethanolamine (Figure 1.23), where $n\Sigma/S = 0.537$. Apparently, the marked mismatch between S and $n\Sigma$ leads to this extreme chain tilt. At S 2$n\Sigma$, i.e., $n\Sigma/S$ 0.5, interdigitation occurs, as is indeed the case for lauroylpropanediol-1-phosphocholine, a lysophosphatidylcholine analog (Figure 1.24). The lipid molecules pack laterally head-to-tail, resulting in a fully interdigitated hydrocarbon chain layer confined on both sides by polar group layers (Figure 1.24). The total thickness of this interdigitated, layered structure is 2.45 nm, which is about half the thickness of the bilayer formed by phosphatidylethanolamine. By chain interdigitation, two hydrocarbon chains are accommodated per polar group (Figure 1.24). If S > 2$n\Sigma$, as is true for the lysophosphatidylcholine analogue (Figure 1.24), the interdigitated hydrocarbon chains will attain a tilt given by $\cos \phi = 2n\Sigma/S$.

The molecular packing of the crystal structure of 2,3-dimyristoyl-D-glycero-1-phosphocholine is shown in Figure 1.25. It differs from a normal bilayer arrangement in that two neighboring molecules are mutually displaced in the direction of the bilayer normal by 0.25 nm or one zigzag unit of their hydrocarbon chains. Such an offset with respect to the bilayer normal, together with an inclined orientation of the phosphocholine group, produces a relatively small molecular area S = 0.389 nm^2 comparable to that of phosphatidylethanolamine. Note that S ~0.50 nm^2 if the polar group is oriented parallel to the layer plane.

TABLE 1.7
Glycerophospholipids: Crystallographic Data

Glycerophospholipid[a]	Stereo Isomer	System	Space Group	S (Å²)	NΣ (Å²)	a (Å)	b (Å)	c (Å)	α (°)	β (°)	γ (°)	φ (°)	Volume (Å³)	Density (g/cm³)	Chain Packing	Ref.[b]
2,3-dilauroyl-DL-PE	DL	monoclinic	$P2_1/c$	38.6	38.6 n = 2	47.73	7.773	9.953	90	92.0	90	0	3690.4	1.19	HS1	1
2,3-dimyristoyl-D-PC	D	monoclinic	$P2_1$			8.72	8.92	55.4				(0)				
3-palmitoyl-DL-glycero-phosphoethanolamine	DL	monoclinic	$P2_1/a$	38.9	38.0 n = 2	7.66	9.08	37.08	90	97.4	90	12 (12.3)	4273.3		HS	2
3-lauroylpropanediol-1-phosphocholine monohydrate	—	monoclinic	$P2_1/c$			24.82	9.53	10.94								
2,3-dilauroyl-DL-glycero-1-phospho-N,N-dimethylethanolamine	DL	triclinic	$P\bar{1}$			5.64	8.20	39.86								
2,3-dimyristoyl-D-glycero-1-phosphate monosodium salt	D	monoclinic	$P2_1$			5.44	7.95	43.98								
3-lauroyl-DL-glycero-1-phosphate dihydrate (disodium salt)	DL	triclinic	$P\bar{1}$	34.8	18.7 n = 1	7.74	5.54	32.87	90	90.2	90	57.5 (57.5)	2579	1.168	T∥	3
3-hexadecylpropanediol-1-phosphoric acid monohydrate	—	triclinic	$P\bar{1}$			4.746	5.720	44.36								
3-hexadecylpropanediol-1-phosphoric acid monohydrate	—	monoclinic	$P2_1$			4.752	5.722	88.72								
2,3-dimyristoyl-D-glycero-1-phospho-DL-glycerol	DL	monoclinic	$P2_1$			10.370	8.482	45.52								
-D-galactosyl-N-(2-D-hydroxyoctadecanoyl)-D-dihydrosphingosine	D	monoclinic	$P2_1$			11.202	9.262	46.46								

90	99.7	99.7	52.1	19.7 (n = 1)	41 (40.9)	2552	1.149	$M_{\parallel}$	4
94.5	90.1	101.9	45.2	37.9 (n = 2)	33 (33)	1796	1.124	$O_{\perp}$	5
90	114.2	90	43.3	37.2 (n = 2)	31 (31)	1735	1.175	$T_{\parallel}$	6
92.6	99.2	128.3	33.6	19.3 (n = 1)	55 (54.9)	1071.6	1.347	$T_{\parallel}$	7
91.0	101.5	100.5	26.69	18.2 (n = 1)	47.1 (47.0)	1159	1.142	$T_{\parallel}$	8
90	90.0	100.8	26.71	18.5 (n = 1)	46.1 (46.2)	2370	1.117	$T_{\parallel}$	8
90	95.22	90	44.0	38.4 (n = 2)	29.0 (2902)	3987	1.146	hybrid subcell[c]	9
90	99.0	90	51.9	38.2 (n = 2)	41 (42.6)	4761	1.073	HS2	10

Note: S = molecular area at the layer interface; Σ = cross-sectional area of the hydrocarbon chains perpendicular to the chain axis; n = number of the hydrocarbon chains; φ = angle of tilt between the hydrocarbon chain axis and the bilayer normal. For a pictorial representation of these parameters the reader is referred to Figure 4 to the review by Hauser et al., 1981. The tilt angle φ is related to S and Σ by cosφ = n Σ/S. The f-values are experimentally determined, the values in parenthesis are calculated from S and n Σ.

[a] PE stands for Phosphatidylethanolamine, PC for Phosphatidylcholine.

[b] References: 1. Hitchcock et al., 1974; 2. Pearson and Pascher, 1979; 3. Pascher et al., 1981b; 4. Hauser et al., 1981a; 5. Pascher and Sundell, 1986; 6. Harlos et al., 1984; 7. Pascher and Sundell, 1985; 8. Pascher et al., 1984; 9. Pascher et al., 1987; 10. Pascher and Sundell, 1977.

[c] The hybrid subcell combines features of both $T_{\parallel}$ and $O_{\perp}$ chain packing modes.

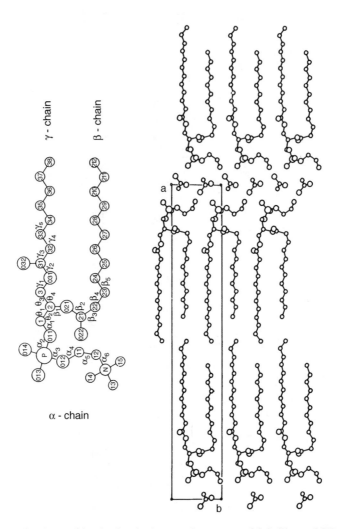

FIGURE 1.22 The molecular packing in the single-crystal structure of 2,3-dilauroyl-DL-glycerol-phospho-rylethanolamine acetic acid projected onto the ab-plane. The rectangle shown represents the a,b axes of the unit cell that contains four molecules. The characteristic phospholipid bilayer arrangement is clearly empha-sized in this projection. A single molecule of the crystal structure is shown expanded on the left: besides the molecular conformation, it gives the atom numbering and notation for torsion angles for glycerophospholipids. (Adapted from Hitchcock PB et al.: *Proc. Natl. Acad. Sci. USA* 1974, 71:3636.)

The molecular packing of the crystal structure of 2,3-dilauroyl-DL-glycero-1-phospho-*N*,*N*-dimethylethanolamine (Figure 1.26) is interesting for two reasons:

- The molecules pack in a typical bilayer. However, one monolayer of the bilayer is made up of the natural D-enantiomer, the other half of the L-enantiomer.
- In contrast to other single-crystal structures of phosphatidylcholines and phosphatidyl-ethanolamines, the headgroup of this molecule extends perpendicular to the layer plane and interacts by interdigitation with the headgroups of the opposition bilayer (Figure 1.26) (cf. Pascher and Sundell, 1986).

The molecular packing of the monosodium salt of 2,3-dimyristoyl-D-glycerophosphate is shown in Figure 1.27. The molecules pack in bilayers which are connected at the interface. The connection is provided by alternate phosphate groups that belong to the upper and lower bilayer and form a

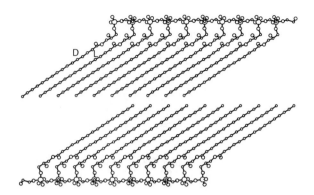

FIGURE 1.23 Molecular packing in the single-crystal structure of 3-palmitoyl-DL-glycero-1-phosphoetha-nolamine, a lysophosphatidylethanolamine, viewed along the b-axis. In this projection, the D and L enantiomers appear to alternate in the a direction. However, the D and L molecules are displaced by half a unit cell axis (4.54 Å) along the b direction. (Adapted from Pascher I et al.: *J. Mol. Biol.* 1981b, 153:807.)

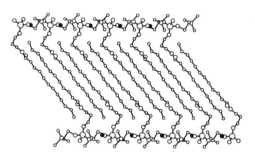

FIGURE 1.24 Molecular packing in the single-crystal structure of 3-lauroylpropanediol-l-phosphocholine monohydrate projected onto the ac-plane. The shaded spheres represent water molecules of hydration. (Adapted from Hauser H et al.: *J. Mol. Biol.* 1980b, 137:249.)

common hydrogen bonded phosphate ribbon extending in the b-direction. In this way, two sodium phosphate groups of opposing phosphatidic acid molecules contribute to the cross-sectional area $S = 0.433$ nm^2 of one phosphatidic acid molecule (Figure 1.27). Because the two hydrocarbon chains of one phosphatidic acid molecule have a total cross-sectional area of $\Sigma = 0.392$ nm^2, they adjust to the molecular area $S = 0.433$ nm^2 by attaining a chain tilt of 31° (Table 1.7). A striking feature of the hydrocarbon chain packing is the difference in orientation of the hydrocarbon chain axes: the parallel alignment of the hydrocarbon chains, i.e., the hydrocarbon chain stacking, is warranted within one molecule as well as within one monolayer of the bilayer. The orientation of the hydrocarbon chain axes is, however, different in the two halves (monolayers) of each bilayer (Figure 1.27).

In the single-crystal structure of sodium 2,3-dimyristoyl-D-glycero-1-phospho-DL-glycerol (Pascher et al., 1987) the lipid molecules are also packed in a typical bilayer (Figure 1.28). Two independent molecules A and B form the asymmetric unit, and two such A–B pairs constitute the unit cell. The two molecules A and B in each pair are diastereomers: their diacylglycerol moiety has the natural D-configuration; the two molecules differ in the configuration of the glycerol headgroup. This group has the natural L-configuration in molecule A; it has the D-configuration in molecule B. The phosphoglycerol headgroup is oriented parallel to the bilayer plane, forming a layer of negatively charged phosphoglycerol groups. A layer of positively charged Na+ ions is sandwiched by two opposing negatively charged phosphoglycerol layers. The cross-sectional area of the phosphoglycerol headgroup $S = 0.44$ nm^2, and the two hydrocarbon chains with a cross-sectional area $2\Sigma = 0.384$ nm^2 adjust to the headgroup area by adopting a tilt of $\phi = 29°$ (cf. Table

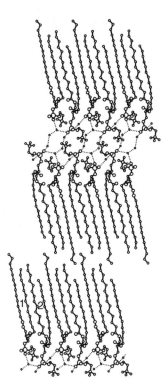

FIGURE 1.25 Molecular packing in the single-crystal structure of 2,3-dimyristoyl-D-glycero-1-phosphocholine dihydrate projected onto the ac-plane. The position of the water molecules is indicated by small circles, and the hydrogen bond system is given by dotted lines. (Adapted from Pearson RH and Pascher I: *Nature* 1979, 281:499.)

1.7). The hydrocarbon chains are tilted with respect to both the a-axis and b-axis (Figure 1.28). Furthermore, the hydrocarbon chain axes exhibit an opposite tilt in the two halves of the bilayer when viewed along the a-axis (Figure 1.28a). A more detailed discussion of this structure is provided in the original publication (Pascher et al., 1987).

1.4.4.2 Molecular Conformation

In the discussion of molecular conformations of glycerophospholipids present in single-crystal structures, the question of a preferred conformation is of prime interest. In order to shed light on this question, torsion angles of the molecule(s) present in the single-crystal structures are summarized in Table 1.8. Inspection of this table reveals that there is a preferred conformation both in the headgroup and in the diacylglycerol part glycerophospholipids.

1.4.4.2.1 Conformation of the Headgroup or α-Chain

The orientation of the headgroup is preferably parallel to the bilayer plane. This is true for the phosphoethanolamine (Figure 1.22 and Figure 1.23), phosphocholine (Figure 1.24 and Figure 1.25), and phosphoglycerol headgroup (Figure 1.28). For instance, the inclination of the P–N vector is 15° in dilauroylphosphatidylethanolamine, 17° and 27° in the two molecules present in the single-crystal structure of dimyristoylphosphatidylcholine, and 7° in the lysophosphatidylcholine analog. The inclinations of the vectors drawn from the center of gravity of the two unesterified phosphate oxygens to the N-atom are even smaller: 8°, 9°, 18°, and −3°, respectively. The parallel orientation of the zwitterionic headgroup of phosphocholine and phosphoethanolamine is apparently the most stable arrangement: the positive and negative charges come to lie in a plane that is nearly layer-

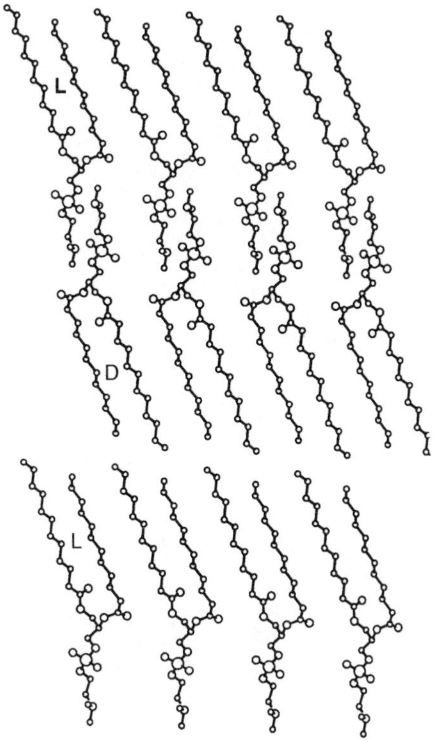

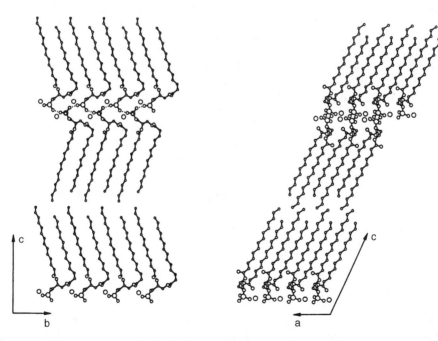

FIGURE 1.26 Molecular packing in the single-crystal structure of 2,3-dilauroyl-DL-glycero-l-phospho-*N, N*-dimethylethanolamine viewed along the a-axis. Enantiomers D and L pack in separate layers. (Adapted from Pascher I and Sundell S: *Biochem Biophys. Acta* 1986, 855:68.)

FIGURE 1.27 Molecular packing in the single-crystal structure of monosodium salt of 2,3-dimyristoyl-D-glycero-l-phosphate viewed along the a-axis (left-hand side) and along the b-axis (right-hand side). The dotted lines represent hydrogen bonds between phosphate groups that belong alternately to the upper and lower bilayer. The hydrogen-bonded phosphate ribbon thus formed extends along the b-axis. (Adapted from Harlos K et al.: *Chem. Phys. Lipids* 1984, 34:115.)

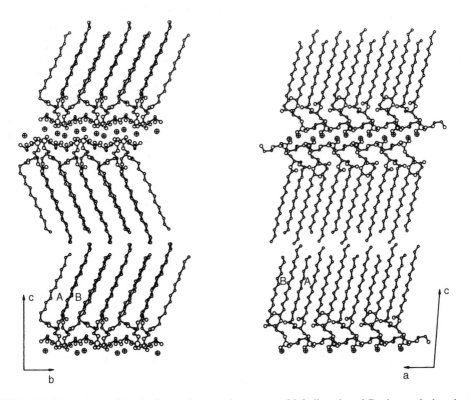

FIGURE 1.28 Molecular packing in the single-crystal structure of 2,3-dimyristoyl-D-glycero-1-phospho-DL-glycerol viewed along the a-axis (left) and along the b-axis (right). The position of the sodium ions is marked by a +. (Adapted from Pascher I et al.: *Biochem. Biophys. Acta* 1984, 896:77.)

parallel and electrostatically neutral. The exception is the headgroup of 2,3-dilauroyl-DL-glycero-1-phospho-*N,N*-dimethylethanolamine (Figure 1.26), which has a layer-perpendicular orientation. Such an orientation leads to the energetically unfavorable charge separation in the direction of the layer normal. The layer-perpendicular orientation in the crystal structure of Figure 1.26 is stabilized by intermolecular interdigitation of the phosphoethanolamine headgroups. In this way, two layer-parallel planes are generated, 0.4 nm apart, each one containing an equal number of positive and negative charges, thus preserving electroneutrality in each of the two planes.

The preferred conformation of the zwitterionic headgroup or α-chain of phosphocholine and phosphoethanolamine is characterized by the following sequence of torsion angles α (cf. Table 1.8):

- Torsion angle α_1 is always ap.
- Torsion angles α_2/α_3 are correlated and are either +sc/+sc or −sc/−sc.

As pointed out before (Hauser et al., 1981), other combinations are unfavorable for steric reasons, because they lead to collisions of the hydrogen atoms at C(1) and C(11) and/or to lone-pair repulsion of the phosphate oxygens. The favored +sc/+sc or −sc/-sc sequence of Θ_2/Θ_3 produces a 90° bend at the phosphorus atom so that the phosphocholine or phosphoethanolamine group is oriented at right angles with respect to the diacylglycerol part (Figure 1.22). Torsion angle α_4 is variable within ±ac, and torsion angle α_5 is always ±sc. The latter is determined by intramolecular, electrostatic interaction between the positively charged ammonium nitrogen and the negatively charged phosphate group.

Although there is a preferred conformation in the headgroup (α-chain), the molecules listed in Table 1.8 vary widely in their orientation of the headgroup with respect to the diacylglycerol moiety.

TABLE 1.8
Torsion Angles Derived From Single-Crystal Structures of Glycerophospholipids and a Cerebroside

Glycerophospholipid		α_1	α_2	α_3	α_4	α_5	Θ_1	Θ_2	Θ_3	Θ_4	Rotamer
2,3-dilauroyl-DL-PE		-154	58	66	106	67	-52	65	-172	69	A_γ
2,3-dimyristoyl-D-PC	Molecule 1	163	62	68	143	-64	58	177	-178	63	A_γ
	Molecule 2	177	-74	-47	-150	54	168	-82	166	51	A_γ
3-palmitoyl-DL-glycero-1-phosphoethanolamine		-154	58	66	106	67	-52	65	-172	69	A_γ
3-lauroylpropanediol-1-phosphocholine monohydrate		-162	-86	45	-129	-84	-28	—	-78	—	B_γ
2,3-dilauroyl-DL-glycero-1-phospho-N,N-dimethylethanolamine		179	65	54	144	-96	176	-66	56	-60	B_γ
2,3-dimyristol-D-glycero-1-phosphate monosodium salt		153	—	—	—	—	-54	62	-178	61	A_β
3-lauroyl-DL-glycero-1-phsphate dihydrate (disodium salt)		-125	—	—	—	—	173	-53	-176	63	A_γ
3-hexadecylpropanediol-1-phosphoric acid monohydrate	Triclinic	164	—	—	—	—	180	—	180	—	A_γ
	Monoclinic A	158	—	—	—	—	180	—	180	—	A_γ
	Monoclinic B	160	—	—	—	—	180	—	180	—	A_γ
2,3-dimyristoyl-D-glycero-1-phospho-DL-glycerol	Molecule A	-146	-76	-86	143	180	151	-78	64	-63	B_β
	Molecule B	116	58	78	-147	-173	71	179	45	-58	B_γ
D-galactosyl-N-(2-D-hydroxyoctadecanoyl)-D-dihydrosphingosine	Molecule A	—	—	—	—	—	-75	48	-176	61	A
	Molecule B	—	—	—	—	—	-57	63	-173	66	A

This variation in headgroup orientation arises from different conformations about the C(1)-C(2) glycerol bond defined by the torsion angles Θ_1 and Θ_2. As is evident from Table 1.8, all three staggered and nearly eclipsed conformations about this bond are possible. In dilauroylphosphatidylethanolmine (Figure 1.22) and dimyristoylphosphatidylcholine (molecule 2, Figure 1.25) the headgroup is oriented such that it is close to its diacylglycerol part; in dimyristoylphosphatidylcholine (molecule 1, Figure 1.25), the lysophosphatidylcholine analog (Figure 1.24); and in dimyristoylphosphatidylglyercol (Figure 1.28), the headgroup points away from the diacylglycerol part. The variability in torsion angle Θ_1 indicates that there is no preferred conformation about the C(1)-C(2) glycerol bond. There is a corollary in the liquid crystalline state of glycerophospholipids: there is no rotational restriction about C(1)-C(2) bond, which has been shown to act as an axis of rotation. It is very likely that the orientation of this bond is almost parallel to the bilayer normal in the liquid crystalline state. Rotation about such a C-C bond makes the headgroup (with fixed torsion angles α) rotate within a cone. Under these conditions, the orientation of the P–N vector with respect to the layer normal remains unchanged.

1.4.4.2.2 Conformation of the Diacylglycerol Moiety

As mentioned earlier, the parallel chain stacking of diaclyglycerophospholipids is determined by the conformation of the C(2)-C(3) glycerol bond. Inspection of Table 1.8 shows that the single-crystal structures can be divided into two classes:

- The first class, denoted by A, is characterized by torsion angles Θ_3/Θ_4 = ap/sc consistent with rotamer A (Figure 1.29).
- The second class, denoted by B, is characterized by torsion angles Θ_3/Θ_4 = +sc/-sc consistent with rotamer B (Figure 1.29).

There are more single-crystal structures in class I with Θ_3/Θ_4 = ap/sc (Table 1.8). Torsion angle Θ_3 is ap, ranging from 166° to −172°, and torsion angle Θ_4 is +sc, ranging from 51° to 69° (see Table 1.8). There are three single-crystal structures belonging to class II with Θ_3 = +sc (range 45° to 78°) and Θ_4 = -sc (range −58° to −63°). In both classes of single-crystal structures, Θ_4 is either +sc or −sc. This torsion angle range apparently facilitates the parallel alignment of the two hydrocarbon chains. In contrast, the conformation with Θ_4 = ap, where the two oxygen atoms on C(2)

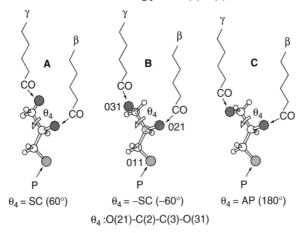

Rotamers about the glycerol C(2)–C(3) bond

θ_4 = SC (60°) θ_4 = −SC (−60°) θ_4 = AP (180°)

θ_4 :O(21)-C(2)-C(3)-O(31)

FIGURE 1.29 The three staggered conformations A-C of the glycerol C(2)-C(3) bond of phospholipids. For the atom numbering, see Figure 1.22; for the notation of torsion angles, see Table 1.6. (From Hauser H et al.: *Biochemistry* 1988, 27:9166. Copyright 1988, Americam Chemical Society, with permission.)

and C(3) are in an ap arrangement, is difficult to reconcile with parallel chain stacking and intramolecular close contacts between the two hydrocarbon chains. Therefore, rotamer C (Figure 1.29) is energetically unfavorable and is not found in diacylglycerophospholipids. There are, however, single-crystal structures with $\Theta_3/\Theta_4 = -$sc/ap. These are crystal structures of constituent molecules of glycerophospholipids lacking hydrocarbon chains. In this case, Θ_4 is not restricted by chain stacking requirements. Examples are the single-crystal structures of D-glycero-1-phosphocholine (Abrahamsson and Pascher, 1966) and D-glycero-1-phosphoethanolamine (DeTitta and Craven, 1971).

From an inspection of Table 1.8, it is clear that in single-crystal structures of lipids, not only are there two types of minimum free energy conformations about the C(2)-C(3) glycerol bond (represented by rotamers A and B), but each rotamer can be subdivided into two groups. These two subgroups differ in the chain stacking mode designated by β and γ. These letters are used as subscripts to capitals A and B. A_β stands for rotamer A with the β-chain extending in an almost perfect, continuous zigzag from the glycerol backbone. All torsion angles along this chain are close to 180°. The second hydrocarbon chain is initially oriented layer-parallel and undergoes an approximately 90° bend at its second C atom. A_γ stands for the conformation of the A rotamer with the γ-chain forming a continuous zigzag and torsion γ angles close to 180°. What is said for A_β and A_γ regarding hydrocarbon chain packing is also applicable to the B rotamer. In the last column of Table 1.8, this notation is used to classify existing single-crystal structures. Representative single-crystal structures of the two subgroups or rotamers A and B are presented in Figure 1.30. The two subgroups of rotamer A are represented by dimyristoylphosphatidylcholine (Pearson and Pascher, 1979) and dilauroylphosphatidic acid (Hauser et al., 1988), corresponding to the A_γ and A_β structures, respectively. In dimyristoylphosphatidylcholine as A_γ, the γ-chain is oriented approximately perpendicular to the bilayer plane, forming a continuous zigzag with the glycerol group. The β-chain of this molecule is initially oriented layer-parallel, and at the second carbon atom C(22) there is a 90° bend. From C(22) onward, the β-chain becomes parallel to the γ-chain (Figure 1.30). The torsion angles in the β-chain responsible for the 90° bend are β_3 and β_4.

The packing mode of dimyristoylphosphatidylcholine contrasts with that of the two hydrocarbon chains of dilauroylphosphatidic acid. In this case, the β-chain extends in a straight zigzag from the glycerol atom C(2). The γ-chain is initially oriented in a layer-parallel way, and at C atom C(32) the γ-chain makes a 90° bend (see Figure 1.30, bottom left). The torsion angles in the -chain producing the 90° bend are γ_3 and γ_4. The values of the torsion angles responsible for the 90° bend have been given and discussed in some detail in a previous publication (Hauser et al., 1988). The two subgroups differ not only in the mode of chain stacking but also in the orientation of the glycerol group with respect to the bilayer plane (Figure 1.30). In the crystal structure of dimyristoylphosphatidylcholine, the orientation of the glycerol group is approximately perpendicular with respect to the bilayer plane; in the crystal structure of dilauroylphosphatidic acid, the glycerol group has a coplanar orientation (cf. the two structures, top and bottom on the left-hand side of Figure 1.30). It should be stressed that there is this difference in the orientation of the glycerol group, although there is no difference in the conformation of the C(2)-C(3) glycerol bond: in both single-crystal structures, the conformation is $\Theta_3/\Theta_4 = $ ap/sc.

The two subgroups of rotamer B are represented by the single-crystal structures of N-methyl substituted dilauroylphosphatidylethanolamine and dimyristoylphosphatidylglycerol (see right-hand side of Figure 1.30). In the single-crystal structure of the former, the -chain extends almost straight from the C(3) atom of glycerol with a minor twist about the C(3)-O(31) bond ($\gamma_1 = 129°$). The β-chain produces a 60° bend at C atom C(22) with torsion angle $\beta_3 = -57°$. The 60° bend ensures the parallel alignment of the two hydrocarbon chains. This single-crystal structure is therefore denoted B_γ, indicating that we are dealing with the B rotamer ($\Theta_3/\Theta_4 = $ sc/-sc) and that the γ-chain forms a straight zigzag.

In contrast, the chain stacking mode of dimyristoylphosphatidylglycerol is such that the β-chain extends straight from the glycerol backbone, whereas the ψ-chain makes a 90° bend at C atom

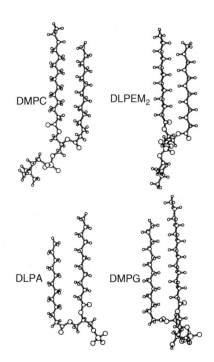

FIGURE 1.30 The four single-crystal structures presented represent possible conformations about the C(2)-C(3) glycerol bond and, related to these conformations, possible chain stacking modes. The molecular conformation of DMPC and DLPA (left-hand side) represent rotamer B. DMPC and DLPA differ in both chain stacking and the orientation of the glycerol group. In contrast, DLPEM$_2$ and DMPG differ only in chain stacking; their glycerol groups have similar orientations. DMPC = 2,3-dimyristoyl-D-glycero-1-phosphocholine. (From Pearson and Pascher, 1979); DLPEM$_2$ = 2,3-dilauroyl-DL-glycero-1-phospho-N,N-dimethylethanolamine (Pascher and Sundell, 1986); DLPA = 2,3-dilauroyl-D-glycero-1-phosphate (Pascher and Sundell, unpublished results); DMPG = 2,3-dimyristoyl-D-glycero-1-phos-DL-glycerol (Pascher et al., 1987). (From Hauser H et al.: *Biochemistry* 1988, 27: 9166. Copyright 1988, American Chemical Society, with permission.)

C(32). This is evident from a comparison of the top and bottom structures on the right-hand side of Figure 1.30. The single-crystal structure of dimyristoylphosphatidylglycerol is therefore denoted B$_\beta$, β indicating that the β-chain forms a straight zigzag. It should be noted that both B rotamers shown in Figure 1.30 have similar orientations of the glycerol group: the glycerol group is tilted by about 45° with respect to the bilayer normal.

From the different single-crystal structures presented in Figure 1.30, it can be concluded that the orientation of the glycerol group, with respect to the bilayer plane, is independent of the conformation of the C(2)-C(3) bond. It can vary from a bilayer-normal to a bilayer-parallel orientation. A detailed discussion of the conformation and motion of the diacylglycerol group in the liquid crystalline state would be beyond the scope of this chapter. Heat is taken up at the crystal-to-liquid crystal transition temperature, and above this temperature the glycerophospholipid molecule undergoes both molecular and segmental motions. The dynamics of the liquid crystalline state have been studied by spectroscopic methods. These methods, particularly NMR and infrared spectroscopy, indicate that the molecule switches between at least two states with rates that are fast on the NMR time scale. ^{1}H-NMR suggests that these states are identical to rotamers A and B (Figure 1.29 and Figure 1.30) observed in single-crystal structures of glycerophospholipids (Hauser et al., 1988). This relationship between single-crystal structures and liquid crystalline state is one example demonstrating the relevance and usefulness of single-crystal structures.

1.4.5 CRYSTAL STRUCTURES OF GLYCOSPHINGOLIPIDS

Due to the efforts of Abrahamsson and Pascher and coworkers in Gothenburg, a number of single-crystal structures of sphingolipids have been solved, culminating in the report of the single-crystal structure of a cerebroside. Because much of the earlier work has been reviewed by Pascher (1976) and Abrahamsson et al. (1977), the discussion here is focused on the cerebroside structure (Pascher and Sundell, 1977). The cerebroside β-D-galactosyl-*N*-(2-D-hydroxyoctadecnoyl)-D-dihydrosphingosine packs in a bilayer with disordered ethanol molecules intercalated between bilayers (Figure 1.31). The most remarkable feature of the bilayer structure is the tight, two-dimensional network of hydrogen bonds in the polar group region involving the galactose ring and the polar groups of the ceramide moiety. Intramolecular hydrogen bonding between the amide N-H group and the glycosidic oxygen of galactose fixes the galactose ring in a position pointing away from the ceramide part. The plane of the sugar ring has an orientation almost parallel to the bilayer plane, giving rise to an L-shaped molecule. The parallel chain stacking is determined by the conformation of the C(2)-C(3) bond of the sphinganine chain. The relevant torsion angles are Θ_3/Θ_4 = ap/sc, indicating that the conformation about the C(2)-C(3) bond is consistent with rotamer A (cf. Figure 1.29). The plane of the amide group has a perpendicular orientation with respect to the initial part of the sphinganine chain (up to C atom 5 and 6 for molecules B and A, respectively). The fatty acyl chain is twisted about the C(1')-C(2') bond (the corresponding torsion angle = −104°) and continues as an antiperiplanar zigzag thereafter. Due to this twist, the fatty acyl chain has an inclination of about 35° with respect to the amide plane and 55° with respect to the axis of the initial part of the sphinganine chain. The parallel chain stacking is then accomplished by a sharp turn in the sphinganine chain occurring at C atom 5 and 6 in molecules B and A, respectively. In molecule A, the turn is produced by +sc/+sc (69°/89°) conformations about the C(5)-C(6) and C(6)-C(7) bonds of sphinganine, in molecule B by −sc/-sc (−57°/−78°) conformations about the C(4)-C(5) and C(5)-C(6) bonds. This is an unusual chain stacking mode that is not observed in the crystal structures of glycerophospholipids.

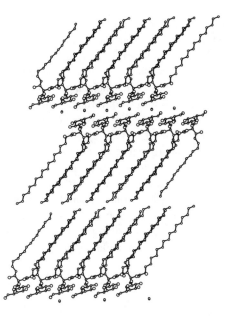

FIGURE 1.31 Molecular packing in the single-crystal structure of the cerebroside β-D-galactosyl-*N*-(2-D-hydroxyocatdecanoyl)-D-dihydrosphingosine viewed along the b-axis. (Adapted from Pascher I and Sundell S: *Chem. Phys. Lipids* 1977, 20:175.)

If hydrogen bonding in the polar group also occurs in the hydrated, liquid crystalline state, it will have important implications regarding the stability and other physicochemical properties of the bilayer. This has been discussed in more detail elsewhere (Pascher, 1976; Abrahamsson et al., 1977). For details of the chain packing mode and the intra- and intermolecular hydrogen bonding network, the interested reader is referred to the original paper by Pascher and Sundell (1977).

1.5 SUMMARY OF SINGLE-CRYSTAL X-RAY CRYSTALLOGRAPHIC ANALYSIS

1. Single-crystal structures of membrane lipids are still scarce due to difficulties in growing single crystals.
2. Single-crystal structures represent minimum energy conformations and serve as a reference in the description of less ordered, dynamic structures.
3. In all existing single crystals, the lipid molecules pack in a layered arrangement. Tail-to-tail packing is the predominant arrangement, leading to the well-known bilayer.
4. Crystal structures of lipids exhibit polymorphism, which is based primarily on different modes of hydrocarbon chain packing.
5. The angle tilt ϕ of the hydrocarbon chains with respect to the bilayer normal is determined by the ratio of cross-sectional areas $n\Sigma/S$ (S = cross-sectional area of the polar group; $n\Sigma$ = sum of the cross-sectional area of the hydrocarbon chains).
6. A fundamental principle of the single-crystal structures of lipids is the intra- and intermolecular alignment or stacking of the hydrocarbon chains. The conformation of the C(2)-C(3) glycerol bond of glycerophospholipids and the C(2)-C(3) bond of sphingosine in glycosphingolipids is intimately related to chain stacking. This bond forms the central link between the two hydrocarbon chains. The conformation of the C(2)-C(3) bond defined by torsion angles Θ_3 and Θ_4 is restricted to two types, represented by rotamer A with Θ_3/Θ_4 = ap/sc and rotamer B with Θ_3/Θ_4 = sc/-sc (cf. Figure 1.29).
7. A characteristic feature of the headgroup is its preferred conformation. This means that in the single-crystal structures of glycerophospholipids, the torsion angles of the headgroup vary within a relatively small range. With few exceptions, the headgroup of phosphatidylcholines and phosphatidylethanolamines is oriented approximately parallel to the plane of the bilayer. This is also true of the phosphoglycerol group of phosphatidylglycerol.
8. The exception to the rule of a preferred conformation in the lipid polar group is the C(1)-C(2) glycerol bond. The conformation of this bond, defined by torsion angles Θ_1 and Θ_2, determines the orientation of the headgroup with respect to the diacylglycerol moiety. All three staggered and also nearly eclipsed conformations are observed about the C(1)-C(2) bond. This result indicates that there is no preferred conformation about this bond and no preferred orientation of the headgroup with respect to the diacylglycerol part.
9. The free energy gained by the formation of a single hydrogen bond is significant. The formation of an intramolecular hydrogen bond produces conformational changes in one or several of the torsion angles of the polar group. The affected torsion angle or angles come to lie outside the normal torsion angle range(s).

ACKNOWLEDGMENTS

We are grateful to Dr. B. Schweizer for useful discussions and help with the computer graphics. This project was supported by the Swiss National Science Foundation (Grant 31-25719.88).

REFERENCES

Abrahamsson S and Pascher I: Crystal and molecular structure of L-α-glycerylphosphorylcholine. *Acta Crystallogr.* 1966, 21:79–87.

Abrahamsson S, Dahlén B, Löfgren H, Pascher I, and Sundell S: Molecular arrangement and conformation of lipids of relevance to membrane structure, in *Structure of Biological Membranes*. Eds. Abrahamsson S and Pascher I, Plenum Press, New York, 1977, pp. 1–23.

Abrahamsson S, Dahlén B, Löfgren H, and Pascher I: Lateral packing of hydrocarbon chains. *Prog. Chem. Fats Other Lipids* 1978, 16:125.

Bell RM: Protein kinase C activation by diacylglycerol second messengers. *Cell* 1986, 45:631.

Blöcher D, Six L, Gutermann R, Henkel B, and Ring K: Physicochemical characterization of tetraether lipids from *Thermoplasma acidophilum*. Calorimetric studies on miscibility with diether model lipids carrying branched or unbranched alkyl chains. *Biochim. Biophys. Acta* 1985, 818:333.

Brade H, Brade L, and Rietschel E Th: Structure–activity relationships of bacterial lipopolysaccharides (endotoxins). *Zbl. Bakt. Hyg. A* 1988, 268:151.

Craven BM: Crystal structure of cholesterol monohydrate. *Nature* 1976, 260:727.

Craven BM: Pseudosymmetry in cholesterol monohydrate. *Acta Crystallogr. B* 1979, 35:1123.

Craven BM: Cholesterol crystal structures: adducts and esters, in *The Physical Chemistry of Lipids: From Alkanes to Phospholipids*. Ed. Small DM, Plenum Press, New York, 1986, pp. 149–182.

Craven BM and DeTitta GT: Cholesteryl myristate: structures of the crystalline solid and mesophases. *J. Chem. Soc. Perkin II*, 1976, 814.

Daum G: Lipids of mitochondria. *Biochim. Biophys. Acta* 1985, 822:1.

DeTitta GT and Craven BM: Conformation of O-(L-α-glycerylphosphoryl)-ethanol-amine in the crystal structure of its monohydrate. *Nature New Biol.* 1971, 233:118.

Fredrickson DS and Ferrans VJ: Acid cholesteryl ester hydrolase deficiency (Wolman's disease and cholesteryl ester storage disease), in *The Metabolic Basis of Inherited Disease, 4th ed*. Eds. Stanbury JB, Wyngaarden JB, and Fredrickson DS, McGraw-Hill, New York, 1978, pp. 670–687.

Fredrickson DS, Goldstein JL, and Brown MS: The familial hyperlipoproteinemias, in *The Metabolic Basis of Inherited Disease, 4th ed*. Eds. Stanbury JB, Wyngaarden JB, and Fredrickson DS, McGraw-Hill, New York, 1978, pp. 604–655.

Gennis RB: *Biomembranes: Molecular Structure and Function*. Springer-Verlag, New York, 1989.

Goto M and Asada E: The crystal structure of the B-form of stearic acid. *Bull. Chem. Soc. Jpn.* 1978, 51:2456.

Gross RW: Identification of plasmalogen as the major phospholipid constituent of cardiac sarcoplasmic reticulum. *Biochemistry* 1985, 24:1662.

Hakomori S: Glycosphingolipids in cellular interaction, differentiation, and oncogenesis. *Annu. Rev. Biochem.* 1981, 50:733.

Harlos K, Eibl H, Pascher I, and Sundell S: Conformation and packing properties of phosphatidic acid: the crystal structure of monosodium dimyristoylphosphatidate. *Chem. Phys. Lipids* 1984, 34:115.

Hauser H, Howell K, Dawson RMC, and Bowyer DE: Rabbit small intestinal brush border membrane. Preparation and lipid composition. *Biochim. Biophys. Acta* 1980(a), 602:567.

Hauser H., Pascher I, and Sundell S: Conformation of phospholipids. Crystal structure of a lysophosphatidylcholine analogue. *J. Mol. Biol.* 1980(b), 137:249.

Hauser H, Pascher I, Pearson RH, and Sundell S: Preferred conformation and molecular packing of phosphatidylethanolamine and phosphatidylcholine. *Biochim. Biophys. Acta* 1981, 650:21.

Hauser H and Gains N: Spontaneous vesiculation of phospholipids: a simple and quick method of forming unilamellar vesicles. *Proc. Natl. Acad. Sci. USA* 1982, 79:1683.

Hauser H, Pascher I, and Sundell S: Preferred conformation and dynamics of the glycerol backbone in phospholipids. An NMR and x-ray single-crystal analysis. *Biochemistry* 1988, 27:9166.

Herbert PN, Gotta AM, and Fredrickson DS: Familial lipoprotein deficiency (abetalipoproteinemia, hypobetalipoproteinemia and Tangier disease), in *The Metabolic Basis of Inherited Disease, 4th ed*. Eds. Stanbury JB, Wyngaarden JB, and Fredrickson DS, McGraw-Hill, New York, 1978, pp. 589–603.

Hirschmann H: The nature of substrate asymmetry in stereoselective reactions. *J. Biol. Chem.* 1960, 235:2762.

Hitchcock PB, Mason R, Thomas KM, and Shipley GG: Structural chemistry of 1,2-dilauroyl-DL-phosphatidylethanolamine: molecular conformation and intermolecular packing of phospholipids. *Proc. Natl. Acad. Sci. USA* 1974, 71:3036.

IUPAC-IUB Commission on Biochemical Nomenclature (CBN): The nomenclature of lipids. *Eur. J. Biochem.* 1977, 79:11.

Jain MK: *Introduction to Biological Membranes, 2nd ed.* John Wiley & Sons, New York, 1988.

Jensen LH and Mabis AJ: Refinement of the structure of -tricaprin. *Acta Crystallogr.* 1966, 21:770.

Klyne W and Prelog V: Description of steric relationships across single bonds. *Experientia* 1960, 16:521.

Labischinski H, Barnickel G, Bradaczek H, Naumann D, Rietschel ET, and Giesbrecht P: High state of order of isolated bacterial lipopolysaccharide and its possible contribution to the permeation barrier property of the outer membrane. *J. Bacteriol.* 1985, 162:9.

Larsson K: On the crystal structure of the forms $_1$ and $_2$ of racemic 1-mono-glycerides. *Ark. Kemi* 1964(a), 23:29.

Larsson K: On the crystal structure of 2-monolaurin. *Ark. Kemi* 1964(b), 23:23.

Larsson K: The crystal structure of the L-1-monoglyceride of 11-bromoundecanoic acid. *Acta Crystallogr.* 1966, 21:267.

Lodish HF and Rothman JE: The assembly of cell membranes. *Sci. Am.* January 1979, 38.

Lüderitz O, Freudenberg MA, Galanos C, Lehmann V, Rietschel E Th, and Shaw DH: Lipopolysaccharides of Gram-negative bacteria, in *Current Topics in Membranes and Transport*, Vol. 17. Eds. Razin S and Rottem S, Academic Press, New York, 1982, pp. 79–151.

Luzzati V, Gambacorta A, DeRosa M, and Gulik A: Polar lipids of thermophilic prokaryotic organisms: chemical and physical structure. *Annu. Rev. Biophys, Biophys. Chem.* 1987, 16:25.

Meissner G and Fleischer S: Characterization of sarcoplasmic reticulum from skeletal muscle. *Biochim. Biophys. Acta* 1971, 241: 346.

Ourisson G: Bigger and better hopanoids. *Nature* 1987, 326:126.

Pascher I: Molecular arrangements in sphingolipids. Conformation and hydrogen bonding of ceramide and their implication on membrane stability and permeability. *Biochim. Biophys. Acta* 1976, 455:433.

Pascher I and Sundell S: Molecular arrangements in sphingolipids. The crystal structure of cerebroside. *Chem. Phys. Lipids* 1977, 20:175.

Pascher I, Sundell S, and Hauser H: Glycerol conformation and molecular packing of membrane lipids. The crystal structure of 2,3-dilauroyl-D-glycerol. *J. Mol Biol.* 1981(a), 153:791.

Pascher I, Sundell S, and Hauser H: Polar group interaction and molecular packing of membrane lipids. The crystal structure of lysophosphatidylethanolamine. *J. Mol. Biol.* 1981(b), 153:807.

Pascher I, Sundell S, Eibl H, and Harlos K: Interactions and space requirement of the phosphate headgroup of membrane lipids: the single-crystal structures of a triclinic and a monoclinic form of hexadecyl-2-deoxyglycerophosphoric acid monohydrate. *Chem. Phys. Lipids* 1984, 35:103.

Pascher I, and Sundell S: Interaction and space requirements of the phosphate headgroup in membrane lipids. The crystal structure of disodium lysophosphatidate dihydrate. *Chem. Phys. Lipids* 1985, 37:241.

Pascher I and Sundell S: Membrane lipids: preferred conformational states and their interplay. The crystal structure of dilauroylphosphatidyl-*N,N*-dimethylethanolamine. *Biochim. Biophys. Acta* 1986, 855:68.

Pascher I, Sundell S, Harlos K, and Eibl H: Conformation and packing properties of membrane lipids: the crystal structure of sodium dimyristoylphosphatidylglycerol. *Biochim. Biophys. Acta* 1987, 896:77.

Pearson RH and Pascher I: The molecular structure of lecithin dihydrate. *Nature* 1979, 281:499.

Perrone JR, Hackney JF, Dixon JF, and Hokin LE: Molecular properties of purified (sodium + potassium)-activated adenosine triphosphatases and their subunits from the rectal gland of *Squalus acanthias* and the electric organ of *Electrophorus electricus*. *J. Biol. Chem.* 1975, 250:4178.

Pind S and Kuksis A: Structure and function of enterocyte membrane lipids, in *Fat Absorption*, Vol. I. Ed. Kuksis A, CRC Press, Boca Raton, FL, 1986, pp. 43–82.

Popot J-L, Demel RA, Sobel A, van Deenen LLM, and Changeux JP: Interaction of the acetylcholine (nicotinic) receptor protein from *Torpedo marmorata* electric organ with monolayers of pure lipids. *Eur. J. Biochem.* 1978, 85:27.

Prince RC: Hopanoids: the world's most abundant biomolecules? *Trends Biochem. Sci.* 1987, 12:455.

Raetz CRH: *Escherichia coli* mutants that allow elucidation of the precursors and biosynthesis of lipid A, in *Handbook of Endotoxin*, Vol. 1. Ed. Rietschel ET, Elsevier, Amsterdam, 1984, pp. 248–268.

Rietschel ET, Galanos C, Lüderitz O, and Westphal O: The chemistry and biology of lipopolysaccharides and their lipid A component, in *Immunopharmacology and the Regulation of Leukocyte Function*. Ed. Webb DR, Marcel Dekker, Basel, 1982, pp. 183–229.

Rietschel E Th, Wollenweber H-W, Brade H, Zähringer U, Lindner B, Seydel U, Bradaczek H, Barnickel G, Labischinski H, and Giesbrecht P: Structure and conformation of the lipid A component of lipopolysaccharides, in *Handbook of Endotoxins*, Vol. 1. Ed. Rietschel ET, Elsevier, Amsterdam, 1984, pp. 187–220.

Roberts JD and Caserio MC: *Basic Principles of Organic Chemistry, 2nd ed.* WA Benjamin, Menlo Park, 1979.

Sawzik P and Craven BM: The crystal structure of cholesteryl acetate at 123 K. *Acta Crystallogr. B* 1979, 35:895.

Shieh HS, Hoard LG, and Nordman CE: Crystal structure of anhydrous cholesterol, *Nature* 1977, 267:287.

Shipley GG: X-ray crystallographic studies of aliphatic lipids, in *The Physical Chemistry of Lipids: From Alkanes to Phospholipids*. Ed. Small DM, Plenum Press, New York, 1986, pp. 97–147.

Small DM: Liquid crystals in living and dying systems. *J. Colloid Interface Sci.* 1977, 58:581.

Small DM: *The Physical Chemistry of Lipids: From Alkanes to Phospholipids*. Plenum Press, New York, 1986, pp. 1–19.

Small DM and Shipley GG: Physical-chemical basis of lipid deposition in atherosclerosis. *Science* 1974; 185:222.

Stults CLM, Sweeley CC, and Macher BA: Glycosphingolipids: structure, biological source, and properties. *Methods Enzymol.* 1989, 179:167.

Sundaralingam M: Discussion paper: molecular structures and conformations of the phospholipids and sphingomyelins. *Ann. N.Y. Acad. Sci.* 1972, 195:324.

White DA: The phospholipid composition of mammalian tissues, in *Form and Function of Phospholipids*, Vol. 3. Eds. Ansell GB, Hawthorne JN, and Dawson RMC, Elsevier, Amsterdam, 1973, pp. 441–482.

2 The Mesomorphic Phase Behavior of Lipid Bilayers

Ruthven N.A.H. Lewis and Ronald N. McElhaney

CONTENTS

0-8493-1403-8/05/$0.00+$1.50

2.1 INTRODUCTION

Membrane lipids are invariably polymorphic, that is, they can exist in a variety of different kinds of organized structures, especially when hydrated. The particular polymorphic form that predominates depends not only on the structure of the lipid molecule itself and on its degree of hydration, but also on such variables as temperature, pressure, ionic strength, and pH [see 1, 2]. X-ray diffraction techniques are usually used to determine the structure of these lipid phases [see 3], and differential scanning calorimetry (DSC) is used to study the transition of one lipid phase to another [see 4]. This combination of a direct structural technique (X-ray diffraction) with a thermodynamic technique (DSC) has proved extremely valuable in studies of lipid polymorphism in both model and biological membranes.

Under physiologically relevant conditions, most (but not all) membrane lipids exist in the lamellar or bilayer phase, usually in the lamellar liquid–crystalline but sometimes in the lamellar gel phase. It is not surprising, therefore, that the lamellar gel to liquid–crystalline or chain-melting phase transition has been the most intensively studied lipid phase transition. This cooperative phase transition involves the conversion of a relatively ordered gel-state bilayer, in which the hydrocarbon chains exist predominantly in their rigid, extended, all-*trans* conformation, to a relatively disordered liquid–crystalline bilayer, in which the hydrocarbon chains contain a number of gauche conformers and the lipid molecules exhibit greatly increased rates of intra- and intermolecular motion. The gel to liquid–crystalline phase transition is accompanied by a pronounced lateral expansion and a concomitant decrease in the thickness of the bilayer, and by a small increase in the total volume occupied by the lipid molecules. There is also evidence that the number of water molecules bound to the surfaces of the lipid bilayer increases upon chain melting. Thermodynamically, the gel to liquid–crystalline phase transition occurs when the entropic reduction in free energy arising from chain isomerism counterbalances the decrease in bilayer cohesive energy arising from the lateral expansion and from the energy cost of creating gauche conformers in the hydrocarbon chains [see 5].

Gel to liquid–crystalline phase transitions can be induced by changes in temperature and hydration, as well as by changes in pressure and in the ionic strength or pH of the aqueous phase. In this chapter, we will concentrate on thermally induced phase transitions, because these have been most extensively studied and are of direct biological relevance, particularly for organisms that cannot regulate their own temperature. However, we shall also deal briefly with hydration-induced (lyotropic) and pressure-induced (barotropic) phase transitions, as these may also be biologically relevant under special environmental circumstances. Finally, phase transitions induced by alterations in pH and in the nature and quantity of ions present in the aqueous phase surrounding the bilayer will be briefly discussed, because these transitions may also be of importance in living cells.

Pure synthetic lipids often exhibit gel-state polymorphism, and phase transitions between various forms of the gel-state bilayer can occur. Although we will illustrate this behavior for a common phospholipid, dipalmitoylphosphatidylcholine (DPPC), gel-state transitions will not be emphasized here because, with only one known exception (see Section 2.4.1), they do not seem to occur in the heterogeneous collection of lipid molecular species found in biological membranes. However, phase transitions between different types of gel-state bilayers, including hydrocarbon chain-interdigitated bilayers, will be discussed in Chapter 3. Moreover, certain synthetic or naturally occurring lipid species can exist in liquid–crystalline nonlamellar phases, especially three-dimensional reversed cubic and hexagonal phases. Although the actual existence of nonbilayer lipid phases in biological membranes has never been demonstrated under physiological conditions, there is evidence to suggest that the relative proportion of bilayer-preferring and nonbilayer-preferring lipids may be biosynthetically regulated in response to variations in temperature and membrane lipid fatty acid composition and cholesterol content in some organisms. Thus lipid species that, in isolation, may form nonlamellar phases may have important roles to play in the liquid–crystalline bilayers found in essentially all biological membranes. The transitions between lamellar and nonlamellar lipid phases have been reviewed by us and others [see 6 and references cited therein] and will also be discussed in Chapter 5.

2.2   DIFFERENTIAL SCANNING CALORIMETRY

As mentioned earlier, the technique of DSC has been of primary importance in studies of lipid phase transitions in model and biological membranes. The principle of DSC is comparatively simple. A sample and an inert reference (i.e., material of comparable thermal mass that does not undergo a phase transition within the temperature range of interest) are simultaneously heated or cooled at a predetermined constant rate (dT/dt) in an instrument configured to measure the differential rate of heat flow (dE/dt) into the sample relative to that of the inert reference. The temperatures of the sample and reference may either be actively varied by independently controlled units (power compensation calorimetry) or passively changed through contact with a common heat sink, which has a thermal mass that greatly exceeds the combined thermal masses of the sample and reference (heat conduction calorimetry). For our purposes, the sample would normally be a suspension of lipid or membrane in water or an aqueous buffer and the reference cell would contain the corresponding solvent alone. At temperatures distant from any thermotropic events, the temperatures of the sample and reference cells change linearly with time and the temperature difference between them remains zero. The instrument thus records a constant difference between the rates of heat flow into the sample and reference cells which, ideally, is reflected by a straight, horizontal baseline. When the sample undergoes a thermotropic phase transition, a temperature differential between the sample and reference occurs, and the instrument either actively changes the power input to the sample cell to negate the temperature differential (power compensation calorimetry) or passively records the resulting changes in the rate of heat flow into the sample cell until the temperature differential eventually dissipates (heat conduction calorimetry). In both instances, there is a change in the differential rates of heat flow into the sample and reference cells, and either an exothermic or endothermic deviation from the baseline condition occurs. Upon completion of the thermal event, the instrument either reestablishes its original baseline condition or establishes a new one if a change in the specific heat of the sample has occurred. The output of the instrument is thus a plot of differential heat flow (dE/dt) as a function of temperture in which the intensity of the signal is directly proportional to the scanning rate (dT/dt).

The variation of excess specific heat (dE/dt) with temperature for a simple two-state, first-order endothermic process, such as the gel to liquid–crystalline phase transition of a single, highly pure phosphatidylcholine, is illustrated schematically in Figure 2.1. From such a DSC trace, a number of important parameters can be directly determined. The phase transition temperature, usually denoted Tm, is that temperature at which the excess specific heat reaches a maximum. For a symmetrical curve, Tm represents the temperature at which the transition from the gel to liquid–crystalline state is one-half complete. However, for asymmetric traces, which are characteristic of certain pure phospholipids and many biological membranes, the Tm does not represent the midpoint of the phase transition, and a T1/2 value may be reported instead. Once normalized with respect to the scan rate, the peak area under the DSC trace is a direct measure of the calorimetrically determined enthalpy of the transition, ΔHcal, usually expressed in kcal/mol. The area of the peak can be determined by planimetry or by the cutting and weighing technique; alternatively, the calorimeter output can be digitized and the Tm and ΔHcal calculated by a computer. At the phase transition midpoint temperature, the change in free energy (ΔG) of the system is zero, so the entropy change associated with the transition can be calculated directly from the equation

$$\Delta S = \Delta H_{cal}/T_m$$

where ΔS is normally expressed in cal/$K^{-1}$$mol^{-1}$.

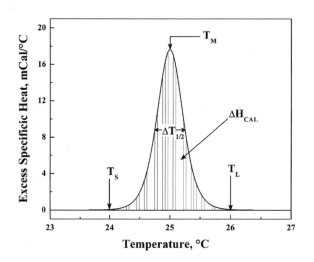

FIGURE 2.1 Idealized scan rate normalized DSC heating thermogram exhibited by a first-order two-state thermotropic process at equilibrium. The process being simulated has a Tm of 25°C, a ΔT1/2 of 0.5°C, and a thermal load (ΔHcal) of 10 mCal.

The sharpness or cooperativity of the gel to liquid–crystalline phase transition can also be evaluated from the DSC trace. The sharpness of the phase transition is often expressed as the temperature width at half-height, ΔT1/2, or as the temperature difference between the onset or lower boundary of the phase transition, Ts, and the completion or upper boundary, Tl, or ΔT = Tl − Ts. The ΔT1/2 values may range from < 0.1°C, for very pure synthetic phospholipids, to as much as 10° to 15°C for biological membranes. From the Tm and ΔT1/2 values determined for a particular phase transition, the van't Hoff enthalpy, ΔHvH, can be approximately determined from the relationship

$$\Delta H_{vH} \cong 6.9\ T_m^2/\Delta T_{1/2}$$

From the ratio $\Delta H_{vH}/\Delta H_{cal}$, the cooperative unit size (CUS) (in molecules) can be determined. The CUS is a measure of the degree of intermolecular cooperation between phospholipid molecules in a bilayer. For a completely cooperative, first-order phase transition of an absolutely pure substance, this ratio should approach infinity; for a completely noncooperative process, this ratio should approach unity. Although the absolute CUS values determined should be regarded as tentative, because this parameter is markedly sensitive to the presence of impurities and may be limited by instrumental parameters, carefully determined CUS values can be useful in assessing the purity of synthetic phospholipids and in quantitating the degree of cooperativity of lipid phase transitions.

It must be stressed that the thermodynamic parameters derived from DSC measurements are valid only if measurements are performed under conditions in which the instrument response is true to the properties of the sample (so-called high-fidelity DSC) and if the thermotropic process being studied is at equilibrium throughout the measurement. In practice, this means that the measurement must be made with a high-sensitivity instrument operating with a modest thermal load at scan rates that are slow relative to the thermal time constant of the instrument, and to both the width and half-life of the thermotropic process under investigation. For processes such as the gel to liquid–crystalline phase transition of certain single, pure synthetic phosphatidylcholines, this is rarely a problem, because such processes are usually rapid enough to be effectively free of kinetic limitations even at moderate scan rates [7, 8]. However, processes such as the pretransition and the subtransition of synthetic disaturated phosphatidylcholines are known to be kinetically limited at all temperatures and scan rates at which calorimetric observation is feasible [9–12]. For such processes, some of their thermodynamic parameters cannot be reliably measured by DSC.

Another aspect of the thermodynamic equilibrium problem is the question of whether or not the system is at equilibrium *before* the calorimetric scan is initiated. In many DSC studies of model and biological membranes, the sample is placed in the calorimeter and cooled fairly rapidly to a low temperature, and a calorimetric heating scan then begun relatively quickly. Because, as mentioned previously, the kinetics of lipid phase transitions in complex systems are not well studied, and because the rates of reversible lipid phase transitions are generally considerably slower upon proceeding from a higher-temperature to a lower-temperature state than the reverse, the possibility exists that the system under study may not be at thermodynamic equilibrium when the calorimetric run is begun. This can be the case even if no exothermic events are observed upon heating. For this reason, it is always advisable to cool the sample slowly and to investigate the effect of variations in the annealing time at low temperatures on the DSC results obtained.

2.3 STUDIES OF MODEL MEMBRANES

2.3.1 THE THERMOTROPIC PHASE BEHAVIOR OF DPPC

A DSC heating scan of a fully hydrated aqueous dispersion of DPPC, which has been annealed at 0°C for 3.5 days, is presented in Figure 2.2. The sample exhibits three endothermic transitions, termed (in order of increasing temperature) the *subtransition, pretransition,* and *main phase transition.* The thermodynamic parameters associated with each of these lipid phase transitions are presented in Table 2.1. The presence of three discrete thermotropic phase transitions indicates that four different phases can exist in annealed, fully hydrated bilayers of this phospholipid, depending on temperature and thermal history. The structures and dynamics of each of these phases are illustrated schematically in Figure 2.3 and are also discussed briefly below (see Reference 5 for a

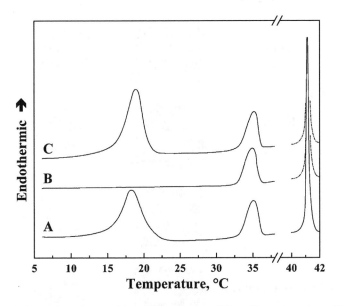

FIGURE 2.2 Smoothed calorimetric transition curves for a multilamellar suspension of high-purity DPPC observed at a scan rate of 0.5°C min-1. Curve A: scan of suspension after being cooled at 0°C for 3.5 days. Curve B: reheating after cooling to 1°C for 3 h. A barely detectable endothermic peak, which was observed at 14°C, does not show up in the figure. Curve C: reheating after cooling to 2°C for three days. Main transitions (only one shown) are plotted on different scales. (Redrawn from [303].)

TABLE 2.1
Thermodynamic Characteristics of the Three Phase Transitions Exhibited by Multilamellar Suspensions of Dipalmitoylphosphatidylcholine Annealed at 0°C for 3.5 Days[a]

Transition Type	T_m (°C)	$\Delta T^{1/2}$ (°C)	ΔH_{cal}[b]	ΔS[c]	CUS[d]
Subtransition	18.4	3.0	3.23	11.1	70[e]
Pretransition	35.1	1.8	1.09	3.5	380[e]
Main Transition	41.1	0.18	6.9	22.0	600

[a] Ref. [303].
[b] Kcal/mol.
[c] cal.K-1.mol-1.
[d] Molecules.
[e] Due to kinetic limitations, the apparent Tm increases slightly and the apparent ΔT1/2 moderately with increases in scan rate. Thus, the CUS values calculated from a linear extrapolation of Tm and ΔT1/2 to zero scan rate, which should approximate the correct equilibrium values for these transitions, are actually somewhat higher than those given here, particularly for the subtransition. Nevertheless, the subtransition remains considerably less cooperative than the pretransition and main transition, which actually exhibit similar intrinsic cooperativities.

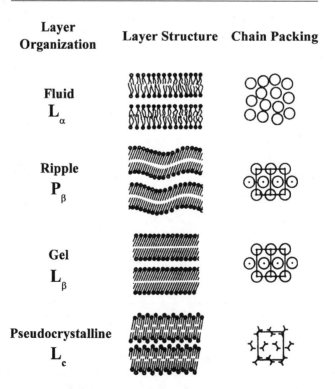

FIGURE 2.3 Organization of the lamellar bilayer phases of DPPC in the fluid (Lα), ripple (Pβ′), gel (Lβ′) and pseudo-crystalline or subgel (Lc′) states. A top view of the packing of the hydrocarbon chains is shown in the last column. (Redrawn from [1].)

more detailed review). All of these phases are lamellar or bilayer phases, differing only in their degrees of organization.

The low-temperature gel phase corresponds closely to that of the crystalline dihydrate and is thus denoted the *Lc′ phase*. The structure of this phase has been discussed in detail in Chapter 1. The Lc′ phase is characterized by extended hydrocarbon chains that are tilted slightly with respect to the bilayer normal. These chains are packed very tightly, and rotation about their long axes is severely restricted. The polar headgroup contains only a few bound water molecules, and its motion is also severely restricted.

With increasing temperature, the steric and van der Waals interchain interactions that favor a crystallinelike packing of the DPPC molecules are progressively overcome by thermally induced rotational excitations of the hydrocarbon chains. Thus, at the subtransition temperature, the Lc′ phase converts to the lamellar gel or Lβ′ phase. In this phase, the extended hydrocarbon chains are tilted more strongly from the bilayer normal, are packed in a distorted orthorhombic lattice, and undergo relatively slow, restricted rotational motion about their long axes. The polar headgroup now contains about 15 to 18 waters of hydration and exhibits slow, hindered rotation on the NMR time scale. The subtransition results in a small increase ($\cong 0.2$ nm^2) in the lipid hydrocarbon chain cross-sectional area and a larger increase ($\cong 0.53$ nm^2) in the interfacial area. Thus, the Lβ′ phase is less tightly packed and much more strongly hydrated than the Lc′ phase.

Further increases in temperature result in a marked increase in the long-axis rotational rates of the hydrocarbon chains, and at the pretransition temperature, the Lβ′ phase converts to the so-called ripple or Pβ′ phase. In the Pβ′ phase, the extended hydrocarbon chains appear to remain tilted with respect to the normal to the local bilayer plane but behave as if they are rotationally symmetric, packing into a hexagonal lattice. The cross-sectional areas of the hydrocarbon chain thus show a small increase at the pretransition temperature. The interfacial area increases much more substantially, however, due to the displacement of each lipid molecule along its long axis with respect to its neighbor. The increased area occupied by the polar headgroups allows them to rotate almost freely, although the degree of hydration does not appear to change. In contrast to the Lc′ and Lβ′ phases, the bilayer is no longer planar but exists as a series of periodic, quasi-lamellar segments.

With increasing temperature, the formation of gauche conformers in the hydrocarbon chain becomes increasingly favorable until, at the gel to liquid–crystalline phase transition temperature, chain melting occurs. Spectroscopic and thermodynamic studies have shown that the hydrocarbon chains of DPPC in the melted or Lα phase contain about 4 to 5 gauche bonds per chain, mostly but not entirely in the form of kink (gauche+-trans-gauche-) sequences [13, 14]. As the melting of the hydrocarbon chains produces a marked increase in cross-sectional area and effectively shortens the length of the chains, the bilayer expands laterally and thins at the main phase transition. Although the hydrocarbon chains exhibit rapid flexing and rotation in the Lα phase, they are, on average, oriented normal to the bilayer plane and pack in a loose hexagonal lattice. This increase in the cross-section area per molecule results in an increase in the area available to the polar headgroup, with the result that rotational motion becomes fast on the nuclear magnetic resonance (NMR) time scale and the hydration at the bilayer interface increases, due in part to the partial exposure of more deeply located polar residues, such as the carbonyl oxygens of the fatty acyl chains, to the aqueous phase.

The pattern of gel-state polymorphism exhibited by an aqueous dispersion of any lipid molecular species will vary considerably, depending on the length of the hydrocarbon chains. For example, phosphatidylcholines (PCs) containing linear saturated fatty acids of 10 to 13 carbon atoms form very stable Lc′ phases, which melt directly to the Lα phase on heating, so that their DSC heating traces show only a single, highly energetic endothermic phase transition. PCs containing linear saturated fatty acids of 14 to 15 carbons exhibit only Lc′/Pb′ and Pb′/Lα phase transitions, whereas very long chain PCs show only Lc′/Lβ′ and Lβ′/Lα transitions. Only PCs containing linear saturated fatty acids of 16 to 21 carbon atoms exhibit all four phases at temperatures between 0 and 100°C [12]. Moreover, there is thermodynamic [12] and spectroscopic [15, 16] evidence indicating that the structures of the most stable Lc phase formed, and the kinetics of their formation, are also markedly dependent on fatty acid chain length. In particular, the Lc phases of shorter-chain PCs are characterized by the relatively rapid formation of a more extensive hydrogen-bonding network in the interfacial region of the bilayer but by a *suboptimal* packing of the hydrocarbon chains. Thus, the type of Lc phase formed by these PCs represents a compromise between the partially incompatible packing requirements of the polar headgroup and hydrocarbon chains, such that the former dominates at shorter chain lengths and the latter at longer chain lengths.

The complexity of the gel-phase polymorphism exhibited by any particular PC, and the structure of the gel phases formed, is markedly dependent on the chemical structure of the hydrocarbon chains, as well. For example, chain-symmetric PCs containing methyl isobranched [17–20] or *cis*-monounsaturated [21] fatty acids exhibit only two stable gel states — an Lc phase and a relatively disordered higher-temperature gel state—whereas methyl anteisobranched PCs may form as many as four or five different gel states, depending on the length of the hydrocarbon chain [22, 23]. In addition, the thermotropic phase behavior of a member of a homologous series of PCs usually depends on whether the phospholipid in question contains an odd or an even number of carbon atoms. A particularly dramatic example of this behavior is afforded by PCs containing two identical ω-cyclohexyl fatty acids. Odd-chain members of this homologous series are characterized by a single DSC endotherm on heating, whereas even-chained members exhibit as many as four overlapping but distinct endothermic events prior to the chain-melting phase transition [24, 25]. The molecular basis for this rather dramatic difference in the thermotropic phase behavior of closely related members in a homologous series of PCs is just beginning to be understood. However, because the gel-phase polymorphism of phospholipids depends on the length and chemical structure of their hydrocarbon chains and also on whether these chains contain an even or an odd number of carbon atoms, there is no such thing as a "representative member" of a homologous series of phospholipids.

2.3.2 THE LYOTROPIC PHASE BEHAVIOR OF DPPC

The number and types of phases formed by a phospholipid, and thus its thermotropic phase behavior, depend strongly on its degree of hydration. This is illustrated in Figure 2.4, which depicts a series of DSC heating scans of DPPC as a function of the water content of the system [26]. Anhydrous DPPC (probably actually the dihydrate) undergoes a single, chain-melting phase transition at $>100°C$. Increasing water content results in a large and progressive decrease in the phase transition temperature until a certain water content is reached, at which point no further decreases in Tm are noted. The decrease in Tm observed upon increasing hydration indicates that the progressive adsorption of water molecules decreases the strength of the interactions of adjacent lipid molecules in the bilayer, primarily by the disruption of polar headgroup–headgroup interactions. Only at the higher water levels is an ice endotherm at $0°C$ noted, indicating that, at the lower levels of hydration, all water molecules are bound to the lipid polar headgroups and do not freeze. Also, the pretransition at about $35°C$ appears only above a certain level of hydration. The exact stoichiometry of water binding is difficult to discern from the DSC traces, because at water contents between 17 and 48 wt% the system is obviously heterogeneous, exhibiting simultaneously a free water peak and a higher Tm shoulder on the major chain-melting endotherm, the latter presumably due to a population of lipid molecules that are not fully hydrated. X-ray diffraction [27] and other studies [28], however, reveal that the Lβ′ and Pβ′ phases of DPPC bind about 30 wt% of water, whereas the Lα phase binds about 40 wt%. These values correspond to 17–18 and 27–28 water molecules bound per DPPC molecule in the Lβ′ or Pβ′ and the Lα phase, respectively. The former value is generally compatible with the DSC results just discussed. Egg phosphatidylserine (PS) and particularly egg phosphatidylethanolamine (PE) bind fewer water molecules in their Lα phases than does egg PC [29].

Clearly, in most living cells sufficient water is present to fully hydrate the membrane lipids, and in fact a large excess water phase is normally present. Therefore, in the remainder of this

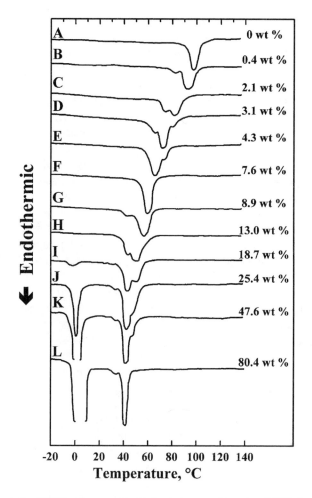

FIGURE 2.4 Representative DSC heating curves obtained at a scanning rate of 1°C/min of DPPC as a function of increasing hydration. (Redrawn from [26].)

chapter, only fully hydrated lipid–water systems will be discussed. One should note, however, that the topic of lyotropic mesomorphism may be of biological as well as physical interest. This is because there is considerable evidence that the often lethal effects of dehydration on living cells, induced either by desiccation or by freezing, are due primarily to damage to cellular membranes, and that this damage is mediated in part by dehydration-induced changes in lipid phase structure. In particular, the dehydration-induced conversion of the lamellar liquid–crystalline lipid phase to the lamellar gel phase or to a nonlamellar phase has been proposed to result in the disruption of the lipid bilayer permeability barrier and to affect adversely the activity and stability of membrane proteins. This view is supported by the fact that dehydration-resistant organisms are often protected by the presence of high amounts of polyols, such as the disaccharide trehalose, and that trehalose and related compounds can prevent these lyotropic phase transitions from occurring by substituting for the waters of hydration normally bound to the lipid polar headgroups [30–33].

2.3.3 THE BAROTROPIC PHASE BEHAVIOR OF DPPC

Phase transitions in anhydrous or in fully hydrated lipid bilayers can be induced by variations in pressure alone when the temperature is held constant. Thus the application of increasing pressure to fully hydrated dispersions of DPPC at a constant temperature above its Tm results in the sequential conversion of the Lα phase, which exists at atmospheric pressure, to one of a series of five different gel phases. The first two gel phases formed by the application of increasing pressure appear to correspond at least approximately in structure to the Pβ′ and Lβ′ phases, which can be induced in DPPC by a reduction in temperature, whereas the final gel phase formed appears to be very similar in structure to the DPPC dihydrate crystal [see 33].

The thermotropic phase behavior of DPPC is also affected by pressure. Thus the Lβ′/Pβ′ and the Pβ′/Lα phase transition temperatures of DPPC both increase linearly as the pressure increases. Specifically, the pretransition temperature rises by about 16°C/kbar and the main transition temperature rises by 20°C/kbar. Qualitatively, this variation is expected from the relative increase in molecular volume as one proceeds from the Lβ′ through the Pβ′ to the Lα phase. Interestingly, however, vibrational spectroscopic and dilatometric results indicate that the volumes occupied by both the Lβ′ and Pβ′ phases are *greater* at higher pressures. This is due to the fact that while pressure initially causes a conformational ordering of the liquid–crystalline lipid hydrocarbon chains by restricting their lateral mobility, at higher pressures compression along the chain axis predominates over lateral compression, leading to a distortion of the methyl end groups and thus to a conformational disordering of the methylene chains. Thus the volume changes at the (elevated) pretransition and main transition temperatures of DPPC actually decrease with increasing pressure [33].

As with lyotropic mesomorphism, lipid barotropic mesomorphism is probably not generally relevant to most living cells, except perhaps to marine organisms living at great depth. Therefore, in the remainder of this chapter we will consider only the thermotropic phase behavior of lipid systems at a constant atmospheric pressure. However, the study of lipid barotropic mesomorphism in model and biological membranes is important from the biophysical viewpoint. This is because variations in temperature at a constant pressure nevertheless result in simultaneous changes in both the molecular volume and the thermal energy of the lipid molecule, making the separation of kinetic and volume effects on the organization of lipid bilayers difficult. On the other hand, variations in pressure at a constant temperature permit investigations of the effects of alterations in volume alone on lipid polymorphism. Thus, by comparing the thermotropic and the barotropic phase behavior of lipid–water systems, greater insights into the molecular mechanisms of phase transitions, and thus into the structures of the lipid phases involved in these transitions, can be obtained [see 33].

2.3.4   THE EFFECT OF VESICLE SIZE ON THE THERMOTROPIC PHASE BEHAVIOR OF DPPC

The thermotropic phase behavior of small, single-bilayer vesicles of DPPC differs significantly from that of the same lipid in a large unilamellar or multilamellar vesicle. The phase behavior of small phospholipid vesicles prepared by ultrasonic irradiation has been studied by DSC, in conjunction with other physical techniques. It was initially reported that the calorimetric characteristics of sonicated and unsonicated disaturated PC dispersions are indistinguishable by conventional DSC [34], and later that sonicated vesicles exhibit a less enthalpic pretransition and a downward-shifted, somewhat broadened main phase transition with only a slightly reduced ΔHcal [35]. However, studies employing high-sensitivity DSC demonstrated that, in freshly sonicated vesicles of disaturated PCs, the pretransition is absent and that the ΔHcal and cooperativity of the main transition are drastically reduced [36–38]. The results obtained with conventional DSC were probably due to a time-dependent aggregation of unilamellar vesicles into larger multilamellar vesicles promoted by the high lipid concentrations required. Also, vesicle fusion was shown to be enhanced by cycling through the lipid phase transition or by the freezing of the aqueous phase, as well as by an increased medium osmolarity, the addition of fatty acids, or the presence of divalent cations such as calcium [35, 37]. A number of other physical techniques have indicated that the small radius of curvature of sonicated phospholipid vesicles leads to less orientational order and to a greater freedom of motion of the phospholipid hydrocarbon chains than are found in larger vesicles, and to marked differences in molecular packing in the inner and outer lipid monolayers, thus explaining the DSC results.

TABLE 2.2
The Thermodynamic Parameters of the Gel to
Liquid–Crystalline Phase Transitions of a
Homologous Series of Symmetric-Chain
Linear Saturated Phosphatidylcholines[a]

PC	T_m (°C)	$\Delta T_{1/2}$ (°C)	ΔH_{cal}[b]	ΔS[c]	CUS[d]
13:0	13.7	0.10	4.4	15.3	1309
14:0	23.9	0.10	5.9	19.9	961
15:0	34.7	0.12	6.9	22.4	872
16:0	41.4	0.13	7.7	24.5	691
17:0	49.8	0.16	8.7	26.9	540
18:0	55.3	0.24	9.8	29.8	321
19:0	61.8	0.24	10.7	31.9	306
20:0	66.4	0.32	11.4	33.6	221
21:0	71.1	0.47	12.2	35.4	145

[a] Ref. [12].
[b] Kcal/mol.
[c] cal.K-1.mol-1.
[d] Molecules.

In the remainder of this chapter, we will consider only the thermotropic phase behavior of lipid vesicles with relatively large radii, as these systems more closely mimic the lipid bilayers found in most biological membranes.

2.3.5 THE EFFECT OF VARIATIONS IN THE LENGTH AND STRUCTURE OF THE HYDROCARBON CHAINS ON LIPID THERMOTROPIC PHASE BEHAVIOR

In this section, we will consider how changes in the length and chemical structure of the hydrocarbon chains affect the gel to liquid–crystalline phase transition of a series of symmetric-chain diacylphosphatidylcholines, because this is the class of lipids that has been most extensively studied in this regard. However, we will also briefly discuss the behavior of other diacylphospho- and glycolipid classes.

2.3.5.1 Variations in Hydrocarbon Chain Length

The Tm, ΔHcal, ΔS, and CUS values for the Pβ' to Lα phase transition of a homologous series of PCs containing two identical linear saturated fatty acyl chains are presented in Table 2.2. As illustrated in Figure 2.5, the Tm increases smoothly but nonlinearly with increases in the number of carbon atoms in the hydrocarbon chain. The nonlinear increase in Tm with increasing chain length supposedly results from the fact that each additional methylene group gives rise to a gradually diminishing, incremental increase in the stability of the gel phase. This is because the chain-length independent "end effects" should vary as some reciprocal of n. Indeed, as shown in Figure 2.6, a plot of Tm versus 1/n−2 gives an almost linear relationship, with a Tm at infinite chain length of 423 K and an n at 0 K of 4.8. (The best linear relationship is obtained with n = 2.316.) Although caution should be used in attaching physical significance to these numerical values, the near linear variation of Tm with n−2 seems reasonable in view of the fact that the number of methylene groups in a linear saturated fatty acyl chain is equal to the total number of carbon atoms minus 2, there being one carbonyl carbon and one methyl terminal group per chain. Although these two "end-group" carbon atoms contribute to the overall stability of the bilayer, they do not participate in the

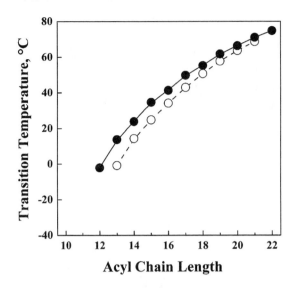

FIGURE 2.5 Effect of chain length of the peak temperature of the Lβ′→Pβ′ transition (-○-) and Pβ′→Lα (-●-) transitions of a homologous series of saturated 1,2-diacyl PCs. (Data obtained from [12].)

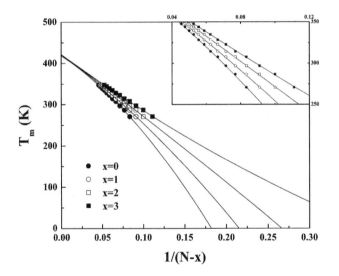

FIGURE 2.6 Plots of the gel to liquid–crystalline phase transition temperatures of a homologous series of linear saturated 1,2-diacyl PCs versus 1/n, 1/(n−1), 1/(n−2), etc., where n is the total number of carbon atoms in the acyl chain. (Data obtained from [12].)

chain-melting process. The extrapolated Tm at infinite chain length of 151°C is near to but somewhat greater than that for the melting point of polymethylene (138°C), indicating that the gel to liquid–crystalline phase transition in this series of PCs involves primarily, but not exclusively, a chain-melting process. The extrapolated n value of 4.8 at 0°K suggests that PCs must have hydrocarbon chains of at least five carbon atoms in length to undergo a cooperative gel to liquid–crystalline phase transition. A similar variation of Tm with hydrocarbon chain length is observed in PCs containing a homologous series of methyl iso- or anteisobranched, ω-cyclohexyl, or *cis*-monounsaturated hydrocarbon chains (see Figure 2.8).

    These empirical observations are also compatible with various thermodynamic, structural, and other empirical rationalizations of the melting behavior of crystalline lipids in general and of hydrated glycerolipid bilayers in particular [see 39–47 and references cited therein]. It should be noted that the effect of hydrocarbon chain length on Tm is relatively modest, being about 7 to 8°C/methylene group at n values of 16 to 18. In contrast, variations in the chemical structure of the lipid hydrocarbon chain can produce more marked changes in Tm when the total number of carbon atoms in the chain or its effective length are held constant (see below). It is probably for this reason that the living cells normally regulate the fluidity and phase state of their membrane lipids, in response to variations in environmental temperature, for example, primarily by changes in the chemical structure rather than the length of the hydrocarbon chains of their membrane lipids [see 48, 49]. Moreover, changes in lipid hydrocarbon chain length would produce changes in bilayer thickness, which could in turn produce undesirable changes in the thermodynamic stability and passive permeability of the lipid bilayer of biological membranes [see 50]. Also, changes in membrane thickness can produce a mismatch between the dimensions of the lipid bilayer hydrocarbon core and the hydrophobic transmembrane α-helical segments of integral membrane proteins, thus inhibiting their function [50, 51]. It is understandable, then, that *homeoviscous* or *homeophasic* adaptation in living cells is normally accomplished primarily by changes in the relative proportion of, for instance, the saturated and unsaturated fatty acids incorporated into the membrane lipids, rather than by changes in the average chain length of these two classes of fatty acids.

The variation in the ΔHcal and ΔS values of the Pβ′/Lα phase transitions of this homologous series of disaturated PCs is illustrated in Figure 2.7. Except for the two shortest chain members of this series, which exhibit anomalous phase behavior, both ΔHcal and ΔS appear to increase in an approximately linear fashion with increases in chain length. Interestingly, a linear extrapolation of ΔHcal to zero yields an intercept on the x-axis between 7 and 8, suggesting that PCs containing fewer than nine carbon atoms would not undergo cooperative gel to liquid–crystalline phase transitions. Very similar behavior is exhibited by a homologous series of linear saturated PEs [52]. In this regard, it may be significant that PCs, phosphatidylserines (PSs), and phosphatidylglycerols (PGs) with acyl chains of eight or fewer carbon atoms form primarily micelles rather than bilayers when dispersed in aqueous media under physiologically relevant conditions [53, 54], whereas with PEs, lamellar and micellar structures coexist at acyl chain lengths of six carbon atoms [54]. It is also noteworthy that vesicles formed from PCs containing 12 or fewer carbon atoms in their hydrocarbon chains are "leaky" to normally rather impermeable solutes trapped within their aqueous cores [55]. Interestingly, however, extrapolation of ΔS to zero yields a value of n near 5, which is similar to the value obtained by extrapolating the Tm to 0°K.

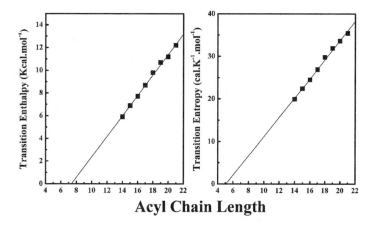

Acyl Chain Length

FIGURE 2.7 A plot of the hydrocarbon chain length dependence of the transition enthalpy (left panel) and the transition entropy (right panel) of the gel to liquid–crystalline (Pβ′→Lα) phase transitions of a homologous series of linear saturated PCs. (Data obtained from [12].)

From Figure 2.7, the incremental ΔS and ΔHcal values can be derived. For both the linear disaturated PC and PE homologous series, an incremental ΔS per CH2 group per chain of ~1.5 cal K-1 mol-1 is obtained. This value is just over half that of the incremental ΔS of fusion of crystalline hydrocarbons. Because the incremental ΔS value is determined largely by the intramolecular conformational disorder of the hydrocarbon chains in the fluid phase, this result indicates that the hydrocarbon chains in liquid–crystalline phospholipid bilayers are considerably more ordered than those of liquid hydrocarbons. Similarly, the incremental transition ΔHcal per CH2 group per chain (~0.5 kcal/mol) is smaller than the incremental ΔH due to the melting of a crystalline hydrocarbon of similar chain length by a comparable factor. The incremental ΔH depends on both the changes in conformational order and the strength of interchain interactions accompanying the gel to liquid–crystalline phase transition.

The cooperativity of the Pβ'/Lα phase transition appears to decrease as the length of the PC hydrocarbon chain increases (see Table 2.2). This could be due to progressively greater levels of impurities in the longer chain members of this series or to small but progressively greater degrees of thermally induced hydrolysis at the increasingly elevated temperatures required to induce chain melting in the longer-chain members of this series. However, these factors do not appear to be of sufficient magnitude to account entirely for the relatively large effects observed. At present, the molecular basis for this decrease in the cooperativity of the Pβ'/Lα phase transition has not been firmly established. However, because membrane lipid hydrocarbon chain length variations in living cells are normally small and the heterogeneous mixture of lipid molecular species in biological membranes exhibit relatively broad gel to liquid–crystalline phase transitions, the intrinsic transition cooperativity of a lipid molecular species is not likely to be a thermodynamic parameter of great biological significance.

2.3.5.2 Variations in Hydrocarbon Chain Structure

A comparison of the calorimetrically determined thermodynamic parameters of the gel to liquid–crystalline phase transitions of a series of PCs having the same total number of carbon atoms in their hydrocarbon chains but different chain structures is presented in Table 2.3. Although the thermotropic events being compared here are "pure" chain-melting phase transitions (transitions from the highest-temperature gel to the liquid–crystalline state), the detailed structures of the gel states from which these lipid bilayers melt are not the same. In fact, it is primarily the differences

TABLE 2.3

Comparison of the Thermodynamic Parameters of the Gel to Liquid–Crystalline Phase Transition of a Series of Phosphatidylcholines Having Different Hydrocarbon Chain Structures but the Same Total Number of Carbon Atoms in the Chain

PC	T_m (°C)	ΔH_{cal}[a]	$\Delta H_{cal}/CH_2$a	ΔS[b]	$\Delta S/CH_2$b	Ref.
di 18:0	55.3	9.8	0.61	29.8	1.86	12
di 18:0i	37.2	~10.0	0.7	32.2	2.30	17
d 18:0ai	18.9	~5	0.36	17.1	1.22	22
di 18:0ch	16.0	~2.7	0.25	9.3	0.85	24
di 18:1tΔ9	9.5	7.3	0.52	25.8	1.84	220
di 18:1cΔ9	−17.3	~3.5–4.0	0.27	14.7	1.05	21

[a] Kcal/mol.

[b] cal.K-1.mol-1.

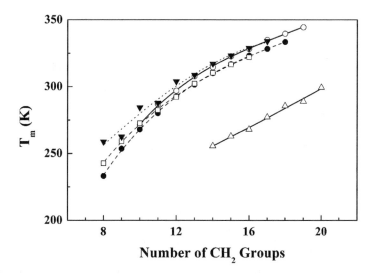

FIGURE 2.8 A plot of the gel to liquid–crystalline phase transition temperatures of a group of homologous series of chain-symmetric PCs containing different classes of fatty acids vs. the number of CH2 groups in the hydrocarbon chain. The symbols are as follows: (○), linear saturated PCs, (●), methyl isobranched PCs, (□), methyl anteisobranched PCs, (▼), ω-cyclohexyl PCs; and (△), *cis*-monounsaturated PCs. (Data obtained from [12, 17, 21, 22, 24].)

in the degree of organization of these different gel states that determine the values of the various thermodynamic parameters being measured.

The Tm values of various 18-carbon PCs found in Table 2.3 are markedly dependent on the chemical structure of their fatty acyl chains. The PC containing two identical linear saturated chains exhibits the highest Tm, and the Tm values decrease progressively in this order: linear saturated > methyl isobranched > methyl anteisobranched > *omega*-cyclohexyl > *trans*-monounsaturated > *cis*-monounsaturated fatty acids. Interestingly, if the Tms of the linear saturated, methyl iso- and anteisobranched, and the *omega*-cyclohexyl PC series are plotted vs. the number of CH2 groups per chain (see Figure 2.8), the curves for these four series of lipids are quite similar, indicating that the primary effect of these chemical modifications on the Tm arise from an effective reduction in the number of CH2 groups that stabilize the high-temperature gel state. However, the fact that the methyl branched PCs exhibit slightly lower Tms than their linear saturated analogues, even when normalized according to the number of CH2 groups present, indicates not only that the isopropyl and sec-butyl groups present at the methyl termini of the methyl iso- and ante-isobranched hydrocarbon chains do not contribute positively to the thermal stability of the gel state, but actually perturb slightly the packing of the polymethylene segments of the chain. This perturbation effect on chain packing is manifested in a much more pronounced form in the *trans*- and *cis*-monounsaturated PCs, which exhibit Tms equivalent to much shorter linear saturated PCs. Dioleoyl PC (DOPC), for example, behaves as if it contains only eight or nine methylene groups, rather than 14. Evidently, the *cis*-double bonds in the center of the hydrocarbon chain severely perturb the interactions of the polymethylene regions on either side of it.

In considering the ability of various chemical modifications of a linear saturated chain to lower the Tm, due cognizance must be taken of the position, as well as the nature, of the substituted group. This is because, as illustrated in Figure 2.9, the Tm-lowering effect of, for example, a *cis*-double bond depends markedly on its location within the hydrocarbon chain and is greatest when located near the center of the chain, as predicted by structural, thermodynamic, and molecular mechanics principles [45, 46, 56] However, although it has been stated that the *cis*-double bond is intrinsically more potent in reducing Tm than a single methyl substitution, the opposite is in fact

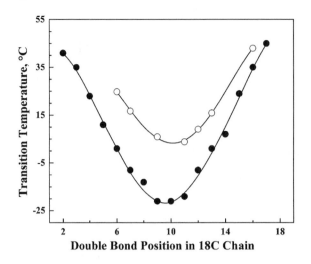

FIGURE 2.9 Plots of the apparent gel to liquid–crystalline phase transition temperature of a series of chain symmetric dioctadecenoyl PCs (-●-) and 1-stearoyl-2-octadecenoyl PCs (-○-) vs. the position of the double bond in the *cis*-monounsaturated hydrocarbon chain. (Data obtained from [56 and 316].)

the case when both of these substituents are compared at comparable positions along the chain. Increasing the number or the size of the alkyl group substituents, or the number of double bonds present in the hydrocarbon chain, results in a further lowering of the Tm of the corresponding PC, but the Tm-lowering effect of each additional substituent becomes progressively less. The lack of an observable cooperative phase transition between −120 and +120°C in aqueous dispersions of diphytanoyl PC [57] is doubtless due to the cumulative perturbing effects of the four methyl group substitutions on each hydrocarbon chain, and to the fact that only ten CH2 groups are present in this highly substituted 20-carbon fatty acid.

The ΔH and ΔS values of the gel to liquid–crystalline phase transition of this series of PCs also vary markedly with the chemical structure of the fatty acyl chains. Although there is a general trend for both the ΔHcal and ΔS values to decrease as the Tm decreases (as one would intuitively expect), these parameters are not highly correlated, even when normalized to the number of CH2 groups in the hydrocarbon chain. In general, this indicates that the various chemical modifications of the linear saturated hydrocarbon chain have rather disparate effects on the changes in the strength of interchain interactions and on the conformational disorder accompanying the chain-melting phase transition. The methyl isobranch substitution is unique in producing an increase in the ΔHcal and ΔS values, compared to a linear saturated chain containing the same total number of carbon atoms. Because there is good evidence that the high-temperature gel state of the methyl isobranched PCs is somewhat less well ordered and the liquid–crystalline state slightly more ordered than that of their linear saturated analogues [18–20, 58–61], the larger ΔHcal and ΔS values observed experimentally are surprising. Apparently, these elevated values arise from the increased cross-bilayer interactions in the center of the model membrane. This is due to the fact that each terminus of an isobranched fatty acid contains two methyl groups positioned such that both can interact with their opposite members in the other monolayer. A similar explanation has been proposed to explain why methyl isobranched fatty acids exhibit melting points only about 0.5°C below those of their linear saturated analogues. All other chemical modifications lead to decreases in ΔHcal and ΔS values, as expected, with the magnitude of these decreases increasing in this order: *trans*-monounsaturated > methyl anteisobranched > *cis*-monounsaturated > *omega*-cyclohexyl. Although to some extent these decreases are due to an effective decrease in the number of CH2 groups present in the hydrocarbon chain, especially in the case of the *omega*-cyclohexyl fatty acids (where none of the carbons in the rigid ring system actually melt at the gel to liquid–crystalline phase transition) [25],

TABLE 2.4

Comparison of the Thermodynamic Parameters of the Gel to Liquid–Crystalline Phase Transitions of a Series of Phosphatidylcholines Having Different Hydrocarbon Structures but the Same Effective Chain Lengths

PC	T_m (°C)	ΔH_{cal}[a]	$\Delta H_{cal}/CH_2$a	ΔS[b]	$\Delta S/CH_2$b	Ref.
di 18:0	55.3	9.8	0.61	29.8	1.86	12
di 19:0i	43.7	11.2	0.75	35.3	2.35	17
di 19:0ai	29.5	7.9	0.53	26.1	1.74	22
di 21:0ch	46.6	~10	0.71	31.3	2.24	24
di 18:1tΔ9	9.5	7.3	0.52	25.8	1.84	220
di 18:1cΔ9	−17.3	3.5–4.0	0.27	14.7	1.05	21

[a] Kcal/mol.
[b] cal.K-1.mol-1.

it is clear that these latter chemical modifications produce substantial reductions in the ΔHcal and ΔS primarily by producing reductions in the tightness of organization of the gel phase.

A comparison of the calorimetrically determined thermodynamic parameters of the gel to liquid–crystalline phase transition of a series of PCs having the same effective main hydrocarbon chain length but different chemical structures (and different numbers of carbon atoms) is presented in Table 2.4. In general, the variations in Tm, ΔHcal, and ΔS are similar to those already discussed for Table 2.3. However, the *omega*-cyclohexyl PC now becomes the second-highest melting PC; the Tm, ΔHcal, and ΔS values of the methyl isobranched PC now exceed those of the linear saturated PC by a greater amount; and the values for the *omega*-cyclohexyl and methyl ante-isobranched PCs are increased and move closer to the comparable values for linear saturated PCs. Comparing a series of phospholipids of the same effective chain length is physically attractive because the thickness of their bilayers, at least in the gel state, should be more or less comparable, as generally seems to be the case with biological membranes from organisms grown at the same temperature [see 50, 62]. However, neither comparison by total number of carbon atoms nor by effective chain length faithfully reflects the situation in real membranes for all classes of fatty acids. This is because, in the eubacteria in which they naturally occur, the total numbers of carbon atoms in the linear saturated and methyl iso- and anteisobranched membrane lipid hydrocarbon chains are generally quite similar, making Table 2.3 the table of choice for comparing the gel to liquid–crystalline phase transition thermodynamic parameters of phospholipids containing these groups of hydrocarbon chains. Moreover, the eubacterial *omega*-cyclohexyl fatty acids contain on average about two additional carbon atoms, making Table 2.4 more appropriate for the comparison of these phospholipids with those containing linear saturated or methyl branched fatty acids. However, the *cis*-monounsaturated fatty acids of both eubacterial and eucaryotic membranes generally contain two additional carbon atoms, in comparison to the linear saturated fatty acyl groups from lipids of the same membrane, making neither table suitable for comparing this class of fatty acids with the others. In this regard, it is noteworthy that the "kink" in the hydrocarbon chain of the monounsaturated fatty acids due to the presence of the *cis*-double bond reduces the thickness of the bilayer, such that the additional carbons found in such fatty acids in nature may result in their real effective lengths in the gel state being similar to those of the apparently shorter chain linear saturated or methyl branched fatty acids. A similar argument could be made for the *omega*-cyclohexyl fatty acids, because three of the six carbon atoms of the cyclohexyl ring do not contribute to the effective length of the hydrocarbon chain.

2.3.5.3 Mixed-Chain Diacyl Phosphatidylcholines

The thermotropic behavior of a number of mixed-chain saturated diacyl PCs has been investigated by conventional DSC [63–70], high-sensitivity DSC [71–74], X-ray diffraction [68, 70, 74], Raman and infrared spectroscopy [74–78] and NMR spectroscopy [78, 79]. A consistent feature of all studies of these compounds is that the Tm and ΔHcal values of each pair of positional isomers of these phospholipids are different. The one in which the longer chain is located at the *sn*-1 position of glycerol always exhibits significantly lower Tm and ΔHcal values than does the reverse isomer. This phenomenon can be rationalized on the basis of the known conformational difference between the *sn*1 and *sn*2 fatty acyl chains of glycerolipids. Both spectroscopic and neutron and X-ray diffraction studies (see Chapter 1) have shown that the acyl chains at the *sn*1- and *sn*2-positions of the glycerol backbone have different conformations. The acyl chain at position 2 begins roughly parallel to the bilayer surface before bending at C2 to orient the hydrocarbon chain perpendicular to the bilayer plane, whereas the fatty acyl chain at position 1 is perpendicular to the bilayer throughout its length. Thus, even when identical fatty acyl chains are esterified to the *sn*1- and *sn*2-positions of glycerol, the penetration of the fatty acyl chain at the 1-position into the bilayer is some 1.8 to 1.9 Å, or about 1.5 C-C bond lengths, greater than that of the *sn*2- fatty acyl chain. Consequently, saturated mixed-acid PCs with the shorter chain acid at position 1 tend to minimize the intrinsic chain length mismatch that arises from the different conformations of the two fatty acyl chains, whereas this intrinsic chain length mismatch is accentuated in the reverse isomer. As a result, the location of the shorter chain at the *sn*1 position of glycerol stabilizes the bilayer by increasing the number of lateral van der Waals contacts between the two chains, and this is manifest in Tm and ΔHcal values, which are significantly higher than those of the corresponding reverse isomer [for a more detailed discussion, see 80–82]. The studies of these saturated mixed-chain diacyl PCs have also shown that the asymmetric substitution of fatty acyl moieties to the glycerol promotes the formation Lc-type gel phases, and of interdigitated gel phases when the effective length mismatch between the *sn*1 and *sn*2 acyl chains is very great. These properties of mixed-chain phospholipids are not the focus of this chapter and will not be discussed any further here. For more details on these aspects of the mixed-chain phospholipids, the reader is referred to Chapter 3 and reviews by Huang and coworkers [80, 82].

The thermotropic behavior of several mixed-acid, saturated–unsaturated PCs containing oleic (O) and elaidic (E) acids has been examined by DSC and differential thermal analysis [83–88]. For 1-palmitoyl-2-oleoyl PC (POPC), reported Tm values range from −5 to +3°C and ΔHcal values from 8.0 to 8.1 kcal/mol; comparable values for the reversed isomer OPPC are −11°C and 6.7 kcal/mol, qualitatively the results expected from a consideration of the differential effective chain length effect just discussed. It is noteworthy that the Tm values for these mixed-acid, saturated–unsaturated PCs fall well below the average for the corresponding disaturated and diunsaturated compounds (Tm ~41°C and −14 to −22°C for DPPC and DOPC, respectively [see 89]). On the other hand, the ΔHcal for POPC (8.0 kcal/mol) falls near the average for DPPC and DOPC (about 8.5 kcal/mol and 7.6 kcal/mol, respectively), whereas the corresponding value for OPPC (6.7 kcal/mol) is below that of DOPC (but note that the higher enthalpy value quoted here for DOPC is due to a concomitant gel/gel and gel/liquid–crystalline phase transition being observed [see 21]).

In contrast, the Tm of 1-palmitoyl-2-elaidoyl PC (PEPC, Tm = 35°C) lies closer to the Tm of DPPC than to that of dielaidoyl PC (DEPC) (9.5 to 13°C). Unfortunately, a ΔHcal value for this lipid is not available, nor has the reverse isomer been studied. For 1-stearoyl-2-oleoyl PC (SOPC), reported Tm values range from 3 to 13°C and ΔHcal values from 4.3 to 5.3 kcal/mol; a Tm value is not available for 1-oleoyl-2-stearoyl PC (OSPC), but a ΔHcal of 6.7 kcal/mol has been reported. Again, the Tm values fall well below the average for the corresponding disaturated and diunsaturated analogs (Tm ~54°C for DSPC), and the ΔHcal values fall well below those of DOPC. Interestingly, the Tm for 1-stearoyl-2-elaidoyl PC (SEPC) (26°C) falls closer to that of DEPC than of DSPC and well below the Tm of PEPC, despite its greater average chain length — again illustrating the effect of the different chain conformations at positions 1 and 2. The ΔHcal for SEPC (8.4 kcal/mol) again falls below the average for DSPC and DEPC (ΔHcal for DEPC = 7.3 to 10.0 kcal/mol). It thus appears that, in the absence of intrinsic chain length mismatch compensation, the gel-state stability of at least the mixed-acid saturated-*cis*-unsaturated PCs more closely resembles that of the corresponding diunsaturated than that of the disaturated analogs. Also, the mixed-acid, saturated–unsaturated PCs, like the diunsaturated species, do not exhibit pretransitions.

2.3.6 The Effect of Variations in the Chemical Structure of the Polar Headgroup Structure on Lipid Thermotropic Phase Behavior

An evaluation of the influence of the polar headgroup on the thermotropic phase properties of hydrated lipid bilayers is not a simple or a straightfoward task. In general, the manner in which the lipid headgroup can influence the properties of a lipid bilayer can be conveniently examined from the perspectives of size, polarity, the number and nature of the charged groups present (if any), the capacity for forming hydrogen bonds, etc. However, such an approach is probably not strictly correct. Because factors such as the size and polarity of the headgroup, as well as its capacity for ionic and/or hydrogen-bonding interactions, largely determine the nature and the strengths of the headgroup–headgroup and headgroup–solvent interactions at the bilayer surface [5, 90], the influence of polar headgroup structure on the properties of any lipid bilayer will be a reflection of the additive and/or synergistic effects of all such factors. Moreover, given the amphipathic nature of the lipid molecules that comprise model and natural membranes, it is even unlikely that a combination of these factors alone can be the sole determinants of the headgroup–headgroup and headgroup–solvent interactions, which actually occur at the bilayer surface. This is because the geometric requirements for the maximization of polar interactions at the bilayer surface and at the polar–apolar interface, and of the van der Waals interactions between the hydrocarbon chains in the bilayer core, can rarely be simultaneously satisfied. Thus, the nature and the strength of the polar and hydrophobic interactions that actually occur in any given bilayer will be the result of a "compromise" between these competing and often incompatible driving forces. One thus expects that the length and the structure of the hydrocarbon chain will also influence the way in which the structure of the polar headgroup affects the thermotropic phase properties of a lipid bilayer. Nevertheless, we will examine the effects of lipid polar headgroup structure on the properties of a lipid bilayer from the same general perspectives described previously. It should be understood, however, that the organization of this chapter along such lines is merely an attempt to make the process of reviewing this area more tractable and is not intended to suggest that any given property of the lipid polar headgroup can exert its effect in isolation from any other.

2.3.6.1 The Effect of Size

The effect of polar headgroup size on lipid thermotropic phase behavior cannot always be easily determined, mainly because it is virtually impossible to affect any variation in the size of that moiety without significantly altering its other properties. Moreover, within the context of a hydrated lipid bilayer, the concept of polar headgroup size is not easy to define, especially because the critical factors determining its effective size should be the intrinsic size of the polar headgroup and its associated waters of solvation, along with its orientation and motion relative to the bilayer surface. Nevertheless, it is intuitively obvious that the physical size of the polar headgroup (however defined) must affect the phase properties of a lipid bilayer from simple geometric considerations alone. Indeed, from a consideration of the geometric constraints involved in assembling lipid molecules, it has been shown that the size of the polar headgroup relative to that of the hydrophobic domain even determines the type the lipid assembly (i.e., whether micellar, lamellar hexagonal, or cubic) that can be formed [91, 92]. Thus, within the context of a lipid bilayer, the probable effect

TABLE 2.5
Gel to Liquid–Crystalline Phase Transition
Temperatures of the Monoalkyl Esters of
Dipalmitoyl Phosphatidic Acid

Alkyl Group	T_m at pH 7	Ref.	T_m at pH 1	Ref.
Methyl	44	94	53	93
Ethyl	41	94	51	93
Propyl	41	94	38	93
Butyl	41	94	34	93
Pentyl	33	94	28	93

of headgroup size can be deduced from such geometric considerations, of which the key factor would be the effective cross-sectional area of the polar headgroup.

From such a perspective one can suggest that, in the absence of other factors, the gel phases of lipids with headgroups of cross-sectional areas that are sufficiently small so as to allow close interactions between the hydrocarbon chains should be stabilized with respect to their liquid–crystalline phases, and this should result in relatively high phase transition temperatures. In addition, once the headgroup becomes large enough to force an increase in the mean separation of the hydrocarbon chains, there should also be a relative destabilization of the gel phase and a progressive lowering of the gel to liquid–crystalline phase transition temperature. This effect is not usually obvious in most studies with natural phospholipid bilayers, where the effect of headgroup size is often obscured by other effects, such as charged-group interaction, hydrogen bonding, etc. However, the predicted effect of polar headgroup size has been observed with phospholipid bilayers under conditions in which interference from charged-group interaction or other effects is suppressed. For example, in experiments with a series of dipalmitoylphosphatidylalkanols at low pH, Eibl [93] demonstrated that a decrease in Tm does occur when the size of the headgroup is increased (Table 2.5). However, such effects are not observed at near neutral pH [94]. In fact, a similar trend is also evident from a comparison of the Tm values of DPPC and DPPE bilayers at neutrality and at very low pH (Table 2.7). Evidently, the protonation of the phosphate moieties of both the normally anionic and the zwitterionic phospholipids suppresses the critical interference attributable either to charged-group repulsion or to hydrogen-bonding interactions and allows the effect of polar headgroup size to be observed.

The effect of polar headgroup size can be more easily demonstrated, however, with nonionic glycolipid bilayers, where interference from charged-group interactions does not exist. Such effects are clearly evident from a survey of the available data on the chain-melting transition temperatures of the monoglycosyl- and diglycosyl glycerolipids (Table 2.6). Here, the observation that the monoglycosyl glycerolipids generally exhibit considerably higher gel/liquid–crystalline phase transition temperatures than their diglycosyl counterparts with similar hydrocarbon chains can be attributed primarily to the larger size of the diglycosyl headgroup.

However, one should be very cautious when considerations of headgroup size are involved. Mannock et al. [95, 96] observed that the gel/liquid–crystalline phase transition temperatures of a homologous series of α-D-glucosyl diacylglycerols were significantly lower than those of comparable β-anomers. In principle, the concept of effective headgroup size can be invoked to explain these observations, if one considers the fact that the change in the anomeric configuration of the glucose headgroup alters the orientation of the headgroup relative to the bilayer surface [97]. NMR spectroscopic studies indicate that the sugar headgroup of the α-linked anomers is aligned nearly parallel to the bilayer surface, whereas that of the β-anomer is extended away from the bilayer surface [97]. Thus, the lower phase transitions of the α-anomers may be the result of an effectively

TABLE 2.6
Gel to Liquid–Crystalline Phase Transition Temperatures of the Monoglycosyl and Diglycosyl Glycerolipids[a]

Monoglycosyl Glycerolipids			Diglycosyl Glycerolipids		
Sample	T_m (°C)	Ref.	Sample	T_m (°C)	Ref.
14:0 dialkyl β-D-Glc	50.8	304	14:0 dialkyl Gentiobiose	27.5	308
16:0 dialkyl β-D-Glc	63.6	304	16:0 dialkyl Cellobiose	54.0	309
18:0 dialkyl β-D-Glc	72.0	304	16:0 dialkyl Maltose	52.0	309
12:0 dialkyl β-D-Gal	31.7	137			
14:0 dialkyl β-D-Gal	52.2	305	18:0 diacyl digalactosyl	50.3	307
14:0 dialkyl α-D-Glc	52.0	97			
14:0 dialkyl α-D-Man	52.0	97			
16:0 diacyl β-D-Gal	51.6	306			
18:0 diacyl β-D-Gal	68.5	306			

[a] See also Table 2.9.

larger headgroup cross-sectional area resulting from the change in the orientation of the sugar moiety. However, as pointed out by Mannock et al. [96], the change in the anomeric configuration of the sugar headgroup also changes parameters, such as the interaction between the sugar and its hydration sphere, the orientation of the hydrophobic surfaces of the sugar to the bilayer surface, and even the penetration of water into the interfacial region of the bilayer. Because all such "subsidiary" effects will undoubtedly have some effect on the phase behavior of the lipid bilayer, one should be particularly cautious when the effects of polar headgroup size are being assessed, especially when the structural changes being considered are relatively small.

2.3.6.2 The Effect of Charge

Lipid molecules that contain polar headgroups with ionized or ionizable moieties are found in all cell membranes. The presence of these charged groups is a significant determinant of the physical properties of the lipid bilayer. Clearly, the presence of charged or ionizable headgroups confers sensitivity to parameters that directly or indirectly alter their capacity to interact with charged or otherwise polar species. Moreover, as illustrated previously, the mere existence of such charged groups can effectively suppress the manifestation of the influence of other physical factors, such as headgroup size. The lipids that exhibit such properties include the overwhelming majority of the common glycerophospholipids and the glycolipid species in which there are sugar moieties esterified to phosphate groups (i.e., phosphorylated glycolipids), sulfate groups (e.g., cerebroside sulfate), or amino acid moieties. With such lipid bilayers, factors such as pH, the overall ionic strength of the medium, and the presence of specific ions tend to influence bilayer physical properties, primarily by their effects on the charged group interactions between the polar headgroups.

This effect is illustrated by the data shown in Table 2.7 and in Figure 2.10, which demonstrate the effect of pH on the gel/liquid–crystalline phase transition temperatures of some common phospholipids. As expected, the significant changes in Tm generally occur over the pH ranges in which there is protonaton/deprotonation of the ionizable groups. This seems reasonable, because protonation and/or deprotonation of ionizable groups would inevitably alter the attractive/repulsive interactions between the polar headgroups and change the relative stability of the gel phase by their effects on the close packing interactions of the lipid molecules. However, the particular effect being expressed is also strongly dependent on the location of the charged group being titrated. Thus, for example, the neutralization of the negatively charged phosphodiester group of DPPC and DMPC at low pH

TABLE 2.7
Effect of pH on the Gel to Liquid–Crystalline Phase Transition Temperatures Dipalmitoyl Glycerophospholipids

Sample	pH 1			pH 7		
	Charge	Tm (°C)	Ref.	Charge	Tm (°C)	Ref.
DPPC	+	49	98	+ −	41	98
DPPM		53	111	−	44	111
DPPG		61	178	−	41	178
DPPE	+	66	93[a]	+ −	64	52
DPPS	+	69	100[b]	− − +	55	100[b]
DPPA				−	71	310[c]

[a] Tm = 41°C at pH 12.
[b] Tm = 32°C at pH 13.
[c] Tm = 45°C at pH 11.

FIGURE 2.10 pH dependence of the chain-melting phase transition temperatures of various dimyristoylglycerophospholipid bilayers. Transitions involving metastable states or little hydrated, highly tilted lipids at low pH are not indicated. The superscripts give the lipid charge and the abbreviations are given in the text. (Redrawn from [5].)

results in a marked increase in Tm [98, 99], despite the fact that the molecular species formed at low pH has a net positive charge. Presumably, this arises because close packing of the PC molecules without short range interactions of the positively charged trimethyl ammonium moieties near the end of the phosphorylcholine group is feasible, whereas short-range interactions between the phosphate ester moieties always occur whenever PC molecules are assembled into a bilayer.

The effect of the binding of specific ions to the bilayer surface and of bulk aqueous phase ionic strength on the thermotropic phase behavior of the ionic lipids can also be rationalized from a consideration of their effects on the interaction between the charged groups. Thus, at near-neutral pH, the elevation of the Tm of PS bilayers by an increase in ionic strength of the bulk aqueous phase can be attributed to increased shielding of the charged moieties of the headgroup by the increased polarity of the solvent phase [100]. In those instances where the specific binding of cations such as Ca^{2+} and Li^+ to anionic lipid bilayers occurs, there is a marked elevation in Tm [see 101 and references cited therein]. This effect is undoubtedly similar to that which occurs when the negative charge of the phosphate moiety is neutralized at low pH. Here, the binding of the counterion promotes closer packing of the lipid molecules by an effective neutralization of the charged moiety on the headgroup, thereby stabilizing the gel phase of the lipid. Moreover, in those instances where divalent cations are bound, there may even be further promotion of lipid close packing by the formation of divalent cation bridges between the polar headgroups [102]. In such instances, the stabilization of the gel phase of the lipid may be so strong that nucleation and growth of highly ordered crystalline phases may even occur [103–105].

2.3.6.3 The Effect of Hydrogen Bonding

It is now generally accepted that the physical properties of a lipid bilayer are influenced by hydrogen-bonding interactions at its surface. With regard to the crystalline phases that many lipids form, single-crystal X-ray studies of glyco- and phospholipids [106, 107, and references cited therein] have established that the formation of a hydrogen-bonding network at the bilayer surface is often a key to the enhanced stability of that structure. However, it is more difficult to evaluate the influence of hydrogen-bonding interactions on the thermotropic properties of fully hydrated lipid bilayers: once fully hydrated, the lipid polar headgroups seem more disposed to hydrogen-bonding interactions with the solvent than with each other. With hydrated lipid bilayers, experimental evidence that hydrogen bonding is a major determinant of its thermotropic phase behavior is often indirect and subject to varied interpretations, mainly because it is often difficult to design experiments that would definitively assign any particular result to the effects of intermoleular hydrogen bonding. Thus, despite the many studies that have directly or indirectly addressed this issue [for reviews, see 108–110], there is yet to be a consensus on the extent to which these forces affect the thermotropic phase properties of fully hydrated lipid bilayers, though there is general agreement that their effects are significant.

For example, in studies of the pH dependence of the gel/liquid–crystalline phase transition of 1,2 di-tetradecyl and 1,2 di-hexadecyl phosphatidic acids (PAs), Eibl [111] observed that the plot of Tm vs. pH showed a maximum in the pH range where partial, but not complete, protonation of the phosphate moiety occurs (Figure 2.11). With these lipids, one can rationalize the progressive increase in Tm that occurs with acidification down to near pH 3 from the perspective of charge-group interaction effects (see Section 2.3.6.2), but not the decrease in Tm as the formal charge of the polar headgroup approaches neutrality. To explain the latter, it was postulated that a stabilizing network of hydrogen bonds forms as the pH of the medium approaches the pK1 of phosphatidic acid. The formation of this network presumably promotes the formation of tight interactions between the lipid molecules (thereby elevating Tm), and the decrease in Tm at lower pH is the result of the destruction of this network by further acidification [111]. The involvement of hydrogen-bonding in this manner is certainly plausible (to date it is the only explanation that has been proposed) and currently forms the basis of the rationalization of similar observations in studies of dimyristoyl- and dipalmitoylphosphatidylmethanol bilayers [see 111].

It is apparent, however, that the type of hydrogen-bonding interaction described previously probably has more in common with that observed in the crystalline phases of lipid bilayers than in hydrated lipid bilayers. With PE bilayers, for example, a different pattern of hydrogen-bonding interaction is often considered in attempts to explain why at near neutral pH, its Tm is unusually high when compared with comparable anionic and zwitterionic phospholipids [see 110]. In this case, a system is envisaged in which there is dynamic equilibrium between "free" phosphate groups, phosphate groups hydrogen-bonded to water, and phosphate groups hydrogen-bonded to the amine protons of neighboring headgroups. This seems plausible, especially because it has been shown theoretically that the enhanced gel phase stability resulting from even such transient interheadgroup

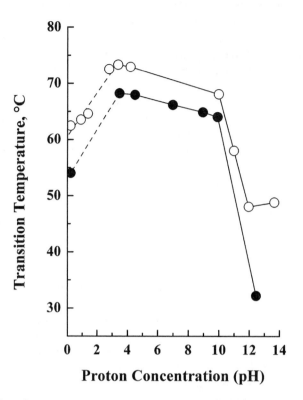

FIGURE 2.11 The effect of proton concentration on the temperature of the main transition of 1,2-dihexadecyl- and 1,2-ditertadecyl-*sn*-glycero-3-phosphoric acid. Dispersions contained 20 mg of lipid in 20 mg of bidistilled water. The pH was adjusted by the addition of dilute NaOH. The dotted lines indicate regions where two transitions are sometimes observed: a lower and an upper transition. (Redrawn from [111].)

hydrogen-bonding interactions is great enough to result in a significant increase in Tm [112]. Moreover, because this model does not require or even suggest any immobilization of the polar headgroups, it is quite compatible with the 31P-NMR spectroscopic data, which clearly show that, on the NMR timescale, the motion of the phosphate headgroup is axially symmetric in both the gel and liquid–crystalline phases [113, 114]. The presence of a dynamic hydrogen-bonding network in the headgroup and polar/apolar interfacial regions of fully hydrated PE bilayers is supported by Fourier transform infrared (FTIR) spectroscopic evidence for the existence of a population of H-bonded ester carbonyl groups, which is not present in comparable PC bilayers [114].

A similar process of dynamic interchanges of headgroup–headgroup and headgroup–solvent hydrogen-bonding interactions may also be occurring with the nonionic glycolipids. Upon examination of the gel/liquid–crystalline phase transition temperatures of such lipids (see Tables 2.7 and 2.9), it becomes apparent that their Tms are still too high to be rationalized by factors that do not include a significant contribution from intermolecular hydrogen bonding. In fact, because there are no charged or ionizable groups present, hydrogen-bonding effects probably form the basis of virtually all aspects of the physical properties of the nonionic glycolipids, either as crystalline solids or hydrated bilayers.

Finally, in any evaluation of the influence of hydrogen-bonding interactions on bilayer physical properties, it should be recognized that in addition to the effects described previously, hydrogen-bonding interactions also influence other properties of the polar headgroup. For example, hydrogen bonding is an important determinant of the capacity of the polar headgroup to interact with and to bind water. In this way, hydrogen bonding also affects the properties of the lipid bilayer by its influence on headgroup hydration number, which in turn determines the effective size of the polar

TABLE 2.8
Gel to Liquid–Crystalline Transition Temperatures of Aqueous Dispersions of the Phosphatidyl Sulfocholines and Thiophosphatidylcholines

Sample	T_m (°C)	Ref.
Dimyristoyl Phosphatidylsulfocholine	26.5	115
Dipalmitoyl Phosphatidylsulfocholine	45.0	115
Distearoyl Phosphatidylsulfocholine	60.5	115
Dioleoyl Phosphatidylsulfocholine	−24.5	115
(Rp) Dipalmitoyl ThioPhosphatidylcholine	44.9	117
(Sp) Dipalmitoyl ThioPhosphatidylcholine	45.0	117
(Rp+Sp) Dipalmitoyl ThioPhosphatidylcholine	44.8	117

headgroup. Moreover, the influence of "subsidiary" effects such as these is often antagonistic to the "primary" factors being considered. As is the case with any of the properties being considered in this review, the effects of hydrogen bonding are very complex; unless great care is exercised and all of the possibilities considered, the conclusions drawn can be misleading.

2.3.6.4 Miscellaneous Effects

It is not unexpected that the phase behavior of a lipid bilayer is sensitive to changes in the chemical structure of the lipid polar headgroup. However, what is remarkable is the fact that significant changes in bilayer physical properties are often observed as a result of seemingly small chemical modifications, which do not appear to alter the essential character of the polar headgroup, whereas in many instances the properties of the bilayer are relatively unresponsive to considerably larger chemical modifications. For example, it has been shown that the replacement of the trimethyl ammonium group of the PCs with a dimethylsulfonium moiety affects both the Tm (Table 2.8) and the structural basis of their gel-phase polymorphism [115, 116]. In this case, the changes in the chemical properties of the headgroup are very subtle, are not close to the polar/apolar interfacial region of the bilayer, and are certainly smaller than those of replacing the phosphorylcholine moiety with phosphorylglycerol, for example — and yet the effect on the thermotropic phase properties of the bilayer is considerably greater. These observations are yet to be explained or satisfactorily rationalized.

A similar trend was also observed in some unrelated studies by Tsai and coworkers (Table 2.8), who showed that the replacement of one of the unesterified phosphate oxygens of DPPC with a sulfur atom results in an increase in Tm [117], changes in the dynamic properties of the polar headgroup at temperatures above Tm [118], and major changes in the basis of the gel-phase polymorphism [119]. These observations may be related to the generation of a chiral phosphorus close to the polar/apolar interfacial region of the lipid bilayer. However, as is the case with the phosphatidylsulfocholines described previously, the reasons why such small changes significantly affect the thermotropic phase properties of the bilayer, whereas other more drastic chemical modifications exert much smaller effects, have not been established. Evidently, there are many parameters to consider, even when seemingly small chemical modifications are made to the lipid polar headgroup.

Finally, we point out that attention is rarely given to hydrophobic interactions in considerations of the influence of the lipid polar headgroup on the thermotropic properties of the lipid bilayer. In principle, these effects can originate from the interchain interactions in the hydrophobic core of the lipid bilayer, as well as hydrophobic components in the polar headgroup itself. Of these, the influence of the former is relatively straightforward and can be easily explained. It has become

TABLE 2.9
Effect of Hydrocarbon Chain Length and Structure on the Gel to Liquid–Crystalline Phase Transition Temperatures of Phosphatidylcholines, Phosphatidylethanolamines, and the Monoglucosyl Diacylglycerols

Chain Structure	T_m (°C)			
	PCs	PEs	α-D-GlcDG[a]	β-D-GlcDG[b]
12:0	−2.1[c]	31.3[d]	19.5	26.0
13:0	13.7[c]	42.1[d]	32.9	35.7
14:0	23.9[c]	50.4[d]	40.5	45.5
15:0	34.7[c]	58.4[d]	50.7	54.2
16:0	41.4[c]	64.4[d]	57.2	61.0
17:0	49.8[c]	70.5[d]	63.4	67.0
18:0	55.3[c]	74.2[d]	68.4	71.7
19:0	61.8[c]	79.2[d]	73.7	76.5
20:0	66.4[c]	83.4[d]	76.8	79.7
18:1cΔ9	−17.3[e]	−16.0[f]		
18:1tΔ9	13.0[g]	38.3[h]		
19:0i	43.7[i]	59.0[j]		
19:0ai	29.5[k]	44.6[i]		

[a] Ref. [96].
[b] Ref. [95].
[c] Ref. [12].
[d] Ref. [114].
[e] Ref. [21].
[f] Ref. [311].
[g] Ref. [86].
[h] Ref. [312].
[i] Ref. [17].
[j] Ref. [113].
[k] Ref. [22].

increasingly apparent that the physical properties of any given lipid bilayer is the result of a "compromise" between the driving forces governed by polar interactions in the headgroup and interfacial regions of the bilayer and hydrophobic interactions in the bilayer core. Moreover, because the magnitude of the interchain hydrophobic interactions is intimately linked to the length and the structure of the hydrocarbon chain, it is logical to infer that the influence of the polar headgroup should also be responsive to these parameters. This expectation is indeed corroborated by an examination of the data shown in Table 2.10, which show that the influence of the polar headgroup diminishes with increasing chain length. Thus, for example, the replacement of the choline moiety of dilauroylphosphatidylcholine with an ethanolamine group increases Tm by some 32°C, whereas the same operation with diarachidoylphosphatidylcholine increases Tm by only 17°C (see Table 2.9). This trend is a logical consequence of the increase in the magnitude of the interchain hydrophobic interactions, coincident with an increase in hydrocarbon chain length, and the tendency of these forces to dominate the properties of the bilayer as hydrocarbon chain length increases. Also, in marked contrast to bilayers composed of saturated fatty acyl chains, the effect of the replacement of a choline moiety with an ethanolamine group is very small when *cis*-monounsaturated fatty acyl chains are present (see Table 2.9). In this case, it has been suggested that with

**TABLE 2.10
Gel to Liquid–Crystalline Phase
Transition Temperatures of Dimyristoyl-
Phosphatidylethanolamine and Its
C-alkylated Analogues[a]**

Amine Moiety	T_m (°C)
2-Aminoethanol (PE)	50.1
2-Methyl, 2-Aminoethanol	43.4
2, 2-Dimethyl, 2-Aminoethanol	37.5
2-Ethyl, 2-Aminoethanol	34.2
3-Amino, 1-Propanol	41.9
4-Amino, 1-Butanol	34.4

[a] Measurements all made at near neutral pH [313].

both lipid classes, the introduction of the *cis* double bond increases the molecular area, resulting in a marked diminution in the strength of the hydrogen-bonding interactions of the polar headgroups, which are believed to stabilize the gel phase of the PE bilayers [110].

Unlike the effects of the hydrocarbon chain, however, the influence of hydrophobic interactions involving the polar headgroup is more difficult to evaluate and/or predict. There is little doubt, however, that such effects must influence bilayer physical properties, because there is a hydrophobic component to the chemical structure of virtually all lipid polar headgroups. The inherent difficulties encountered here are typified by the examples shown in Tables 2.5 and 2.10. Although the data seem to suggest that there is sensitivity to the size of the hydrophobic component of the polar headgroup, unambiguous interpretation is difficult because the chemical modifications of the head-group moieties alter not only the hydrophilic/hydrophobic balance of the moiety, but also its effective size, its hydrogen-bonding capacity, its solubility in water, and many other factors. Thus, it would be virtually impossible to assign specifically any single effect to the influence of hydro-phobic interaction, though there is little doubt that these effects are important. This provides a timely reminder that the effects of any property of the polar headgroup (or indeed of any part of the lipid molecule) on the properties of a lipid bilayer are probably best evaluated with due regard to the interactions of all of the moieties that comprise the given lipid molecule.

2.3.7 THE EFFECT OF VARIATIONS IN THE CHEMICAL STRUCTURE OF THE GLYCEROL BACKBONE REGION ON LIPID THERMOTROPIC PHASE BEHAVIOR

In this section, we shall examine how chemical variations in the structure of the *glycerol backbone region,* and the linkages between this region and the hydrocarbon chains, affect lipid thermotropic phase behavior. That the structure of this portion of the lipid molecule should influence the physicochemical properties of the lipid bilayer is not surprising, because the glycerol backbone region determines many of the properties of the polar/apolar interfacial region of the bilayer. In particular, the chemical structure of the interfacial region can influence the overall conformation of the lipid molecule and thus the nature of the mesomorphic phases that it can form. Moreover, it can also influence the degree of hydration of the interfacial region and the depth of water penetration into the bilayer, which in turn influence the properties and conformation of the lipid polar headgroup [120]. Our evaluation of the influence of the glycerol backbone region's chemical structure on bilayer physical properties will be largely comparative, and the intensively studied 1,2-diacyl-*sn*-glycerolipids will serve as the reference.

2.3.7.1 The Effect of Chirality

On account of the high stereospecificity of virtually all biosynthetic pathways, the overwhelming majority of naturally occurring lipids contain at least one optically active center in the interfacial region of the molecule. In the case of the phosphoglycerolipids, for example, a single chiral center at C2 of the glycerol backbone is present. To date, little emphasis has been placed on a comparison of the physical properties of such D- and L-stereoisomers, probably because their physical properties are expected to be identical. The few studies performed with such lipids have concentrated on a comparison of one of the optically active enantiomers with the racemic mixture. Such studies have found that the Tms of phospholipid bilayers are relatively insensitive to the loss of optical activity at the glycerol, but have indicated that their gel-phase behavior is radically altered [121, 122]. However, in a recent study of a polymerizable phosphatidylcholine, a marked difference in the thermotropic phase properties of the racemic and optically active isomers was observed [123]. In this case, the sensitivity of the chain-melting transition to the loss of optical activity at the glycerol is probably attributable to the fact that these lipids readily form highly ordered crystalline phases at temperatures below Tm, as inferred from infrared spectroscopic studies [124].

The relative insensitivity of bilayer physical properties to the stereochemistry at a single optically active center, as observed in phosphoglycerides, is probably unlikely in molecules such as the glycoglycerolipids and the sphingolipids, for which additional chiral centers exist either in the headgroup or the polar/apolar interface, respectively. With such molecules, the inversion of configuration at any one chiral center results in the formation of diastereomers, as opposed to enantiomers. Some recent studies have shown that there can be differences in the physical properties of diastereomeric pairs of lipids in the liquid–crystalline phase [118] and especially in the gel phase of such bilayers [119, 125] but report little difference between the chain-melting phase transition temperatures of diastereomeric lipids [see 125 and references cited therein]. However, because the stereochemistry at the various chiral centers on such molecules will undoubtedly affect the orientation of the polar groups relative to each other, parameters such as intermolecular hydrogen bonding, headgroup, and interfacial hydration should be affected. Given this, one would expect that more detailed investigations into this area would indeed find that there are major differences in the phase behavior of these lipids.

2.3.7.2 The 1,3-Diacyl Glycerolipids

Relatively few studies have been directed at the influence of the position of the acyl and headgroup moieties on the glycerol backbone on the physical properties of the lipid bilayer. A examination of the thermodynamic data on the few compounds that have been studied so far (see Table 2.11) indicates that these compounds have lower chain-melting phase transition temperatures than their 1,2-diacyl counterparts. It is possible that this may be related to the effect of conformational differences at the glycerol backbone [126]. It has also been deduced from 2H-NMR spectroscopic studies of 1,3 DPPC that, unlike its 1,2-diacyl counterpart, its glycerol backbone lies parallel to the bilayer surface [127]. As a result, the attachment of both acyl chains to the glycerol is parallel to the bilayer surface, and they are both bent perpendicular to the bilayer at their respective C2 carbons. Presumably, the lowering of the chain-melting transition temperature is the result of small changes in the interchain van der Waals contacts that result from this conformational change. The available data also suggest that the altered conformation of the glycerol backbone promotes the formation an interdigitated gel phase at temperatures below Tm [128] and of highly ordered crystalline structures at still lower temperatures [128, 129].

2.3.7.3 Glycerolipids with Ether-Linked Hydrocarbon Chains

Glycerolipids with ether-linked hydrocarbon chains are found as major constituents of the membrane lipids of thermoacidophylic microorganisms, where, presumably, the resistance of ether

TABLE 2.11
Thermtropic Transitions of 1,3 Diacyl Glycerolipids

Sample	Subtransition		Chain-Melting Transition	
	Tm (°C)	$\Delta H_{cal}{}^a$	T_m (°C)	$\Delta H_{cal}{}^a$
1,3 DPPE[b]	42.8	4.5	53.1	9.3
1,3 DMPC[c]	15.0	4.3	19.0	6.1
1,3 DPPC[d]	27.0	9.1	37.0	10.5
1-O-14:0,3-O-18:0 PC[c]	16.0	5.9	30.0	7.1
1-O-16:0,3-O-18:0 PC[c]	26.0	7.3	46.0	10.4
1,3 DMPA[e]			42.0	

[a] Values are in Kcal/mol.
[b] Ref. [129].
[c] Ref. [66].
[d] Ref. [128].
[e] Ref. [310].

linkages to hydrolytic cleavage would enhance survival. Interest in ether-linked glucerolipids also stems from the observation that their levels appear to be elevated in neoplastic tissue [130, 131]. The Tms of all ether-linked glycerolipids are higher than those of the corresponding ester-linked analogues (see Table 2.12 for some representative examples). The stabilization of glycerolipid gel phases by ether-linked hydrocarbon chains varies somewhat with the nature of the polar headgroup but is a consistent feature of all of the data published so far. Available data also show that the lamellar/nonlamellar phase transition temperatures of all ether-linked glycerolipids are substantially lower than those of the corresponding ester-linked analogues [see 6], indicating that ether-linked hydrocarbon chains also destabilize liquid–crystalline glycerolipid bilayers with respect to inverted nonlamellar phases.

In principle, both of these effects can be attributed to the combination of the smaller size and reduced polarity of the ether bonds, which normally favor tighter packing and reduced hydration at bilayer polar/apolar interfaces. Interestingly, however, a comparison of properties of DPPC with those of its dialkyl and acyl–alkyl analogues suggests that ether-linked hydrocarbon chains may promote the formation of chain interdigitated bilayers at temperatures below the Tm [132–134], facilitate greater hydration of the headgroup and polar/apolar interfacial regions of the bilayer [135], and cause a change in the preferred conformation and/or orientation of the glycerol backbone [135]. For the most part, these effects seem incompatible with the smaller size and lower polarity of the ether linkage, and to our knowledge such behavior has not been observed with any of the PEs and monoglycosyl glycerolipids examined [see 136–141]. However, relatively little information on this aspect of the glycerolipid behavior is available, and it is therefore difficult to make any definitive conclusions on this topic. Nevertheless, the fact that even these small changes in the chemical structure of lipid polar/apolar interfaces can have such wide-ranging effects on glycerolipid structure and organization underscores the critical importance of this region of glycerolipid molecules.

2.3.7.4 Lipids with Interfacial Amide Groups

A distinguishing feature of the sphingolipids is the presence of an amide rather than an ester group in the polar/apolar interfacial region. This class of lipids includes the sphingomyelins (SMs) and the glycosphingolipids, both of which are important components of the cell membranes of higher animals. There is relatively little physical data available on these lipids, mainly because they have not been studied as intensively as the diacylglycerolipids. The SMs and their synthetic analogues

TABLE 2.12
Gel to Liquid–Crystalline Phase Transition
Temperatures of Diacyl and Dialkyl Glycerolipids

Sample	T_m (Diacyl)	T_m (Dialkyl)
14:0 PC	23.9[a]	27.0[b]
16:0 PC	41.4[a]	44.2[c]
12:0 PE	31.3[b]	35.0[b]
14:0 PE	50.5[b]	55.5[b]
16:0 PE	64.4[b]	68.5[b]
18:0 PE	74.2[b]	77.0[b]
14:0 PA	50.0[d]	61.5[e]
16:0 PA	67.0[f]	75.0[e]
14:0 N-Methyl PE	42.7[g]	46.0[b]
14:0 N,N-Dimethyl PE	31.4[g]	34.0[b]
14:0 Phosphoryl 3-amino-propane	41.9[g]	46.0[b]
14:0 Phosphoryl 4-amino-butane	34.4[g]	38.0[b]
14:0 PG	23.7[d]	26.0[h]
14:0 β-D-GlcDG	45.5[i]	50.0[j]
16:0 β-D-GlcDG	61.0[i]	63.6[j]
18:0 β-D-GlcDG	71.7[i]	72.0[j]
14:0 α-D-GlcDG	40.5[k]	52.0[l]

[a] Ref. [12].
[b] Ref. [114].
[c] Ref. [133].
[d] Ref. [178].
[e] Ref. [310].
[f] Ref. [165].
[g] Ref. [313].
[h] Ref. [314].
[i] Ref. [95].
[j] Ref. [96].
[k] Ref. [97].
[l] Ref. [304].

tend to form highly ordered crystalline phases [125, 142–146], presumably because of the stabilization of such structures by hydrogen-bonding interactions involving the interfacial amide and hydroxyl moieties. The gel to liquid–crystalline phase transition temperatures of the SMs are also higher than those of comparable PCs (the Tm of the DL N-16:0, dihydrosphingosine compound is some 6°C higher than that of DPPC, see Table 2.13), possibly due to stronger hydrogen-bonding interactions involving the interfacial amide bond. Presumably, this is the result of increased interfacial hydration caused by the greater polarity of the amide groups. Given this, it is not clear why the melting temperatures of the SMs should be so different from those of the diacyl PCs.

The pattern that emerges from the available data on the glycosphingolipids is remarkably similar to that of the SMs. Like the SMs, the glycosphingolipids also form very stable crystalline phases [148–152 and references cited therein], and in this instance the involvement of the interfacial amide in hydrogen bonding interactions has been confirmed by single-crystal X-ray studies [106, 107, and references cited therein]. Their crystalline phases usually melt directly to the liquid–crystalline phase at fairly high temperatures. With the overwhelming majority of these lipids, the Lβ-type of lamellar gel phase is very difficult to characterize because it is unstable with respect to one or more of the crystalline phases. As a result, data on the typical gel/liquid–crystalline phase transition of

---

TABLE 2.13
Gel to Liquid–Crystalline Phase Transition Temperatures of Sphingomyelins

Sample	T_m (°C)
DL N-16:0	41.3[a]
DL N-16:0, dihydrosphingosine	47.8[a]
DL N-18:0	46.0[b]
D N-18:0	44.7[c]
L N-18:0	44.2[c]
DL N-24:0	48.8[d]

[a] Ref. [142].
[b] Ref. [144].
[c] Ref. [125].
[d] Ref. [143].

---

these lipids are rarely reported. However, the DSC cooling thermograms that have been published do show that typical Lβ/Lα type transitions occur at temperatures lower than those of the Lc/Lα melting transitions that are commonly observed [for examples, see 148–152]. Our estimates of the gel to liquid–crystalline phase temperatures of these lipids (estimated from the published DSC cooling thermograms) tend to be lower than those available for comparable monoglycosyl diacyl-glycerols, and in this respect the glycosphingolipids appear to differ from the SMs (see above).

TABLE 2.14
Gel to Liquid–Crystalline Phase Transition Temperatures of Amide and Carbamyl Containing PC Analogues of Sphingomyelin

Sample	T_m (°C)
1-O-Octadecyl, 2-N-16:0 PC	47.8[a]
1-O-18:0, 2-N-18:0 PC	52.9[a]
1-O-18:0, 2-N-(Octadecylamino)carbonyl PC	57.7[a]
1-O-18:0, 2-N-(Octyloxy)carbonyl PC	6.8[a]
1,2 Ditridecanylcarbamyloxy PC	33[b,c]
1,2 Dipentadecanylcarbamyloxy PC	46g
1,2 Diheptadecanylcarbamyloxy PC	55g
1-O-16:0, 2-O-Tridecanylcarbamyl PC	38g
1-O-16:0, 2-O-Pentadecanylcarbamyl PC	47g
1-O-16:0, 2-O Heptadecanylcarbamyl PC	53g

[a] Ref. [147].
[b] Ref. [145].
[c] Ref. [146]. The transition temperatures reported for these lipids are those of the melting of their crystalline phases. The authors reported that these lipids exhibited two transitions (most probably the Lα/Lβ and Lβ/Lc transitions) at lower temperatures upon cooling. Their data suggest that the normal gel to liquid–crystalline phase transition temperatures of 1-O-16:0, 2-O-Tridecanylcarbamyl PC and 1,2 Dipentadecan-ylcarbamyloxy PC should be 27°C and 36°C, respectively.

Interestingly, however, this latter observation is compatible with published work on amide-containing PC analogues, which indicates that interfacial amide groups may actually cause a decrease in Tm [147]. Thus the relatively high Tms of SMs may be the result of special inter- or intramolecular interactions that are not possible with the synthetic amide-containing PC analogues of SM.

2.3.7.5 Lipids with Conformationally Restricted Glycerol Backbones

In this section, we shall briefly review the effect of chemical modifications that alter the conformation in the polar/apolar interfacial region of the lipid bilayer. We recognize that any chemical modification of the lipid polar/apolar interface will have some effect on the conformation of its constituent moieties. Thus we will concentrate here on those experimental studies in which conformational changes were induced without major changes in the chemical character of the molecule.

Of the various approaches adopted in these studies, the simplest involves substitution of methyl groups for hydrogens at the C1 and C3 positions of the glycerol moiety [153]. Both of these operations result in a 4 to 6°C reduction of Tm, presumably because of expansion and/or conformational changes in the interfacial region to accommodate the methyl group. Another, more novel approach involved the synthesis of 1,2- and 1,3-DPPC analogues, in which the 1,2,3-triol residues normally supplied by glycerol were supplied by the 1,2,3-triol groups of cyclopentane-1,2,3 triol [154]. The physical data of two of the analogues of 1,2 DPPC that were synthesized were somewhat unusual, and it was later demonstrated that they form highly ordered crystalline bilayers that melt directly to the liquid–crystalline phase at relatively high temperatures [155]. Moreover, liposomes composed of these two lipids were also permeable to sodium ions at all temperatures [154]. One can easily demonstrate with molecular models that the orientation of the acyl chains of these two lipids is such that it is difficult to align the chains to form a bilayer without their adopting a very unusual packing. Their properties may be a reflection of such conformational restrictions. However, for those analogues in which the orientation of the acyl chains on the cyclopentane ring was more favorable to bilayer formation, it was found that the cyclopentano analogues of 1,2-DPPC exhibited lower (3 to 5°C) phase transition temperatures than that of the reference glycerolipid, whereas the Tms of the cyclopentano analogues of 1,3-DPPC were higher (5 to 7°C) than those of 1,3-DPPC [154].

These changes in the thermotropic properties of these bilayers seem remarkably small when one takes into account the nature of the chemical change affected. Perhaps this is a reflection of the fact that the conformation of the glycerol backbone at the polar/apolar interface of normal glycerolipids is normally very restricted and that the effect of further restrictions is not as large as one would imagine. Interestingly, the type of the conformational restrictions imposed by this particular cyclopentane ring system affects the positional isomers of DPPC differently. Evidently, the conformational freedoms of the glycerol moieties of the 1,2 and 1,3 phospholipids are very different.

The effect of any conformational restriction of components of the polar/apolar interfacial region of the lipid bilayer also depends on the part of the molecule upon which such "restrictions" are imposed. This effect is vividly demonstrated by the dipentadecylmethylidine analogues of dihexadecyl phospholipids developed by Blume and Eibl [156]. In contrast to the analogues described previously, the linkage of the hydrocarbon chains to the interfacial region of the dipentadecylmethylidine analogues imposes severe conformational restraints on the hydrocarbon chains themselves. Evidently, irrespective of the chemical structure of the of polar headgroup, this has a major destabilizing effect on the gel phase of the lipid bilayer, as evidenced by the substantial (20 to 30°C) decreases in Tm that result (Table 2.15). One can show, with molecular models of these analogues, that the hydrocarbon chains will not have the conformational freedom to maximize van der Waals contacts in the gel phase. This is in marked contrast to the other work described previously, where the chemical changes restricted the mobility of the glycerol backbone while leaving the acyl chains relatively free to adopt an optimal conformation.

TABLE 2.15
Gel to Liquid–Crystalline Phase Transition Temperatures of Dipentadecylmethylidene Phospholipids and Their Dihexadecyl Analogues

Polar Headgroup	T_m (Dihexadecyl)	T_m (Dipentadecylmethylidene)
PA	75.0[a]	38.5[b]
PE	68.5[c]	37.5[b]
N-Methyl PE	62.0[d]	31.0[b]
N,N-Dimethyl PE	51.0[d]	24.0[b]
PC	44.2[e]	17.9[b]

[a] Ref. [310].
[b] Ref. [156].
[c] Ref. [52].
[d] Ref. [315].
[e] Ref. [133].

2.3.8 THE THERMOTROPIC PHASE BEHAVIOR OF LIPID MIXTURES

Although studies of the thermotropic phase behavior of single-component multilamellar phospholipid vesicles are necessary and valuable, these systems are not realistic models for biological membranes, which normally contain at least several different types of phospholipids and a variety of fatty acyl chains. As a first step toward understanding the interactions of both the polar and apolar portions of different lipids present in mixtures, DSC studies of various binary phospholipid systems have been carried out. Phase diagrams can be constructed by specifying the onset and completion temperatures for the phase transition of a series of mixtures and by an inspection of the shapes of the calorimetric traces. A comparison of the observed transition curves with the theoretical curves supports a literal interpretation of the phase diagrams obtained by DSC [157].

The thermotropic behavior of mixtures of two linear saturated PCs differing in the nature of their fatty acyl constituents has been investigated. Disaturated PCs differing by only two carbons in the length of their hydrocarbon chains exhibit almost ideal behavior in binary phospholipid–water dispersions (see Figure 2.12). Although a difference in chain length of four carbons presents a system considerably removed from ideality, isothermal melting of the shorter-chain lipid is not observed and a significant degree of lateral phase separation does not occur (see 158]. A difference in hydrocarbon chain length of six carbons results in monotectic behavior, with the chain-melting onset temperature remaining constant over most of the concentration range; a region of pronounced lateral phase separation is evident from the DSC traces (see Figure 2.13) [86, 158–161]. Monotectic behavior is also observed for the systems DOPC plus DMPC, DPPC, or DSPC, indicating that di-cis-monounsaturated PCs and disaturated PCs of whatever chain length are largely immiscible in the gel state [159]. In contrast, DEPC and DMPC are nearly perfectly miscible in all proportions, while DEPC and DPPC mix less ideally and exhibit a solid–solid immiscibility gap at mole fractions of DPPC from about 0.30 to 0.55 [158]. In general, the smaller the difference in the Tm values of these simple diacyl PCs, the more nearly ideal is their mixing behavior. Phase diagrams for the binary systems POPC:DPPC and SOPC:DPPC have also been constructed using DSC [88]. The POPC–DPPC system exhibits behavior that is far from ideal, but there is little gel-state immiscibility at any composition, whereas the SOPC–DPPC system exhibits appreciable gel state immiscibility, particularly at DPPC concentrations of less than 50 mol%. Of these latter two binary systems, the more nearly ideal behavior is exhibited by the PC pair with the largest difference in Tm values, indicating that the miscibility of PCs in bilayers may be influenced by more subtle structural variations as well as by differences in chain-melting temperatures.

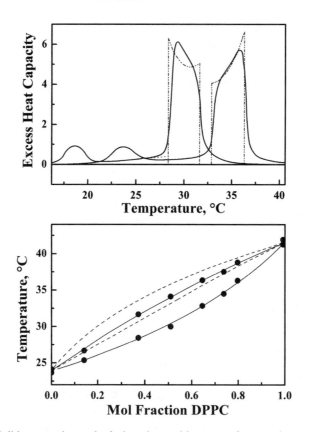

FIGURE 2.12 (A) Solid curves: observed calorimetric transition curves for two mixtures of DMPC and DPPC. (Redrawn from [161].) Dashed curves: transition curves calculated [161] on the basis of the phase diagram in B. (B) Solid curves: phase diagram constructed from initiation and completion temperatures read from observed transition curves. Dashed lines: ideal phase diagram. (Calculations are described in [161 and 157].)

A number of calorimetric studies have been done of binary mixtures of two different phospholipids containing either similar or dissimilar fatty acids. For binary mixtures of the sodium salts of disaturated phosphatidylglycerols (PG-Na+) and PCs with identical acyl chains, nearly ideal mixing is observed. Mixing simple PGs and PCs with increasingly large differences in the chain lengths of their saturated fatty acyl groups, or mixing disaturated with diunsaturated lipids, produces increasingly nonideal behavior, but no more so than for a binary mixture of PCs containing the same fatty acids. These results indicate a high degree of miscibility of the PG and PC headgroups, either in the presence or absence of Ca2+. Interestingly, small amounts of PC were found to abolish the formation of high-melting metastable PG-Ca2+ or PG-Mg2+ complexes observed with the pure PGs alone [162].

Several groups have studied binary mixtures of PC and PE [160, 163, 164]. Mixtures of these two phospholipids having identical fatty acyl groups, such as DMPC–DMPE, exhibit quite nonideal behavior. Although these lipids seem miscible in the liquid–crystalline phase, the solidus curves of these phase diagrams show a minimum at about 20 mol%, indicating gel-state immiscibility in this composition range. Some gel-state immiscibility is also observed in DMPE–DPPC and DMPE–DSPC mixtures; however, those mixtures, particularly the latter one, exhibit more nearly ideal behavior than the DMPE–DMPC system. Because the ΔTm values also decrease as the chain length of the PC component of the DMPE–PC mixtures increases from 14 to 18 carbon atoms, these results suggest that the relative chain-melting temperatures, rather than the absolute relative chain lengths, are of primary importance in determining PE–PC miscibility. Nevertheless, even when the Tm values of

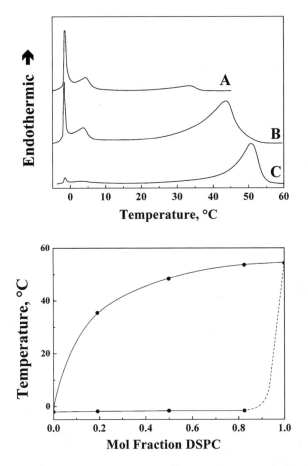

FIGURE 2.13 (A) Solid curves, dashed baselines: transition curves for DLPC–DSPC mixtures with mole fraction DSPC (X_{DSPC}) = 0.191 and 0.819. Dashed curve, solid baseline: transition curve for X_{DSPC} = 0.498. (B) Phase diagram constructed from calorimetric transition curves. (Redrawn from [161].)

both components are closely matched, PE–PC mixtures display considerably more nonideality than comparable PC–PC or PC–PG mixtures, indicative of some polar headgroup immiscibility.

The phase behavior of mixtures of DPPC with bovine brain PS [165, 166] has been studied by DSC in conjunction with other techniques. Complex and somewhat nonideal phase mixing occurs, but more quantitative conclusions about the miscibility of polar headgroups are difficult because the natural PS is already a mixture of molecules differing in the length and degree of unsaturation of their fatty acyl chains. Calorimetric studies of egg PC–cerebroside [167] and synthetic PC–ganglioside [87, 168] mixtures have also been published. For a summary of calorimetric and other studies of the thermotropic phase behavior of natural and synthetic SMs and of their mixing properties with cholesterol and phospholipids, the reader is referred to a comprehensive review [169].

Theoretical analyses of the phase diagrams of binary mixtures of disaturated PCs of different chain lengths and of PC–PE mixtures indicate that microscopic mixing may be significantly nonideal in the liquid–crystalline phase as well as the gel phase, even though no macroscopic lateral phase separations occur [170, 171]. Thus, microscopic clusters of like lipid molecules may exist in the gel and liquid–crystalline phases, although generally more random mixing will occur in the liquid–crystalline phase. Interestingly, the propensity for the self-association of two different liquids in the gel and liquid–crystalline states is not always related; neither is this propensity always correlated with the macroscopic phase behavior of that particular system. Some indirect noncalorimetric evidence for phospholipid clustering in the liquid–crystalline state is available [108].

Because alterations in pH and ionic strength can cause significant changes in the Tm, particularly of acidic phospholipids, isothermal phase transitions can be induced in pure phospholipids by variation in pH or divalent cation concentration. Similarly, an isothermal lateral phase separation in liquid–crystalline mixtures of a zwitterionic lipid, such as PC, and an acidic lipid, such as PG, PS, or phosphatidic acid (PA), can also be induced by changes in pH or by the addition of Ca2+. In this case, domains of gel or quasi-crystalline acidic phospholipids can presumably be formed, leaving the neutral phospholipid in separate fluid domains. Calorimetric and other evidence for Ca2+-induced phase separations in PS–PC [165, 172–174,], PA–PC [175–178], and PG–PC [162, 178, 179] mixtures has been published. Apparently, Mg2+ is not as effective at inducing lateral phase separations in such binary mixtures [165, 175, 176]. Currently, it is not clear whether or not such divalent cation-induced isothermal phase transitions occur in biological membranes [see 1–2].

2.3.9   THE EFFECT OF CHOLESTEROL ON THE THERMOTROPIC PHASE BEHAVIOR OF PHOSPHOLIPIDS

The occurrence of cholesterol and related sterols in the membranes of eukaryotic cells has prompted many investigations of the effect of cholesterol on the thermotropic phase behavior of phospholipids [see 4]. Studies utilizing calorimetric and other physical techniques have established that cholesterol can have profound effects on the physical properties of phospholipid bilayers and probably plays an important role in controlling the fluidity of biological membranes. Cholesterol induces an "intermediate state" in phospholipid molecules with which it interacts, increasing the fluidity of the hydrocarbon chains below and decreasing the fluidity above the gel to liquid–crystalline phase transition temperature [see 180, 181]. See Chapter 7 for a detailed discussion of the interactions of cholesterol with phospholipids and sphingolipids in model and biological membranes.

A considerable number of DSC studies of cholesterol–phospholipid binary mixtures have been carried out, by far the majority on cholesterol–DPPC or, to a lesser extent, on cholesterol–DMPC systems. Although all earlier studies, which used relatively low-sensitivity calorimeters, agreed that the progressive addition of cholesterol broadens the gel to liquid–crystalline phase transition of these phospholipids and progressively reduces the transition enthalpy, there was considerable disagreement about the details of this interaction. For example, the addition of cholesterol was reported either to considerably reduce [182], to have little effect on [183, 184], or to increase [185] the phospholipid phase transition temperature. Also, the cooperative phospholipid phase transition was reported to be completely abolished at cholesterol concentrations ranging from 33 to 43 mol%. Similar results were reported on the interaction of cholesterol with various binary mixtures of linear saturated PCs [186]. Moreover, only a single endothermic transition was detected in all of these studies at all cholesterol concentrations examined.

More recent high-sensitivity DSC studies of cholesterol–PC interactions, however, have revealed a more consistent but also more complex picture of cholesterol–DPPC and cholesterol–DMPC interactions [187–190]. At cholesterol concentrations from 0 to 20–25 mol%, the DSC endotherm consists of two components. The sharp component exhibits a phase transition temperature and cooperativity only slightly reduced from those of the pure phospholipid, and the enthalpy of this component decreases linearly with increasing cholesterol content, becoming zero at 20–25 mol%. In contrast, the broad component exhibits a progressively increasing phase transition temperature and enthalpy, with a progressively decreasing cooperativity over this same range of cholesterol content. Above cholesterol levels of 20–25 mol%, the broad component becomes progressively less cooperative, the phase transition midpoint temperature continues to increase, and the transition enthalpy continues to decrease, eventually approaching zero only at cholesterol concentrations near 50 mol%. Moreover, these observations have been confirmed by a high-sensitivity dilatometric analysis [191]. These results suggest that at low cholesterol concentrations, cholesterol-poor and cholesterol-rich domains coexist, with the former decreasing in proportion to the latter as cholesterol concentrations increase. In fact, a cardinal point in the cholesterol–DPPC

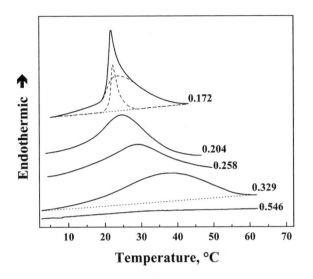

FIGURE 2.14 DSC heating scans of four mixtures of cholesterol with DMPC (DML) in aqueous suspension at pH 7. (Redrawn from [157].)

phase diagram at about 22 mol% had been predicted from the earlier model-building study of Engleman and Rothman [192, 193], who calculated that the cholesterol molecule could interact with a maximum of seven adjacent phospholipid hydrocarbon chains (or 3.5 phospholipid molecules) and thus that free phospholipid would exist only at cholesterol concentrations below this value. This model also explains the decreasing enthalpy of the broad component observed above 22 mol% cholesterol, because an increasing proportion of phospholipid molecules would interact with more than one cholesterol molecule, rather than with the more flexible hydrocarbon chains of adjacent phospholipids, thus progressively decreasing and eventually abolishing the cooperative chain-melting phase transition.

McMullen and coworkers [194] have studied the effects of cholesterol on the thermotropic phase behavior of aqueous dispersions of a homologous series of linear saturated PCs, using high-sensitivity DSC and an experimental protocol which ensures that the broad, low-enthalpy phase transitions present at high cholesterol concentrations are accurately monitored. They found that the incorporation of small amounts of cholesterol progressively decreases the temperature and the enthalpy, but not the cooperativity, of the pretransition of all PCs exhibiting such a pretransition and that the pretransition is completely abolished at cholesterol concentrations above 5 mol% in all cases. Moreover, the incorporation of increasing quantities of cholesterol was found to alter the main or chain-melting phase transition of these phospholipid bilayers in both hydrocarbon chain length–dependent and hydrocarbon chain length–independent ways. The temperature and cooperativity of the sharp component are reduced only slightly and in a chain length–independent manner with increasing cholesterol concentration, an observation ascribed to the colligative effect of the presence of small quantities of cholesterol at the domain boundaries. Moreover, the enthalpy of the sharp component decreases and becomes zero at 20 to 25 mol% cholesterol for all of the PCs examined. In contrast, the broad component exhibits a chain length–dependent shift in temperature and a chain length–dependent decrease in cooperativity, but a chain length–independent relative increase in enthalpy over the same range of cholesterol concentrations. Specifically, cholesterol incorporation progressively increases the phase transition temperature of the broad component in PCs having hydrocarbon chains of 16 or fewer carbon atoms, while decreasing the broad-component phase transition temperature in PCs having hydrocarbon chains of 18 or more carbon atoms, an effect attributed to hydrophobic mismatch between the cholesterol molecule and its host PC bilayer [194]. The best match between the effective length of the cholesterol molecule and the mean hydrophobic thickness of the PC bilayers is obtained with the diheptadecanoyl PC molecule. Moreover, cholesterol decreases the cooperativity of the broad component more rapidly and to a greater extent in the shorter-chain, as compared to the longer-chain, PCs. At cholesterol concentrations above 20 to 25 mol%, the sharp component is abolished, and the broad component continues to manifest the chain length–dependent effects on the temperature and cooperativity described previously. However, the enthalpy of the broad component decreases linearly and reaches zero at about 50 mol% cholesterol, regardless of the chain length of the phosphatidylcholine.

This latter finding does not agree with previous studies of PC and PE bilayers [195, 196], which found that the cholesterol concentration required to reduce the enthalpy of the main phase transition to zero appeared to increase steeply and approximately linearly with phosphatidylcholine hydrocarbon chain length. These previous results were ascribed to an experimental artifact arising from the use of a low-sensitivity calorimeter and an experimental protocol not optimized to detect broad, low-enthalpy phase transitions. Moreover, subsequent studies [197] showed that an apparent increase in the cooperativity of the DPPC endotherm at about 7 mol% cholesterol was due to a crossing over of the sharp and broad components of this endotherm and not to the existence of a triple point in the phase diagram, as had been postulated earlier [187]. Interestingly, hydrophobic mismatch effects can also explain the limited miscibility of androsterol [198] and alkyl chain–truncated cholesterol derivatives in DPPC and SOPC bilayers [199].

The effect of cholesterol on the thermotropic phase behavior of PCs appears to depend on their fatty acid composition, in particular on the number and type of unsaturated fatty acyl chains present, at least according to the early low-sensitivity DSC studies [200–202]. For example, the amount of cholesterol required to completely abolish a calorimetrically detectable phase transition in SOPC and OSPC has been reported to be 40 and 30 mol%, respectively, considerably less than the 50 mole% required for DSPC. Moreover, when the oleoyl chain is replaced with a linoleoyl chain, only 17 mole% cholesterol is required to remove the phase transition. Thus, the amount of cholesterol necessary to abolish cooperative chain melting seems to decrease with increasing unsaturation in saturated/unsaturated mixed-acid PCs and to depend on the position of the unsaturated chain on the glycerol backbone. However, in the mixed-chain saturated/unsaturated phospholipids AOPC and OAPC, the phase transition enthalpy did not decrease to zero until 50 mol% cholesterol. Moreover, with DOPC a calorimetrically detectable phase transition apparently remains even at 50 mole% cholesterol, and cholesterol has been reported to have no effect at all on the broad, low-temperature transitions detected in aqueous dispersions of PC containing two identical polyunsaturated fatty acids.

It is very difficult to rationalize the complex and apparently inconsistent results reported in these early low-sensitivity DSC studies of the interactions of cholesterol with PCs containing one or two unsaturated fatty acyl chains. It is particularly difficult to explain how the presence of a single unsaturated fatty acyl chain in a PC molecule could make these phospholipids more sensitive to the presence of cholesterol but the presence of two unsaturated fatty acids could make them less sensitive, or how a small change in the chain length of the saturated fatty acid in mixed-chain PCs could have such a profound effect on its interactions with cholesterol. We therefore believe it likely that the low sensitivity of the calorimeter employed and high scan rates utilized in these studies obscured the true thermotropic phase behavior of the phospholipid–cholesterol systems being investigated. In support of this contention, we note that recent high-sensitivity DSC studies of SOPC–cholesterol [199] and DEPC–cholesterol [203] mixtures indicate that their thermotropic phase behavior is essentially identical to that of DSPC and demonstrate clearly that the transition enthalpy in both cases decreases progressively with increasing cholesterol concentration and approaches zero at 50 mol% cholesterol. These latter results indicate that the presence of a single *cis*-monounsaturated or of two *trans*-monounsaturated fatty acids in a PC does not significantly alter the nature or stoichiometry of cholesterol–phospholipid interactions, bringing into question many of the findings reported earlier. However, several recent studies indicate that PCs containing two highly polyunsaturated fatty acids do exhibit a markedly reduced miscibility with cholesterol [204–206].

McMullen et al. [207] also studied aqueous dispersions of cholesterol-containing PE bilayers by high-sensitivity DSC and NMR spectroscopy. Regardless of hydrocarbon chain length, the incorporation of low levels of cholesterol into these bilayers causes progressive reductions in the temperature, enthalpy, and overall cooperativity of the lipid hydrocarbon chain-melting phase transition. Moreover, at low cholesterol levels, the heating and cooling thermograms observed for the cholesterol–PE binary mixtures are similar, indicating comparable levels of lateral miscibility of cholesterol with PE bilayers in the gel and liquid–crystalline states. However, at higher levels of cholesterol incorporation, marked differences between the heating and cooling thermograms are noted. Upon heating, complex multicomponent thermograms are observed in PE bilayers containing large amounts of cholesterol, and the temperature and overall enthalpy values *increase* discontinuously from the pattern of monotonic decrease observed at lower cholesterol levels. Moreover, these discontinuities begin to emerge at progressively lower cholesterol concentrations as PE hydrocarbon chain length increases. Upon cooling, a simpler pattern of thermotropic behavior is observed, and the measured temperature enthalpy values continue to decrease monotonically with increases in cholesterol content. These results suggest that at higher concentrations, cholesterol exhibits a decreased degree of lateral miscibility in the gel or crystalline as compared to the liquid–crystalline states of PE bilayers, particularly in the case of the longer-chain PEs. Upon subsequent heating, the melting of these crystalline phases gives rise to the complex thermograms detected by DSC and to the discontinuities in the phase transition temperature and enthalpy noted previously. This pattern of behavior differs markedly from that observed with the corresponding PCs, where comparable degrees of cholesterol miscibility are observed in the gel and liquid–crystalline states even at high cholesterol concentrations, and where cholesterol inhibits rather than facilitates the formation of lamellar crystalline phases. These workers also found that the presence of cholesterol does not result in the hydrophobic mismatch-dependent shifts in the phase transition temperature in PE bilayers previously observed in PC bilayers of varying thickness. These differences in the effects of cholesterol on phospholipid thermotropic phase behavior were attributed to stronger electrostatic and hydrogen bonding interactions at the surfaces of PE and compared to PC bilayers.

The effect of cholesterol on the thermotropic phase behavior of negatively charged phospholipids appears to differ somewhat from that observed for the zwitterionic phospholipids PC, PE, and SM [see 208–210]. The addition of cholesterol to natural or synthetic PS, PG, and PA bilayers, respectively, results in a slight and marked reduction in Tm, as is also observed for the strongly hydrogen-bonded PE bilayers. However, at cholesterol levels above 30 to 40 mol%, a phase separation of cholesterol appears to occur, so that these phospholipids continue to exhibit a chain-melting transition even at cholesterol concentrations of 50 mol% or more. Moreover, the formation of a separate cholesterol crystallite phase at higher cholesterol concentrations in PS bilayers has been reported [211–214]. A limited solubility of cholesterol of the gel phase of the uncharged sphingolipid galactocerebroside has also been reported [208, 215].

McMullen and coworkers [216] have recently reinvestigated the thermotropic phase behavior of mixtures of cholesterol with a homologous series of linear saturated PSes. In contrast to previous reports, these workers found that the temperature, enthalpy, and cooperativity of the gel to liquid–crystalline phase transition of the host PS bilayer were progressively reduced by the incorporation of increasing quantities of cholesterol, such that a cooperative phase transition is abolished at 50 mol% sterol. Moreover, a separate anhydrous cholesterol or cholesterol monohydrate phase was not detected in these binary mixtures of PS and cholesterol, even at the higher cholesterol concentrations examined. These workers thus concluded the cholesterol is fully miscible in PS bilayers up to at least 50 mol% and ascribed the apparent limited solubility of cholesterol in PS bilayers reported previously to a fractional crystallization of the cholesterol and phospholipid phases during the removal of organic solvent prior to the hydration of the sample. Indeed, Feigenson and coworkers have recently demonstrated that when such problems are avoided, the maximum solubility of cholesterol in PC and PE bilayers is 67 and 50 mol%, respectively [217, 218], and approaches 67 mol% in PS bilayers [G.W. Feigenson, personal communication]. These latter results argue against the idea that anionic phospholipids in general exhibit a reduced ability to interact with the uncharged cholesterol molecule when present in high concentrations. This conclusion is further supported by earlier work (discussed below) showing that limiting amounts of cholesterol show comparable or higher affinities for the anionic phospholipids PS and PG than for the zwitterionic phospholipids PC and PE, respectively, and that cholesterol exhibits a greater miscibility in anionic PG than in uncharged glycolipid bilayers [219].

Several conventional DSC studies of binary mixtures of two phospholipids exhibiting gel phase immiscibility have indicated that cholesterol may preferentially associate with different lipid classes in such mixtures [220, 221]. When cholesterol is added to monotectic mixtures of PC–PC, PC–PG, PC–PS or PS–PE, cholesterol preferentially interacts with the lower-melting lipid, as indicated by a decrease in ΔHcal and a broadening of the phase transition of that component. In other binary systems, however, cholesterol exhibits a preference for a particular phospholipid, whether or not that lipid is the lower- or higher-melting component. The order of preference of cholesterol thus established was SM > PS ~ PG > PC > PE. Cholesterol does not appear, however, to show any calorimetrically detectable preferential affinity in several PC-PE-cholesterol [222] and PC-SM-cholesterol [185] systems which, although nonideal, were not monotectic. Because the extracted phospholipids from eukaryotic membranes without cholesterol show single, relatively broad transitions and no monotectic behavior, it seems unlikely that in biological membranes preferential cholesterol–phospholipid interactions of sufficient strength and specificity occur, inducing the formation of large cholesterol-free and cholesterol-rich domains as apparently occurs in certain noncocrystallizing binary mixtures of synthetic phospholipids. Cholesterol-poor or cholesterol-rich microdomains or clusters that are also enriched or depleted of certain phospholipid classes may, however, occur, and there is evidence for the formation of cholesterol- and SM-enriched lipid domains in both model and biological membranes (see Chapter 7).

2.3.10 THE EFFECT OF SMALL MOLECULES ON THE THERMOTROPIC PHASE BEHAVIOR OF PHOSPHOLIPIDS

A number of lipid-soluble small molecules, including drugs like tranquilizers, antidepressants, narcotics, and anesthetics, produce biological effects in living cells. Although some of these compounds are known to produce their characteristic effects by interacting with specific membrane proteins, others seem to interact rather nonspecifically with the lipid bilayer of many biological membranes. The effect on the Tms of synthetic PCs of over 100 hydrophobic small molecules producing biological effects has now been studied by DSC [223]. At least four different types of modified transition profiles can be distinguished:

> In so-called type C profiles, the addition of the additive shifts Tm usually (but not always) to a lower temperature, while having little or no effect on the cooperativity (ΔT1/2) or ΔHcal of the transition. Other physical evidence suggests that additives producing this behavior are usually localized in the central region of the bilayer, interacting primarily with the C9-C16 methylene region of the phospholipid hydrocarbon chains.
> Type A profiles are characterized by a shift in Tm usually to a lower temperature, an increase in ΔT1/2, and a relatively unaffected ΔHcal upon the addition of the appropriate small molecules. These additives appear to be partially buried in the hydrocarbon core of the bilayer, interacting primarily with the C2-C8 methylene region of the hydrocarbon chains.
> In type B profiles, a shoulder emerges on the main transition, the area of which increases in conjunction with a corresponding decrease in the area of the original peak as the concentration of additive increases. The total area of both peaks is relatively unchanged, at least at low additive concentrations. Additives that produce type B profiles are generally present at the hydrophobic/hydrophilic interface of the bilayer and interact primarily with the glycerol backbone of the phospholipid molecules.

Finally, type D profiles exhibit a discrete new peak, which grows in area at the expense of the parent peak as the additive concentration increases. Normally, however, the final ΔHcal and ΔT1/2 values of the new and original peaks are not greatly different. Type D additives usually seem to be located at the bilayer surface and to interact with the phosphorylcholine headgroup.

Although this classification is useful, not all small molecules produce one of these four types of DSC profiles. It remains to be determined whether or not a consistent relationship exists between the type of transition profile produced by a small molecule and its physiological effects.

Free fatty acids occur as minor components of many biological membranes, and these compounds can alter the permeability properties of model membranes and the activity of certain enzymes in biomembranes. Moreover, free fatty acids can promote the fusion or lysis of phospholipid vesicles and cells. DSC and other physical techniques have been utilized to monitor the effect of different fatty acids on the gel to liquid–crystalline phase transition of synthetic PCs [37, 224–228]. In general, the addition of small amounts of saturated free fatty acids of 12 to 18 carbons to DPPC multilamellar dispersions results in increased Tm, ΔT1/2, and ΔHcal values for the main transition, while abolishing the pretransition. Palmitic acid has the largest effect on DMPC bilayers, probably because the location of the carboxyl group at the hydrophobic/hydrophilic interface results in the palmitic acid hydrocarbon chain having the same effective length as the myristoyl chains in DMPC. Saturated fatty acids with ten or fewer carbons and unsaturated fatty acids also increase ΔT1/2 but, in contrast, lower the Tm and ΔHcal values for chain melting. Very long chain saturated fatty acids (20 to 22 carbons) increase ΔT1/2 and decrease ΔHcal without affecting Tm, as does the addition of cholesterol. Interestingly, the presence of saturated or unsaturated fatty acids does not appear to alter lipid fluidity in the liquid–crystalline state as monitored by pyrene eximer fluorescence [227]. The addition of high levels of palmitic acid to DPPC (at a mole ratio of 2:1) produces a sharp, asymmetric melting profile with a Tm of 61.5°C, only a few degrees below the Tm of DPPE [226]. This observation was taken to support the concept that the lower Tm values characteristic of PCs, as compared to PEs, may be the result of a destabilizing crowding of the relatively bulky PC headgroups. However, it has now been shown that the sharp endothermic phase transition at 61.5°C exhibited by the palmitic acid:DPPC (2:1) complex is actually a lamellar gel to a nonlamellar reversed hexagonal (Lβ to HII) phase transition, not the Pβ'/Lα phase transition exhibited by DPPC alone [229]. The formation of the thermally stable Lβ phase by the palmitic acid–DPPC complex may be due in part to the formation of hydrogen bonds between the largely protonated fatty acids and the oxygens of the phosphate headgroups [230].

Lysophospholipids also occur as minor components of most biological membranes and, like free fatty acids, can affect membrane permeability, promote membrane fusion, and modulate the activity of some membrane enzymes. The effect of increasing concentrations of 1-palmitoyl-*sn*-glycerol-3-phosphorylcholine (lysoPPC) on the thermotropic phase behavior of DPPC has been studied by DSC [231]. LysoPPC, which alone exhibits an endothermic transition at 3.4°C, causes a nonlinear decrease in the Tm of DPPC. LysoPPC also causes an initial, slight increase in the ΔHcal of DPPC, followed by a gradual decrease at higher lysoPPC concentrations, but at low concentrations the ΔT1/2 is unaffected. No phase separation can be detected, and the lamellar phase persists up to 50 mol% lysoPPC. Mixtures of lysoPPC with DOPC or DSPC, or of lysoOPC with either DPPC or DOPC, however, exhibit immiscibility in the PC gel state [232, 233]. Small amounts of lysoPPC also abolish the pretransition of DPPC. Cholesterol was also found to decrease the ΔHcal of the pure lysoPPC transition, eliminating it at a concentration of about 50 mol%, as for DPPC. Moreover, cholesterol increases the Tm of the lysoPPC transition [231]. The ether analog of lysoPPC was also found to lower the Tm of DPPE, but in contrast to DPPC, the ΔT1/2 is markedly increased, as is the ΔHcal, from 8.2 to 18.0 kcal/mol. This very high ΔHcal was ascribed to the breaking of the network of intermolecular hydrogen bonds between the PE molecules caused by the insertion of lysoPPC. The addition of lysoPPC to DPPC–DPPE mixtures (which, in the absence of the lysophospholipid, exhibit almost complete gel phase miscibility) induces a separation into three different phases. In contrast, the addition of the ether analog of lysoPPC to DMPE–DSPC mixtures abolishes the miscibility gap normally found in this particular binary system [232]. Thus, the effect of lysoPPC on phospholipid mixtures is complex and depends on the nature of both the lysophospholipid and the diacyl phospholipid being studied.

2.3.11 The Effect of Transmembrane Peptides on Lipid Thermotropic Phase Behavior

The synthetic peptide acetyl-K_2-G-L_{24}-K_2-A-amide (P_{24}) and its analogs have been successfully utilized as a model of the hydrophobic transmembrane α-helical segments of integral membrane proteins [see 234 and referenes cited therein]. These peptides contain a long sequence of hydrophobic leucine residues capped at both the N- and C-termini with two positively charged, relatively polar lysine residues. Moreover, the normally positively charged N-terminus and the negatively charged C-terminus have both been blocked in order to provide a symmetrical tetracationic peptide, which will more faithfully mimic the transbilayer region of natural membrane proteins. The central polyleucine region of these peptides was designed to form a maximally stable α-helix, which will partition strongly into the hydrophobic environment of the lipid bilayer core; the dilysine caps were designed to anchor the ends of these peptides to the polar surface of the lipid bilayer and to inhibit the lateral aggregation of these peptides. In fact, circular dichroism (CD) and FTIR spectroscopic studies of P_{24} have shown that it adopts a very stable α-helical conformation both in solution and in lipid bilayers [235, 236]. X-ray diffraction [237], fluorescence quenching [238], FTIR [236], and ^{2}H-NMR [239] spectroscopic studies have confirmed that P_{24} and its analogs assume a transbilayer orientation with the N- and C-termini exposed to the aqueous environment and the hydrophobic polyleucine core embedded in hydrocarbon core of the lipid bilayer when reconstituted with PCs. ^{2}H-NMR and electron spin resonance (ESR) spectroscopic studies have shown that the rotational diffusion of P_{24} about its long axis perpendicular to the membrane plane is rapid in the liquid–crystalline state of the bilayer and that the closely related peptide acetyl-K_2-L_{24}-K_2-amide (L_{24}) exists at least primarily as a monomer in the liquid–crystalline PC bilayers, even at relatively high peptide concentrations [for a review, see 234].

High-sensitivity DSC and FTIR spectroscopy were used to study the interaction of P_{24} and members of the homologous series of n-saturated PCs [240, 241]. In the low range of the peptide mole fractions, the DSC thermograms exhibited by the lipid–peptide mixtures are resolvable into two components (see Figure 2.15). One of these components is fairly narrow and highly cooperative, and exhibits properties that are similar to but not identical with those of the pure lipid. In addition, the fractional contribution of this component to the total enthalpy change, the peak transition temperature, and cooperativity decrease with an increase in the peptide concentration, more or less independently of the acyl chain length. The other component is very broad and predominates in the high range of peptide concentration. These two components were assigned to the chain-melting phase transitions of populations of bulk lipid and peptide-associated lipid, respectively. Moreover, when the mean hydrophobic thickness of the PC bilayer is less than the peptide hydrophobic length, the peptide-associated lipid melts at higher temperatures than the bulk lipid and vice versa. In addition, the chain-melting enthalpy of the broad endotherm does not decrease to zero, even at high peptide concentrations, suggesting that this peptide reduces but does not abolish the cooperative gel/liquid–crystalline phase transition of the lipids with which it is in contact. The DSC results indicate that the width of the phase transition observed at high peptide concentration is inversely but discontinuously related to the hydrocarbon chain length and that gel phase immiscibility occurs when the hydrophobic thickness of the bilayer greatly exceeds the hydrophobic length of the peptide. The FTIR spectroscopic data again indicate that the peptide forms a very stable α-helix but that small distortions of its α-helical conformation are induced in response to any mismatch between peptide hydrophobic length and bilayer hydrophobic thickness (see Figure 2.16). These results also

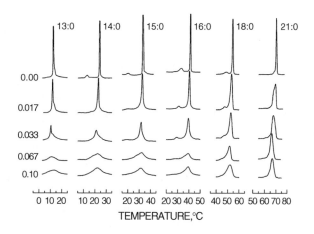

FIGURE 2.15 The effect of increasing quantities of peptide P_{24} on the DSC thermograms of a series of n-saturated diacyl PCs. Thermograms are shown as a function of the acyl chain length (N:0) of the lipids, and the approximate mol fraction of P_{24}, as indicated on the column of numbers on the left side of the figure. (Redrawn from [241].)

indicate that the peptide alters the conformational disposition of the acyl chains in contact with it and that the resultant conformational changes in the lipid hydrocarbon chains tend to minimize the extent of mismatch of the peptide hydrophobic length and bilayer hydrophobic thickness. Interestingly, ESR and ^{2}H-NMR spectroscopic studies have shown that, when incorporated into liquid–crystalline DPPC and DOPC bilayers, the closely related peptide L_{24} increases the orientational order of the phospholipid hydrocarbon chains, in contrast to natural transmembrane proteins [242, 243].

High-sensitivity DSC and FTIR spectroscopy were also used to study the interaction of P_{24} and members of a homologous series of n-saturated PEs [244]. In the lower range of peptide molecular fractions, the DSC endotherms exhibited by the lipid–peptide mixtures again consist of two components, which can be attributed to the chain-melting phase transitions of peptide-nonassociated and peptide-associated PE molecules, respectively. Although the temperature at which the peptide-associated PE molecules melt is progressively decreased by increases in peptide concentration, the

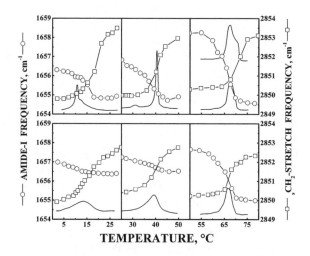

FIGURE 2.16 Combined plots of the lipid CH_2 symmetric stretching frequency (-□-), the peptide amide I band maxima (-○-), and the calorimetric thermograms as a function of temperature, for mixtures of P_{24} with 13:0 PC (left), 16:0 PC (middle), and 21:0 PC (right). The peptide mol fractions are 0.03 (top panels) and 0.1 (bottom panels). (Redrawn from [241].)

magnitude of this shift is *independent* of the length of the PE hydrocarbon chain. In addition, the width of the phase transition observed at higher peptide concentrations is also relatively insensitive to PE hydrocarbon chain length, except that peptide gel-phase immiscibility occurs in very short or very long chain PE bilayers. Again, the enthalpy of the chain-melting transition of the peptide-associated PE does not decrease to zero even at high peptide concentrations, suggesting that this peptide does not abolish the cooperative gel/liquid–crystalline phase transition of the lipids with which it is in contact. The FTIR spectroscopic data indicate that the peptide remains in a predominantly α-helical conformation but that the peptide α-helix is subject to small distortions coincident with the changes in the hydrophobic thickness that accompany the chain-melting phase transition of the PE bilayer. These data also indicate that the peptide significantly disorders the hydrocarbon chains of the adjacent PE molecules in *both* the gel and liquid–crystalline states relatively *independently* of the lipid hydrocarbon chain length. The relative independence of many aspects of PE-peptide interactions on the hydrophobic thickness of the host bilayer is in marked contrast to the results of the previous study of peptide/PC model membranes [241], where strong hydrocarbon chain length–dependent effects were observed. The differing effects of peptide incorporation on PE and PC bilayers were ascribed to the much stronger lipid polar headgroup interactions in the former system. The primary effect of transmembrane peptide incorporation into PE bilayers was postulated to be the disruption of the relatively strong electrostatic and hydrogen-bonding interactions at the bilayer surface, this effect being sufficiently large to mask the effect of the hydrophobic mismatch between the lengths of the hydrophobic core of the peptide and its host bilayer. Interestingly, as discussed in Section 2.3.9, the effect of cholesterol on the gel/liquid–crystalline phase transition temperature exhibits a similar differential dependence of phospholipid bilayer thickness in PC and PE systems.

The interactions of the hydrophobic helical transmembrane peptide Ac-K_2-(LA)$_{12}$-K_2-amide [(LA)$_{12}$] with a series of *n*-saturated PCs and PEs were also studied by high-sensitivity DSC and FTIR spectroscopy [245–247]. In general, the effects of (LA)$_{12}$ on phospholipid thermotropic phase behavior are similar to those observed with P_{24}, except that (LA)$_{12}$ decreases the temperature and enthalpy of the gel to liquid–crystalline phase transition of the host phospholipid to a greater extent, indicating a more substantial reduction in the organization of the gel-state bilayers. The FTIR spectra of (LA)$_{12}$ in these phospholipid bilayers indicate that this peptide also retains a predominantly α-helical conformation in both the gel and liquid–crystalline phases of the short- to medium-chain phospholipids studied. However, when incorporated into the bilayers composed of the longer-chain phospholipids, (LA)$_{12}$ undergoes a reversible conformational distortion at the gel/liquid–crystalline phase transition of the mixture. In the liquid–crystalline phase, the amide I regions of the FTIR spectra of these mixtures indicate a predominantly α-helical peptide conformation. However, upon freezing of the lipid hydrocarbon chains, populations of (LA)$_{12}$ giving rise to a sharp, conformationally unassigned band near 1665 cm^{-1} are formed, indicating a greater degree of conformational flexibility than for P_{24}. A comparison of the results of these calorimetric and FTIR spectroscopic studies with similar studies of a polyleucine-based analogue of (LA)$_{12}$ suggests that the thermodynamics of the interaction of hydrophobic transmembrane helices with lipid bilayers can be influenced by factors such as the polarity and topology of the helical surface, factors that are dependent on the amino acid sequence of the helix. Also, possible adjustments to a hydrophobic mismatch between the peptide and its host lipid bilayer covers a spectrum of possibilities, which can include changes in the degree of conformational disorder in the lipid chains and/or significant conformational changes on the part of the peptide.

The conformation and amide proton exchangeability of the model transmembrane peptide acetyl-K_2-A_{24}-K_2-amide (A_{24}) and its interaction with PC bilayers have also been studied by a variety of physical techniques [248]. A combination of the CD and FTIR spectroscopic results indicate that, when dissolved in methanol or deposited from methanol as a dried film, A_{24} is predominantly α-helical. Upon dissolution in aqueous media, rapid H-D exchange of A_{24} amide protons occurs, indicating that the peptide is sufficiently conformationally dynamic that all amide protons are fully exposed to the solvent. CD and FTIR spectroscopic techniques also reveal that A_{24} exists primarily as a mixture of helical (but probably not α-helical) and β-sheet structures in aqueous media at room temperature. Upon heating, A_{24} converts reversibly primarily to unordered structures in unbuffered water but irreversibly to antiparallel β-sheet structures in phosphate-buffered saline. These studies also indicate that although A_{24} exists primarily as a membrane α-helix when incorporated into phospholipids in the absence of water, the hydration of that system results in a quick exchange of all amide protons. Also, when dispersed with PC in the aqueous media, the conformation and thermal stability of A_{24} are not significantly altered by the presence of the phospholipid, its phase state, or its gel/liquid–crystalline phase transition. The DSC and ESR spectroscopic studies also indicate that A_{24} has relatively minor effects on the thermodynamic properties of the lipid hydrocarbon chain-melting phase transition, that it does not abolish the lipid pretransition, and that its presence has no significant effect on the orientational order or rates of motion of the phospholipid hydrocarbon chains. These results indicate that A_{24} has sufficient α-helical propensity, but insufficient hydrophobicity, to maintain a stable transmembrane association with the phospholipid bilayers in the presence of water. Instead, A_{24} exists primarily as a dynamic mixture of helices, β-sheets, and other conformers and resides mostly in the aqueous phase, where it interacts weakly with the bilayer surface or the interfacial regions of phosphatidylcholine bilayers. These results are consistent with other studies, which show that the insertion of alanine-rich peptides into lipid membranes is neither kinetically nor energetically favored [249–252], and clearly show that polyalanine-based peptides are not good models for the transmembrane α-helical segments of natural membrane proteins.

High-sensitivity DSC and FTIR spectroscopy were also used to study the interaction of L_{24} and odd-chain members of the homologous series of n-saturated PCs. An analog of L_{24}, in which the lysine residues were all replaced by 2,3-diaminopropionic acid, and another, in which a leucine residue at each end of the polyLeu sequence was replaced by a tryptophan, were also studied [253]. FTIR spectroscopic data indicate that these peptides form very stable α-helices under all experimental conditions but that small distortions of their α-helical conformations are induced in response to mismatch between peptide hydrophobic length and gel-state bilayer hydrophobic thickness. Evidence was also presented that these distortions are localized to the N- and C-terminal regions of these peptides. Interestingly, replacing the terminal Lys residues of L_{24} by 2,3-diami-nopropionic acid residues actually attenuates the hydrophobic mismatch effects of the peptide on the thermotropic phase behavior of the host PC bilayer, in contrast to the predictions of the snorkel hypothesis [see 254, 255]. This attenuated hydrophobic mismatch effect was rationalized by postulating that the 2,3-diaminopropionic acid residues are too short to engage in significant electrostatic and hydrogen-bonding interactions with the polar headgroups of the host phospholipid bilayer, even in the absence of any hydrophobic mismatch between incorporated peptide and the bilayer. Similarly, the reduced hydrophobic mismatch effect, also observed when the two terminal Leu residues of L_{24} are replaced by Trp residues, is rationalized by considering the lower energetic cost of exposing the Trp (as opposed to the Leu residues) to the aqueous phase in thin PC bilayers and the higher cost of inserting the Trp (as opposed to the Leu residues) into the hydrophobic cores of thick PC bilayers.

The effects of the model α-helical transmembrane peptide L_{24} on the thermotropic phase behavior of aqueous dispersions of 1,2-dielaidoylphosphatidylethanolamine (DEPE) have also been studied [256]. In particular, the effect of L_{24} and three derivatives thereof on the liquid–crystalline lamellar (L_α)-reversed hexagonal (H_{II}) phase transition of DEPE model membranes was investigated by DSC and ^{31}P-NMR spectroscopy. The incorporation of L_{24} was found to decrease progressively the temperature, enthalpy, and cooperativity of the L_α-H_{II} phase transition, as well as to induce the formation of an inverted cubic phase. This indicates that this transmembrane peptide promotes the formation of the inverted nonlamellar phases, despite the fact that the hydrophobic length of this peptide exceeds the hydrophobic thickness of the host lipid bilayer. These characteristic effects are not altered by truncation of the side chains of the terminal lysine residues or by replacing each of the leucine residues at the end of the polyleucine core of L_{24} by a tryptophan residue. Thus, the characteristic effects of these transmembrane peptides on the DEPE thermotropic phase behavior are independent of their detailed chemical structure. Importantly, significantly *shortening* the polyleucine core of L_{24} results in a *smaller* decrease in the L_α-H_{II} phase transition temperature of the DEPE matrix into which it is incorporated, and *reducing* the thickness of the host phosphatidylethanolamine bilayer results in a *larger* reduction in the L_α-H_{II} phase transition temperature. These results are inconsistent with those predicted by the hydrophobic mismatch effects reported in studies of model α-helical transmembrane peptides composed of an alternating sequence of leucine and alanine residues capped with either di-tryptophan (WALP peptides) or di-lysine (KALP peptides) sequences at their N- and C-termini [257–261]. Evidence was presented that both the WALP and KALP peptides lowered the L_α/H_{II} phase transition temperatures of PEs [257–259] and induced nonlamellar phase formation in PCs [260], under conditions where bilayer hydrophobic thickness significantly exceeds peptide hydrophobic length. Also, the magnitudes of these effects were proportional to the degree of hydrophobic mismatch, and the WALP peptides were more effective at promoting nonlamellar phase formation than the KALP analogues [257]. The results obtained in studies of the WALP and KALP peptides were remarkably similar to those obtained in comparable studies of the interactions of gramicidin A with phospholipid bilayers [262–266]. It thus seems that the effect of transmembrane peptides on the nonlamellar phase-forming propensity of its host lipid membrane is a very complex phenomenon that is subject to many influences other than hydrophobic mismatch considerations.

2.3.12 THE EFFECT OF PROTEINS ON THE THERMOTROPIC PHASE BEHAVIOR OF PHOSPHOLIPIDS

Because of their obvious relevance to biological membranes, the effect of a number of peptides and proteins on the thermotropic phase behavior of single synthetic phospholipids or phospholipid mixtures has been studied by many groups [see 267]. It was originally proposed by Papahadjopoulos et al. [268] that polypeptides and proteins can be considered as belonging to one of three types according to their characteristic effects on phospholipid gel to liquid crystalline phase transitions:

Type 1 proteins typically produce no change or a modest increase in Tm, a slight increase or no change in ΔT1/2, and an appreciable and progressive increase in ΔHcal as the amount of protein added is increased. These proteins normally do not expand phospholipid monolayers or alter the permeability of phospholipid vesicles into which they are incorporated. Type 1 proteins are hydrophilic proteins which are thought to interact with the phospholipid bilayer exclusively by electrostatic forces and, as such, normally show stronger effects on the phase transitions of charged rather than zwitterionic phospholipids.

Type 2 proteins produce a decrease in Tm, an increase in ΔT1/2, and a considerable and progressive decrease in ΔHcal. Phospholipid monolayers are typically expanded by such proteins, and these proteins normally increase the permeability of phospholipid vesicles. These proteins, which are also hydrophilic, are believed to interact with phospholipid bilayers by a combination of electrostatic and hydrophobic forces, initially adsorbing to the charged polar headgroups of the phospholipids and subsequently partially penetrating the hydrophilic/hydrophobic interface of the bilayer to interact with a portion of the hydrocarbon chains.

Type 3 proteins usually have little effect on the Tm or ΔT1/2 of the phospholipid phase transition, but ΔHcal decreases linearly with protein concentration. Type 3 proteins are hydrophobic proteins that markedly expand phospholipid monolayers and increase the permeability of phospholipid vesicles. These proteins are thought to penetrate deeply into or through the hydrophobic core of anionic or zwitterionic lipid bilayers, interacting strongly with the phospholipid fatty acyl chains and essentially removing them from participation in the cooperative chain-melting transition. It should be noted, however, that some type 3 proteins may also interact electrostatically with phospholipid polar head-groups, particularly with those bearing a net negative charge. For example, the hydrophobic integral protein of the myelin membrane, lipophilin, exhibits preferential binding to acidic phospholipids [269], even though it behaves as a type 3 protein calorimetrically and immobilizes and disorders the hydrocarbon chains of its boundary lipid [270]. Similarly glycophorin, a membrane-spanning glycoprotein of the erythrocyte membrane, immobi-lizes about nine negatively charged phospholipid molecules per molecule of protein via strong electrostatic interactions with the phosphate headgroup [271].

The results of more recent DSC and other studies of lipid–protein model membranes clearly indicate that the classification scheme originally proposed is not completely appropriate for naturally occurring membrane proteins. Thus, none of the water-soluble, peripheral membrane-associated proteins studied thus far exhibit classical type 1 behavior (no change or a modest increase in Tm, a slight increase in $\Delta T1/2$, and an increase in the ΔH of the phospholipid phase transition). For example, although the vesicular stomatitis virus (VSV) M protein does increase the Tm of phos-pholipid chain-melting transitions, it also *markedly* increases the $\Delta T1/2$ and does not change the ΔH. Similarly, cytochrome c, another peripheral membrane protein, actually exhibits type 2 behavior when reconstituted with most anionic phospholipids, indicating that both hydrophobic and electro-static lipid–protein interactions are important in this system. Even poly(L-lysine), a polypeptide model for type 1 proteins, only exhibits classical type 1 behavior when interacting with PGs and then only under certain conditions; with other anionic phospholipids, the $\Delta Hcal$ of the phospholipid gel to liquid–crystalline phase transition is actually reduced rather than increased. Therefore, it seems doubtful that natural membrane proteins ever interact with phospholipid bilayers *exclusively* by electrostatic interactions.

On the other hand, there are a few examples of membrane proteins that exhibit more or less classical type 2 behavior. These include the myelin basic protein and cytochrome c, all of which usually reduce the Tm, increase the $\Delta T1/2$, and substantially reduce the ΔH of the chain-melting transition of anionic phospholipids. Strictly speaking, few if any membrane proteins actually exhibit classical type behavior as originally defined (no change in the Tm or $\Delta T1/2$ and a progressive linear reduction in the ΔH of both neutral and anionic phospholipid phase transitions with increasing protein concentration). This is because, with the advent of high-sensitivity calorimeters and the availability of pure phospholipids, it has become clear that all integral membrane proteins reduce the cooperativity of gel to liquid–crystalline phase transitions, as indeed would be expected from basic thermodynamic principles.

Moreover, some type 3 proteins exhibit a nonlinear decrease in ΔH with changes in protein levels, whereas others can produce at least moderate shifts in the Tm of phospholipid phase transitions. However, if we relax the original type 3 criteria somewhat, then a number of integral, transmembrane proteins can be said to exhibit modified type 3 behavior. These include the VSV G protein, myelin proteolipid protein, glycophorin, bacteriorhodopsin, cytochrome oxidase, cyto-chrome P-450, and the (Ca2++Mg2+)-ATPase, as well as several lysine-hydrophobic amino acid copolymers and the membrane-spanning polyleucine model polypeptides. Finally, the concanavalin A receptor, the *Acholeplasma laidlawii* B (Na++Mg2+)-ATPase, and poly(L-lysine) at high con-centrations do not fit into even this slightly modified classification scheme, because they exhibit a mixture of type 1, 2, and 3 characteristics, depending on the protein concentration range examined and, in the case of poly(L-lysine) at least, also on the particular phospholipid studied.

Thus, the classification scheme of Papahadjopoulos et al. [268], appropriately modified for type 3 proteins, is still of some use in studies of lipid–protein interactions, although some proteins, at least under certain conditions, do not fall neatly into any of these three categories. It seems that all naturally occurring membrane proteins studied to date interact with lipid bilayers by both hydrophobic and electrostatic interactions, and that different membrane proteins differ only in the specific types and relative magnitudes of these two general classes of interactions. It is also clear that the behavior exhibited by any particular membrane protein can depend on its conformation, method of reconstitution, and relative concentration, as well as on the polar headgroup and fatty acid composition of the lipid bilayer with which it is interacting [see 267].

Although DSC and other physical techniques have made considerable contributions to the elucidation of the nature of lipid–protein interactions, a number of outstanding questions remain. For example, it remains to be definitively determined whether some integral, transmembrane proteins completely abolish the cooperative gel to liquid–crystalline phase transition of lipids with which they are in direct contact, or whether there is only a partial abolition of this transition, as suggested by the studies of the interactions of the model transmembrane peptides with phospholipid bilayers (see Section 2.3.11). The mechanism by which some integral, transmembrane proteins perturb the phase behavior of very large numbers of phospholipids also remains to be determined. Finally, the molecular basis of the complex and unusual behavior of proteins such as the concanavalin A receptor and the *A. laidlawii* B ATPase is still obscure.

A plot of the ΔHcal of a phospholipid gel to liquid–crystalline phase transition vs. the protein/phospholipid molar ratio can yield the number of phospholipid molecules withdrawn from the cooperative chain-melting transition by each type 3 protein, when ΔHcal is extrapolated to zero. These values have ranged from 6 and 10 (for the small hydrophobic peptides gramicidin and melittin, respectively) to about 15 (for the relatively small membrane protein lipophilin). For these proteins, it appears that only one layer of phospholipid molecules — that is, those phospholipids interacting directly with the surface of the protein hydrophobic region — are withdrawn from the cooperative phase transition. On the other hand, the membrane-spanning and somewhat larger bacteriophage M-13 coat protein appears to remove 70 to 100 phospholipid molecules, whereas the considerably larger hydrophobic protein glycophorin also removes 80 to 100 molecules from the transition; the membrane-spanning regions of these proteins would appear to withdraw roughly three phospholipid layers. The VSV hydrophobic glycoprotein, whose incorporation into DPPC vesicles decreases the Tm as well as the ΔHcal and also increases $\Delta T1/2$, bound 270±150 phospholipid molecules, corresponding to the removal of five to six concentric shells of phospholipid per glycoprotein molecule. Moreover, human erythrocyte concanavalin A receptor and the *A. laidlawii* B (Na++Mg2+)-ATPase, which are both large, integral transmembrane proteins, appear to remove about 685 and 1000 phospholipid molecules, respectively, from participating in a cooperative phase transition [see 267, 272]. The basis for this differing behavior is not presently understood but may be related in part to differential interactions, both electrostatic and hydrophobic, between the non-membrane-spanning regions of these larger proteins and the lipid bilayer.

As previously mentioned, proteins can induce an isothermal lateral phase separation of negatively charged lipids in a somewhat similar manner to Ca2+. A number of intrinsic membrane proteins, including lipophilin, several ATPases, and rhodopsin, preferentially bind acidic phospholipids in their boundary lipid regions in preference to PC or PE. Also, a number of water-soluble proteins, such as polylysine, cytochrome c, and myelin basic protein, can act as polycations and separate out acidic lipids, even in binary mixtures where these lipids would normally be nearly ideally mixed. Moreover, the relative strengths of the interactions between a type 2 protein, such as the myelin basic protein, and various negatively charged phospholipids can vary markedly with the nature of the lipid headgroup [108]. These findings have important implications for biological membranes, because the conformation and activity of membrane proteins may be determined by the properties of the lipids in their own microenvironments rather than by the properties of the bulk phase lipids. There is considerable calorimetric and noncalorimetric evidence that the conformation and activity of membrane proteins can be altered by varying the nature of the phospholipid headgroup, the amount of cholesterol present, or the fluidity of the lipid bilayer [see 2, 50, 51, 108, 273].

2.4 STUDIES OF BIOLOGICAL MEMBRANES

2.4.1 MYCOPLASMA AND BACTERIAL MEMBRANES

Acholeplasma laidlawii is a member of the mycoplasmas, a diverse group of prokaryotic microorganisms that lack a cell wall. Because the mycoplasmas are genetically and morphologically the simplest organisms capable of autonomous replication, they provide useful models for the study of a number of problems in molecular and cellular biology. Mycoplasmas are particularly valuable for studies of the structure and function of cell membranes. Being nonphotosynthetic prokaryotes and lacking a cell wall or outer membrane, mycoplasma cells possess only a single membrane: the limiting, or plasma, membrane. This membrane contains essentially all the cellular lipid and, because these cells are small, a substantial fraction of the total cellular protein, as well. Due to the absence of a cell wall, substantial quantities of highly pure membranes can easily be prepared by gentle osmotic lysis followed by differential centrifugation, a practical advantage not offered by other prokaryotic microorganisms [274, 275].

Another useful property of mycoplasmas in general, and of *A. laidlawii* in particular, is the ability to induce dramatic yet controlled variations in the fatty acid composition of their membrane lipids. Thus, relatively large quantities of a number of exogenous saturated, unsaturated, branched-chain, or alicyclic fatty acids can be biosynthetically incorporated into the membrane phospho- and glycolipids of these organisms. In cases where *de novo* fatty acid biosynthesis is either inhibited or absent, fatty acid–homogeneous membranes (membranes whose glycerolipids contain only a single species of fatty acyl chain) can be produced. Moreover, by growing mycoplasmas in the presence or absence of various quantities of cholesterol or other sterols, the amount of these compounds present in the membrane can be altered dramatically. The ability to manipulate membrane lipid fatty acid composition and cholesterol content, and thus to alter the phase state and fluidity of the membrane lipid bilayer, makes these organisms ideal for studying the roles of lipids in biological membranes [274, 275].

The unique properties of the *A. laidlawii* membrane were utilized by Steim et al. [276] to show for the first time that biological membranes can undergo a gel to liquid–crystalline lipid phase transition similar to that previously reported for lamellar phospholipid–water systems. These workers demonstrated that when whole cells or isolated membranes are analyzed by DSC, two relatively broad endothermic transitions are observed on the initial heating scan. The lower-temperature transition is fully reversible, varies markedly in position with changes in the length and degree of unsaturation of the membrane lipid fatty acyl chains, is broadened and eventually abolished by cholesterol incorporation, and exhibits a transition enthalpy characteristic of the mixed-acid synthetic phospholipids. Moreover, an endothermic transition having essentially identical properties is observed for the protein-free total membrane lipid extract dispersed in excess water or aqueous buffer, indicating that the presence of membrane proteins has little effect on the thermotropic phase behavior of most of the membrane lipids. The higher-temperature transition, in contrast, is irreversible, independent of membrane lipid fatty acid composition or cholesterol content, and absent in total membrane lipid extracts, indicating that it is due to an irreversible thermal denaturation of the membrane proteins. A comparison of the enthalpies of transition of the lipids in the membrane and in water dispersions indicates that at least 75% of the total membrane lipids participate in this transition. Evidence was also presented that the lipids must be predominantly in the fluid state to support normal growth. These results were later confirmed and extended by Reinert and Steim [277] and by Melchior et al. [278], who showed that the gel to liquid–crystalline lipid phase transition is a property of living cells and that about 90% of the lipid participates in the gel to liquid–crystalline phase transition. These studies provided strong, direct experimental evidence for the hypothesis that lipids are organized as a liquid–crystalline bilayer in biological membranes, a basic feature of the currently well accepted fluid-mosaic model of membrane structure.

In these early DSC studies of the lipid thermotropic phase behavior in *A. laidlawii* membranes, cells whose membrane lipids were only moderately enriched in various exogenous fatty acids and low-sensitivity calorimeters were employed. The resultant broad lipid phase transitions (due primarily to fatty acid compositional heterogeneity) and the relative poor quality of the DSC traces obtained (due to baseline instability and noise) could have obscured subtle differences in lipid thermotropic phase behavior in intact cells, isolated membranes, and total membrane lipid dispersions. Indeed, Mantsch and coworkers, using FTIR spectroscopy, reported that the gel to liquid–crystalline phase transition in intact cells highly enriched in saturated fatty acids occurs some 5 to 10°C below that of isolated membranes derived from them, suggesting that the organization of the lipids in the membranes of living cells differs from that of the isolated membranes [279, 280]. This result is in contrast to the finding of the earlier DSC studies, which showed that the lipid chain-melting transitions in living cells, isolated membranes, and lipid dispersions is essentially identical, except that in the former systems about 10% of the lipid is prevented from participating in this cooperative phase transition by their interaction with membrane proteins.

In order to resolve this apparent discrepancy in results and to confirm or refute the original DSC findings, the experiments of Steim et al. [276] were repeated, using fatty acid-homogeneous *A. laidlawii* B cells (to remove fatty acid compositional heterogeneity) and a modern, high-sensitivity calorimeter (to improve the quality of the DSC traces obtained) [281]. The three exogenous fatty acids employed in this study (elaidic, isopalmitic, and myristic acid) were selected because they produce sharp phase transitions at temperatures above that of the ice-melting endotherms but below that of the protein denaturation endotherms. Thus, the thermotropic phase behavior of the lipids and proteins are well separated, and the use of potentially perturbing antifreeze additives, such as ethylene glycol, can be avoided.

Representative high-sensitivity DSC initial heating scans of viable cells, isolated membranes, and total membrane lipid dispersions are shown in Figure 2.17. In this instance, cells, membranes, and lipids were made nearly homogeneous (97 to 99 mol%) in elaidic acid. The fully reversible gel to liquid–crystalline lipid phase transitions observed in cells and membranes have essentially identical phase transition temperatures, enthalpies, and degrees of cooperativity, suggesting that membrane lipid organization in these two samples is very similar or identical. In contrast, the midpoint of the chain-melting transition of the membrane lipid dispersion is shifted to a higher temperature, exhibits a greater enthalpy, and is considerably less cooperative than in cells or membranes, suggesting that native membrane lipid organization has been perturbed during extraction and resuspension of the membrane lipids in water. The thermal denaturation of the proteins in the cells and membranes has absolutely no effect on the peak temperature or cooperativity of the lipid phase transition. However, about 15% of the lipids do not participate in the cooperative gel to liquid–crystalline phase transition in either the native or heated-denatured membranes, presumably because their cooperative phase behavior is abolished by interaction with the transmembrane regions of integral membrane proteins. Alternatively, a larger proportion of the membrane lipids may interact with the membrane proteins but have their cooperative melting behavior only partially perturbed, thereby leading to the 15% reduction in the transition enthalpy observed. The fact that the gel to liquid–crystalline lipid phase transition in cells and membranes exhibits a similar temperature maximum and a *higher* cooperativity than the membrane lipid dispersion favors the former interpretation. In general, the results obtained with intact cells and membranes support the earlier studies of Steim and coworkers, who reported a nearly identical lipid thermotropic phase behavior in both systems, and not the IR spectroscopic results of Mantsch and coworkers, who reported significantly different phase behavior in these systems. This difference in results may be due at least in part to the existence of a thermal history–dependent gel-state lipid polymorphism (see below).

When similar calorimetric experiments are performed with isopalmitic acid-homogeneous *A. laidlawii* B cells, membranes, and lipids, two well-resolved endotherms are observed in all three systems, as indicated in Figure 2.18. The properties of the lower enthalpy lipid transition centered

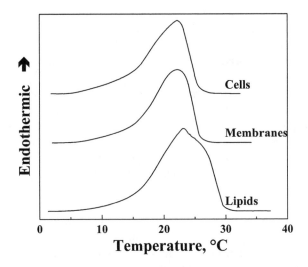

FIGURE 2.17 DSC heating scans of *A. laidlawii* B elaidic acid-homogenous intact cells, isolated membranes, and extracted total membrane lipids dispersed as multilamellar vesicles in water. The heating scan rate is 30°C/hour, at which the sample and reference are in thermal equilibrium throughout the DSC run. Redrawn from [281].)

at 8 to 9°C are dependent on the heating scan rate and the thermal history of the sample. In particular, the apparent transition temperature increases with increasing scan rate, and annealing the sample at 0°C for 24 h before beginning the DSC run results in a two- to threefold increase in the observed calorimetric enthalpy. Because similar hysteresis is typically observed in the formation and interconversions of highly ordered gel phases in bilayers of synthetic phospholipids, the lower temperature endotherm was tentatively identified as a phase transition between a more highly ordered and a less highly ordered gel state. In contrast, the properties of the higher enthalpy transition centered at 21 to 22°C exhibit no dependence on heating scan rate or on thermal history, indicating that this is the typical gel to liquid–crystalline or chain-melting transition previously observed in this organism by a variety of techniques.

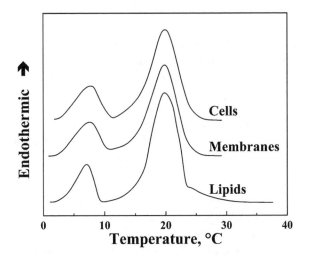

FIGURE 2.18 DSC scans of unannealed *A. laidlawii* B isopalmitic acid-homogenous intact cells, isolated membranes, and extracted total membrane lipids dispersed in water. The heating scan rate is 30°C/hour. (Redrawn from [281].)

The structural changes associated with each of the two lipid phase transitions detected by DSC were investigated by FTIR and 31P NMR spectroscopy. These spectroscopic techniques confirm that the lower-temperature endotherm is due to a transition from a highly ordered gel phase (in which the all-*trans* lipid hydrocarbon chains are very closely packed, the bilayer interfacial region is partly dehydrated, and the phospholipid polar headgroups are undergoing slow axially asymmetric motion) to a disordered gel phase (in which the lipid hydrocarbon chains, although still largely extended, are loosely packed, the interfacial region is fully hydrated, and the phospholipid polar headgroups are undergoing fast, axially symmetric motion). These spectroscopic techniques also confirm that the higher-temperature transition corresponds to a conversion from a loosely packed gel state to the liquid–crystalline state, in which the lipid hydrocarbon chains are conformationally disordered and contain a number of gauche conformers. All three physical techniques indicate that at least 80% of the total membrane lipid participates in both the gel/gel and gel/liquid–crystalline phase transitions [281].

The finding that gel-phase polymorphism can exist in *A. laidlawii* B membranes is quite surprising, in view of the fact that most binary mixtures of synthetic phospholipids do not exhibit multiple gel states, even when they contain identical fatty acyl chains. Thus, the ability of the *A. laidlawii* membrane lipids to form a highly ordered gel phase seems all the more remarkable, because this organism contains three major and two minor lipid classes, including both phospho- and glycoglycerolipids. These results thus imply that the *A. laidlawii* B membrane lipid classes are highly miscible in all three lipid phase states detected, a result compatible with an earlier differential thermal analysis study of mixtures of the individual membrane lipid classes [282]. Moreover, gel-state polymorphism in this organism is not restricted to membranes containing a single methyl isobranched fatty acid: *A. laidlawii* B membranes made homogeneous with members of most fatty acid classes tested also exhibit multiple gel-state phase transitions. In fact, we have unpublished calorimetric and spectroscopic evidence for partially resolved gel/gel and gel/liquid–crystalline lipid phase transitions in membranes containing various proportions of two different classes of fatty acids. It thus seems clear the gel-state polymorphism is not restricted to single-component lipid model membranes but can occur in lipid bilayers and in biological membranes containing appreciable polar headgroup and fatty acyl chain compositional heterogeneity, as well. However, to our knowledge, this phenomenon has not been reported in any other biological membrane.

By utilizing *A. laidlawii* B membranes containing essentially only a single, exogenous fatty acid, it is possible to assess the relative contributions of polar headgroup and fatty acyl group heterogeneity to the relatively broad phase transitions characteristic of native or fatty acid–enriched membranes. It was found that a strong correlation existed between fatty acid heterogeneity and the sharpness of the gel to liquid–crystalline membrane lipid phase transition, as detected by differential thermal analysis (DTA), with the transition width decreasing from a normal value of 25 to 30°C to a limiting value of about 7°C, indicating that fatty acid heterogeneity was the primary contributor to the broad transitions observed in native membranes [283]. Additional DTA studies confirmed that the five major membrane lipids of this organism (two glycolipids, two phospholipids, and a phosphorylated glycolipid) show a high degree of miscibility in both the gel and liquid–crystalline states and that Mg2+ does not induce a lateral phase separation of the acidic lipid components [282]. Also, the Tm values of the fatty acid–homogeneous membranes correlated very well with the Tm values of synthetic PCs containing the same acyl chains. Finally, the $\Delta T1/2$ of the membrane transitions depended on the nature and chain length of the fatty acyl group present, even with comparable levels of enrichment (98 to 99 mol%). This may reflect the fact that overlapping gel/gel and gel/liquid–crystalline phase transitions may have been observed in some instances. Subsequent work with high-sensitivity DSC and many other physical techniques have confirmed that the general findings for *A. laidlawii* B also apply to a number of bacterial membranes, in particular for unsaturated fatty acid auxotrophs of *Escherichia coli* [see 284, 285]. The only exception to this general behavior reported so far appears to be the membranes and membrane lipid extracts from the halophilic bacterium *Halobacterium halobium*, where no gel to liquid–crystalline lipid phase transition could be detected by high-sensitivity DSC [286, 287]. This is presumably due to the large amounts of the dihydrophytanyl hydrocarbon chains present in the membrane lipids of this organism. As discussed earlier, the highly branched synthetic phospholipid diphytanoyl PC does not undergo a discrete cooperative chain-melting transition in the temperature range −120 to +120°C [57].

2.4.2 Eukaryotic Cell Membranes

The presence of high levels of cholesterol in many eukaryotic membranes, particularly plasma membranes, abolishes a discrete cooperative gel to liquid–crystalline membrane lipid phase transition in these systems. Thus, no lipid phase transitions could be detected by DSC or DTA in the cholesterol-rich erythrocyte [288] or myelin [182] membranes. Cholesterol-free lipid extracts of these membranes did, however, exhibit a single, broad phase transition: the former centered near 0°C and the latter extended from 25 to 60°C. The high Tm of the cholesterol-free myelin extract is due primarily to its high content of sphingolipids. Similar reversible thermal transitions were observed, however, by DSC and fluorescence polarization in the microvillus and basolateral regions of rat small intestinal enterocyte plasma membrane and in hydrated lipid extracts [289]. These transitions occur over a temperature range of about 25 to 40°C and exhibit very low ΔHcal values of about 0.10 to 0.15 cal/g for the intact membranes and about 0.40 to 0.55 cal/g for the extracted lipids. The generally low enthalpies observed were attributed to the large amounts of cholesterol present and to the comparatively lower enthalpy observed in the intact membranes to lipid–protein interactions. The nature of this lipid endotherm remains unclear, because the high levels of cholesterol and polyunsaturated fatty acyl groups present make it unlikely that it represents a bulk-phase gel to liquid–crystalline transition.

The thermotropic behavior of rat liver microsomal membranes, which contain moderate levels of cholesterol, has been studied by DSC. An early study using conventional DSC revealed a single, reversible, broad phase transition occurring between −15 and +5°C in both intact membranes and isolated lipids [290]. A more recent high-sensitivity DSC study confirmed the absence of a reversible phase transition above 0°C [291]. However, rats fed a fat-free diet, which increased the degree of saturation of the membrane lipid fatty acids, also exhibit two reversible membrane lipid phase transitions centered at 3°C and 14°C; after protein denaturation, the lower-temperature peak increases in area and shifts to a higher temperature, while the higher-temperature lipid peak decreases in area. It thus appears that normal microsomal membrane lipids exist entirely in the fluid state above 0°C and that the organization of at least a portion of the membrane lipid is dependent on the state of the membrane protein, in contrast to the situation in mycoplasma and bacterial membranes. The existence of a second reversible, higher-temperature transition in rat liver microsomal membranes, detected by conventional DSC, has been reported. This transition, which occurs between 18 and 40°C in intact membranes and between 10 and 20°C in extracted lipids, is of relatively low enthalpy and is not affected by protein denaturation [292]. It is not clear why this transition could not be detected by high-sensitivity DSC. The molecular basis for this higher-temperature transition, if real, is unknown.

Rat liver mitochondrial membranes, which are low in cholesterol, have been studied by several groups using DSC and other techniques. The earliest work with whole mitochondrial membranes revealed a reversible, broad gel to liquid–crystalline phase transition centered at 0°C in mitochondrial membranes and in extracted lipids [290]. A later study of both intact mitochondria and isolated inner and outer membranes confirmed these results, except that the outer membrane transition seemed to occur at a slightly lower temperature than the inner membrane transition [293]. However, a more recent study of the rat liver inner mitochondrial membrane reported a narrower membrane lipid transition centered near 10°C; by artificially increasing cholesterol content some tenfold to about 30 mol%, the inner membrane gel to liquid–crystalline phase transition could be lowered and broadened, and its ΔHcal reduced to less than one-tenth that of the native membrane [294]. It has also been reported that in beef heart mitochondrial inner membranes, a broad reversible endothermic phase transition centered at $-10°C$ occurs; after protein thermal denaturation, a new reversible transition of low enthalpy is observed at about $20°C$. The extracted lipids exhibit thermal behavior nearly identical to that of the intact protein-denatured membrane, indicating that a latent pool of higher-melting lipids may exist in this membrane [295]. In liver mitochondrial inner membranes from hibernating and nonhibernating ground squirrels, lipid phase transitions centered at $-9°C$ and $-5°C$, respectively, and occurring over a range of 16 to $18°C$, have been reported [296]. However, protein denaturation does not affect lipid thermotropic phase behavior in this case. It is clear from all these studies that the lipids of both mitochondrial membranes exist exclusively in the fluid state at physiological temperatures.

DSC has been used to study the individual protein components of biological membranes of relatively simple protein composition, and the interaction of several of these components with lipids and with other proteins. The red blood cell membrane, which has been most intensively studied, exhibits five discrete protein transitions, each of which has been assigned to a specific membrane protein. The response of each of these thermal transitions to variations in temperature and pH, as well as to treatment with proteases, phospholipases, specific labeling reagents, and modifiers and inhibitors of selected membrane activities, has provided much useful information about the interactions and functions of these components in the intact erythrocyte membrane [297–300]. Similar approaches have been applied to the bovine rod outer segment membrane [301] and to the spinach chloroplast thylakoid membrane [302].

2.5 CONCLUDING REMARKS

A considerable body of knowledge about the thermotropic phase behavior of lipid bilayers has been developed over the past 30 years. One should note, however, that by far the majority of calorimetric, spectroscopic, and X-ray diffraction studies of lipid bilayers have employed PCs containing two identical, saturated fatty acyl groups, particularly DMPC and DPPC, as the only lipid species investigated. The major reasons for this choice are probably practical, because these simple, disaturated PCs are relatively easy to synthesize and purify chemically, are stable to oxidation, hydrate readily, and, in excess water, form only lamellar phases at physiological temperatures. Moreover, DMPC and DPPC exhibit single, highly cooperative gel to liquid–crystalline phase transitions at about $23°C$ and $41°C$, respectively, temperatures well above the freezing point of water but below the thermal denaturation temperature of most proteins. These compounds would thus seem ideal choices for studying the effect of membrane proteins on membrane lipid thermotropic phase behavior.

It is important to realize, however, that simple, disaturated glycerolipids are not major molecular species in any biological membrane and that no individual phospholipid class is found universally in nature. Thus, although PCs are the major zwitterionic phospholipid class in most eukaryotic cellular membranes, PCs are not abundant or widely distributed in the eubacteria and are absent entirely from the archaebacteria. Moreover, although normal saturated fatty acids are universal (but not exclusive) constituents of the membrane lipids of eukaryotic cell membranes, normal saturated fatty acids are at least largely replaced by single methyl branched or alicyclic fatty acyl groups in some eubacteria and are absent entirely from the archaebacteria, where the lipid hydrocarbon chains are multiple methyl branched phytanyl ethers or derivatives thereof. Finally, in eukaryotic and eubacterial membrane glycerolipids, the predominant molecular species almost always contain two different classes of fatty acids. Thus, the glycerol carbon 1 normally bears a relatively high-melting normal saturated or methyl isobranched fatty acyl group, whereas a low-melting unsaturated, methyl anteisobranched or alicyclic fatty acid is esterified to glycerol carbon 2. Although many of the major findings obtained in calorimetric studies of DMPC or DPPC bilayers are likely to be generally valid, at least qualitatively, one should bear in mind that these simple, disaturated PCs, although convenient to use, are not entirely typical membrane lipids. Moreover, the interaction of cholesterol with phospholipids, for example, depends markedly on the nature of the polar headgroup and on the chemical structure and chain length of the hydrocarbon chains, as well as on the positional distribution of these chains on the glycerol backbone. Clearly, in future studies of lipid thermotropic phase behavior, a much wider range of lipids of the type actually found in biological membranes should be studied.

ACKNOWLEDGMENTS

Work performed in the authors' laboratory was supported by operating, major equipment, and personnel support grants from both the Canadian Institutes of Health Research (formerly the Medical Research Council of Canada) and the Alberta Heritage Foundation for Medical Research.

REFERENCES

1. Jain, M., *Introduction to Biological Membranes,* 2nd ed., John Wiley & Sons, New York, 1988.
2. Gennis, R.G., *Biomembranes. Molecular Structure and Function*, Springer-Verlag, New York, 1989.
3. Luzatti, V. and Tardieu, A., Lipid phases: structure and structural transitions, *Annu. Rev. Phys. Chem.*, 25, 79, 1974.
4. McElhaney, R.N., The use of differential scanning calorimetry and differential thermal analysis in studies of model and biological membranes, *Chem. Phys. Lipids*, 30, 229, 1982.
5. Cevc, G. and Marsh, D., *Phospholipid Bilayer: Physical Principles and Models*, John Wiley & Sons, New York, 1987.
6. Lewis, R.N.A.H., Mannock, D.A., and McElhaney, R.N., Membrane lipid molecular structure and polymorphism, in *Lipid Polymorphism and Membrane Properties*, Epand, R.M., ed., Academic Press, San Diego, CA, 1996, 25.
7. Tsong, T.Y., Kinetics of the crystalline-liquid-crystalline phase transition of dimyristoyl L-α-lecithin bilayers, in *Proc. Natl. Acad. Sci., U.S.A.,* 71, 2684, 1974.
8. Albon, N. and Sturtevant, J.M., Nature of the gel to liquid crystalline transition of synthetic phosphatidylcholines, *Proc. Natl. Acad. Sci. U.S.A.*, 75, 2258, 1978.
9. Cho, K.C., Choy, C.L., and Young, K., Kinetics of the pretransition of synthetic phospholipids. A calorimetric study, *Biochim. Biophys. Acta*, 663, 14, 1981.
10. Lentz, B.R., Friere, E., and Biltonen, R.L., Fluorescence and calorimetric studies of phase transitions in phosphatidylcholine multilayers: kinetics of the pretransition, *Biochemistry*, 17, 4475, 1978.
11. Tristram-Nagle, S. et al., Kinetics of the subtransition in dipalmitoyl phosphatidylcholine, *Biochemistry*, 26, 4288, 1987.
12. Lewis, R.N.A.H., Mak, N., and McElhaney, R.N., Differential scanning calorimetric study of the thermotropic phase behavior of model membranes composed of phosphatidylcholines containing linear saturated fatty acyl chains, *Biochemistry*, 26, 6118, 1987.
13. Mendelsohn, R. et al., Quantitative determination of conformational disorder in the acyl chains of phospholipid bilayers by infrared spectroscopy, *Biochemistry*, 28, 8934, 1989.
14. Casal, H.L. and McElhaney, R.N., Quantitative determination of hydrocarbon chain conformational order in bilayers of saturated phosphatidylcholines of various chain lengths by Fourier transform infrared spectroscopy, *Biochemistry*, 29, 5423, 1990.
15. Lewis, R.N.A.H. and McElhaney, R.N., The subgel phases of the *n*-saturated diacyl phosphatidylcholines. A Fourier transform infrared spectroscopic study, *Biochemistry*, 29, 7946, 1990.
16. Lewis, R.N.A.H. and McElhaney, R.N., Structures of the subgel phases of the *n*-saturated diacyl phosphatidylcholine bilayers. FTIR spectroscopic studies of ^{13}C=O and ^{2}H labeled lipids, *Biophys. J.,* 61, 63, 1992.
17. Lewis, R.N.A.H. and McElhaney, R.N., Thermotropic phase behavior of model membranes composed of phosphatidtylcholines containing isobranched phosphatidylcholines. 1. Differential scanning calorimetric studies, *Biochemistry*, 24, 2431, 1985.

18. Mantsch, H.H. et al., Thermotropic phase behavior of model membranes composed of phosphatidtyl-cholines containing isobranched fatty acids. 2. Infrared and 31P-NMR spectroscopic studies, *Biochemistry*, 24, 2440, 1985.

19. Church, S.E. et al., X-ray structure study of thermotropic phases in isoacylphosphatidyl-choline multibilayers, *Biophys. J.*, 49, 597, 1986.

20. Yang, C.P. et al., Dilatometric studies of isobranched phosphatidylcholines, *Biochim. Biophys. Acta*, 863, 33, 1986.

21. Lewis, R.N.A.H., Sykes, B.D., and McElhaney, R.N., Thermotropic phase behavior of model membranes composed of phosphatidylcholines containing cis-monounsaturaed acyl chain homologues of oleic acid. Differential scanning calorimetric and 31P-NMR spectroscopic studies, *Biochemistry*, 27, 880, 1988.

22. Lewis, R.N.A.H., Sykes, B.D., and McElhaney, R.N., Thermotropic phase behavior of model membranes composed of phosphatidylcholines containing *dl*-methyl anteisobranched fatty acids. 1. Differential scanning calorimetric and 31P-NMR spectroscopic studies, *Biochemistry*, 26, 4036, 1987.

23. Mantsch, H.H. et al., Thermotropic phase behavior of model membranes composed of phosphatidyl-cholines containing *dl*-methyl anteisosobranched fatty acids. 2. An infrared spectroscopy study, *Biochemistry*, 26, 4045, 1987.

24. Lewis, R.N.A.H. and McElhaney, R.N., Thermotropic phase behavior of model membranes composed of phosphatidylcholines containing ω-cycloheyl fatty acids. Differential scanning calorimetric and 31P-NMR spectroscopic studies, *Biochemistry*, 24, 4903, 1985.

25. Mantsch, H.H. et al., An infrared spectroscopic study of the thermotropic phase behavior of phosphatidylcholines containing ω-cyclohexyl fatty acyl chains, *Biochim. Biophys Acta,* 980, 42, 1989.

26. Kodama, M., Kuwabara, M., and Seki, S., Successive phase-transition phenomena and phase diagram of the phoshatidylcholine-water system as revealed by differential scanning calorimetry, *Biochim. Biophys. Acta,* 689, 567, 1982.

27. Janiak, M.J., Small, D.M., and Shipley, G.G., Temperature and compositional dependence of the structure of hydrated dimyristoyl lecithin, *J. Biol. Chem.*, 254, 6068, 1979.

28. Jendrasiak, G.L. and Mendible, J.C., The effect of the phase transition on the hydration and electrical conductivity of phospholipids, *Biochim. Biophys. Acta*, 424, 133, 1976.

29. Jendrasiak, G.L. and Mendible, J.C., The phospholipid headgroup orientation: effect on hydration and electrical conductivity, *Biochim. Biophys. Acta*, 424, 149, 1976.

30. Crowe, J.H. et al., Interactions of sugars with membranes, *Biochim. Biophys. Acta*, 947, 367, 1988.

31. Crowe, J.H., Hoekstra, F.A., and Crowe, L.M., Anhydrobiosis, *Annu. Rev. Physiol.*, 54, 579, 1992.

32. Crowe, L.M., Reid, D.S., and Crowe, J.H., Is trehalose special for preserving dry biomaterials?, *Biophys. J.*, 71, 2087, 1996.

33. Wong, P.T.T., Siminovitch, D.J., and Mantsch, H.H., Structure and properties of model membranes: new knowledge from high-pressure infrared spectroscopy, *Biochim. Biophys. Acta*, 947, 139, 1988.

34. de Kruijff, B., Cullis, P.R., and Radda, G.K., Differential scanning calorimetry and 31P NMR studies on sonicated and unsonicated phosphatidylcholine liposomes, *Biochim. Biophys. Acta*, 406, 6, 1975.

35. van Dijck, P.W.M. et al., Phase transitions in phospholipid model membranes of different curvature, *Biochim. Biophys. Acta,* 506, 183, 1978.

36. Suurkuush, J. et al., A calorimetric and fluorescent probe study of the gel-liquid crystalline phase transition in small single-lamellar dipalmitoylphosphatidylcholine vesicles, *Biochemistry*, 15, 1393, 1976.

37. Gruenwald, B., Stankowski, S., and Blume, A., Curvature influence on the cooperativity and the phase transition enthalpy of lecithin vesicles, *FEBS Lett.*, 102, 227, 1979.

38. Kantor, H.L. et al., A calorimetric examination of stable and fusing lipid bilayer vesicles, *Biochim. Biophys. Acta*, 466, 402, 1977.

39. Zacharis, H.M., A linear function for the melting behavior of lipids, *Chem. Phys. Lipids*, 18, 221, 1977.

40. Mason, J.T. and Huang, C., Chain length dependent thermodynamics of saturated symmetric-chain phosphatidylcholine bilayers, *Lipids*, 16, 604, 1981.

41. Davis, P.J. and Keough, K.M.W., Chain arrangements in the gel state and the transition temperatures of phosphatidylcholines, *Biophys. J.*, 48, 915, 1985.

42. Marsh, D., Analysis of the chainlength dependence of lipid phase transition temperatures: main and pretransitions of phosphatidylcholines: main and non-lamellar transitions of phosphatidylethanolamines, *Biochim. Biophys. Acta*, 1062, 1, 1991.

43. Marsh, D., Analysis of the bilayer phase transition temperatures of phosphatidylcholines with mixed chains, *Biophys. J.*, 61, 1036, 1992.

44. Huang, C. et al., Dependence of of the bilayer phase transition temperatures on the structural parameters of phosphatidylcholines, *Lipids*, 28, 365, 1993.

45. Wang, Z. et al., Phase transition behavior and molecular structure of monounsaturated phosphatidylcholines, *J. Biol. Chem.*, 270, 2014, 1995.

46. Marsh, D., Thermodynamic analysis of chain-melting transition temperatures for monounsaturated phospholipids membranes: Dependence on *cis*-monoenoic double bond position, *Biophys. J.*, 77, 953, 1999.

47. Dollhpf, W., Grossmann, H.P., and Leute, U., Some thermodynamic quantities of n-alkanes as a function of chain length, *Colloid & Polymer Sci.*, 259, 267, 1981.

48. Cronan, J.E., Molecular biology of bacterial membrane lipids, *Annu. Rev. Biochem.*, 47, 163, 1978.

49. Fulco, A.J., Regulation and pathways of membrane lipid biosynthesis in bacilli, in *Membrane Fluidity, Biomembranes*, Kates, M. and Manson, L.A., eds., Plenum Press, New York, 1984, 303.

50. McElhaney, R.N., The relationship between membrane lipid fluidity and phase state and the ability of bacteria and mycoplasmas to grow and survive at various temperatures, in *Membrane Fluidity*, Kates, M. and Manson, L.A., eds., Plenum Press, New York, 1984, 249.

51. McElhaney, R.N., Effects of membrane lipids on transport and enzymic activities, in *Current Topics in Membranes and Transport*, Razin, S. and Rottem, S., eds., Academic Press, New York, 1982, 317.

52. Seddon, J.M., Cevc, G., and Marsh, D., Calorimetric studies of the gel-fluid (Lβ-Lα) and lamellar-inverted hexagonal (Lα-HII) phase transitions in dialkyl- and diacylphosphatidylethanolamines, *Biochemistry*, 22, 1280, 1983.

53. de Haas, G.H. et al., Studies on phospholipase A and its zymogen from porcine pancreas. III. Action of the enzyme on short-chain lecithins, *Biochim. Biophys. Acta*, 239, 252, 1971.

54. Kleinschmidt, J.H. and Tamm, L.K., Structural transitions in short-chain lipid assemblies studied by ^{31}P-NMR spectroscopy, *Biophys. J.*, 83, 994, 2002.

55. de Gier, J., Mandersloot, J.G., and van Deenen, L.L.M., Lipid composition and permeability of liposomes, *Biochim. Biophys. Acta*, 150, 666, 1968.

56. Barton, P.G. and Gunstone, F.D., Hydrocarbon chain packing and molecular motion in phospholipid Bilayers formed from unsaturated lecithins, *J. Biol. Chem.*, 250, 4470, 1975.

57. Lindsey, H., Petersen, N.O., and Chan, S.I., Physicochemical characterization of 1,2-diphytanoyl-*sn*-glycero-3-phosphocholine in model membrane systems, *Biochim. Biophys. Acta*, 555, 147, 1979.

58. Macdonald, P.M. et al., 19F-nuclear magnetic resonance studies of lipid fatty acyl chain order and dynamics in *Acholeplasma laidlawii* B membranes. The effect of methyl branch substitution and of trans-unsaturation upon membrane chain orientational order, *Biochemistry*, 22, 5103, 1983.

59. Macdonald, P.M., Sykes, B.D., and McElhaney, R.N., Fatty acyl chain structure, orientational order and the lipid phase transition in *Acholeplasma laidlawii* B membranes. A review of recent 19F-NMR studies, *Can. J. Biochem. Cell. Biol.*, 62, 1134, 1984.

60. Macdonald, P.M., Sykes, B.D., and McElhaney, R.N., 19F-Nuclear magnetic resonance studies of lipid fatty acyl chain order and dynamics in *Acholeplasma laidlawii* B membranes. Gel state disorder in the presence of methyl iso- and anteisobranched chain substituents, *Biochemistry*, 24, 2412, 1985.

61. Rice, D.K. et al., A comparative monomolecular film study of a straight chain phosphatidylcholine with three isobranched phosphatidylcholines (diisoheptadecanoyl-phosphatidylcholine, diisooctadecanoylphosphatidylcholine, and diisoeicosanoyl-phosphatidylcholine, *Biochemistry*, 26, 3205, 1987.

62. McElhaney, R.N., The biological significance of alterations in the fatty acid composition of membrane lipids in response to changes on environmental temperature, in *Proceedings of the 1974 NASA Conference on Extreme Environments: Mechanisms of Microbial Adaptation*, Heinfrich, M., ed., Academic Press, New York, 1975, 255.

63. Keough, K.M.W. and Davis, P.J., Gel to liquid-crystalline phase transitions in water dispersions of saturated mixed-acid phosphatidylcholines, *Biochemistry*, 18, 1453, 1979.

64. Davis, P.J. et al., Gel to liquid crystalline transition temperatures of water dispersions of two pairs of positional isomers of unsaturated mixed-acid phosphatidylcholines, *Biochemistry*, 20, 3613, 1981.

65. Stumpel, J., Niksch A., and Eibl, H., Calorimetric studies on saturated mixed-chain lecithin-water systems. Nonequivalence of acyl chains in the thermotropic phase transition, *Biochemistry*, 20, 662, 1981.

66. Stumpel, J., Eibl, H., and Niksch A., X-ray analysis and calorimetry on phosphatidylcholine model membranes. The influence of length and position of acyl chains upon structure and phase behaviour, *Biochim. Biophys. Acta,* 727, 246, 1983.

67. Coolbear, K.P., Berde, C.B., and Keough, K.M.W., Gel to liquid crystalline phase transitions of aqueous dispersions of polyunsaturated mixed acid phosphatidylcholines, *Biochemistry*, 22, 1466, 1983.

68. Serrallach, E.N., de Haas, G.H., and Shipley, G.G., Structure and thermotropic properties of mixed-chain phosphatidylcholine bilayer membranes, *Biochemistry*, 23, 713, 1984.

69. Keough, K.M.W., Giffin, B., and Kariel, N., The influence of unsaturation on the phase transition temperatures of a series of heteroacid phosphatidylcholines containing twenty carbon chains, *Biochim. Biophys. Acta*, 902, 1, 1987.

70. Mattai, J., Sripada, P.K., and Shipley, G.G., Mixed-chain phosphatidylcholine bilayers: structure and properties, *Biochemistry*, 26, 3287, 1987.

71. Mason, J.T., Huang, C.-H., and Biltonen, R.L., Calorimetric investigations of saturated mixed-chain phosphatidylcholine bilayer dispersions, *Biochemistry*, 20, 6086, 1981.

72. Chen, S.C. and Sturtevant, J.M., Thermodynamic behavior of bilayers formed from mixed-chain phosphatidylcholines, *Biochemistry*, 20, 713, 1981.

73. Xu, H., Stephenson, F.A., and Huang, C.-H., Binary mixtures of phosphatidylcholines with one acyl chain twice as long as the other, *Biochemistry*, 26, 5448, 1987.

74. Lewis, R.N.A.H. et al., Enigmatic thermotropic phase behavior of highly asymmetric mixed chain phosphatidylcholines which form mixed interdigitated gel phases, *Biophys. J.*, 66, 207, 1994.

75. Huang, C., Mason, J.T., and Levin, I.W., Raman spectroscopic study of saturated mixed-chain phosphatidylcholine multilamellar dispersions, *Biochemistry*, 22, 2775, 1983.

76. Wong, P.T.T. and Huang, C.-H., Structural aspects of pressure effects on infrared spectra of mixed-chain phosphatidylcholine assemblies in D2O, *Biochemistry,* 28, 1259, 1989.

77. Lewis, R.N.A.H. and McElhaney, R.N., Studies of mixed chain diacyl phosphatidylcholines with highly asymmmetric acyl chains. A Fourier transform infrared spectroscopic study of interfacial hydration and hydrocarbon chain packing chain in the mixed interdigitated gel phase, *Biophys. J.*, 65, 1866, 1993.

78. Lewis, R.N.A.H. et al., Studies of highly asymmetric mixed chain diacyl phosphatidylcholines which form mixed-interdigitated gel phases. Fourier transform infrared and ^{2}H-NMR spectroscopic studies of hydrocarbon chain conformation and order in the liquid-crystalline state, *Biophys. J.*, 67, 197, 1994.

79. Lewis, B.A., Das Gupta, S.K., and Griffin, G.G., Solid-state NMR studies of the molecular dynamics and phase behavior of mixed-chain phosphatidylcholines, *Biochemistry*, 23, 1988, 1984.

80. Huang, C., Mixed-chain phospholipids and interdigitated bilayer systems, *Klin. Wochenschr.,* 68, 145, 1990.

81. Davis, P.J. and Keough, K.M.W., Chain arrangements in the gel state and the transition temperatures of phosphatidylcholines, *Biophys. J.*, 48, 915, 1985.

82. Huang, C. and Mason, J.T., Structure and properties of mixed-chain phospholipid assemblies, *Biochim. Biophys. Acta*, 864, 423, 1986.

83. Ladbrooke, B.D. and Chapman, D., Thermal analysis of lipids, proteins and biological membranes. A review and summary of some recent studies, *Chem. Phys. Lipids*, 3, 304, 1969.

84. de Kruijff, B., Demel, R.A., and van Deenen, L.L.M., The effect of cholesterol and epicholesterol incorporation on the permeability and on the phase transition of intact *Acholeplasma laidlawii* cell membranes and derived liposomes, *Biochim. Biophys. Acta*, 255, 331, 1972.

85. de Kruijff, B. et al., The effect of the polar headgroup on the lipid-cholesterol interaction: A monolayer and differential scanning calorimetry study, *Biochim. Biophys. Acta*, 307, 1, 1973.

86. Op den Kamp, J.A.F., Kaurez, M.T., and van Deenen, L.L.M., Action of pancreatic phospholipase A2 on phosphatidylcholine bilayers in different physical states, *Biochim. Biophys. Acta*, 406, 169, 1975.

87. Bunow, M.R. and Bunow, B., Phase behavior of ganglioside-lecithin mixtures. Relation to dispersion of gangliosides in membranes, *Biophys. J.*, 27, 325, 1979.

88. Davis, P.J., Coolbear, K.P., and Keough, K.M.W., Differential scanning calorimetric studies of the thermotropic phase behavior of membranes composed of dipalmitoyl lecithin and mixed-acid unsaturated lecithins, *Can. J. Biochem.*, 58, 851, 1980.

89. Silvius, J.R., Thermotropic phase transitions of pure lipids and their modification by membrane proteins, in *Lipid Protein Interactions*, vol. 2, Jost, P.C. and Griffith, O.H., eds., John Wiley & Sons, New York, 1982, 239.

90. Cevc, G., How membrane chain melting properties are regulated by the polar surface of the lipid bilayer, *Biochemistry*, 26, 6305, 1987.

91. Israelachvilli, J.N., Mitchell, D.J., and Ninham, B.W., Theory of self assembly of lipid bilayers and vesicles, *Biochim. Biophys. Acta*, 470, 185, 1977.

92. Israelachvilli, J.N., Marcelja, S., and Horn R.G., Physical principles of membrane organization, *Q. Rev. Biophys.*, 13, 121, 1980.

93. Eibl, H., Phospholipid bilayers: Influence of Structure and Charge, in *Polyunsaturated Fatty Acids*, Kunau, W.H. and Holman, R.T., eds., American Oil Chemists Soc., 1977, 229.

94. Browning, J.L., Motions and interactions of phospholipid headgroups at the membrane surface. I. Simple alkyl headgroups, *Biochemistry*, 20, 7123, 1981.

95. Mannock, D.A. et al., The physical properties of glycosyl diacylglycerols. Calorimetric studies of a homologous series of 1,2-Di-O-acyl-3-O-(β-D-glucopyranosyl)-*sn*-glycerols, *Biochemistry*, 27, 6852, 1988.

96. Mannock, D.A., Lewis, R.N.A.H., and McElhaney, R.N., The physical properties of glycosyl diacyl-glycerols. 1. Calorimetric studies of a homologous series of 1,2-Di-O-acyl-3-O-(α-D-glucopyranosyl)-*sn*-glycerols, *Biochemistry*, 29, 7790, 1990.

97. Jarrell, H.C. et al., The dependence of glyceroglycolipid orientation and dynamics on head-group structure, *Biochim. Biophys. Acta*, 897, 69, 1987.

98. Träuble, H. and Eibl, H., Electrostatic effects on lipid phase transitions: membrane structure and ionic environment, *Proc. Natl. Acad. Sci. U.S.A.*, 71, 214, 1974.

99. Blume, A. and Eibl, H., The influence of charge on bilayer membranes: calorimetric investigations of phosphatidic acid bilayers, *Biochim. Biophys. Acta*, 558, 13, 1980.

100. Cevc, G. Watts, A., and Marsh, D., Titration of the phase transition of phosphatidylserine bilayer membranes. Effects of pH, surface electrostatics, ion binding and headgroup hydration, *Biochemistry*, 20, 4955, 1981.

101. Demel, R.A., Paltauf, F., and Hauser, H., Monolayer characteristics and thermal behavior of natural and synthetic phosphatidylserines, *Biochemistry*, 26, 8659, 1987.

102. Feigenson, G.W., On the nature of ion inding between phosphatidylserine lamellae, *Biochemistry*, 25, 5819, 1986.

103. Hauser, H. and Shipley, G.G., Interactions of divalent cations with phosphatidylserine bilayer mem-branes, *Biochemistry*, 23, 34, 1984.

104. Casal, H.L. et al., Infrared studies of fully hydrated unsaturated phosphatidylserine bilayers. Effect of Li+ and Ca2+, *Biochemistry*, 26, 7395, 1987.

105. Casal, H.L., Mantsch, H.H., and Gauser, H., Infrared studies of fully hydrated phosphatidylserine bilayers. Effect of lithium and calcium, *Biochemistry*, 26, 4408, 1987.

106. Nyholm, P.-G., Pascher, I., and Sundell, S., The effect of hydrogen bonds on the conformation of glycosphingolipids. Methylated and unmethylated cerebroside studied by x-ray single crystal analysis and model calculations, *Chem. Phys. Lipids*, 52, 1, 1990.

107. Shipley, G.G., X-ray crystallographic studies of aliphatic lipids, in *Handbook of Lipid Research*, vol. 4, Hanahan, D.J. ed., Plenum Press, New York, 1986, 97.

108. Boggs, J.M., Intermolecular hydrogen bonding between lipids: influence on organization and function of lipids in membranes, *Can. J. Biochem.*, 58, 755, 1980.

109. Boggs, J.M., Effect of lipid structural modifications on their intermolecular hydrogen bonding inter-actions and membrane function, *Biochem. Cell. Biol.*, 64, 50, 1986.

110. Boggs, J.M., Lipid intermolecular hydrogen bonding: influence on structural organization and mem-brane function, *Biochim. Biophys. Acta*, 906, 353, 1987.

111. Eibl, H., The effect of the proton and of monovalent cations on membrane fluidity, in *Membrane Fluidity in Biology*, Aloia, R.C., ed., Academic Press, New York, 1983, 217.

112. Nagle, J.F., Theory of lipid monolayer and bilayer phase transitions: effect of headgroup interactions, *J. Memb. Biol.*, 27, 233, 1976.

113. Lewis, R.N.A.H. et al., The effect of fatty acyl chain length and structure on the lamellar gel to liquid-crystalline and lamellar to reversed hexagonal phase transitions of aqueous phosphatidylethanolamine dispersions, *Biochemistry*, 28, 541, 1989.

114. Lewis, R.N.A.H. and McElhaney, R.N., Calorimetric and spectroscopic studies of the polymorphic phase behavior of a homologous series of *n*-saturated 1,2-diacyl phosphatidylethanolamines, *Biophys. J.* 64, 1081, 1993.

115. Tremblay, P.A. and Kates, M., Comparative physical studies of phosphatidylsulfocholine and phosphatidylcholine. Calorimetry, fluorescence polarization and electron paramagnetic resonance spectroscopy, *Chem. Phys. Lipids,* 28, 307, 1981.

116. Mantsch, H.H. et al., Phosphatidylsulfocholine bilayers. An infrared spectroscopic characterization of the polymorphic phase behavior, *Biochim. Biophys. Acta,* 698, 63, 1982.

117. Wisner, D.A., Rosario-Jensen, T., and Tsai, M.-D., Phospholipids chiral at phosphorous. Configurational effect on the thermotropic properties of chiral dipalmitoyl phosphatidylcholine, *J. Am. Chem. Soc.,* 108, 8064, 1986.

118. Tsai, M.-D., Jiang, R.-T., and Bruzik, K., Phospholipids chiral at phosphorous. 4 Could membranes be chiral at phosphorous?, *J. Am. Chem. Soc.,* 105, 2478, 1983.

119. Sarvis, H.E. et al., Phospholipids chiral at phosphorous. Characterization of the subgel phase of thiophosphatidylcholines by use of x-ray diffraction, phosphorous-31 nuclear magnetic resonance, and Fourier transform infrared spectroscopy, *Biochemistry,* 27, 4625, 1988.

120. Scherer, J.R., On the position of the hydrophobic/hydrophylic boundary in lipid bilayers, *Biophys. J,* 55, 957, 1989.

121. Boyanov, A.I. et al., Absence of subtransition in racemic dipalmitoylphosphatidylcholine vesicles, *Biochim. Biophys. Acta,* 732, 711, 1983.

122. Kodama, M., Hashigami, H., and Seki, S., Static and dynamic calorimetric studies of three kinds of phase transition in the systems of L- an DL-dipalmitoylphosphatidylcholine/water, *Biochim. Biophys. Acta,* 814, 300, 1985.

123. Singh, A. et al., Lateral phase separation based on chirality in a polymerizable lipid and its influence on formation of tubular structures, *Chem. Phys. Lipids,* 47, 135, 1988.

124. Rudolph, A.S. and Burke, T.G., A Fourier-transform spectroscopic study of the polymorphic phase behavior of 1,2-bis(tricosa-10,12-diynoyl)-*sn*-glycero-3-phosphocholine; A polymerizable lipid which forms novel microstructures, *Biochim. Biophys. Acta,* 902, 349, 1987.

125. Bruzik., K.S. and Tsai, M.D., A calorimetric study of the thermotropic behavior of pure sphingomyelin diastereomers, *Biochemistry,* 26, 5364, 1987.

126. Buldt, G. and de Haas, G.H., Conformational differences between *sn*-3-phospholipids and *sn*-2-phospholipids, *J. Mol. Biol.,* 158, 55, 1982.

127. Seelig, J., Dijkman, R., and de Haas, G., Thermodynamic and conformational studies on *sn*-2-phosphatidylcholines in monolayers and bilayers, *Biochemistry,* 19, 2215, 1980.

128. Serrallach, E.N., et al., Structure and thermotropic properties of 1,3-dipalmitoyl-glycero-2-phosphocholine, *J. Mol. Biol.,* 170, 155, 1983.

129. Chowdhry, B.Z., Dalziel, A.W., and Sturtevant, J.M., Phase transition properties of 1,3-dipalmitoylphosphatidylethanolamine, *J. Phys. Chem.,* 88, 5397, 1984.

130. Howard, B.V. et al., Lipid metabolism in normal and tumor cells in culture, in *Tumor Lipids: Biochemistry and Metabolism,* Wood, R., ed., American Oil Chem. Soc., 1973, 200.

131. Spencer, F., Ether lipids in clinical diagnosos and medical research, in *Ether Lipids, Biochemical and Biomedical Aspects,* Mangold, H.K. and Paltauf, F., eds., Academic Press, New York, 1983, 239.

132. Ruocco, M.J., Siminovitch, D.J., and Griffin, R.G., Comparative study of ether- and ester-linked phosphatidylcholines, *Biochemistry,* 24, 2406, 1985.

133. Kim, J.T., Mattai, J., and Shipley, G.G., Bilayer interactions of ether- and ester-linked phosphlipids: dihexadecyl- and dipalmitoylphosphatidylcholines, *Biochemistry,* 26, 6599, 1987.

134. Kim, J.T., Mattai, J., and Shipley, G.G., Gel phase polymorphism in ether linked dihexadecylphosphatidylcholine bilayers. *Biochemistry* 26, 6592, 1987.

135. Lewis, R.N.A.H., Pohle, W., and McElhaney, R.N. The interfacial structure and the phospholipid bilayers: Differential scanning calorimetry and Fourier transform infrared spectroscopic of 1,2-dipalmitoyl-*sn*-glycero-3-phosphocholine and its dialkyl and acyl-alkyl analogs. *Biophys. J.,* 70, 2736, 1996.

136. Seddon, J.M. et al., X-ray diffraction study of the polymorphism of hydrated diacyl- and dialkyk-phosphatidylethanolamines, *Biochemistry,* 23, 2364, 1984.

137. Kuttenreich, H. et al., Polymorphism of synthetic 1,2-dialkyl-3-O-β-D-galactosyl-*sn*-glycerols of different chain lengths, *Chem. Phys. Lipids,* 47, 245, 1988.

138. Hing, F.S., Maulik, P.R., and Shipley, G.G. Structure and interactions of ether- and ester-linked phosphatidylethanolamines, *Biochemistry,* 30, 9007, 1991.

139. Hinz, H.-J. et al., Stereochemistry and size of sugarheadgroups determine structure and phase behavior of sugar membranes: densitometric, calorimetric and x-ray studies, *Biochemistry*, 30, 5125, 1991.

140. Mannock, D.A. et al., Synthesis and thermotropic characterization of an homologous series of racemic β-D-glucosyl dialkyl-glycerols, *Biochim. Biophys. Acta*, 1509, 203, 2000.

141. Mannock, D.A. et al., The physical properties of glycosyl diacylglycerols. Calorimetric, x-ray diffraction and Fourier transform infrared spectroscopic studies of a homologous series of 1,2-di-O-acyl-3-O-(β-D-galactopyranosyl)-*sn*-glycerols, *Chem. Phys. Lipids*, 111, 139, 2001.

142. Barenholz, Y. et al., A calorimetric study of the thermotropic behavior of aqueous dispersions of natural and synthetic sphingomyelins, *Biochemistry*, 15, 2441, 1976.

143. Estep, T.N. et al., Thermal behavior of synthetic sphingomyelin-cholesterol dispersions, *Biochemistry*, 18, 2112, 1979.

144. Estep, T.N. et al., Evidence for metastability in stearoyl sphingomyelin bilayers, *Biochemistry*, 19, 20, 1980.

145. Curatolo, W., Bali, A., and Gupta, C.M., Metastable phase behavior of a sphingolipid analogue, *Biochim. Biophys. Acta*, 690, 89, 1982.

146. Curatolo, W., Bali, A., and Gupta, C.M., Phase behavior of carbamyloxyphosphatidylcholine, a sphingolipid analogue, *J. Pharmatect. Sci.*, 74, 1255, 1985.

147. Chowdhry, B.Z. et al., Phase transition properties of 1-alkyl and 1-acyl-2-amidophosphatidylcholines and related derivatives, *Biochemistry*, 23, 2044, 1984.

148. Reed, R.A. and Shipley, G.G., Structure and metastability of N-lignocerylgalactosylsphingosine (cerebroside) bilayers, *Biochim. Biophys. Acta,* 896, 153, 1988.

149. Reed, R.A. and Shipley, G.G., Effect of chain length on the structure and thermotropic properties of galactocerebroside, *Biophys. J.*, 55, 281, 1989.

150. Boggs, J.M., Koshy, K.M., and Rangaraj, C., Effect of fatty acid chain length and hydroxylation and various cations on phase behavior of synthetic cerebroside Sulfate, *Chem. Phys. Lipids*, 36, 65, 1984.

151. Boggs, J.M., Koshy, K.M., and Rangaraj, C., Influence of structural modifications on the phase behavior of semi-synthetic cerebroside sulfate, *Biochim. Biophys. Acta*, 938, 361, 1988.

152. Curatolo, W. and Jungalwala, F.B., Phase behavior of galactocerebrosides from bovine brain, *Biochemistry*, 24, 6608, 1985.

153. Lewis, E.N., Bittman, R., and Levin, I.W., Methyl group substitution at C(1) or C(3) of the glycerol backbone of a diether phosphocholine. A comparative study of the bilayer chain disorder in the gel and liquid crystalline phase, *Biochim. Biophys. Acta*, 861, 44, 1986.

154. Singer, M.A. et al., The properties of membranes formed from cyclopentanoid analogues of phosphatidylcholine, *Biochim. Biophys. Acta,* 731, 373, 1983.

155. Jain, M.K. et al., Phase transition properties of aqueous dispersions of homologues of all trans 2,3-dipalmitoylcyclopentano-1-phosphocholine, *Biochim. Biophys. Acta*, 774, 199, 1984.

156. Blume, A. and Eibl, H., A calorimetric study of the thermotropic behaviour of 1,2-dipentadecylmethylidene phospholipids, *Biochim Biophys. Acta*, 640, 609, 1981.

157. Mabrey, S. and Sturtevant, J.M., High-sensitivity differential scanning calorimetry in the study of biomembranes and related model systems, in *Methods in Membrane Biology*, Korn, E.D., ed., Plenum Press, New York, 1978, 237.

158. van Dijck, P.W.M. et al., Miscibility properties of binary phosphatidylcholine mixtures. A calorimetric study, *Biochim. Biophys. Acta,* 470, 58, 1977.

159. Phillips, M.C., Ladbroke, B.D., and Chapman, D., Molecular interactions in mixed lecithin systems, *Biochim. Biophys. Acta,* 196, 35, 1970.

160. Chapman, D., Urbina, J., and Keough, K.M., Biomembrane phase transitions. Studies of lipid-water systems using differential scanning calorimetry, *J. Biol. Chem.*, 249, 2512, 1974.

161. Mabrey, S. and Sturtevant, J.M., Investigation of phase transitions of lipids and lipid mixtures by high sensitivity differential scanning calorimetry, *Proc. Natl. Acad. Sci. U.S.A.*, 73, 3862, 1976.

162. Findlay, E.J. and Barton, P.G., Phase behavior of synthetic phosphatidylglycerols and binary mixtures with phosphatidylcholines in the presence and absence of calcium ions, *Biochemistry*, 17, 2400, 1978.

163. Blume, A. and Ackerman, T., A calorimetric study of the lipid phase transitions in aqueous dispersions of phosphorylcholine-phosphorylethanolamine mixtures, *FEBS Lett.*, 43, 71, 1974.

164. Lee, A.G., Lipid phase transitions and phase diagrams. II. Mixtures involving lipids, *Biochim. Biophys. Acta*, 472, 285, 1977.

165. Jacobson, K. and Paphadjopoulos, D., Phase transitions and phase separations in phospholipid membranes induced by changes in temperature, pH, and concentration of bivalent cations, *Biochemistry*, 14, 152, 1975.

166. Stewart, T.P. et al., Complex phase mixing of phosphatidylcholine and phosphatidylserine in multilamellar vesicles, *Biochim. Biophys. Acta*, 556, 1, 1979.

167. Clowes, A.W., Cherry, R.J., and Chapman, D., Physical properties of lecithin-cerebroside bilayers, *Biochim. Biophys. Acta*, 249, 301, 1971.

168. Sillerud, L.O. et al., Calorimetric properties of mixtures of ganglioside G_M1 and dipalmitoylphosphatidylcholine, *J. Biol. Chem.*, 254, 10876, 1979.

169. Barenholz, Y. and Thompson, T.E., Sphingomyelins in bilayers and biological membranes, *Biochim. Biophys. Acta*, 604, 129, 1980.

170. Lee, A.G., Calculation of phase diagrams for non-ideal mixtures of lipids, and a possible non-random distribution of lipids in lipid mixtures in the liquid crystalline phase, *Biochim. Biophys. Acta*, 507, 433, 1978.

171. Von Dreele, P.H., Estimation of lateral species separation from phase transitions in nonideal two-dimensional lipid mixtures, *Biochemistry*, 17, 3939, 1978.

172. Ito, T. and Ohnishi, S.-I., Ca2+ Induced lateral phase separations in phosphatidic acid-phosphatidylcholine membranes, *Biochim. Biophys. Acta*, 352, 29, 1974.

173. Galla, H.-J. and Sackman, E., Chemically induced lipid phase separation in mixed vesicles containing phosphatidic acid. An optical study, *J. Am. Chem. Soc.*, 97, 4114, 1975.

174. Galla, H.-J. and Sackman, E., Chemically induced lipid phase separation in model membranes containing charged lipids: A spin label study, *Biochim. Biophys. Acta*, 401, 509, 1975.

175. Ohnishi, S. and Ito, T., Clustering of lecithin molecules in phosphatidylserine membranes induced by calcium ion binding to phosphatidylserine, *Biochem. Biophys. Res. Commun.*, 51, 132, 1973.

176. Papahadjopoulos, D. et al., Membrane fusion and molecular segregation in phospholipid vesicles, *Biochim. Biophys. Acta*, 352, 10, 1974.

177. Lee, A.G., Fluorescence studies of chlorophyll a incorporated into lipid mixtures, and the interpretation of "phase" diagrams, *Biochim. Biophys. Acta*, 413, 11, 1975.

178. van Dijck, P.W.M. et al., Comparative studies of the effect of pH and Ca2+ on bilayers of various negatively charged phospholipids, and their mixtures with phosphatidylcholine, *Biochim. Biophys. Acta*, 512, 84, 1978.

179. van Dijck, P.W.M. et al., Influence of Ca2+ and Mg2+ on the thermotropic phase behavior and permeability properties of liposomes prepared from dimyristoylphosphatidylglycerol and mixtures of dimyristoylphosphatidylglycerol and dimyristoylphosphatidylcholine, *Biochim. Biophys. Acta*, 406, 465, 1975.

180. Demel, R.A. and de Kruijff, B., The function of sterols in membranes, *Biochim. Biophys. Acta*, 457, 109, 1976.

181. Yeagle, P.L., Cholesterol and the cell membrane, in *Biology of Cholesterol*, Yeagle, P.L., ed., CRC Press, Boca Raton, FL, 1988, 122.

182. Ladbrooke, B.D. et al., Physical studies of myelin. I. Thermal analysis. *Biochim. Biophys. Acta*, 164, 101, 1968.

183. Hinz, H.-J. and Sturtevant, J.M., Calorimetric investigation of the influence of cholesterol on the transition properties of bilayers formed from synthetic L-α-lecithins in aqueous suspensions, *J. Biol. Chem.*, 247, 3697, 1972.

184. Gershfeld, N.L., Equilibrium studies of lecithin-cholesterol interactions. I. Stoichiometry of lecithin-cholesterol complexes in bulk systems, *Biophys. J.*, 22, 469, 1978.

185. Calhoun, W.I. and Shipley, G.G., Sphingomyelin-lecithin bilayers and their interaction with cholesterol, *Biochemistry*, 18, 1717, 1979.

186. De Kruijff, B. et al., Non-random distribution of cholesterol in phosphatidylcholine bilayers, *Biochim. Biophys. Acta*, 356, 1, 1974.

187. Vist, M.R. and Davis, J.H., Equilibria of cholesterol/dipalmitoylphosphatidylcholine mixtures: 2H Nuclear magnetic resonance and differential scanning calorimetry, *Biochemistry*, 29, 451, 1990.

188. Mabrey, S., Mateo, P.L., and Sturtevant, J.M., High-sensitivity scanning calorimetry study of mixtures of cholesterol with dimyristoyl- and dipalmitoylphosphatidylcholine, *Biochemistry*, 17, 2464, 1978.

189. Estep, T.N. et al., Studies on the anomalous thermotropic behavior of aqueous dispersions of dipalmitoylphosphatidylcholine-cholesterol mixtures, *Biochemistry*, 17, 1984, 1978.

190. Genz, A., Holzwarth, J.F., and Tsong, T.Y., The influence of cholesterol on the main phase transition of unilamellar dipalmitoylphosphatidylcholine vesicles, *Biophys. J.*, 50, 1043, 1986.

191. Melchoir, D.L., Scavitto, F.J., and Steim, J.M., Dilatometry of dipalmitoyllecithin-cholesterol bilayers, *Biochemistry*, 19, 4828, 1980.

192. Engelman D.M. and Rothman, J.E., The planar organization of lecithin-cholesterol bilayers, *J. Biol. Chem.*, 247, 3694, 1972.

193. Rothman, J.E. and Engelman, D.M., Molecular mechanism for the interaction of phospholipid with cholesterol, *Nature — New Biol.*, 237, 42, 1972.

194. McMullen, T.P.W., Lewis, R.N.A.H., and McElhaney, R.N., Differential scanning calorimetric study of the effect of cholesterol on the thermotropic phase behavior of a homologous series of linear saturated phosphatidylcholines, *Biochemistry*, 32, 516, 1993.

195. Singer, M.A. and Finegold, L., Cholesterol interacts with all of the lipid in bilayer membranes, *Biophys. J.*, 57, 153, 1990.

196. Singer, M.A. and Finegold, L., Interaction of cholesterol with saturated phospholipids: role of the C(17) side chain, *Chem. Phys. Lipids*, 56, 217, 1990.

197. McMullen, T.P.W. and McElhaney, R.N., New aspects of the interaction of cholesterol with dipalmitoylphosphatidylcholine bilayers as revealed by high-sensitivity differential scanning calorimetry, *Biochim. Biophys. Acta*, 1234, 90, 1995.

198. McMullen, T.P.W., Lewis, R.N.A.H., and McElhaney, R.N., Comparative differential scanning calorimetric and FTIR and ^{31}P-NMR spectroscopic studies of the effects of cholesterol and androstenol on the thermotropic phase behavior and organization of phosphatidylcholine bilayers, *Biophys. J.*, 66, 741, 1994.

199. Vilcheze, C. et al., The effect of side chain analogues of cholesterol on the thermotropic phase behavior of 1-stearoyl-2-oleoyl-phosphatidylcholine bilayers: a differential scanning calorimetric study, *Biochim. Biophys. Acta*, 1279, 235, 1996.

200. Davis, P.J. and Keough, K.M.W., Differential scanning calorimetric studies of aqueous dispersions of mixtures of cholesterol with some mixed-acid and single-acid phosphatidylcholines, *Biochemistry*, 22, 6334, 1983.

201. Davis, P.J. and Keough, K.M.W., Scanning calorimetric studies of aqueous dispersions of bilayers made with cholesterol and a pair of positional isomers of 3-*sn*-phosphatidylcholine, *Biochim. Biophys. Acta*, 778, 305, 1987.

202. Keough, K.M.W., Giffin, B., and Matthews, P.L.J., Phosphatidylcholine-cholesterol interactions: bilayers of heteroacid lipids containing linoleate lose calorimetric transitions at low cholesterol concentration, *Biochim. Biophys. Acta*, 983, 51, 1989.

203. McMullen, T.P.W. and McElhaney, R.N., Differential scanning calorimetric studies of the interaction of cholesterol with distearoyl and dielaidoyl molecular species of phosphatidylcholine, phosphatidylethanolamine and phosphatidylserine, *Biochemistry*, 36, 4979, 1997.

204. Shaikh, S.R. et al., Lipid phase separation in phospholipid bilayers and monolayers modeling the plasma membrane, *Biochim. Biophys. Acta*, 1512, 317, 2001.

205. Brzustowicz, M.R. et al., Molecular organization of cholesterol in polyunsaturated membranes: microdomain formation, *Biophys. J.*, 82, 285, 2002.

206. Brzustowicz, M.R. et al., Controlling membrane cholesterol content. A role for polyunsaturated (docosahexaenoate) phospholipids, *Biochemistry* 41, 12509, 2002.

207. McMullen, T.P.W., Lewis, R.N.A.H., and McElhaney, R.N., Calorimetric and spectroscopic studies of the effects of cholesterol on the thermotropic phase behavior and organization of a homologous series of linear saturated phosphatidylethanolamine bilayers, *Biochim. Biophys. Acta*, 1416, 119, 1999.

208. Bach, D., Differential scanning calorimetric study of mixtures of cholesterol with phosphatidylserine or galactocerebroside, *Chem. Phys. Lipids*, 35, 385, 1984.

209. Bach, D. and Wachtel, E., Thermotropic properties of mixtures of negatively charged phospholipids with cholesterol in the presence and absence of Li+ or Ca2+ ions, *Biochim. Biophys. Acta*, 979, 11, 1989.

210. Bach, D., Borochov, N., and Wachtel, E., Phase separation of cholesterol in dimyristoylphosphatidylserine-cholesterol mixtures, *Chem. Phys. Lipids*, 92, 71, 1998.

211. Epand, R.M. et al., Cholesterol crystalline polymorphism and the solubility of cholesterol in phosphatidylserine, *Biophys. J.*, 78, 866, 2000.

212. Epand, R.M. et al., A new high-temperature transition of crystalline cholesterol in mixtures with phosphatidylserine, *Biophys. J.*, 81, 1511, 2001.

213. Bach, D. and Miller, I.R., Attenuated total reflection (ATR) Fourier transform infrared of dimyristoyl phosphatidylcholine-cholesterol mixtures. *Biochim. Biophys. Acta.*, 1514, 318, 2002.

214. Epand, R.M. et al., Properties of mixtures of cholesterol with phosphatidylcholine or with phosphatidylserine studied by C-13 magic angle spinning nuclear magnetic resonance, *Biophys. J.*, 83, 2053, 2002.

215. Johnston, D.S. and Chapman, D., A calorimetric study of the thermotropic behaviour of mixtures of brain cerebrosides with other brain lipids, *Biochim. Biophys. Acta*, 939, 603, 1988.

216. McMullen, T.P.W., Lewis, R.N.A.H., and McElhaney, R.N., Differential scanning calorimetric and Fourier transform infrared spectroscopic studies of the effects of cholesterol on the thermotropic phase behavior and organization of a homologous series of linear saturated phosphatidylserine bilayer membranes, *Biophys. J.*, 79, 2056, 2000.

217. Bublotz, J.T. and Feigenson, G.W., A novel strategy for the preparation of liposomes: rapid solvent exchange, *Biochim. Biophys. Acta*, 1417, 233, 1999.

218. Huang, J., Bublotz, J.T., and Feigenson, G.W., Maximum solubility of cholesterol in phosphatidylcholine and phosphatidylethanolamine bilayers, *Biochim. Biophys. Acta*, 1417, 89, 1999.

219. McMullen, T.P.W. et al., Differential scanning calorimetric study of the interaction of cholesterol with the major lipids of the *Acholeplasma laidiawii* B membrane, *Biochemistry*, 35, 16789, 1996.

220. van Dijck, P.W.M. et al., The preference of cholesterol for phosphatidylcholine in mixed phosphatidylcholine-phosphatidylethanolamine bilayers, *Biochim. Biophys. Acta*, 455, 576, 1976.

221. Demel, R.A. et al., The preferential interaction of cholesterol with different classes of phospholipids, *Biochim. Biophys. Acta*, 465, 1, 1977.

222. Blume, A., Thermotropic behavior of phosphatidylethanolamine-cholesterol and phosphatidylethanolamine-phosphatidylcholine-cholesterol, *Biochemistry*, 19, 4908, 1980.

223. Jain, M.K. and Wu, N.M., Effect of small molecules on the dipalmitoyl lecithin liposomal bilayer. III. Phase transitions in lipid bilayers, *J. Membrane Biol.*, 34, 157, 1977.

224. Eliasz, A.W., Chapman, D., and Ewing, D.F., Phospholipid phase transitions. Effects of *n*-alcohols, *n*-monocarboxylic acids, phenylalkyl alcohols and quaternary ammonium compounds, *Biochim. Biophys. Acta*, 448, 220, 1976.

225. Kremer, J.M.H. and Wiersema, P.H., Exchange and aggregation in dispersions of dimyristoylphosphatidylcholine vesicles containing myristic acid, *Biochim. Biophys. Acta*, 471, 348, 1977.

226. Mabrey, S. and Sturtevant, J.M., Incorporation of saturated fatty acids into phosphatidylcholine bilayers, *Biochim. Biophys. Acta*, 486, 444, 1977.

227. Usher, J.R., Epand, R.M., and Papahadjopoulos, D., The effect of free fatty acids on the thermotropic phase transition of dimyristoyl glycerophosphocholine, *Chem. Phys. Lipids*, 22, 245, 1978.

228. Ortiz, A. and Gomez-Fernandez, J.C., A differential scanning calorimetry study of the interaction of free fatty acids with phospholipid membranes, *Chem. Phys. Lipids*, 45, 75, 1987.

229. Marsh, D. and Seddon, J.M., Gel to inverted hexagonal (Lβ-Hʜ) phase transitions in phosphatidylethanolamines and fatty acid-phosphatidylcholine mixtures demonstrated by 31P-NMR spectroscopy and x-ray diffraction, *Biochim. Biophys. Acta*, 690, 117, 1982.

230. Cevc, G. et al., Phosphatidylcholine-fatty acid membranes. I. Effects of protonation, salt concentration, temperature and chain-length on the colloidal and phase properties of mixed vesicles, bilayers and nonlamellar structures, *Biochim. Biophys. Acta*, 940, 219, 1988.

231. Klopfenstein, W.E. et al., Differential scanning calorimetry of mixtures of lecithin, lysolecithin and cholesterol, *Chem Phys. Lipids*, 13, 215, 1974.

232. Blume, A., Arnold, B., and Weltzien, H.U., Effects of a synthetic lysolecithin analog on the phase transition of mixtures of phosphatidylethanolamine and phosphatidylcholine, *FEBS Lett.*, 61, 199, 1976.

233. van Echteld, C.J.A., de Kruijff, B., and De Gier, J., Differential miscibility properties of various phosphatidylcholine/lysophosphatidylcholine mixtures, *Biochim. Biophys. Acta*, 595, 71, 1980.

234. Lewis, R.N.A.H. et al., Mechanisms of the interaction of α-helical transmembrane peptides with phospholipid bilayers, *Bioelectrochem.*, 56, 135, 2002.

235. Davis, J.M. et al., Interaction of a synthetic amphiphilic polypeptide and lipids in a bilayer structure, *Biochemistry* 22, 5298, 1983.

236. Axelsen, P.H. et al., The infrared dichroism of transmembrane helical polypeptides, *Biophys. J.*, 69, 2770, 1995.

237. Huschilt, J.C., Millman, B.M., and Davis, J.H., Orientation of α-helical peptides in a lipid bilayer, *Biochim. Biophys. Acta*, 979, 139, 1989.

238. Bolen, E.J. and Holloway, P.W., Quenching of tryptophan fluorescence by brominated phospholipid, *Biochemistry*, 29, 9638, 1990.

239. Huschilt, J.C., Hodges, R.S., and Davis, J.H., Phase equilibria in an amphiphilic peptide–phospholipid membrane by deuterium nuclear magnetic resonance difference spectroscopy, *Biochemistry*, 24, 1377, 1985.

240. Zhang, Y.-P. et al., FTIR spectroscopic studies of the conformation and amide hydrogen exchange of a peptide model of the hydrophobic transmembrane α-helices of membrane proteins, *Biochemistry*, 31, 11572, 1992.

241. Zhang, Y.-P. et al., Interaction of a peptide model of a hydrophobic transmembrane α-helical segment of a membrane protein with phosphatidylcholine bilayers: DSC and FTIR spectroscopic studies, *Biochemistry*, 31, 11579, 1992.

242. Pare, C. et al., Differential scanning calorimetry and ^{2}H-NMR and FTIR spectroscopic studies of the effects of transmembrane α-helical peptides on the organization of phosphatidylcholine bilayers, *Biochim. Biophys. Acta*, 1511, 60, 2001.

243. Subczynski, W.K. et al., Molecular organization and dynamics of 1-palmitoyl-2-oleoyl-phosphatidyl-choline bilayers containing a transmembrane α-helical peptide, *Biochemistry*, 37, 3156, 1998.

244. Zhang, Y.-P. et al., Interaction of a peptide model of a hydrophobic α-helical segment of a membrane protein with phosphatidylethanolamine bilayers: DSC and FTIR spectroscopic studies, Biophys. J., 68, 847, 1995.

245. Zhang, Y.-P. et al., Peptide models of helical hydrophobic transmembrane segments of membrane proteins. 1. Studies of the conformation, intrabilayer orientation and amide hydrogen exchangeability of Ac-K$_2$-(LA)$_{12}$-K$_2$-amide, *Biochemistry*, 34, 2348, 1995.

246. Zhang, Y.-P. et al., Peptide models of helical hydrophobic transmembrane segments of membrane proteins. 2. DSC and FTIR spectroscopic studies of the interaction of Ac-K$_2$-(LA)$_{12}$-K$_2$-amide with phosphatidylcholine bilayers, *Biochemistry*, 34, 2362, 1995.

247. Zhang, Y.-P. et al., Peptide models of the helical transmembrane segments of membrane proteins: interactions of acetyl-K$_2$-(LA)$_{12}$-K$_2$-amide with phosphatidylethanolamine bilayer membranes, *Biochemistry*, 40, 474, 2001.

248. Lewis, R.N.A.H. et al., A polyalanine-based peptide cannot form a stable transmembrane α-helical in fully hydrated phospholipid bilayer membranes, *Biochemistry,* 40, 12103, 2001.

249. Chung, L.A. and Thompson, T.E., Design of membrane-inserting peptides: spectroscopic characterization with and without lipid bilayers, *Biochemistry*, 35, 11343, 1996.

250. White, S.H. and Wimley, W.C., Membrane protein folding and stability: physical principles, *Annu. Rev. Biophys. Biomol. Struct.*, 28, 319, 1999.

251. Percot, H., Zhu, X.X., and Lafleur, M., Design and characterization of anchoring amphiphilic peptides and their interactions with lipid vesicles, *Biopolymers*, 50, 647, 1999.

252. Moll, T.S. and Thompson, T.E., Semisynthetic proteins: model systems for the study of the insertion of hydrophobic peptides into preformed lipid bilayers, *Biochemistry*, 33, 15469, 1994.

253. Liu, F. et al., Effect of variations in the structure of a polyleucine-based α-helical transmembrane peptide on its interactions with phosphatidylcholine bilayers, *Biochemistry*, 41, 9197, 2002.

254. Segrest, J.P. et al., Amphipathic helix motif — classes and properties, *Proteins: Struct., Funct. Genet.*, 8, 103, 1990.

255. Monne, M. et al., Positively and negatively charged residues have different effects on the position in the membrane of a model transmembrane helix, *J. Mol. Biol.*, 284, 1177, 1998.

256. Liu, F. et al., A DSC and ^{31}P-NMR spectroscopic study of the effects of transmembrane α-helical peptides on the lamellar/reversed hexagonal phase transition of phosphatidylethanolamine model membranes, *Biochemistry*, 40, 760, 2001.

257. de Planque, M.R. et al., Different membrane anchoring positions of tryptophan and lysine in synthetic transmembrane alpha-helical peptides, *J. Biol. Chem.*, 274, 20839, 1999.

258. Van der Wel, P.C.A. et al., Tryptophan-anchored transmembrane peptides promote formation of nonlamellar phases in phosphatidylethanolamine model membranes in a mismatch-dependent manner, *Biochemistry*, 39, 3124, 2000.

259. Killian, J.A. et al., Induction of nonbilayer structures in diacylphosphatidylcholine model membranes by transmembrane alpha-helical peptides: importance of hydrophobic mismatch and proposed role of tryptophans, *Biochemistry*, 35, 1037, 1996.

260. Morein, S. et al., Influence of membrane-spanning alpha-helical peptides on the phase behavior of the dioleoylphosphatidylcholine/water system, *Biophys. J.*, 73, 3078, 1997.

261. Morein, S. et al., The effect of peptide/lipid hydrophobic mismatch on the phase behavior of model membranes mimicking the lipid composition of *Escherichia coli* membranes, *Biophys. J.*, 78, 2745, 2000.

262. Killian, J.A. et al., The tryptophans of gramicidin are essential for the lipid structure modulating effect of the peptide, *Biochim. Biophys. Acta,* 820, 154, 1985.

263. Killian, J.A. and de Kruijff, B., Thermodynamic, motional, and structural aspects of gramicidin-induced hexagonal H_{II} phase formation in phosphatidylethanolamine, *Biochemistry*, 24, 7881, 1985.

264. Killian, J.A. et al., Gramicidin-induced hexagonal H_{II} phase formation in negatively charged phospholipids and the effect of N- and C-terminal modification of gramicidin on its interaction with zwitterionic phospholipids, *Biochim. Biophys. Acta*, 857, 13, 1986.

265. Killian, J.A. et al., A mismatch between the length of gramicidin and the lipid acyl chains is a prerequisite for H_{II} phase formation in phosphatidylcholine model membranes, *Biochim. Biophys. Acta*, 978, 341, 1989.

266. Killian, J.A., Gramicidin and gramicidin-lipid interactions, *Biochim. Biophys. Acta*, 1113, 391, 1992.

267. McElhaney, R.N., Differential scanning calorimetric studies of lipid-protein interactions in model membrane systems, *Biochim. Biophys. Acta,* 864, 361, 1986.

268. Papahadjopoulos, D. et al., Effects of proteins on thermotropic phase transitions of phospholipid membranes, *Biochim. Biophys. Acta*, 401, 317, 1975.

269. Boggs, J.M. et al., Lipid phase separation induced by a hydrophobic protein in phosphatidylserine-phosphatidylcholine vesicles, *Biochemistry*, 16, 2325, 1977.

270. Boggs, J.M., Vail, W.J., and Moscarello, M.A., Preparation and properties of vesicles of a purified myelin hydrophobic protein and phospholipid. A spin label study, *Biochim. Biophys. Acta*, 448, 517, 1976.

271. van Zolen, E.J.J. et al., Effect of glycophorin incorporation on the physico-chemical properties of phospholipid bilayers, *Biochim. Biophys. Acta,* 514, 9, 1978.

272. George, R., Lewis, R.N.A.H., and McElhaney, R.N., Studies on the purified Na+,Mg2+-ATPase from *Acholeplasma laidlawii* B membranes: a differential scanning calorimetric study of the protein-phospholipid interactions, *Biochem. Cell Biol.*, 68, 161, 1990.

273. Sanderman, H., Regulation of membrane enzymes by lipids, *Biochim. Biophys. Acta*, 515, 209, 1978.

274. McElhaney, R.N., The structure and function of the *Acholeplasma laidlawii* plasma membrane, *Biochim. Biophys. Acta,* 779, 1, 1984.

275. McElhaney, R.N., The influence of membrane lipid composition and physical properties on membrane structure and function in *Acholeplasma laidlawii, CRC Critical Rev. Microbiol.*, 17, 1, 1989.

276. Steim, J.M. et al., Calorimetric evidence for the liquid crystalline state of lipids in a biomembrane, *Proc. Natl. Acad. Sci. U.S.A.*, 63, 104, 1969.

277. Reinert, J.C. and Steim, J.M., Calorimetric detection of a membrane lipid phase transition in living cells, *Science,* 168, 1580, 1970.

278. Melchoir, D.L. et al., Characterization of the plasma membrane of *Mycoplasma laidlawii*. VIII. Phase transitions of membrane lipids, *Biochim. Biophys. Acta*, 219, 114, 1970.

279. Cameron, D.G., Martin, A., and Mantsch, H.H., Membrane isolation alters the gel to liquid-crystalline transition of *Acholeplasma laidlawii, Science*, 219, 180, 1983.

280. Cameron, D.G. et al., Infrared spectroscopic study of the gel to liquid-crystal phase transition in live *Acholeplasma laidlawii* cells, *Biochemistry*, 24, 4355, 1985.

281. Seguin, C. et al., Calorimetric studies of the thermotropic phase behavior of cells, membranes and lipids from fatty acid-homogenous *Acholeplasma laidlawii* B, *Israel J. Med. Sci.*, 23, 403, 1987.

282. Silvius, J.R., Mak, N., and McElhaney, R.N., Lipid and protein composition and thermotropic lipid phase transitions in fatty acid-homogenous membranes of *Acholeplasma laidlawii* B, *Biochim. Biophys. Acta*, 597, 199, 1980.

283. Silvius, J.R. and McElhaney, R.N., Growth and membrane lipid properties of *Acholeplasma laidlawii* B lacking fatty acid homogeneity, *Nature*, 272, 645, 1978.
284. Melchoir, D.L. and Steim, J.M., Thermotropic transitions in biomembranes, *Annu. Rev. Biophys. Bioeng.*, 5, 205, 1976.
285. Melchoir, D.L. and Steim, J.M., Lipid-associated thermal events in biomembranes, in *Progress in Surface and Membrane Science,* Cadenhead, D.A. and Danielli, J.F., eds., Academic Press, New York, 1979, 211.
286. Jackson, M.D. and Sturtevant, J.M., Phase behavior of lipids from *Halobacterium halobium, Biochemistry*, 17, 4470, 1978.
287. Jackson, M.D. and Sturtevant, J.M., Phase transitions of the purple membrane of *Halobacterium halobium, Biochemistry*, 17, 911, 1978.
288. Ladbrooke, B.D., Williams, R.M., and Chapman, D., Studies on lecithin-cholesterol-water interactions by differential scanning calorimetry and x-ray diffraction, *Biochim. Biophys. Acta*, 150, 333, 1968.
289. Brasitus, T.A., Tall, A.R., and Schachter, D., Thermotropic transitions in rat intestinal plasma membranes studied by differential scanning calorimetry and fluorescence polarization, *Biochemistry*, 19, 1256, 1980.
290. Blazyk, J.F. and Steim, J.M., Phase transitions in mammalian membranes, *Biochim. Biophys. Acta*, 266, 737, 1972.
291. Mabrey, S. et al., Calorimetric study of microsomal membranes, *J. Biol. Chem.*, 252, 2929, 1977.
292. Bach, D., Bursuker, I., and Goldman, R., Differential scanning calorimetry and enzymic activity of rat liver microsomes in the presence and absence of Δ1-tetrahydrocannabinol, *Biochim. Biophys. Acta*, 469, 171, 1977.
293. Hackenbrock, C.R., Hochli, M., and Chau, R.M., Calorimetric and freeze fracture analysis of lipid phase transitions and lateral translational motion of intramembrane particles in mitochondrial membranes, *Biochim. Biophys. Acta*, 455, 466, 1976.
294. Madden, T.D. et al., The incorporation of cholesterol into mitochondrial membranes and its effect on lipid phase transition, *Biochim. Biophys. Acta*, 599, 528, 1980.
295. Blazyk, J.F. and Newman, J.L., Calorimetric studies of lipid phase transitions in native and heat-denatured membranes of beef heart submitochondrial particles, *Biochim. Biophys. Acta*, 600, 1007, 1980.
296. Pehowich, D.J. et al., Calorimetric and spectroscopic studies of lipid thermotropic phase behavior in liver inner mitochondrial membranes from a mammalian hibernator, *Biochemistry*, 27, 4632, 1988.
297. Brandts, J.F. et al., Calorimetric studies of the structural transitions of the human erythrocyte membrane. The involvement of spectrin in the A transition, *Biochemistry*, 16, 3450, 1977.
298. Brandts, J.F. et al., Calorimetric studies of the structural transitions of the human erythrocyte membrane. Studies of the B and C transitions, *Biochim. Biophys. Acta*, 512, 566, 1978.
299. Snow, J.W., Brandts, J.F., and Low, P.S., The effects of anion transport inhibitors on structural transitions in erythrocyte membranes, *Biochim. Biophys. Acta,* 512, 579, 1978.
300. Snow, J.W., Vincentelli, J., and Brandts, J.F., A relationship between anion transport and a structural transition of the human erythrocyte membrane, *Biochim. Biophys. Acta*, 642, 418, 1978.
301. Miljanich, G. et al., Calorimetric studies of the retinal rod outer segment membrane, *Biophys. J.*, 16, 37a, 1976.
302. Cramer, W.A., Whitmarsh, J., and Low, P.S., Differential scanning calorimetry of chloroplast membranes: identification of an endothermic transition association with the water-splitting complex of photosystem II, *Biochemistry*, 20, 157, 1981.
303. Chen, S.C., Sturtevant, J.M., and Gaffney, B.J., Scanning calorimetric evidence for a third phase transition in phosphatidylcholine bilayers, *Proc. Natl. Acad. Sci U.S.A.,* 77, 5060, 1980.
304. Hinz, H.-J. et al., Head-group contributions to bilayer stability: monolayer and calorimetric studies on synthetic, stereochemically uniform glucolipids, *Biochemistry*, 24, 806, 1985.
305. Koynova, R.D. et al., Influence of head-group interactions on the miscibility of synthetic stereochemically pure glycolipids and phospholipids, *Biochemistry*, 27, 4612, 1988.
306. Mannock, D.A. and McElhaney, R.N., Differential scanning calorimetry and x-ray diffraction studies of a series of synthetic β-D-galactosyl diacylglycerols, *Biochem. Cell Biol.*, 69, 863, 1991.
307. Sen, A. et al., Thermotropic phase properties and structure of 1,2-distearoylgalactosylglycerols in aqueous systems, *Proc. R. Soc. Lond. B.*, 218, 349, 1983.

308. Carrier, D. et al., Dynamics and orientation of glycolipid headgroups by 2H-NMR: gentiobiose, *Biochim. Biophys. Acta*, 983, 100, 1989.

309. Iwamoto, K. et al., Lipisomal membranes XV. Importance of surface structure in liposomal membranes of glyceroglycolipids, *Biochim. Biophys. Acta*, 691, 44, 1982.

310. Eibl, H. and Blume, A., The influence of charge on phosphatidic acid bilayer membranes, *Biochim. Biophys. Acta*, 553, 476, 1979.

311. Cullis, P.R. and de Kruijff, B., The polymorphic phase behavior of phosphatidylethanolamines of natural and synthetic origin, *Biochim. Biophys. Acta*, 513, 31, 1978.

312. Gagne, J. et al., Physical properties and surface interactions of N-methylated phosphatidylethanolamines, *Biochemistry*, 24, 4400, 1985.

313. Silvius, J.R., Brown, P.M., and O'Leary, T.J., Role of headgroup structure in the phase behavior of amino phospholipids. I. Hydrated and dehydrated lamellar phases of saturated phosphatidylethanolamine analogues, *Biochemistry*, 25, 4249, 1986.

314. Harlos, K. and Eibl, H., Influence of calcium ion on phosphatidyl glycerol. Two separate lamellar structures, *Biochemistry*, 19, 895, 1980.

315. Vaughan, D.J. and Keough, K.M., Changes in the phase transitions of phosphatidylethanolamine- and phosphatidylcholine-water dispersions induced by small modifications in the headgroup and backbone regions, *FEBS Lett.*, 47, 158, 1974.

316. Wang, Z-Q. et al., Phase transition behavior and molecular structures of monounsaturated phosphatidylcholines. Calorimetric studies and molecular mechanics simulations, *J. Biol. Chem.*, 270, 2014, 1995.

3 Lipid Bilayer Interdigitation

James L. Slater and Ching-hsien Huang

CONTENTS

ABBREVIATIONS

BP	myelin basic protein
C(X):C(Y)PC	phosphatidylcholine with X carbons in the *sn*-1 acyl chain and Y carbons in the *sn*-2 acyl chain
C(18)PAF	C(18)-platelet-activating factor
DHPC	dihexadecylphosphatidylcholine
DMPC	dimyristoylphosphatidylcholine
DPH	1,6-diphenyl-1,3,5-hexatriene
DPPC	dipalmitoylphosphatidylcholine
DPPE	dipalmitoylphosphatidylethanolamine
DPPG	dipalmitoylphosphatidylglycerol
DSPC	distearoylphosphatidylcholine
DTA	differential thermal analysis
FTIR	Fourier-transform infrared spectroscopy
G(I)	Ripple phase of DPPC
G(II)	noninterdigitated gel phase of DPPC
L	lamellar liquid crystalline phase
NMR	nuclear magnetic resonance
^{2}H NMR	deuterium NMR
^{13}C NMR	carbon-13 NMR
^{31}P NMR	phosphorus-31 NMR
PC	phosphatidylcholine
PE	phosphatidylethanolamine
PMB	polymyxin-B
PMBN	polymyxin-B nonapeptide
ΔC/CL	chain inequivalence parameter

3.1 INTRODUCTION

The lipid bilayer has long been recognized as a fundamental unit providing the permeability barrier in the fluid-mosaic model (Singer, 1974) of cell membranes. In this highly dynamic structure, many specific tasks of the cell membrane are accomplished by the variety of proteins, whose functioning, in principle, may depend on the physical state of this two-dimensional lipid bilayer fluid matrix, which solvates the transmembrane domains of the integral membrane proteins. The organization and physical characteristics of the lipid domains surrounding these transmembrane segments of the proteins may be able to influence the functioning of these proteins. It is natural to wonder whether, in the biological membrane, different proteins are surrounded by their own "local" lipid composition, providing optimal conditions for each protein. Such lipid domain formation is observed in a variety of model systems when the lipid headgroup composition is varied. Lipid domains also result as a consequence of miscibility properties attributable to the acyl chain composition and packing arrangements.

Although typical textbook drawings depict the arrangement of the lipid components as two opposing lipid monolayers with the acyl chain termini meeting in a well-defined bilayer midplane, other characteristic arrangements are possible. Lipids may also form interdigitated domains, whereupon one or both of the hydrocarbon chains extend beyond the region of the bilayer midplane, interpenetrating into the opposing monolayer, allowing portions of any given acyl chain to reside in different monolayers. Interdigitation, therefore, is yet another means of varying the structure and properties of the lipid bilayer component of cell membranes.

3.1.1 THREE MODES OF INTERDIGITATION

Bilayer interdigitation occurs when the terminal methyl groups of the acyl chains extend beyond the bilayer midplane, effectively interpenetrating into the opposing monolayer. Therefore, in addition to the classic, or noninterdigitated, arrangement of acyl chains, three classes of interdigitation are diagrammed in Figure 3.1.

Fully interdigitated bilayers occur in two distinct situations, each at the opposite end of the spectrum with regard to acyl chain-length asymmetry. Acyl chain asymmetry exists both in phospholipids containing *sn*-1 and *sn*-2 acyl chains that are dissimilar in length, and in sphingolipids bearing an acyl chain dissimilar in length to the sphingosine base. In bilayer systems composed of lipids exhibiting little or no chain-length asymmetry, a fully interdigitated system may be induced by a variety of perturbations to the bilayer. The resulting bilayer state exists with four acyl chain cross-sectional areas subtended by a single headgroup, as the acyl chains span the entire width of the bilayer upon interdigitation (Figure 3.1D).

In contrast, lipid systems exhibiting the highest degree of asymmetry possible, including the lysolipids and their analogs, may also form fully interdigitated bilayers characterized by the interpenetration of acyl chains across the entire width of the bilayer. However, because these systems predominantly consist of a single acyl chain, the resulting interdigitated bilayer state exists with two acyl chain cross-sectional areas subtended by the headgroup cross-sectional area.

Mixed interdigitation occurs when the effective length of the shorter acyl chain in the nearly extended conformation is half the length of the long acyl chain in its all *trans*, zigzag chain conformation. The longer acyl chain extends completely across the bilayer span; two of the shorter acyl chains, one from each lipid in opposing monolayers, meet end-to-end in the bilayer midplane. This packing arrangement results with three acyl chain cross-sectional areas per headgroup cross-section (Figure 3.1C).

Finally, when the acyl chain-length asymmetry cannot be accommodated by either noninterdigitation, full interdigitation, or mixed interdigitation, the partially interdigitated bilayer packing mode exists, with the short chain from one monolayer pairing with the longer acyl chain counterpart contributed by the opposing monolayer. This interdigitated phase is characterized by two acyl chain cross-sectional areas subtended by a single headgroup cross section (Figure 3.1B).

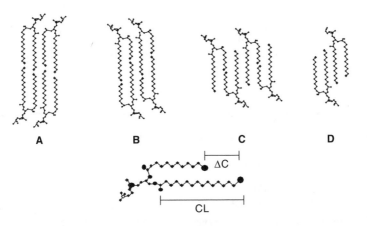

FIGURE 3.1 A schematic representation of the different acyl chain arrangements possible in bilayers: (A) C(16):C(16) PC noninterdigitated bilayer; (B) C(16):C(10) PC partially interdgitated bilayer; (C) C(16):C(10) PC mixed interdigitated bilayer; (D) C(16):C(16) PC fully interdigitated bilayer; (E) diagrammatic representation of the quantities used to calculate the chain inequivalence parameter shown for C(16):C(10) PC. (From Slater JL and Huang C: *Prog. Lipid Res.* 1988, 27:328. With permission from Pergamon Press.)

3.1.2 THE BALANCE OF FORCES THAT DETERMINE INTERDIGITATION

Self-assembly of lipids into bilayers is primarily driven by the requirement to minimize the ordering of water molecules at the hydrophobic interface between the acyl chain region of nonassembled monomeric lipids and the water; bilayer or micelle formation minimizes the hydrophobic surface, which is solvent-accessible, thus minimizing the amount of oriented water molecules. Van der Waals interaction between water and hydrocarbon chains are comparable in magnitude to van der Waals interaction between acyl chains packed within the bilayer interior (Cevc and Marsh, 1987). However, acyl chains within the self-assembled structure must pack without leaving voids. This excluded volume effect helps determine the arrangement of the acyl chains. Therefore, although van der Waals forces between the acyl chains is not the primary driving force for bilayer formation, the lowest-energy packing arrangement tends to maximize the van der Waals interactions in the bilayer interior. At the same time, orderly packing requires that the number of gauche rotamers per chain must be kept at a minimum. Furthermore, the lipid headgroup surface area at the interface region and the acyl chain cross-sectional area must match one another.

Because the transition from noninterdigitated to interdigitated bilayers results in altering the bilayer thickness, diffraction techniques are perhaps the most direct method for determining the structure of interdigitated bilayers. Yet just about every physical technique, in the proper hands, is capable of characterizing interdigitation, and each one has its own strengths and weaknesses. Slater and Huang (1988) reviewed the ways in which many of these techniques may characterize interdigitation, including X-ray and neutron diffraction, differential scanning calorimetry, Raman and Fourier-transform infrared spectroscopy, nuclear magnetic resonance, electron spin resonance, fluorescence, and electron microscopy. The interested reader is encouraged to see this reference and the original literature cited within for more details.

The ability of bilayer lipids to form interdigitated states has come into focus as a means to alter membrane bilayer properties significantly. Interdigitated bilayers, as a consequence of the unusual packing arrangement of the acyl chains, provide for the possible formation of discrete domains within the plane of the bilayer. Besides changing the character of the hydrophobic region of the bilayer, interdigitation alters the electrostatic properties of the bilayer. Furthermore, the bilayer thickness and corresponding elastic properties of the membrane may be altered by interdigitation. Although the primary emphasis of the present chapter is to outline both the forces governing the self-assembly of lipid molecules into interdigitated bilayers and the physical properties distinctive of this interdigitated bilayer state, we might speculate for a moment on the ways in which interdigitation may influence the behavior of biological membranes.

3.2 SPECULATION CONCERNING THE POSSIBLE BIOLOGICAL CONSEQUENCES OF THE INTERDIGITATED STATE

3.2.1 COUPLING OF MONOLAYERS

For both partially and mixed interdigitated systems, as well as for fully interdigitated systems of lysolipids and their analogs, the individual monolayers become progressively more tightly coupled with each other. This has been directly observed for the mixed interdigitated system as a broadening of the ^{31}P-NMR lineshape, indicating a transition to a unit of rotation consisting of two coupled lipid molecules, one from each monolayer (Xu et al., 1987). This coupling influences both the ability of a perturbation on one monolayer to be transmitted across the bilayer and the ability of a particular lipid to form laterally segregated domains within the bilayer plane.

3.2.2 MODULATION OF THE SURFACE CHARGE DENSITY

The transition to the interdigitated phase often results in a change in the surface area subtended by the lipid headgroup. The resulting increase in this surface area necessarily reduces the charge density per unit area (Simon et al., 1988).

The hydration repulsive pressure (McIntosh and Simon, 1986; LeNeveu et al., 1976; Parsegian et al., 1979; Marra and Israelachvilli, 1985; for a review, see Rand and Parsegian, 1989) is a dominant interaction preventing bilayer–bilayer contact in fully hydrated phosphatidylcholines. This hydration pressure is related to the work required to remove solvent molecules from between the two bilayer surfaces that are being brought into close proximity with each other, and is the predominant interaction preventing bilayer–bilayer contact and subsequent fusion.

If this hydration repulsive pressure is measured as a function of the interfacial zwitterionic density by making measurements on the gel, fluid, and interdigitated gel phases (Simon et al., 1988), the results indicate that this hydration pressure decreases with increasing headgroup surface area. Furthermore, the addition of cholesterol to DPPC bilayer dispersions separates the headgroups and reorganizes the interfacial water, thereby reducing short-range repulsive (steric) interactions acting between bilayers and allowing phosphatidylcholine headgroups of opposing bilayer surfaces to interpenetrate at high applied osmotic pressures (McIntosh et al., 1989).

3.2.3 ALTERATION OF BILAYER THICKNESS

Several recent studies of lipid–protein interactions have observed that the optimal functioning of certain integral membrane proteins, including rhodopsin, gramicidin, and bacterial photosynthetic reaction centers, may require the phospholipid bilayer to be a certain minimum thickness. Interdigitation may influence the activity of integral membrane proteins by modulating the bilayer thickness. In particular, lipids that may exist either in mixed interdigitated or in partially interdigitated bilayer states are likely candidates for this role, especially because interactions with inducer molecules may bring about this transition isothermally.

A study of the photocycle following the absorption of light in reconstituted preparations of rhodopsin indicates that, if the acyl chain length of the reconstituting lipid is C(14) or shorter, the normal sequence of conformational transitions no longer occurs. The spectroscopic evidence indicates that the production of metarhodopsin II intermediate from metarhodopsin I in these short-chain lipid bilayers is interrupted as a direct consequence of the lipid environment, suggesting an optimal bilayer thickness is required for this integral membrane protein (Baldwin and Hubbell, 1985).

The measured lifetime of a conducting channel of dimeric gramicidin appears to depend on the bilayer thickness within which it is incorporated (Elliott et al., 1983). The channel is destabilized by the restoring force arising from the deformation of the bilayer as it attempts to accommodate locally the mismatch between the hydrophobic region of the peptide and the hydrophobic region contributed by the acyl chains. These elastic interactions, which influence the lifetime of the conducting state of the gramicidin channel, have been successfully modeled in terms of a deformation free energy (Huang, 1986).

Furthermore, an elastic contribution to lipid–protein interactions may arise because the bilayer becomes distorted by the presence of protein. When a mismatch exists between the membrane spanning hydrophobic alpha helical segments of incorporated integral membrane proteins and the optimal bilayer thickness of the lipid matrix, a departure from this optimal bilayer thickness results (Mouritsen and Bloom, 1984). The experiments of Peschke et al. (1987) and Riegler and Mohwald (1986) represent the first experimental verification of the theoretical predictions. Depending on the length of the acyl chains relative to the protein hydrophobic region, either a stretching or a compression of the acyl chains results. If this mismatch is great enough, the elastic interactions may lead to the clustering of the protein into domains (Riegler and Mohwald, 1986).

The lipid phase transition in these reconstituted systems becomes broadened by the presence of the integral membrane proteins. However, the phase transition temperature may either decrease, remain unchanged, or increase, relative to the value for the pure lipid, depending on the chain length of the lipid used in the reconstitution (Riegler and Mohwald, 1986). This transition temperature shift increases with increasing protein concentration. However, even small amounts of incorporated protein shift the entire transition. The apparent absence of a two-component phase transition suggests that these elastic interactions mediating the change in lipid phase transition properties exert long-range influences (Peschke et al., 1987).

3.2.4 Loss of the Bilayer Midplane

The well-defined bilayer midplane, originally consisting of the terminal methyl groups of the acyl chains, no longer exists in interdigitated bilayers. This manifests itself in the absence of a preferred fracture plane in freeze fracture studies. The bilayer midplane is a region of lower density than the rest of the hydrocarbon interior of the bilayer. Fluorescence anisotropy measurements using DPH indicate that this midplane provides a favorable environment for the bulky fluorophore. This concept of internal partitioning has been advanced to draw a distinction between those classes of molecules that prefer intermonolayer locations and those that prefer intramonolayer locations (MacDonald and MacDonald, 1988). If this low-density region in the bilayer midplane were to provide a niche for the bulkier side groups of membrane-spanning peptides in biological membranes, the loss of this bilayer midplane, which results in the transition from a noninterdigitated to interdigitated bilayer state, could certainly trigger a conformational transition in the protein, as displaced peptide side chains seek a new environment.

3.3 INTERDIGITATION RESULTING FROM ACYL CHAIN-LENGTH ASYMMETRY

3.3.1 Acyl Chain Inequivalence

A tremendous amount of research has been devoted to determining the effect of acyl chain-length asymmetry and its role in influencing the formation of interdigitated bilayer states. This topic has been reviewed (Huang and Mason, 1986, and references contained therein). Because much of the wealth of information and coherence of thought existing in the original article would be lost by attempting to condense this review, only a summary of pertinent results will be presented here. Readers are referred to the original work for greater detail.

A systematic evaluation of the influence of acyl chain inequivalence began by examining the phase behavior of a series of synthetic phospholipids ranging from C(18):C(18)-PC to C(18):C(0)-PC, where the *sn*-2 acyl chain was successively varied by two methylene units (Huang and Mason, 1986). As the variable-length acyl chain is made shorter, the acyl chain packing arrangements progress from noninterdigitated packing through the full spectrum of intermediate possibilities, ending with full interdigitation for lysoPC, a lipid exhibiting the maximum chain asymmetry. These conclusions have been successfully reproduced by calorimetric and X-ray diffraction investigation (Mattai et al., 1987).

X-ray diffraction data on single crystals for dilauroyl phosphatidylethanolamine and dimyristoyl phosphatidylcholine have determined that the *sn*-1 and *sn*-2 acyl chains in these identical-chain phospholipids are conformationally inequivalent (Elden et al., 1977; Pearson and Pascher, 1979). In particular, the fatty acyl chain linked to the *sn*-1 position of the glycerol backbone continues from the *trans* configuration of the glycerol moiety, with the primary ester group being planar. However, the initial segment of the *sn*-2 chain projects out normal to the direction of the *sn*-1 acyl chain, with the plane of the secondary ester approximately perpendicular to that of the primary ester. Furthermore, the *sn*-2 acyl chain bends over at the bond between carbon-2 and carbon-3;

consequently, the rest of the chain runs parallel to the *sn*-1 acyl chain. As a result of this sharp bend on the *sn*-2 acyl chain, the terminal methyl groups of the *sn*-1 and *sn*-2 chains are out of register, being separated along the long molecular axis by approximately 3.1 Å. In addition to the single-crystal studies, this conformational inequivalency has also been observed for the identical chain phospholipids in the gel-state bilayer, based on neutron diffraction data (Zaccai et al., 1979). However, a separation of 1.8 Å or 1.5 carbon–carbon bond lengths is observed for the two terminal methyl groups in the gel-state bilayer. Therefore, the acyl chain-length asymmetry may be quantitated by the following expression (Figure 3.1E):

$$\text{chain inequivalence} = \Delta C/CL$$

In this expression, CL is the length of the longer of the two acyl chains. The parameter ΔC represents the absolute difference in the chain lengths of the two hydrocarbon chains and is given by

$$\Delta C = n_1 - n_2 + 1.5$$

where n_1 and n_2 are the number of carbons in the *sn*-1 and *sn*-2 acyl chains. In this expression for ΔC, a factor of 1.5 is included to account for the 1.5 carbon–carbon bond lengths, representing the effective chain-length difference present in identical chain phospholipids in the gel state. The chain inequivalence parameter therefore represents the magnitude of the chain-length inequivalence normalized by the length of the hydrocarbon region of the bilayer (Mason et al., 1981; Mason and Huang, 1981). This chain asymmetry exerts its greatest influence on the chain packing interactions in the gel phase.

3.3.2 MIXED INTERDIGITATED SYSTEMS

When the chain inequivalence parameter is approximately 0.5, conditions are optimal for mixed interdigitation, where the long *sn*-1 acyl chain spans the entire bilayer width, while the shorter *sn*-2 chains meet end-to-end at the bilayer center. This results in the formation of a gel-phase bilayer consisting of three acyl chain cross-sectional areas per unit headgroup cross-sectional area (Hui et al., 1984; McIntosh et al., 1984).

3.3.2.1 Mixed-Chain Phosphatidylcholines

3.3.2.1.1 Single-Component Phospholipid Systems

Raman investigation of the two asymmetric lipids C(18):C(10)PC and C(18):C(12)PC (Huang et al., 1983), both of which have a value of the chain inequivalence parameter close to 0.5, reveals that these dispersions undergo a sharp cooperative phase transition near 17°C. The intensity ratios from the C-H stretch region indicate that, at temperatures below the phase transition, these two lipids have interchain interactions and intrachain order similar to that of di-C(18)PC. This ordered packing must result from the formation of interdigitated bilayers for these asymmetric lipids.

X-ray diffraction measurements of these fully hydrated dispersions reveal that, in the gel phase, both lipids possess a reduced bilayer thickness in comparison to the di-C(14)PC. Furthermore, the headgroup area subtends three acyl chain cross sections, indicating the formation of mixed interdigitated bilayers in which the long *sn*-1 acyl chain spans the entire hydrocarbon width, while the shorter *sn*-2 acyl chains pack end-to-end, meeting in the bilayer center.

Freeze-fracture studies of these mixed interdigitated gel phase bilayers indicate a discontinuous fracture pattern, where the fracture planes are interrupted by up and down steps. This behavior is predicted for a system such as this mixed interdigitated bilayer, for which there is no preferred weak bonding plane in the bilayer center.

Upon undergoing a thermotropic phase transition, these mixed interdigitated gel phases transform to partially interdigitated bilayers with two acyl chain cross-sectional areas subtended by the average headgroup cross section. This liquid–crystalline phase, at least for C(18):C(10)PC, appears to retain some degree of partial interdigitation. It should be emphasized that the C(18):C(10)PC bilayer in the presence of excess water is the first known example in which the acyl chain interdigitation is implied for bilayers in both the gel and liquid–crystalline states.

Further experimental work has been carried out with respect to these mixed-chain lipids whose chain inequivalence parameter lies close to 0.5. First, the data on this series of asymmetric lipids with chain inequivalence close to 0.5 have been extended for various chain lengths, and high-resolution differential scanning calorimetry indicates that the formation of this mixed interdigitated phase is not unique to C(18):C(10)PC, but is a rather general phenomenon that depends on the chain-length inequivalence between the *sn*-1 and *sn*-2 acyl chains (Xu and Huang, 1987). Particularly noteworthy is the observation that these mixed interdigitated bilayers form regardless of whether the long chain resides in the *sn*-1 position or the *sn*-2 position, provided that the chain inequivalence parameter has a value close to 0.5.

The barotropic behavior of these bilayers, determined by using high-pressure Fourier-transform infrared spectroscopy (FTIR), has revealed that these highly ordered, mixed interdigitated gel phase bilayers of C(18):C(10)PC exist with the zigzag chain planes of neighboring acyl chains intramolecularly and intermolecularly nearly perpendicular to each other in the two-dimensional packing lattice (Wong and Huang, 1989).

3.3.2.1.2 Binary Phospholipid Systems

The first example of studies involving a binary mixture of two highly asymmetric phospholipids, which form mixed interdigitated lamellae, involves calorimetry measurements of the C(10):C(22)PC/C(22):C(12)PC binary system (Xu et al., 1987). Each component lipid in this mixture has a value of the chain inequivalence parameter close to 0.5, except that the positions of the long and short chains are reversed with respect to the glycerol backbone. The temperature–composition phase diagram for C(10):C(22)PC/C(22):C(12)PC has been constructed on the basis of the onset and completion temperatures of a series of phase transition curves. The phase boundaries (the solidus and liquidus lines) do not display either a flat region or a point of sharp reflection, but they do show smooth and continuous changes over the entire composition range, resulting in a binary phase diagram of cigar shape. This indicates that, in the same plane of the bilayer, C(10):C(22)PC and C(22):C(12)PC are mixed nearly ideally in both the gel and liquid–crystalline states over the entire composition range. It should be noted that this particular system studied involved a mixture of two lipids: one with the longer acyl chain in the *sn*-1 position and the other with the longer acyl chain in the *sn*-2 position. This provides a system composed of a single headgroup, with which the phase transition temperature may be continuously varied between two limits while maintaining a relatively constant bilayer thickness.

Another study of a binary system composed of lipids forming mixed interdigitated bilayers involves the determination of the packing behavior in C(18):C(11:1)PC, a synthetic lipid whose *sn*-2 chain is similar in length to the two species known to form mixed interdigitated bilayers [C(18):C(10)- and C(18):C(12)-PC], but which contains a double bond at the *sn*-2 chain terminus (Ali et al., 1989).

If the phase behavior for binary mixtures of this lipid with C(18):C(10)-PC and C(18):C(11)-PC is determined by calorimetry, each of these binary mixtures exhibits a cigar-shaped phase diagram, indicating complete miscibility in all proportions, both in the gel phase and the liquid–crystalline phase. These results suggest that domain formation does not occur even when unsaturation is present at the chain terminus. This near-ideal mixing behavior indicates a similar mode of chain packing for these three species, whose chain inequivalence parameter is approximately 0.5.

However, in binary systems composed of a symmetric chain species di-C(14)PC mixed with each of these three chain asymmetric species, the calorimetric behavior indicates that lateral gel

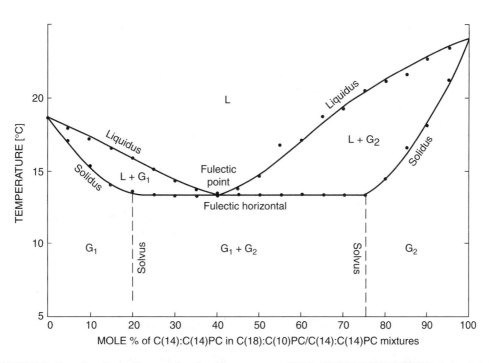

FIGURE 3.2 Eutectic phase diagram for the binary system C(14):c(14)PC/C(18):C(20)PC derived from differential scanning calorimetry. (From Lin H and Huang C: *Biochem. Biophys. Acta,* 946, 182, 1988. With permission from Elsevier Science Publishers.)

phase separation into two-dimensional spatial domains occurs. Despite the small difference in phase transition temperatures between the components of these mixtures, the packing differences are sufficient to prevent the constituent species from mixing appreciably in the gel state. Therefore, the mode of acyl chain packing appears to be a more important determinant of miscibility than the mismatch in acyl chain lengths.

When the two component lipids of a binary system are only partially miscible in the gel phase, but completely miscible in the liquid crystalline phase, the temperature–composition phase diagrams derived from calorimetry indicate eutectic behavior (Figure 3.2). In this system, there are three one-phase regions (G_1, G_2, L), three two-phase regions (G_1 + L, G_2 +L, and G_1 + G_2), and one degenerate three-phase region, the eutectic point at fixed temperature and composition (Lin and Huang, 1985).

At $T < T_m$, C(18):C(10)PC molecules are known to pack into mixed interdigitated bilayers, and consequently they are expected to be miscible only partially with the noninterdigitated C(14):C(14)PC molecules in the bilayer (Lin and Huang, 1988). Because the packing properties of the two components are quite similar in the liquid–crystalline phase, they are completely miscible at temperatures above the phase transition.

An interesting consequence of the eutectic phase diagram is that near the eutectic point, the thermodynamic state of the lipid components in the bilayer can be converted reversibly between a one-phase region (the liquid–crystalline state) and a two-phase region in which the two phases are laterally phase segregated. This reversible interconversion can, in principle, be readily achieved by only a slight modification in either the local temperature or the lipid composition. The dynamics and functions of bilayer-spanning proteins can thus be potentially modulated by such reversible changes in the local lipid-bilayer composition generated from these multiple lipid-phase regions in the membrane (Huang, 1990).

When the acyl chain inequivalence parameter is close to one half, the gel phase exists as a mixed interdigitated system, whereas the fluid phase exists with partial interdigitation. Therefore, the phase transition results in a substantial change in bilayer thickness. When the chain symmetric species in the mixture is chosen with its chain length equal to the longer acyl chain of the chain asymmetric lipid in the mixture, another interesting phenomenon arises.

For example, the phase diagram for the mixture of di-C(18)PC with C(18):C(10)PC has been constructed on the basis of differential scanning calorimetry (Mason, 1989). The proposed model suggests that these two components are partially immiscible in both the gel and the liquid–crystalline bilayer phases. In addition to exhibiting lateral phase segregation of interdigitated and noninter-digitated gel phases, of particular interest is a region on this phase diagram representing laterally segregated, immiscible fluid phases. This is the first demonstration of liquid–liquid phase immis-cibility in a two-component system of lipids bearing the same headgroup.

Differential scanning calorimetry studies of the C(18):C(10)PC/cholesterol binary system indi-cates that cholesterol decreases both the temperature and the enthalpy of the mixed interdigitated gel to partially interdigitated fluid phase transition (Chong and Choate, 1989). The phase transition is completely absent above 25 mol% cholesterol. The disordering effect of cholesterol on this asymmetric lipid may prevent the formation of these interdigitated phases.

3.3.2.2 Sphingomyelin

Raman spectroscopic investigation (Levin et al., 1985) of synthetic D-erythro-N-lignoceroylsphin-gophosphocholine (C(24):SPM) indicates that this fully hydrated lipid undergoes two thermotropic phase transitions at 48.5°C and 54.5°C.

At temperatures below the lowest transition, the Raman peak height intensity ratios derived from the C-H stretch region indicate that this sphingomyelin has strong lateral chain–chain inter-actions, with a high degree of intrachain conformational order, suggesting the likelihood of a mixed interdigitated phase. At temperatures between the two phase transitions, both intra- and interchain disorder increases, but the order parameters are consistent with partial interdigitation of the acyl chains, where the short chain from one monolayer is packed end-to-end with a longer chain from the opposing monolayer, resulting in two acyl chain cross sections subtended by a single headgroup area. The high-temperature transition involves a conversion from the partially interdigitated state to the liquid–crystalline phase.

Therefore, this particular sphingomyelin is proposed to undergo the transition sequence of mixed interdigitated gel to partially interdigitated gel to liquid–crystalline phase. The observation of a mixed interdigitated gel state at temperatures below 48.5°C implies that such interdigitation may also occur between sphingomyelin molecules packed in the opposite faces of a biological membrane at physiological temperatures.

3.3.2.3 Glycosphingolipids

Spin-labeled analogs of glycosphingolipids have been examined to determine the relative order parameter of a chain methylene segment containing a covalently bound spin label (Grant et al., 1987; Mehlhorn et al., 1988). When these glycosphingolipid analogs are immersed in a host bilayer matrix consisting of noninterdigitating bilayer lipids, they provide a measure of the order parameter existing near the bilayer midplane of the host. Two different analogs were synthesized; both contained the spin label covalently attached to the 16 position of the acyl chain. These two analogs had different chain lengths: one had an 18-carbon acyl chain, whereas the other had a much longer 24-carbon acyl chain. Although in both cases the spin label resides at the same depth within the host bilayer, the 24-carbon acyl chain species exhibits considerable chain-length inequivalence.

When these spin-labeled glycosphingolipids are incorporated at low molar ratios into host matrices composed of either symmetric chain synthetic lipids or egg PC, the order exhibited by the segment at the C16 position is always greater for the 24-carbon acyl chain than for the 18-carbon acyl chain. The spin-labeled segment resides near the bilayer midplane; because the longer chain exhibits greater order, this observation is consistent with the interpenetration of the long acyl chain past the depth of the bilayer midplane of the host lipid matrix. Yet it appears that, at this low concentration of glycosphingolipid, these labels are homogeneously dispersed, because they show no sign of spin-exchange broadening, indicating a high local concentration of probe resulting when lateral phase separation of the glycosphingolipid analogs occurs. Therefore, the long interpenetrated chain may not necessarily have a pairing partner in the opposite monolayer. Although the original result was obtained on lactosylceramide (Grant et al., 1989), these observations have been extended to include galactosyl ceramide, globoside, and GM1 (Mehlhorn et al., 1988), indicating that *trans*-bilayer interdigitation may be a general property of these glycosphingolipids containing a long acyl chain.

A slightly different approach to using spin labels to study interdigitation involves observing the behavior of a spin-labeled fatty acid incorporated into the bilayer of interest. The polar carboxylic acid residue ensures that this end of the probe is anchored at the polar interface region of the bilayer; the order parameter at different depths in the bilayer may be examined by placing the spin label at different positions along the fatty acyl chain.

This approach has yielded the following information on systems whose packing behavior was already suggested by other methods:

For asymmetric PC systems exhibiting mixed interdigitation [C(18):C(10)- and C(18):C(12)-PC], the flexibility gradient normally existing in noninterdigitated systems is altered to the extent that these 16-DS-SL and 5-DS-SL probe molecules exhibit comparable values of the order parameter. This result implies that the chain terminus of the fatty acid spin-label probe exists in an environment similar to that found at the interface region of bilayers formed from symmetric chain noninterdigitating lipids (Boggs and Mason, 1986). These asymmetric lipids are also capable of forming a highly ordered phase below their main phase transition temperature: the spin-label probe becomes immobilized in the low-temperature phase, suggesting that these phases are similar to crystalline phases observed in symmetric chain phosphatidylcholines. This identity of the low-temperature phase has been confirmed by X-ray diffraction and differential scanning calorimetry (Mattai et al., 1987).

The phase behavior of cerebroside sulfates formed semisynthetically by substituting the naturally derived heterogeneous acyl chains with a long acyl chain of known composition has been investigated by differential scanning calorimetry (Boggs et al., 1988). These lipids exhibit complex phase behavior composed of several stable and metastable forms, whose appearance and interconversion is documented by scanning calorimetry. Investigation of these cerebrosides using the spin-labeled fatty acid probes (Boggs et al., 1988) reveals that motional restriction of the 16-DS-SL probe is similar to the motional restriction observed previously in C(18):C(10)-PC, which is known to exist with mixed interdigitated packing. This motional restriction occurred only for the long-chain semisynthetic cerebroside sulfates, but not in cerebrosides containing shorter C(18) and C(16) acyl chains, indicating that the highly asymmetric cerebroside sulfates containing long acyl chains are arranged in mixed interdigitated bilayers. A model is proposed that this mixed interdigitated gel phase may arise from a partially interdigitated metastable gel phase, which is formed initially upon cooling the liquid–crystalline bilayers. Hydroxylation of the fatty acid appears to increase the kinetic barrier for the conversion from the metastable (partially interdigitated) phase to the stable (mixed interdigitated) phase.

The mixing behavior of these highly asymmetric sphingolipids with C(16):C(16)PC has been investigated utilizing differential scanning calorimetry (Gardam and Silvius, 1989). Although these asymmetric lipids should interdigitate in single-component systems, these calorimetry results are interpreted to suggest that acyl chain length asymmetry in these sphingolipids does not automatically increase their propensity to phase segregate laterally in these binary systems. Therefore, in addition to exhibiting their interdigitated lifestyle, these asymmetric sphingolipids may successfully integrate into a noninterdigitated neighborhood.

3.3.3 PARTIALLY INTERDIGITATED SYSTEMS

3.3.3.1 The Intermediate Situations between Zero Chain Inequivalence and Mixed Interdigitation

When the transition entropy derived from differential scanning calorimetry is normalized to the number of carbon units undergoing the transition, the trend indicates that this normalized transition entropy decreases in this order: C(18):C(18) > C(18):C(16) > C(18):C(14) (Mason et al., 1981). In order to maintain optimal van der Waals contacts with the ensuing chain length asymmetry, portions of the acyl chains must remain highly disordered in the gel phase. This is required, because the region of the longer chain near its methyl terminus must compensate to fill the void in the bilayer resulting from the shorter sn-2 acyl chain (Mason et al., 1981).

Raman spectroscopy indicates that C(18):C(14)PC bilayers exhibit the least amount of gel-state order measured by the C-H stretching mode peak height intensity ratio parameters (Huang et al., 1983). This results from a significant number of residual gauche conformers present in the gel phase. If this intramolecular disorder is compared for the series C(18):C(18)PC to C(18):C(10)PC by observing spectral band intensities in the C-C stretching region, it is clear that C(18):C(14)PC exhibits the greatest amount of intramolecular disorder in the gel state; C(18):C(14)-PC has a structure intermediate to those lipids that form partially interdigitated gel states and those that form mixed interdigitated gel states.

X-ray diffraction results indicate that the C(18):C(14)PC does not pack into mixed interdigitated bilayers. Instead, the portion of the longer sn-1 acyl chain extending beyond the bilayer depth corresponding to the terminal methyl group of the shorter sn-2 acyl chain remains highly disordered in the gel phase (Hui et al., 1984). It appears that a pretransition to a ripple phase ($T_m = 19°C$) precedes the main phase transition ($T_m = 30°C$) for the C(18):C(14)PC. Furthermore, as a result of the chain mismatch, the transition to this ripple phase is accompanied by a fragmentation of the multilamellar liposomes as they form smaller vesicles with a higher radius of curvature (Hui et al., 1984).

The phase behavior of C(18):C(16)PC, C(18):C(14)PC, C(18):C(12)PC has also been examined in binary mixtures with C(18):C(18)PC. Like the C(18):C(10)PC/C(18):C(18)PC binary system described earlier, C(18):C(12)PC is only partially miscible with C(18):C(18)PC. On the other hand, both C(18):C(16)PC and C(18):C(14)PC appear to be completely miscible with C(18):C(18)PC over the entire composition range. However, the mixing behavior exhibited by this binary system is highly nonideal; this asymmetric lipid species may remain uninterdigitated throughout the entire composition range (Mason, 1988).

3.3.4 FULLY INTERDIGITATED SYSTEMS

The maximum possible chain asymmetry exists for the lysolipids and other analogs in which the sn-2 acyl chain consists of either a proton or an acetyl ester. These species form a fully interdigitated phase in which the sn-1 acyl chain spans the entire width of the bilayer, resulting in the equivalent of two acyl chain cross-sectional areas per unit headgroup cross-sectional area.

3.3.4.1 Lysolipids and Analogs Exhibiting Chain Interdigitation

An X-ray crystallography study of 1-lauroylpropanediol-3-phosphocholine, a lysophosphatidylcholine analog, reveals a structure with the single acyl chain spanning the entire hydrocarbon width of the bilayer, forming a fully interdigitated bilayer in the crystalline nonhydrated state (Hauser et al.,

1980). This fully interdigitated state may also form in fully hydrated dispersions of lysolipids and their analogs. For example, fully hydrated dispersions of C(16):C(0) and C(18):C(0)-PC are capable of self-assembling into a highly ordered two-dimensional organization with fully interdigitated acyl chains. This structural organization was deduced initially using Raman spectroscopy to observe that the interchain conformational order and intrachain order of the C(18)-lysoPC was greater than that of the di-C(18)-PC below the phase transition temperature; however, the C(18)lysoPC was more disordered than the C(18):C(18)PC above their respective phase transitions (Wu et al., 1982). Since the headgroup area for the lysoPC is greater than twice the acyl chain cross-sectional area, these greater lateral chain–chain interactions in the lysoPC gel phase are possible only with fully inter-digitated acyl chains. Later confirmation of this result was obtained by X-ray diffraction measurements on these fully hydrated lamellar dispersions (Hui and Huang, 1986).

This fully interdigitated gel phase undergoes a thermotropic phase transition to become a micellar phase. Within these micelles, the lyso-PC molecules are characterized by an increase in chain disorder and the loss of motional anisotropy (Wu et al., 1982). Supercooling of the micelles is required to form this interdigitated lamellar gel state; the kinetic mechanism has been analyzed, and the formation of the interdigitated gel-phase bilayer requires nucleation followed by two-dimensional growth (Wu and Huang, 1983).

A lysolipid analog missing the glycerol backbone (N-octadecylphosphocholine) behaves similarly with regard to the kinetics of formation of the lamellar phase from the micellar phase (Jain et al., 1985). Although the transformation from the interdigitated lamellar phase to micelles occurs instantaneously, the reverse process requires considerable time. Additional experiments using high-pressure FTIR techniques reveal that increasing hydrostatic pressure facilitates the transformation of micelles to this highly organized interdigitated gel phase (Wong et al., 1988).

Lysophosphatidylethanolamine (lyso-PE) containing an unsaturated acyl chain in the sn-1 position may also form a lamellar structure, presumably with fully interdigitated chains (Tilcock et al., 1986). These lamellar structures are relatively impermeable to ion gradients. ^{31}P NMR measurements suggest that the interdigitated bilayers are transformed to micelles as the monounsaturated lyso-PE undergoes the thermotropic phase transition. If the acyl chain unsaturation is increased, inverted micelles may form instead of micelles, as these interdigitated structures undergo the phase transition.

When the acyl chain in lyso-PE is saturated, entirely different phase behavior results. The micellar phase, upon cooling, transforms immediately to an interdigitated lamellar gel phase. However, in contrast to the interdigitated lamellar gel phase of lysoPC and lysoPC analogs, this lyso-PE interdigitated lamellar gel state is metastable and ultimately undergoes slow conversion to a crystalline phase. Upon heating, however, this micellar phase is approached by a different pathway, whereby the crystalline phase entirely bypasses the metastable interdigitated lamellar gel phase from which it was originally derived and transforms directly into micelles (Slater et al., 1989).

3.3.4.2 Simultaneous Headgroup and Acyl Chain Interdigitation

An X-ray crystal structure of octadecyl-2-methyl-glycerophosphocholine has recently become available (Pascher et al., 1986). This molecule is a lysolipid analog with the glycerol backbone ether-linked for both the long sn-1 acyl chain and the short sn-2 methyl group. This crystal structure reveals some distinctly unusual features, and the bilayer arrangement indicates both the acyl chains and the headgroups are interdigitated in the stacked lamellar organization.

3.3.4.3 Platelet-Activating Factor and C(18):C(2)PC

Although the lysolipids exhibit the ultimate degree of chain asymmetry, two related compounds with acetoyl esters at the sn-2 position have been characterized. C(18):platelet-activating factor [C(18)PAF] contains an 18-membered saturated acyl chain ether-linked to the sn-1 position. In contrast, C(18):C(2)PC has the corresponding long chain attached by an ester link to the *sn*-1 position. Both of these lipids exhibit two thermotropic phase transitions; as expected, a great degree of similarity exists between the phase behavior of these lipids (Huang et al., 1984; Huang et al., 1986).

At temperatures below the lower-temperature phase transition, Raman spectroscopy indicates that these highly asymmetric lipid species have greater lateral chain–chain interactions than the symmetric chain species, suggesting that they are organized in an interdigitated gel phase. The short *sn*-2 acyl chain is arranged parallel to the bilayer surface; this segment retains the conformation normally present in the interface region, where the *sn*-2 chain initially extends parallel to the bilayer surface prior to extending into the bilayer interior. Raman measurements indicate two unique environments for the carbonyl group of the C(18):C(2)PC. In contrast, one of these environments corresponding to the *sn*-1 carbonyl group of the ester linkage is absent for the C(18)PAF.

The primary difference between the phase behavior of these two species results from loss of structural rigidity in the interface region when the ether linkage replaces the ester linkage to form C(18)PAF (Huang and Mason, 1986). A greater rotational disorder of the short *sn*-2 acetoyl segment in C(18)PAF is observed by Raman spectroscopy. In addition, enhanced rotational motion of the headgroup region for gel-phase C(18)PAF is evident by ^{31}P NMR measurements.

On the basis of Raman and ^{31}P NMR spectroscopic results, the low-temperature transition has characteristics of a gel-to-gel phase transition, whereas the higher-temperature transition involves a gel-to-micellar phase transition. In addition, differential scanning calorimetry indicates that a pretransition occurs prior to the gel-to-micellar phase transition. The pretransition is often not detected using Raman spectroscopy to evaluate the C-H stretching modes; the pretransition appears not to involve significant changes in the lateral chain–chain interactions and intrachain disorder.

3.4 INTERDIGITATION IN SYMMETRIC-CHAIN PHOSPHOLIPIDS

3.4.1 PRESSURE-INDUCED INTERDIGITATION

In comparison with the G(II) gel phase, the acyl chain cross-sectional area is slightly reduced for the pressure-induced fully interdigitated bilayer phase. An increase in hydrostatic pressure drives the equilibrium toward a decrease in total system volume. Under conditions of high hydrostatic pressure, DPPC and DSPC may form fully interdigitated phases (Braganza and Worcester, 1986).

High-pressure neutron diffraction of DMPC, DPPC, and DSPC detected this pressure-induced fully interdigitated gel phase. Lipids were prepared with perdeuterated acyl chains and nondeuterated headgroups, allowing these two regions of the bilayer to be distinguished in the Fourier profiles obtained from the diffraction patterns.

The temperature–pressure phase diagrams constructed for these identical chain diacylphosphatidylcholines indicate a region of existence of a fully interdigitated phase with untilted chains. This fully interdigitated gel phase is characterized by an abrupt decrease in the lamellar repeat spacing relative to the noninterdigitated gel phase. The phase boundary between the noninterdigitated and interdigitated gel phases exhibits unusual curvature, making it possible for a thermotropic transition to occur from the noninterdigitated gel phase to the interdigitated gel phase, back into the noninterdigitated gel phase, as the temperature is raised. Furthermore, the pressure required for interdigitation is chain length-dependent, with lower pressure inducing interdigitation in the longer-chain lipids.

Independent experimental evidence supports the possibility of this pressure-induced interdigitation. A study of the hydrostatic pressure-induced phase behavior in DMPC and DPPC, monitored by optical transmission, determined the temperature and pressure dependence of phase changes detected as changes in the intensity of transmitted light. With this technique, a pressure-induced phase, called the *X phase,* is observed at pressures above 0.93 kbar (Prasad et al., 1987). This X phase appears on the phase diagram between the G(II) phase and the G(I) phase, resulting in an additional phase transition between the subtransition and the pretransition temperatures. At pressures sufficient to observe this X phase, the transition sequence becomes S to G(II) to X to G(I) to L.

Increasing the hydrostatic pressure increases the temperature range of existence for this pressure-induced X phase, at the expense of decreasing the temperature range of existence for the G(I) ripple phase. At a pressure of 2.87 kbar and above, the G(I) to L phase transition is replaced by the X to L phase transition; this X phase may coincide with the pressure-induced phase observed by the neutron diffraction studies outlined previously. The existence of this second triple point in the temperature–pressure phase diagram for phosphatidylcholine has been determined independently utilizing high-pressure DTA (Utoh and Takemura, 1985), high-pressure dilatometry (Utoh and Takemura, 1985), and high-pressure X-ray diffraction (Utoh and Takemura, 1985) techniques.

3.4.2 Chemically Induced Interdigitation Model Systems

A variety of molecules are capable of inducing the formation of a fully interdigitated gel phase in symmetric chain phospholipids. Despite the apparently large diversity among the various inducer molecules, the properties required to induce interdigitation have been reviewed by Simon and coworkers (1986).

A formal charge is not required, but the inducer molecule must be amphipathic, not hydrophobic. Furthermore, the inducer must localize at the interface region of the bilayer, because alkanes and benzene, both of which localize in the bilayer interior, cannot induce the interdigitated phase. There is a size limit to the amphiphile; long-chain fatty acids and cholesterol, both of which reside at or near the interface, do not cause interdigitation, suggesting that the inducer must displace some headgroup-associated water molecules and increase the headgroup surface area.

3.4.2.1 Induction by Polyols, Short-Chain Alcohols, and Anesthetics

3.4.2.1.1 Single-Lipid Component Systems

Perhaps the first observation of induced interdigitation in DPPC bilayers evolved from a study of the effect of concentrated glycerol, ethylene glycol, and methanol solutions on the phase behavior of DPPC, determined both by differential scanning calorimetry and by X-ray diffraction (McDaniel et al., 1987).

These results indicate that glycerol may substitute for water in the interlamellar fluid space. X-ray diffraction of these systems reveals that the bilayer thickness abruptly decreases at about mole fraction 0.5 glycerol. Accompanying this change in bilayer thickness is a transformation in acyl chain packing, from the tilted quasihexagonal arrangement characteristic of the noninterdigitated gel phase to an untilted arrangement. Electron-density profiles also indicate that the bilayer width has decreased, and the deep trough in electron density associated with the G(II) gel phase bilayer midplane is replaced with two shallower troughs, separated by approximately 12 Å, perhaps corresponding to the methyl groups for interpenetrated acyl chains. In this interdigitated phase, the terminal methyl groups interpenetrate to the level corresponding with the first or second methylene group of the lipid in the opposing monolayer, indicating a fully interdigitated arrangement. Raman spectroscopy of the DPPC/glycerol system also reveals greater acyl chain order in this fully interdigitated gel phase relative to the noninterdigitated gel phase (O'Leary and Levin, 1984).

Further confirmation of the fully interdigitated structure results from comparing the calculated ratio of surface area per lipid headgroup to the cross-sectional area per acyl chain for these various phases. For the interdigitated phase, this ratio is approximately 4, indicating that full interdigitation results when the mole fraction of glycerol present exceeds 0.5.

Two necessary criteria for induction of full interdigitation in these symmetric chain lipids are proposed:

- First, water must be removed from certain locations at the interface region.
- This displaced water must be replaced by a larger, surface-active amphiphilic molecule, creating an increase in the interfacial area. This surface-area increase allows the positioning of the acyl chain terminal methyl groups at the bilayer interface region.

Other surface-active molecules have been examined for their ability to induce bilayer interdigitation (McIntosh et al., 1984). Fully interdigitated bilayer phases result from the introduction of chlorpromazine, tetracaine, benzyl alcohol, phenylethanol, and phenylbutanol to DPPC bilayers. In contrast, it appears that phosphatidylethanolamine bilayer dispersions are not readily induced to interdigitate, at least in the presence of chlorpromazine (Suwalsky et al., 1988) or alcohol (Rowe, 1985).

Much of our knowledge concerning the effect of alcohols on model lipid bilayer systems and their ability to induce interdigitation is contributed by Rowe and coworkers. When these short-chain alcohols are introduced into lipid dispersions, the phase transition midpoint (measured by optical light scattering) initially decreases relative to its value in the absence of any alcohol. This transition midpoint decreases linearly with increasing concentration of alcohol until a particular alcohol concentration, beyond which the trend is reversed. Any further increase in the alcohol beyond this reversal concentration results in a subsequent rise in the transition temperature. Therefore, alcohols are said to exhibit a *biphasic effect* on the lipid main phase transition temperature (Rowe, 1983).

X-ray diffraction of DPPC multilayer dispersions in the presence of alcohol has determined that this biphasic behavior is accompanied by a change in lipid bilayer organization of the gel phase (Simon and McIntosh, 1984). Below the reversal concentration of ethanol, the usual noninterdigitated (G(II)) gel phase persists. At ethanol concentrations exceeding this reversal concentration, a fully interdigitated gel phase appears.

At ethanol concentrations below the reversal point, the main phase transition temperature decreases with increasing ethanol, because ethanol preferentially partitions into the fluid bilayer phase. At the reversal concentration, the fully interdigitated bilayer phase is induced. This sudden increase in the number of acyl chains subtended per headgroup at the bilayer interface yields additional binding sites for ethanol at the interfacial region. Consequently, ethanol partitions preferentially into this interdigitated gel phase, elevating the phase transition temperature as the ethanol concentration increases above the reversal concentration.

Because the van der Waals interaction is greater for these more closely packed, interdigitated acyl chains than in the G(II) gel phase, interdigitated bilayer formation becomes favored. This favorable interaction is counterbalanced by the unfavorable exposure of the terminal methyl groups to the polar environment of the aqueous phase at the bilayer interfacial region. Because the energy cost of exposing these terminal methyl groups is independent of chain length, longer-chain lipids are expected to interdigitate more readily (Simon and McIntosh, 1984). As predicted, progressively lower concentrations of ethanol are required to induce interdigitation as the acyl chain length of the lipid increases. The increased order in the hydrocarbon region of the interdigitated phase is reflected in a narrowing of the phase transition width.

Using previously established absorbance methods to monitor the main phase transition, the ethanol effect on the reversibility of the main phase transition was compared for symmetric chain PC and PE (Rowe, 1985). This reversibility is manifested in a different apparent phase transition temperature, depending on whether heating scans or cooling scans are employed. In contrast to phosphatidylcholine, phosphatidylethanolamine does not exhibit a biphasic dependence of the phase transition temperature on alcohol concentration. Instead, only a lowering of the phase transition temperature results with increasing alcohol concentration. Furthermore, in PE, the main transition appears to be completely reversible.

This reversibility of the main transition and a lack of biphasic dependence on alcohol suggests that PE does not readily interdigitate when treated with alcohol. This is not surprising, because the interdigitated phase may replace the ripple phase, which is not a characteristic phase displayed by phosphatidylethanolamines.

Because the concentration of membrane-associated alcohol required to induce the biphasic behavior is relatively independent of the size of the *n*-alkyl group of the alcohol or the acyl chain length of the PC, the important determinant governing the transition to the interdigitated phase must be the actual membrane concentration of the alcohol. For a given acyl chain length, the partition coefficients for alcohol in PC and PE dispersions are very similar, indicating that the ability to form the interdigitated bilayer depends on properties of the lipid class and not on different amounts of membrane-associated alcohol present as a result of different partitioning behaviors.

Alcohols also affect the pretransition behavior of DPPC bilayers (Veiro et al., 1987). As the alcohol concentration is increased, the pretransition temperature begins to decrease linearly; above a threshold alcohol content, the pretransition becomes extinguished. At this threshold concentration, the absorbance change associated with the main phase transition suddenly increases significantly. The alcohol concentration at which the pretransition abruptly disappears correlates well with the concentration at which all of the other previously observed alcohol-induced phenomena take place.

Therefore, in summary, the induction of the fully interdigitated gel phase in the presence of short chain alcohols may be monitored by the following three parameters:

- The appearance of biphasic behavior in the main phase transition temperature
- The appearance of nonreversibility manifested in a temperature hysteresis associated with the main phase transition
- The shift to lower temperatures and the ultimate disappearance of the pretransition

Although the previous observations of the concentration dependence of the alcohol effects on DPPC bilayers have yielded insights into the threshold concentration at which the alcohol-induced, fully interdigitated gel phase occurs, a direct observation of the transition between the noninterdigitated gel and the fully interdigitated gel phase was not possible with these previous methods. However, a fluorescence method, calibrated by direct measurements on the same preparations with X-ray diffraction, may monitor this transformation between the two gel phases (Nambi et al., 1988).

For a particular ethanol concentration (1.2 M), the temperature dependence of the DPH fluorescence intensity during ascending temperature scans indicates a phase transition signaled by an abrupt decrease in fluorescence intensity as the temperature is raised. X-ray diffraction using identical DPPC dispersions indicates that the lamellar repeat period undergoes an abrupt decrease at a temperature coinciding with the phase transition observed by the fluorescence technique. Electron density profiles derived from these X-ray measurements confirm that this fluorescence method directly monitors a phase transition from the noninterdigitated gel phase to the fully interdigitated gel phase with ascending temperature when sufficient ethanol is present.

When the ethanol concentration dependence of this fluorescence quenching is determined isothermally by adding aliquots of ethanol to dispersions maintained at fixed temperatures, the results indicate that at higher temperature, less ethanol is required to induce this transition from the noninterdigitated gel phase to the fully interdigitated gel phase. Furthermore, this temperature dependence indicates that a thermotropic transition must exist between the noninterdigitated G(II) gel phase and this fully interdigitated gel phase.

In light of this observation, the biphasic effect of ethanol on the main phase transition may be interpreted by postulating that, at the reversal concentration of ethanol, the G(I) ripple phase to L liquid–crystalline main phase transition is replaced by the fully interdigitated gel (G_1) to liquid–crystalline (L) phase transition. In addition, at the reversal alcohol concentration, the pretransition involving the transformation from the G(II) noninterdigitated gel to (GI) ripple phase is replaced by the G(II) noninterdigitated gel to G_1 fully interdigitated gel phase transition.

This thermotropic phase transition from the noninterdigitated gel phase to the fully interdigitated gel phase is not immediately reversible simply by lowering the temperature. Instead, the formation of the noninterdigitated gel phase from the fully interdigitated gel phase exhibits significant hysteresis, requiring an overnight incubation in order to regenerate the noninterdigitated state.

3.4.2.1.2 Induced Interdigitation in a Binary Lipid System

Because ethanol interacts differently with PC and PE, the thermotropic behavior of a PC–PE mixture was examined for this ethanol effect (Rowe, 1987). If interdigitation is induced only for the PC within the mixture, lateral phase separation is expected from the resultant mismatch of the hydrophobic regions contributed by these two components. The particular species C(12)-PE and C(16)-PC were chosen for the mixture so that the higher-melting species of the pair was the PC. These two components are miscible in both the fluid and gel phases; a single phase transition occurs in the absence of ethanol. In contrast, when sufficient ethanol is present, the thermotropic behavior of the PE–PC mixtures containing a low mole fraction of PE (between 0.11 and 0.40) exhibits two separate transitions, indicating that lateral phase separation occurs for these compositions. When the mole fraction of PE exceeds 0.4, a single transition is observed again.

These observations are interpreted by proposing that for the solid phases at low PC concentrations (below 0.1 mole fraction), the interdigitated DPPC phase accommodates the PE molecules in increasing amounts until it becomes saturated with PE. Increasing the PE beyond this solubility limit results in the formation of a separate noninterdigitated gel phase, which consists of PE–PC. This gel phase coexists with the interdigitated gel phase composed of DPPC saturated with PE.

Beyond 0.4 mole fraction PE, the interdigitated phase no longer coexists with the regular gel phase, and again a single gel-phase region appears, consisting of the noninterdigitated PE–PC. Because these gel phases transform to the fluid phases upon raising the temperature, the phase diagram exhibits a line of coexistence of three phases: the noninterdigitated gel phase, the interdigitated gel phase, and the (noninterdigitated) fluid phase.

3.4.2.1.3   Ethanol Effect on Biological Membranes

Vibrational infrared spectroscopic studies have been performed to suggest an animal model of alcohol-induced physiological changes in membrane structure (Lewis et al., 1989). These studies suggest that liver plasma membranes of ethanol-treated rats may respond by increasing the membrane order. These adaptive responses are fully reversible upon alcohol withdrawal. Furthermore, if the membrane preparations isolated from the alcohol-treated animals are treated with additional ethanol, these membranes respond by significantly increasing their membrane order. These observations are interpreted by suggesting that membranes from alcohol-treated animals are partially interdigitated.

3.4.2.2   Induction by Polymyxin and Myelin Basic Protein

Until now, our focus has been on the interaction of small amphipathic molecules with zwitterionic lipid bilayers. Ample evidence exists that lipids containing negatively charged headgroups may also interdigitate under the proper circumstances. For example, the interaction of myelin basic protein (BP) with negatively charged lipids has been well studied. This protein, by virtue of its net positive charges, is capable of both an electrostatic interaction with the negatively charged headgroup region, and a hydrophobic interaction whereby portions of the protein are intercalated into the bilayer interior. Investigation of the requirements of bilayer fluidity on the interaction of BP with DPPG bilayers (Boggs et al., 1981) indicates that the protein must first interact with the fluid-phase bilayer before it gains the ability to perturb the gel-to-fluid phase transition, indicating that the protein may only intercalate significantly into the fluid-phase bilayers. If a calorimetry sample is taken above the phase transition (thereby allowing the protein to interact first with the fluid phase) then cooled prior to subsequent rescanning, two endothermic components appear in the thermogram, replacing the single endotherm that appears in the absence of protein. These endothermic transitions may be separated by an additional exothermic phase transition.

Because X-ray diffraction measurements were already available for both the glycerol-induced interdigitated gel phase of DPPC (McDaniel et al., 1987) and the polymyxin-induced interdigitated gel phase in DPPG (Ranck and Toccane, 1982), a unique opportunity existed to determine the feasibility of the spin-label technique for detecting interdigitation (Boggs and Rangaraj, 1985). This technique utilizes spin-labeled fatty acid bilayer probes by comparing the behavior of probes labeled at different positions along the fatty acyl chain. In a noninterdigitated bilayer, the 5-DS-

SL samples are environment near the bilayer interface while the 16-DS-SL probes a region near the bilayer midplane. However, in fully interdigitated bilayers, the 16-DS-SL spin label should interpenetrate past the bilayer midplane to reside at the bilayer interface region of the opposing monolayer. This environment should be similar to the surroundings encountered by the 5-DS-SL spin label. Therefore, the basic premise behind this technique is that if the 5-DS-SL and 16-DS-SL probes both show similar values of the order parameter, these probes must be sampling similarly ordered environments within the bilayer, indicating bilayer interdigitation.

Below the main transition, DPPG samples prepared in water display the usual flexibility gradient characteristic of noninterdigitated bilayers, with disordering of the acyl chain segments being more pronounced as one advances away from the interface toward the bilayer center. When glycerol is added to the system, the order parameters for both 5-DS-SL and 16-DS-SL are similar and slightly greater than the value for 5-DS-SL probed in DPPG bilayers without glycerol. Since the 16-DS-SL is not immobilized in this phase, these results are consistent with the abolishment of the flexibility gradient when glycerol is present, indicating interdigitation. Above the main transition, both probes show isotropic motion, indicating fluid-phase bilayers.

When the spin-label technique was applied toward characterizing the interaction between DPPG and BP, the immobilizing effect of the protein was probed at different depths within the bilayer interior. The ordinary flexibility gradient characteristic of noninterdigitated bilayers appears to be absent in either of the low-temperature gel phases that appear when BP is present.

The greatest immobilizing effect for a 16-DS-SL label, which ordinarily localizes near the relatively disordered bilayer midplane, is associated with the phase exhibiting the higher melting endothermic transition. Because the mobility of this end-labeled probe molecule is comparable in this low-temperature phase to the mobility characteristic for probes localized near the interface region, this greatly decreased amplitude of motion for the 16-DS-SL probes may result because its surroundings are interdigitated. As a consequence of interdigitation, both the 16-DS-SL and 5-DS-SL probes may reside in comparable bilayer environments.

The sequence of thermal events leading to interdigitation in DPPG is interpreted as an initial transition involving melting of the partially interdigitated state, followed by an exothermic transition to the fully interdigitated phase. The fully interdigitated gel phase is subsequently converted to noninterdigitated fluid phase during the main transition.

However, in contrast to the DPPG/BP and DPPG/glycerol, the phase behavior of DPPG/poly-myxin appears to involve a sequential transformation of the fully interdigitated gel phase to the noninterdigitated gel phase prior to the formation of the noninterdigitated fluid bilayer phase. The DPPG/polymyxin-B (DPPG/PMB) system may be compared with DPPG/polymyxin-B nonapeptide (DPPG/PMBN) system, a derivative of PMB lacking the fatty acyl chain (Kubesch et al., 1987). Although PMB induces two endothermic components to appear in the calorimetric thermogram when mixed with DPPG, an equivalent amount of PMBN results in only a single endotherm, indicating that the fatty acyl tail of this peptide antibiotic interacts with bilayers.

PMBN induces a fully interdigitated gel phase in DPPG, similar to the phase formed when DPPG is mixed with the parent compound PMB. A conspicuous absence of a calorimetric transition from the fully interdigitated gel phase to noninterdigitated gel phase suggests that this transition does not involve any significant cooperative intrachain order/disorder events.

Vibrational spectroscopy investigating the DPPG/PMB complex probed its structure in greater detail (Babin et al., 1987). These results indicate some disagreement between the published literature documenting the thermotropic behavior of the DPPG-polymyxin complex. Analysis of the C-C stretching of the resulting spectra utilized the peak height intensity ratio of 1090 cm^{-1} to the 1125 cm^{-1} as a parameter to monitor the proportion of gauche conformers in the acyl chain region, thereby indicating the relative intramolecular order. This ratio indicates that, with respect to the relative number of gauche conformers, the gel phase with polymyxin present is not significantly different from the gel phase that forms without polymyxin. However, polymyxin does appear to disorder the acyl chains in the fluid phase.

The C-H stretching region of the spectrum provides two peak height ratio parameters that aid in the characterization of interchain interactions and intrachain order (Levin, 1984; Wong, 1984). The 2880 cm^{-1}/2845 cm^{-1} peak height intensity ratio monitors the intermolecular interactions; greater intermolecular order is indicated for the gel phase which forms with polymyxin. The 2930 cm^{-1}/2880 cm^{-1} peak height intensity ratio may monitor the overall disorder in the hydrocarbon region; this parameter indicates that DPPG exhibits less overall disorder in the gel phase formed with polymyxin. Because both noninterdigitated and interdigitated gel phases possess similar acyl chain conformation, the polymyxin acyl chain must not significantly perturb the hydrophobic bilayer interior of the fully interdigitated gel phase.

3.4.2.3 Other Inducers

Additional X-ray diffraction studies concerning the fully interdigitated gel phase seem to suggest that Tris buffer may induce its formation (Wilkinson et al., 1987). Tris is a relatively large cation; its association with the negatively charged DPPG headgroup may promote a larger surface area per headgroup at the interface, resulting in bilayer interdigitation.

Differential scanning calorimetry indicates that, although the pretransition occurs in the presence of several different salt solutions containing small cationic counterions, no pretransition is ever observed when Tris is in the bathing medium. The main transition temperature and enthalpy are both increased when Tris is present; these calorimetry trends share similarities with the phase behavior of other induced, fully interdigitated bilayer systems. Dilatometry measurements indicate that the partial specific volume of these bilayer lipids is smaller with Tris than with the other solutions studied, consistent with observed closer packing of the acyl chains in other fully interdigitated gel phases. The calculated values of the van der Waals interaction energy between acyl chains are greater with Tris than with phosphate buffer, providing additional support for Tris-induced gel-phase interdigitation.

Thiocyanate, a small amphipathic anion, adsorbs to lipid bilayers (McLaughlin et al., 1975). X-ray diffraction measurements reveal that the lamellar repeat spacing of DPPC bilayers becomes significantly diminished in the presence 1 M thiocyanate, indicating possible bilayer interdigitation (Cunningham and Lis, 1986). This interdigitation is proposed to result from the perturbation that the thiocyanate causes in the structured water or in the polarizability of water associated with the headgroup region. Furthermore, differential scanning calorimetry detected a slight increase in the main transition temperature and a disappearance of the pretransition. Therefore, in the presence of thiocyanate, the main transition of DPPC most likely involved the direct transformation from the fully interdigitated gel phase to the liquid–crystalline bilayer phase.

3.4.3 STRUCTURAL MODIFICATIONS RESULTING IN INTERDIGITATION

3.4.3.1 Interdigitation for Ether-Linked Lipids

3.4.3.1.1 Comparing DHPC with DPPC

Several studies document the idea that when the ester linkages normally found in symmetric diacylphosphatidylcholines are replaced by ether linkages to form dialkylphosphatidylcholines, the gel-phase bilayers spontaneously interdigitate without any inducers. Differential scanning calorimetry, X-ray diffraction, and ^{31}P-NMR were used to compare the thermotropic behavior of DPPC and DHPC in order to assess any change in phase behavior resulting as the ester linkage in DPPC was replaced by ether linkages (Ruocco et al., 1985).

Both lipids behave very similarly in the fluid phase, and the major distinctions in the phase behavior are observed for the gel phases. The principal calorimetric difference between these two species is that the subtransition in DHPC has a much lower transition enthalpy and is immediately reversible. In contrast, this subtransition for DPPC is conditionally reversible and requires an extended, low-temperature preincubation period prior to its appearance.

X-ray diffraction indicates that, for all phases below the temperature of the pretransition in DHPC, the acyl chains are arranged in a fully interdigitated bilayer, in contrast with the noninterdigitated gel phase for DPPC in the absence of inducers or high pressure. Unlike in DPPC, the lamellar repeat (d spacing) does not change significantly during the DHPC subtransition; rather, the subtransition in DHPC is associated only with a rearrangement in acyl chain packing and does not appear to involve substantial hydration changes in the headgroup region. The wide-angle scattering suggests that the DHPC subtransition is a transformation from orthorhombic to hexagonal packing of the acyl chains. NMR measurements suggest that the rapid, whole-molecule axial diffusion for DHPC persists to much lower temperatures, indicating that a fundamental difference lies in the degree of restriction of the headgroup motion for this low-temperature crystalline phase. Phosphorus NMR results indicate a lack of difference in the magnitude of motional averaging between the DHPC crystalline and the interdigitated gel phases, supporting the notion that the subtransition in DHPC does not significantly affect the rate of reorientational motions of the headgroup phosphate moiety.

Deuterium NMR may potentially determine both the average orientation of a deuterated segment with respect to a reference axis, and the dynamics of motion of this segment. In an effort to detail the molecular differences in phase behavior between these ester- and ether-linked symmetric phosphatidylcholines, a deuterium NMR study using specifically headgroup- and chain-labeled phospholipids was performed (Serrellach et al., 1983).

The alpha methylene segment of the choline group was deuterated for both lipid species, while the chain labeling involved the alpha methylene units of both chains for each lipid. Little difference in headgroup behavior was ascertained between the two lipid species in the liquid–crystalline phase, in agreement with the previous phosphorus NMR results.

These deuterium NMR data for the headgroup-labeled species also suggest that the axial diffusion persists in DHPC to a much lower temperature than in DPPC. Undoubtedly, the increased surface area available to the headgroups for the interdigitated gel phase contributes to the persistence of the axial diffusion in these bilayers.

However, a distinction arises when comparing the properties of the liquid–crystalline phase between these two classes of lipids. DPPC ordinarily exhibits three quadrupolar splittings for the chain alpha methylene segment. This has been shown by specific labeling studies to arise from a single splitting for both deuterons of the sn-1 chain, and two separate quadrupole splittings for the sn-2 chain deuterons, as a consequence of the magnetic inequivalence of these two atoms on this methylene segment in the sn-2 acyl chain. Each chain at the interface is inequivalent, and the individual deuterons of the sn-2 chain are magnetically inequivalent with each other.

In comparison with the results for DPPC, four components can be distinguished in the deuterium NMR spectra of the liquid–crystalline phase DHPC, and these are attributed to the magnetic inequivalence of all four deuterons in the acyl chain region.

A high-pressure FTIR spectroscopic study compares the structure and dynamics of the acyl chain region for DPPC and DHPC (Siminovitch et al., 1987). The pressure-induced correlation field splitting of the methylene rocking and scissoring modes of the acyl chains is examined in detail, because these vibrational modes are potentially useful for characterizing differences in interchain packing and interchain interactions arising from bilayer interdigitation.

Generally, the pressure at which the correlation field component becomes distinct is lower for DHPC than for DPPC. The methylene rocking mode is examined in detail; the quantitation of the intensities of the scissoring mode is difficult because of interface by an overlapping band contributed by the quartz pressure calibrant. Because previous experimental work has established interdigitation in the DHPC gel phase, the authors proposed a new peak height intensity ratio for diagnosing interdigitation. This ratio consists of the intensities of the methylene scissoring mode and its correlation field component; the study of its pressure dependence has been further utilized in characterizing other interdigitated bilayer systems (Siminovitch et al., 1987; Wong and Huang, 1989).

Temperature-scanning densitometry data indicate that the partial specific volume for DHPC is smaller than for DPPC at all temperatures (Laggner et al., 1987). The greatest difference between these two lipids arises during the pretransition, where the associated volume change for DHPC is approximately three times greater than for DPPC. This is not surprising, because the pretransition involves the deinterdigitation of the DHPC acyl chains. The subtransition volume change is only a fraction (approximately one-tenth) of the volume change observed in DPPC, indicating once again that the subtransitions must have fundamentally different molecular origins in these two species.

3.4.3.1.2 Binary System of DPPC/DHPC

Examining the phase diagram constructed for the fully hydrated binary mixtures of DPPC and DHPC (Lohner et al., 1987) reveals that the addition of DHPC to DPPC attenuates the DPPC subtransition and leads to its disappearance at a molar ratio of approximately 1:6. However, the DHPC subtransition is only slightly affected by the presence of DPPC up to equimolar concentrations.

Adding the second component to either of the pure species results in a depression of the pretransition enthalpy and temperature in a symmetrical manner, with a minimum occurring at the equimolar mixture. The main transition shows a linear composition dependence, consistent with the view that the ripple phase of each component is similar in character and therefore should exhibit ideal mixing.

These X-ray data on this DPPC/DHPC binary system suggest that, below the pretransition temperature, the chain packing arrangement changes from noninterdigitated to fully interdigitated gel phase when the DHPC exceeds equimolar concentrations.

Below the subtransition temperature, the presence of DHPC increases the repeat spacing, suggesting that the DPPC crystalline phase is unable to persist and becomes transformed into the more hydrated lamellar gel phase when minor amounts of DHPC are present. Yet the low-temperature subphase of DHPC can accommodate up to equimolar amounts of DPPC without becoming noninterdigitated. These results suggest that the carbonyl groups in DPPC contribute to the molecular interactions that govern the packing arrangement in the low-temperature gel phases (Lohner et al., 1987). These interactions are disrupted by even minor amounts of DHPC within this DPPC lattice. The main determinant that governs the mode of packing in DHPC appears to be chain–chain interactions in the hydrocarbon region.

A study of the DPPC/DHPC phase behavior using Raman spectroscopy to investigate a mixture of perdeuterated DPPC with nondeuterated DHPC has allowed a determination of the packing and conformational order of each lipid within the mixture (Devlin and Levin, 1989). The conclusion is drawn that, relative to the lipid–lipid interactions observed in the single component systems, gel-phase DHPC is more perturbed by the addition of DPPC, whereas gel-phase DPPC appears to retain much of its single-component character even with DHPC present.

3.4.3.1.3 Effect of Cholesterol on DHPC Phase Behavior

In contrast to the thermotropic behavior of DPPC/cholesterol mixtures, the presence of cholesterol causes a dramatic increase in the main phase transition for DHPC dispersions (Levin et al., 1985). An X-ray diffraction study indicates that the presence of equimolar cholesterol increases the lamellar repeat spacing relative to pure DHPC, suggesting that these normally fully interdigitated gel-phase bilayers have become noninterdigitated because of the cholesterol (Siminovitch et al., 1987). The diffraction patterns obtained for equimolar DPPC–cholesterol mixtures and DHPC–cholesterol mixtures appear to be identical. Because cholesterol occupies a location in the acyl chain region near the interface, its presence in the DHPC bilayer must act to reduce the mismatch between the headgroup cross section and the acyl chain cross section.

3.4.3.1.4 Ethanol and DHPC

What happens when short chain alcohols, which induce the formation of the interdigitated phase in DPPC, are allowed to interact with a gel phase that is ordinarily interdigitated? One might expect that the acyl chain terminal methyl groups at the interface region could serve as binding sites for these amphipathic molecules (Simon et al., 1983; Veiro et al., 1988), resulting in preferential interaction of the alcohol for the interdigitated DHPC phase. This preferential interaction would yield an increase in the gel-to-ripple phase transition temperature.

The DHPC pretransition, which involves the conversion from a fully interdigitated gel to the noninterdigitated ripple phase, may be observed using optical light scattering the DPH fluorescence. If this pretransition is monitored as a function of increasing alcohol concentration (Veiro et al., 1988), the transition temperature increases. When sufficient alcohol is available (0.25 M), the pretransition coalesces with the main phase transition and is no longer distinguishable from the main phase transition. Consistent with previous results for DPPC, the longer-chain alcohols are more effective at shifting the transition temperature. A biphasic effect of alcohol on the main transition temperature of DHPC is observed. Because this threshold alcohol concentration corresponds with that concentration required to abolish the pretransition, the DHPC interdigitated gel phase must transform directly into the liquid–crystalline phase, bypassing the ripple phase in analogy with the events occurring when DPPC is under the influence of alcohol. Therefore, the alcohol effect on the main phase transition of DHPC is similar to the effect on DPPC, in agreement with the proposed similar structure of the ripple phases for these two lipids.

The threshold concentrations of alcohol required for interdigitation are consistently lower for DHPC than for DPPC, suggesting perhaps that the intermolecular interactions at the interface are weaker for the ether-linked lipids (Veiro et al., 1988).

3.4.3.2 Bilayers of Thermophilic Bacteria

The thermophilic bacteria contain several unusual classes of lipids, as one might expect for adaptation to the extreme conditions of their environment (Luzzati et al., 1987). These organisms have evolved a unique mechanism to couple their separate bilayer surfaces.

Lipids from the thermophilic archaebacteria appear to consist exclusively of ether linkages; one category of these lipids is the diglyceryl tetraether lipids (Luzzati et al., 1987). These tetraether lipids exist with two membrane-spanning branched acyl chains, covalently attached to a polar group at each surface. Their membranes superficially resemble an interdigitated system, except the arrangement results from a monolayer of these bipolar lipids. These lamellar systems undergo order/disorder transitions; furthermore, thermodynamic parameters of these phase transitions are influenced by the content of cyclopentane rings in these branched acyl chains, perhaps superficially analogous to the influence of sterols in animal cell membrane systems.

3.4.3.3 Positional Isomers Exhibiting Interdigitation

Although most studies of symmetric chain phospholipids reviewed here involve the naturally occurring isomer with acyl chains occupying the *sn*-1 and *sn*-2 positions, the thermotropic behavior of the L-β-DPPC has been investigated (Serrellach et al., 1983). This isomer has its acyl chains attached at the *sn*-1 and *sn*-3 positions of the glycerol backbone, and this alteration at the interface region profoundly influences the acyl chain packing arrangements. An immediately reversible high-temperature endothermic phase transition and a conditionally reversible low-temperature endothermic phase transition are observed. This lower endotherm involves a transformation from the crystalline phase with tilted, noninterdigitated acyl chains to a gel phase with fully interdigitated acyl chains. Therefore, this positional isomer may spontaneously form a fully interdigitated gel phase by virtue of its altered structure at the interface. The high-temperature endothermic phase transition represents the transformation from this interdigitated gel phase to a liquid–crystalline phase.

3.4.3.4 Headgroup Interdigitation in the Crystalline Phase

A recent X-ray crystallographic investigation of dilauroyl-*N,N*-dimethyl-ethanolamine reveals that, under some circumstances, the lipid headgroups may adopt a bilayer perpendicular organization. Although this orientation of the headgroups is normally energetically unfavorable, it may occur if the headgroup from one bilayer surface interdigitates with the headgroup from an opposing bilayer surface (Pascher and Sundell, 1986). The interdigitation of the headgroups across the interbilayer space results in a structure without charge separation in the headgroup plane.

ACKNOWLEDGMENTS

We wish to thank those countless researchers whose work has paved the way for studies on bilayer interdigitation. J.L. Slater is also indebted to Dr. Ira W. Levin and an Intramural Research Training Award at the National Institutes of Health for providing support during preparation of this manuscript.

REFERENCES

Ali S, Lin H, Bittman R, and Huang C: *Biochemistry* 1989, 28:522.
Babin Y, D'Amour J, Pigeon M, and Pezolet M: *Biochim. Biophys. Acta* 1987, 903:78.
Baldwin PA and Hubbell WL: *Biochemistry* 1985, 24:2624.
Boggs JM, Stamp D, and Moscarello MA: *Biochemistry* 1981, 20:6066.
Boggs JM and Rangaraj G: *Biochim. Biophys. Acta* 1985, 816:221.
Boggs JM and Mason JT: *Biochim. Biophys. Acta* 1986, 863:231.
Boggs JM, Koshy KM, and Rangaraj G: *Biochim. Biophys. Acta* 1988, 938:373.
Boggs JM, Koshy KM, and Rangaraj G: *Biochim. Biophys. Acta* 1988, 938:361.
Braganza LF and Worcester DL: *Biochemistry* 1986, 25:2591.
Cevc G and Marsh D: in *Phospholipid Bilayers. Physical Principles and Models.* Wiley-Interscience, New York, 1987.
Chong PL-G and Choate D: *Biophys. J.* 1989, 55:551.
Cunningham BA and Lis LJ: *Biochim. Biophys. Acta* 1986, 861:237.
Devlin MT and Levin IW: *Biochemistry* 1989, 28:8912.
Elder M, Hitchcock P, Mason R, Shipley GG: *Proc. R. Soc. London A* 1977, 354:157.
Elliott JR, Needham D, Dilger JP, and Hayden DA: *Biochim. Biophys. Acta* 1983, 735:95.
Gardam M and Silvius JR: *Biochim. Biophys. Acta* 1989, 980:319.
Grant CWM, Mehlhorn IE, Florio E, and Barber KR: *Biochim. Biophys. Acta* 1987, 939:151.
Hauser H, Pascher I, and Sundell S: *J. Mol. Biol.* 1980, 137:249.
Huang C: Heinrich-Wieland Prize lecture delivered on October 27, 1989. *Klin. Wochenschr.* 1990, 68:149–165.
Huang C, Mason JT, Levin IW: *Biochemistry* 1983, 22:2775.
Huang C, Mason JT, Stephenson FA, and Levin IW: *J. Phys. Chem.* 1984, 88:6454.
Huang C, Mason JT, Stephenson FA, and Levin IW: *Biophys. J.* 1986, 49:587.
Huang C and Mason JT: *Biochim. Biophys. Acta* 1986, 864:423.
Huang HW: *Biophys. J.* 1986, 50:1061.
Hui SW, Mason JT, and Huang C: *Biochemistry* 1984, 23:5570.
Hui SW and Huang C: *Biochemistry* 1986, 25:1330.
Jain MK, Crecely RW, Hille JDR, de Haas GH, and Gruner SM: *Biochim. Biophys. Acta* 1985, 813:68.
Kubesch P, Boggs J, Luciano L, Maass G, and Tummler B: *Biochemistry* 1987, 26:2139.
Laggner P, Lohner K, Degovics G, Muller K, and Schuster A: *Chem. Phys. Lipids* 1987, 44:31.
LeNeveu DM, Rand RP, and Parsegian VA: *Nature* 1976, 259:601.
Levin IW: *Advances in Infrared and Raman Spectroscopy,* vol. 11. Eds. Clark RJH and Hester RE, John Wiley & Sons, New York, 1984, pp. 1–48.
Levin IW, Keihn E, and Harris WC: *Biochim. Biophys. Acta* 1985, 820:40.
Levin IW, Thompson TE, Barenholtz Y, and Huang C: *Biochemistry* 1985, 24:6282.
Lewis EN, Levin IW, and Steer CJ: *Biochim. Biophys. Acta* 1989 (in press).
Lin H and Huang C: *Biochim. Biophys. Acta* 1988, 946:178.
Lohner K, Schuster A, Degovics G, Muller K, and Laggner P: *Chem. Phys. Lipids* 1987, 44:61.
Luzzati V, Gambacorta A, DeRosa M, and Gulik A: *Annu. Rev. Biophys. Biophys. Chem.* 1987, 16:25.
MacDonald RC and MacDonald RI: *J. Biol. Chem.* 1988, 263(21):10052.
Marra J and Israelachvilli J: *Biochemistry* 1985, 24:4608.

Mason JT: *Biochemistry* 1988, 27:4421.

Mason JT, Huang C, and Biltonen RL: *Biochemistry* 1981, 20:6086.

Mason JT and Huang C: *Lipids* 1981, 168:604.

Mattai J, Sripada PK, and Shipley GG: *Biochemistry* 1987, 26:3287.

McDaniel RV, McIntosh TJ, and Simon SA: *Biochim. Biophys. Acta* 1987, 731:97.

McIntosh TJ, Magid AD, and Simon SA: *Biochemistry* 1989, 28:17.

McIntosh TJ, McDaniel RV, and Simon SA: *Biochim. Biophys. Acta* 1983, 731:109.

McIntosh TJ, Simon SA, Ellington JC, Jr., and Porter NA: *Biochemistry* 1984, 23:4038.

McIntosh TJ and Simon SA: *Biochemistry* 1986, 25:4058.

McLaughlin S, Bruder A, Chen S, and Maser C: *Biochim. Biophys. Acta* 1975, 394:304.

Mehlhorn IE, Florio E, Barber KR, Lordo C, and Grant CWM: *Biochim. Biophys. Acta* 1988, 939:151.

Mouritsen OG and Bloom M: *Biophys. J.* 1984, 36:141.

Nambi P, Rowe ES, and McIntosh TJ: *Biochemistry* 1988, 27:9175.

O'Leary TJ and Levin IW: *Biochim. Biophys. Acta* 1984, 776:185.

Parsegian VA, Fuller N, and Rand RP: *Proc. Natl. Acad. Sci. U.S.A.* 1979, 76:2750.

Pascher I, Sundell S, Eibl H, and Harlos K: *Chem. Phys. Lipids* 1986, 39:53.

Pascher I and Sundell S: *Biochim. Biophys. Acta* 1986, 855:68.

Pearson RH and Pascher I: *Nature* 1979, 281:499.

Peschke J, Riegler J, and Mohwald H: *Eur. J. Biophys.* 1987, 14:385.

Prasad SK, Shashidhar R, Gaber BP, and Chandrasekhar SC: *Chem. Phys. Lipids* 1987, 143:227.

Ranck JL and Toccanne JF: *FEBS Lett.* 1982, 143:171.

Rand RP and Parsegian VA: *Biochim. Biophys. Acta* 1989, 988:351.

Riegler J and Mohwald H: *Biophys. J.* 1986, 49:1111.

Rowe ES: *Biochemistry* 1983, 22:3299.

Rowe ES: *Biochim. Biophys. Acta* 1985, 813:321.

Rowe ES: *Biochemistry* 1987, 261:46.

Ruocco MJ, Siminovitch DJ, and Griffin RG: *Biochemistry* 1985, 24:4844.

Ruocco MJ, Siminovitch DJ, and Griffin RG: *Biochemistry* 1985, 24:2406.

Serrellach EN, Dijkman R, de Haas GH, and Shipley GG: *J. Mol. Biol.* 1983, 170:155.

Siminovitch DJ, Ruocco MJ, Makriyannis A, and Griffin RG: *Biochim. Biophys. Acta* 1987, 901:191.

Siminovitch DJ, Wong PTT, and Mantsch HH: *Biophys. J.* 1987, 51:465.

Siminovitch DJ, Wong PTT, and Mantsch HH: *Biochim. Biophys. Acta* 1987, 900:163.

Simon SA and McIntosh TJ: *Biochim. Biophys. Acta* 1984, 773:169.

Simon SA, McIntosh TJ, and Hines ML: *Molecular and Cellular Mechanisms of Anesthetics.* Eds. Roth SH and Miller KW, Plenum Press, New York, 1986, pp. 297–308.

Simon SA, McIntosh TJ, and Magid AD: *J. Colloid Interface Sci.,* 1988 (in press).

Singer SJ: *Annu. Rev. Biochem.* 1974, 43:805.

Slater JL and Huang C: *Prog. Lipid Res.* 1988, 27:325.

Slater JL, Huang C, Adams RG, and Levin IW: *Biophys. J.* 1989, 56:243.

Suwalsky M, Giminez V, Saenger V, and Neira F: *Z. Naturforsch.* 1988, 43c:742.

Tilcock CPS, Cullis PR, Hope MJ, and Gruner SM: *Biochemistry* 1986, 25:816.

Utoh S and Takemura T: *Jpn. J. Appl. Phys.* 1985, 24:356.

Utoh S and Takemura T: *Jpn. J. Appl. Phys.* 1985, 24:1404.

Veiro JA, Nambi P, Herold LL, and Rowe ES: *Biochim. Biophys. Acta* 1987, 900:230.

Veiro J, Nambi P, and Rowe ES: *Biochim. Biophys. Acta* 1988, 943:108.

Wilkinson DA, Tirrell DA, Turek AB, and McIntosh TJ: *Biochim. Biophys. Acta* 1987, 905:447.

Wong PTT: *Annu. Rev. Biophys. Bioeng.* 1984, 13:1.

Wong PTT, Siminovitch DJ, and Mantsch HH: *Biochim. Biophys. Acta* 1988, 947:139.

Wong PTT and Huang C: *Biochemistry* 1989, 28:1259.

Wu W, Huang C, Conley TG, Martin RB, and Levin IW: *Biochemistry* 1982, 21:5957.

Wu W and Huang C: *Biochemistry* 1983, 22:5068.

Xu H and Huang C: *Biochemistry* 1987, 26:1036.

Xu H, Stephenson FA, and Huang C: *Biochemistry* 1987, 26:5448.

Zaccai G, Buldt G, Seelig A, and Seelig J: *J. Mol. Biol.* 1979, 134:693.

4 The Dynamics of Membrane Lipids

Klaus Gawrisch

CONTENTS

4.1 INTRODUCTION

The lipid matrix of almost all biomembranes is in a liquid–crystalline lamellar state characterized by the formation of a stable bilayer of mostly phospholipids that maintain liquidlike properties in their molecular motions [1, 2]. The lipid motional degrees of freedom are inherently linked to bilayer elasticity but also to membrane functions, like permeability for solvents and solutes, and the action of peripheral and integral membrane proteins. Maintenance of the liquid–crystalline nature of the lipid matrix requires a particular membrane composition, temperature range, and state of hydration [3, 4].

The essentiality of lipid dynamics for membrane function was recognized more than three decades ago with the introduction of the fluid-mosaic model of biomembranes by Singer and Nicholson in 1972 [5]. The model recognized the liquid–crystalline nature of the lipid bilayer as the matrix of biological membranes. Interaction of peripheral and integral membrane proteins with the matrix, as well as protein function, were directly linked to an image of membranes as "two-dimensional oriented viscous solution."

Lipid order and dynamics are studied by nuclear magnetic resonance (NMR), infrared and Raman spectroscopy, time-resolved fluorescence spectroscopy on labels located in different regions of the bilayer, electron paramagnetic resonance (EPR) on spin labels, differential scanning calorimetry (DSC), X-ray and neutron diffraction experiments, and ever more sophisticated molecular simulations.

Molecular simulations provide the time-dependent location of all atoms in the bilayer, permitting the most detailed description of molecular motions [6, 7]. But to achieve this kind of resolution, the simulations rely on effective interaction potentials that are adjusted to reproduce experimental results [8]. Also, depending on molecular detail and computational power, the length of simulations is still limited to the range from nano- to microseconds. Simulated bilayer patches are relatively small in size, of the order of 100 lipids only. With increasing computer power, those limitations may be eased. Perhaps in the future simulations will be able to substitute for experiments. However, at present the experiments are guiding simulations.

Spectroscopic data on lipid order and dynamics are typically analyzed in terms of order parameters [9–11] and motional correlation times [12]. Order parameters report the orientation of vectors representing orientation of chemical bonds, molecular segments, molecules, or assemblies of molecules. Motional correlation times report the characteristic time scale for reorientation of those vectors. In fluid bilayers, the time scale of motions covers 15 orders of magnitude, from femtoseconds to seconds. In general, the correlation times increase with increasing size of the moving segments. Movements of individual bonds are fastest, followed by motions of lipid segments, motions of the entire lipid molecule, undulatory motions of patches of the bilayer, etc. This chapter will address the following lipid motions and their approximate correlation times (see Figure 4.1):

- Vibrations of bond length and bond torsional oscillations (subpicosecond range)
- Librational motions of bond orientation (picoseconds)
- Rotations about chemical bonds, gauche/*trans* isomerization (pico- to nanoseconds)
- Diffusional reorientation of lipid molecules about their long axis (nanoseconds)
- Wobble of lipids (nanoseconds)
- Lateral diffusion in the plane of the bilayer (nano- to microseconds for movement over a distance of 0.8 nm, the approximate lateral lattice spacing between two lipids)
- Undulatory motions of bilayer patches (microseconds to seconds)
- Flip-flop of lipids between the two monolayers of a bilayer (milliseconds to hours)

4.1.1 MEMBRANE FLUIDITY

The widespread recognition of the fluid nature of the lipid matrix is reflected in the use of the term *membrane fluidity*. In textbooks on fluid dynamics, fluidity is introduced as an inverse value of viscosity [13]. Therefore, membrane fluidity may be interpreted as a measure of the viscous drag acting on membrane constituents (e.g., integral membrane proteins) in dynamic processes like rotational and lateral diffusion. Such drag may also be important for the dynamics of protein structural transitions. In the 1980s, this term became very popular in biomembrane research. The outcome of complex experiments that reported order parameters and motional correlation times of membrane-imbedded labels and lipid molecular bonds, lateral diffusion rates, membrane permeability to solutes and solvents, and phase transition temperatures of the lipid matrix were summarized as increased or decreased fluidity. The intention was to give results of experiments conducted at the molecular level a predictive value for biomembrane function. But the link between those measurements and viscous drag is obscure. Definition of viscous drag in spatially anisotropic media like membranes is operational, and parameters for specific membrane constituents may not be transferable to other compounds without major correction. Also, viscous drag is just one of many properties that control membrane function. It may not have such central importance.

There is also growing awareness that one parameter may not adequately describe membrane dynamic properties. Modern spectroscopic approaches measure dozens of parameters per mem-

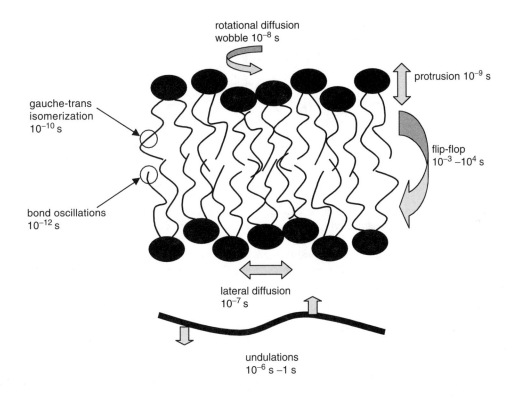

FIGURE 4.1 Lipid motions in biomembranes and their approximate correlation times.

brane. Order parameters and motional correlation times are routinely determined with resolution for most carbon atoms of lipid molecules. Recently published results show that every type of perturbation to membrane structure leaves a characteristic fingerprint. For example, the influences on membrane order and dynamics from changes in phospholipid headgroup structure, hydrocarbon chain length and unsaturation, cholesterol content of membranes, temperature, state of hydration — but also from interaction with solvents and solutes like alcohols — have distinct characteristics that cannot be summarized as increasing or decreasing fluidity [*14*]. Furthermore, functional studies on membrane proteins reconstituted into different lipid environments discriminate between the type of membrane perturbation (see, e.g., sensitivity of rhodopsin to the chain length of membrane-incorporated alcohol [*15*]).

Another negative aspect of the use of "fluidity" is the confusion it may create when comparing results from different experiments. The same type of perturbation may result in an increase or decrease of order parameters, depending on the time scale of the method [*16, 17*]. For example, an integral protein may slow diffusional rotation of spin-labeled hydrocarbon chains in its lipid annulus. Certain motions may no longer contribute effectively to averaging of magnetic interactions, because EPR reports all motions with correlation times longer than 10 ns as "frozen out." The result is an apparent increase of order parameters of the spin-labeled hydrocarbon chain. In contrast, a method with a time base that is longer by three orders of magnitude, like the measurement of C-^{2}H bond order parameters by ^{2}H NMR, does not sense the increase in correlation times if they remain shorter than the time scale of ^{2}H NMR, which is 10 μs. Instead, NMR measurements may report a decrease in order from additional modes of motion with correlation times in the range from 10 ns to 10 μs that lipid hydrocarbon chains will experience when interacting with the rugged surface of the protein. A careful analysis of chain order parameters and motional correlation times reported by the two methods would recognize both the increase of motional correlation times and the presence of additional motions with correlation times in the submicrosecond range. In contrast,

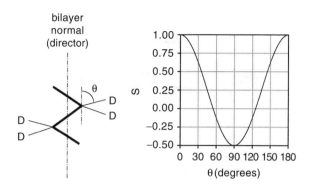

FIGURE 4.2 Angular dependence of the bond order parameter S of hydrocarbon chains. The value of S depends on the angle θ between the bilayer normal and the C-D bond.

the reduction of results to "membrane fluidity" generates confusion. Therefore, the term "membrane fluidity" has been avoided in this chapter. Instead, I will introduce the motions by their influence on order parameters and their correlation times.

4.2 EXPERIMENTAL STUDIES

4.2.1 LABELING OF LIPID BILAYERS

Spectroscopic approaches frequently employ labels for enhancement of sensitivity and resolution. The influence of isotopic labeling (e.g., ^{2}H- or ^{13}C labels) on membrane structure and dynamics is benign, except for known minor shifts in phase transition temperatures from deuteration (http://www.lipidat.chemistry.ohio-state.edu). However, the data from the much larger spin and fluorescence labels must be analyzed with caution. Those labels, even if added to the membrane at very low concentration, tend to perturb their immediate environment. It is the duty of the investigator to interpret responsibly results obtained by perturbing labels. For example, the use of perturbing labels to detect the molecular motions of lipid segments on short time scales could be inappropriate. However, such labels may be well suited for reporting collective properties of lipid packing in bilayers. Three decades' worth of research with those potentially perturbing labels has demonstrated their usefulness for membrane studies.

4.2.2 LIPID ORDER AND MOTIONAL CORRELATION TIMES

Lipid order is typically expressed as a set of order parameters, in the simplest case a Legendre polynomial function of second order:

$$S_i = \frac{1}{2}(3 \cos^2(\theta_i) - 1)$$

where θ_i is the angle between the frame of reference e.g., the normal to the bilayer surface, and a chemical bond labeled i. According to this definition, the order parameter $S_i = +1$ if the bond is oriented parallel to the axis of reference ($\theta_i = 0$), $S_i = 0$ at the "magic" angle ($\theta_i = 54.7°$), and $S_i = -0.5$ if the bond is oriented perpendicular ($\theta_i = 90°$) (Figure 4.2).

4.2.2.1 Order Parameters and Motions

A frequent misconception is that low order parameters are always indicative of the presence of rapid, more isotropic motions. The value of an order parameter *per se* does not distinguish between being low because of a particular fixed orientation of the vector to the axis of reference and being low from rapid averaging over many orientations. For example, the order parameter may be zero because a bond is oriented at a magic angle to the bilayer normal or because motions are entirely random.

However, order parameters are often defined such that immobilized bonds have highest order S = 1. Under those conditions, motional averaging always reduces order. If analyzed in terms of increasing length of correlation times, order reduction takes place in increments from faster to slower motions. Every consecutive motion reduces order by a certain amount, depending on geometry and motional amplitudes. Consequently, lipid order parameters are always lowest on an infinite time scale that covers all motions.

It can be shown that order parameters of two or more superimposed motions may be well approximated by a product of order parameters according to

$$S(tot)_i = S(\tau_1)_i \cdot S(\tau_2),....,\cdot S(\tau_n)$$

if the correlation times τ of those motions differ by at least one order of magnitude [17]. The symbol $S(tot)_i$ is the order parameter of a vector (i) (e.g., the orientation of a chemical bond after averaging by motions with correlation times $\tau_1,..., \tau_n$); $S(\tau_1)_i$ is the order parameter of this bond after averaging over the fastest motion (e.g., bond vibrational fluctuations and gauche/*trans* isomerization). The parameter $S(\tau_2)$ reflects averaging by a slower motion with $\tau_1 \leq 10\ \tau_2$ (e.g., lipid rotational diffusion and wobble, etc.). The index (i) at the order parameter for all but the fastest motion $S(\tau_1)_i$ has been omitted, because slower motions are collective and the same order parameters $S(\tau_2),....,S(\tau_n)$ do apply to many bonds.

It is straightforward to show that the effective symmetry axis of motions is always defined by the symmetry of the slowest motion. In liquid–crystalline theory, this axis is typically called the *director* [18]. For example, the symmetry axis for motions in a fluid lipid bilayer is the bilayer normal, because reorientational diffusions are symmetric about the bilayer normal and have longer correlation times than gauche/*trans* isomerization.

4.2.2.2   Ensemble and Time Averages of Order Parameters

Experimental order parameters and motional correlation times are almost always measured as ensemble averages over a large number of molecules. For example, even under very favorable experimental conditions it takes at least 10^{15} atoms with the same resonance frequency to detect an NMR resonance. Other methods, like EPR and optical spectroscopy, require fewer molecules but are ensemble averages, as well.

The concept of time averaging of order parameters was introduced previously. For a method that operates on the time scale τ, the order parameter is

$$S(\tau) = \int_0^{\tau} \frac{1}{\tau} S(t)\,dt$$

If lipid bilayers are in equilibrium, then conformational transitions of lipids are truly statistical events. In such systems, according to the ergodic theorem, ensemble and time averages converge over sufficiently long observation times. For lipids in a fluid matrix, this is the case on a time scale of 10 μs or longer, as in ^{2}H NMR. This explains the exceptional resolution of ^{2}H resonances of labeled lipids: properties of all superimposed resonances are exactly the same.

According to the ergodic theorem, lipid order parameters can be calculated from simulations on a shorter time scale by averaging over a large number of molecules in the data set. The error limits of calculated order depend on both the simulation time and the number of molecules over which the result is averaged.

4.2.3 Spectroscopic Studies of Lipid Order and Dynamics

Spectroscopies employing electromagnetic radiation (X-rays, light, radiowaves), and also diffraction with elementary particles like neutrons, operate on specific time scales. All motions with correlation times shorter than such a scale reduce the order parameter to one effective value, which is the weighted average over all conformers. In contrast, motions with longer time scales result in superposition of signals from states with different order. For example, the time scale of ^2H NMR is 10 μs. On this scale, the gauche/*trans* isomerization of hydrocarbon chains with a correlation time of 100 ps is fast, and a single, time-averaged order parameter per bond is measured. It is a weighted average of order from contributing isomers. In contrast, infrared spectroscopy operates on a subpicosecond time scale that is much shorter than the 100 ps correlation time of gauche/*trans* isomerization. Therefore, infrared spectra report the isomers as superposition of resonances. The intensity of those resonances is proportional to their concentration. Approximate correlation times of motions may be determined by exploiting differences in time scales between both methods.

4.2.3.1 Solid-State NMR

NMR has contributed the most to our understanding of internal lipid dynamics. Solid-state NMR of membranes measures the magnitude of angularly dependent (anisotropic) interactions. This includes chemical shifts that depend on the orientation of a lipid in the magnetic field (e.g., the ^{31}P NMR resonance of phospholipids), magnetic dipole–dipole interactions between nuclei (e.g., interactions between directly bonded ^1H and ^{13}C nuclei), and the quadrupole interaction of lipids labeled with ^2H nuclei. The last approach is particularly popular because it provides information from labeled sites only.

4.2.3.1.1 ^2H NMR

The resonance signal of ^2H NMR nuclei is split into a doublet by a quadrupolar interaction between the electric quadrupolar moment of the deuterium nucleus and the electric field gradient in C-^2H bonds [19]. The resonance frequencies of the doublet lines depend on the order parameter $S(\theta)$, according to

$$\nu_{1/2} = \nu(iso) \pm \frac{3}{4} \frac{eqQ}{h} S(\theta),$$

where $\nu(iso)$ is the chemical shift of the resonance, and the constant eqQ/h characterizes the strength of quadrupole interactions (C-^2H bonds in saturated carbons 168 kHz [20]; olefinic carbons 175 kHz [21]). The difference in resonance frequencies between both lines, called the *quadrupole splitting*, is

$$\Delta\nu = \frac{3}{2} \frac{eqQ}{h} S(\theta)$$

The time scale τ of the experiment depends on the fraction of the frequency range of anisotropic interactions that is averaged out by motions. For simplicity, let us assume that this range is one-tenth of eqQ/h, $\Delta \approx 17$ kHz. This yields a time scale $\tau = (2\pi \cdot \Delta\nu)^{-1} \approx 10\,\mu s$. Bond librational motions ($\tau \approx 1 - 10$ ps), gauche/*trans* isomerization ($\tau \approx 100$ ps), diffusional motions about the long axis of the lipid, and molecular wobble ($\tau \approx 1$–10 ns) have shorter correlation times. Those motions reduce the quadrupole splitting to a new effective value

$$\Delta\nu_{eff}(\tau) = \frac{3}{2} \frac{eqQ}{h} \int_0^\tau \frac{1}{\tau} S(\theta(t)) dt$$

where the integral is the time average of the order parameter over the 10 μs.

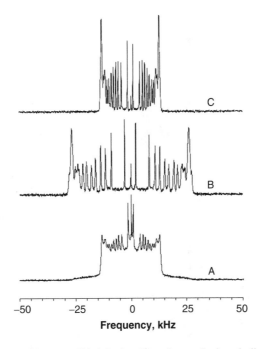

FIGURE 4.3 ^{2}H NMR spectra of 1-stearoyl(d_{35})-2-oleoyl-*sn*-glycero-3-phospholine. (A) Large, multilamellar liposomes with random orientation of bilayer normals to the magnetic field. (B) Oriented membranes with bilayer normal parallel to the magnetic field. (C) Oriented membranes with bilayer normal perpendicular to the field.

The motions also result in a new effective symmetry axis of quadrupolar interactions. Rotational diffusion and wobble have the longest correlation times. They are axially symmetric about the bilayer normal. This designates the bilayer normal as the new symmetry axis of fast molecular reorientation (director). An internal molecular order parameter $S(mol)_i$ for those motions can be defined such that $S(total)_I = S(mol)_i\,S(_{DL})$, where $S(mol)_i$ is the order parameter of a specific lipid C-^{2}H bond with respect to the bilayer normal. The order parameter $S(_{DL})$ reflects the orientation of the bilayer normal to the magnetic field. The latter order parameter is not related to motions. It only reflects the static distribution of bilayer orientations. The quadrupole splitting is maximal if the bilayer normal is oriented parallel to the magnetic field ($\theta_{DL} = 0°$). The splitting is zero at the magic angle ($\theta_{DL} = 54.7°$) and has half the maximal value if the bilayer normal is perpendicular to the field ($\theta_{DL} = 90°$).

In large multilamellar liposomes, the bilayers adopt a random orientation with respect to the magnetic field, resulting in a superposition of resonance lines with different quadrupolar splittings, a so-called powder pattern. The probability $p(\theta_{DL})$ to find a vector at θ_{DL} is

$$p(\theta_{DL}) = \sin(\theta_{DL})$$

indicating that it is much more likely in a random distribution to find a vector that is oriented perpendicular to the field than parallel. Therefore, quadrupole splittings for $\theta_{DL} = 90°$, corresponding to half the quadrupole splitting, have the highest intensity. In contrast to multilamellar liposomes, spectra of well-oriented membranes correspond to very narrow distribution functions. The resultant spectra are better resolved and have higher signal intensity (Figure 4.3).

4.2.3.1.2 NOESY Cross-Relaxation
Rates of relaxation are determined by the time dependence of interactions between the nuclei and their environment. Therefore, they are a direct measure of motional correlation times within a

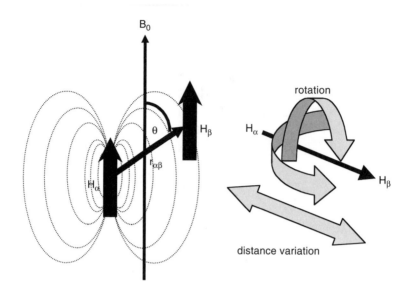

FIGURE 4.4 ¹H-¹H NOESY cross-relaxation rates between the hydrogens H_α and H_β depend on reorientation of the vector $r_{\alpha\beta}$ connecting both hydrogens and also on variation of vector length. The vector B_0 represents the magnetic field of the NMR instrument.

particular range. As an example, I will describe the principles of ¹H-¹H cross-relaxation in lipid bilayers as measured by nuclear Overhauser enhancement spectroscopy (NOESY) [22, 23].

NOESY cross-relaxation reports the rate of magnetization exchange between neighboring protons. The rates are a function of time-dependent interactions between the proton magnetic dipoles. All protons j surrounding the proton i contribute to a magnetic field strength at the site of i. This field is fluctuating if protons are moving. A fluctuating field of proper frequency triggers transitions between spin states, changing their lifetime.

The effects of magnetic dipole–dipole interactions on spins are expressed as dipolar correlation function

$$C_{ij}(t) = \frac{4}{5} \sum_i \sum_j \left\langle \frac{Y_{20}\left(\vec{r}_{ij}(0)\right)}{r_{ij}^3(0)} \frac{Y_{20}\left(\vec{r}_{ij}(t)\right)}{r_{ij}^3(t)} \right\rangle$$

where the angular dependent term $Y_{20}(r) = [5/16\pi]^{1/2}(3\ cos^2\theta - 1)$, and θ is the angle between the internuclear vector $\vec{r}_{ij}(t)$ and the normal to the membrane (see [24, 25]). The term $Y_{20}(r)$ contributes a time-dependent decay to the correlation function from the reorientation of the vector $\vec{r}_{ij}(t)$, whereas the expression in the denominator reflects the variation of vector length. In fluid bilayers, the order parameter and vector length are time-dependent, resulting in a decay of the correlation function with a time τ (Figure 4.4).

The spectral density of those fluctuations $J_{ij}(\omega)$ is the Fourier transform of the autocorrelation function

$$J_{ij}(\omega) = \int_0^\infty C_{ij}(t) \cdot \cos(\omega t) dt$$

It can be shown that the rates of NOESY cross-relaxation Γ_{ij} depend on the spectral densities $J(2\omega_0)$ and $J(0)$ according to $\Gamma_{ij} = \xi\left[3J_{ij}(2\omega_0) - 0.5J_{ij}(0)\right]$, where ξ is a constant and $\omega_0 = 2\pi\nu_0.$

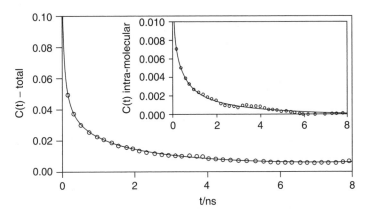

FIGURE 4.5 Correlation function of intra- and intermolecular dipolar interactions between the hydrogens at carbon C_2 of lipid hydrocarbon chains and the methyl protons of the choline headgroup. The insert shows the contribution from intramolecular interactions only. Correlation functions were calculated from a molecular simulation of a 1,2-dipalmitoyl-*sn*-glycero-3-phosphocholine bilayer. The solid line is a multiexponential fit to the MD data. (Reproduced from [23] with permission.)

The frequency v_0 is the Larmor precession frequency of protons at a magnetic field strength B_0 (e.g., 500 MHz at a magnetic field strength of 11.74 Tesla for protons). The value of the spectral density $J(2\omega_0)$ is dominated by fast motions with correlation times $\tau_1 = (2\omega_0)^{-1} \approx 320\,ps$ at 500 MHz. The spectral density $J(0)$ is determined by motions with correlation times much longer than the inverse of the Larmor frequency $\tau_2 >> \omega_0^{-1}$.

A typical correlation function of a dipolar interaction between two lipid resonances is shown in Figure 4.5. It was obtained from a molecular dynamics simulation. The function was approximated by superposition of exponentially decaying functions according to

$$C_{ij}(t) = \sum_{n=1}^{4} A_n \exp\left[-\frac{t}{\tau_n}\right]$$

where the coefficient A_n is the amplitude of the relaxation process (n) and τ_n its correlation time. The correlation times reflect bond librational motions (τ_1, picoseconds), gauche/*trans* isomerization ($\tau_2 = 50$ to 100 ps), diffusional rotation and wobble of lipids (τ_3, nanoseconds), and lateral diffusion (τ_4, hundreds of nanoseconds) [23, 26].

Note that the correlation function does not decay to zero over the observation time. This is because the distance between interacting protons $r_{ij}(t)$ is finite and order parameter $S(\theta_{ij}(t))$ in bilayers is not averaged to zero.

Approximation of the correlation function by a superposition of exponentials not only provides insight into motional correlation times, it also permits the evaluation of contributions from every motion to cross-relaxation rates. The Fourier transform of the correlation function

$$C(t) = \sum_{n} A_n \exp\left[-\frac{t}{\tau_n}\right]$$

yields the spectral density functions

$$J(2\omega_0)_n = \frac{2A_n\tau_n}{1 + 4\tau_n^2\omega_0^2}$$

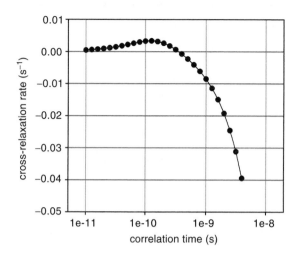

FIGURE 4.6 ^{1}H-^{1}H NOESY cross-relaxation rates as a function of motional correlation time at a proton resonance frequency of 500 MHz. The model assumes isotropic motions of the vector connecting the interacting protons. Cross-relaxation rates are positive at short correlation times of less than 300 ps and negative at longer correlation times. The absolute value of rates is scaled by the sixth power of the distance between interacting hydrogen atoms.

$$J(0)_n = 2A_n\tau_n$$

and, after substitution, a cross-relaxation rate

$$\Gamma = \sum_n \Gamma_n = \sum_n \xi \left[\frac{6A_n\tau_n}{1 + 4\tau_n^2\omega_0^2} - A_n\tau_n \right]$$

A plot of cross-relaxation rates as a function of the correlation time τ for an NMR instrument with a proton resonance frequency of 500 MHz is shown in Figure 4.6. The graph indicates that contributions to Γ from fast motions with correlations in the ps-range are positive, whereas contributions with correlation times of nanoseconds and longer are negative. Homonuclear NOESY experiments have the advantage that they distinguish between fast and slow motions by a change in the sign of cross-relaxation rates. Under favorable conditions, the sign change may be detected in experiments conducted as a function of temperature. If cross-relaxation is dominated by one motion, this change in sign is a very precise measure of the motional correlation time.

4.2.3.1.3 ^{13}C Spin-Lattice and Spin-Spin Relaxation

The spin-lattice relaxation times of ^{13}C nuclei are primarily determined by dipole interactions between carbons and directly bonded protons. Because the ^{13}C-^{1}H bond length is constant in good approximation, the decay of the correlation function is entirely determined by time-dependent changes of the bond order parameter $S(\theta_{CH}(t))$.

The spin lattice relaxation rates R_I depend on spectral density functions according to

$$R_1 = K\left[J(\omega_C - \omega_H) + 3J(\omega_C) + 6J(\omega_C + \omega_H) \right],$$

where K is a constant; $J(\omega_C - \omega_H)$, $J(\omega_C)$, and $J(\omega_C + \omega_H)$ are spectral density functions; and ω_C and ω_H are the Larmor precession frequencies of ^{13}C and ^{1}H nuclei, respectively [27]. On an NMR instrument with a proton resonance frequency of 500 MHz, ^{13}C spin-lattice relaxation is most sensitive to motions with correlation times of $\tau_{C-H} = |(\omega_C - \omega_H)^{-1}| \approx 420 ps$, $\tau_C = (\omega_C)^{-1} \approx 1.3 ns$, and $\tau_{C+H} = (\omega_C + \omega_H)^{-1} \approx 250 ps$, which covers the range of gauche/*trans* isomerization, rotational

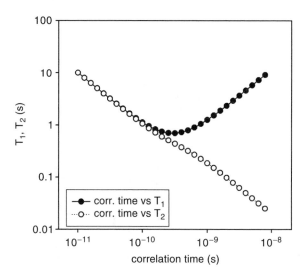

FIGURE 4.7 Spin-lattice (T_1) and spin-spin (T_2) relaxation times of ^{13}C nuclei as a function of motional correlation time at a carbon resonance frequency of 125 MHz (proton resonance frequency 500 MHz). It is assumed that the ^1H-^{13}C bond moves isotropically. Relaxation times are scaled by the bond length and the number of hydrogens directly bonded to the ^{13}C nucleus.

diffusion of lipids, and wobble. Experimental results are frequently reported as relaxation times T_1, which are the inverse of relaxation rates $T_1 = (R_1)^{-1}$.

The spin-spin relaxation rates R_2 cover not only motions with correlation times in the nanosecond range but also much slower motions represented by the spectral density function $J(0)$:

$$R_2 = \frac{K}{2}\left[4J(0) + J(\omega_C - \omega_H) + 3J(\omega_H) + 6J(\omega_C) + 6J(\omega_C + \omega_H)\right]$$

$$T_2 = (R_2)^{-1}$$

The correlation time dependence of T_1 and T_2 is shown in Figure 4.7. Spin-lattice and spin-spin relaxation times are identical if motional correlation times are dominated by very rapid motions like gauche/*trans* isomerization. However, contributions from slower motions with correlation times of nanoseconds and longer further decrease T_2, while T_1 values increase. The difference between T_1 and T_2 is a good measure of motional correlation times.

Similar derivations may be provided for other spin relaxation processes and for other nuclei. In particular, ^2N NMR relaxation of deuterated lipids has been extensively studied [28]. The resonance frequency of ^2H nuclei is 6.5-fold lower than for protons. Furthermore, most experiments have been conducted at a magnetic field strength of 7 Tesla or lower, resulting in highest sensitivity to lipid rotational diffusion and wobble with correlation times of the order of nanoseconds.

Relaxation measurements are particularly useful for the study of membrane dynamics, because they may cover a wide range of correlation times.

4.3 LIPID ORDER AND MOTIONS

4.3.1 PHOSPHOLIPID GLYCEROL AND HEADGROUP REGIONS

Order and dynamics of the glycerol and headgroup regions have been studied extensively by measurements of ^2H NMR quadrupolar interactions on specifically deuterated lipids and of the ^{31}P

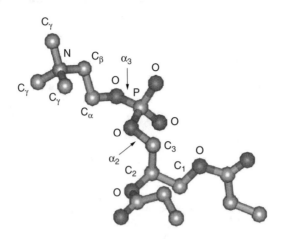

FIGURE 4.8 Schematic presentation of the phosphatidylglycerol polar region.

NMR anisotropy of chemical shift of lipid phosphate groups. The glycerol backbone is the least flexible region of phospholipids. The dihedral bond angle C_1-C_2 of glycerol has a preference for gauche+/gauche− isomers. Gauche/*trans* isomerization is severely restricted, as reflected by non-equivalence of chemical shifts and order parameters of the two hydrogens/deuterons at C_1 [29]. The relative rigidity of this glycerol bond is also seen in the difference between the orientation of *sn*-1 and *sn*-2 chain carbonyl groups measured by ^{13}C NMR [30].

The orientation of the glycerol C_2-C_3 bond is close to parallel to the lipid bilayer normal, and gauche/*trans* isomerization about this bond is not as constrained [31, 32]. It has been suggested that rapid isomerization about this bond is responsible for the decoupling of lipid headgroup and glycerol/hydrocarbon chain motions in gel-phase lipids. At temperatures below the fluid–gel phase transitions the headgroup motions continue, despite the crystalline packing of lipid hydrocarbon chains, as reflected in the ^{31}P NMR spectra of lipid headgroups and ^{2}H NMR spectra of glycerol backbone resonances [33–35].

The internal rigidity of the glycerol group of phospholipids is also reflected by the ^{2}H spin-lattice relaxation times of glycerol deuterons. Relaxation times are dominated by motions with correlation times of the order of nanoseconds, which are typical for rotational diffusion and wobble of the entire lipid molecule [33, 34] (Figure 4.8).

Order and motions of headgroups have been extensively studied for phosphatidylcholines (PC) and phosphatidylethanolamines (PE). The orientation of the headgroup electric dipole, formed by the negative charge at the phosphate group and the positive charge of choline/ethanolamine, is primarily determined by the dihedral bond angles α_2 and α_3 at the phosphorus atom. NMR and neutron diffraction results indicate that the headgroups have a preference for orienting parallel to the plane of the bilayer (perpendicular to the bilayer normal) [36–40]. This suggests a gauche-gauche conformation for α_2 and α_3, similar to the conformation of the phosphocholine headgroup in glycerophosphocholine crystals [41]. But headgroups in fluid membranes are highly flexible and other orientations occur as well, albeit with lower probability [42]. This relative orientational freedom of headgroups was confirmed by ^{2}H NMR on membranes with electric surface potentials. The order parameters of deuterons on α- and β-methylene groups of choline are in agreement with a headgroup model that has the positively charged choline group closer to the bilayer core when the potential is negative, and farther away for positive potentials. The effect was explained by an interaction of surface electric fields with the electric dipole moment of PC and PE headgroups [43]. Changes in the state of lipid hydration alter the headgroup orientation, as well [44]. NMR relaxation measurements on headgroups conducted as a function of water concentration indicated that the headgroup moves in the first approximation as one unit [45, 46]. Relaxation times of ^{2}H nuclei increase insignificantly from $^2H_2$-C_α to $^2H_2$-C_β but are much longer for $^2H_3$-C_γ. The small increase from C_α to C_β may indicate a tendency of faster bond isomerization/flickering motions at C_β, whereas the much longer relaxation times at C_γ reflect the rotational freedom that is inherent to all methyl groups.

4.3.2 LIPID HYDROCARBON CHAINS

^2H NMR experiments on phospholipids with specifically ^2H-labeled hydrocarbon chains revealed the existence of a distinct chain order profile. In chains linked to the *sn*-1 position of glycerol, order parameters are high for carbon atoms up to the middle of the chain and then decay rapidly toward the terminal methyl group [47–49]. The region of high chain order is referred to as the *order parameter plateau*. In contrast, chains linked to the *sn*-2 position have lower order parameters of C-^2H bonds at carbon C_2, and order of the two deuterons at C_2 is nonequivalent. The order profile of the remainder of the *sn*-2 chain resembles *sn*-1 chain order. The lower order at C_2 is a reflection of differences in chain orientation near the glycerol [50]. Whereas the entire *sn*-1 chain is, on average, parallel to the bilayer normal, the first segment of the *sn*-2 chain is tilted, similar to *sn*-2 chain packing in crystalline lipid [51].

Chain order profiles of saturated chains in all model and biomembranes are similar (see [28, 52] for a summary of results). The higher order of chains near the lipid/water interface is the consequence of anchoring one end of the flexible chains to the rigid glycerol group [53]. The methyl end of chains is permitted to move freely, resulting in the characteristic order parameter profile. Diffraction experiments on fluid membranes indicate that the projection of chain length on the bilayer normal is significantly less than expected for an extended chain because of gauche/*trans* isomerization and/or tilt [54].

In a fully extended, saturated chain with all C-C bonds in *trans* configuration, the C-^2H bonds are oriented perpendicular to the bilayer normal ($\theta_{CD} = 90°$, where θ_{CD} is the angle between orientation of the C-^2H bond and the bilayer normal). The corresponding chain order parameter is $S_{CD} = -0.5$ (see Figure 4.2). For comparison, experimentally measured order parameters in the order parameter plateau are $|S_{CD}| \approx 0.25$. The negative sign of the order parameter is usually not reported, because it cannot be determined from the symmetric ^2H NMR spectra. Chain order near the methyl terminal end is much lower, with values of $|S_{CD}| \approx 0.05$. It has been suggested that a higher density of gauche isomers in the lower half of the chain is primarily responsible for the decay in lipid order parameters.

But the decline in order could also reflect increasing chain tilt. Indeed, molecular simulations suggest that small deviations in dihedral bond angles from the preferred *trans* (180°), gauche$^+$ (+60°), or gauche$^-$ (-60°) orientations, as well as gauche conformers, may result in substantial tilt of chains with respect to the bilayer normal [42].

Not only do order parameters decrease toward the methyl groups of hydrocarbon chains, ^1H-^2H- and ^{13}C spin-lattice relaxation times increase significantly along the chain, with shorter values near the carbonyl and longest relaxation times at the terminal methyl group [17, 55–62]. This is a reflection of a gradient in chain dynamics with higher density of gauche conformers, larger amplitudes of bond librational motions, and, perhaps, shorter correlation times of those motions at the terminal methyl end. Those faster motions result in less effective spin-lattice relaxation. The data yield correlation times for gauche/*trans* isomerization in the range of 100 ps, as well as correlation times of lipid rotational diffusion about the long axis and wobble of the lipid in the nanosecond range. Data have been analyzed with much greater precision than the correlation time ranges given here. However, since these relaxation rates are the result of superimposed motions, interpretation is somewhat model-dependent (see, e.g., [63]).

4.3.2.1 Mono- and Polyunsaturated Hydrocarbon Chains

Most phospholipids in biomembranes carry an unsaturated chain at position *sn*-2 of glycerol. The fluid–gel phase transition temperature of lipids with unsaturated chains is about 50°C lower than transitions of saturated phospholipids with a comparable number of carbons per hydrocarbon chain [64, 65]. This has been frequently labeled as an indication of "higher fluidity" of membranes with unsaturated hydrocarbon chains. But it is doubtful that the phase transition temperature truly reflects order and motions of the biologically relevant fluid state. Most likely, phase transition temperatures are determined by differences in interactions between crystalline hydrocarbon chains that are much stronger than interactions in the fluid phase. Therefore, phase transition temperatures are a poor measure of lipid dynamics.

But order parameters in unsaturated, fluid membranes are indeed lower [49, 64]. As discussed earlier, lower order parameters are not necessarily the result of higher motional disorder. Indeed, the value of C-^{2}H bond order parameters of unsaturated carbons may be dominated by differences in bond geometry between saturated and unsaturated hydrocarbon chains. If a double bond is oriented parallel to the bilayer normal, then the two C-^{2}H bonds are at the magic angle to the normal, yielding an order parameter $S_{CD} = 0$, even without any motions.

^{2}H NMR experiments on deuterated lipids have shown that other chain order parameters of unsaturated lipids are lower, as well. This includes order parameters of methylene groups in the sn-2 chain near the double bond and order of saturated chains in position sn-1 [49, 64]. EPR studies with spin-labeled hydrocarbon chains [66] and fluorescence studies on membrane-incorporated labels [67], as well as infrared and Raman studies [68], also reported lower order and higher content of gauche isomers. The lower chain order was related to packing disorder from the bulky cis–double bonds of unsaturated chains. Lipid order parameters have a tendency to decrease with increasing unsaturation of the bilayer [64].

Among the polyunsaturated fatty acids, highest membrane concentrations are found for docosa-hexaenoic acid (DHA, 22:6n3), a polyunsaturated fatty acid with 22 carbons and six double bonds, separated by methylene groups (see Figure 4.9) [69]. DHA is found at low concentrations (several mol%) in membranes of most organs, but concentration is much higher in the brain frontal cortex (15 to 20 mol%, [70]). In mammalian synapses and the retina, but also in spermatozoids, 30 to 50 mol% of all phospholipid hydrocarbon chains are DHA. Insufficient supply of DHA or its ω-3 fatty acid precursors to the developing brain results in replacement of DHA by docosapentaenoic acid (DPA), a ω-6 fatty acid with the same number of carbon atoms that lacks the last double bond at the methyl terminal end (22:5n6) [71].

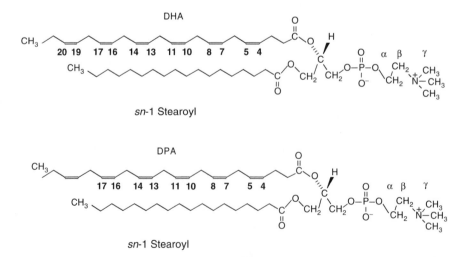

FIGURE 4.9 Chemical structure of polyunsaturated lipids. DHA: 1-stearoyl-2-docosahexaenoyl-sn-glycero-3-phosphocholine (18:0-22:6n3PC); DPA: 1-stearoyl-2-docosapentaenoyl-sn-glycero-3-phosphocholine (18:0-22:5n6PC).

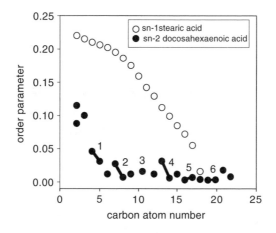

FIGURE 4.10 Order parameter profiles of stearic acid (18:0) and docosahexaenoic acid (DHA, 22:6n3) in 18:0-22:6n3PC. The carbons linked by the six double bonds of DHA are indicated by bars and numbers. The graph was prepared using data published in [72].

^{2}H- and ^{13}C NMR experiments on the phosphatidylcholines 18:0-22:6n3 PC and 18:0-22:5n6 PC with saturated stearic acid (18:0) in position *sn*-1, paired with the polyunsaturated acid DHA (22:6n3) or DPA (22:5n6) in *sn*-2, revealed surprising motional properties of polyunsaturated chains [72, 73]. Most of the ^{13}C resonances of unsaturated chains were resolved by magic-angle spinning (MAS) NMR [74]. Assignment of resonances was achieved by one- and two-dimensional MAS NMR experiments, and by comparison with results from quantum chemical and molecular simulations [72].

Order parameters of both DHA and DPA are much lower than order in saturated hydrocarbon chains (see Figure 4.10). Although the lower order of hydrogens/deuterons in double bonds may be related to double-bond geometry, the very low order parameters of the methylene groups between double bonds suggest that chain dynamics are a contributing factor. This was confirmed by ^{13}C NMR relaxation time studies (Figure 4.11).

The shortest relaxation times were measured for the carbons C_2 and C_3 next to the carbonyl/glycerol moiety. Relaxation times increase stepwise from one double bond to the next. The T_1 values

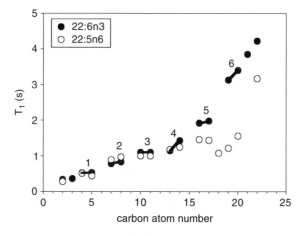

FIGURE 4.11 Spin-lattice relaxation times T_1 of docosahexaenoic acid (DHA, 22:6n3) and docosapentaenoic acid (DPA, 22:5n6) in 18:0-22:6n3PC and 18:0-22:5n6PC, respectively. The graph was prepared using data published in [72].

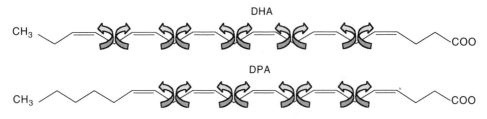

FIGURE 4.12 Location of C-C bonds in DHA and DPA polyunsaturated chains with low potential energy barriers for bond rotation.

of carbons at the methyl terminal end, including double bond number 6, are comparable to values of hydrocarbon chains dissolved in liquids. The relaxation results clearly point to the presence of rapid motions, in particular near the methyl group of the polyunsaturated chains. This is surprising, because double bonds in polyunsaturated chains reduce the number of degrees of freedom. However, the stepwise increase in T_1 from double bond to double bond suggests that isomerization about the C-C bonds to the methylene carbons, sandwiched by double bonds, is very rapid.

Quantum chemical calculations revealed the cause of those high isomerization rates [75]. Although the C-C bonds of saturated chains have lowest energy in *trans*/gauche+/gauche- conformations, the two dihedral bond angles sandwiched between double bonds (=CH-CH$_2$-CH=) prefer skew+ (120°)/skew- (120°) (see Figure 4.13) [73]. Transitions between *trans*/gauche+/ gauche- require passing energy barriers of ≈ 3 kcal/mol, which is sixfold higher than a unit of thermal energy (kT ≈ 0.6 kcal/mol) at ambient temperature. In contrast, the energy barrier between skew+/skew- is only 0.8 kcal/mole, permitting vary rapid isomerization about the C-C bonds between the olefinic carbons of DHA and DPA (see Figure 4.12). Therefore, it can be concluded that polyunsaturated chains themselves are exceptionally flexible, with rapid structural transitions between large numbers of conformations.

The significant difference in chain dynamics between DHA and DPA at the methyl terminal end is related to the distribution of flexible bonds over the polyunsaturated chains. DPA is an ω-6 fatty acid with a stretch of four methylene segments near the terminal methyl group, in contrast to DHA, an ω-3 fatty acid with a single methylene group. The C-C bonds between methylene segments have much higher potential barriers for chain isomerization than C-C bonds linking double bonds. This influences chain dynamics not only at the terminal methyl end, but also higher up in the polyunsaturated chains. Remarkably, we not only observed differences in motional correlation times but also a significant difference in chain density distribution along the membrane normal between DHA and DPA. We speculate that this has consequences for integral membrane protein function [72, 73] (Figure 4.13).

4.3.3 PACKING DISORDER IN THE LIPID MATRIX

The lipid matrix of biomembranes is usually depicted as a bilayer of well-aligned molecules. Hydrocarbon chains are drawn as wavy lines with higher disorder in the bilayer center to account for the fluid nature of lipids. Unfortunately, most cartoons grossly underrepresent the level of packing disorder in fluid bilayers. The results from NOESY experiments, but also X-ray and neutron diffraction data, indicate much higher levels of packing disorder. A model that correctly reproduces experimental observations is the snapshot of a 1,2-dipalmitoyl-*sn*-glycero-3-phosphocholine bilayer obtained from a molecular simulation in Figure 4.14 [76].

The image shows that hydrocarbon chains bend and tilt. Terminal methyl groups of chains are widely distributed over the hydrophobic core of the bilayer. The lipid water interface is a ≈10 Å-wide band with rugged edges on both sides of the hydrophobic core. It comprises lipid headgroups, glycerol, and chain carbonyls, as well as water molecules interdispersed between lipid segments.

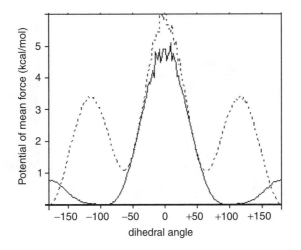

FIGURE 4.13 Potential of mean force for rotation about single bonds located between double bonds (solid line) and rotation about bonds in saturated hydrocarbon chains (dashed line). The energy minima correspond to conformations skew⁺ (120°)/skew⁻ (−120°) for the solid line and gauche⁺ (60°)/gauche⁻ (−60°)/*trans*(180°) for the dashed line.

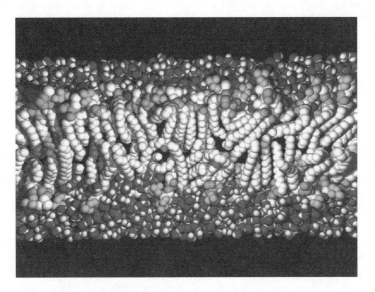

FIGURE 4.14 Snapshot from a molecular dynamics trajectory of a 1,2-dipalmitoyl-*sn*-glycero-3-phospho-choline bilayer. The image shows disordered hydrocarbon chains in the center, flanked by layers of polar headgroups and water on both sides. The black areas in the bilayer center are visualization artifacts and should not be interpreted as void volumes. (Reproduced from [76] with permission.)

The ¹H MAS NOESY experiments reflect packing disorder in membranes by unexpected magnetization transfer over what appear to be excessive distances [22]. The rates of magnetization transfer between lipid protons vary by up to two orders of magnitude, but rarely are too weak to be detected. Even protons that are, on average, at distances of 15 Å from each other (e.g., the choline of PC and methyl groups of hydrocarbon chains) exchange measurable amounts of magnetization. This is surprising, because the strength of magnetic dipole interactions depends on the sixth power of the distance between protons. Crosspeaks usually vanish if distances exceed 5 Å. In experiments on deuterated lipids, we established that these "long-range" transfers of magnetization take place between neighboring lipid molecules. Transfer of magnetization in multiple steps,

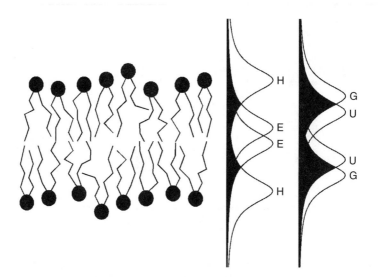

FIGURE 4.15 The probability of interactions between protons depends on the overlap of the spatial distribution functions of lipid segments. Two hypothetical distributions are shown on the right side. Because of intermolecular disorder, the glycerol (G) and upper chain (U) regions of lipids significantly overlap in distribution functions, which results in high cross-relaxation rates between hydrogens. Much less overlap in location occurs between the headgroup hydrogens (H) and the hydrogens at the end of hydrocarbon chains (E), which translates into low cross-relaxation rates. (Reproduced from [*23*] with permission.)

called *spin diffusion*, has low efficiency in fluid bilayers and does not account for the unexpected result [*77*].

Therefore, cross-relaxation rates in lipid bilayers do reflect a low but finite probability of close approach between normally distant protons. Such encounters are the result of packing disorder in combination with rapid motions. Protons that are located at the same depth in the bilayer, like protons of glycerol and protons of upper chain segments, meet each other more frequently. The corresponding rates of magnetization exchange are high. In contrast, protons of lipid headgroups and terminal methyl groups of hydrocarbon chains have a very low probability of interaction (see Figure 4.15). The corresponding cross-relaxation rates are low [*22*].

It is worth mentioning that this cross-relaxation model is radically different from cross relaxation in the more rigid proteins, where rates reflect the different but fixed distances between protons. Both models, encounter by chance vs. fixed distances, are an oversimplification. In reality, cross relaxation in lipids is modulated not only by distance variation of the vector connecting protons but also by reorientation of this vector, and cross relaxation in proteins is not solely a function of the correlation time of vector reorientation, but may also reflect some distance variation between protons.

4.3.4 LATERAL DIFFUSION

Lipid diffusion is studied by NMR [*78*] and fluorescence techniques [*79*]. Rapid lateral diffusion with rates in the range of 10^{-12} to 10^{-11} m²/s is one of the hallmarks of a fluid lipid bilayer. If diffusion in the two-dimensional membranes is unrestricted, then lipids travel, on average, over the distance $x = \sqrt{4Dt}$ from their origin during the observation time t. The symbol D is the diffusion constant. Lipids travel over distances of ≈10 μm within 1 s, equivalent to the typical dimension of cells in mammalian tissue. However, this estimate is based on the assumption that lipids do not encounter any barriers during their journey over the cell surface. Because membranes compartmentalize into domains with different lipid/protein composition, this is very unlikely. Barriers for diffusion may be formed by integral and peripheral membrane proteins, as well as an inherent tendency for lipid mixtures to separate, particularly in the presence of cholesterol [80].

The lateral heterogeneity of membranes introduces an apparent distance dependence of diffusion rates. Diffusion over short distances is less likely to be impeded by barriers than diffusion over larger distances. Therefore, diffusion rates have a tendency to be smaller if measured for travel of lipids over larger distances. Every experiment operates on a specific distance scale: measurement of lateral diffusion via analysis of fluorescence excimer formation or exchange broadening of the signal of EPR labels requires molecular collision between labels; NOESY NMR experiments report magnetization transfer over distances < 1 nm; fluorescence resonance energy transfer (FRET) is measured over distances < 10 nm; NMR experiments with pulsed magnetic field gradients (PFG-NMR) are sensitive to lipid diffusion over distances in the range of 100 nm to 10 μm, depending on gradient strength; and measurements by fluorescence recovery after photo-bleaching (FRAP) are sensitive to distances ≈1 μm. The difference between rates measured as a function of distance contains information about domain size and shape, as well as domain properties.

4.3.5 LIPID FLIP-FLOP

The movement of lipids between the two monolayers of a bilayer requires a translational movement along the bilayer normal and a rotation by 180°. The latter was the reason for naming such movement *lipid flip-flop*. Its rates have been measured by EPR [81], NMR [82], and fluorescence techniques [83]. The rate is reported as the time for movement of 50% of lipids from one monolayer to the other. In fluid model membranes composed of lipids with saturated hydrocarbon chains and no protein, this half-time of flip-flop is days to hours. This is not surprising, because such transbilayer movement of phospholipids requires that the hydrophilic headgroup traverses the hydrophobic bilayer core, a very unlikely event. In contrast, the energy barrier of molecules with less distinct polarity, like cholesterol and fatty acids, is much lower. They have half-times of flip-flop from seconds to milliseconds [84].

Typical rates of phospholipid flip-flop in biomembranes are much higher than in model membranes. The differences have been assigned to the action of flippases, proteins that actively or passively facilitate transbilayer movement of lipids [85].

4.3.6 MEMBRANE UNDULATIONS AND PROTRUSION OF LIPIDS

The study of repulsive interactions between lipid bilayers [86] — but also quantitative flicker analysis of liposome shape changes [87], X-ray diffraction experiments [88], and NMR relaxation studies [89, 90] — indicates that bilayers undergo thermally driven undulations with correlation times that cover more than six orders of magnitude from a fraction of a microsecond to seconds. The corresponding wavelength of such undulations covers nano- to micrometers. Amplitudes and frequencies of membrane fluctuations are directly related to membrane elastic properties, viscosity of the surrounding solution, and the distance to other membranes or a solid support. A variety of cellular processes are related to membrane undulations, such as membrane fusion, exo- and endocytosis, cell–cell interaction, adhesion of cells to solid surfaces.

Membrane undulations are collective movements of the entire bilayer that are distinct from the movement of individual lipids or lipid segments along the bilayer normal. The latter is called *protrusion*. There is evidence from incoherent quasielastic neutron scattering [91] that some protrusion of molecules and/or molecular segments takes place. Protrusion of individual lipids would make a large contribution to disjoining pressure between bilayers if movement of neighboring lipid molecules along the bilayer normal is statistically independent [92]. However, analysis of repulsive interactions between membranes conducted as a function of osmotic stress suggests that such interactions may be important only when the headgroup layers of apposing membranes start to penetrate each other [93]. Therefore, it is reasonable to assume that the amplitude of protrusions is small. Most likely, these are lipid headgroup movements. Protrusion of the entire lipid could be restricted because of the necessity to correlate movements between neighboring lipids in the membrane.

4.3.7 DYNAMICS OF WATER OF HYDRATION

The stability of the lipid bilayer as a matrix of biomembranes is intrinsically linked to surface hydration. It has many functional and morphological consequences [94]. The first layer of water at the membrane/water interface forms hydrogen bonds with phosphate-, carbonyl-, carboxyl, amino-, and hydroxyl groups. Similar to hydrogen bonds between water molecules, the lifetime of those bonds is very short.

Nevertheless, those temporary interactions with the lipid matrix introduce some degree of anisotropy into the motions of water molecules. This anisotropy is easily detected as ^{2}H-NMR quadrupole splitting of deuterated water in large multilamellar liposomes [95, 96]. The quadrupole splitting decreases with increasing water content. In homogeneous samples, only one splitting is measured for all water molecules, indicating that on the characteristic time scale of the experiment $\approx 10^{-4}$ s all water molecules are in rapid exchange between the different environments. At water concentrations of ≈ 5 waters/lipid or less, all water molecules form hydrogen bonds directly with the polar groups of the lipid, and quadrupolar splittings are in the range from 1 to 3 kHz. Considering the quadrupolar coupling constant of water, $e^2qQ/h = 220kHz$, this corresponds to rather low order parameters, S = 0.006 – 0.018. Such low order appears to be in conflict with the well-documented contribution of the first hydration shell to the electrical membrane dipole potential [97]. Consequently, we suggested that low apparent order may result from rapid motions of water molecules about an axis that is the bisection of the ^{2}H-O-^{2}H bond angle of water (109°). The O-^{2}H bonds of water are oriented at an angle of 54.5° to the bisection, almost the magic angle at which all quadrupolar splittings vanish. Quadrupole splittings of a second layer of water molecules, which forms hydrogen bonds with the first water layer, are ≈ 0.5 kHz, corresponding to order parameters S ≈ 0.003. Quadrupole splittings of all consecutive water layers are indistinguishable from zero, indicating that motional properties correspond to those of free water.

I would like to emphasize that separate water layers exist only on the picosecond time scale. Water molecules exchange rapidly between environments on the time scale of nanoseconds. The residence time of membrane-associated water is measured by spin-lattice and NOESY cross-relaxation experiments. Relaxation times are determined by two motions: the fast rotational reorientation of water molecules (with correlation times of picoseconds) and the slower distance variation between protons from diffusion of water molecules. Rates of diffusion near the membrane surface are controlled by formation of hydrogen bonds between water molecules and polar groups of lipids. The slower motions near the membrane surface reduce T_1 relaxation times of water hydrogens or deuterons. It has been estimated that the lifetime of hydrogen bonds at interfaces is of the order of 100 ps [98].

The magnitude and sign of NOESY crosspeaks between lipid and water protons is a good measure of water residence time in the first hydration shell, as well. NOESY crosspeaks between protons of 1-palmitoyl-2-oleoyl-*sn*-glycero-3-phosphocholine (POPC) and water are weak and negative [99, 99], in contrast to the stronger, positive crosspeaks between lipid protons. According to derivations in Section 4.2.3.1.2, negative crosspeaks (positive cross-relaxation rates) correspond to motional correlation times of the order of 100 ps, in good agreement with estimates derived from spin-lattice relaxation times of membrane-bound water (Figure 4.16).

4.4 SUMMARY

The fluid nature of the lipid matrix of biomembranes is reflected by motions beginning with bond vibrations on the subpicosecond range and ending with slow, undulatory motions of the membrane

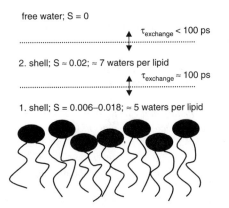

free water; S = 0

$\tau_{exchange}$ < 100 ps

2. shell; S ≈ 0.02; ≈ 7 waters per lipid

$\tau_{exchange}$ ≈ 100 ps

1. shell; S = 0.006–0.018; ≈ 5 waters per lipid

FIGURE 4.16 Schematic presentation of membrane hydration shells. Water molecules have highest order parameters in the first hydration shell. Order in the second shell is significantly lower. Water molecules exchange rapidly between the hydration shells and free water.

with correlation times of up to seconds. The fast dynamics of lipid molecules in the correlation time range from femto- to nanoseconds are now reasonably well understood. Recent molecular simulations are in good agreement with spectroscopic observations [23, 72, 75, 100]. The principal modes of slower motions with correlation times from a fraction of a microsecond to seconds have been determined, as well. Much has been learned about those slower motions from ^{2}H NMR order parameter studies that operate on a time scale of 10^{-5} to 10^{-4} s. However, a lack of knowledge about slower motions still challenges both experimenters and theoreticians.

REFERENCES

1. Luzzati, V., Reiss-Husson, F., Rivas, E., and Gulik-Krzywicki, T., Structure and polymorphism in lipid-water systems, and their possible biological implications. *Ann. N. Y. Acad. Sci.* 137, 409, 1966.
2. Luzzati, V., Gulik-Krzywicki, T., and Tardieu, A., Polymorphism of lecithins. *Nature* 218, 1031, 1968.
3. Luzzati, V., Gulik-Krzywicki, T., Rivas, E., Reiss-Husson, F., and Rand, R.P., X-ray study of model systems: structure of the lipid-water phases in correlation with the chemical composition of the lipids. *J. Gen. Physiol.* 51, Suppl: 37S+, 1968.
4. Reiss-Husson, F. and Luzzati, V., Phase transitions in lipids in relation to the structure of membranes. *Adv. Biol. Med. Phys.* 11, 87, 1967.
5. Singer, S.J. and Nicolson, G.L., The fluid mosaic model of the structure of cell membranes. *Science* 175, 720, 1972.
6. Pastor, R.W., Venable, R.M., and Karplus, M., Model for the structure of the lipid bilayer. *Proc. Natl. Acad. Sci. U. S. A.* 88, 892, 1991.
7. Pastor, R.W. and Feller, S.E., Time scales of lipid dynamics and molecular dynamics. In *Biological Membranes: A Molecular Perspective from Computation and Experiment.* Eds. Merz, K.M. and Roux, B. Birkhauser, Boston. 1996, 3.
8. Feller, S.E., Molecular dynamics simulation of phospholipid bilayers. In *Lipid Bilayers.* Eds. Katsaras, J. and Gutberlet, T. Springer, New York. 2000, 89.
9. Maier, E. and Saupe, A., *Z. Naturforsch.* 13a, 564, 1958.
10. Maier, E. and Saupe, A., *Z. Naturforsch.* 14a, 882, 1959.
11. Maier, E. and Saupe, A., *Z. Naturforsch.* 15a, 187, 1960.
12. Emsley, J.W., *Nuclear Magnetic Resonance of Liquid Crystals.* Reidel Publishing Company, Dordrecht, Holland. 1985.
13. Batchelor, G. K., *An Introduction to Fluid Dynamics.* Cambridge University Press, Cambridge, U.K. 1967.
14. Gawrisch, K. and Holte, L.L., NMR investigations of non-lamellar phase promoters in the lamellar phase state. *Chem. Phys. Lipids* 81, 105, 1996.

15. Mitchell, D.C., Lawrence, J.T.R., and Litman, B.J., Primary alcohols modulate the activation of the G-protein-coupled receptor rhodopsin by a lipid-mediated mechanism. *J. Biol. Chem.* 271, 19033, 1996.

16. Gaffney, B.J. and McConnell, H.M., Paramagnetic-resonance spectra of spin labels in phospholipid membranes. *J. Magn. Reson.* 16, 1, 1974.

17. Petersen, N.O. and Chan, S.I., More on the motional state of lipid bilayer membranes: interpretation of order parameters obtained from nuclear magnetic resonance experiments. *Biochemistry* 16, 2657, 1977.

18. Luckhurst, G.R. In *Liquid Crystals and Plastic Crystals*. Ellis Horwood, Chichester, U.K. 1974.

19. Abragam, A. *The Principles of Nuclear Magnetism*. Oxford University Press, New York. 1961.

20. Burnett, L.J. and Muller, B.H., Deuteron quadrupole coupling constants in three solid deuterated paraffin hydrocarbons: $C_{2D6, C4D10, C6D14}$. *J. Chem. Phys.* 55, 5829, 1971.

21. Gall, C.M., Diverdi, J.A., and Opella, S.J., Phenylalanine ring dynamics by solid-state ^{2}H NMR. *J. Am. Chem. Soc.* 103, 5039, 1981.

22. Huster, D., Arnold, K., and Gawrisch, K., Investigation of lipid organization in biological membranes by two-dimensional nuclear Overhauser enhancement spectroscopy. *J. Phys. Chem. B* 103, 243, 1999.

23. Feller, S.E., Huster, D., and Gawrisch, K., Interpretation of NOESY cross-relaxation rates from molecular dynamics simulation of a lipid bilayer. *J. Am. Chem. Soc.* 121, 8963, 1999.

24. Brüschweiler, R. and Wright, P.E., Water self-diffusion model for protein-water NMR cross-relaxation. *Chem. Phys. Lett.* 229, 75, 1994.

25. Brüschweiler, R. and Case, D.A., Characterization of biomolecular structure and dynamics by NMR cross-relaxation. *Prog. NMR Spectrosc.* 26, 27, 1994.

26. Yau, W.M. and Gawrisch, K., Lateral lipid diffusion dominates NOESY cross-relaxation in membranes. *J. Am. Chem. Soc.* 122, 3971, 2000.

27. Harris, R.K., *Nuclear Magnetic Resonance Spectroscopy. A Physicochemical View.* Pitman Publishing, Marshfield, MA. 1983.

28. Davis, J.H., The description of membrane lipid conformation, order and dynamics by ^{2}H-NMR. *Biochim. Biophys. Acta* 737, 117, 1983.

29. Allegrini, P.R., Pluschke, G., and Seelig, J., Cardiolipin conformation and dynamics in bilayer-membranes as seen by deuterium magnetic-resonance. *Biochemistry* 23, 6452, 1984.

30. Braachmaksvytis, V.L.B. and Cornell, B.A., Chemical-shift anisotropies obtained from aligned egg-yolk phosphatidylcholine by solid-state C-13 nuclear magnetic resonance. *Biophys. J.* 53, 839, 1988.

31. Seelig, J., Gally, G.U., and Wohlgemuth, R., Orientation and flexibility of the choline head group in phosphatidylcholine bilayers. *Biochim. Biophys. Acta* 467, 109, 1977.

32. Kohler, S.J. and Klein, M.P., Orientation and dynamics of phospholipid head groups in bilayers and membranes determined from 31P nuclear magnetic resonance chemical shielding tensors. *Biochemistry* 16, 519, 1977.

33. Auger, M., Vancalsteren, M.R., Smith, I.C.P., and Jarrell, H.C., Glycerolipids — common features of molecular motion in bilayers. *Biochemistry* 29, 5815, 1990.

34. Auger, M., Carrier, D., Smith, I.C.P., and Jarrell, H.C., Elucidation of motional modes in glycoglycerolipid bilayers — an H-2 NMR relaxation and line-shape study. *J. Am. Chem. Soc.* 112, 1373, 1990.

35. Frye, J., Albert, A.D., Selinsky, B.S., and Yeagle, P.L., Cross polarization P-31 nuclear magnetic resonance of phospholipids. *Biophys. J.* 48, 547, 1985.

36. Shepherd, J.C. and Büldt, G., Zwitterionic dipoles as a dielectric probe for investigating head group mobility in phospholipid membranes. *Biochim. Biophys. Acta* 514, 83, 1978.

37. Seelig, J. and Gally, H., Investigation of phosphatidylethanolamine bilayers by deuterium and phosphorus-31 nuclear magnetic resonance. *Biochemistry* 15, 5199, 1976.

38. Yeagle, P.L., Hutton, W.C., Huang, C.H., and Martin, R.B., Structure in the polar head region of phospholipid bilayers: a 31P [1H] nuclear Overhauser effect study. *Biochemistry* 15, 2121, 1976.

39. Griffin, R.G., Powers, L., and Pershan, P.S., Head-group conformation in phospholipids: a phosphorus-31 nuclear magnetic resonance study of oriented monodomain dipalmitoylphosphatidylcholine bilayers. *Biochemistry* 17, 2718, 1978.

40. Buldt, G., Gally, H.U., Seelig, J., and Zaccai, G., Neutron diffraction studies on phosphatidylcholine model membranes. I. Head group conformation. *J. Mol. Biol.* 134, 673, 1979.

41. Sundaralingam, M., Discussion paper: molecular structures and conformations of the phospholipids and sphingomyelins. *Ann. N. Y. Acad. Sci.* 195, 324, 1972.

42. Frischleder, H. and Peinel, G., Quantum-chemical and statistical calculations on phospholipids. *Chem. Phys. Lipids* 30, 121, 1982.

43. Scherer, P.G. and Seelig, J., Electric charge effects on phospholipid headgroups. Phosphatidylcholine in mixtures with cationic and anionic amphiphiles. *Biochemistry* 28, 7720, 1989.

44. Bechinger, B. and Seelig, J., Conformational changes of the phosphatidylcholine headgroup due to membrane dehydration. A 2H-NMR study. *Chem. Phys. Lipids* 58, 1, 1991.

45. Ulrich, A.S. and Watts, A., Lipid headgroup hydration studied by H-2-NMR — a link between spectroscopy and thermodynamics. *Biophys. Chem.* 49, 39, 1994.

46. Ulrich, A.S. and Watts, A., Molecular response of the lipid headgroup to bilayer hydration monitored by 2H-NMR. *Biophys. J.* 66, 1441, 1994.

47. Seelig, A. and Seelig, J., The dynamic structure of fatty acyl chains in a phospholipid bilayer measured by deuterium magnetic resonance. *Biochemistry* 13, 4839, 1974.

48. Seelig, A. and Seelig, J., Effect of a single cis double bond on the structures of a phospholipid bilayer. *Biochemistry* 16, 45, 1977.

49. Seelig, J. and Waespe-Sarcevic, N., Molecular order in cis and trans unsaturated phospholipid bilayers. *Biochemistry* 17, 3310, 1978.

50. Seelig, A. and Seelig, J., Bilayers of dipalmitoyl-3-sn-phosphatidylcholine. Conformational differences between the fatty acyl chains. *Biochim. Biophys. Acta* 406, 1, 1975.

51. Hitchcock, P.B., Mason, R., Thomas, K.M., and Shipley, G.G., Structural chemistry of 1,2 dilauroyl-dl-phosphatidylethanolamine: molecular conformation and intermolecular packing of phospholipids. *Proc. Natl. Acad. Sci. U. S. A.* 71, 3036, 1974.

52. Seelig, J. and Seelig, A., Lipid conformation in model membranes and biological membranes. *Q. Rev. Biophys.* 13, 19, 1980.

53. de Gennes, P.G., General features of lipid organization. *Phys. Lett.* 47A, 123, 1974.

54. Nagle, J.F. and Tristram-Nagle, S., Structure of lipid bilayers. *Biochim. Biophys Acta* 1469, 159, 2000.

55. Brown, M.F., Seelig, J., and Häberlen, U., Structural dynamics in phospholipid bilayers from deuterium spin-lattice relaxation time measurements. *J. Chem. Phys.* 70, 5045, 1979.

56. Brown, M.F. and Davis, J.H., Orientation and frequency-dependence of the deuterium spin-lattice relaxation in multilamellar phospholipid dispersions — implications for dynamic models of membrane structure. *Chem. Phys. Lett.* 79, 431, 1981.

57. Brown, M.F., Ribeiro, A.A., and Williams, G.D., New view of lipid bilayer dynamics from 2H and 13C NMR relaxation time measurements. *Proc. Natl. Acad. Sci. U. S. A.* 80, 4325, 1983.

58. Meier, P., Ohmes, E., and Kothe, G., Multipulse dynamic nuclear-magnetic-resonance of phospholipid membranes. *J. Chem. Phys.* 85, 3598, 1986.

59. Morrison, C. and Bloom, M., Orientation dependence of ^{2}H nuclear magnetic resonance spin-lattice relaxation in phospholipid and phospholipid-cholesterol systems. *J. Chem. Phys.* 101, 749, 1994.

60. Rommel, E., Noack, F., Meier, P., and Kothe, G., Proton spin relaxation dispersion studies of phospholipid membranes. *J. Phys. Chem.* 92, 2981, 1988.

61. Mayer, C., Grobner, G., Muller, K., Weisz, K., and Kothe, G., Orientation-dependent deuteron spin-lattice relaxation times in bilayer-membranes — characterization of the overall lipid motion. *Chem. Phys. Lett.* 165, 155, 1990.

62. Mayer, C., Muller, K., Weisz, K., and Kothe, G., Deuteron NMR relaxation studies of phospholipid membranes. *Liquid Crystals* 3, 797, 1988.

63. Brown, M.F. and Chan, S I., Bilayer membranes: deuterium & carbon-13 NMR. In *Encyclopedia of Nuclear Magnetic Resonance*. Eds. Gran, D.M. and Harris, R.K. Wiley, New York. 1995, 871.

64. Holte, L.L., Peter, S.A., Sinnwell, T.M., and Gawrisch, K., ^{2}H nuclear magnetic resonance order parameter profiles suggest a change of molecular shape for phosphatidylcholines containing a polyunsaturated acyl chain. *Biophys. J.* 68, 2396, 1995.

65. Niebylski, C.D. and Salem, N., Jr., A calorimetric investigation of a series of mixed-chain polyunsaturated phosphatidylcholines: effect of sn-2 chain length and degree of unsaturation. *Biophys. J.* 67, 2387, 1994.

66. Marsh, D. Electron spin resonance: spin labels. In *Membrane Spectroscopy*. Ed. Grell, E. Springer, Berlin. 1981, 51.

67. Heyn, M.P., Order and viscosity of membranes — analysis by time-resolved fluorescence depolarization. *Methods Enzymol.* 172, 462, 1989.

68. Mantsch, H.H., Casal, H.L, and Jones, R.N. In *Spectroscopy of Biological Systems*. Eds. Clark, R.J.H. and Hester, R.E. Wiley, New York. 1986, 1.

69. Salem, N. Omega-3 fatty acids: molecular and biochemical aspects. In *New Protective Roles for Selected Nutrients*. Eds. Spiller, G.A. and Scala, A. Alan R. Liss, Inc., New York. 1989, 109.

70. Moriguchi, T. and Salem, N., Recovery of brain docosahexaenoate leads to recovery of spatial task performance. *J. Neurochem.* 87, 297, 2003.

71. Mohrhauer, H. and Holman, R.T., Alteration of fatty acid composition of brain lipids by varying levels of dietary essential fatty acids. *J. Neurochem.* 10, 523, 1963.

72. Eldho, N.V., Feller, S.E., Tristram-Nagle, S., Polozov, I.V., and Gawrisch, K., Polyunsaturated docosahexaenoic vs. docosapentaenoic acid — differences in lipid matrix properties from the loss of one double bond. *J. Am. Chem. Soc.* 125, 6409, 2003.

73. Gawrisch, K., Eldho, N.V., and Holte, L.L., The structure of DHA in phospholipid membranes. *Lipids* 38, 445, 2003.

74. Yeagle, P.L. and Frye, J., Effects of unsaturation on ^{2}H-NMR quadrupole splittings and ^{13}C-NMR relaxation in phospholipid bilayers. *Biochim. Biophys. Acta* 899, 137, 1987.

75. Feller, S.E., Gawrisch, K., and MacKerell, A.D., Polyunsaturated fatty acids in lipid bilayers: intrinsic and environmental contributions to their unique physical properties. *J. Am. Chem. Soc.* 124, 318, 2002.

76. Feller, S.E., Venable, R.M., and Pastor, R.W., Computer simulation of a DPPC phospholipid bilayer: structural changes as a function of molecular surface area. *Langmuir* 13, 6555, 1997.

77. Huster, D. and Gawrisch, K., NOESY NMR crosspeaks between lipid headgroups and hydrocarbon chains: spin diffusion or molecular disorder? *J. Am. Chem. Soc.* 121, 1992, 1999.

78. Lindblom, G. and Orädd, G., NMR studies of translational diffusion in lyotropic liquid crystals and lipid membranes. *Prog. NMR Spectrosc.* 26, 483, 1994.

79. Jovin, T.M. and Vaz, W.L.C., Rotational and translational diffusion in membranes measured by fluorescence and phosphorescence methods. *Methods in Enzymology* 172, 471, 1989.

80. Anderson, R.G.W. and Jacobson, K., Cell biology — a role for lipid shells in targeting proteins to caveolae, rafts, and other lipid domains. *Science* 296, 1821, 2002.

81. Kornberg, R.D. and McConnell, H.M., Inside-outside transitions of phospholipids in vesicle membranes. *Biochemistry* 10, 1111, 1971.

82. de Kruijff, B. and Wirtz, K.W.A., Induction of a relatively fast transbilayer movement of phosphatidylcholine in vesicles. *Biochim. Biophys Acta* 255, 311, 1977.

83. Doody, M.C., Pownall, H.J., Kao, Y.J., and Smith, L.C., Mechanism and kinetics of transfer of a fluorescent fatty-acid between single-walled phosphatidylcholine vesicles. *Biochemistry* 19, 108, 1980.

84. Hamilton, J.A., Fast flip-flop of cholesterol and fatty acids in membranes: implications for membrane transport proteins. *Curr. Opin. Lipidol.* 14, 263, 2003.

85. Devaux, P.F., Lipid transmembrane asymmetry and flip-flop in biological membranes and in lipid bilayers. *Curr. Opin. Struct. Biol.* 3, 489, 1993.

86. Helfrich, W., Steric interaction of fluid membranes in multilayer systems. *Z. Naturforsch.* 33a, 305, 1978.

87. Faucon, J.F., Mitov, M.D., Méléard, P., Bivas, I., and Bothorel, P., Bending elasticity and thermal fluctuations of lipid membranes. Theoretical and experimental requirements. *J. Phys. France* 50, 2389, 1989.

88. Nagle, J.F. and Tristram-Nagle, S. Structure and interactions of lipid bilayers: role of fluctuations. In *Lipid Bilayers*. Eds. Katsaras, J. and Gutberlet, T. Springer-Verlag, New York. 2001, 1.

89. Bloom, M. and Evans, E. Observation of surface undulations on the mesoscopic length scale by NMR. In *Biologically Inspired Physics*. Ed. Peliti, L. Plenum Press, New York. 1991, 137.

90. Stohrer, J., Grobner, G., Reimer, D., Weisz, K., Mayer, C., and Kothe, G., Collective lipid motions in bilayer-membranes studied by transverse deuteron spin relaxation. *J. Chem. Phys.* 95, 672, 1991.

91. Konig, S., Bayerl, T.M., Coddens, G., Richter, D., and Sackmann, E., Hydration dependence of chain dynamics and local diffusion in L-alpha-dipalmitoylphosphtidylcholine multilayers studied by incoherent quasi-elastic neutron-scattering. *Biophys. J.* 68, 1871, 1995.

92. Israelachvili, J.N. and Wennerstrom, H., Hydration or steric forces between amphiphilic surfaces. *Langmuir* 6, 873, 1990.

93. McIntosh, T.J. and Simon, S.A., Contributions of hydration and steric (entropic) pressures to the interactions between phosphatidylcholine bilayers: experiments with the subgel phase. *Biochemistry* 32, 8374, 1993.

94. Gawrisch, K., Parsegian, V.A, and Rand, R.P. Membrane hydration. In *Biophysics of the Cell Surface*. Eds. Glaser, R. and Gingell, D. Springer-Verlag, New York. 1990, 61.

95. Finer, E.G. and Darke, A., Phospholipid hydration studied by deuteron magnetic resonace spectroscopy. *Chem. Phys. Lipids* 12, 1, 1974.

96. Gawrisch, K., Richter, W., Möps, A., Balgavy, P., Arnold, K., and Klose, G., The influence of water concentration on the structure of egg yolk phospholipid/water dispersions. *studia biophysica* 108, 5, 1985.

97. Gawrisch, K., Ruston, D., Zimmerberg, J., Parsegian, V.A., Rand, R.P., and Fuller, N., Membrane dipole potentials, hydration forces, and the ordering of water at membrane surfaces. *Biophys. J.* 61, 1213, 1992.

98. Volke, F., Eisenblatter, S., Galle, J., and Klose, G., Dynamic properties of water at phosphatidylcholine lipid bilayer surfaces as seen by deuterium and pulsed-field gradient proton NMR. *Chem. Phys. Lipids* 70, 121, 1994.

99. Volke, F. and Pampel, A., Membrane hydration and structure on a subnanometer scale as seen by high-resolution solid-state nuclear magnetic resonance — POPC and POPC/$C_{12}EO_4$ model membranes. *Biophys. J.* 68, 1960, 1995.

100. Feller, S.E., Brown, C.A., Nizza, D.T., and Gawrisch, K., Nuclear Overhauser enhancement spectroscopy cross-relaxation rates and ethanol distribution across membranes. *Biophys. J.* 82, 1396, 2002.

5 Nonlamellar Lipid Phases

Sol M. Gruner

CONTENTS

5.1 INTRODUCTION

5.1.1 WHY STUDY LIPID POLYMORPHISM?

The many integral membrane proteins found in biomembranes are subject to a complex array of molecular interactions arising from the close proximity of oillike and aqueous environments. Some of these interactions are quite specific in nature, such as the binding of particular membrane molecules to recognition sites of proteins. Other interactions, such as those that couple to the membrane surface charge, are nonspecific in that they arise from the collective interactions of many molecules. The collective, nonspecific interactions generally relate to colligative physical properties of the lipid bilayer, such as the bilayer thickness, surface charge, dielectric constant profile, permeability, etc., which vary in systematic ways as the polar lipid composition of the bilayer is changed. Ideally, one would like to understand how these colligative physical properties affect membrane function, their relative importance, and the manner in which these properties arise from different lipid constituents.

The concern of this chapter is the physical basis of nonlamellar phases and the way in which an understanding of this basis lends insight into the forces at play in lipid bilayers. The structural forms assumed by polar biomembrane lipids when mixed with water considerably exceed the range of structures typically found in cells. In addition to bilayer lamellae, polar lipid–water dispersions form a remarkable variety of structures, including tubes, rods, and three-dimensionally periodic assemblies with cubic symmetries (Figure 5.1). The ability of a given mixture of lipids to form crystallographically diverse structures is called *polymorphism* or, equivalently, *mesomorphism*. The term polymorphism (from *poly* = many and *morph* = form) is a general term in the chemical

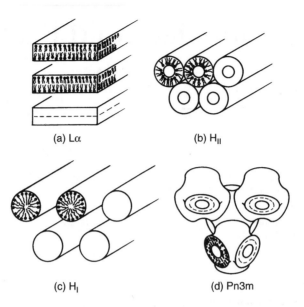

(a) Lα   (b) H$_{II}$

(c) H$_I$ (d) Pn3m

FIGURE 5.1 Several liquid–crystalline lipid mesomorphs are shown. (a) L$_\alpha$ phase; (b) H$_{II}$ phase; (c) H$_I$ phase; (d) a cubic phase with Pn3m crystallographic symmetry.

literature. An example of polymorphism is that carbon can exist in either of the crystallographically distinct phases of graphite or diamond. The term mesomorphism refers particularly to the polymorphism of liquid crystals, including the liquid–crystalline, i.e., melted chain, polymorphs of lipids. The word comes from the many forms (= *morph*) or phases that occur in the middle (= *meso*) of a thermal sequence between the low-temperature solid phases and the high-temperature isotropic liquid phase. In practice, lipid mesomorphism is often loosely used to refer to the full range of lipid phases, including the frozen (gel) state phases.

Why study lipid polymorphism? The existence of the numerous lipid polymorphs that are seen when lipids are extracted from biomembranes are structural manifestations of the release of constraints on the intermolecular forces that exist in the biomembrane. As shall be discussed in this chapter, by studying the structural polymorphism observed with isolated lipids, one gains an understanding of the forces that are locked up in biomembranes and that affect the organization and function of proteins, even though the predominant lipid organization in biomembranes is in the form of bilayers.

Another good reason for studying lipid polymorphism is intrinsic interest: as a group, lipids exhibit what is arguably the broadest range of polymorphic structures of any known class of molecules. The polymorphic behavior of lipids has been key to understanding the basis of amphiphilic and surfactant action in general and thus has enormous technological importance. The historical significance of this aspect of technology may be appreciated by comparison to other technological advances, such as the invention of the transistor: imagine what life would be like in the absence of transistors compared to the absence of soaps, detergents, and food emulsifiers.

The virtues of studying lipid behavior for reasons of intrinsic interest are often not fully appreciated. Studies of the physical properties of isolated membrane components, such as lipids, have been enormously important toward elucidating the properties of biomembranes. It is important to bear in mind that biomembranes are incredibly diverse and complex mixtures of molecules and that any isolated constituent system has properties peculiar to the isolated system, as well as to biomembranes from which the constituents are derived. Historically, it has often been the case that a property of a synthetic lipid system which initially had little obvious relevance to biology, upon further study provided information that was, indeed, important toward understanding biomembranes. In other instances, study of the properties of isolated systems led to technology that is medically important. For example, an understanding of the behavior of pure lipid vesicles is proving to be essential toward the development of liposomal drug delivery systems. It is important to maintain a broad perspective.

Polymorphism of membrane lipids is far too large a subject to be covered comprehensively in a single chapter; in consequence, the emphasis here will be largely restricted to polar membrane lipids that have two chains of roughly 8 to 24 carbons, and to polymorphs in which the chains are liquid–crystalline, i.e., in a melted, fluidlike state. This covers the lipids that make up the bulk of the bilayer matrix of most eukaryotic biomembranes and includes many phospholipids that predominate in animals and the carbohydrate-headed lipids that predominate in plants. It is important to recognize, however, that by restricting attention to these molecules, many important lipids, such as sterols, polyisoprenes, glycoproteins, soaps, fatty acids, lysolipids, etc., will not be adequately treated. By considering only liquid–crystalline phases, many important frozen (i.e., gel) chain phases also will not be discussed. Readers interested in these subjects are referred to the literature (Small, 1986) and the other chapters of this volume.

The central concern of this chapter is the set of polymorphic phase transitions that involve changes in the curvature of the lipid–water interface. Although curvature-altering phase transitions and interfacially curved lipid mesomorphs have been known for many years (see Luzatti, 1968, for a classic review), it is only recently that a qualitative understanding of the competition of free energies that control curvature-altering phase transitions has been developed. This has been the result of an explosive growth of interest in the biological ramifications of nonlamellar phases since Cullis and colleagues, in the late 1970s, began using ^{31}P NMR as a probe of phase behavior in biomembrane lipids (e.g., see Cullis and de Kruyff, 1976; Cullis et al., 1985). It is now realized that most biomembranes contain large lipid fractions, which in isolation do not readily form lamellar bilayers. Because nature could perfectly well have chosen biomembrane lipid compositions that do not contain such "nonlamellar-prone" lipid constituents and still have accounted for all the physical properties of liquid–crystalline bilayers that were believed to be biologically important, researchers began to question why nonlamellar-prone lipids were so prevalent in biomembranes. This led to studies on the properties of lipid mixtures containing nonlamellar-prone lipids and advances in the understanding of the physics of curvature-altering phase transitions. It has also led to a number of hypotheses about how the presence of nonlamellar-prone lipid affects biomembrane function.

The emphasis in this chapter will be to provide an intuitive understanding of the forces that drive the formation of curvature-altering phase transitions. No attempt will be made to cover comprehensively all aspects of such phases or to go into formal detail of the physics involved; readers interested in more detail are referred to various reviews in the literature (Cullis et al., 1985; Gruner et al., 1985; Caffrey, 1989; Lindblom and Rilfors, 1989; Gruner, 1989; and Seddon, 1989).

5.1.2 Terminology

Before continuing, it is important to clarify a few ambiguities about the terminology in common use in the field. The term *nonbilayer phase* really is meant to refer to nonlamellar phases, i.e., liquid–crystalline phases that are not L_α phases. In fact, certain "nonbilayer" phases, such as some bicontinuous cubic phases (e.g., Figure 5.1d), are composed entirely of multiplied connected bilayers. An *inverted* or *water-in-oil* phase refers to one in which the lipid/water interface has the same sign of curvature as an H_{II} phase, i.e., a net concave curvature when viewed from the water domain. Thus, the H_I (Figure 5.1c) phase is a noninverted or oil-in-water phase. This terminology may be understood by reference to the curvature and majority constituent of oil-water-surfactant micelles: at a high water-to-oil ratio, micelles of surfactant-coated oil in water are commonly found. As the water-to-oil ratio decreases to the point where the oil is the majority constituent, the system often "inverts" and results in surfactant-coated water droplets in oil. Another ambiguous term is *liquid–crystalline,* which in the lipid literature refers to systems in which the chains are in a fluidlike, melted state. In the more general liquid crystal literature, liquid–crystalline specifically refers to phases that are intermediate to the rigorously crystalline solids and true isotopic liquids, including systems that do not have long flexible chains at all.

A more fundamental ambiguity relates to the distinction between a chemical phase and an enduring nonequilibrium state. Because many lipids have very low monomer solubilities in water (typically 10^{-10} M for diacyl phospholipids), kinetic relaxation of a lipid assembly into a state of lower free energy may be so severely hindered that the system may, for all practical experiments, be unable to come to true equilibrium. An example of this is the still-debated question as to whether the bilayers in an aqueous dispersion of unilamellar vesicles should be considered to be in a distinct phase. A state of lower free energy exists, namely, the L_α phase, for the same composition and temperature if the system were reassembled in the form of stacked bilayer lamellae in coexistence with excess, bulk water. This is because there are stabilizing interaction energies between bilayers that are closely opposed (at least for uncharged bilayers, which are sufficiently rigid). In fact, the individual bilayers in the L_α phase differ little from the bilayer wall of the unilamellar vesicle (the thickness of the stacked and isolated bilayers are thought to be very slightly different), so isolated liquid–crystalline bilayers are usually said to be in an L_α *phase*. A similar ambiguity exists for other systems when the geometry of the system is locally similar to a crystallographically well-defined phase, but the density of geometric defects is sufficiently high to destroy crystallographic periodicity. Note that this may be thought about in a continuum fashion. As an example, the bilayers of bicontinuous cubic phases overtly differ from L_α phases by the way in which the lipid/water interface is curved. Defects in this periodically curved surface may be imagined as being progressively introduced until obvious semblance of the periodic cubic symmetry is erased. The system now looks like a random "sponge" of multiplied connected bilayer walls.

When should one stop calling the system a cubic phase and start calling it a nonequilibrium state, especially given that the defect-ridden state can endure practically indefinitely? This is not an idle academic exercise, because such ambiguous lipid states are very common. For example, a lot of literature exists on the "lipidic particles" (Verkleij, 1984) sometimes thought to be defect states on the way to true cubic phases (Siegel, 1984, 1986a–c, 1987). It is frequently very difficult to produce the crystallographically well-defined assembly. An interesting example whereby cubic phases are produced by the gradual introduction of defect structures is given by Shyamsunder et al. (1988).

5.2   EXPERIMENTAL TECHNIQUES

Historically, advances in the understanding of lipid mesomorphism have been closely tied to the development and use of techniques for probing molecular structure. Many different physical techniques — X-ray and neutron diffraction, NMR, electron spin resonance, electron microscopy, infrared and visible light spectroscopies, calorimetry, etc. — have been useful in studying lipid mesomorphism. Of these, X-ray diffraction, NMR, and calorimetry have been most central to our understanding of mesomorphism. X-ray diffraction is directly sensitive to the relative positions of the atoms in a mesomorph and is, therefore, a fundamental tool for the identification of the structural rearrangements that characterize different lipid phases. Unfortunately, X-ray diffraction patterns typically require long exposure times (although this is changing with the advent of intense synchrotron X-ray sources), so the information derived is time-averaged over the exposure interval. NMR, on the other hand, probes the local environment of the excited nuclei over time scales of, typically, tens of microseconds and may be used to gain information on the dynamics of the mesomorph. The drawback of NMR is that long-range structure, which is very readily determined by X-ray diffraction, can only be indirectly inferred via NMR. Calorimetry has been most important as a simple identifier of the temperature at which phase transitions occur, because many lipid phase transitions involve heat changes that are relatively straightforwardly measured. Calorimetry does not, itself, yield structural information.

It is well beyond the scope of this chapter to discuss thoroughly the principles of any of the major physical methods used to study lipid phases. However, an understanding of X-ray diffraction and NMR techniques is especially important for much of the discussion of this chapter. Therefore, these techniques will be summarily described. The emphasis will be on understanding the uses and limitations of these methods. Considerably more attention will be given to X-ray diffraction, because NMR has already been discussed in other chapters.

5.2.1 X-Ray Diffraction

X-ray diffraction is the most important and least ambiguous method of structural determination of lipid phases (Luzzati, 1968). X-ray diffraction involves the placement of a thick (typically 1 mm) specimen of the lipid material in a well-collimated beam of X-rays and the subsequent recording of the scattered X-ray pattern. The scattered X-ray pattern is fundamentally related to the electron density distribution within the specimen in that, up to geometrical factors, the scattered intensity is a projection of the Fourier transform of the electron density distribution of the specimen. Two kinds of information are contained in the diffraction patterns: the positions of discrete X-ray diffraction reflections relate directly to the lattice structure that may be present in the sample, whereas the intensity of the reflections yields the modulus of the Fourier transform of the electron density distribution of the repeating unit, or unit cell.

In single-crystal X-ray diffraction, the intensities of many pointlike X-ray diffraction spots are recorded and used to reconstruct the unit cell electron density distribution, from which atomic positions are determined. This is the case, for example, in protein crystallography. In liquid crystals, however, the near-perfect translationally repetitive molecular positions representative of an ideal crystal are replaced by various degrees of molecular disorder. The result is that the liquid crystal structure is better represented by a statistical description of the relative placement of the molecules and the exact relative placement of the molecules is replaced by time-averaged positions. For example, in the L_α phase (Figure 5.1a), the bilayers are stacked atop one another with intervening water widths. The stacking is periodic in a direction perpendicular to the bilayer planes, but within the planes of the bilayers, the lipids are packed next to one another with liquidlike disorder. Thus, the appropriate electron density description is the average electron density as a function of distance along the stacking axis, appropriately called the *lamellar electron density profile* (e.g., see Levine, 1973). Identification of the peaks and dips in this profile, corresponding to the electron-rich phosphorus atoms and electron-light terminal chain methyls, were first used to prove the existence of the phospholipid bilayer structure (Wilkins et al., 1971). For the H_{II} phase (Figure 5.1b), in which the molecules are free to diffuse around and along the water-cored lipid cylinders, the appropriate electron density distribution is a description of the average electron density as a function of the radius from the tube axis and the angle about this axis.

Even given the positional averaging inherent in liquid crystals, it is frequently difficult to solve the structure for the appropriate electron density distribution because of three problems:

First, ordered liquid–crystalline arrays, such as macroscopically aligned H_{II} tubes, are difficult to obtain. As a result, the X-ray patterns are indicative of the superposition of an ensemble of microspecimens, each with its own orientation. Most specimens, in fact, are unoriented dispersions, giving rise to pure "powder patterns." It is sometimes difficult to determine the intensity of the individual lattice reflections from unoriented specimens.

A second impediment to obtaining electron density distributions is that the phase information required to invert the Fourier transform and obtain the real-space electron density distribution is lost when the intensity is recorded. It must be determined by alternative means (see, for example, Stamatoff and Krimm, 1976; Mariani et al., 1988). For this reason, lamellar electron densities are rarely determined. The determination of nonlamellar electron densities is very rare, indeed, but possible. H_{II} phases have been reconstructed (Caron et al., 1974; Gulik et al., 1985, 1988; Turner and Gruner, 1989), as have cubic phases (for a review of cubic phase reconstructions, see Luzzati et al., 1987; Mariani et al., 1988). A third problem in performing reconstructions is that the small number of X-ray reflections yielded by liquid–crystalline specimens limits the resolution of the reconstruction.

Despite the limitations and difficulties of obtaining electron density reconstructions, this method is still the most practical way known to obtain detailed static structural information on lipid mesomorphs and to verify the basic structural schemes inferred from considerations of lattice symmetry and other probes, such as NMR and electron microscopy.

The lattice structure that is present in the lipid mesomorph directly determines the symmetry of the diffraction pattern observed (they are related by a Fourier transformation). The most frequent use of X-ray diffraction of lipids is as an unambiguous probe of lattice symmetry, which can frequently constrain the determination of the lipid phase. For instance, the presence of a one-dimensional periodic lattice, in which the low-angle X-ray reflections occur in the distance ratio of 1:2:3..., is indicative of a lamellar stacking of bilayers. Although it is possible to conceive of nonlamellar one-dimensional structures (e.g., a stack of monolayers, all oriented the same way so that water contacts both the headgroup and terminal methyl surfaces of the layer), the energetic constraints of exposing hydrocarbon to water eliminate most such alternative structures. Likewise, a two-dimensional hexagonal lattice has the signature of low-angle reflections in the distance ratio of A:3:2:7.... This symmetry is present in both the H_{II} and H_I phases (Figure 5.1b and Figure 5.1c). Determination of which of the two is involved requires additional information, such as the observation that all known H_I phases swell indefinitely upon the continuous addition of water, whereas the lattices of all known H_{II} phases swell to a limit and then coexist with excess bulk water. (The point at which this happens is called the *excess water point.*)

Whereas the presence of a given lattice is usually taken as compelling proof of a given set of phases, the absence of lattice structure in the diffraction pattern simply means the absence of a long-range ordered lattice within the specimen. For example, a dispersion of large, unilamellar vesicles of liquid–crystalline bilayers precludes a lamellar stacking. In this case, the bilayers yield weak, very diffuse (so-called unsampled) diffraction, indicative of the absence of a lattice. The vesicle bilayers are usually stated to be in an L_α phase. As noted earlier, the vesicle bilayers are in a different state than in a true L_α phase, because the presence of nearby lamellae affects the bilayer thickness. But this effect is small at high water contents and is usually ignored.

Deviations from perfect lattices, either in the form of topological defects of the surfaces or distance fluctuations within the mesomorphs, limit the resolution and the number of X-ray reflections observed. Definitive lattice assignments, especially among the many possible cubic lattice symmetries, cannot be unambiguously performed with a limited number of X-ray reflections. Rigorously speaking, an X-ray lattice assignment cannot *a priori* prove the presence of a lattice; it can only exclude the presence of alternative lattices. This is because a given reflection allowed by the lattice symmetry may have zero intensity due to the electron density of the unit cell. On the other hand, the presence of a reflection not allowed by the lattice symmetry rigorously excludes that lattice. The practical consequences of this are that five or six uniformly spaced, low-angle reflections are generally taken as compelling evidence of a lamellar phase assignment, whereas more than 12 reflections are needed to be confident of a cubic phase assignment.

While on the subject of diffraction techniques, mention should be made of the potential of neutron diffraction. Neutron diffraction relies on the different scattering lengths of different nuclei to low-energy neutrons. Most notably, the large scattering-length differences between the proton and the deuteron can be used to control contrast variation in biological materials by selective deuteration. Further, normal and deuterated water can be mixed to contrast-match the proton concentration of regions of the mesomorphs, such as the hydrocarbon regions, thereby effectively contrast-enhancing regions that differ in proton concentration (see Worcester, 1976, for a review). Although neutron diffraction studies on biological lipids have been predominately confined to studies of bilayers, the broad experiences of neutron diffraction on surfactant systems (e.g., Alperine et al., 1985; Hendrikx et al., 1987) have demonstrated the power of the neutron diffraction technique. Because neutron diffraction requires an appropriate neutron source, such as a reactor, the most important practical limitation of neutron diffraction is the paucity of suitable experimental facilities.

5.2.2 NMR

NMR has proved to be a very powerful tool in the study of lipid mesomorphism. NMR methods often complement X-ray diffraction techniques: whereas X-ray diffraction sees time-averaged structure, is lattice sensitive, and suffers from a limited number of available facilities, NMR can be sensitive to motion on much longer time scales, is lattice blind, and, most importantly, benefits from the existence of many facilities. Because the use of NMR for lipid studies is well described in the literature (for reviews, see Seelig, 1977; Cullis et al., 1985; Bloom, 1988; Lindblom and Rilfors, 1989) and the basic principles have been described by Gawrisch in Chapter 4, the discussion here will be confined to the uses and limitations of NMR for mesomorphic studies.

The hindered molecular diffusion of a liquid crystal broadens the width of NMR lines. This is a consequence of correlations that exist between the initial and final orientations of the molecule containing the nucleus being probed over the roughly 10^{-5} s time scale of the NMR experiment. The diffusional motions sampled by the atom in question determine the shape of the NMR lines. Moreover, the types of motions experienced by lipids in mesomorphs are, in certain cases, sufficiently different that the line shape can be used to distinguish between the characteristic geometries of different phases. The two most important resonances in this regard are due to phosphorus 31, which is naturally present in adequate concentration in most phospholipids, and the quadrupole signal of deuterium, which typically requires synthetically deuterated lipids. Examples of these signals are shown in Figure 5.2.

The use of ^{31}P NMR was important in stimulating interest in lipid mesomorphism in the late 1970s in that it was used to demonstrate that H_{II} phases are very common with biomembrane phospholipids. This also led to a certain amount of controversy, because the shape of the ^{31}P NMR signal is not only dependent on the diffusional motion of the phospholipid but also on the particular conformation of the headgroup. Although, in principle, this can result in misleading identifications (Thayer and Kohler, 1981), the consistency of phase identification between X-ray and NMR methods has, in fact, been excellent (Tilcock et al., 1986). But it must always be borne in mind that a phase identification based on the shape of the ^{31}P NMR resonance is strong, but not irrefutable, evidence toward identifying lamellar and H_{II} phases.

Three canonical ^{31}P NMR line shapes are typically seen (Figure 5.2), indicative of motion in which the director of the molecule (i.e., the long axis) is free to diffuse only parallel to itself, as in lamellae, or is additionally free to diffuse in other directions. If the director is free to diffuse around the surface of a tube, as in the H_{II} phase, the shape shown in Figure 5.2b is typically seen. Note that this degree of motional freedom is also present in any extended tubular lipid organization, such as the H_I phase. The lipids need not be organized in a lattice; isolated, extended lipid tubes would also yield the H_{II} ^{31}P NMR signature if the diffusion times along and around the tube were rapid enough. The third line shape, which is simply a narrow symmetrical spike (Figure 5.2c), occurs if the lipid director samples all directions over the NMR time scale. This occurs in cubic phases, with small vesicles, and with a surprisingly large variety of disorganized lipid structures.

The reader should appreciate that the three ^{31}P NMR line shapes are commonly occurring manifestations of a continuous variation in molecular diffusional capability and that ambiguous line shapes do occur. For example, imagine the signal expected from a tubular bilayer vesicle. In the limit that the vesicle is very long and the tube diameter small, H_{II} type line shapes are expected. In the limit that the tube length and diameter are very large, lamellar-type signals will be seen. In the limit that the length and tube diameter both become very small, an isotropic resonance is

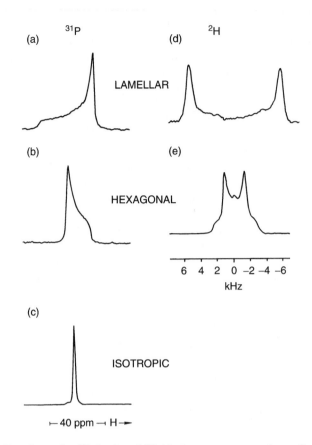

FIGURE 5.2 NMR line shapes for ^{31}P (a–c) and ^2H (d–e) resonances are shown for several mesomorphic phases. (a) L_α phase; (b) H_{II} phase; (c) isotropic resonance characteristic, for example, of cubic phases; (d) ^2H quadrupole resonance for an H_{II} phase. The quadrupole resonance for a cubic phase (not shown) is typically a single, sharp spike.

expected. As the tube shape is changed continuously between these shapes, the shape of the ^{31}P NMR signal obtained will, likewise, pass continuously from one expected type to another.

In recent years, ^2H NMR has grown in importance as a lipid diagnostic tool. The utility of this nucleus is that hydrogen may be selectively replaced with deuterium at many sites in a lipid molecule without appreciably perturbing the molecular structure. The deuteron is a spin 1 nucleus and exhibits a spectrum characterized by a quadrupole splitting (Figure 5.2d and Figure 5.2e). The magnitude of the splitting, denoted by ΔQ, is a function of the diffusive reorientations of the deuteron over the time scale of the NMR experiment; thus, it should not be too surprising that it varies with the lipid phase.

The use of ^2H NMR in mesomorphic studies is not confined to phase identification. Because lipids contain so many sites where hydrogen may be replaced by deuterium, ^2H NMR can be used to examine the relative motions of different parts of the lipid molecule. For example, deuteration of the lipid chains can be used to extract the order parameter profile of the chains, which is a measure of the degree to which the orientation of the carbon–carbon bonds are correlated with the normal to the lipid/water interface. The correlation is very high near the headgroups and decreases to being essentially uncorrelated at the terminal methyl groups of the chain ends. This variation of orientational freedom manifests as slightly different values of ΔQ for deuterons at different positions along the chain. A procedure called *dePakeing* (Sternin et al., 1983) can be used to determine the ΔQ value as a function of carbon number along the chains, thereby allowing the order parameter profile to be extracted from single ^2H NMR spectra of totally perdeuterated chains. The alternative

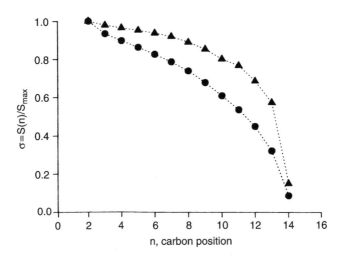

FIGURE 5.3 The hydrocarbon order parameter profiles are shown for an L_α (triangles) and an H_{II} phase (circles). This figure was derived from the ^{2}H NMR spectra of perdeuterated tetradecanol mixed into the lipid mesomorph; however, perdeuterated lipid chains show very similar order parameter profiles (Lafleur et al., 1989). (From Sternin E et al.: *Biophys. J.* 1988, 54:689. With permission.)

procedure of obtaining the order parameter profile, namely, using samples in which specific carbons on the chain are deuterated, involves many difficult lipid syntheses and many NMR spectra. Using the dePakeing procedure, Sternin et al. (1988) and, more recently, Lafleur et al. (1989) have examined the changes that occur in the order parameter profile when lipids undergo the L_α and H_{II} transition. The profiles are distinctly different (Figure 5.3).

5.3 PHYSICAL BASIS OF NONLAMELLAR PHASES

5.3.1 H_{II} Phases

The central concern of this chapter is to understand the physical forces that determine the liquid–crystalline mesomorphic behavior of lipid–water systems, with the aim of gaining insight into how these forces affect the physical properties of biomembranes. Attention will be focused on the liquid–crystalline phases that occur at temperatures above crystalline gel-chain phases and below the temperature at which the system totally melts into an isotropic fluid (this upper temperature is frequently above 100°C, in which case high pressures must be used to study it). As opposed to the low-temperature gel-chain phases, in which different phases are identified by the relative positions of essentially rigid, relatively immobile molecules, the liquid–crystalline phases are distinguished by changes in the shape of the lipid/water interface. Within a given liquid–crystalline phase, the molecules diffuse about rapidly, subject to the constraint that the headgroups move within the lipid/water interface. Conceptually, these phases are well characterized as assemblies of two-dimensional liquid interfaces, and attention must be focused on the forces that give the interfaces their shape.

Study of the interfaces begins with determination of the phase diagram of the system, which catalogs the conditions under which a given interfacial shape occurs. In general, the exact transition temperatures and, indeed, the sequence of phases observed, vary both with the temperature and with the relative concentrations of the constituents. The complete phase diagram of a given system is a multidimensional plot of phase behavior as a function of variation of the relative amounts of each of the constituents, the temperature, and any other thermodynamic variables of interest, such as the pressure. The terms *thermotropism* and *lyotropism* are used to refer to phase variation as a function of temperature and of composition, respectively. Once the phase boundaries have been identified, one attempts to understand which of the many components of the free energy of the system change most markedly at the phase transition boundary under the assumption that the change in the interfacial boundary and the free energy component are causally linked.

The phase diagram catalogs the conditions under which a given interfacial shape occurs, but it does not generally contain information about the dimensions of the structure of the mesomorph. The work of our group at Princeton University (Gruner, 1989) has illustrated the importance of also knowing how the dimensions of the shapes change at phase boundaries. This information is primarily obtained by X-ray diffraction. The reason the spatial dimensions are important, as detailed later, is that the magnitudes of the energies involved in the phase transitions are very sensitive to the dimensions.

How is the phase behavior of lipid liquid crystals to be understood? To date, the most insightful approaches have been phenomenological. A phenomenological approach is based on the recognition that a prediction of phase behavior attained by a detailed summing of the interactions of a complex, multiatom system is beyond present capabilities. Moreover, a focus on the complexity of the system, as is required for such a detailed microscopic understanding, can obscure unifying themes and quantities that result from the colligative interactions of large numbers of atoms. A simple example of a phenomenological rule is Hook's law, namely, that the force required to extend an elastic body is linearly proportional to the degree of extension. This simple rule describes the force required to deform a remarkably diverse variety of objects, from steel springs to glass rods to rubber bands, without ever having to consider in detail the diverse and complicated molecular interactions that give the law validity. As a phenomenological rule, Hook's "law" is approximately true only over a limited regime of extension; indeed, the limits to the "law" yield information about the nature of the molecular interactions. As an empirical rule, the law is useful because the relationship of force to the extension may be quantitatively measured and applied, again, without a detailed understanding of how the fundamental molecular interactions sum to yield the force constants.

The phenomenological approach taken at Princeton University (see Gruner, 1989, for a review) begins with the recognition that the fundamental unit of all lipid mesomorphs is the lipid monolayer and that this monolayer may be endowed with a spontaneous tendency to curl. The monolayers pair tails-to-tails to form bilayers, curl into water-cored rods in H_{II} phases, and into hydrocarbon chain–cored tubes in H_I phases. Many asymmetric material systems exhibit a spontaneous curvature. A simple example is a bimetallic strip or a strip of rubber cut from the side of a tennis ball. A strip of one of these materials will curl to a well-defined radius when relaxed. We call this the *spontaneous radius of curvature*, R_0, and the inverse of this radius the *spontaneous curvature*, $C_0 = 1/R_0$. The energy per unit area required to bend the strip to a different radius, R, at least for a limited regime of bend away from the relaxed radius, is given by

$$\Delta E = (K_c/2)(1/R - 1/R_0)^2 \qquad (5.1)$$

This equation simply quantifies the fact that the deformation energy is zero when $R = R_0$ and rises parabolically for deviations, either larger or smaller, of R from R_0. The coefficient K_c is called the *rigidity* of the object because it is a measure of the stiffness of the strip, i.e., the energy required to bend it for a given deviation of R from R_0. K_c may be defined for a unit area of the strip, in which case Equation 5.1 is energy per unit area, and the energy for an arbitrary area of strip may be found by integrating Equation 5.1 over the area. Helfrich (1973) first considered a form of Equation 5.1 in attempting to explain the shapes of bilayer vesicles. In what follows, it shall be assumed that similar description applies to lipid monolayer tubes or planes. Equation 5.1 may be taken as a hypothesis about the elastic properties of lipid monolayers and, as an empirical equation, is subject to experimental verification.

The fundamental reason that a bimetallic strip bends is a mismatch between the surface areas of the two metals of which the strip is composed, coupled with a constraint of connection between the two metals that prevents them from sliding past one another. Another example of such a bending moment is the curling of paper when it is exposed to moist air. In this case, the curling arises because the fibers on the side of the paper exposed to the moisture swell more than the fibers on the other side. Another example of an object with a spontaneous curvature, although it may seem strange to think of it as an elastic object, is an arch composed of trapezoidal blocks (the tremendous rigidity of the arch is the reason for its great strength and resistance to deformation). Much of the literature dealing with the basis of lipid mesomorphism deals with a "shape concept" (Tartar, 1955; Israelachvili et al., 1980) and may be understood by analogy with the arch. This concept holds that the mesomorphic tendency of a lipid layer is a consequence of the side-by-side packing of molecules whose cross-sectional areas vary systematically along the length of the molecules. Thus, lipids that maintain an overall uniform cross-sectional area along their length tend to form planar mesomorphs, such as bilayers (i.e., have zero spontaneous curvature), molecules with smaller headgroup than tail areas form H_{II} structures (i.e., have a finite spontaneous curvature, to which we may arbitrarily assign a negative sign), and those that have larger headgroups than tails tend to form H_{II} structures (i.e., exhibit a positive spontaneous curvature).

A shape concept formulation of lipid mesomorphism is simple, appealing, and useful, provided that it is not taken too literally. It must be remembered that lipids are flexible molecules and do not have a single, simple shape. Further, the "shape" must include energetic contributions, which bias the cross-sectional areas. For example, increasing the net headgroup charge of the lipids, as may often be accomplished by changing the pH, increases headgroup–headgroup repulsion and the area per headgroup without really changing the steric shape of the headgroups. The shape to be considered is more subtle: it is not the shape of the molecule *per se,* but the shape of the average volume which packs to fill the lipid volume and which minimizes the free energy of packing the molecules.

A measure of molecular shape that is frequently used (Israelachvili et al., 1980) is the dimensionless quantity v/al, where v is the volume per molecule, a is the area per headgroup, and l is the approximate thickness of the lipid hydrocarbon layer. Values of v/al that are significantly less than one are associated with monolayers with a positive spontaneous curvatures (H_I phases), values approximately equal to one lead to zero spontaneous curvatures (L_α phases), and values greater than one lead to negative spontaneous curvatures (inverted phases). Although this conceptual measure has much appeal, it is of limited practical utility in helping to understand phase transitions in which the curvature of the lipid/water interface changes sharply at the transition. Consider, for example, the two common lipids, dioleoylphosphatidylcholine (DOPC) and dioleoylphosphatidylethanolamine (DOPE), in excess water. At 2°C, both lipids are in an L_α phase and measurement of the bilayers in both cases would yield v/al values of roughly unity. If the temperature is raised above about 6°C, DOPE forms an H_{II} phase, with a very different v/al value, whereas DOPC remains lamellar to above 100°C. There was little in the v/al values at 2°C to indicate that one system is very H_{II}-prone while the other is not. This is because v/al is really a measure of the expressed shape of the mesomorph, not its deviation from a relaxed shape. By contrast, an important distinction between the actual and spontaneous curvatures of the interface is inherent in Equation 5.1. As shall be seen, it is straightforward to measure R on a system and infer R_0 from the measured values.

It is important to recognize that the spontaneous curvature is *assumed* to be a phase-independent quantity that promotes a given curvature of the interface. Consider, for example, the L_α-H_{II} transition of DOPE. It is assumed, of course, that at high temperatures, the spontaneous curvature is primarily responsible for the actual curvature seen in the H_{II} phase. Below the phase transition temperature, the expressed curvature is approximately zero. (It will turn out that R_0 values relevant to H_{II} phases are of the order of magnitude of the diameter of H_{II} tube water cores, i.e., typically less than 50 Å. The curvatures encountered with L_α bilayers are much larger than this, so on the length scales of interest, L_α bilayers are practically flat.) If one assumes that H_{II} phases are elastically relaxed such that $R \cong R_0$, then in undergoing the H_{II}-to-L_α transition, R deviates sharply from R_0. Then, by Equation 5.1, the L_α phase is a geometry in which the bending energy of the layer is very high, i.e., the layer is very far from being relaxed with respect to bend. This is possible if there exists another temperature- and shape-dependent component of the free energy, which is in competition with the layer bending energy such that both free energies cannot be simultaneously minimized in either the flat or the tubular geometry. Physicists call a system with such competing free energies *frustrated*. Because it is the net free energy that is important in determining the equilibrium phase of a system, it is possible for the L_α phase to be in a high free energy state with respect to bend if the contribution of the other free energy components results in a reduction of the net free energy.

A competing free energy may be identified with spatial inhomogeneities in the hydrocarbon chains that inevitably accompany any geometry in which opposed lipid/water interfaces are not parallel (Kirk et al., 1984; Charvolin, 1985). Consider the chains of a L_α phase, as shown in Figure 5.4. The geometry of a bilayer is such that a translation of a molecule along the surface of the bilayer does not result in a change in the average hydrocarbon environment. By contrast, in the H_{II} phase shown in Figure 5.4b, the environment varies as molecules diffuse around the tubes. For example, the distance labeled d_{HII} is shorter than the distance labeled d_{max}. One way to accommodate this is if the chains stretch a bit farther in reaching to fill the volume indicated by d_{max} than the volume indicated by d_{HII}. Lipid chains act effectively as short polymer strands and resist changes in the mean extension, much in the same way that the polymer strands of a rubber band resist being straightened out. Liquid–crystalline chains also form a fluidlike environment and will entropically resist the introduction of inhomogeneities, such as variations in the volume density or mean direction of the chains as a function of position around the cylinder. The requirement that the chains fill the hydrophobic volume as uniformly as possible, and the impossibility of doing so without variation around the tubes in an H_{II}-like geometry, results in a rise in the free energy of the hydrocarbon in undergoing the L_α-to-H_{II} transition. Although there is some coupling of hydrocarbon packing and bending energies, the coupling is, in fact, small.

The competition between monolayer bending energies and hydrocarbon packing energies in the H_{II} and the L_α phase is now clear. If the spontaneous curvature of the system is such that it tends to curl strongly, then in the L_α phase, the bending energy is high but the hydrocarbon packing energy is low. In the H_{II} phase, the bending energy may be low, but the hydrocarbon packing energy is high. Although both of these energies may be expected to be a function of temperature, there is no *a priori* reason to expect that they have the same functional dependence. The L_α-to-H_{II} phase transition is then largely determined by the point at which the drop in free energy incurred by curling exceeds the magnitude of rise in free energy, which is thereby required to pack the chains most uniformly.

Insofar as the thickness of the lipid monolayer varies little with temperature, the relative fraction of the hydrocarbon volume outside inscribed circles that reach to the edges of the hexagon shown in Figure 5.4 (which is an indication of the degree of hydrocarbon stress) increases with the radius of the water core (Gruner, 1985). Indeed, in the limit of very large cylinders, the distance to corners of the hexagon exceeds the length of a fully extended lipid chain. This is never observed experimentally, so it may be assumed that such large hydrocarbon extensions are intolerably energetic. One way to allow very large water cores without requiring overly extended chains would be to have water cores that were not perfectly circular but rather more polygonal in shape. It is not clear whether this would be energetically favorable because it involves bending of the monolayers. More importantly, X-ray studies of DOPE have shown the water cores to be almost (but not perfectly) circular in cross section (Turner and Gruner, 1989).

The actual change in the dimensions of the water cores of H_{II} phases as the temperature is varied is consistent with the fact that lamellar phases always occur at lower temperatures than H_{II} phases. X-ray studies have shown that the core size increases as the temperature drops (Figure 5.5; Tate, 1987; Tate and Gruner, 1989. Note that in these experiments, the lipid was dispersed in excess bulk water, which readily diffuses through the sample, so the amount of available water was not a constraint.) Simple geometrical arguments can be used to show that the amount of hydrocarbon stress, for monolayers of approximately constant thickness, increases with the size of the water core (Gruner, 1985). Thus, the energetic price of packing the hydrocarbon in the H_{II} phase increases as the temperature decreases. At some point it has increased sufficiently that the corresponding

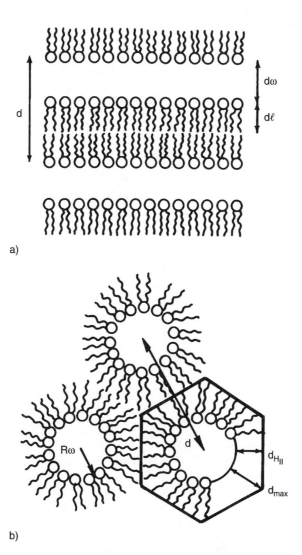

a)

b)

FIGURE 5.4 Idealized cross sections and dimensions of L_α and H_{II} phases are shown. The unit cell repeat spacing is indicated by d. In the L_α phase, the mean monolayer and water thicknesses are indicated by d_L and d_w, respectively. In the H_{II} phase, R_w indicates the mean water core radius. The lipid chains must extend at least over the distances d_H and d_{max}, as shown in (b), if the hydrocarbon volume is to be filled. The resultant sixfold periodic axial symmetry in the hydrocarbon environment is the source of the hydrocarbon packing stress.

curvature cost of going into a flat, lamellar geometry, while reducing the hydrocarbon inhomogeneity, leads to an overall lower free energy.

The model described here has been confirmed by many experimental tests and has proved to have significant predictive power. For example, if it were possible to remove the hydrocarbon packing price associated with the H_{II} phase, one would expect that H_{II} phases would extend to much lower temperatures. A simple method for removing most of the hydrocarbon packing stress is to add a few percent of a light oil, such as dodecane or tetradecane. Because these alkanes are not anchored to the lipid/water interface, one might imagine that they would diffuse about in the hydrocarbon and preferentially partition into zones where the lipid chains are most stressed, thereby reducing the stress. Figure 5.5 shows the H_{II} phases that occur if a few percent of dodecane is added to DOPE. Note that the repeat spacings of the H_{II} phases, indicative of the size of the water cores (Kirk and Gruner, 1985), are similar with and without dodecane. This size of the lattice in

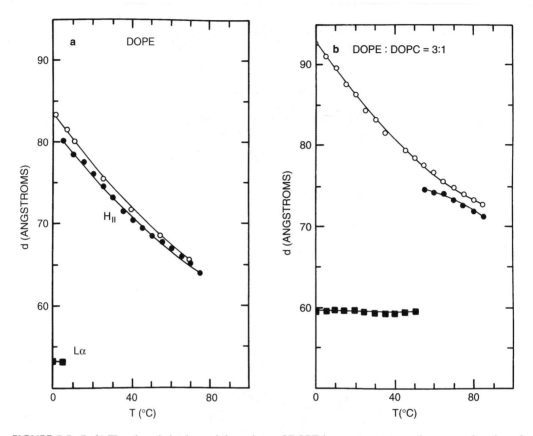

FIGURE 5.5 (Left) The phase behavior and d-spacings of DOPE in excess water are shown as a function of temperature. Squares represent L_α phases, and circles are for H_{II} phases. In the absence of alkane (solid symbols), the phase transition occurs at about 5°C. In the presence of several percent of alkane (open symbols), the transition is below 0°C. (Right) The phase behavior of DOPE:DOPC = 3:1 is similar, except the larger spontaneous radius of curvature characteristic of the mixed-lipid system increases the d-space and causes the transition temperature in the absence of alkane to be raised to the vicinity of 55°C. Note that the d-spacings of H_{II} phases with alkane are close to those without alkane over temperatures where H_{II} phases occur without alkane. This suggests that the alkane does little to modify the spontaneous radius of curvature. (Adapted from Kirk GL and Gruner SM: *J. Phys. Paris* 1985, 46:761.)

the presence of dodecane has also been shown to be fairly insensitive to the exact amount of the oil or, within limits, the type of oil. These data suggest that the oil is not strongly affecting the spontaneous curvature that controls the lattice size. However, the dodecane has a strong effect on the L_α-H_{II} transition temperature, effectively eliminating the L_α phase, because it is removing the hydrocarbon packing stress that otherwise prevents low-temperature H_{II} phases.

One predicts that if the spontaneous radius of curvature of a system were increased, then the L_α-H_{II} transition temperature would also increase. This is because a larger radius means larger tubes, which, in turn, involve an increased hydrocarbon stress. One would have to go to higher temperatures before the spontaneous radius of curvature would drop sufficiently that H_{II} tubes could actually form. A simple way of continuously adjusting the spontaneous curvature is to mix two different lipids with different spontaneous curvatures in different proportions. Although there is no reason to expect that the curvatures will add linearly, they may be expected to add systematically. Earlier, it was noted that DOPE readily undergoes the L_α-H_{II} transition, whereas DOPC, which differs only slightly from DOPE, exhibits L_α phases at least to 90°C. In terms of the model, DOPC does not readily form H_{II} phases because it has a very large spontaneous radius of curvature, and this precludes overwhelming the hydrocarbon stress at low temperatures. By mixing DOPE and DOPC, one observes that the size of the H_{II} lattice that forms and the L_α-H_{II} transition temperature, T_{bh}, both systematically increase with the fraction of DOPC (Kirk and Gruner, 1985). Again, adding a light oil mitigates the increase in T_{bh} but hardly affects the size of the lattice (Figure 5.5).

The spontaneous curvature value, $C_0 = 1/R_0$, and the rigidity, K_c, of Equation 5.1 have been measured by the method of osmotic dehydration (Gruner et al., 1986; Rand et al., 1989). X-ray diffraction was used to measure the H_{II} phase dimensions for a series of samples placed in osmotic equilibrium with a series of hydrophilic polymer solutions of known osmotic pressure. An increase in osmotic pressure corresponds to an increased competition for the water in the H_{II} cores and a corresponding decrease in the radius of the cores. By plotting the osmotic pressure vs. the volume of core water per lipid molecule, one effectively can examine the pressure–volume relationship of the system. The integral under this curve is the chemical work required to achieve a given radius, which can be compared against Equation 5.1. A least-squares fit of such data to Equation 5.1 not only indicates the range of validity of that equation but also provides values for R_0 and K_c.

Four main conclusions emerged from the osmotic studies on DOPE and DOPE-DOPC systems:

First, Equation 5.1 fits the data well for extraction of most of the water. It fails only for extremely dehydrated systems, in which case there are presumably other significant energetic terms associated with chemical dehydration of the headgroups. This is reasonable because the diameters of the H_{II} water cores were in the range of 40 to 50 Å, so most of the water in the cores was several water diameters removed from a lipid headgroup and is, therefore, not strongly involved in chemical hydration of the headgroup. This suggests that much of the H_{II} water is sucked into the core simply to fill the volume, as demanded by the interfacial curvature.

Second, the fully hydrated core size corresponds closely to $R = R_0$, i.e., at full hydration, the lipid curvature stress is almost fully relaxed.

Third, the rigidity values are similar (when doubled) to the rigidity values of bilayers as determined by bilayer fluctuation and mechanical bending measurements (e.g., see Bo and Waugh, 1989, and references therein).

Finally, the osmotic studies helped to remove an ambiguity in the definition of R. Because the thickness of the monolayer is of the same order of magnitude as the radius of the water core, it was initially unclear to what depth in the monolayer R should be measured. As an H_{II} tube is dehydrated, the headgroup area/molecule shrinks and the area/molecule at the chain terminal methyl increases. There exists a zone near the middle of the chain where the change in area varies least with changes in the water content. R is most unambiguously defined to this zone.

An important consequence of the curvature vs. packing model is that it has focused attention on the need to consider structural dimensions as well as transition temperatures. Previously, T_{bh} alone was taken as the indication of the tendency for a lipid system to go into the H_{II} phase. The curvature vs. packing model has made it clear that spontaneous curvature is a better measure in that it more directly relates to the curvature stress accumulated in the L_α phase. The model also clearly distinguishes between two (not necessarily exclusive) classes of molecules:

- Those that primarily affect curvature, as assayed by the size of H_{II} lattices
- Those that primarily affect hydrocarbon packing, as assayed by T_{bh} for a given H_{II} lattice size at a given temperature (Tate and Gruner, 1987)

As examples, DOPC mixed into a DOPE system primarily affects curvature, whereas dodecane mostly affects packing. In general, molecules can contribute to both, as in the case of adding an extra-long chain PC to a DOPE system (Tate and Gruner, 1987).

NMR studies have supported the primary features of the curvature vs. packing model. Sjolund et al. (1987, 1989) demonstrated that very large H_{II} phases can be formed with PCs in the presence of relatively large amounts of normal alkanes and water, even though PCs do not normally form H_{II} phases at the temperatures used. They showed that increasing amounts of dodecane and water are required to form H_{II} phases as the degree of unsaturation of the PC chains increases. This was interpreted in terms of increasing amounts of alkane required to reduce the hydrocarbon packing stress in inverted lipid cylinders of increasing size. It was also found that, although the number of carbon atoms added per DOPC was kept constant, the efficacy of the alkanes in inducing the H_{II} phase decreased from octane to eicosane. Siegel et al. (1989b) used NMR, calorimetry, and X-ray diffraction to investigate the H_{II}-inducing abilities of alkanes and diglycerides on DOPC and N-methylated DOPE (DOPE-Me). The data were consistent with the alkanes' being preferentially located in the periphery of the H_{II} tubes with relatively little effect on the size of the H_{II} lattice. By contrast, 1,2-dipalmitoylglycerol, which also lowers T_{bh}, is preferentially located with the glycerol backbones at the lipid/water interface and the acyl chains roughly parallel to the host phospholipid chains. The lowering of T_{bh} by the diglyceride was accompanied by a decrease in the H_{II} lattice dimension, consistent with a decrease in R_0. As was the case with the Sjolund et al. (1987, 1989) studies, the ability of hydrocarbon to lower T_{bh} was a function of the hydrocarbon used.

5.3.2 OTHER NONLAMELLAR PHASES

Thus far, the discussion has been mainly concerned with the L_α-H_{II} transition. What about other nonlamellar phases, such as the bicontinuous cubic phases (Figure 5.1d)? As reviewed by Lindblom and Rilfors (1989), bicontinuous cubic lipid phases are not uncommon. However, fewer studies have been performed on cubic mesomorphs and, therefore, these phases are less well understood than H_{II} phases or lamellar phases. Cubic phases are difficult to study both because of the structural complexity and because crystallographically well-formed samples are difficult to obtain. The latter feature is probably related to the very low enthalpies associated with bicontinuous cubic transitions, suggesting that formation of these phases is not strongly favored energetically. There is also evidence that the formation of bicontinuous cubic phases involves large energy activation barriers (Shyam-sunder et al., 1988), leading to difficult specimen equilibration problems.

Although an understanding of lipid cubic phases is just beginning to evolve, several general features of these fascinating phases have emerged. In what follows, these features will only be summarized because several detailed reviews of cubic phases, which reference essentially all of the relevant literature, have recently appeared (Luzzati et al., 1987; Mariani et al., 1988; Lindblom and Rilfors, 1989). Crystallographically, there are many different lattices with cubic symmetry, and many diffracted orders are needed to distinguish between them. For many years it was believed that cubic mesomorphs typically consisted of closely packed micelles of either the inverted or noninverted variety. Such phases are, in fact, relatively common in the ternary phase diagrams of many surfactants with high monomer solubilities in water (high critical micelle concentrations) and represent a natural progression of increasing magnitude of curvature from lamellar, to hexagonally packed cylindrical, to spherical or globular surfactant/water interfaces. However, it was recognized early on that many cubic phases appear *between* lamellar and hexagonal phases, both lyotropically and thermotropically, and this phase sequence is inconsistent with a monotonic progression of interfacial curvatures tending toward globular micelles. Kirk et al. (1984) also pointed out that the hydrocarbon packing energy associated with packed micelles was expected to be higher than with hexagonal phases and, therefore, on energetic ground packed micelle phases are unlikely structures between lamellar and hexagonal phases. Cubic mesomorphs that appear between lamellar and hexagonal phases in the phase diagrams are especially common with biomembrane lipids.

NMR investigations of lipid and water diffusion in cubic phases indicated macroscopic diffusion on time scales that were inconsistent with the slow macroscopic diffusion expected of captive micelles (see Lindblom and Rilfors, 1989, for a review). The diffusion coefficients, if extrapolated from L_α phases, are suggestive of lipid/water interfaces that extend macroscopically through the sample in a bicontinuous manner. Seminal X-ray diffraction work by the Luzzati group (e.g., Luzzati et al., 1968), followed by detailed structural solutions of X-ray diffraction patterns by workers in Sweden, France, and the U.S., has demonstrated the bicontinuous nature of cubic mesomorphs that appear between lamellar and hexagonal phases in the phase diagrams. Scriven (1976) suggested that bicontinuous cubic phases might be described by the periodic minimal surfaces (PMS) known from differential geometry. Many researchers now believe that most bicontinuous cubic mesomorphs have the topology of PMS.

A periodic minimal surface is a surface that is three-dimensionally periodic and has zero mean curvature everywhere (see Andersson et al., 1984; Lindblom and Rilfors, 1989). The interest here is on surfaces that are smooth, i.e., without intersections corresponding to discontinuous gradients of the surface. In differential geometry, it is taught that the curvature of a surface at a given point may be uniquely described by two principal curvatures, denoted by C_1 and C_2. These are the inverse radii of the algebraically (signed) extremal best fitting circles tangent to the point in question. Zero mean curvature, $0 = H \equiv 1/2\,(C_1 + C_2)$, corresponds to either the trivial case of a plane (in which case $C_1 = C_2 = 0$) or a saddle surface, in which $C_1 = -C_2$. The PMS cubic surfaces being considered here are such that every point on the surface is a saddle point for which $H = 0$ and the surface structure is periodic in 3-space. An example is the surface traced out by the terminal chain methyls of Figure 5.1d. A number of PMSes are known. At present it is not known as to how many PMSes exist; new ones are being discovered all the time, including ones without cubic symmetry.

At first glance, a surface topology with minimal mean curvature seems oddly placed between the mean curvatures of L_α ($H = 0$) and H_{II} (H large) phases. However, as pointed out by both Gruner and S. Leibler (Gruner et al., 1988), most cubic PMS structures consist of lipid monolayers that drape both sides of PMS-like surfaces such that the PMS-like surface is traced out by the mean position of the terminal chain methyl groups of the multiplied connected bilayers (see Figure 5.1d). In this case, there are two lipid/water interfaces about equally removed by a monolayer thickness from the PMS-like surface. These surfaces do not have zero mean curvature. In fact, they are similar to the surfaces of *constant, nonzero* mean curvature examined by Anderson (1986). Helfrich (1973, 1978) had shown that the energy density of bending a surface with spontaneous curvature, C_0, to the principal curvatures, C_1 and C_2, is given approximately by

$$\Delta E = (K_c/2)(C_1 + C_2 - C_0)^2 + K_g C_1 C_2 \qquad (5.2)$$

This reduces to Equation 5.1 if one of the principal curvatures is zero, as for L_α or H_{II} phases. Note that the middle term of Equation 5.2 is zero if a surface in question has a constant mean curvature such that $0 = C_1 + C_2 - C_0$. Recall that the surface appropriate to the area integration of Equation 5.1 and Equation 5.2 is a bit below the headgroups (Gruner et al., 1986; Rand et al., 1989). If these surfaces conform to a constant mean curvature surface, then the curvature (middle) term of Equation 5.2 integrates to zero, corresponding to a minimum value of the curvature stress. Further, the distance between these opposed near-headgroup surfaces, on either side of the PMS-like surface, is nearly constant, corresponding to a nearly uniform hydrocarbon environment. Thus, for appropriate values of C_0 and the mean hydrocarbon thickness, the PMS-like cubic structures may represent an excellent compromise between curvature stress and packing. This was demonstrated formally by Anderson et al. (1988). Moreover, the progression of the mean curvature of the energetically important near-headgroup surface in a phase sequence from L_α to cubic to H_{II} is now monotonically increasing in magnitude, i.e., bicontinuous cubic structures fit naturally into the observed phase sequence. The same curvature vs. packing argument used to explain the L_α-H_{II} transition may be extended to explain bicontinuous cubic phases.

The physical significance of the rightmost term of Equation 5.2 is unclear. Note that for L_α or H_{II} phases, for which at least one of the principal curvatures is zero, this term is identically zero. The quantitative significance for phases in which neither C_1 nor C_2 is zero is being actively explored.

This term, which is called the *Gaussian curvature term,* could conceivably be involved in more subtle contributions, such as the lattice size or transitions between three-dimensionally periodic structures. The Gaussian term and transitions between different bicontinuous mesomorphs are just two of many unresolved questions about nonlamellar phases.

5.3.3 OTHER FORCES

The spontaneous curvature is the net result of many fundamental interactions that combine in an ill understood manner to yield a quantitative relationship, such as described by Equation 5.2. How do other well-described membrane interactions affect the mesomorphic behavior, either directly or indirectly? An example of a direct interaction is that the surface charge of the monolayer, in low ionic strength solutions, contributes an electrostatic energy to the water volume which is different for the opposed planes of the L_α phase than for the cylindrical cavity of the H_{II} phase. This certainly is expected to affect the phase transition directly. But membrane surface charge may also be expected to have indirect effects. For example, if the lipid headgroups become charged, as may be done with many zwitterionic lipids by simply changing the pH, then the repulsion between neighboring headgroups within the monolayer surface can affect the spontaneous curvature by increasing the net headgroup area. Because spontaneous curvature is a phenomenological interaction, it is unclear where to draw the dividing line between direct and indirect effects due to electrical charge, or, indeed, if the spontaneous curvature is a well-defined, phase-invariant property in the presence of many charged headgroups.

Questions of how specific interactions, such as electrical charge, affect the spontaneous curvature and modulate the phase behavior for a given spontaneous curvature value are just beginning to be explored, both experimentally and theoretically. A number of fundamental interactions that have not been explicitly considered previously in this chapter are known to affect lipid mesomorphic behavior. Besides surface charge, these include hydration effects, van der Waals interactions, geometry-dependent double layer effects of ions in solution, specific adsorption of multivalent ions, steric and chemically specific effects resulting from certain molecules imbedded in the monolayers, and molecular and morphological features that dominate the transition kinetics. Kirk et al. (1984), in first presenting the curvature vs. packing model, attempted to include electrostatic and hydration interactions. This was explicitly a first attempt that led to some qualitative insight, even though the model was relatively unsophisticated. Enough is known about van der Waals, hydration, and electrostatic interactions to begin the theoretical work of examining better models in which these interactions are explicitly included. Specific adsorption effects are more difficult to deal with, because these interactions fall outside the realm of continuum and mean field models. Examples of chemically specific effects include the modulation of phase behavior by gramicidin (e.g., Killian and de Kruijff, 1988) and the interaction of cholesterol with phosphatidylcholine. Molecular and morphological features refer not to equilibrium phase phenomena but rather to features that can so dominate the kinetic pathways of mesomorphic behavior as to completely change it on any reasonable human time scale. For example, there is evidence that the L_α-to-H_{II} transition can only occur at the interface between opposed lipid monolayers (Gruner et al., 1985b). This means that an isolated lipid bilayer, such as an isolated large unilamellar vesicle, can remain in the L_α phase above T_{bh} observed with multilamellae. Any other molecular or morphological feature that simply keeps lamellae apart, such as surface charge in the presence of ample water, bulky "spacer" molecules, or very tightly curved small unilamellar vesicles that resist monolayer opposition over sufficiently large areas, can likewise suppress mesomorphism (Gruner, 1987). Some or all of the effects considered in this paragraph may be operative in the complex environment of real biomembranes. Obviously, much work remains to be done.

5.3.4 KINETICS

It has been proposed by many researchers that one of the reasons that H_{II}-prone lipids constitute such a large fraction of the lipids in biomembranes is that this endows the bilayers with a mesomorphic plasticity, which facilitates the disruption of the lipid/water interface that must necessarily accompany events such as membrane fusion, endocytosis, and large protein transport across membranes. The L_α-H_{II} phase transition is a model for such events in two senses:

- Small R_0 values energetically favor tight local bends of the monolayers.
- The respective topologies of L_α and H_{II} phases are such that disruption of the lipid/water interface necessarily occurs during the phase transition.

What are the exact shapes of the lipid/water interfaces during mesomorphic transitions? This turns out to be a very difficult question to answer, because the rate-limiting shapes are likely to be short lived and do not often form a periodic structure amenable to diffraction analysis. As a result, much of the information about transitory shapes involved during the transitions have been derived from rapid-freeze electron microscope techniques and from NMR, both of which frequently involve controversial interpretation of the data. Prior to discussing kinetic effects, however, it is important to understand how lipid mesomorphic transitions differ from most of the phase transitions studied, for example, in solid-state physics.

A typical polymorphic phase transition in the solid state involves the relative movement of atoms past one another and concomitant breaking and reforming of covalent bonds or redistributions of electron orbitals. The distinguishing feature of lipid mesomorphic transitions is that they are really transitions between the shapes of interfaces, subject to the constraint that a stiff energetic cost is involved in stretching, strongly bending, or tearing the interfaces. Interfacial volumes enclosing hydrocarbon chains are also subject to stiff constraints of separation because of the limited lengths and statistics of the lipid chains. Because one is dealing with a liquid crystal, the lipid headgroups can flow along the interfaces. This implies that once there is a disruption and resealing of an interface that involves a change in the topology, then flow along the new surface pathways can enlarge the change with relatively little energetic cost. The topology of the interfaces is all-important in mesomorphic transitions. Small numbers of defects, such as occur in all real lattices, are likely to dominate the transition behavior by providing low-energy pathways for new, gross geometries. Note also that interfaces cannot readily cross one another. This implies that many pathways derived by flow along random defects will be blocked. The question of kinetic pathways of the phase transitions is, thus, complicated because defects play a large role in determining the minimum energy routes toward forming new geometries.

Evidence for the pathway constraints set by the interfaces is the observation that when an H_{II} phase grows in a multilamellar bilayer array, it always grows in such a manner that the H_{II} tubes are coplanar with the surrounding lamellae (Gruner et al., 1982, 1985b; Caffrey, 1985). Information about the interfacial constraints may be derived by analysis of the shapes of isolated single domains of one phase growing in a matrix of another phase (Gruner et al., 1985b), but very little work, has, in fact, been done on liquid crystal growth habits. Part of the problem in performing such work is the difficulty of growing isolated, single domains. Another provocative experiment, which suggested the importance of interfacial constraints, was the observation of the formation of a crystallographically well-defined cubic phase in DOPE in a region of the phase diagram where a cubic phase had never before been seen (Shyamsunder et al., 1988). It has been noticed that, after performing a temperature jump across the L_α-H_{II} phase transition boundary, there inevitably were traces in the X-ray diffraction patterns of the former phase. These traces disappeared very slowly, as though there were regions of the specimen where the phase transition activation barrier was anomalously high. If rapid, successive jumps back and forth across the transition were taken, the magnitude of the remnant phase, as assayed by the strengths of the diffraction lines, could be increased, as though the "blocked" phase domains were being accumulated. After many jumps through the transition, the lines regrouped, sharpened, and eventually coalesced into a well-defined pattern with cubic symmetry (Figure 5.6).

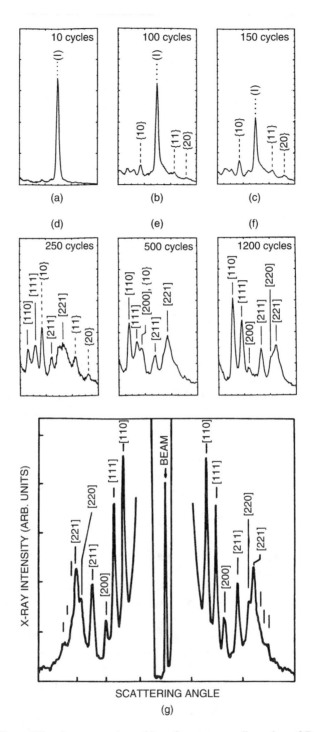

FIGURE 5.6 (a–f) X-ray diffraction pattern intensities of an aqueous dispersion of DOPE after successive thermal cycles across the L_α-H_{II} transition. The primary, unscattered X-ray beam is off the left edge of each pattern. The patterns shown were all recorded at $-5°C$ after the number of cycles indicated in the figures. The dotted line indicates the position of the first order L_α peak. Dashed lines indicate the position of an undercooled H_{II} phase, and solid lines indicate the positions of peaks that index to a cubic phase. The fully evolved cubic phase pattern is shown in (g). (From Shyamsunder E et al.: *Biochemistry* 1988, 27:2332. With permission.)

This formation of a cubic mesomorph is consistent with the model presented earlier, whereby bicontinuous cubic geometries are better compromises between curvature and packing than either L_α-H_{II} phases. Could it be that there are actually equilibrium cubic phases between all L_α and H_{II} phases, but the high degree of topological change (i.e., the number of rips that must be introduced into the lipid/water interface to form the cubics) is so large and energetically restrictive that these equilibrium phases are difficult to form? The Shyamsunder et al. (1988) experiment suggests that this may be the case.

Siegel (1984, 1986a–c, 1987) has presented detailed theoretical models of pathways for topologically disruptive transitions between phases (see Chapter 8). According to this theory, two types of lamellar/nonlamellar transitions may be distinguished. The first is characterized by facile transitions as, for example, seen with L_α-H_{II} phase transitions with DOPE. In these cases, the bulk of the sample undergoes the phase transition within a few degrees of the extremum in the enthalpy profile, as determined by normal scanning differential calorimetry. Further, the temperatures of the enthalpy extremae differ by only few degrees upon heating and cooling. Lipids of this type have relatively small R_0 values. A second type of transition, characterized by N-monomethylated-DOPE (DOPE-Me) exhibits very different behavior upon heating and cooling, typically has sluggish H_{II}-to-L_α transitions, and is identified with relatively large values of R_0 (e.g., see Gruner et al., 1988; Ellens et al., 1989). Siegel has proposed that the transition kinetics of the two types are different because of the relative rate of accumulation of topological features that disrupt the lamellar interfacial integrity. In particular, in the second (sluggish) type of transition, there is an accumulation of grommetlike "interlamellar attachments" that can rearrange to form systems that yield isotropic NMR resonances. Support for the existence of interlamellar attachmentlike structures has been obtained via time-resolved cryotransmission electron microscopy (Siegel et al., 1989).

Kinetically limited topological defects in the lipid/water interface are frequently observed in freeze-fracture electron micrographs of lipid–water systems in the vicinity of mesomorphic transition conditions. These defects, collectively known as *lipidic particles* (Verkleij, 1984), probably span a range of shapes that are dependent on the system in question and the history of the specimen. The defects extend the range of diffusional reorientation of the lipid molecules and, consequently, give rise to sharp, isotropic resonances in ^{31}P and ^{2}H NMR.

5.4 BIOMEMBRANES

The factors that determine the lipid compositions of biomembranes are not known. Biomembranes differ from one organelle to another in specific ways. For example, one type of membrane might always contain certain lipids that are required because they bind in a highly specific manner to the membrane proteins. It is believed that many, if not most, of the lipid species in biomembranes are not in this category because experiments in which the lipid composition of biomembranes is chemically or genetically limited have shown that deletion of many of the normally occurring lipid species does not have serious effects on the test cells. Typically, deletion of certain lipids results in changes in the relative amounts of the remaining lipids, as if several physical parameters were being optimized within the range of compositions available to the cell. Physical parameters that are believed to be optimized, depending on the membrane in question, include the surface charge, bilayer thickness, and the state of liquid crystallinity of the chains. There is no reason to believe that this is the full set of parameters optimized by the cell. An understanding of the lipid compositions that result in a given circumstance would involve a much better understanding than presently exists of the parameters important to a given membrane and the mechanism whereby the lipid composition is actually achieved.

As stated at the beginning of this chapter, most biomembranes contain large amounts of lipids that, in isolation, are prone to form nonlamellar phases. Why are these lipids present? All of the known physical parameters usually thought to be optimized by cells (e.g., bilayer thickness, surface charge, "fluidity," etc.) can be varied without the need for nonlamellar-prone lipid. A premise that underlies much of the research on the biological importance of nonlamellar-prone lipid is that these lipids are prevalent in biomembranes because they perform some function(s) that cannot be performed in their absence.

Wieslander and coworkers (e.g., see Rilfors, 1984; Lindblom et al., 1986; Wieslander et al., 1986) have performed experiments that suggest that the nonlamellar-prone compositions of biomembranes are carefully regulated by living cells. The experiments involved growing *Acholeplasma laidlawii* A in the presence of exogenous fatty acid. This organism, which is a mycoplasma that normally inhabits other bacteria, lacks the cell wall of typical bacteria and has only a single membrane, so pure single-membrane fractions are readily obtained. The organism can be grown so that it incorporates the exogenous fatty acid into the lipids out of which the membrane is made. This still leaves the organism with considerable choice as to the membrane lipid composition, because there is still freedom of selection of the headgroup attached to the fatty acid. It has been found that the resulting lamellar- vs. nonlamellar-prone lipid compositions are carefully regulated. More direct evidence of the regulation of R_0 comes from experiments in which the organisms were grown on various ratios of palmitic and oleic acid. The polar neutral membrane lipids were extracted, and the phase behaviors were examined by NMR (Lindblom et al., 1986). Although the ratio of incorporated oleic acid varied from 20 to 95% and the headgroup ratios to which these chains were attached varied widely, all the samples exhibited transitions out of the lamellar phase over a relatively narrow temperature range. This suggests that the extracted lipid mixtures had similar R_0 values.

The generality of the results obtained with *A. laidlawii* A are not known. Experiments with other bacterial systems (e.g., Goldfine et al., 1987) also point to the importance of regulating the lamellar- vs. nonlamellar-prone lipid composition, although the experimental system was more difficult to work with and the results are, consequently, not as clear cut. Similar experiments will be much more difficult to perform with higher organisms because the fatty acid compositions are not as easily externally varied. However, demonstration of a nonlamellar-prone lipid regulation in any one species is very important, despite the experimental difficulties of extension to other organisms, because it would seem odd if nature chose to devise a specific lipid regulation scheme exclusively for one species.

Four rationales for nonlamellar-prone lipid have been proposed. These four rationales are not mutually exclusive; each might have some measure of validity.

   The first is that there is functional importance to the occurrence of nonlamellar phases in cells. Although nonlamellar phases have been observed in cells, they are, in general, rare. Consequently, this explanation is not likely to account for the compositions of most membranes.

   A second proposition, advocated, for example, by Pieter Cullis (Cullis et al., 1985), is that nonlamellar-prone lipids modulate the permeability properties of biomembranes. Relatively little information either for or against this hypothesis is known.

   A third reason, which has been proposed by many researchers, is that the presence of nonlamellar-prone lipid endows the membrane with mesomorphic plasticity, which is needed during the rare, localized, but inevitably necessary events during which the integrity of the lipid/water interface is disrupted. Biomembranes are assembled at specific sites in cells from which they are blebed off. Many biomembranes are transported across cells as vesicles and eventually fuse with other membranes. The processes of membrane fission and fusion necessarily involve disruption of the lipid/water interface. Although these processes are certainly under protein control, the presence of lipid mesomorphic plasticity may still be required.

   A fourth reason, which has been advocated by Gruner (1985) and more recently by Hui (1987), is that the presence of nonlamellar-prone lipid in bilayers results in the buildup of elastic strain in the bilayers and that this strain modulates the function of certain membrane proteins.

The rationale that lipid mesomorphic plasticity plays a role in membrane fusion has been demonstrated in several pure lipid vesicle systems (see Ellens et al., 1989, and references cited therein). An understanding of how mesomorphic plasticity can be manipulated to control vesicle fusion may have technological importance for applications such as drug delivery and chemical processing (Gruner, 1987). The role mesomorphic plasticity may play in living cells is uncertain, largely because of lack of knowledge of the mechanism of the proteins involved in processes such as biomembrane fusion. Because biomembrane fusion occurs under protein control, an understanding of the role mesomorphic plasticity must ultimately involve a better understanding of the interactions of protein with lipid. Consequently, the rationale of mesomorphic plasticity is really a subset of the notion that lipid curvature strain affects the function of specific proteins.

Nonlamellar phases, as explained earlier in this chapter, may be understood on the basis of a competition between curvature elasticity and the constraints of uniform packing of the hydrocarbon chains. In the bilayer phase, lipid mixtures characterized by small, spontaneous curvature radii result in the buildup of a curvature strain, which is released upon undergoing nonlamellar phase transitions, such as the L_α-H_{II} transition. Other mixtures, characterized by larger R_0 values at the given temperature, are more elastically relaxed in the bilayer phase. What are the biological implications of the buildup of a bilayer curvature strain energy? To choose a concrete example, bilayers rich in unsaturated PEs tend to have small R_0 values, and those rich in PCs tend to have larger R_0 values. Thus, at physiological temperature, bilayers of the former composition will have a built-in strain, whereas the PC-rich bilayers will be more relaxed. How will this physical difference affect, for instance, the functioning of imbedded membrane proteins?

One can conceive of experiments that directly test the hypothesis that protein function is modulated by the spontaneous curvature of the membrane lipid monolayers (Gruner, 1985). It would be necessary to reconstitute the protein, typically a membrane protein, in lipid bilayers in which the composition is varied in a way so as to attain a given value of the spontaneous curvature. This is possible because the spontaneous curvature is a colligative quantity that can be adjusted to a given value by many different lipid compositions. If the activity of the protein were correlated most directly with the spontaneous curvature (and not the compositions used to achieve the spontaneous curvature), then the experiment would be very strong evidence that the spontaneous curvature modulates lipid activity. In practice, such experiments are very difficult. Simply purifying and reconstituting integral membrane proteins is often experimentally difficult. When lipid mixtures are used for reconstitutions, some of the lipid does not end up in the reconstituted vesicles, so the final lipid compositions must be determined by assay. Also, membrane proteins are complex conformational engines that may be subject to many simultaneous requirements, such as the bilayer thickness and charge, as well as to specific auxiliary lipid requirements. In varying the compositions, it will be important to keep all the other requirements constant as well, if the dependence on R_0 is to be unambiguously measured. Even given these difficulties, it will be important to perform experiments whereby correlations with R_0 are examined if the interactions between protein and nonlamellar-prone lipid are to be examined. Investigation of such correlations is an important area for future research.

5.5 CONCLUSION

Although nonlamellar lipid phases have been known for decades, modern investigations that have sought to elucidate the relationship between nonlamellar-prone lipid and biomembranes have mostly been performed since the late 1970s. There was a time when research on lipid mesomorphism was regarded by many biologically oriented scientists as a peripheral activity to the study of biomembranes because nonlamellar lipid phases were rare in living systems. It is now understood that study of the appearance of nonlamellar phases under nonphysiological conditions is not divorced from biological concern. This is because the phase boundaries are frequently very close to the physiological conditions of biomembranes, suggesting that biomembranes regulate their compositions near the edge of instability (Gruner, 1985). There must be advantages to doing so, because it is not hard to select lipid compositions that are far removed from the phase boundaries. Attention is now being focused on understanding what these advantages might be.

Interest in the biological roles of nonlamellar-prone lipid has resulted from an improved understanding of the physical basis of lipid mesomorphism. In particular, it is now understood that a competition between monolayer spontaneous curvature and hydrocarbon packing dominates the mesomorphic behavior of lipid systems. Significantly, it is now recognized that the forces that drive lipid mesomorphism are present in both lamellar and nonlamellar phases and that the focus of study of the biological importance of lipid mesomorphism should be on understanding these forces. It is especially important to investigate the interaction of these forces with biomembrane proteins. Recent experiments have suggested that the activities of at least some important proteins are strongly modulated by the nonlamellar-prone characteristics of the imbedding lipid bilayer.

Interest in nonlamellar-prone lipid has also been catalyzed by studies that have shown that the presence of such lipid is carefully regulated in at least some cellular systems. A major difficulty in performing studies of this kind has been in identifying physical properties that relate to, indeed, define the nonlamellar tendency of a given lipid composition. In the past, the temperature of the L_{α}-nonlamellar phase transition has most commonly been used as a measure of the nonlamellar tendency. Physical studies on isolated lipid systems have demonstrated that the transition temperature is a result of the balance between curvature and other competing forces and that the temperature can be shifted by very small levels of specific impurity molecules. The spontaneous curvature is, itself, a better measure of the nonlamellar tendencies of a given mixture, but few studies have yet correlated biological behavior with this measure. Another problem that limits progress is the fact that most measures of the nonlamellar tendencies of a system require extraction of the lipid. Physical probes are needed that can measure curvature stress in bilayers in the presence of protein.

ACKNOWLEDGMENTS

Lipid research in the author's laboratory is supported by DOE grant DE-FG02-87ER60522-A000, NIH grant GM32614, and ONR contract N00014-86-K-0396P00001.

REFERENCES

Alperine S, Hendrikx Y, and Charvoline J: Internal structure of aggregates of two ampiphilic species in a lyotropic liquid crystal. *J. Phys. Lett.* 1985, 46:L27.
Anderson DM: Studies in the Microstructure of Microemulsion. Ph.D. thesis, University of Minnesota, Minneapolis, MN, 1986.
Anderson D, Gruner SM, and Leibler S: Geometrical aspects of the frustration in cubic phases of lyotropic liquid crystals. *Proc. Natl. Acad. Sci. U.S.A.* 1988, 85:5364.
Andersson S, Hyde ST, and von Schnering HG: The intrinsic curvature of solids. *Z. Kristallogr.* 1984, 168:1.
Bloom M: NMR studies of membrane and whole cells, in *Physics of NMR Spectroscopy in Biology and Medicine.* C. Corso, Italy, 1988.
Bo L and Waugh RE: Determination of bilayer membrane bending stiffness by tether formation from giant, thin-walled vesicles. *Biophys. J.* 1989, 55:509.
Caffrey M: Kinetics and mechanism of the lamellar gel/lamellar liquid crystal and lamellar/inverted hexagonal phase transition in phosphatidylethanolamine: a real-time x-ray diffraction study using synchrotron radiation. *Biochemistry* 1985, 24:4826.

Caffrey M: The study of lipid phase transition kinetics by time-resolved x-ray diffraction. *Annu. Rev. Biophys. Biophys. Chem.* 1989, 18:159.

Caron F, Mateu L, Rigny P, and Azerad R: Chain motions in lipid-water and protein-lipid-water phase: a spin-label and x-ray diffraction study. *J. Mol. Biol.* 1974, 85:279.

Charvolin J: Crystals of interfaces: the cubic phases of ampiphile/water systems. *J. Phys. Paris* 1985, 46:C3–173.

Cullis PR and de Kruijff B: ^{31}P NMR studies of unsonicated aqueous dispersion of neutral and acidic phospholipids: effects of phase transitions, p^{2}H and divalent cations on the motion in the phosphate region of the polar headgroup. *Biochim. Biophys. Acta* 1976, 436:523.

Cullis PR, Hope MJ, de Kruijff B, Verkleij AJ, and Tilcock CPS: Structural properties and functional roles of phospholipids in biological membranes, in *Phospholipids and Cellular Regulations* Vol. 1. Ed. Kuo JF, CRC Press, Boca Raton, FL, 1985.

Ellens H, Siegel DP, Alford D, Yeagle PL, Boni L, Lis LJ, Quinn PJ, and Bentz J: Membrane fusion and inverted phases. *Biochemistry* 1989, 28:3692.

Goldfine H, Johnston NC, Mattai S, and Shipley GG: Regulation of bilayer stability in *Clostridium butyricum:* studies on the polymorphic phase behavior of ether lipids. *Biochemistry* 1987, 26:2814.

Gruner SM, Rothschild KJ, and Clark NA: X-ray diffraction and electron microscope study of phase separation in rod outer segment photoreceptor membrane multilayers. *Biophys. J.* 1982, 39:241.

Gruner SM: Curvature hypothesis: does the intrinsic curvature determine biomembrane lipid composition? A role for non-bilayer lipids. *Proc. Natl. Acad. Sci. U.S.A.* 1985, 82:3665.

Gruner SM, Cullis PR, Hope MJ, and Tilcock CPS: Lipid polymorphism: the molecular basis of non-bilayer phases. *Annu. Rev. Biophys. Biophys. Chem.* 1985, 14:211.

Gruner SM, Rothschild KJ, de Grip WJ, and Clark NA: Co-existing lyotropic liquid crystals: commensurate, faceted and co-planar single hexagonal (H_{II}) domains in lamellar photoreceptor membranes. *J. Phys.* 1985(b), 46:193.

Gruner SM, Parsegian VA, and Rand RP: Directly measured deformation energy of phospholipid H_{II} hexagonal phases. *Faraday Disc.* 1986, 81:29.

Gruner SM: Materials properties of liposomal bilayers, in *Liposomes*, Vol. 2. Ed. Ostro M, Marcel Dekker, New York, 1987.

Gruner SM, Tate M, Kirk GL, So PTC, Turner DC, Keane DT, Tilcock CPS, and Cullis PR: X-ray diffraction study of the polymorphic behavior of N-methylated dioleoylphosphatidylethanolamine. *Biochemistry* 1988, 27:2853.

Gruner SM: Stability of lyotropic phases with curved interfaces. *J. Phys. Chem.* 1989, 93:7562.

Gulik A, Luzzati V, DeRosh M, and Gambacorta A: Structure and polymorphism of bipolar isopranyl ether lipids from archaebacteria. *J. Mol. Biol.* 1985, 182:131.

Gulik A, Luzzati V, DeRosh M, and Gambacorta A: Tetraether lipid components from a thermoacidophilic archaebacterium chemical structure and physical polymorphism. *J. Mol. Biol.* 1988, 201:429.

Helfrich W: Elastic properties of lipid bilayers: theory and possible experiments. *Z. Naturforsch.* 1973, 28c:693.

Helfrich W: Steric interaction of fluid membranes in multilayer systems. *X. Naturforsch.* 1978, 33a:305.

Hendrikx Y, Charvolin J, Kekicheff P, and Roth M: Structural fluctuations in the lamellar phase of sodium decyl sulphate/decanol/water. *Liquid Cryst.* 1987, 2:677.

Hui SW: Non-bilayer-forming lipids: why are they necessary in biomembranes? *Comments Mol. Cell. Biophys.* 1987, 4:233.

Israelachvili JN, Marcelja S, and Horn R: Physical principles of membrane organization. *Q. Rev. Biophys.* 1980, 13:121.

Jensen JW and Schutzbach JS: Modulation of dolichyl-phosphomannose synthase activity by changes in the lipid environment of the enzyme. *Biochemistry* 1988, 27:6315.

Killian JA and de Kruijff B: Proposed mechanism for H_{II} phase induction by gramicidin in model membranes and its relation to channel formation. *Biophys. J.* 1988, 53:111.

Kirk GL, Gruner SM, and Stein DL: A thermodynamic model of the lamellar (L_α) to inverse hexagonal (H_{II}) phase transition of lipid membrane-water systems. *Biochemistry* 1984, 23:1093.

Kirk GL and Gruner SM: Lyotropic effects of alkanes and headgroup composition on the L_α-H_{II} lipid liquid crystal phase transition: hydrocarbon packing versus intrinsic curvature. *J. Phys. Paris* 1985, 46:761.

Lafleur M, Bloom M, and Cullis PR: Lipid polymorphism and hydrocarbon order. *Biochem. Cell. Biol.* 1989 (in press).

Levine YK: X-ray diffraction studies of membranes. *Prog. Surface Sci.* 1973, 3:279.

Lindblom G, Brental I, Sjoland M, Wikander G, and Wieslander A: Phase equilibria of membrane lipids from *Acholeplasma laidlawii*: importance of a single lipid forming nonlamellar phases. *Biochemistry* 1986, 25:7502.

Lindblom G and Rilfors L: Cubic phases and isotropic structures formed by membrane lipids — possible biological relevance. *Biochim. Biophys. Acta* 1989, 988:221.

Luzzati V: X-ray diffraction studies of lipid-water systems, in *Biological Membranes*, Vol. 1. Ed. Chapman D, Academic Press, New York, 1968, pp. 71–123.

Luzzati V, Mariani P, and Gulik-Krzywicki T: The cubic phases of lipid-containing systems: physical structure and biological implications, in *Physics of Amphiphilic Layers*. Eds. Meunier J, Langluin D, and Boccara N, Springer, Berlin, 1987.

Mariani P, Luzzati V, and Delacroix H: Cubic phases of lipid-containing systems. Structure analysis and biological implications. *J. Mol. Biol.* 1988, 204:165.

Navarro J, Toivio-Kinnucan M, and Racker E: Effect of lipid composition on the calcium/adenosine 5'-triphosphate coupling ratio of the Ca^{2+}-ATPase of sarcoplasmic reticulum. *Biochemistry* 1984, 23:130.

Rand RP, Fuller NL, Gruner SM, and Parsegian VA: Membrane curvature, lipid segregation, and structural transitions for phospholipids under dual solvent stress. *Biochemistry* 1989 (in press).

Rilfors L, Lindblom G, Wieslander A, and Christiansson A: Lipid bilayer stability in biological membranes, in *Membrane Fluidity*, Vol. 12. Eds. Kates M and Manson LA, Plenum Press, New York, 1984.

Scriven LE: Equilibrium bicontinuous structure. *Nature* 1976, 263:123.

Seddon JM: Structure of the inverted hexagonal (H_{II}) phase, and non-lamellar phase transitions of lipids. 1989 (in preparation).

Seelig J: Deuterium magnetic resonance: theory and application to lipid membranes. *Q. Rev. Biophys.* 1977, 10:353.

Shyamsunder E, Gruner SM, Tate MW, Turner DC, and So PTC: Observations of inverted cubic phase in hydrated dioleoylphosphatidylethanolamine membranes. *Biochemistry* 1988. 27:2332.

Siegel DP: Inverted micellar structures in bilayer membranes. *Biophys. J.* 1984, 45:399.

Siegel DP: Inverted micellar intermediates and the transitions between lamellar, cubic and inverted hexagonal lipid phases. I. Mechanism of the L_{α} H_{II} phase transitions. *Biophys. J.* 1986(a), 49:1155.

Siegel DP: Inverted micellar intermediates and the transitions between lamellar, cubic and inverted hexagonal phases. II. Implications for membrane-membrane interactions and membrane fusion. *Biophys. J.* 1986(b), 49:1171.

Siegel DP: Membrane-membrane interactions in lamellar-to-inverted hexagonal phase transitions, in *Membrane Fusion*. Ed. Sowers AE, Plenum Press, New York, 1986(c).

Siegel DP: Inverted micellar intermediates and the transitions between lamellar, cubic, and inverted hexagonal amphiphilic phases. II. Isotropic and inverted cubic state formation via intermediates in transitions between L_{α} and H_{II} phases. *Chem. Phys. Lipids* 1987, 42:279.

Siegel DP, Burns JL, Chestnut MH, and Talmon Y: Intermediates in membrane fusion and bilayer/nonbilayer phase transitions imaged by time-resolved cryotransmission electron microscopy. *Biophys. J.* 1989, 56:161.

Siegel DP, Banschbach J, and Yeagle PL: Stabilization of H_{II} phases by low levels of diglycerides and alkanes: an NMR, calorimetric and x-ray diffraction study. *Biochemistry* 1989(b), 28:5010.

Sjolund M, Lindblom G, Rilfors L, and Arvidson G: Hydrophobic molecules in lecithin-water systems. I. Formation of reversed hexagonal phases at high and low water contents. *Biophys. J.* 1987, 52:145.

Sjolund M, Rilfors L, and Lindblom M: Reversed hexagonal phase formation in lecithin-alkane-water systems with different acyl chain unsaturation and alkane length. *Biochemistry* 1989, 28:1323.

Small DM: *Physical Chemistry of Lipids from Alkanes to Phospholipids*, Handbook of Lipid Research Series, Vol. 3. Ed. Hanahan D, Plenum Press, New York, 1986.

Stamatoff JB and Krimm S: Phase determination of x-ray reflections for membrane-type systems with constant fluid density. *Biophys. J.* 1976, 16:503.

Sternin E, Bloom M, and MacKay AL: De-pake-ing of NMR spectra. *J. Magn. Reson.* 1983, 55:274.

Sternin E, Fine B, Bloom M, Tilcock CPS, Wong KF, and Cullis PR: Acyl chain orientational order in the hexagonal H_{II} phase of phospholipid-water dispersions. *Biophys. J.* 1988, 54:689.

Tartar HV: A theory of the structure of the micelles of normal paraffin chain salts in aqueous solution. *J. Phys. Chem.* 1955, 59:1195.

Tate MW: Equilibrium and Kinetic States of the L_α-H_{II} Transition. Ph.D. thesis, Princeton University, Princeton, NJ, 1987.

Tate MW and Gruner SM: Lipid polymorphism of mixtures of dioleoylphosphatidylethanolamine and saturated and mono-unsaturated phosphatidylcholines of various chain lengths. *Biochemistry* 1987, 26:231.

Tate MW and Gruner SM: Temperature dependence of the structural dimensions of the inverted hexagonal (H_{II}) phase of phosphatidylethanolamine containing membranes. *Biochemistry* 1989, 28:4245.

Thayer AM and Kohler SJ: Phosphorus-31 nuclear magnetic resonance spectra characteristic of hexagonal and isotropic phospholipid phases generated from phosphatidylethanolamine in the bilayer phase. *Biochemistry* 1981, 20:6831.

Tilcock CPS, Cullis PR, and Gruner SM: On the validity of ^{31}P NMR determinations of phospholipid polymorphic phase behavior. *J. Chem. Phys. Lipid* 1986, 40:47.

Turner DC and Gruner SM: Electron density reconstruction of the inverted hexagonal (H_{II}) phase in phospholipid-water membranes. *Biophys. J.* 1989, 55:116a.

Verkleij AJ: Lipidic intermembraneous particles. *Biochim. Biophys. Acta* 1984, 779:43.

Wiedmann TS, Pates RD, Beach JM, Salmon A, and Brown MF: Lipid-protein interactions mediate the photochemical function of rhodopsin. *Biochemistry* 1988, 27:6469.

Wieslander A, Rilfors L, and Lindblom G: Metabolic changes of membrane lipid composition in *Acholeplasma laidlawii* by hydrocarbons, alcohols and detergents: arguments for effects on lipid packing. *Biochemistry* 1986, 25:7511.

Wilkins MHF, Blaurock AE, and Engelman DM: Bilayer structure in membranes. *Nature* 1971, 230:72.

Worcester DL: Neutron beam studies of biological membranes and membrane components, in *Biological Membranes*, Vol. 3. Eds. Chapman D and Wallach DFH, Academic Press, New York, 1976.

Wuthrich K: Protein structure determination in solution by nuclear magnetic resonance spectroscopy. *Science* 1989, 243:45.

6 The Forces between Interacting Bilayer Membranes and the Hydration of Phospholipid Assemblies*

R.P. Rand and V.A. Parsegian

CONTENTS

* Much of this work appeared as a major review in *Biochimica et Biophysica Acta*, 1989, 988:351–376. Reprinted with permission.

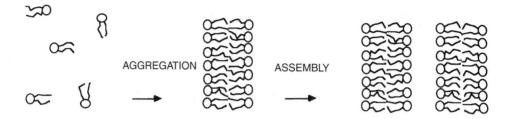

FIGURE 6.1 The hierarchy of lipid aggregation and assembly into bilayers: the short-range driving force of hydrocarbon aggregation to form bilayers covered with polar groups; the assembly of bilayers into multilayer arrays whose spacing reflects long-range interlamellar forces.

6.1 INTRODUCTION

Hydration forces, now known to dominate the interaction of all phospholipid membranes as they approach within 10 to 20 Å, remained unrecognized until relatively recent times. This is surprising, because the phospholipids that merge their hydrocarbon tails create bilayers covered with a phalanx of polar groups that must hold on to the solvent into which they would otherwise dissolve. The tenacity of holding this water is part of a natural tension in amphiphilic aggregates, a balance between the high energy of a hydrocarbon/water interface and the energy lowering adsorption of solvent. That water costs energy to remove. So, when two such surfaces come together, that cost translates into a strong force of repulsion. It has now been recognized as the dominant force between all hydrophilic surfaces.

But why does that force extend 10 or 20 Å and prevent the aggregates from making molecular contact? Most probes of water near bilayer surfaces indicate little perturbation beyond the first hydration layer. The answer appears to be in the very size of the membrane, in the fact that the displacement of a bilayer entails the displacement of hundreds or thousands of water molecules. Even the tiniest energetic perturbation per water molecule is multiplied by hundreds or thousands to be an important energy on the scale of the bilayer membrane (see Figure 6.1).

Our purpose here is to collect information rather strictly on phospholipid hydration, information that has become available from different experimental methods. It would be wrong, though, not to mention for reference what has been learned in other systems. All modern studies of solvation and hydration follow the major achievements of Derjaguin and his school. These people built and designed the first successful surface force apparatus, developed much of the physical theory of long-range forces, and recognized the importance of the "structural component of the disjoining force" (for which read *solvation* or *hydration* repulsion). This work is the subject of a book and several reviews (Churaev and Derjaguin, 1985; Barclay and Ottewill, 1970; Derjaguin and Churaev, 1986; Derjaguin et al., 1987). The swelling of clays, by the action of both electrostatic and hydration forces, has been recognized for several decades. Studies during the past two decades, particularly those of Low and collaborators (e.g., Viani et al., 1982, from which references to the very large earlier literature may be traced) have shown, exponentially, varying forces measured by osmotic stress. We have written elsewhere of the relevance of hydration repulsion to bilayer fusion processes (Rand and Parsegian, 1986; Parsegian and Rand, 1988). Forces measured between natural nerve myelins strongly resemble those seen between phospholipids, although the cell surface is likely to be a far more complicated structure (Rand et al., 1979). We will forgo the temptation here to list the many biological phenomena that may relate to the hydration properties of molecular and membrane surfaces.

Although there were the early experimental signs (Clunie et al., 1967; Churaev and Derjaguin, 1985; Barclay and Ottewill, 1970; Derjaguin and Churaev, 1986; Derjaguin et al., 1987; Luzzati and Husson, 1962; Small and Bourges, 1966; Parsegian, 1967), only during the past decade or so have hydration forces been characterized systematically in terms of their exponential decay, compared with other operative forces (such as van der Waals, electrostatic double layer and steric interactions), and examined in terms of the molecular features of the bilayer surface that regulates them. Even now, these interactions are just beginning to be utilized in analyses of membrane fusion and in theories of phospholipid polymorphism.

We begin with a summary and comparison of various direct methods of force or energy measurement and then attempt to summarize the large body of experimental results to allow their comparison and to attempt to discover their origin. We next describe the ways in which hydration forces combine with other interactions occurring at 10 to 30 Å separations, and the ways in which these forces are expected to act in situations other than those in which they were measured, in phenomena such as unilamellar vesicle interaction and phospholipid phase transitions. In this way, we hope to stimulate systematic work and further examination of outstanding questions relating these important forces to microscopic properties of phospholipid and other molecular assemblies.

6.2 MEASURING HYDRATION REPULSION

Figure 6.2 schematically illustrates three complementary methods of force or energy measurement that contribute to our understanding of these interactions: osmotic stress (OS), surface force apparatus (SFA), and pipette aspiration (PA). Each provides direct measure of some quantities, but each is limited by inference or extrapolation for estimating others. Our comparison shows their complementarity. The best strategy is to use all three to the extent possible.

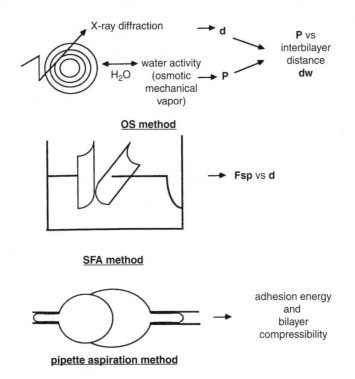

FIGURE 6.2 Three ways of measuring forces between bilayers. Osmotic stress (S) gives the dehydration or desolvation free energy of thermodynamically well-defined phospholipid phases equilibrated with water at different activities; the structural consequences of solvent removal are usually monitored by X-ray diffraction (LeNeveu et al., 1976, 1977; Parsegian et al., 1979, 1986). The surface force apparatus (SFA) is designed to measure attractive and repulsive interactions between crossed cylinders of mica coated with lipid bilayers and immersed in different solutions (Israelachvili and Adams, 1978; Israelachvili and Marra, 1986). Pipette aspiration (PA) measures energies of adhesive contact between large unilamellar bilayer vesicles in solution, as well as bilayer strength or compressibility under lateral deformation (Kwok and Evans, 1981; Evans and Metcalfe, 1984; Evans and Needham, 1987).

6.3 OSMOTIC STRESS

The osmotic stress (OS) method of measuring interbilayer forces in multilamellar systems and between macromolecules in ordered assemblies has been reviewed in detail (Parsegian et al., 1986). It is shown schematically in Figure 6.2. The water in a multilayer array is brought to thermodynamic equilibrium with a second phase of known water activity. This equilibration can be achieved three ways:

- The multilayer is equilibrated with a polymer solution of osmotic pressure P, whose large solutes cannot penetrate between bilayers.
- The multilayer is physically squeezed under a pressure P in a chamber with a semiper-meable membrane to allow exchange with a reservoir of pure water.
- The multilayer is brought to equilibrium with a vapor of known relative humidity (p/p_o) to create an effective osmotic pressure $P = (kT/v_w)\ln(p_o/p)$, where k = Boltzmann's constant, T the absolute temperature, and v_w the partial molar volume of water.

The chemical potential of the water with which the lipid is equilibrated, whether controlled osmotically, by mechanical means, or through the vapor phase, gives the net repulsive pressure P between bilayers. X-ray diffraction of that equilibrated phase gives the repeat distance d of the lipid plus water layers, often to better than angstrom accuracy. The method is of general application and has also been applied to inverted hexagonal lipid phases (Gruner et al., 1986) and to a variety of macromolecular solutions (Evans and Needham, 1987; Gruner et al., 1986). The amount of water per lipid molecule V_w, at any repeat spacing, is determined from gravimetrically mixed lipid and water samples. The amount of water removed under pressure P, ΔV_w, yields the work of dehydration, P, ΔV_w, which is a change in the chemical free energy of the lipids. This work is independent of any model of hydration and of any assumptions about the structure of the phospholipid phase. Using the three methods of applying osmotic stress, it is possible to bring structures from full hydration in water to virtually complete dehydration at pressures corresponding to over 1000 atmospheres, or 10^9 dynes/cm². Attractive forces cannot be measured directly but are inferred from the point of balance between repulsive and attractive forces.

A relation of interbilayer pressure vs. bilayer separation, d_w, could automatically be constructed from measured pressure P vs. repeat spacing d, if d could simply be reduced by a constant bilayer thickness d_1. But bilayers are laterally compressible (Kwok and Evans, 1981; Evans and Needham, 1986, 1987; Evans and Kwok, 1982; Luzzati, 1968). The same isotropic osmotic stress that pushes bilayers together also acts to deform them laterally (Parsegian et al., 1979; Evans and Skalak, 1980), causing the bilayer thickness, d_1, to increase, accompanied by a decrease in cross-sectional area A. Consequently, estimates of these structural changes are required in order to estimate bilayer separation d_w.

There are two traditional ways of gauging the bilayer thickness, as shown in Figure 6.3. The first, the *gravimetric* traditional procedure of Luzzati, divides the repeat spacing d into a lipid layer thickness d_1 that contains all the lipid and none of the water plus a water layer d_w that contains only water (Luzzati and Husson, 1962). This division requires a knowledge of specific volumes of the lipid molecules and their parts and of the intervening water. The procedure works well except near limiting, or saturating, amounts of water where slight changes in d with added water make difficult an accurate determination of water content. Although this approach provides clear evidence of bilayer deformation, it circumvents the difficult issue of interfacial structure. In the second method, the *electron density* profile, determined from low-resolution X-ray diffraction analysis, is used to estimate bilayer thickness. The low resolution requires assumptions about the structure of the interface and takes the bilayer thickness as the distance between two electron density maxima, corresponding to the center of the polar groups, plus a 5 Å width on each side of the bilayer to include the (hydrated) headgroup. By this method it is concluded that there is essentially negligible

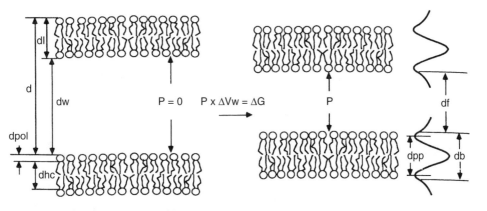

FIGURE 6.3 Geometric parameters for describing bilayer thickness and separation as a function of applied pressure. The repeat spacing d is traditionally divided into a pure lipid layer of thickness d_l and a pure water layer thickness d_w. The volume of water per lipid molecule can be written as $V_w = A.d_w/2$, where A is a mean cross-sectional area projected onto the average plane of the bilayer. A further division of d_l is to imagine a pure hydrocarbon interior of thickness, d_{hc}, between two polar layers, d_{pol}, which contain the lipid polar groups. The work of removal of water $V_w = G$ in the lipid bilayer free energy. Such pressures cause decrease in area A as well as in separation d_w. It is sometimes possible to describe the bilayer dimension by low-resolution electron density distributions, whose peaks correspond roughly to the location of the lipid/water interfaces. The peak-to-peak distance d_{pp} plus a constant to include the width of the polar group layer is defined to be a bilayer thickness d_b. The remaining space, d_f (= d − d_b) is another measure of separation.

change in bilayer thickness. A comparative study (Janiak et al., 1979) showed differences between thicknesses measured in this way and by the gravimetric method.

Each of these methods suffers from allowing only an insensitive measure of bilayer compressibility, i.e., a measure of the changes in bilayer thickness or molecular area with changes in hydration. However, this difficulty can be circumvented by using the independent, very sensitive measurements of bilayer compressibility itself (Evans and Needham, 1987). One may begin by using either the gravimetric estimates of d_1 and d_w at low enough hydration to be quite accurate or by using the estimates from electron densities. Then, using measured lateral compressibilities, it is possible to compute bilayer deformation and changes in d_1 and d_w over a range of deformation for which compressibility is known. We now consider this procedure to be the best available way to determine the variation of bilayer thickness and bilayer separation with hydration. Consequently it is the method of choice to determine most accurately not only the pressure vs. separation relationships but also the structural parameters of the multilamellar phases. In what follows, therefore, we describe this procedure in detail.

6.3.1 MEASURING BILAYER THICKNESS AND SEPARATION

Figure 6.4 shows an example of the relation between the experimentally measured X-ray repeat spacing of the spontaneously formed multilamellar structure, and the weight percent lipid, c, in the samples determined by gravimetrically adding water to SOPC.

On the basis of the densities of lipid and of water, c can be converted to the volume of water per lipid molecule V_w, the volume fraction of lipid in the sample ϕ, and the area A available per lipid molecule on one plane perpendicular to the axis of the lamellar repeat.

$$\phi = 1/(1 + (1 - c) \cdot v_w/c \cdot v_1)$$

$$A = 2 \cdot 10^{24} \cdot MW_1 \cdot v_1/\phi \cdot d \cdot N_o$$

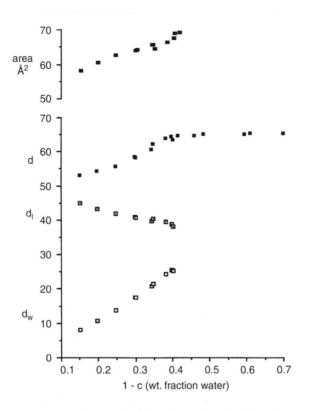

FIGURE 6.4 Structural parameters of the lamellar phase formed by SOPC as they vary with water content and determined by X-ray diffraction; d, lamellar repeat spacing; d_l, bilayer thickness determined gravimetrically; d_w, bilayer separation. Area available per molecule on a plane perpendicular to the direction of lamellar repeat.

$$V_w = (1 - \phi) \cdot A \cdot d/2$$

MW_1 is the molecular weight of the lipid, N_o Avogadro's number, and v_1 and v_w are the partial specific volumes of water and phospholipid, respectively.

These structural parameters are independent of the distribution of the water and lipid within the lamellar repeat distance d, i.e., are model independent. They are dependent, however, on a good knowledge of the partial specific volumes of the lipid. We have listed in Table 6.1 the values used for the indicated lipids. These are taken from, and are consistent with, the measurements and derivations from a number of references listed in Table 6.1.

Interesting observations emerge from these data:

- v_1 for CH_2 chains in bilayers are 1.05 cm³/gm if frozen and 1.17 if melted. Hydrocarbon solutions are 1.07 and 1.29, respectively. This suggests that, when constrained to bilayers, the melted hydrocarbon chains have a lower partial volume than when free in solution.
- The lower value for melted chains in bilayers is required to give sensible polar group partial volumes. For example, from the measured value for DAG, 1.07, if $v_{hc} = 1.17$, v_{pol} = .745, comparable to the published value of .793 for glycerol. This correlation also requires that the C=O groups be included in the polar group molecular weight.
- White et al. (1987) have shown that the global partial specific volume of phospholipids does not change over the range of dehydration used to study interbilayer forces. This lends credence to the structural parameters and their changes derived using the Luzzati formalism.

TABLE 6.1
Partial Specific Volumes for Lipids and Their Parts (cm³/gm)

Liquid paraffin	1.29	*Handbook of Physics and Chemistry*, 1967; Templin, 1956
Melted hc	1.17	Nagle and Wilkinson, 1978
Melted hc, PC, PE	1.17	Nagle and Wilkinson, 1978
Crystalline hc	0.998	Nagle and Wilkinson, 1978
	1.07	Templin, 1956
Crystalline hc PC	1.05	Templin, 1956
	1.008	
	1.006	
gly-Pc+C=O	0.668	Nagle and Wilkinson, 1978
	0.758	Abrahamsson and Pascher, 1966
	0.713 DOPC	
	0.768 DPPCm	
	0.751 DPPCf	
Serine-P	0.660	Putkey and Sundaralingan, 1970
2-amino-eth PO4	0.640	Kraut, 1961
Glycerol	0.793	*Handbook of Physics and Chemistry*, 1967
	0.745 DAG	
gly-PE	0.693	*Handbook of Physics and Chemistry*, 1967
	0.67 DMPE	*Handbook of Physics and Chemistry*, 1967
DPPC, DMPC, DSPC (gel)	0.94	Nagle and Wilkinson, 1978
DPPC (45°C)	1.005	Nagle and Wilkinson, 1978
dmpc (30°C)	0.98	Nagle and Wilkinson, 1978
DSPC (55°C)	1.02	Nagle and Wilkinson, 1978
DOPC (20°C)	0.990	
DMPE	0.96	Nagle and Wilkinson, 1978
DAG	1.07	
Synthetic PEs	0.96–1.02	Seddon et al., 1984

To proceed to define bilayer thickness and separation, assumptions are required about the distribution of water and lipid within the repeat distance d of the multilamellar phase. Using the Luzzati method, which assumes that the lipid and water pack into completely separate layers containing all and only the single component, d can be partitioned into a layer of lipid of thickness $d_1 = \phi.d$ and a layer of water of thickness $d_w = d - d_1$. Further, a knowledge of the molecular weights and densities of the hydrocarbon and polar parts of the lipid molecule (see Table 6.2) allows the bilayer itself to be divided into hydrocarbon, $d_{hc} = \phi_{hc} = \cdot \phi \cdot d$, and polar group layer, $d_p = d_1 = d_{hc}$, thicknesses.

$$\phi_{hc} = MW_{hc} \cdot v_{hc}/(MW_1 \cdot v_1)$$

where MW_{hc} is the molecular weight of the hydrocarbon portion of the lipid molecule and v_{hc} is the partial specific volume of that hydrocarbon.

An alternative definition of bilayer thickness is to use the electron density distribution of the bilayers, shown schematically in Figure 6.3.

Because each of these methods using X-ray dimensions suffers from too low structural resolution to define bilayer thickness and separation adequately, and how they change with dehydration, the independently measured bilayer compressibility modulus, K (Evans and Needham, 1987), is applied to the bilayer thickness measured by X-ray diffraction. This is illustrated for the gravimetric data using the following procedure, but it could as well be applied to the dimensions derived using electron density profiles.

TABLE 6.2

	MW	MW hc	MW p sv.	v	v_{hc}	K	d_l @ log P = 7
POPE	712	433	269	1	1.17	233	42.0@7.04
DOPS	832	471	361	1	1.17		
SOPC	786	475	311	1	1.17	200	41.2@7.08
DGDG	933	462	471	1	1.17	200	39.1
POPE/SOPC							
19/1	715	444	271	1	1.17		
9/1	719	446	273	1	1.17	233	42.0@7.06
4/1	729	449	277	1	1.17		
2/1	731	454	277	1	1.17		
3/2	741	455	286	1	1.17		
1/1	749	459	290	1	1.17		
DGDG/SOPC 45/55	852	469	384	1	1.17	200	39.3@7.06
DGDG/POPE 1/1	822	452	370	1	1.17	216	39.2@7.05
DOPE/DOPC 3/1	750	471	279	0.99	1.17	200	39
eggPE	733	464	269	1	1.17	200	34.1@7.01
eggPEt	733	464	269	1	1.17	200	37.7
eggPEt-Me	747	464	283	1	1.17	200	41.3@7.03
eggPEt-Me$_2$	761	464	297	1	1.17	200	40.8@6.99
eggPC	775	464	311	1	1.17	145	37.7@7.03
PC 16-22	800	490	310	1	1.17	145	39
eggPC/CHOL 1/1	1177	851	326	1.03	1.12	1000	42.1
DPPC/CHOL 1/1	1120	810	310	1.02	1.12	600	43.3
eggPC/DAG-12.5	787	455	332	1	1.17	145	37.3
DLPC	621	311	310	0.98	1.17	145	32.2
DMPC-27	677	367	310	0.98	1.17	145	36.4
DPPC-50	733	423	310	1.005	1.17	145	36.7
DOPC	787	471	316	0.99	1.17	145	36.6
DPPC-25	733	423	310	0.94	1.05	1000	47.2
DSPC	789	473	316	0.94	1.05	1000	47.8
DPPC/CHOL 8/1	772	462	310	0.94	1.05	1000	50.9

Note: Molecular weights of the total lipid (MW), its hydrocarbon (MWhc) and polar (MWP) parts; lipid specific volume (v, cm^3/gm) and specific volume of the hydrocarbon part of the molecule (vhc); bilayer compressibility, K, and bilayer thickness, d_l, at the osmotic stress of log P = 7 used to calculate the structural parameters.

Table 6.2 then provides molecular weights, partial volumes, and compressibilities used to calculate the structural parameters of the lamellar phases of phospholipids.

Figure 6.5 shows the experimentally determined relation between the net interbilayer pressure P and the repeat spacing d of the resultant lamellar phase, again for SOPC. The linear part of the curve can be described by

$$P = P_o \exp(-d/\lambda_d)$$

By reference to the gravimetric data the repeat spacing, d can be translated into the volume of water per lipid molecule, V_w. The relation between P and V_w, also shown in Figure 6.5, can then be described by

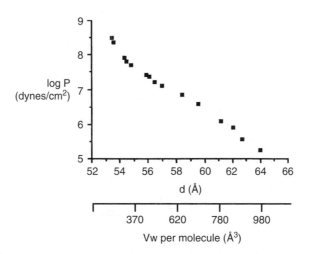

FIGURE 6.5 Net interbilayer pressure P, for SOPC, as it varies either with water content V_w, the volume of water per lipid molecule or with lamellar repeat distance d.

$$P = P_o \exp(-V_w/\mu)$$

The structural parameters d_l and d_w for the osmotic stress data, which make use of the independently measured compressibility of the bilayers, have then been derived the following way. First, one particular bilayer thickness, d^*_l, corresponding to a water content at log P* ~7, is chosen because the water content is known accurately and the compressibility of the bilayer, K, is in the linear range. Second, the compressibility modulus, K dynes/cm, measured by Evans and Needham (1987) is used to calculate bilayer thickness, d_l, and separation, $d_w = d - d_l$, for all the osmotic stress experimental points where log P < 8. The actual values of d^*_l and K used for a variety of lipids are shown in Table 6.2.

K is the fractional change in area for a change in bilayer tension T, and is equal to $\Delta T/\Delta A/A_o$.

For osmotic stress, changes from P* to P cause changes in lateral tension $\Delta T = (P - P^*).d_w$. The fractional change in area $\Delta A/A_o = -\Delta d_l/d^*_l = (d^*_l - d_l)/d^*_l$ for constant lipid molecular volume.

Hence

$$d_l/d^*_l = 1 + (P - P^*/K)d_w$$

and, since $d = d_l + d_w$, then

$$d_l/d^*_l = (K + (P - P^*) \cdot d)/(K + (P - P^*) \cdot d^*_l)$$

from which can be derived from the new d and P, the new d_l and other structural parameters.

We have shown (Rand et al., 1988) that the derived parameters are independent of the chosen osmotic pressure for log P* < 7.5.

Figure 6.6 shows the relation between log P and d_w derived this way. As a descriptor of the data, the linear part of the log P vs. d_w curves is then best fitted to $P = P_o \exp(-d_w/\lambda_c)$. These fits are shown in Table 6.3 for a number of lipid species.

By extrapolating to low stress, the limiting value of d_l can be determined, and, therefrom, all the structural parameters describing the lamellar phase in excess water. They are within error of the gravimetrically derived data. Because they come from the preferred method for deriving lamellar

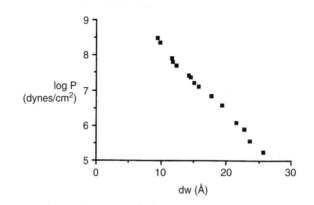

FIGURE 6.6 Net interbilayer pressure P, for SOPC, as it varies with interbilayer separation d_w, determined gravimetrically.

phase dimensions and hydration force parameters, these results are shown for a wide variety of lipids in Table 6.3.

In Table 6.3, repeat spacings of lamellar lattices in excess water, d_o, are directly measured. Water content expressed as dry weight fraction, c_o, cross-sectional areas, A_o, and bilayer separation, d_{wo}, again in excess water, are derived either by the gravimetric method or, where osmotic stress data have been measured, by the new "compressibility" method (see text) using the compressibility K (dyne/cm) either directly measured or inferred (see Table 6.2). Volume of water per lipid molecule $V_{wo} = A_o \times d_{wo}/2$ for comparison among different species; this volume has been renormalized as V_{wo}/PE, the volume of water per mass of PE headgroup. Hydration force parameters P_o and are fitted to data in the high-pressure region where log P vs. separation is a straight line. The G_{min} (ergs/cm^2) are absolute values of negative quantities; these estimates are based on extrapolation of exponentially decaying P to separation d_{wo} with the assumption of a van der Waals attraction vs. exponential hydration repulsion. Data without reference are our unpublished results.

6.3.2 Extracting Hydration Pressure P_h from P

Dissection of the measured net pressure P into its physically distinct components is a problem almost as difficult as the theoretical explanation of those components themselves. For convenience, we imagine that the net force per unit area is made up of four kinds of interactions:

- P_h — repulsion due to surface hydration
- P_{es} — electrostatic double layer repulsion due to surface charge
- P_{vdw} — attractive van der Waals or dispersion forces
- P_{fl} — repulsion from the structural undulations of the thermally excited bilayers

Because a proper theoretical formulation of any of these components poses severe problems at the observed distances of bilayer encounter (Rand et al., 1988), we emphasize here an empirical description of P vs. d_w rather than forcing the data to fit into an arbitrary formalism. Still, some assumptions must be made.

Except near the limit of swelling in unlimited amounts of water, the nature of the force observed between neutral bilayers is essentially exponential with decay distance of some 1 to 3 Å. We therefore describe this force in form

$$P_o \exp(-d_w/\lambda) \qquad (6.6)$$

by fitting to points that are well away from the region where attractive forces create a deviation from exponentiality.

TABLE 6.3

	d_o (Å)	c_o	Å (Å²)	d_{lo} (Å)	d_{wo} (Å)	V_{wo} (Å³)	V_{wo}/PE (Å³)	K	λ	$\log P_o$ (Å)	G_{min} (Å)	References
DDPE	45.8	0.72	55	32.5	13.3	365	365					Seddon et al. 1984
DAPE	57.3	0.79	58	47.3	10	290	290					Seddon et al. 1984
DLPE	46.1					270	270					Seddon et al. 1984
POPE (30°C)	53.2	0.79	56.6	41.8	11.4	323	323	233	0.8	12.5	0.14	Rand et al. 1988
DOPS (.8M)	53.5	0.74	70	39.6	13.9	485	361					Rand et al. 1988
SOPC (30°C)	64.6	0.63	64.3	40.6	24	771	667	200	2.0	10.5	0.02	Rand et al. 1988
DGDG	53.2	0.73	79.8	38.8	14.4	574	328	200	1.7	10.3	0.24	Rand et al. in press
POPE/SOPC												
19/1	54.5											Rand et al. 1988
9/1	56.4	0.74	57.3	41.7	14.7	421	415	233	1.3	11.2	0.09	Rand et al. 1988
4/1	59.9	0.68	58.5	41.5	19.7	576	559	222	2.1	10.0	0.08	Rand et al. 1988
2/1	61.2											Rand et al. 1988
3/2	63.3											Rand et al. 1988
1/1	63.8											Rand et al. 1988
DGDG/SOPC 45/55	57.2	0.68	72.8	38.9	18.3	666	467	200	1.8	10.6	0.18	Rand et al. in preparation
DGDG/POPE 1/1	54	0.72	70.2	38.9	15.1	530	385	216	1.7	10.3	0.23	Rand et al. in preparation
DOPE/DOPC 3/1	58	0.67	63.8	38.6	19.4	619	597	200	1.8	10.2	0.03	Rand et al., in preparation
eggPE	52.9	0.64	72.1	33.8	19.1	690	690	200	1.3	12.5	0.14	Lis et al., 1982
DOPE (-2°C)												Gruner et al., 1988
DOPE-Me (-2°C)	52	0.70	65	37	15	487	487					Gruner et al., 1988
DOPE-(Me)₂	61	0.63	62	39	22	682	648					Gruner et al., 1988
(-2°C)	62	0.60	66	38	25	825	747					Gruner et al., 1988
DOPC (-2°C)	61	0.59	70	36	24	840	727					Gruner et al., 1988
eggPEt	52	0.72	65	37.4	14.6	474	474	200	1.1	12.3	0.20	Rand et al., 1988
eggPEt-Me	61.8	0.66	60.7	40.8	21	637	605	200	1.8	10.3	0.01	Rand et al., 1988
eggPEt-(Me)₂	63.1	0.64	62.6	40.4	22.7	713	646	200	1.8	10.4	0.01	Rand et al., 1988
eggPC	61.9	0.60	69.5	37	24.9	866	749	145	2.1	10.6	0.03	Rand et al., 1988
PC 16-22	63.5	0.60	69.3	38.3	25.2	873	758	145	2.1	10.1	0.01	Rand et al., 1988
eggPC/CHOL 1/1	65.5	0.64	95.6	42	23.5	1126	929	1000	1.1	13.8	0.003	Lis et al., 1982

TABLE 6.3 (Continued)

	d_o (Å)	c_o	A_o (Å²)	d_{1o} (Å)	d_{wo} (Å)	V_{wo} (Å³)	V_{wo}/PE (Å³)	K	λ	$logP_o$ (Å)	G_{min} (Å)	References
DPPC/CHOL 1/1	66	0.65	87.9	43.1	22.9	1005	872	600	1.5	11.5	0.01	Lis et al., 1982
eggPC/DAG-12.5	63	0.58	81.2	36.6	26.4	1070	829	145	2.4	10.4	0.05	Das and Rand, 1986
DLPC	59	0.54	64	31.6	27.4	877	761	145	2.0	10.6	0.01	Lis et al. 1982
DMPC (27°C)	62.2	0.57	61.7	35.7	26.5	816	708	145	2.2	10.5	0.02	Lis et al. 1982
DPPC (50°C)	67	0.54	68.1	35.9	31.1	1059	919	145	2.1	11	0.01	Lis et al. 1982
DOPC	64	0.56	72.1	35.9	28.1	1013	862	145	2.1	10.6	0.01	Lis et al., 1982
DPPC (25°C)	63.8	0.74	48.6	47.1	16.7	405	351	1000	1.2	12.3	0.03	Lis et al., 1982
DSPC	67.3	0.71	51.6	47.7	19.6	506	431	1000	1.3	12.9	0.15	Lis et al., 1982
DPPC/CHOL 8/1	80	0.64	47.5	50.8	29.2	694	602	1000	2.0	10.7	0.004	Lis et al. 1982

Note: Repeat spacings of lamellar lattices in excess water, d_o, are directly measured. Water content, expressed as dry weight fraction, c_o, cross-sectional areas, A_o, and bilayer separation, d_{wo}, again in excess water, is derived either by the gravimetric method or, where osmotic stress data have been measured, by the "compressibility" method (see text) using the compressibility K (dyne/cm) either directly measured or inferred. Volume of water per lipid molecule $V_{so} = A_o \cdot d_{wo}/2$. For comparison among different species, this volume has been renormalized as V_{wo}/PE, the volume of water per mass of PE headgroup. Hydration force parameters P_o and are fitted to data in the high pressure region where log P vs. separation is a straight line. The G_{min} (ergs/cm²) are absolute values of negative quantities; these estimates are based on extrapolation of exponentially decaying P to separation d_{wo} with the assumption of a van der Waals attraction vs. exponential hydration repulsion. Data without references are our unpublished results.

6.4 SURFACE FORCE APPARATUS

A second method of force measurement (Figure 6.2) applied to phospholipid bilayer interactions is by means of a *surface force apparatus* (SFA) (Israelachvili and Adams, 1978; Israelachvili and Marra, 1986). Here, one coats lipids, either by adsorption from suspension (Horn, 1984) or by passage through monolayers (Marra, 1985), onto mica sheets glued onto cylindrical surfaces. One measures the distance between the crossed cylinders by means of interference fringes that are set up between the silvered backs of the mica sheets. Forces between the surfaces of the crossed cylinders are read from the deflection of a cantilever spring system of variable tension that can be moved to bring the surfaces to a given separation. Repulsive forces are seen as a continuous deflection away from contact and are limited by the onset of deformation of the mica surface. Attractive forces are seen either from the position of a jump into "contact," as surfaces are brought together with springs of different thickness, or from the position of a jump away from a spontaneously assumed minimum energy position. Relative changes in position can be measured to an accuracy of 1 Å. The "zero" of separation is computed by subtracting from the measured distance of contact between half-bilayers in air and subtracting again the thickness of a bilayer based on estimated phospholipid volume and the lipid cross-sectional area of the source monolayer.

Two important differences between this method and the osmotic stress method are

- The immobilization of bilayers that comes from attachment to the mica surface
- The cylindrical vs. parallel geometry

To correct for the geometry, the mica surface measurements routinely assume the validity of a transformation, due to Derjaguin (1934), that the force between crossed cylinders of equal radius R is the same as the force F_{sp} between a sphere of radius R and a plane flat surface. Further, this *force* F_{sp} is equivalent to the *energy* E_{pp} between plane parallel surfaces of the same material. Specifically

$$E_{pp} = F_{sp}/2\,R \qquad (6.7)$$

For this reason, forces F_{sp} measured with the SFA are routinely plotted as F_{sp}/R and are therefore implicitly related to the *energy* rather than the *force* between parallel surfaces.

The position of a spontaneously assumed minimum energy (zero force) position between bilayers in the multilayer system will occur at a greater separation than that seen as a point of force balance in the mica cylinder system (see Figure 6.7). So, to compare forces measured on multilayers with those between crossed cylinders, it is necessary either to differentiate the cylinder-cylinder forces or to integrate multilayer forces from a hypothetical infinity.

The observation (Marra, 1986) that mica surfaces bend at F_{sp}/R 10 dyne/cm places an upper limit on the equivalent pressure between planar surfaces to which the SFA method can be used. We can say that an exponential pressure P of decay rate corresponds to an energy, $\lambda*P = E_{pp}$ with a maximum value 10/2 2 dyne/cm. Then for λ 2 Å, typical of phospholipid hydration repulsion, the maximum measurable pressure will be P 10^8 dynes/cm^2.

6.5 PIPETTE ASPIRATION

Evans (Kwok and Evans, 1981; Evans, 1980; Evans and Metcalfe, 1984; Evans and Needham, 1987) has developed a procedure (schematically shown in Figure 6.2) for manipulating vesicles, aspirated into the ends of pipettes, in order to determine mechanical properties of isolated vesicles and contact energies of adhering vesicles. On an isolated vesicle one measures the tongue length inside the pipette as a function of applied suction pressure ΔP. At first, small pressures have a large effect on tongue length because of the removal of bends and folds; then the bilayer becomes taut with the subsequent length/pressure relation reflecting the bilayer area elasticity. We follow Evans's terminology in referring to the modulus of this elasticity as a "compressibility."

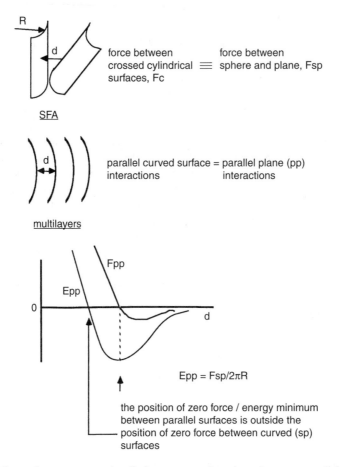

FIGURE 6.7 Forces between crossed cylinders compared to those between parallel surfaces of the same material. By the Derjaguin approximation, the force between crossed cylinders F_c is equivalent to that between a sphere and plane F_{sp}; it corresponds to the energy E_{pp} between parallel surfaces when their separation, d, is much less than the radius of curvature R. The relation between them is $E_{pp} = F_{sp}/2\pi R$. The separation at the point of force balance (or minimum energy) between long-range attraction and shorter-range repulsion will always occur at smaller separations for oppositely curved surfaces than for parallel surfaces. Remarkably, the equilibrium separation between oppositely curved surfaces is independent of their size.

For bilayer adhesion measurements, two vesicles are drawn taut and brought to contact on the ends of approximately coaxial pipettes. Measured diameters and tongue lengths are used to determine bilayer area. Then, keeping both pipettes fixed and maintaining tension on one vesicle, tension is relaxed on the other allowing it to spread over its taut neighbor. Measurement of the diameter of the contact area, together with monitored pressure and tongue length, allows determination of contact angle and lateral tension in the bilayer. These then combine using Young's equation to give the adhesive contact energy, G_{min}, as described later. These are used as a standard to compare with energies derived from the integrated force curves from the osmotic and SFA techniques.

6.6 MAXIMUM HYDRATION AND STRUCTURAL DIMENSIONS OF LIPID MULTILAYERS

The most unassuming measure of the strength of hydration of bilayers is the amount of water multibilayers imbibe from excess solution. Whether determined from the simpler gravimetric method or further refined by adjusting for compressibility makes little difference to the measured

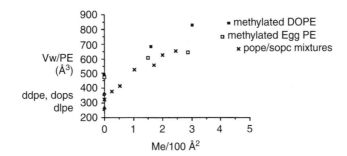

FIGURE 6.8 The uptake of water by multilayers under zero osmotic stress correlates strongly with the density of polar surface methyl groups. The correlation seems to hold whether methyls are successively bound to polar groups (squares) or are added by mixing methylated (SOPC) and unmethylated (POPE) species (X symbols). Pure DDPE and DLPE, as well as a nonmethylated DOPS in solutions of high salt concentration, hydrate in the same low range as DOPE and POPE. DOPE data from Gruner et al. (1988); POPE from Seddon et al. (1984); the remainder from Rand et al., 1988.

maximum volume of water per lipid molecule. Polar group identity, polar group methylation, the physical state of the hydrocarbon chain, chain heterogeneity, mixing of lipid species all appear to affect total hydration. We have grouped the entries in Table 6.3 to facilitate recognition of these factors without intending to obfuscate other comparisons that might occur to the reader. This is not a comprehensive list of lipids that have been studied but is selected to highlight the major differences in maximum hydration of neutral lipids, or of charged lipids in high ionic strength. Also, we have used the preferred compressibility adjusted values where available, otherwise the gravimetric. Still, it is worth noting that qualitative comparisons using the gravimetric and compressibility derivations are little different. This updates an earlier review of phospholipid hydration (Rand, 1981).

In order to make comparisons among lipids that differ in size of polar group, the maximum volume of water per molecule, V_{wo}, has been normalized to V_{wo}/PE, a volume of water per polar group mass equal to that of PE. Note that differences in the amount of water usually correlate with comparable differences in maximum bilayer separation in excess water, d_{wo}.

The most striking factor that increases maximum hydration is methylation of the polar group layer. This summarized in Figure 6.8, where maximum hydration V_{wo}/PE is plotted as it varies with the number of methyl groups per 100 $Å^2$ of polar group surface. The dramatic effect of methylation is seen among the following factors which affect maximum hydration.

6.6.1 HYDRATION OF LIPIDS WITH DIFFERENT POLAR GROUPS

In the comparisons of the homogeneous synthetic lipids the PEs, DOPS, SOPC, and DGDG, the methylated species hydrate nearly twice as much as the other lipids. As a class, phosphatidylcholines (PCs) hydrate more than PEs, even though there is a wide range of sorption within each class. Compare, for example, palmitoyloleylPE (POPE) with steroyloleylPC (SOPC), whose hydrocarbon chains differ by only a –CH2-CH2-link in one chain. Their cross-sectional areas Å differ by less than 15%, bilayer thicknesses d_{lo} by less than 3%, yet the volumes taken up and bilayer separations differ by more than a factor or two.

Dioleylphosphatidylserine (DOPS), a charged lipid, put in .8 M NaCl to screen out electrostatic repulsion, and digalactosyldiglyceride (DGDG), a neutral species, swell only as much as POPE. However, the swelling of melted chain PCs, egg and dilauryl at room temperature, dimyristoyl at 27°C, dipalmitoyl at 50°C, are much like SOPC. To us, the higher hydration suggests the action of polar group methylation, the defining difference between the PCs and their unmethylated sisters.

6.6.2 POLAR GROUP METHYLATION

In the methylation of eggPEt to eggPC, large but disproportionate increases in hydration result with each methylation. Beginning with eggPEt, a PE created by replacement of the polar groups of eggPC, then creating singly (eggPEt-Me) and doubly (eggPEt-Me^2) methylated derivatives, one may systematically examine the effect of methylation alone. A single methylation results in a 28% increase in hydration, whereas successive methylations give 7%, then 16%, increases for the fully methylated PC. This is seen as well with the successive methylations of DOPE (Gruner et al., 1988) and DMPE (not shown in Table 6.3) (Cevc, 1988). (See also Vaughan and Keough, 1974; Mulukutla and Shipley, 1984).

6.6.3 METHYLATED LIPIDS ADDED TO BILAYERS

These last effects of methylation hold also when the methylation of the polar layer is varied by mixing, in the bilayer, methylated and unmethylated species, either SOPC and POPE, shown here, or egg PC and eggPE (Jendrasiak and Mendible, 1976). Figure 6.8 shows a remarkable parallel between these mixed bilayers and the methylated series just described.

   These studies of the systematic methylation of bilayers show that there is a disproportionate effect of the first methyl groups. A single methylation of the PE polar group results in a larger increase in hydration, with smaller increases on successive methylations. SOPC/POPE mixtures 2/3 hydrate to the same extent as pure SOPC. These disproportionate effects suggest that, beyond bringing their complement of water to these mixtures, lipids with methyl groups induce a structural change in PE bilayers that results in further hydration. We suggest later in the text that this change is a disruption of hydrogen bonding that appears as an attractive force between (PE) bilayers (Rand et al., 1988).

6.6.4 CHAIN MELTING AND HETEROGENEITY

There are differences in hydration that appear to reflect effects of the hydrocarbon chains. Gel-phase lipids hydrate less than their melted counterparts; DPPC-25°C < DPPC-50°C. Among the PEs, hydration increases with chain heterogeneity and degree of polyunsaturation: POPE < eggPEt < egg PE. Between SOPC and DOPC, SOPC with one unsaturated bond seems to hydrate less than DOPC, which has two.

6.6.5 ADDITION OF NONPOLAR LIPIDS TO BILAYERS

Cholesterol or diacylglycerol (DAG), which can be considered to act as a lateral spacer between polar groups, causes large increases in water uptake per polar group mass. Thus, cholesterol added to DPPC at low levels, so that most of the hydrocarbon chains are still in the gel state, results in a large increase in hydration (Rand et al., 1980). Addition to the extent of disordering the chains at room temperature gives hydration levels equal to that of the melted state. Equimolar levels of cholesterol (38) or 12 mol% DAG (Das and Rand, 1986) added to egg PC result in hydration levels somewhat larger than pure egg PC. To pressures of 10^8 dynes/cm^2, McIntosh et al. (1988) report little effect in pressure vs. separation for egg PCs to which cholesterol has been added up to 1:1 molar ratios. Given the lateral dilution of polar groups by cholesterol (Rand and Luzzati, 1986), this again shows increased hydration normalized per polar group.

6.7 FORCES BETWEEN APPROACHING LIPID BILAYERS

Even to choose the mathematical form for describing pressure vs. separation, one must be aware of at least five different kinds of interactions expected to occur between bilayers: electrostatic double layers forces, the hydration force due to perturbations of water by the polar surface, van

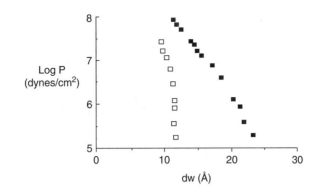

FIGURE 6.9 Comparison of pressure vs. separation for POPE (open squares) and SOPC (solid squares) both at 30°C. Note difference in range and slope (cf. Table 6.1). Both lipids undergo phase transitions at pressures above those shown here.

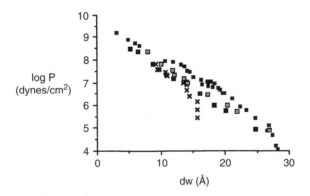

FIGURE 6.10 Pressure vs. distance measured for eggPEt and its methylated derivatives. (x) eggPEt, (open square) singly, (square with dot) doubly, and (solid square) fully methylated. The greatest change is wrought by the first methylation.

der Waals attraction that limits multilayer hydration, repulsion due to thermal undulations of the whole bilayer, and possibly steric interactions of polar groups whose conformations are confined by an approaching surface. In the sense that these all involve a positive or negative work of removal of water between bilayers, they are all "hydration forces" of some kind. The challenge is to estimate the relative contribution of each to the total force or energy. Each of these interactions will be considered in more detail later. The problem with any empirical description is to decide how to fit an experimental curve, which can be fit with a minimum of parameters, with a set of postulated interactions that involve many more. Little can be learned using more than the minimum required parameters.

We emphasize here a minimum parameter description of the measured force curves. Plots of pressure as log P vs. separation d_w, Figure 6.9 through Figure 6.12, all suggest exponential decay of the net total force at high pressures, then a drop to a limiting separation, d_{wo}. A minimum description of the exponential part is given by

$$P_o \exp(-d_w/\lambda) \tag{6.8}$$

Whether we fit to the full curve using this exponential plus an attractive van der Waals potential to enforce the "hydration minimum" at d_{wo}, or whether we fit an exponential to the upper part alone, there is no qualitative effect on the extracted λ and P_o. These parameters do give, respectively,

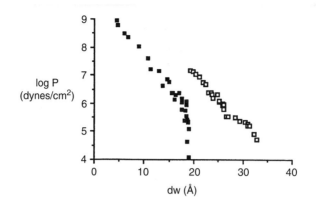

FIGURE 6.11 Effect of chain melting on hydration repulsion. DPPC at 50°C (open squares) and at 25°C (solid squares). The 50°C data are limited to log P 7, because further dehydration causes acyl chain freezing.

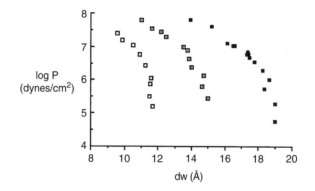

FIGURE 6.12 Comparison of forces between bilayers with identical (PE) polar groups but different hydrocarbon chains; (open square) synthetic POPE, (solid square) natural eggPE, (square with dot) a derivative eggPEt made by transphosphatidylation of eggPC. Chain heterogeneity and degree of polyunsaturation increase tendency for PE hydration (Rand et al., 1988).

a good comparative empirical measure of the range of the repulsive force, as well as the strength it is expected to reach at a given separation. From these, one knows the energy per unit area (or per molecule) that is encountered when two bilayers approach. Much more difficult to determine is the contribution of each separate, underlying force with an aim to understanding its physical origin (Rand et al., 1988).

Using compressibility-adjusted estimates, decay constants, λ, do correlate in a systematic way with polar group identity and state of the hydrocarbon chain (Table 6.3). Statistical tests of the decay lengths (Rand et al., 1988) show that with a probability >98% of these decay lengths are in the sequence

$$POPE < eggPEt = eggPE < eggPEt\text{-}Me = eggPEt\text{-}Me_2 = egg\ PC = SOPC \qquad (6.9)$$

One sees that all the PEs (POPE, eggPE, eggPEt) have λs from 0.8 to 1.3 Å, a range that includes decay rates for the two frozen-chain PCs (DPPC-25°C and DSPC). Decay constants for the melted chain PCs, with the exception of eggPC/cholesterol 1/1, are from 1.5 to 2.4 Å. λ for DGDG is somewhat closer to those for PCs.

As with total hydration V_{wo}, when compared within related lipid species, there is an effect of methylation on λ (Figure 6.10, Table 6.3). A single methylation changes the value from 1.1 Å for eggPEt to 1.8 Å for the monomethyl eggPEt-Me and the dimethyl eggPEt-Me$_2$, compared to 2.1

for the fully methylated eggPC. And a 9/1 POPE/SOPC mixture shows a λ of 1.3 Å compared to 2.1 Å for a 2/1 mixture. Chain melting (Figure 6.11) and increased chain heterogeneity (Figure 6.12) increase bilayer hydration, seen in terms of force curves, just as they do in terms of maximum water adsorption.

6.8 HYDRATION FREE ENERGY

It is important to recognize that the dehydration measurements made under osmotic stress are in fact a direct measure of the free energy of the lipids as a function of the amount of water. Because lipid phase transitions usually involve significant changes in water content, these measured free energies can be a useful source of information in examining the phospholipid–water phase diagram and in testing various models of phase transitions. Guldbrand et al. (1982) were the first to recognize this possibility in a model of the gel to liquid–crystal transition. Leibler and Goldstein have recently developed an order parameter formalism to include hydration energies in this same transition (Goldstein and Leibler, 1988). Cevc and coworkers have performed practical and imaginative measurements of the temperature and entropy of this same transition as a function of water content (Cevc and Marsh, 1985; Cevc et al., 1986). Because we believe that this kind of analysis is just the beginning of many possible uses of dehydration/phase transition data, we have codified the data on bilayer dehydration in terms of osmotic stress vs. water volume parameters (Rand and Parsegian, 1989). The enormous energies of bilayer dehydration may be appreciated by examining one case, eggPC say, where forces have been measured virtually to zero water. Taking $\lambda^* P_o$ as a measure of the integrated work, one sees (Table 6.3) that at a pressure of 10^9 dynes/cm^2, say, dehydration has involved a work of some 20 erg/cm^2, and that for bilayers approaching zero-water contact, this energy can grow to the order of 100 erg/cm^2. Translated into chemical units, this amounts to 2 to 10 kcal/mol. These energies are of the magnitude known for oil/water or vapor/liquid contact.

6.9 MEASURED BILAYER ADHESION ENERGIES

Phospholipid hydration repulsion, preventing anhydrous contact, is an important factor affecting the strength of adhesion between electrically neutral bilayers. For this reason, the measured strength of adhesion between bilayers can be a useful inverse indicator of the strength of bilayer hydration. The three methods described provide estimates of adhesion.

First, the pipette aspiration method provides the most direct measure of the adhesion energy per unit area, G_{min} of spontaneous interaction between bilayers. The measured contact angle Φ and applied bilayer tension T in the bilayer give, by Young's equation (Evans and Parsegian, 1986),

$$G_{min} = 2T(\cos \Phi - 1) \qquad (6.10)$$

One can combine this G_{min} with an estimate of average bilayer separation d_{wo} measured by X-ray diffraction. This combination allows one to test various models for bilayer attraction forces and to correlate strength of adhesion with hydration force measurements (Evans and Needham, 1987; Rand et al., in preparation).

The osmotic stress measurements alone allow a second estimate of contact energy, but it relies on an extrapolation of the exponential repulsive force

$$P_o \exp(-d_w / \lambda) \qquad (6.11)$$

to the position, d_{wo}, where this force is equal and opposite to a longer-range attractive force. For example, if one assumes the distance dependence of van der Waals attraction in its simplest form, one has (LeNeveu et al., 1976)

$$F_{vdw} = A_h/(6d_w^3) \tag{6.12}$$

as indicated from SFA measurements (Marra, 1985, 1988; Marra and Israelachvili, 1985). Then, integrating these two forces from infinity, one infers

$$G_{min}(d_{wo}) = F_{vdw}(d_{wo})[(d_{wo}/2) - \lambda] \tag{6.13}$$

At d_{wo}

$$F_{vdw}(d_{wo}) = P_o \exp(-d_{wo}/\lambda) \tag{6.14}$$

so that

$$G_{min}(d_{wo}) = [d_{wo}/2) - \lambda]P_o \exp(-d_{wo}/\lambda) \tag{6.15}$$

which can be evaluated from measured P_o, d_{wo}, and λ.

As long as the attractive force is of much longer range than the repulsive, the order of magnitude of G_{min} extracted by this procedure is not very sensitive to the form of attraction. For example, if one includes the finite bilayer thickness or even subdivides the bilayer into regions of different polarizability (LeNeveu et al., 1977), estimates of G_{min} will not be qualitatively affected.

The force F_o at the position of maximum attraction where two phospholipid coated mica surfaces jump together in the SFA gives yet a third way to measure G_{min} (Marra and Israelachvili, 1985; Marra, 1988). By the Derjaguin approximation (see Figure 6.7 and related text)

$$G_{min} = F_o/2\,R \tag{6.16}$$

In Table 6.4 we have compared estimates from these three methods. What is puzzling is the much greater estimate of G_{min} from the coated mica surface measurements compared to those between unsupported bilayers. In any case, all these minima are erg/cm², relatively weak on the scale of oil/water or vapor/liquid interfacial energies.

6.10 HYDRATION OF CHARGED PHOSPHOLIPID BILAYERS

In general, charged bilayers separate indefinitely in excess solution unless sufficiently screened. But at close range, because of the high pressures produced by combined hydration and electrostatic double layer forces, hydration interactions between charged phospholipids are not always easy to see. With the OS method, it is necessary to apply high stress and go to small spacings, especially in solutions of low salt concentration, in order to see deviation from pure electrostatic repulsion. Often, such pressures cannot be attained with the SFA before there is bending of the supporting mica surfaces. For example, measurements of forces between disteroylphosphatidylglycerol (DSPG) in NaCl solutions, using the surface force apparatus, were limited to separations greater than 20 Å, a separation too large and at pressures too low to see hydration repulsion (Marra, 1986).

It is worth considering where one should see a transition from electrostatic double layer-dominated repulsion to a regime of hydration force dominance.

Consider, for simplicity, an electrostatic repulsion, P_{es}, between parallel surfaces, separated by a distance d, of the form

$$P_{es}(d) = P_{eo} \exp(-d/\lambda_e) \tag{6.17}$$

TABLE 6.4
Measured Adhesion Energies Compared

Lipid	SFA	PA	OS
eggPC			0.01
DLPC	0.1 (22°) (Marra and Israelachvili, 1985)	0.01–0.015 (Evans and Metcalfe, 1984)	0.01 (25°)
SOPC		0.012[a] (Evans and Needham, 1987)	0.02 (25°)
DMPC			0.02 (27°)
DPPC	0.15 (21°) (Marra, 1988)		0.03 (25°)
DPPE	0.80 (Lβ) (Marra, 1988)		
POPE		0.12–0.15 (Lα) (Evans and Needham, 1987)	0.14
DGDG	0.29 (Marra, 1988)	0.25 (Rand et al. in preparation)	0.24
MGDG	0.48 (Marra, 1985)		

Note: Bilayer–bilayer adhesion energies derived by three methods. Osmotic stress (OS) extrapolations assume a simple $1/d^3$ van der Waals attraction. Agreement from all three methods is excellent for digalactosyldiglycerides but not for PEs and PCs. Osmotic stress (OS) and pipette aspiration (PA) measurements are on unsupported films. Undulations of these films might explain some of the difference from surface force apparatus (SFA) measurements for the PCs and PEs, but most of these differences are not understood. (Source references are in brackets. OS values are from Table 6.3.) Temperature is taken to be at 25°C unless stated otherwise. G_{min} tabulated are absolute values of negative quantities (ergs/cm²).

[a] When accounting for undulation forces, present in the OS and possibly in the PA and absent in the SFA bilayers, 0.012 erg/cm² becomes 00165 erg/cm² (E. Evans, personal communication).

where the coefficient P_{eo} depends on surface charge, and the decay distance $_e$ is the Debye–Huckel decay length. Add to this a hydration repulsion of the form

$$P_{hyd}(d) = P_o \exp(-d/\lambda) \tag{6.18}$$

Between *parallel* surfaces, then, there will be a transition from electrostatic to hydration forms around a position d_p, where these two quantities are of comparable magnitude:

$$P_{eo} \exp(-d_p/\lambda_e) = P_o \exp(-d_p/\lambda) \tag{6.19}$$

or

$$d_p = [\lambda \times \lambda_e)/(\lambda_e - \lambda)] \ln(P_o/P_{eo}) \tag{6.20}$$

This separation d_p will typically be less than the maximum distance assumed by neutral bilayers of comparable hydration tendency.

Between oppositely *curved* surfaces, such as the cross-mica cylinders, this transition will occur at a separation d_c, where the integrated *energies* $_e \times P_{es}$ (d) and $\times P_{hyd}$ (d) are comparable. That is

$$\lambda_e P_{eo} \exp(-d_c/\lambda_e) = \lambda \times P_o \exp(-d_c/\lambda) \tag{6.21}$$

so that

$$d_c = [(\lambda \times \lambda_e)/\lambda_e - \lambda)] \ln[(\lambda P_o)/(\lambda_e P_{eo})] = d_p - [(\lambda - \lambda_e)/(\lambda_e - \lambda)] \ln(\lambda_e/\lambda) \tag{6.22}$$

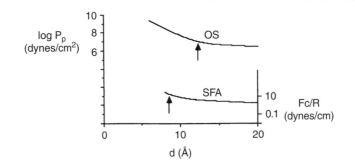

FIGURE 6.13 Difficulty of detecting hydration forces between charged bilayers. The sum of electrostatic and hydration repulsion is plotted as a pressure, P_p, between parallel surfaces, or its integrated equivalent, F_c/R, between crossed cylinders of radius R; $F_c/R = 2\pi E_{pp}$ where E_{pp} is the energy between parallel surfaces. Parameters used are, for hydration repulsion similar to that between PEs, $\lambda = 1$ Å, $P_o = 10^{12}$ dynes/cm^2, and for electrostatic double layer, made so the $P_{es} = P_{hyd}$ at 12 Å with Debye length 10 Å. Lines are drawn over regions accessible to osmotic stress, P_p, and to the surface force apparatus, F_c/R, without deformation. Note shift in position of switchover from electrostatic double layer to hydration force (arrows). Theoretical curves omit contribution of undulatory fluctuations.

The effect of opposite curvatures is to shift in toward contact the place where hydration forces "take off." Except for solutions of very high (i.e., molar) salt concentrations, $\lambda_e \gg \lambda$, and this shift is approximately

$$d_c - d_p = \lambda \times \ln(\lambda_e/\lambda) \qquad (6.23)$$

This is some 2 to 5 hydration decay lengths in solutions of 100 to 1 mM ionic strength, respectively. This represents a nontrivial difference in the stress required to see hydration forces between oppositely curved surfaces, as illustrated in Figure 6.13.

Although hydration repulsion between charged bilayers has not been as thoroughly examined as between neutral bilayers, there is clear evidence of an extra, nonelectrostatic repulsion. Early measurements of eggPG/eggPC and erythrocytePl/eggPC mixtures under osmotic stress in zero-salt solutions showed an extra repulsion at small separations that was taken to indicate hydration repulsion (Cowley et al., 1978). The deviations from electrostatic repulsion at 5 and 10 mol% PG or Pl almost exactly followed the curve for pure eggPC at the same separations.

The very high forces encountered with pure PG and egg phosphatidylserine (egg PS), in 0.01 to 1.0 M univalent salt solutions at close separation, are similar to those seen with eggPC hydration, where evidence of an extra, nonelectrostatic force is much less ambiguous (Loosley-Millman et al., 1982). The sudden onset of a repulsion at some 20 Å separation, of much steeper slope than expected from electrostatic decay and apparently unscreened even by high salt concentrations. The results with PG have been confirmed recently using the OS method (McIntosh et al., 1990) (see Figure 6.14).

In retrospect, these data suggest a nonelectrostatic double layer repulsion more like that between eggPC than between eggPC bilayers. In fact, we have recently found that DOPS in 0.8 M NaCl swells like eggPE: a repulsion that becomes important at about 14 Å separation and varies with the ~1 to 2 Å decay rate of eggPE. The opportunity exists for further measurement in these and related systems.

We have measured forces between bilayers of the nonphospholipid, dihexadecyldimethylamine acetate (DHDAA) in 5 to 500 mM acetate solutions. In 5 mM acetate solutions, for example, there is a clear break away from electrostatic double layer repulsion at a 15 to 17 Å separation and at a pressure of 10^7 dynes/cm^2. Below 15 to 17 Å separation, there is a clear break and transition to a region with an exponentially varying force with a 2.7 to 2.9 Å decay constant. That these

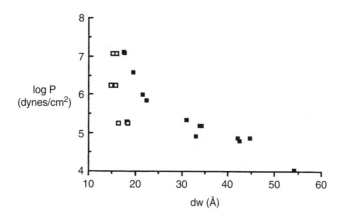

FIGURE 6.14 Interbilayer pressure between PS bilayers as it varies with separation in $1.0\ M$ NaCl (open squares) and $0.4\ M$ NaCl (closed squares). For $0.4\ M$ NaCl the sudden onset of a repulsion at about 20 Å is of much steeper slope than expected from electrostatic decay.

shorter range interactions were not seen in previous measurements with lipids adsorbed to the crossed cylindrical mica sheets of a "surface force apparatus" (Pashley et al., 1986) can be explained by the opposite bilayer curvatures enforced in that system and the stress limitations caused by mica bending as outlined earlier. That the forces measured by SFA were accounted for by electrostatic double layer interactions alone emphasize that critical examination of forces between contact and 20 Å separations should, where possible, be made between parallel rather than oppositely curved surfaces.

Qualitatively different kinds of interactions occur between acidic phospholipid bilayers exposed to divalent cation solutions. The many studies of such systems show that the interactions are strong enough to break through the hydration barrier and allow very close contact. Much use has been made of this in attempts to model membrane fusion. An example of such a remarkable change occurs in PS bilayers exposed to Ca ions. Even at micromolar Ca concentrations, these bilayers precipitate to virtually anhydrous contact, often with crystallization of hydrocarbon chains (Portis et al., 1979; Feigenson, 1986; Florine and Feigenson, 1987). By comparing the binding constant of Ca to the outer surface of multilayers (Eisenberg et al., 1979) with the strength of binding between bilayers (Feigenson, 1986) we estimate an energy of contact on the order of 100 ergs/cm^2 in these Ca-collapsed PS multilayers, quite enough to completely overcome hydration repulsion. [Method of Parsegian and Rand (1983) updated with the binding constants of Feigenson (1986).] The fact that this precipitate contains no detectable water argues against an attractive force based on ionic fluctuations (Guldbrand et al., 1984; Kjellander and Marcelja, 1985) and suggests rather the kind of dehydration characteristic of insoluble ionic crystals.

6.11 AMPLIFICATION OF BILAYER REPULSION BY THERMAL BILAYER UNDULATIONS

For some time, since the pioneering work of Helrich (1978) on steric repulsion between lipids, there has been the sense of a dilemma about deciding whether lipid bilayers repelled because of actual forces between them or because of collisions that occurred when they experienced normal thermal undulations. It appears now (Sornette and Ostrovsky, 1985) that there is no dilemma. Bilayers do undulate. These undulations are suppressed by long-range interactions rather than the hard collisions originally imagined. And the loss of undulatory entropy, suppressed by membrane repulsion, is an important part of bilayer packing energy.

There appear to be two limiting regimes:

- One where bilayers are so close that undulations are effectively suppressed and bilayers interact only through the underlying or direct interbilayer repulsive force
- Another where bilayers are sufficiently far apart that forces between them are weak enough and of relatively short enough range for them to repel as predicted by the original Helfrich model (Safinya et al., 1986)

In between, there is a coupling of the undulatory and the underlying or bare interactions, a coupling that results in behavior different from either taken alone.

To clarify the relative strength of bending undulatory and direct interaction forces, it is worth examining the form of the undulatory fluctuation force, P_{fl}, in a regime where the underlying interaction is dominated by a single exponentially decaying force

$$P_o \exp(-d_w/\lambda) \tag{6.24}$$

In that case

$$P_{fl} = (\ kT/32\,\lambda)[(P_o/B\,\lambda)\ \exp(-d_w/2\,\lambda)] \tag{6.25}$$

(To derive this result, see Evans and Parsegian, 1986. Introduce Equation 6.18 or 6.19 into Equation 16 or into the derivative of Equation 14 of that paper.) Here, B is the bilayer bending modulus (usually about 25 kT) and λ and P_o are the decay rate and coefficient of the underlying repulsion. For distances d_w much bigger than λ, this fluctuation component will dominate to give a force that decays half as fast as the underlying force. It is possible at these larger distances to infer the actual bilayer–bilayer interaction only through a theoretical construct that takes into account the undulatory force.

It is instructive to compute the point of crossover between the dominance of a direct exponential force

$$P_o \exp(-d_w/\lambda) \tag{6.26}$$

and the undulatory fluctuation force. Set

$$P_o \exp(-d_w/\lambda) = (\ kT/32\,\lambda)\ [(P_o/B\,\lambda)\ \exp(-d_w/2\,\lambda)] \tag{6.27}$$

For = 2 Å P_o = 10^{10} dyne/cm^2, B = 25 kT = 10^{-12} erg, equality is satisfied for d_w 17 Å. Below this distance, one would not expect appreciable contributions from fluctuations. At greater distances, one may see expanded exponential decay due to fluctuations.

Indeed, recent measurements of forces between parallel DNA double helical linear polyelectrolytes (Podgornik et al., 1988) show precisely this halving of the decay rate. In salt solutions of low concentration, but at separations much greater than the Debye length, forces vary with half the classical Debye decay rate. In very high salt concentrations, where charge interactions are screened, there is an exponentially varying hydration force at separations less than 10 Å and an extended region of half the decay rate at greater separations. Simultaneous measurement of molecular motion indicated by progressive broadening of the X-ray reflections confirms that the region of extended decay corresponds to a regime of steadily increasing molecular motion.

In general, the interplay of direct forces and undulatory fluctuation forces will not always result in cleanly visible behavior of one or the other type. Between phospholipid bilayers, which enjoy undulatory freedom near the position of force balance between van der Waals attraction and hydration repulsion, the action of fluctuations seems to amplify hydration repulsion near the limit of swelling. Fluctuations shift the force balance outward (Evans and Parsegian, 1986).

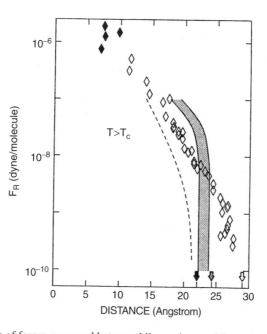

FIGURE 6.15 Comparison of forces measured between bilayers in a multilayer (points), using osmotic stress and between bilayers immobilized onto the crossed mica cylinders of the surface force apparatus (shaded band). The data points are for DLPC at 25°C, where hydrocarbon chains are melted except at high pressures (solid points). Data from Lis et al., 1982. The SFA curves are a set of melted-chain PCs (Horn et al., 1988; Marra and Israelachvili, 1985). The dashed line is the underlying interbilayer force after subtraction of undulatory fluctuation forces in the multilayer system (Evans and Parsegian, 1986). Arrows indicate limiting spacing at zero force. This plot shows (a) the expansive power of undulatory steric fluctuations in the regime of small pressures (lower third of figure), (b) the suppression of these fluctuations at higher pressures, (c) the remarkable agreement between SFA and OS measurements once one takes account of the difference in apparent zero separation. (F_R is a force per molecule. For details, see Horn et al., 1988.)

Fortunately, it is possible to compare experimentally measured forces between bilayers undulating within a multilayer array with those between bilayers immobilized onto rigid mica cylinders, where undulations are presumably impossible. Figure 6.15 shows the force vs. distance between bilayers on crossed mica cylinders, differentiated to give the equivalent force per molecule Fr (shaded band), together with measurements of repulsion between bilayers in a multilayer array, also as a force per molecule (points). Both data sets are for PCs with melted hydrocarbon chains. It is clear that in a region of strong repulsion the two show similar forces, with only a small horizontal shift due probably to differences in the defined "zero" of separation. But at low pressures, there is a distinct divergence between the two data sets: the limiting spacing of the multilayers is considerably greater than that between adsorbed bilayers. If, though, one subtracts undulatory entropic contributions from these data using the theory of Evans and Parsegian (1986), one obtains the dashed line that is remarkably parallel to the fixed-bilayer shaded band of the SFA measurements (Horn et al., 1988).

This comparison actually teaches us at least two things. First, undulations act to enhance the hydration force giving it a greater apparent range. Second, at higher pressures undulations are effectively suppressed, suggesting that one can use measurements in this range to estimate the underlying hydration force.

There are cases in which fluctuations probably always dominate the repulsion of weakly hydrating bilayers, such as the case of the nonionic alkylpoly(oxyethylene) "PEO" surfactants. Tiddy and coworkers have used controlled vapor pressure to measure forces between bilayers of compounds of various hydrocarbon and ethylenoxide lengths. Carvell et al. (1986) and Adam et al. (1984) argue that the polyethylene oxide chain polar groups are extended and probably hydrate with only one layer of water, and within their residence space form a PEO/water mash. Melted bilayers separate to greater extents than when they are frozen. Melted bilayers of the shorter-chain compounds swell appreciably more than the longer-chain species, which are presumably less flexible, and achieve separations greater than the maximum length of the fully extended amphiphile molecule. There is good reason then to think that these long spacings occur from undulatory fluctuations confined by collision between the hydrated polar regions of facing bilayers (Carvell et al., 1986; Adam et al., 1984).

6.12 THE VAPOR PRESSURE PARADOX

Widely recognized among phospholipid physical chemists, and even more widely ignored among those who prepare lipids for laboratory study, is the fact that lipids exposed to a water vapor of 100% humidity will not take up as much water as will the same sample put into contact with liquid water (Jendrasiak and Mandible, 1976; Jendrasiak and Hasty, 1974, compare to Table 6.3 this text). Typically, for example, a phosphatidylcholine multilayer will imbibe some 45 to 55% by weight water from the pure liquid but only some 30% from a water "saturated" vapor (Marra and Israelachvili, 1985). What is more, a sample equilibrated against liquid will actually give up water to a 100% rh vapor and then reversibly regain water from a liquid when given an opportunity to do so (R.P. Rand, unpublished; S. Gruner and R. Templar, personal communication).

Worse, charged phospholipids, e.g., phosphatidylserine (Jendrasiak and Hasty, 1974), form bilayers that separate indefinitely in liquid water (Loosley-Millman et al. , 1982) and will spontaneously form vesicles. But in 100% relative humidity will actually stop swelling at a water contents far less than that taken up by phosphatidylcholine under similar conditions (Jendrasiak and Hasty, 1974). The limit of swelling of multilayers on solid substrates (White et al. 1987; S. Gruner and Richard Templar, unpublished) seems to resemble that of lipids in vapors.

What is going on? Is the activity of a 100% rh vapor not the same as that of the liquid water with which it is supposed to be in equilibrium?

One's first thought is that perhaps, because of slight thermal gradients, the vapor activity is somewhat less than that of its mother liquid. It is very instructive to estimate the large effects that small thermal fluctuations have on the hydration of such systems. Consider the osmotic stress Π equivalent of a vapor of relative humidity p/p_o

$$\Pi = -(kT/v)\ \ln(p/p_o) \tag{6.28}$$

where v is the 30 Å^3 volume of a water molecule and $(kT/v) = 1.4 \times 10^9$ dyne/cm^2. For p near p_o, we may write $p/p_o = 1-\Delta$ and

$$\Pi = (kT/v) \times \Delta = 1.4 \times 10^9 \times \Delta. \tag{6.29}$$

An osmotic pressure of 10^6 dynes/cm^2, enough to remove one-third to one-half of the water from a multilayer, results when $\Delta = 0.00075$, or when the relative humidity is lowered from 100% to more than 99.9%, which could come from a 0.01°C rise in the temperature! Figure 6.16 shows the large effects that small changes in relative humidity, resulting from tiny temperature fluctuations, can have.

Temperature fluctuations will explain the escape of water from bilayers in liquid to vapor. But thermal fluctuations do not explain the observation that water is lost to a vapor maintained at 110% relative humidity. In that experiment, water-saturated air was cooled before being blown at a hydrated sample (S. Gruner, personal communication).

A second possibility, therefore, is that the action of a vapor/multilayer or solid/multilayer interface is to suppress the bilayer undulations that enhance hydration or electrostatic repulsions.

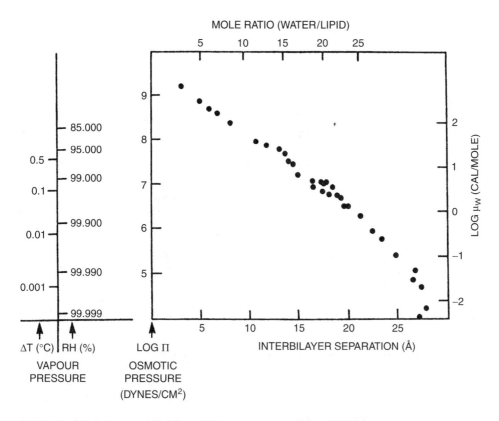

FIGURE 6.16 Mole ratio (water/lipid) and bilayer separation for eggPC bilayers plotted as a function of pressure expressed four different ways: (i) osmotic stress dynes/cm²; (ii) chemical potential relative to bulk water; (iii) equivalent relative humidity; (iv) temperature increase that would correspond to the same changes in equivalent relative humidity and osmotic stress.

Quantitative comparison of water loss in vapor with the predicted shift in equilibrium spacing using the model of Evans and Parsegian (1986) suggests that, at least given present theories, one can account for about half of the observed effect with this explanation.

Other possible explanations might lie in the restraining effects of high surface tensions. The results of Smith et al. (1987) on planar lipid films stretched over a hole and exposed to vapor show that the geometry of the overall multilayer is not critical.

The inability of charged lipids to swell in vapor suggests to us that, at the very least, the phenomenon is a practical problem. One knows that charged lipids must repel. If one is unwittingly preventing this swelling, then that is certainly a problem in handling. People who prepare samples in vapor should be suitably aware that, under these conditions, lipids will not go to full hydration.

6.13 INTERACTION BETWEEN MEMBRANE VESICLES

One of the most interesting applications of these quantitative measurements of the forces between bilayer membranes is to the interaction between bilayers that are in the form of closed unilamellar vesicles or curved surfaces. In that application, two important considerations superimpose on the basic forces themselves, considerations which are often ignored but which affect the outcome of such interactions. First, we emphasize that the form of interaction will depend on whether the curved surfaces are parallel to each other or whether they curve away from each other (have the opposite curvature), as we have discussed in the foregoing. Two cells or vesicles approaching one another, or a small vesicle approaching a plasma membrane, necessarily curve away from each other. Large intracellular vesicles closely apposed to the plasma membrane are more nearly parallel to each other. Second, it is also significant that the forces encountered can be strong enough to deform interacting bilayers, either to restrain thermal undulation or to flatten neighboring vesicles. Deformation itself can then result in a change in regime from one of surfaces that curve away from each other to one of parallel surfaces.

It is instructive to see how hydration repulsion and the adhesion energy, G_{min}, measured between parallel surfaces at a position of force balance, show up in the interaction between curved surfaces. These phenomena have been examined rigorously by Evans and coworkers (1980, 1984, 1986, 1987, 1982).

As described earlier, a convenient approximation due to Derjaguin (1934) allows one to transform forces measured between parallel planar layers (pp) to interactions expected between spherical vesicles (ss) or spherical vesicles and flat layers (sp). Between crossed cylinders of radius R or between a sphere of radius R and a plane, the force F_{sp} is related to the energy E_{pp} by

$$F_{sp}/R = 2 E_{pp} \tag{6.30}$$

Between two spheres, the transform is

$$F_{ss}/R = E_{pp} \tag{6.31}$$

It is remarkable that the point d_o of zero net force or minimum energy is the same between two spheres, between a sphere and a flat, or between two crossed cylinders, *and is quite independent of radius R*. Further, this d_o between oppositely curved surfaces will always be expected to occur at a smaller separation than between parallel surfaces (cf. Figure 6.7, $E_{pp} = 0$ at a smaller separation than where $F_{pp} = 0$). The interaction between surfaces that curve away from each other is a sum of individual interactions at different separations. In the Derjaguin approximation, some parts of the surface may feel net *attraction*, some *repulsion*. The longer-range force will be felt over a greater area of the surface than the repulsive. It will have a proportionally larger "say" in determining the final position of force balance. (For an illustration of the result of mixed attractive and repulsive electrostatic double layer forces between spheres, and a rigorous examination of the accuracy of the Derjaguin approximation for such interactions, see Barouch et al., 1976.) But this same combination of attraction and repulsion will create a torque to deform a curved surface. One must therefore recognize surface deformability in any problem involving curved surfaces.

What can one say at the level of vesicles of 200 Å radius? First, between rigid spheres compared to parallel layers, there will be an inward shift in the position of force balance between long-range attraction and short-range repulsion. Second, because vesicles are in fact not rigid, they will flatten to create regions of planar adhesion having the energy, G_{min}, per unit area described previously.

6.13.1 Rigid Spheres

For simplicity, consider a repulsive force of the form $P = P_o \times \exp(-d/\lambda)$ and attraction of the van der Waals form $F_{vdw} = -(A_h/6\ d^3)$. (Here we use d as the distance between the surfaces. By virtue of the assumption of rigidity, one ignores any action of undulatory repulsion.) The corresponding energy between two planar surfaces experiencing these forces is

$$E_{pp} = \lambda \times P_o \times \exp(-d/\lambda) - A_h/(12\ d^2) = (\lambda \times P) - (d/2) \times F_{vdw} \tag{6.32}$$

Sketches of force and energy per unit area for typical parameters, Figure 6.17, show the inward shift in zero-force position for spheres or cylinders from that for parallel planes.

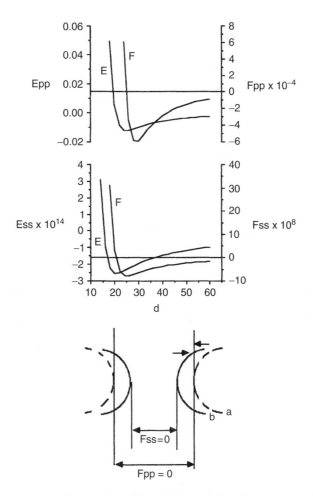

FIGURE 6.17 Simultaneous attraction and repulsion between rigid spheres occurs at a minimum energy position that is less than that of force balance $F_{pp} = 0$ between parallel planes of the same material. (a) Spheres at separation corresponding to maximum attraction; (b) spheres at their minimum energy separation (where $E_{pp} = 0$). Little arrows show conflicting repulsive and attractive pressures creating a torque on curved surfaces.

Because the force between spheres goes as the radius R, the depth of the energy minimum for interacting spheres is proportional to sphere radius. By $F_{ss} = R\,E_{pp}$, the energy of interaction between two spheres goes as

$$E_{ss} = (\lambda^2 \times P_o \times \exp(-d/\lambda) - A_H/(12\,d))R = [(\lambda^2 \times P) - (d^2/2) \times F_v]R \qquad (6.33)$$

The fact that the minimum energy position of two spheres is at a separation less than that of two parallel planes means that the closest parts of the spheres are actually being pushed to a separation where they repel (Figure 6.17). The simultaneous attraction and repulsion on different parts of a vesicle create a torque that can be relaxed by vesicle deformation.

6.13.2 DEFORMABLE VESICLES

The stress of hydration repulsion and even weak van der Waals attraction is such that virtually any curved bilayer surfaces must deform to some extent when in adhesive contact (Evans and Parsegian, 1983). Lateral tension T within the bilayer surface develops against the drive to create a flattened

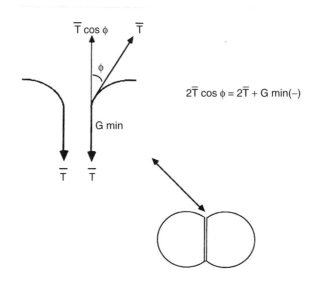

FIGURE 6.18 The balance of line tension T and attractive energy G_{min} (-) to create a deformed region of phospholipid vesicle interaction. G_{min} (-) between neutral phospholipids is usually pictured as a balance between van der Waals attraction and hydration repulsion. Despite the tension developed, there may be some repulsion also from undulatory fluctuations. Between some phospholipids, there may also be hydration attraction or H-bonding across a water layer due to complementary surface polar groups.

area of contact of adhesive energy, G_{min}, in such a way as to satisfy Young's equation and to make a contact angle Φ (Figure 6.18).

$$\cos \phi = 1 + [G_{min}(-)/2 \times T] \tag{6.34}$$

For small contact angles $\cos \phi = 1 - \phi^2/2$ and

$$\phi^2 = -G_{min}(-)/T \tag{6.35}$$

The area of flattening $R^2\phi^2 = R^2 G_{min}$ (-)/T, the fractional area of flattening is

$$R^2 G_{min} (-)/(T \times 4 R^2) = G_{min} (-)/4T \tag{6.36}$$

and the energy of interaction over this flattened area is

$$R^2G_{min}(-)^2/T \tag{6.37}$$

The factor of 10 difference in G_{min} between PE and PC leads to a factor of 100 difference in the contact energy between deformable vesicles. For example, consider an R = 200 Å vesicle under tension T = 1 dyne/cm. The adhesive interaction of G_{min} ~-.01 erg/cm² will be only 1/5 of the thermal energy kT = 4.2 x 10⁻¹⁴ erg.

But for G_{min} ~-.1, the interaction energy will be some 20 kT. Depending on tension, the contact energy will often be dominated by G_{min} x area of contact.

Other contributions, such as the residual attraction between nonflattened areas and the work of deformation, will very often be small by comparison. When vesicles are deformable, as is the usual case with phospholipids, their interaction is more characteristic of forces between parallel planes than between curved surfaces. It is puzzling to us why most models of vesicular aggregation neglect this important feature of interaction.

6.14 HYDRATION IN NONBILAYER SYSTEMS

Phospholipids, generally expected to form bilayer structures, have been shown ever since the early studies by Reiss-Husson and Luzzati to form nonbilayer structures in a variety of aqueous conditions. This seminal work launched many studies on the rich polymorphism of these lipids in aqueous environments (for an early review see Luzzati, 1968).

For some time it has been known that one very common nonbilayer structure that many phospholipids will take on is an inverted hexagonal structure of parallel cylinders of water bounded by lipid polar groups with the space between cylinders filled by nonpolar components (see Chapter 5). An enormous literature now exists of studies examining the role of temperature, polar group and hydrocarbon chain species, nonphospholipid molecules, and water content in the bilayer-hexagonal structural transition. Only relatively recently, however, has it been recognized that the formation of nonbilayer structures can be stimulated by addition of alkane (Kirk and Gruner, 1985) or, more significantly, by small quantities of lipids endowed with particularly long hydrocarbon chains (Tate and Gruner, 1987). This has made it possible to see the balance of structural tendencies in these lipids and to understand how shifting that balance creates phase transitions or phase separations. While this has usually been explained in terms of molecular shape or excluded volume, Gruner and coworkers have utilized the idea of a spontaneous monolayer curvature, $1/R_o$, for each phospholipid or mixture of phospholipids, akin to the ideas originally developed by Helfrich (Helfrich, 1978) for bilayers. The spontaneous monolayer curvature is used to explain the tendency to form nonbilayer structures, a tendency that might be expressed within each monolayer making up a bilayer.

The osmotic stress method described here and originally developed to measure forces between bilayers in multilayer arrays, has been used to measure the work of dehydrating and deforming the aqueous cavities in the hexagonal phase. As one then stresses the H_{II} phase osmotically, removing water and bending the lipid monolayers enforcing deviations from R_o, it has become possible to map the structural dimensions of the lipid/water system at the same time as one determines the free energy of the components, just as has been described for the lamellar phase.

The first detailed analysis of such a system, that of DOPE and DOPE/DOPC mixtures, has been published (Rand et al., 1990). The following is a summary of a number of remarkable features that emerge:

> Lipid allowed to form the hexagonal phase can hydrate to a much greater extent than it does in its lamellar form. This is understood in terms of the monolayer curvature energy, whereby in addition to the water required for direct hydration of the polar groups, additional quantities are required to fill large aqueous cavities in order for R_o to be expressed (see Figure 6.19).
>
> Figure 6.20 shows that as water is removed and the monolayer is bent around an ever smaller aqueous cylinder, the molecular area is compressed at the polar end of the molecule and expanded at the hydrocarbon end. So there is a neutral or pivotal position along the length of the phospholipid molecule, shown schematically in Figure 6.21, each side of which molecular areas change in opposite directions as the monolayer is bent.
>
> By defining R_o at the pivotal position, the measured energies are well fit by a quadratic bending energy $K_c/2 \, (1/R - 1/R_o)^2$; the fit yields bilayer bending moduli of $K_c = 1.2 - 1.7 \times 10^{-12}$ ergs, in good agreement with measurements from bilayer mechanics. On the other hand, we emphasize that even though the measured energies fit so well with a quadratic energy of curvature, this does not exclude the possibility that other kinds of interaction or functional forms might give an equally good description of the data. Further, we (Gawrisch et al., submitted) have recently discovered a facile, 0.1 kT per DOPE molecule, reentrant hexagonal–lamellar–hexagonal sequence of phase transitions with osmotic stress. The low energy of the hexagonal–lamellar transition belies the notion that uncurling the

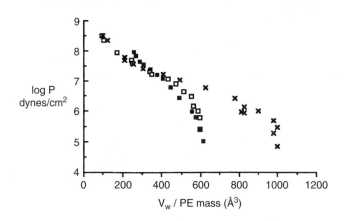

FIGURE 6.19 Variation in the volume of water per DOPE polar mass with osmotic stress for DOPE alone (open squares), for DOPE/DOPC 3/1 lamellar phase (closed squares), and for the hexagonal phase formed by DOPC/DOPE 3/1 with 20% added td (crosses). (Reprinted with permission from Rand et al. *Biochemistry,* 29:76–87. Copyright (1990) American Chemical Society).

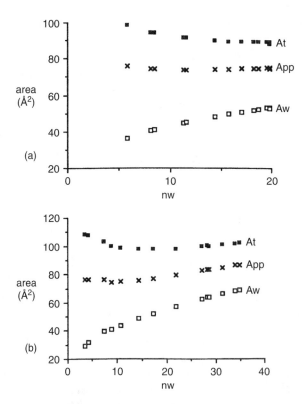

FIGURE 6.20 Molecular areas calculated at the lipid water interface, (A_w), the pivotal position along the length of the phospholipid molecule (see text), (A_{pp}), and the end of the monolayer annulus (A_t), as they vary with n_w, the numbers of water per lip molecule. (a) DOPE (b) DOPE/DOPC 3/1 with added tetradecane assumed to 12 wt% of the lipid. Note that in both cases, the area at the polar group end of the molecule increases, whereas that at the ends of the hydrocarbon chains decreases as water is added and the overall dimension of the hexagonal lattice increases. A_{pp}, on the other hand, remains constant. (Reprinted with permission from Rand et al. *Biochemistry,* 29:76–87. Copyright (1990) American Chemical Society.)

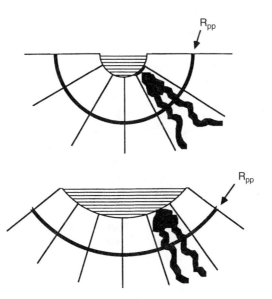

FIGURE 6.21 Scale drawing of monolayers in the minimally and the maximally hydrated DOPE hexagonal phase. The outlined radial sectors, surrounding the central water cores, represent the cross sections of constant-thickness slabs, in the direction of the cylinder axis, which delimit the excluded volumes of single DOPE molecules. The changes in sector dimensions illustrate that, when the monolayer is bent, the areas of the polar and hydrocarbon ends of the molecule change in opposite directions as the molecule pivots around its position of constant area, R_{pp}. That position, the neutral plane of the monolayer, is indicated by the solid line and is one-third of the distance along the hydrocarbon chain from the polar group. (Changes in molecular area can result from changes in dimension parallel to the cylinder axis, along the cylinder circumference, or both. Whichever way area changes are attributed does not change the neutral plane. In this drawing, the slab thickness at the pivotal position has been taken as constant during monolayer bending.) (Reprinted with permission from Rand et al. *Biochemistry*, 29:76–87. Copyright (1990) American Chemical Society.)

hexagonal monolayer, against its measured bending modulus, to fit into a bilayer, which requires 1.6 kT per molecule, dominates that transition. Rather, it clearly suggests that factors other than curvature are working to provide the negative energy favoring the hexagonal-to-lamellar transition.

For lipid mixtures, DOPE and DOPC, for example, enforced deviation of the H_{II} monolayer from R_o is sufficiently powerful to cause demixing of the phospholipids. And this demixing is done in a way suggesting that the DOPE/DOPC ratio self-adjusts its R_o so that the spontaneous curvature energy is minimized.

An understanding of the energies of the forming and deforming of such nonlamellar phases should help in gauging the magnitude of any elastic strains within lipid bilayers. Such strains are perceived as affecting the incorporation and function of proteins in membranes and membrane destabilization in processes of membrane disruption and fusion.

6.15 THEORETICAL QUESTIONS

Much effort has been spent to develop a satisfactory theory of hydration. In lipid systems, difficulties are compounded by the several phenomena affecting surface hydration. Bilayer undulation, lateral compressibility and deformation, and polar group packing and rearrangement all no doubt contribute to the wide range of force decay rates and works of dehydration that emerge in the comparisons presented here and debated in the current literature.

Why "hydration" at all? Essentially because of the work encountered in bringing together neutral bilayers in distilled water or low salt buffers, added salt seems to make relatively little difference, except at molar concentrations (Afzal et al., 1984). Charged bilayers do interact in ways that suggest salt-screened electrostatic double layers, but only at distances greater than where strong forces are encountered between neutral bilayers. At shorter distances (~20 Å) charged bilayer repulsion usually resembles that between neutral phospholipid bilayers (Loosley-Millman et al., 1982).

So, at the root level, a "hydration" or "dehydration" force implies a work of removal of water from between membrane or molecular surfaces. One could include in that work any steric forces of polar groups or of entire undulating bilayers, specific arrangements of polar groups that enable attraction as well as repulsion, and actual adsorption of water to the membrane polar groups. It is not clear when any of these factors stands out so clearly as to be distinctly identified. What is clear, though, is that it has not been possible to rationalize measured forces with any theory that neglects the structure of the intervening solvent.

For simplicity, we still favor the approach originally proposed by Marcelja and coworkers (1976). A polar surface will perturb aqueous solvent just next to it; and the propagation of this perturbation by solvent–solvent interactions mediates a force that extends, with a solvent-characteristic length, over many solvent layers. The strength of interaction was seen as a function of the perturbing strength of the surface, while its exponential decay was a characteristic of the intervening solvent. The original formulation spoke in terms of an order parameter of undetermined type. Gruen and Marcelja (1983) emphasized the importance of water polarization, whereas later studies by Kjellander and Marcelja (1985) examined the possibility of H-bond rearrangement within the water solvent or of the coordination of water molecules. The formulation of Schiby and Ruckenstein (1983) emphasized the importance of surface polarization and water dipole interactions.

Cevc and coworkers (1982, 1985) have argued for recognition of a hydration potential, a measure of the polarizing or hydrating power of polar groups, rather than a fixed value for the operative order parameter at the hydrating surface. This idea of a potential has been developed into an effective surface polarity, a function of polar group ionization, methylation, or other surface parameters, that is responsible for reorganizing boundary water (Cevc and Marsh, 1987). It has been possible to create a self-consistent model allowing a near-quantitative explanation of a large body of data on phase transitions (Cevc, 1987; Cevc and Marsh, 1985). Much of this material has been reviewed in some detail (Cevc and Marsh, 1987). Simon et al. (1988) have argued that the surface potential, measured across phospholipid monolayers in presumed equilibrium with free multilayers, is the organizing potential that would file into the Cevc et al. formalism. Because dipole potentials can be inferred from bilayer transport measurements, it seems worthwhile to investigate this claim by direct comparison.

Kornyshev et al. (1985) use a continuum dielectric formalism with a nonlocal response to rationalize the decay of hydration forces. In recent work, they have succeeded in coupling solvent correlation length with the lattice constant of interacting surfaces. The result is a net decay length that can be different for different surfaces interacting across the same solvent material (Dzhava-khidze et al., 1987). Attard and Batchelor (1988) have proposed a model that recognizes the progressive entropy loss (or enthalpy gain) of the surface-perturbed water H-bond network. Decay lengths reflect surface boundary conditions, as well as solvent lattice lengths, to allow some variation in decay rate. This approach emphasizes the nonelectrostatic nature of the solvent parameter that mediates hydration forces.

Indeed, the continuous 1 to 3 Å range of measured decay rates for forces between phospholipid bilayers suggests that a single-decay picture is either inadequate or results from being combined with other forces. Forces decaying exponentially with ~3 Å decay constants have been seen much more consistently between linear polyelectrolytes (Rau et al., 1984). Perhaps in these systems they can be more fruitfully analyzed theoretically.

Computer simulations of H-bonding water near polar surfaces (Kjellander and Marcelja, 1985) do not show the kind of extended decay of perturbation expected from the original Marcelja formalism or presumably from later H-bond models. But this may be due simply to the fact that these simulations are accurate to some 0.5 kcal/mol of solvent (e.g., Jorgensen et al., 1985) while the perturbations of water that seem to be important are as small as ~1 cal/mol (Figure 6.16). Indeed, the essence of these forces, and the reason they were not expected from probes of water itself, is that they come from virtually undetectable perturbations of solvent summed over large numbers of water molecules. One can get some idea of the difficulty of modeling hydration forces by looking at the force distance curve in chemical, rather than physical, units. The right-hand scale in Figure 6.16 shows the applied osmotic stress in units of small calories. It is immediately clear that the pressures over which forces are observed correspond to perturbations that are less than thermal energy (~600 cal/mol) on most of the intervening water molecules.

McIntosh and coworkers (1988) have suggested from data on eggPC that there is an additional upward break in the pressure vs. spacing curve that is due to steric repulsion between bilayers that have less than ~10 water molecules per PC. Such an upward break is not seen in the PC data processed as we have done here. Its appearance depends heavily on the definition of bilayer thickness. Should one expect interactions between hydrated polar species to be separable into hydration and steric components when the polar group conformations will always involve their associated water? Or does hydration repulsion combine both such interactions, inasmuch as one expects continuously increased polar group restrictions from the very first steps of dehydration?

6.16   HYDRATION ATTRACTION?

In their review of structure in ordered phospholipid phases, Hauser et al. (1981) made clear the intricate pattern of hydrogen bonds among the phospholipid polar groups and their hydrating waters. These patterns are expected among phospholipids such as certain PEs that take up relatively little water compared to PCs or to charged species (Kolber and Haynes, 1979; McIntosh and Simon, 1986). Can the great strength of adhesion in these systems (Table 6.4) be explained solely in terms of weakened hydration repulsion to allow a relatively strong van der Waals attraction? Or can solvent restructuring, thought to cause repulsion between hydrating surfaces, also mediate attraction between laterally ordered surfaces?

Crystal structures have provided direct observation of intermolecular ammonium-phosphate linkages having the dual character of both salt bridge and hydrogen bond in DLPE bilayers (for a review, see Hauser et al., 1981). A variety of water-mediated H-bonded linkages between the phosphate oxygens both within and between apposing DMPC bilayer planes has also been described in DMPC crystals. One expects such arrangements when the number of water molecules per polar group is small enough that discrete polar and water layers are less likely. Kolber and Haynes (1979) and McIntosh et al. (1988) have suggested that such arrangements occur at the low hydration levels seen in PE bilayers. Water networks with both attractive and repulsive orientations exist in other crystalline structures (Savage, 1986). Although bilayers with melted hydrocarbon chains are molecularly more dynamic than the crystals, one would still expect such networks to exist between molecules whose mobility is less than that of water itself. Equilibrium hydration always represents a balance of all attractive and repulsive components. What we propose here is that the observed range of hydration results, in part, from changes in that balance resulting from both attractive and repulsive orientations of water between bilayers.

More specifically, the attractive force may be thought of as resulting from a complementarity between phosphate and amine polar charges on facing surfaces, interacting by structurally perturbing the intervening water. Specifically, if simplistically, one expects that the amine and the phosphate regions on the PE headgroup polarize or otherwise perturb water in opposite directions. The nature of this perturbation and of its propagation into the aqueous space is thought of as a restructuring of solvent molecules, as formulated earlier by Marcelja and Radic (1982). There are two extremes in the relative orientations of the polar groups sitting on apposing surfaces: amine-to-amine and phosphate-to-phosphate, noncomplementary and repulsive; or amine-to-phosphate, complementary

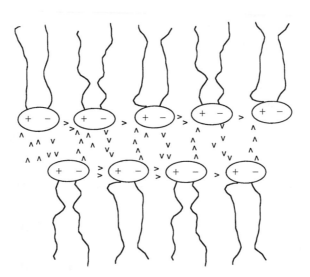

FIGURE 6.22 Schematic diagram showing spatial correlations of the polar heads of the phospholipid molecules, each of which has groups capable of polarizing water in opposite directions. Hydrogen-bonded water bridges can then span the space between complementary groups, which may not themselves hydrogen-bond because of constraints imposed by the packing of the other parts of the molecule.

and expected to be attractive, as shown in Figure 6.22. Totally random orientation, unless the strengths of amine and phosphate are closely matched, will lead to net repulsion. We suggest that all orientations can exist between bilayers, their balance being determined by the correlations that are allowed under the constraints of molecular packing within the bilayers. Such obvious complementarity as between amine and phosphate groups is not required to achieve such water bridges between groups. For example, water-mediated H-bonded linkages between phospholipid oxygens are seen between apposing DMPC bilayer planes in DMPC crystals. We emphasize that any complementarity is likely to be dynamic, in that the polar groups on both surfaces are in continual motion, although much slower than the intervening water molecules, while interacting with each other. In some ways, the attraction thus resembles the mutual motion and perturbation that is the basis of very low frequency van der Waals forces (Mahantly and Ninham, 1976; Parsegian, 1975). These will be opposed by noncomplementing polar groups, the original hydration repulsion.

As noted earlier, multilayers of PE as a class tend to imbibe less water than PCs. Still, there is a larger range in the amount of uptake; natural PEs with heterogeneous chains such as eggPE swell the most. Remarkably, even one methylation of a PE or the addition of methylated species to a bilayer will lead to sudden swelling of the multilayer lattice (Figure 6.8). If one relies only on average properties of the bilayer surface, it is hard to see how a generalized hydration repulsion and van der Waals force cause such different results with chemically similar surfaces.

The same Marcelja order-parameter formulation that describes hydration repulsion between like surfaces will also predict attraction between surfaces of opposite polarizing tendencies. Indeed, this formalism has been applied to the long-range attraction observed between DNA double helices. There is a characteristic 3 Å exponentially decaying repulsion, much more cleanly defined than what is seen among lipids, that becomes a 1.5 Å decay when DNA binds certain polyvalent cations. The shift in decay length is characteristic of the Marcelja model rederived for ordered heterogeneous surfaces.

We have suggested elsewhere (Rand et al., 1988) that the same kind of combined hydration repulsion and attraction might also explain the shortened decay rate of repulsion between poorly swelling PEs, such as POPE, and that a solvent-structure-mediated attraction might explain the anomalously high attraction seen in these cases (Rand et al., 1988; Marra and Israelachvili, 1985). We suggest that the full range of degrees of hydration and of interbilayer spacings observed for different neutral bilayers results in part from variable contributions of the attractive and repulsive hydration components. The extremes are exemplified by the dominance of attraction between PE bilayers with crystalline hydrocarbon chains and the dominance of repulsion and full hydration of melted PC bilayers. Within the variable hydration of the PEs with melted chains, one sees evidence of variable hydration attraction. With or without the postulated attraction, though, good estimates of adhesion energy can be obtained. It is necessary, therefore, to make other critical tests to distinguish the alternatives.

One should also be aware of a growing literature on attractive forces that will occur between surfaces of mutually orienting dipoles. Several models have appeared (Kjellander, 1984; Podgornik, 1988; Atford et al., 1988) using classical electrostatics to formulate a dipolar fluctuation force appropriate to polar zwitterions attached near a dielectric interface.

6.17 CONCLUSIONS

In a broad sense, any work to remove water is a form of hydration (or dehydration) force. One has grown accustomed to using "hydration force" for that part of the work due to perturbation of water structure by the membrane or molecular surface. From this perspective, it is puzzling why neutral bilayers will repel at ~20 Å separation. But membrane undulation, polar group steric factors, electrostatic forces, and even van der Waals attraction might also contribute to the work of dehydration. The real problem is to determine the distinguishing features, relative importance, and interplay of these factors in what we have empirically called "hydration forces." Solution of the problem is impeded by many experimental uncertainties intrinsic to the structural disorder of most phospholipid liquid–crystalline systems.

In phospholipid crystals where such disorder does not exist, X-ray crystallographic determinations show an intricate arrangement of polar groups with positive and negative charges neatly matched between facing bilayers, an intricate set of hydrogen bonds among lipid zwitterions and the few included water molecules (see, e.g., Hauser et al., 1981; Small, 1986). But most lipid species under most aqueous conditions do not form such dehydrated crystals. Apparently, these precise arrangements of polar groups do not occur with such low energy as to create a significant driving force for crystal formation; a preexisting order in the rest of the bilayer seems to be necessary. In fact, most systems are driven to hydrate.

The simplest measure of hydration, maximum water uptake by neutral phospholipids, shows that disruptions of bilayer order (from double bonds in the hydrocarbon chain, from heterogeneity of chain type, from chain melting, from bulky methyl groups on polar amines) all seem to increase the uptake of water, to drive bilayers from crystalline arrangements. There is a kind of synergy in this drive. Disorder in the lipid causes water uptake by disrupted polar groups that concomitantly loosens bilayer structure. We find it remarkable that the equivalent molar concentration of zwitterions, even at full bilayer swelling, is far greater than the saturating solution concentration of these same polar groups existing as pure unattached solutes. But when they are attached to nonpolar hydrocarbon chains, these closely packed polar groups virtually never precipitate into crystalline ordered arrays. The natural tension between the hydrophilic and hydrophobic parts keeps the polar groups themselves posed between precipitation and dissolution.

However, to go further, to describe the physical forces that swell the space between bilayers, one must rely heavily on definitions of bilayer thickness and separation. Difficulties are especially severe at low levels of hydration. Just as there is no mathematically ideal interface to define the progressive change from lipid to water regions, so there is no way to state boundaries without some idealization or ancillary construction. The progressive work of hydration, a pressure times a change in volume, is thermodynamically well defined. Molecular dimensions are not. Different definitions/constructions based on very low-resolution electron density maps with additional assumptions from models, or definitions based primarily on the water to lipid mass, lead to different comparative hydration strengths and even different features of the pressure vs. distance curves.

One such qualitative difference in feature has evoked the proposal of molecular steric interactions of polar groups (McIntosh et al., 1987). These are suggested by an upward break in the pressure vs. separation at less than 10 Å separation. But such a break is far more evident from low-resolution electron density construction than from a mass average construction, which so far suggests hardly any break at all. Clearly, this region of high osmotic stress and very limited hydration merits far more study using both forms of data analysis. If this thermodynamic-structural work could be coupled with better probes of changes in polar-group order — by neutron diffraction or by nuclear magnetic resonance, or perhaps by more precise analysis of thermal transitions — then the action of polar group steric forces would be more systematically understood.

At greater separations between bilayers, where polar group crowding between bilayers is no longer expected, there is still a question how much of the observed hydration repulsion is due to straight affinity of water for the polar interface and to what extent that underlying force is enhanced by forces of membrane undulation. If means could be devised to monitor membrane disorder, as has been possible for forces between polyelectrolytes, then one would have a clearer idea of the magnitude of the underlying hydration itself.

In any event, there are questions about the molecular basis of hydration, about the mechanism by which water will be perturbed some layers from the surface. And of all the possible contributions to total hydration, do differences in total water uptake reflect differences in mechanisms of attraction between different kinds of polar layers? Or are there simply differences in strengths of repulsion with broadly similar van der Waals attractive forces? Does the presence of a net electrostatic charge that can drive bilayers to indefinitely large separation also affect surface-bound water to change surface hydration forces? Does the exclusion of solutes — small sugars or large polymers — that are unable to compete for water near the bilayer interface create thermodynamically different conditions for bilayer stability?

ACKNOWLEDGMENTS

One of the unexpected pleasures of working on this subject is our interaction with many people who are now drawn to it and whose studies figured so strongly in the work cited here. We would especially like to thank the following for their help and comments while preparing this: Gregor Cevc, Sol Gruner, Roger Horn, Tom McIntosh, Sid Simon, and Gordon Tiddy. It could not have been done without the expert support of Nola Fuller and the financial backing, for RPR, of the Natural Sciences and Engineering Research Council of Canada.

REFERENCES

Abrahamsson S and Pascher I: *Acta Crystallogr.* 1966, 21:79.
Adam CG, Durrant JA, Lowry MR, and Tiddy GJT: *J. Chem. Soc. Faraday Trans.* 1 1984, 80:789.
Afzal S, Tesler WJ, Blessing SK, Collins SK, and Lis LJ: *J. Colloid Interface Sci.* 1984, 97:303.
Atford P and Batchelor MT: *Chem. Phys. Lett.* 1988, 149:206.
Atford P, Mitchell DJ, and Ninham B: *Biophys. J.* 1988, 53:457.
Barclay LM and Ottewill RH: *Spec. Discuss. Faraday Soc.* 1970, 1:138.
Barouch E, Matijevic E, and Parsegian VA: *J. Chem. Soc.* 1986, 82:2801.
Bradley WF, Grim RE, and Clark GL: *Z. Kristallogr.* 1937, 97:216.
Carvell M, Hall DG, Lyle IG, and Tiddy GJ: *Faraday Discuss. Chem. Soc.* 1986, 81:223.
Cevc G: *Ber Bunsenjesellschaft Phys. Chem.* 1988 (in press).
Cevc G: *Biochemistry* 1987, 26(20):6305.
Cevc G and Marsh D: in *Phospholipid Bilayers: Physical Principles and Models.* John Wiley & Sons, New York, 1987.
Cevc G, Seddon JM, and Marsh D: *Discussions Chem. Soc.* 1986, 81:179.
Cevc G: *Chemica Scripta* 1985, 25:96.

Cevc G, Podgornik R, and Zeks B: *Chem. Phys. Lett.* 1982, 91:193.

Christenson HK and Claesson PM: *Science* 1988, 239:390.

Churaev NV and Derjaguin BV: *J. Colloid Interface Sci.* 1985, 103:542.

Clunie JS, Goodman JF, and Symons PC: *Nature* 1967, 216:1203.

Cowley AC, Fuller NL, Rand RP, Parsegian VA: *Biochemistry* 1978, 17:3163.

Das S and Rand RP: *Biochemistry* 1986, 25:2882.

Derjaguin BV, Churaev NV, and Miller VM: in *Surface Forces,* Consultants Bureau, 1987.

Derjaguin BV and Churaev NV: in *Fluid Interfacial Phenomena.* Ed. Croxton CA, John Wiley & Sons, Chichester, U.K., 1986, pp. 663–738.

Derjaguin BV: *Kolloid-Z.* 1934, 69:155.

Dzhavakhidze PG, Kornyshev AA, and Levadny VG: International Atomic Energy Agency and UNESCO International Center for Theoretical Physics, Trieste, 1987.

Evans DF and Ninham BW: *J. Phys. Chem.* 1986, 90:226.

Evans DG: *Langmuir* 1988, 4:3.

Evans E and Needham D: *J. Phys. Chem.* 1987, 91:4219.

Evans E and Needham D: *Faraday Discuss. Chem. Soc.* 1986, 81:267.

Evans E and Parsegian VA: *Proc. Natl. Acad. Sci. U.S.A.* 1986, 83:7132.

Evans E and Metcalfe M: *Biophys. J.* 1984, 46:423–425.

Evans E and Kwok R: *Biochemistry* 1982, 21:4874.

Evans EA and Needham D: in *Physics of Amphiphiles.* Eds. Mennier J and Langevin D, Springer-Verlag, Berlin, 1987.

Evans EA and Parsegian VA: *Ann. N.Y. Acad. Sci.* 1983, 416:13.

Evans EA: *Biophys. J.* 1980, 31:425.

Evans EA and Skalak R: in *Mechanics and Thermodynamics of Biomembranes.* CRC Press, Boca Raton, FL, 1980.

Eisenberg M, Gresalfi T, Riccio T, and McLaughlin S: *Biochemistry* 1979, 18:5213.

Feigenson GW: *Biochemistry* 1986, 25:5819.

Florine KI and Feigenson GW: *Biochemistry* 1987, 26:1757.

Gawrisch K, Parsegian VA, Hajduk DA, Tate MW, Gruner SM, Fuller NL, and Rand RP: 1991 (submitted to *Biochemistry.*

Goldstein RE and Leibler S: *Phys. Rev. Lett.* 1988, 16:2113.

Gruen DWR and Marcelja S: *J. Chem. Soc. Faraday Trans. II* 1983, 79:211.

Gruen DWR and Marcelja S: *J. Chem. Soc. Faraday Trans. II* 1983, 79:225.

Gruner SM, Tate MW, Kirk GL, So PTC, Turner DC, Keane DT, Tilcock CPS, and Cullis PR: *Biochemistry* 1988, 27:2853.

Gruner SM, Parsegian VA, and Rand RP: *Discuss. Chem. Soc.* 1986, 81:29.

Guldbrand L, Jonsson B, Wennerstrom H, and Linse P: *J. Chem. Phys.* 1984, 80:2221.

Guldbrand L, Jonsson B, and Wennerstrom H: *J. Colloid Interface Sci.* 1982, 89:532.

Handbook of Physics and Chemistry 1967, 47:c75.

Hauser H, Pasher I, Pearson RH, and Sundell S: *Biochim. Biophys. Acta* 1981, 650:21.

Helfrich W: *Z. Naturforsch.* 1978, 33a:305.

Horn R: *Biochim Biophys. Acta* 1984, 778:224.

Horn RG, Israelachvili JN, Marra J, Parsegian VA, and Rand RP: *Biophys. J.* 1988.

Israelachvili J and Marra J: *Methods Enzymol.* 1986, 127:353.

Israelachvili J and Pashley R: *Nature (London)* 1982, 300:341.

Israelachvili JN: *Acc. Chem. Res.* 1988.

Israelachvili JN and Adams GE: *J. Chem. Soc. Faraday Trans.* 1, 1978, 74:975.

Janiak MJ, Small DM, and Shipley GG: *J. Biol. Chem.* 1979, 254:6068.

Jendrasiak GL and Mendible JC: *Biochim. Biophys. Acta* 1976, 424:149.

Jendrasiak GL and Hasty JH: *Biochim. Biophys. Acta* 1974, 337:79-91.

Jorgensen WL, Gao J, and Ravinohan C: *J. Phys. Chem.* 1985, 89:3470.

Kirk GL and Gruner SM: *J. Phys. Paris* 1985, 46:761.

Kjellander R and Marcelja S: *Chem. Phys. Lett.* 1985, 120:393.

Kjellander R and Marcelja S: *Chemica Scripta* 1985, 25:112.

Kjellander R: *J. Chem. Soc. Faraday Trans. II* 1984, 80:1323.

Kolber MA and Haynes DH: *J. Membr. Biol.* 1979, 48:95.

Kornyshev AA: *Chemica Scripta* 1985, 25:63.

Kraut J: *Acta Crystallogr.* 1961, 14:1146.

Kwok R and Evans E: *Biophys. J* 1981, 35:637.

LeNeveu DM, Rand RP, Parsegian VA, and Gingell D: *Biophys. J.* 1977, 18:209.

LeNeveu DM, Rand RP, and Parsegian VA: *Nature* 1976, 259:601.

Lewis BA and Engelman DM: *Mol. Biol.* 1983, 166;211.

Lis LJ, McAlister M, Fuller NL, Rand RP, and Parsegian VA: *Biophys. J.* 1982, 37:657.

Loosley-Millman ME, Rand RP, and Parsegian VA: *Biophys. J.* 1982, 40:221.

Luzzati V: in *Biological Membranes.* Ed. Chapman D, Academic Press, New York, 1968, pp. 71–123.

Luzzati V and Husson F: *J. Cell. Biol.* 1962, 12:207.

Mahanty J and Ninham B: in *Dispersion Forces.* Academic Press, London, 1976.

Marcelja S and Radic N: *Chem. Phys. Lett.* 1976, 42:129.

Marra J: *J. Colloid Interface Sci.* 1988, 109:11–20.

Marra J: *Biophys. J.* 1986, 50:815.

Marra J: *J. Colloid Interface Sci.* 1985, 107:446.

Marra J and Israelachvili JN: *Biochemistry* 1985, 24:4608.

McIntosh TJ, Magid AD, and Simon SA: *Biophys. J.* 1990, 57:1187.

McIntosh TJ, Magid AD, and Simon SA: *Biochemistry* 1987, 26:7325.

McIntosh TJ and Simon SA: *Biochemistry* 1986, 25:4948.

McIntosh TJ and Simon SA: *Biochemistry* 1986, 25:4058.

Mulukutla S and Shipley GG: *Biochemistry* 1984, 23:2514.

Nagle JF and Wilkinson DA: *Biophys. J.* 1978, 23:159.

Parsegian VA and Rand RP: in *Cellular Membrane Fusion.* Eds. Wilschut J and Hoeckstra D, Marcel Dekker, New York, 1988.

Parsegian VA, Rand RP, Fuller NL, and Rau DC: in *Methods in Enzymology.* Vol. 127. Ed. Packer L, Academic Press, New York, 1986, pp. 400–416.

Parsegian VA, Rand RP, and Rau DC: *Chemica Scripta* 1985, 25:28.

Parsegian VA and Rand RP: *Ann. N.Y. Acad. Sci.* 1983, 416:1.

Parsegian VA, Fuller NL, and Rand RP: *Proc. Natl. Acad. Sci. U.S.A.* 1979, 76:2750.

Parsegian VA: in *Physical Chemistry: Enriching Topics for Colloid Surface Science.* Eds. van Olphen and Mysels J, 1975, pp. 27–72.

Parsegian VA: *J. Theor. Biol.* 1967, 15:70.

Pashley RM, McGuiggan PM, Ninham BW, Brady J, and Evans DF: *J. Phys. Chem.* 1986, 90:1637.

Pashley RM: *J. Colloid Interface Sci.* 1981, 80:153.

Pashley RM: *J. Colloid Interface Sci.* 1981, 83:531.

Podgornik R: *J. Chem. Soc. Faraday Trans. II* 1988, 84:611.

Podgornik R, Rau DC, and Parsegian VA: *Macromolecules* 1988 (in press).

Portis A, Newton C, Pangborn W, and Papahadjopolous D: *Biochemistry* 1979, 18:780.

Putkey E and Sundaralingan M: *Acta Crystallogr. B.* 1970, 26:782.

Rand RP, Needham D, and Evans E: (in preparation).

Rand RP, Fuller NL, Gruner SM, and Parsegian VA: *Biochemistry* 1990, 29:76.

Rand RP and Parsegian VA: *Biochim. Biophys. Acta* 1989, 988:351.

Rand RP, Fuller NL, Parsegian VA, and Rau DC: *Biochemistry* 1988, 27:7711.

Rand RP and Luzzati V: *Biophys. J.* 1986, 8:125.

Rand RP and Parsegian VA: *Annu. Rev Physiol.* 1986, 48:201.

Rand RP and Parsegian VA: *Biochim. Biophys. Acta* 1989, 988:351.

Rand RP: *Annu. Rev. Biophys. Bioeng.* 1981, 10:277.

Rand RP, Parsegian VA, Henry JAC, Lis LJ, and McAlister M: *Can. U. Biochem.* 1980, 58:959.

Rand RP, Fuller NL, and Lis LJ: *Nature* 1979, 279:258.

Rau DC and Parsegian VA: *Biophys. J.* 1987, 51:503.

Rau DC, Lee BK, and Parsegian VA: *Proc. Natl. Acad. Sci. U.S.A.* 1984, 81:2621.

Safinya CR, Roux D, Smith GS, Sinha SK, Dimon P, Clark NA, and Bellocq AM: *Phys. Rev. Lett.* 1986, 57:2718.

Savage H: *Biophys. J.* 1986, 50:967.

Schiby D and Ruckenstein: *Chem. Phys. Lett.* 1983, 95:435.

Seddon JM, Cevc G, Kaye RD, and Marsh D: *Biochemistry* 1984, 23:2634.

Seddon JM, Harlos K, and Marsh D: *J. Biol. Chem.* 1983, 258:3850.

Simon SA, McIntosh J, and Magid AD: *J. Colloid Interface Sci.* 1988 (in press).

Small PM: *The Physical Chemistry of Lipids. Handbook of Lipid Research.* Plenum Press, New York, 1986.

Small DM and Bourges M: *Mol. Cryst.* 1966, 1:541.

Smith GS, Safinya CR, Roux D, and Clark N: *Mol. Cryst. Liquid Cryst.* 1987, 144:235.

Sornette D and Ostrovsky N: *J. Chem. Phys.* 1985, 45:265.

Tate MW and Gruner SM: *Biochemistry* 1987, 26:231.

Templin PR: *Ind. Eng. Chem.* 1956, 48:154.

Torbet J and Wilkins MHF: *J. Theor. Biol.* 1976, 62:447.

Vaughan DJ and Keough KM: *FEBS Lett.* 1974, 47:158.

Viani BE, Low PF and Roth CB: *J. Colloid Interface Sci.* 1982, 96:229.

White SH, Jacobs RE and King GI: *Biophys. J.* 1987, 52:663.

Wilkinson DA and Nagle JF: *Biochemistry* 1981, 20:187.

7 The Roles of Cholesterol in the Biology of Cells

Philip L. Yeagle

CONTENTS

7.1 INTRODUCTION

Cholesterol has long been a subject of intense interest because of the critical roles it plays in normal cell biology and in the pathogenesis of disease. Cholesterol plays these roles from its location in the membranes of cells, where it modulates membrane structure, dynamics, and function. Sufficient information is now available to support at least a partial understanding of the molecular basis for these roles of cholesterol in normal cell biology.

Cholesterol modulates membrane structure, dynamics, and/or function through three major molecular mechanisms in cholesterol-requiring cells:

1. Cholesterol modulates membrane protein function through sterol–protein interactions within the membrane.
2. Cholesterol modulates the internal properties of the lipid bilayer of the cell membrane and through that process affects membrane protein function.
3. Cholesterol alters the lateral distribution of components in the cell membrane.

The first two mechanisms will be discussed in this chapter. The third concept is today called *rafts* and is the subject of Chapter 9. A previous review contains more details from the older literature (Yeagle 1985a).

7.2 FUNCTIONAL EVOLUTION OF CHOLESTEROL

Cholesterol is the end product of a series of over three dozen enzyme-catalyzed steps. Eighteen enzyme-catalyzed steps are required just to transform lanosterol (Figure 7.1), a sterol seemingly rather similar to cholesterol, into cholesterol (Faust and others 1988). Production of cholesterol for cholesterol-requiring cells is therefore complicated and energy intensive. Bloch suggested that evolutionary pressure for a more biologically competent sterol led to the development of the enzymatic pathway from lanosterol to cholesterol (Bloch 1976). Presumably the modest chemical

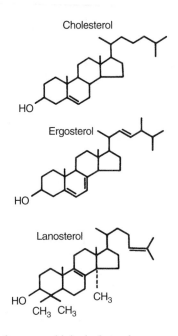

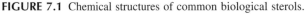

FIGURE 7.1 Chemical structures of common biological sterols.

differences between lanosterol and cholesterol are sufficiently critical in cellular function to justify this enormous investment in metabolic energy and complexity required to achieve cholesterol synthesis. Such considerations suggest that details of sterol chemical structure present important clues to the molecular mechanisms for cholesterol modulation of membrane function. This discussion will follow that lead. These clues will direct us to the first two molecular mechanisms introduced previously.

Mammalian cells require cholesterol for cell growth. This subject has been difficult to investigate experimentally because sterol auxotrophy is normally lethal to mammalian cells. However, inhibition of cholesterol biosynthesis (for example, by inhibition of HMGCoA reductase, the enzyme catalyzing the rate-limiting step in cholesterol biosynthesis) inhibits cell growth (Chen and others 1974). Although this inhibition could result from the loss of other products made intracellularly from mevalonate, addition of cholesterol to the culture medium restored cell growth (Brown and Goldstein 1974). Work on a human macrophagelike cell line further expands these observations. Cell growth demanded a highly specific sterol requirement; specific features of the chemical structure of cholesterol were required to support growth (Esfahani and others 1984). These results illustrate the essential requirement of cholesterol for mammalian cells.

Mycoplasma mycoides have a similar requirement for cholesterol to sustain cell growth. *M. mycoides* do not synthesize cholesterol and so must take up this sterol from the medium. Deprivation of cholesterol inhibits cell growth; addition of cholesterol to the medium restores cell growth. Addition of lanosterol, in the absence of any cholesterol, does not support cell growth (Dahl and Dahl 1988). Thus, the cell is able to recognize the modest differences in chemical structure between lanosterol and cholesterol, such that growth is supported by cholesterol and not by lanosterol.

Saccharomyces cerevisiae have an analogous and specific requirement for a different sterol: ergosterol (see Figure 7.1). Anaerobic growth (when the absence of oxygen inhibits oxidative reactions leading to sterol production) requires the presence of ergosterol in the medium. Addition of cholesterol, only, to the medium will not support cell growth (Andreason and Stier 1953). *S. cerevisiae* therefore have a specific requirement for ergosterol for cell growth and can distinguish between the chemical structure of cholesterol and the chemical structure of ergosterol.

These examples provide a view into the remarkable achievements of functional evolution of sterols. *S. cerevisiae* have evolved a requirement for ergosterol, a requirement that cannot be met by cholesterol. *M. mycoides* have evolved a requirement for cholesterol that cannot be met by lanosterol. Mammalian cells have evolved a requirement for cholesterol that cannot be met by other sterols. By what molecular mechanism *M. mycoides* could distinguish between lanosterol and cholesterol, and *S. cerevisiae* could distinguish between ergosterol and cholesterol, and mammalian cells can differentiate cholesterol from other sterols, is the subject of the next section. This discussion of sterol specificity serves to focus attention on the central question of the critical role of sterols in biology.

7.3 MOLECULAR BASIS FOR SPECIFICITY OF STEROLS IN SUPPORT OF CELL GROWTH

To understand the sterol specificities described earlier requires molecular mechanisms that can distinguish small differences in the chemical structures of the sterols. The ability of cells to distinguish sterol structure is quite dramatic: one sterol cannot substitute for another in supporting cell growth. Such exquisite molecular selectivity is generally achieved in biology through protein recognition of ligands. For example, enzymes can be extraordinarily specific for details of chemical structure in selecting those compounds that will function as substrate. For reasons that will be described later, general sterol effects on the lipid bilayer of membranes cannot explain the sterol requirements for cell growth for mammalian cells, ergosterol support of *S. cerevisiae* growth, or cholesterol support of *M. mycoides* growth. Therefore, one should look for evidence of sterol–protein interactions in membranes as an explanation of this sterol specificity in support of growth.

Cholesterol modulates critical membrane transport processes in an interesting way. In particular, some transport processes appear to require cholesterol in the membrane to operate. For example, consider the $Na^+K^+ATPase$. The $Na^+K^+ATPase$ is found in the plasma membrane and maintains the transmembrane Na^+ and K^+ gradients across the plasma membrane and the correlated transmembrane electrical potential. This enzyme transport system is the single greatest user of ATP in many cells; thus, this protein represents a key component of cell viability. By altering cholesterol content in isolated plasma membranes containing the $Na^+K^+ATPase$, cholesterol was found to stimulate ATPase activity of this membrane enzyme at low to moderate cholesterol contents. From a cholesterol/phospholipid mole ratio of 0 to a ratio of about 0.35, an increase in membrane cholesterol content resulted in an increase in ATP hydrolyzing activity. Extrapolation to a cholesterol/phospholipid mole ratio of 0 suggests that this enzyme is not active (or exhibits very low activity) in the absence of cholesterol in the membrane (Sotomayor and others 2000; Yeagle and others 1988). Even depletion *in vivo* in human subjects of erythrocyte cholesterol led to a significant loss of $Na^+K^+ATPase$ activity (Shanmugasundaram and others 1992).

Similar sensitivity of other transport functions to membrane cholesterol content has been reported. At low to moderate membrane cholesterol contents, cholesterol stimulated activity of Na^+-Ca^{2+} exchange (Vemuri and Philipson 1989), ATP-ADP exchange (Kramer 1982), glutamate transport (Shouffani and Kanner 1990), GABA transport (Shouffani and Kanner 1990), and carrier-mediated lactate transport (Grunze and others 1980). Cholesterol is required for activity of the acetyl choline receptor (Craido and others 1982). Cholesterol is also required for galanin binding to the galanin receptor (Pang and others 1999). In all of these examples, the data suggest that little or no activity is observed in the absence of cholesterol in the membrane.

Some evidence has been reported that suggests binding of cholesterol to membrane proteins. Binding of cholesterol by a protein could readily explain the specificity of the sterol-dependent growth phenomenon discussed earlier for details in sterol structure. Older literature contains data suggestive of cholesterol binding to human erythrocyte glycophorin (Yeagle 1984) and band 3 (anion transport protein) (Klappauf and Schubert 1977; Schubert and Boss 1982). Subsequently, cholesterol binding to the fusion protein of the Sendai virus envelope (Asano and Asano 1988) and to rhodopsin was reported (Albert and others 1996), as well as other receptors (Gimpl and others 1997).

In recent studies with direct relevance to this question, cholesterol binding to proteins involved in cholesterol homeostasis, such as the SCAP protein, has been observed (Brown and others 2002). Cholesterol apparently interacts with P-glycoprotein in a manner that may contribute to raft stability (Garrigues and others 2002). Cholesterol was found to interact specifically with the serotonin transporter (Scanlon and others 2001).

Therefore, cholesterol may bind to membrane proteins. Such binding offers a possible explanation for the specificity observed from the sterol requirements for cell growth. However, the question remains whether the enzymes, noted previously, that require cholesterol for full expression of activity have a specificity for cholesterol relative to other sterols. This question was addressed in experiments with the Na^+K^+ATPase. After removal of cholesterol from the membrane, the ability of sterols to restimulate activity was examined. The Na^+K^+ATPase exhibited considerable sensitivity to sterol structure. Lanosterol and ergosterol were much less effective in supporting activity of this enzyme than was cholesterol (Yeagle and others 1988). This specificity mimicked what was observed for *M. mycoides*; cholesterol supported cell growth, whereas lanosterol did not. This specificity also is consistent with the response to sterols of the human macrophagelike cell line described earlier. If membrane enzymes critical to cell viability, and thus to cell growth, have a specificity for cholesterol in the membrane, then it is reasonable to expect that overall cell growth characteristics would also have a specific requirement for cholesterol.

These data and arguments support an hypothesis that cholesterol modulates some membrane protein function through sterol–protein interactions within the membrane and that these interactions are specific for the chemical structure of the sterol. Thus, this hypothesis suggests that *S. cerevisiae* have evolved to require the particular structure of ergosterol for cell growth through the development of ergosterol–protein interactions in membranes, and mammalian cells have evolved an analogous specificity for cholesterol through cholesterol–protein interactions in membranes.

An alternative hypothesis for the specific requirements of cholesterol invokes the ability of cholesterol (and not other sterols) to alter the physical properties of the lipid bilayer in the membrane. One measure of this ability of cholesterol is the condensing effect of cholesterol on the bilayer (more of this will be discussed later). This is a measure of the ability of the sterol to reduce the average cross-sectional area of the phospholipids in the membrane, proportional to the sterol content of the bilayer. Cholesterol is more effective at condensing the membrane than many other sterols. Cholesterol is the required sterol for mammalian cells, so the superior ability of cholesterol to condense lipid bilayers might be consistent with the requirement of cholesterol by mammalian cells. However, ergosterol is significantly less effective than cholesterol at condensing lipid bilayers, and yet it is the sterol specifically required by *S. cerevisiae*. Analysis of the relationship between the condensing ability of sterols and their ability to support membrane enzyme activities shows no simple correlation (Yeagle 1991). Furthermore, the ability of sterols to condense bilayers is insignificant at low sterol content, where small increases in sterol content (at low overall sterol content) lead to significant increases in activity of enzymes dependent on sterol for function.

Therefore, the latter hypothesis can be rejected. The most reasonable explanation for the specific requirement of cholesterol for mammalian cells and ergosterol for yeast is a direct interaction between the sterol and critical membrane enzymes within the membrane. In this model, the required sterol acts through the membrane analogously to the modulation of water-soluble enzymes by effectors. The site for interaction would be expected to exhibit an affinity for the appropriate sterol, such that the extent of binding would be governed by the sterol content of the membrane: higher sterol content would result in higher occupancy of the site and a greater response by the enzyme. Because the sterol is binding to the enzyme at a site, the chemical nature of that site would best match one particular sterol, giving rise to the observed specificity for sterols. As in the case of water-soluble enzymes, effectors can either stimulate or inhibit enzymes. The Na^+K^+ATPase is an example of specific cholesterol stimulation of a membrane enzyme. In principle, enzyme inhibition could also be achieved by sterol's binding to other membrane enzymes.

7.4 FUNDAMENTAL PROPERTIES OF CHOLESTEROL IN MEMBRANES

Cholesterol consists of a planar ring system of four fused carbon rings (Figure 7.1). The ring system is normally considered to be flat and conformationally restricted. However, X-ray crystallography has revealed that the A ring can adopt an alternate conformation and that the hydrocarbon tail of the molecule has even more possible conformations (Duax and others 1988). Thus, the fused ring system is considered to be largely rigid, whereas the hydrocarbon tail is much more disordered.

Cholesterol is an amphipathic molecule. The A ring contains a 3β-hydroxyl group. Although small, this polar group is sufficient to orient cholesterol in the membrane. In a lipid bilayer, the fused-ring system of cholesterol is oriented parallel to the hydrocarbon chains of the lipids. The hydroxyl makes the sterol surface active and is located in the interfacial region of the bilayer. Cholesterol is very hydrophobic and exhibits very low solubility in aqueous media (nM). Cholesterol is readily accommodated in membranes at high levels, up to 1:1 mole ratio with the phospholipids, or even higher under some circumstances. Later, these features will be discussed more fully. The β-configuration of the hydroxyl is important; epicholesterol, an otherwise identical sterol, has a 3α-hydroxyl group. Epicholesterol has very different physical properties in the membrane from cholesterol.

Cholesterol may form homodimers. Evidence from partitioning experiments (Feher and others 1974; Gilbert and Reynolds 1976) and from X-ray crystallography (Duax and others 1988) suggests such a dimer may form. Observations of cholesterol dimer formation have assisted the development of models for cholesterol organization in lipid bilayers (Martin and Yeagle 1978).

Because of the hydrophobic effect, free cholesterol in cells is located in the membranes. The distribution of cholesterol among the membranes of the cell is inhomogeneous. Membranes with the highest cholesterol content are plasma membranes of cells. For example, the human erythrocyte plasma membrane contains approximately 45 mole% cholesterol with respect to the other membrane lipids (Cooper 1978). In contrast, intracellular membranes contain signficantly less cholesterol. For example, the endoplasmic reticulum contains 10 to 12 mole% cholesterol. Golgi are intermediate in cholesterol content between endoplasmic reticulum and plasma membrane. Typical mitochondrial membranes contain vanishingly small amounts of cholesterol. Cholesterol has been suggested to be asymmetrically distributed across the plasma membrane of cells with an enrichment on the cytoplasmic face (Hale and Schroeder 1982; Schroeder 1981; Schroeder and others 1995).

An interesting example of the inhomogeneity of cholesterol contents among cell membranes can be found in the mammalian retinal rod cells. The outer segment of these cells contains a stack of disk membranes holding the photopigment rhodopsin. These disks are formed at the base of the outer segment from the plasma membrane and, with time, move up the outer segment until the old disks are phagocytosed from the apical tip. New disks are high in cholesterol, and old disks are much lower in cholesterol (Boesze-Battaglia and others 1990). In this dynamic system, the new disks reflect the cholesterol content of the plasma membrane from which they were formed. As the disks age, there is a time-dependent migration of cholesterol out of the disk membranes.

If cell membranes have different cholesterol contents, the question arises whether cholesterol can move between membranes. Lipid vesicles without cholesterol, when incubated with biological membranes, can deplete the biological membrane of cholesterol (Lange and others 1980). Therefore, cholesterol can exchange between membranes. Two general mechanisms are extant. Cholesterol can exchange between two membranes following close approach or collision of the membranes (McLean and Phillips 1981). Cholesterol can also exchange through the aqueous phase without close approach of the membranes, although this is much slower than the first mechanism (Backer and Dawidowicz 1981a). It is thought that, in both mechanisms, cholesterol has some exposure to the aqueous phase. Although this is thermodynamically unfavorable, nevertheless, cholesterol is soluble at nM concentrations, and thus a monomer or dimer in the aqueous phase is a possibility as part of a kinetic mechanism.

Some information is available about cholesterol exchange rates between membranes. The most interesting observation is that sphingomyelin decreases dramatically the exchange rate of cholesterol between membranes (Lange and others 1979; Yeagle and Young 1986). This observation may be relevant to more recent observations of raft formations, where the rafts are believed to consist of sphingomyelin and cholesterol.

From these cholesterol exchange experiments, data on factors affecting the distribution of cholesterol among membranes has been obtained (Yeagle and Young 1986). The most dramatic effect on cholesterol distribution among membranes arises from phosphatidylethanolamine (PE). Membranes rich in phosphatidylethanolamine provide a thermodynamically unfavorable environment for cholesterol. This observation is particularly interesting in the context of the rod outer segment disk membranes noted previously. When new disks are formed, PE content is very high (relative to other membranes) and is the most abundant of phospholipid species in the membrane. Thus it is not surprising that, after formation of the new disk, disk membrane cholesterol is rapidly lost. Recent experiments have identified highly unsaturated phospholipids, such as those containing 22:6 (docosahexenoic acid), as an unfavorable environment for cholesterol, as well (Mitchell and Litman 1998). Exacerbating the problems for cholesterol in the high PE disks is the high content of 22:6 in that membrane.

Cholesterol can undergo flip-flop across lipid bilayers and biological membranes. *Flip-flop* refers to the transfer of cholesterol from one leaflet of the bilayer to the other leaflet of the bilayer, or from exposure of the hydroxyl on one side of the bilayer to exposure of the hydroxyl on the opposite side of the bilayer. In small vesicles, flip-flop of cholesterol is rapid (Backer and Dawidowicz 1981b; Lange and others 1981). Similar measurements in cellular membranes have given mixed results. For example, in erythrocyte membranes, cholesterol flip-flop is rapid (Backer and Dawidowicz 1981b; Lange and others 1981), whereas in mycoplasma cholesterol flip-flop is much slower (Clejan and Bittman 1984).

The dynamics of cholesterol in membranes have been examined in some detail. Cholesterol is quite rigid in the membrane, exhibiting only axial diffusion about an axis perpendicular to the membrane surface and a time dependence of the orientation of the director for axial diffusion (or *wobble*). Spin-label experiments showed a rotational correlation time about the long axis of cholesterol of about 0.1 ns (Schindler and Seelig 1974). Time-resolved fluorescence measurements revealed a correlation time of wobble of the long axis of the sterol of about 1 ns (Yeagle and others 1990).

Sterol dynamics in the membrane show considerable sensitivity to the structure of the sterol. Lanosterol is not as ordered in the membrane as cholesterol (Yeagle 1985b; Yeagle and others 1977b). Epicholesterol (with the 3α hydroxyl) is also less ordered in the membrane than cholesterol (Brainard and Szabo 1981).

Cholesterol has been localized in the membrane through the use of neutron diffraction experiments. The data indicate that the 3β-hydroxyl of the cholesterol molecule is in the immediate vicinity of the ester carbonyl of the phospholipid (Franks 1976; Worcester and Franks 1976). Huang has suggested that the sterol hydroxyl is hydrogen-bonded to the ester carbonyl of the phospholipid (Huang 1976). Even at low cholesterol contents, cholesterol is likely segregated into sterol-rich regions in the membrane (Mabrey and others 1978; Smutzer and Yeagle 1985).

7.4.1 STEROL ORDERING OF MEMBRANES

The most widely recognized property of cholesterol in lipid bilayers and biological membranes is the ordering of the membrane lipids due to the incorporation of cholesterol in the membrane. This phenomenon is now well understood and is the subject of this section.

One of the earliest studies on cholesterol in phospholipid bilayers reported that cholesterol altered the gel to liquid–crystalline phase transition of the phospholipid (Oldfield and Chapman 1972). Cholesterol was suggested to induce a new physical state, separate from the gel state and from the liquid–crystalline state. Today, this state is referred to as the *liquid-ordered state*. The dynamics of this state are best described using the concept of motional order.

Motional order can be readily determined using ^{2}H NMR (Seelig and Seelig 1974). Motional order refers to the degrees of freedom of motion a molecule or a segment of a molecule can experience. Motional order is largely determined by the extent of isomerizations about the carbon–carbon single bonds in the lipids and to overall molecular motion. Motional order is not a measure of the rates of motion (though the rates of motion may limit the sensitivity of a particular probing method to degrees of disorder). For the lipid hydrocarbon chains in a bilayer, increasing motional order corresponds to a decrease in the *trans*-gauche isomerizations of the carbon–carbon single bonds in the hydrocarbon chain. Disordering corresponds to an increase in the incidence of "kinks" in the lipid hydrocarbon chain. A kink results from the conversion of a *trans* configuration into a *cis* configuration. When this occurs, the overall direction of the hydrocarbon chain is altered. A complementary isomerization at a neighboring carbon–carbon bond produces a kink in the chain, with the overall direction of the chain unchanged but with a displacement of the director of the hydrocarbon chain. The introduction of a kink leads to a packing defect in the bilayer, a transient void on the molecular level.

These experiments have provided a view of the hydrophobic interior of the phospholipid bilayer. It is highly anisotropic and is characterized by an ordered region of the hydrocarbon chains near the headgroup and a disordered region in the bilayer center. About half the chain is characterized by relatively high order. When cholesterol is introduced into the lipid bilayer, the rigid fused-ring system of the sterol lies in the highly ordered region and increases its average motional order. The hydrocarbon tail of the cholesterol is substantially more ordered than the corresponding region of the lipid hydrocarbon chains (DuFourc and others 1984).

A decrease in the *trans*-gauche isomerizations of the carbon–carbon single bonds would increase the effective length of the hydrocarbon chain. Therefore, one would expect that cholesterol would increase the thickness of the hydrocarbon portion of a phospholipid bilayer. For phospholipids with up to 16 carbon atoms in the hydrocarbon chain, cholesterol does increase the bilayer thickness (McIntosh 1978). However, for a phospholipid with an 18-carbon chain, an interesting deviation from this behavior is noted. Cholesterol decreases the bilayer width of these longer-chain phospholipids (McIntosh 1978). This result can be readily understood in the context of the motional ordering just discussed. Cholesterol is not as long as a phospholipid with an 18-carbon chain. This leads to a packing defect in the center of the bilayer. The phospholipid hydrocarbon chain must fill this packing defect, and to do so the end of the phospholipid hydrocarbon chain must become more disordered. This should lead to an effective shortening of the chain, as is observed. The ordering of the membrane lipid hydrocarbon chains by cholesterol is observed in biological membranes, as well (Butler and others 1978; Kelusky and others 1983; Owen and others 1982; Pal and others 1983; Rintoul and others 1979). Cholesterol does not interact as well with unsaturated lipids, showing more disordering of the sterol in polyunsaturated bilayers than in less unsaturated bilayers (Brzustowicz and others 1999).

Another measure of the ordering of membrane lipids by cholesterol is the condensing effect observed in lipid monolayers. Cholesterol decreases the surface area per molecule occupied by phospholipids at the air/water interface (Demel and others 1972a). Highly unsaturated phospholipids are not condensed by cholesterol (Demel and others 1972b), consistent with the observations mentioned earlier from exchange experiments that unsaturated phospholipid bilayers provide an unfavorable environment for cholesterol. The 3β-hydroxyl and the planar ring are the most important features of cholesterol for the condensing effect (Cadenhead and Landau 1979; Cadenhead and Landau 1984; Landau and Cadenhead 1979a; Landau and Cadenhead 1979b).

A third measure of the ordering of membrane lipids by cholesterol is membrane permeability to small molecules. Passive permeability of phospholipid bilayers to glucose is inhibited by cholesterol (Demel and others 1972a; Papahadjopoulos and others 1971). Because of the correspondence between the effects of cholesterol on phospholipid chain ordering in bilayers and on bilayer permeability, a model of the permeability of bilayers to small molecules can be advanced. Disordering of the lipid bilayer occurs when *trans*-gauche isomerizations produce local packing defects in the bilayer structure (kink in one chain next to a kink of a noncomplementary type in another chain, or kink against an all-*trans* configuration in a neighboring chain). Permeation can be modeled to occur by small molecules jumping between instantaneous defects in the bilayer structure resulting from the transient existence of voids between hydrocarbon chains that arise from kinks within chains. Small molecules can transiently occupy these voids. The volume of these voids is similar to the volume of a water molecule. Therefore, small molecules can permeate across a lipid bilayer in a liquid–crystal state by moving from defect to defect, but as the size of the molecule increases, the rate of diffusion decreases dramatically. Glucose flux across a lipid bilayer is much less than water, and molecules larger than glucose have vanishingly small permeabilities. Ordering of the bilayer by cholesterol reduces the incidence of those defects and thus reduces the permeability of the bilayer. Reduction in bilayer permeability by cholesterol has structural requirements in the sterol like those required for the condensing effect described previously.

Litman and coworkers (Straume and Litman 1987a; Straume and Litman 1987b) have developed an alternative but complementary means to detect these properties of sterols. Time-resolved fluorescence techniques and the fluorescent probe, diphenylhexatriene (DPH), have revealed that although the widely used steady-state fluorescence anisotropy measurements on DPH have no specific physical meaning, it is possible to extract information on partitioning of the DPH between the bilayer center and an orientation parallel to the lipid hydrocarbon chains. Litman and coworkers interpreted their work in terms of packing voids in the bilayer into which the probe could partition. As with other measurements, the presence of cholesterol reduced the incidence of those voids, or "free volume," in the bilayer.

Although cholesterol orders the hydrocarbon chains of lipids, cholesterol does not induce a true solid state. As noted previously, cholesterol induces a liquid-ordered state in the lipid bilayer. Cholesterol, however, inhibits the formation of a true gel or solid state, much as an impurity in the bilayer (Estep and others 1978; Mabrey and others 1978). This phenomenon has been much studied and has been suggested to be an important role for cholesterol in cell membranes. However, mammalian cell plasma membranes do not form a gel at physiological temperatures even if cholesterol is removed from the membrane (Dijck and others 1976). Therefore, the inhibition of formation of a gel state in the membrane is not a true biological role for cholesterol.

One final comment on the physical effects of cholesterol on the lipid bilayer: cholesterol interferes with naturally occuring phospholipid headgroup–headgroup interactions in membranes. Normally, headgroups like those of phosphatidylcholine and phosphatidylethanolamine lie parallel to the bilayer surface and interact with each other by hydrogen bonds or by weak electrostatic interactions (Yeagle and others 1975). These intermolecular headgroup interactions are disrupted by the insertion of cholesterol among the lipids in the bilayer (Yeagle and others 1977a).

7.5 CHOLESTEROL ORDERING OF MEMBRANES AND ALTERATION OF MEMBRANE FUNCTION

With such profound physical effects on the lipid bilayer in model and biological membranes, cholesterol is likely to modulate membrane protein function through the bilayer. The clearest mechanistic studies of such inhibition have been performed with rhodopsin, the visual pigment sensitive to light that initiates visual signal transduction upon absorption of a photon. Reconstitution of rhodopsin into a lipid bilayer reduces the free volume, as defined earlier (Straume and Litman

1988). Furthermore, rhodopsin is known to expand upon activation by light (Lamola and others 1974). Therefore, one would expect that a protein like rhodopsin would be sensitive to the availability of free volume in the membrane bilayer. The activation of this protein to the metarhodopsin II state has been shown to be enhanced by polyunsaturated fatty acids that increase the incidence of defects in the bilayer and thus the free volume (Mitchell and others 1992). As expected, therefore, when cholesterol is introduced into the bilayer, the free volume falls and the activation of this receptor is inhibited (Boesze-Battaglia and Albert 1990; Mitchell and others 1990).

Experiments such as these provide the second model for cholesterol modulation of membrane protein function. High cholesterol levels in the lipid bilayer reduce free volume and reduce the conformational motility of membrane-bound enzymes. A number of examples of cholesterol inhibition of membrane protein function, including cholesterol inhibition of ion transport, fit this model (Brasitus and others 1988; Gregg and Reithmeier 1983; Rotenberg and Zakim 1991; Saito and Silbert 1979). Recent data show that *in vivo* reductions in cholesterol content in erythrocyte membranes led to an increase in activity of the $Na^+ K^+$ pump, consistent with this same model (Lijnen 1997).

REFERENCES

Albert AD, Young JE, Yeagle PL. 1996. Rhodopsin-cholesterol interactions in bovine rod outer segment disk membranes. *Biochim. Biophys. Acta* 1285:47–55.

Andreason AA, Stier TJB. 1953. Anaerobic nutrition of *Saccharomyces cerevisiae*. Ergosterol requirements for growth in a defined medium. *J. Cell. Comp. Physiol.* 41:23.

Asano K, Asano A. 1988. Binding of cholesterol and inhibitory peptide derivatives with the fusogenic hydrophobic sequence of F-glycoprotein of Sendai virus: possible implication in the fusion reaction. *Biochemistry* 27:1321–1329.

Backer JM, Dawidowicz EA. 1981a. Mechanism of cholesterol exchange between phospholipid vesicles. *Biochemistry* 20:3805–3810.

Backer JM, Dawidowicz EA. 1981b. Transmembrane movement of cholesterol in small unilamellar vesicles detected by cholesterol oxidase. *J. Biol. Chem.* 256:586–588.

Bloch K. 1976. On the evolution of a biosynthetic pathway. In: Kornberg A, HKorecker BL, Cornudella Z, Oro J, editors. *Reflections in Biochemistry*. Oxford: Pergamon Press. p 143.

Boesze-Battaglia K, Albert A. 1990. Cholesterol modulation of photoreceptor function in bovine rod outer segments. *J. Biol. Chem.* 265:20727–20730.

Boesze-Battaglia K, Fliesler SJ, Albert AD. 1990. Relationship of cholesterol content to spatial distribution and age of disk membranes in retinal rod outer segments. *J. Biol. Chem.* 265:18867–18870.

Brainard JR, Szabo A. 1981. Theory for nuclear magnetic relaxation of probes in anisotropic systems. *Biochemistry* 20:4615–4628.

Brasitus TA, Dahiya R, Dudeja PK, Bissonnette BM. 1988. Cholesterol modulates alkaline phosphatase activity of rat intestinal microvillus membranes. *J. Biol. Chem.* 263:8592–8597.

Brown AJ, Sun L, Feramisco JD, Brown MS, Goldstein JL. 2002. Cholesterol addition to ER membranes alters conformation of SCAP, the SREBP escort protein that regulates cholesterol metabolism. *Mol Cell* 10(2):237–245.

Brown MS, Goldstein JL. 1974. Suppression of 3-hydroxy-3-methylglytaryl-coenzyme A reductase activity and inhibition of growth of human fibroblasts of 7-ketocholesterol. *J. Biol. Chem.* 249:7306.

Brzustowicz MR, Stillwell W, Wassall SR. 1999. Molecular organization of cholesterol in polyunsaturated phospholipid membranes: a solid state 2H NMR investigation. *FEBS Lett.* 451(2):197–202.

Butler KW, Johnson KG, Smith ICP. 1978. *Acholeplasma laidlawii* membranes: an electron spin resonance study of the influence on molecular order of fatty acid composition and cholesterol. *Arch. Biochem. Biophys.* 191:289–297.

Cadenhead DA, Landau FM. 1979. Molecular packing in steroid-lecithin monolayers. III: Mixed films of 3-doxyl cholestane and 3-doxyl-17-hydroxyl-androstane with dipalmitoylphosphatidylcholine. *Chem. Phys. Lipids* 25:329–343.

Cadenhead DA, Landau FM. 1984. Molecular packing in steroid-lecithin monolayers. IV. Mixed films of epicoprostanol with dipalmitoyl phosphatidylcholine. *Can. J. Biochem. Cell. Biol.* 62:732–737.

Chen HW, Kandutsch AA, Waymouth C. 1974. Inhibition of cell growth by oxygenated derivatives of cholesterol. *Nature* 251:419.

Clejan S, Bittman R. 1984. Distribution and movement of sterols with different side chain structures between the two leaflets of the membrane bilayer of Mycoplasma cells. *J. Biol. Chem.* 259:449–455.

Cooper RA. 1978. Influence of increased membrane cholesterol on membrane fluidity and cell function in human red blood cells. *J. Supramol. Struct.* 8:413–430.

Craido M, Eibl H, Barrantes FJ. 1982. Effects of lipids on acetylcholine receptor. Essential need of cholesterol for maintenance of agonist-induced state transitions in lipid vesicles. *Biochemistry* 21:3622–3627.

Dahl C, Dahl J. 1988. Cholesterol and cell function. In: Yeagle PL, editor. *Biology of Cholesterol.* Boca Raton, FL: CRC Press. p 147–172.

Demel RA, Bruckdorfer KR, Deenen LLMv. 1972a. The effect of sterol structure on the permeability of liposomes to glucose, glycerol and Rb+. *Biochim. Biophys. Acta* 255:321–330.

Demel RA, Kessel WSMGv, Deenen LLMv. 1972b. The properties of polyunsaturated lecithins in monolayers and liposomes and the interactions of these lecithins with cholesterol. *Biochim. Biophys. Acta* 266:26–40.

Dijck PWMv, Zoelen EJJv, Seldenrijk LR, Deenen LLMv, Gier Jd. 1976. Calorimetric behaviour of individual phospholipid classes from human and bovine erythrocyte membranes. *Chem. Phys. Lipids* 17:336–343.

Duax WL, Wawrzak Z, Griffin JF, Cheer C. 1988. Sterol conformation and molecular properties. In: Yeagle PL, editor. *Biology of Cholesterol.* Boca Raton, FL: CRC Press.

DuFourc EJ, Parish EJ, Chitrakorn S, Smith ICP. 1984. Structural and dynamical details of the cholesterol-lipid interaction as revealed by deuterium NMR. *Biochemistry* 23:6062–6071.

Esfahani M, Scerbo L, Devlin TM. 1984. A requirement for cholesterol and its structural features for a human macrophage-like cell line. *J. Cell. Biochem.* 25:87–97.

Estep TN, Mountcastle DB, Biltonen RL, Thompson TE. 1978. Studies on the anomalous thermotropic behavior of aqueous dispersions of dipalmitoylphosphatidylcholine-cholesterol mixtures. *Biochemistry* 17:1984–1989.

Faust JR, Trzaskos JM, Gaylor JL. 1988. Cholesterol biosynthesis. In: Yeagle PL, editor. *Biology of Cholesterol.* Boca Raton, FL: CRC Press. p 19–38.

Feher JJ, Wright LD, McCormick DB. 1974. Studies of the self-association and solvent-association of cholesterol and other 3β-hydroxysterols in nonpolar media. *J. Phys. Chem.* 78:250–255.

Franks NP. 1976. Structural analysis of hydrated egg lecithin and cholesterol bilayers. I. x-ray diffraction. *J. Mol. Biol.* 100:345–358.

Garrigues A, Escargueil AE, Orlowski S. 2002. The multidrug transporter, P-glycoprotein, actively mediates cholesterol redistribution in the cell membrane. *Proc Natl Acad Sci USA* 99(16):10347–10352.

Gilbert DB, Reynolds JA. 1976. Thermodynamic equilibria of cholesterol-detergent-water. *Biochemistry* 15:71–74.

Gimpl G, Burger K, Fahrenholz F. 1997. Cholesterol as modulator of receptor function. *Biochemistry* 36(36):10959–10974.

Gregg VA, Reithmeier RAF. 1983. Effect of cholesterol on phosphate uptake by human red blood cells. *FEBS Lett.* 157:159–164.

Grunze M, Forst B, Deuticke B. 1980. Duel effect of membrane cholesterol on simple and mediated transport process in human erythrocytes. *Biochim. Biophys. Acta* 600:860–868.

Hale JE, Schroeder F. 1982. Asymmetric transbilayer distribution of sterol across plasma membranes determined by fluorescence quenching of dehydroergosterol. *Eur. J. Biochem.* 122:649–661.

Huang C-h. 1976. Roles of carbonyl oxygens at the bilayer interface in phospholipid-sterol interaction. *Nature* 259:242–244.

Kelusky EC, Dufourc EJ, Smith ICP. 1983. Direct observation of molecular ordering of cholesterol in human erythrocyte membranes. *Biochim. Biophys. Acta* 735:302–304.

Klappauf E, Schubert D. 1977. Band 3 from human erythrocyte membranes strongly interacts with cholesterol. *FEBS Lett.* 80:423–425.

Kramer R. 1982. Cholesterol as activator of ADP-ATP exchange in reconstituted liposomes and in mitochondria. *Biochim. Biophys. Acta* 693:296–304.

Lamola AA, Yamane T, Zipp A. 1974. Effects of detergents and high pressures upon the metarhodopsin I to metarhodopsin II equilibrium. *Biochemistry* 13:738–745.

Landau FM, Cadenhead DA. 1979a. Molecular packing in steroid-lecithin monolayers. I: Pure films of cholesterol, 3-doxyl-cholestane, 3-doxyl-17-hydroxyl-androstane, tetradecanoic acid and dipalmitoyl-phosphatidylcholine. *Chem. Phys. Lipids* 25:299–314.

Landau FM, Cadenhead DA. 1979b. Molecular packing in steroid-lecithin monolayers. II: Mixed films of cholesterol with dipalmitoylphosphatidylcholine and tetradecanoic acid. *Chem. Phys. Lipids* 25:315–328.

Lange Y, Cutler HB, Steck TL. 1980. The effect of cholesterol and other intercalated amphipaths on the contour and stability of the isolated red cell membrane. *J. Biol. Chem.* 255:9331–9336.

Lange Y, D'Alessandro JS, Small DM. 1979. The affinity of cholesterol for phosphatidylcholine and sphingomyelin. *Biochim. Biophys. Acta* 556:388–398.

Lange Y, Dolde J, Steck TL. 1981. The rate of transmembrane movement of cholesterol in the human erythrocyte. *J. Biol. Chem.* 256:5321–5323.

Lijnen P. 1997. The effect of membrane cholesterol content on ion transport processes in plasma membranes. *Cardiovasc. Res.* 35(2):384–386.

Mabrey S, Mateo PL, Strutevant JM. 1978. High-sensitivity scanning calorimetric study of mixtures of cholesterol with dimyristoyl- and dimyristoylphosphatidylcholines. *Biochemistry* 17:2464–2468.

Martin RB, Yeagle PL. 1978. Models for lipid organization in cholesterol-phospholipid bilayers. *Lipids* 13:594–597.

McIntosh TJ. 1978. The effect of cholesterol on the structure of phosphatidylcholine bilayers. *Biochim. Biophys. Acta* 1978:43–58.

McLean LR, Phillips MC. 1981. Mechanism of cholesterol and phosphatidylcholine exchange or transfer between unilamellar vesicles. *Biochemistry* 20:2893–2900.

Mitchell D, Straume M, Miller J, Litman BJ. 1990. Modulation of metarhodopsin formation by cholesterol-induced ordering of bilayers. *Biochemistry* 29:9143–9149.

Mitchell DC, Litman BJ. 1998. Effect of cholesterol on molecular order and dynamics in highly polyunsaturated phospholipid bilayers. *Biophys J.* 75(2):896–908.

Mitchell DC, Straume M, Litman BJ. 1992. Role of sn-1-saturated, sn-2-polyunsaturated phospholipids in control of membrane receptor conformational equilibrium: effects of cholesterol and acyl chain unsaturation on the metarhodopsin I-metarhodopsin II equilibrium. *Biochemistry* 31:662–670.

Oldfield E, Chapman D. 1972. Molecular dynamics of cerebroside-cholesterol and sphingomyelin-cholesterol interactions: implications for myelin membrane structure. *FEBS Lett.* 21:303–306.

Owen JS, Bruckdorfer R, Day RC, McIntyre N. 1982. Decreased erythrocyte membrane fluidity and altered lipid composition in human liver disease. *J. Lipid Res.* 23:124–132.

Pal R, WA, Petri J, Barenholz Y, Wagner RR. 1983. Lipid and protein contributions to the membrane surface potential of vesicular stomatitis virus probed by a fluorescent pH indicator, 4-heptadecyl-7-hydroxy-coumarin. *Biochim. Biophys. Acta* 729:185–192.

Pang L, Graziano M, Wang S. 1999. Membrane cholesterol modulates galanin-GalR2 interaction. *Biochemistry* 38(37):12003-12011.

Papahadjopoulos D, Nir S, Ohki S. 1971. Permeability properties of phospholipid membranes: effect of cholesterol and temperature. *Biochim. Biophys. Acta* 266:561–583.

Rintoul DA, Chou SM, Silbert DF. 1979. Physical characterization of sterol-depleted LM-cell plasma membranes. *J. Biol. Chem.* 254:10070–10077.

Rotenberg M, Zakim D. 1991. Effects of cholesterol on the function and thermotropic properties of pure UDP-glucuronosyltransferase. *J. Biol. Chem.* 266:4159–4161.

Saito Y, Silbert DF. 1979. Selective effects of membrane sterol depletion on surface function thymidine and 3-O-methyl-D-glucose transport in a sterol auxotroph. *J. Biol. Chem.* 255:1102–1107.

Scanlon SM, Williams DC, Schloss P. 2001. Membrane cholesterol modulates serotonin transporter activity. *Biochemistry* 40(35):10507–10513.

Schindler H, Seelig J. 1974. EPR spectra of spin labels in lipid bilayers. II. Rotation of steroid spin probes. *J. Chem. Phys.* 61:2946–2951.

Schroeder F. 1981. Use of a fluorescent sterol to probe the transbilayer distribution of sterols in biological membranes. *FEBS Lett.* 135:127–130.

Schroeder F, Woodford JK, Kavecansky J, Wood WG, Joiner C. 1995. Cholesterol domains in biological membranes. *Mol. Membr. Biol.* 12(1):113–119.

Schubert D, Boss K. 1982. Band 3 protein-cholesterol interactions in erythrocyte membranes. *FEBS Lett.* 150:4–8.

Seelig A, Seelig J. 1974. The dynamic structure of fatty acyl chains in a phospholipid bilayer measured by deuterium magnetic resonance. *Biochemistry* 13:4839–4845.

Shanmugasundaram KR, Padmavathi C, Acharya S, Vidhyalakshmi N, Vijayan VK. 1992. Exercise-induced cholesterol depletion and Na+,K(+)-ATPase activities in human red cell membrane. *Exp. Physiol.* 77(6):933–936.

Shouffani A, Kanner BI. 1990. Cholesterol is required for reconstitution of the sodium- and chloride-coupled GABA transporter from rat brain. *J. Biol. Chem.* 265:6002–6008.

Smutzer G, Yeagle PL. 1985. Phase behavior of DMPC-cholesterol mixtures; a fluorescence anisotrophy study. *Biochim. Biophys. Acta* 814:274–280.

Sotomayor CP, Aguilar LF, Cuevas FJ, Helms MK, Jameson DM. 2000. Modulation of pig kidney Na+/K+-ATPase activity by cholesterol: role of hydration. *Biochemistry* 39(35):10928–10935.

Straume M, Litman BJ. 1987a. Equilibrium and dynamic structure of large, unilamellar, unsaturated acyl chain phosphatidylcholine vesicles. Higher order analysis of 1,6-diphenyl-1,3,5-hexatriene and 1-[4-trime-thylammonio]-6-phenyl-1,3,5-hexatriene anisotropy decay. *Biochemistry* 26:5113–5120.

Straume M, Litman BJ. 1987b. Influence of cholesterol on equilibrium and dynamic bilayer structure of unsaturated acyl chain phosphatidylcholine vesicles as determined from higher order analysis of fluorescence anisotropy decay. *Biochemistry* 26:5121–5126.

Straume M, Litman BJ. 1988. Equilibrium and dynamic bilayer structural properties of unsaturated acyl chain phosphatidylcholine-cholesterol-rhodopsin recombinant vesicles and rod outer segment disk membranes as determined from higher order analysis of fluorescence anisotropy decay. *Biochemistry* 27:7723–7733.

Vemuri R, Philipson KD. 1989. Influence of sterols and phospholipids on sarcolemmal and sarcoplasmic reticular cation transporters. *J. Biol. Chem.* 264:8680–8685.

Worcester DL, Franks NP. 1976. Structural analysis of hydrated egg lecithin and cholesterol bilayers II. Neutron diffraction. *J. Mol. Biol.* 100:359–378.

Yeagle PL. 1984. Incorporation of the human erythrocyte sialglycoprotein into recombined membranes containing cholesterol. *Membr. Biol.* 78:201–210.

Yeagle PL. 1985a. Cholesterol and the cell membrane. *Biochim. Biophys. Acta Biomembrane Rev.* 822:267–287.

Yeagle PL. 1985b. Lanosterol and cholesterol have different effects on phospholipid acyl chain orientation. *Biochim. Biophys. Acta* 815:33–36.

Yeagle PL. 1991. Modulation of membrane function by cholesterol. *Biochemie* 73:1303–1310.

Yeagle PL, Albert AD, Boesze-Battaglia K, Young J, Frye J. 1990. Cholesterol dynamics in phosphatidylcholine bilayers. *Biophys. J.* 57:413–424.

Yeagle PL, Hutton WC, Huang C-h, Martin RB. 1975. Headgroup conformation and lipid-cholesterol association in phosphatidylcholine vesicles: a P-31 {H-1} nuclear Overhauser effect study. *Proc. Natl. Acad. Sci. USA* 72:3477–3481.

Yeagle PL, Hutton WC, Huang C-h, Martin RB. 1977a. Phospholipid headgroup conformations; intermolecular interactions and cholesterol effects. *Biochemistry* 16:4344–4349.

Yeagle PL, Martin RB, Lala AK, Lin H, Bloch K. 1977b. Differential effects of cholesterol and lanosterol on artificial membranes. *Proc. Natl. Acad. Sci. USA* 74:4924–4926.

Yeagle PL, Rice D, Young J. 1988. Effects of cholesterol on (Na,K)-ATPase ATP hydrolyzing activity in bovine kidney. *Biochemistry* 27:6449–6452.

Yeagle PL, Young J. 1986. Factors contributing to the distribution of cholesterol among phospholipid vesicles. *J. Biol. Chem.* 261:8175–8181.

8 Lipid Membrane Fusion

David P. Siegel

CONTENTS

8.1 INTRODUCTION

Membrane fusion is a ubiquitous and critical process in biology; it is central in the everyday economy of cellular life and the function of multicellular organisms. Numerous proteins have evolved to control and catalyze this process [1–6]. However, the basic mechanism by which these proteins act is still indistinct. It seems clear that fusion occurs by a rearrangement of lipid moieties and not by creation of a protein channel between two membranes, which later turns into a membrane-lined channel. The fusion-catalyzing proteins act by making transient changes in the physical chemical environment of small patches of membrane. The changes induce the lipids to collectively form intermembrane structures that eventually form a membrane connection, turning two opposed membranes into one continuous membrane. There are still big gaps in our understanding of what must be done to the lipids of the membranes to make them fuse. Until we know how lipid membranes fuse, it will be difficult to make progress in understanding how proteins catalyze this process. We must know the types of intermediate structures that form between lipid membranes, and how the energies and formation rates of these structures respond to various permutations in lipid composition, the presence of different types of peptides and adsorbed materials, membrane tension, and other parameters. Otherwise, it would be as if one were trying to understand how an enzyme worked without knowing the chemical reaction that it catalyzes. Fairly recent reviews on membrane fusion that emphasize the role of lipids are available [6–9, 9a].

In the past decade, considerable insight has been gained into at least one mechanism of lipid membrane fusion, and promising new techniques have been applied to studies in this area. The purpose of this chapter is to review our present concepts of how lipid membrane fusion occurs, with emphasis on recent developments. Some outstanding questions and prospects for future progress will also be outlined.

8.2 PRINCIPAL LIPID FUSION MECHANISMS

Before the late 1980s, most research on lipid membrane fusion focused on anionic lipid fusion triggered by multivalent cations (for reviews, see [10–13]). Most attention in the last 15 years has been focused on fusion and lipid mixing between membranes in systems where these interactions occur via a mechanism related to inverted phase formation. This mechanism rationalizes the behavior of membranes with no charge or with small amounts of charged lipids in the absence of large divalent cation concentrations, situations often encountered in biomembrane fusion. Some intermediates in this fusion mechanism have been demonstrated, and many observations of protein-mediated fusion are compatible with the predicted intermediate structures. Hence, most of this review will be devoted to these developments. However, the fusion mechanism in the charged lipid systems is still of considerable interest. Some recent results in studies of charged lipid fusion will be discussed in the context of speculations concerning the fusion mechanism at work in Section 8.2.3.2.

8.2.1 Correlation of Membrane Fusion with Inverted Phase Formation

Work in the last 15 years has shown that there is a close association between membrane fusion and inverted phase formation. Fusion is potentiated when the tendency of the lipid composition of the membranes to form inverted phases (Chapter 5) is increased. Depending on the bulk lipid composition of the membranes, this tendency is increased either by heating the system toward the lamellar/inverted phase transition temperature, by addition of small amounts of lipids that lower the transition temperature, by changing the pH, or by addition of certain cations. Moreover, there is now compelling evidence that one membrane fusion mechanism proceeds via the same intermediate structures that form in the course of lamellar/nonlamellar phase transitions. However, it is important to note that there may be other mechanisms of membrane fusion that do not share exactly the same intermediate structures (see Section 8.2.3).

Some phospholipids can adopt so-called inverted phases, or nonlamellar phases, under conditions that are physiological or close to physiological (see Chapter 5, references therein, and [14]). Lipids in lamellar phases have the same planar bilayer structure found in biomembranes (Figure 8.1A). Inverted hexagonal (H_{II}) or inverted cubic (Q_{II}) phases are more complicated structures. The H_{II} phase is a hexagonal, close-packed array of inverted rod micelles (Figure 8.1D, lower right). The Q_{II} phases are labyrinths formed by tubular networks of lipid bilayers (Figure 8.1D, lower left) [15–17]. (Some Q_{II} phases are composed of inverted micellar units [18, 18a], but the phases associated with fusion are the so-called bicontinuous phases just described.) Long ago it was observed that liposomes of a given lipid composition tend to fuse spontaneously under conditions in which the equilibrium phase of the lipids is one of these inverted phases, especially the "isotropic" or Q_{II} phase [19–21]. In early literature, the existence of a so-called isotropic phase was inferred due to the formation of isotropic ^{31}P-NMR resonances and ILAs, as detected by freeze-fracture electron microscopy (see [20] and references therein). Subsequent work showed that these isotropic phases were actually Q_{II} phase precursors (ILAs and arrays of ILAs) or Q_{II} phases [20, 22]. The fusion rate of a given composition increases as the lamellar/Q_{II} phase boundary is approached [20, 21]. In membranes that form only H_{II} phases, some fusion is observed as the temperature approaches

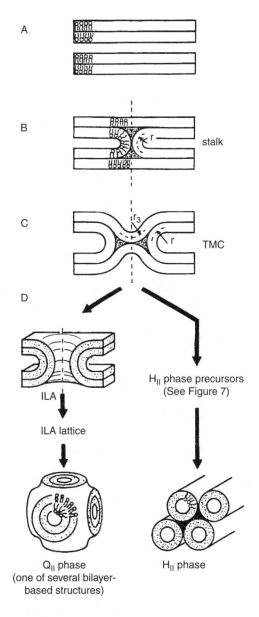

FIGURE 8.1 The modified stalk mechanism of membrane fusion and inverted phase formation, showing the interconversion of lamellar (L_α), inverted cubic Q_{II}), and inverted hexagonal (H_{II}) phases. (A) Planar L_α phase bilayers. (B) The stalk intermediate. This structure is cylindrically symmetrical about the dashed vertical axis. (C) The TMC (*trans* monolayer contact) or hemifusion structure. The diaphragm in the center of the TMC can rupture to form a fusion pore (also referred to as an ILA, or interlamellar attachment). A cross section through a perspective view of a fusion pore is shown at left in D. If fusion pores accumulate in sufficient numbers in a stack of planar bilayers, they form lattices that can rearrange to form Q_{II} phases. For systems close to the L_α/H_{II} phase boundary, TMCs can also aggregate to form H_{II} precursors (see Figure 8.8), which grow directly into domains of H_{II} phase (lower right in D). (Reprinted from [70], with permission.)

the lamellar/H_{II} transition temperature, T_H [23], but predominantly lipid mixing and leakage occur near and above T_H [23–25], and the rate increases rapidly as a function of temperature as T_H is approached [26]. In retrospect, the association of fusion and liposome disruption is not surprising. The lamellar/nonlamellar phase transitions require extensive interactions between opposed lamellar phase bilayers, because the topology changes so drastically. Indeed, when liposomes are made of H_{II} phase-preferring lipids at temperatures near T_H, liposomes do not leak until they come into contact [25], and nonlamellar structures are not observed via cryoelectron microscopy in the membranes of nonaggregated unilamellar liposomes [22, 27–29, 31–33]. Like fusion, the transitions start with formation of intermembrane structures.

The mechanistic relationship between lamellar/inverted phase transitions and membrane fusion was sensed even in very early work (e.g., [30, 32, 34, 35]). More recent work has established this relationship. The intermediates in lamellar/inverted phase transitions have been observed using time-resolved cryoelectron microscopy [22, 27–29, 31, 33] and freeze-fracture electron microscopy (e.g., [30, 32, 35]). One intermediate structure, the interlamellar attachment (or so-called ILA), mediates the lamellar/Q_{II} transition (e.g., [20, 22, 27, 28, 30]). This intermediate constitutes a fusion pore that forms between two opposed membranes (Figure 8.1). Thus, membrane fusion is an obligatory step in the lamellar/Q_{II} phase transition. Although lipid compositions that form only H_{II} phases tend to exhibit more leakage than fusion, fusion is observed to some extent [23], and formation of at least a small number of ILAs in such systems is demonstrated both by cryo-TEM [29, 31] and by the ability to increase the number of ILAs and form Q_{II} phases by repeated temperature cycling through T_H [36–39]. ILAs are thought to arise from earlier intermembrane intermediates, which will be discussed in detail in Section 8.2.3. For now, it will only be noted that recent experiments have also provided compelling evidence for the existence of these earlier intermediate structures.

On first consideration, the association of membrane fusion with lamellar/H_{II} or lamellar/Q_{II} phase transitions may seem peculiar. The structures of these phases are not consistent with the integrity of a membrane around a single, closed aqueous compartment. If unilamellar liposomes are incubated under conditions in which Q_{II} or H_{II} phases are the equilibrium phases, the liposomes will eventually rearrange into bulk inverted phase, with leakage of contents. How can fusion (retention of contents) be associated with this process?

First, intermediates in the phase transitions (and membrane fusion) can begin forming when the system is close to the lamellar/nonlamellar phase boundary, but when the equilibrium phase is still the lamellar phase. For example, transition intermediates are observed by time-resolved cryo-electron microscopy at temperatures as much as 20°C below T_H in DOPE and DOPE-Me [28, 29, 31], and many authors have shown that ILAs (fusion pores) begin to form slowly, as detected via ^{31}P-NMR, tens of degrees below T_H and T_Q in phospholipid systems (e.g., [20, 21, 40]). Second, the lamellar/Q_{II} phase transition can be extremely slow and hysteretic in phospholipids. In DOPE-Me, for example, the Q_{II} phase can be the equilibrium phase at temperatures where no detectable Q_{II} phase forms from the lamellar phase even after many hours [40–43]. Third, even if the equilibrium phase of the liposomal lipids is the H_{II} or Q_{II} phase, the initial interactions between unilamellar liposomes may still result in fusion. Formation of bulk inverted phases requires formation of arrays of intermediates [22, 27–30] within a multilamellar stack. Such arrays cannot form when only two liposomes are opposed to each other. Production of larger aggregates of liposomes produces mul-tilamellar stacks and then inverted phases [28, 29, 31]. This same process is also inferred from studying the kinetics of fusion in systems that form Q_{II} phases. In studying fusion via fluorescent assays, Ellens and coworkers characterized what was probably the onset of bulk Q_{II} phase formation from aggregates of fusing liposomes [20]. They termed it "collapse" and observed it after extensive aggregation and fusion had occurred. In biology, fusion is a process that usually occurs between membranes that are allowed to touch over small areas; hence, the same restriction applies.

Some care must be taken in reading the literature on membrane fusion and inverted phases: many authors implicitly assume that lipid mixing between membranes is due to membrane fusion. However, lipid mixing can occur in the absence of contents mixing of the two original liposomes (e.g., via stalk formation without subsequent fusion pore formation; see Section 8.2.3). Fusion is the mixing of the lipids of two membrane-bounded compartments accompanied by the nonleaky mixing of their enclosed aqueous contents. This is the definition more relevant to fusion events in cells where extensive leakage does not occur (in exocytosis, for example).

Biomembrane lipid compositions may be close to lamellar/inverted phase boundaries under some circumstances. Several classes of lipids that occur frequently in biomembranes adopt inverted phases under physiological conditions when the acyl chains are unsaturated ([14–16, 44, 45]; see also Chapter 5). The principal ones are phosphatidylethanolamine (PE) and monogalactosyldiglyceride, which is prominent in chloroplast thylakoid membranes. In addition, cardiolipin (CL) and phosphatidic acid (PA) can form inverted phases in the presence of divalent cations, and phosphatidylserine (PS) and PA both form inverted phases at low pH.

Lamellar, H_{II}, and Q_{II} phases are composed of lipid monolayers of different topologies. In excess water, the relative tendency of a lipid composition to form inverted phases from planar bilayers is determined primarily by two components of the lipid free energy. These are the *curvature elastic energy* and the *chain-packing energy* (also referred to as the interstitial energy). These energies, discussed in detail in Chapter 5, depend on the lipid composition. Changes in either of these energies that stabilize inverted phases are also observed to increase the rate of membrane fusion and contact-mediated lipid mixing and leakage. The curvature elastic energy of a lipid monolayer (G_c) is given by [46]

$$G_c = (k_m / 2) \int_A \left[C_1 + C_2 - C_0 \right]^2 dA + \kappa \int_A C_1 C_2 dA \qquad (8.1)$$

where k_m is a bending modulus specific to the lipid composition, and the integral is over the area of the monolayer. C_1 and C_2 are the two principal curvatures at each point on the interface (evaluated at a specified depth in the monolayer that corresponds to the plane at which the area per lipid molecule is constant despite changes in curvature [47]). C_1 and C_2 are the inverse of a principal radii of curvature R_1 and R_2, which are defined in Figure 8.2. The principal radii and the principal curvatures are signed quantities. By convention, if the curvature tends to bend the headgroup layer around water, it is negative, and it is positive if the headgroup layer bends to favor water surrounding lipid. C_0 is a very important quantity called the *spontaneous curvature*, which is also a signed quantity. The value of k_m is not very sensitive to lipid composition when the lipids are in the liquid–crystalline (L_α) phase, but the spontaneous curvature can be very sensitive to the lipid composition.

The last term on the right-hand side of Equation 8.1 is the Gaussian curvature energy of the monolayer, and κ is referred to as the Gaussian curvature elastic modulus or the saddle splay modulus [46]. The Gaussian curvature energy of closed surface is a function only of the topology of the membrane and not the area or local variations in shape [16, 46]. There is currently no direct means of measuring κ, but on theoretical grounds its value is believed to be smaller than k_m [16]. The Gaussian curvature energy has usually been neglected in calculations of the bending elastic energy of lipid monolayers.

Effectively, the spontaneous curvature is the principal factor determining differences in G_c between different lipid compositions. Changes in this quantity have important effects on the ability of a lipid composition to form inverted phases and to undergo membrane fusion.

8.2.1.1 Spontaneous Curvature

Briefly, the spontaneous curvature (C_0) is a measure of the tendency of monolayers of amphiphilic lipids to bend into nonplanar geometries. If the headgroup layer prefers to be concave and curl around water, C_0 is assigned as negative. If it prefers to be convex, and bow outward into a water phase (as in a normal detergent micelle), it is assigned as positive curvature. H_{II} and Q_{II} phases form at the expense of lamellar phases when C_0 adopts sufficiently negative values. This is because

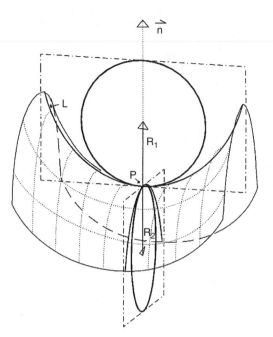

FIGURE 8.2 The curvature of a surface at point P is uniquely determined by the two principal curvatures at that point. If one draws a unit vector n normal to the surface at P and chooses an arbitrary plane that contains n, the plane will intersect the plane on a contour, L. The first and second derivatives of this contour at P specify the best-fitting circle of radius R that is tangent to the surface at P. The radius is taken as opposite in sign to n if it is opposite in direction to n. By rotating the plane about n, an ensemble of best-fitting circles is obtained, along with corresponding ensembles of their radii R and inverses $1/R$. The two principal curvatures are the two extreme values of the set of $1/R$, with values $C_1 = 1/R_1$ and $C_2 = 1/R_2$. Examples: the principal curvatures of a sphere are both the inverse of the sphere radius; the principal curvatures of a circular cylinder are the inverse of the cylinder radius and zero; and the principal curvatures of a plane are both zero. Note that the principal radii of the saddle surface shown in the figure have opposite signs. If the principal curvatures have equal magnitudes, the mean curvature of the surface can be zero. For lipid/water interfaces, by convention, the sign of the curvature is negative if that curvature of the interface surrounds a water-filled space. Hence, the monolayers in the H_{II} phase (Figure 8.1) have negative curvature. (Reprinted from [17], with permission.)

the monolayers in the inverted phases have net negative curvature, where the curvature is defined as the area average of the sum of the two principal curvatures at each point on the lipid/water interface. The curvature energy per unit area of lipid monolayers is determined by the degree to which the curvature deviates from the spontaneous curvature (Equation 8.1): the closer the area-averaged curvature to the C_0, the lower the free energy. The C_0 of a lipid mixture can be measured by X-ray diffraction experiments in the presence of excess water and long-chain alkanes [48, 49].

As discussed in Chapter 5, to a first approximation, spontaneous curvature is determined by the "shape" of the molecule. If the headgroup area has a smaller cross-sectional area than the hydrophobic chains, then the spontaneous curvature will tend to be negative, favoring H_{II} and Q_{II} phase formation. If the headgroup has roughly the same cross-sectional area as the hydrophobic chains, the monolayers will tend to be planar, favoring lamellar phase formation.

8.2.1.1.1 Spontaneous Curvature Is Determined by the Major Lipid Components

The spontaneous curvature of a lipid mixture is determined by the major lipid constituents. It can be decreased by using lipids that form inverted phases (have negative spontaneous curvatures) when they are pure. It has been shown that the spontaneous curvature of a lipid mixture is the mole fraction-weighted average of the spontaneous curvatures of the components [50, 50a, 62a, 81].

Increasing the amount of phosphatidylethanolamine (PE) in a mixture of PE and phosphatidylcholine (PC) decreases the spontaneous curvature, as measured by X-ray diffraction [48, 51], and lowers T_H (e.g., [20]). Addition of more PE also increases the rate of fusion and liposomal destabilization, as expected [19, 20]. Exchange of a PE with a lower T_H for one with a higher T_H also speeds up interliposome interactions at a given temperature [26]. Generally, the spontaneous curvature decreases with increasing temperature and decreasing water content of the lipid–water systems [52, 53]. Thus, a system can be brought closer to the L_α/H_{II} phase boundary by increasing T and/or increasing the mole fraction of lipids like PE. Sometimes, the spontaneous curvature can be reduced by a change in pH or by addition of particular cations, which bind to the headgroups, decrease interheadgroup repulsion, partially dehydrate the headgroups, and decrease their effective cross-sectional areas. PE is anionic at high pH, and liposomes made at pH 9.5 are stable until the pH is dropped to neutral or acidic values (e.g., [20, 21, 26]). Another example is the formation of inverted phases in mixtures of PE and anionic lipids like CHEMS. A mixture of PE and CHEMS (cholesteryl hemisuccinate) is in the lamellar phase at neutral pH, where CHEMS is negatively charged. At low pH, CHEMS is protonated, and the mixture forms an H_{II} phase [54]. Accordingly, PE-CHEMS liposomes are stable at neutral pH but undergo rapid contact-mediated leakage at low pH [24, 26, 54]. Mixtures of other lipids that are anionic at neutral or low pH with PE produce liposomes that also undergo pH-induced fusion and lipid mixing (e.g., [55, 56]).

An example of a system in which interactions are induced by divalent cations is cardiolipin (CL). At neutral pH and physiological salt concentrations, CL forms only lamellar phases. However, addition of divalent cations induces formation of inverted phases, especially the H_{II} phase [57–59]. Ortiz et al. [60] showed that addition of divalent cations (Ca^{2+}, Mg^{2+}, Sr^{2+}, Ba^{2+}) to CL liposomes induces fusion and leakage. Moreover, for Sr^{2+} and Ba^{2+}, the rates of the liposome–liposome interactions are strongly correlated with T_H for each of the respective cation–cardiolipin complexes. The rates of leakage and fusion increase rapidly with temperatures at and above T_H, with leakage being more rapid than fusion. CL is a particularly interesting case because there may be more than one mechanism of membrane interaction and fusion involved. Ca^{2+} and Mg^{2+} induce much larger initial extents of fusion (as much as 80%) than do the other two cations [60]. The fusion rates also increase monotonically as a function of temperature for Ca^{2+} and Mg^{2+} (T_H for these complexes is less than 0°C). Fluorescence and electron microscopic data indicate that the lamellar/H_{II} transitions are slower in the case of Ca^{2+} and Mg^{2+} than for the other divalent cations. The difference in fusion/leakage rate dependence for the two sets of cations suggests either that there is some peculiar hysteresis in intermediate formation that is different for different cations, or that Ca^{2+} and Mg^{2+} induced fusion may occur via a mechanism unrelated to the lamellar/inverted phase transition. The second mechanism may be related to the mechanism presumed to operate for lipids like phosphatidylserine in the presence of Ca^{2+} (see Section 8.2.3). Fusion and leakage of liposomes made of mixtures of anionic lipids with PE can also be triggered by addition of Ca^{2+} at high pH (e.g., [23, 26]).

8.2.1.1.2 Spontaneous Curvature Can Be Altered by Addition of Minor Lipid Components

The spontaneous curvature can also be lowered by addition of small mol fractions of lipids that do not necessarily form H_{II} or Q_{II} phases by themselves, but that nevertheless lower the C_0 when added to phospholipid mixtures. These effects are also rationalized in terms of the molecular shape of the additives. The sign of the effect of an additive on C_0 can be determined by measuring whether or not it lowers T_H, the lamellar/H_{II} phase transition temperature, of a lipid like PE. The size of the effect on C_0 of different additives can be judged by comparing the magnitudes of their effectiveness in lowering T_H per mole percent of additive. Janes assembled a table of such effects for many different lipid additives [61].

One important example of such an additive is the diglycerides (DAGs). DAGs form oils or crystalline phases by themselves [62], yet they lower the spontaneous curvature [21, 62a] and the lamellar/inverted phase transition temperatures of phospholipids (e.g., [21, 63–65]), while stimulating membrane fusion and contact-mediated leakage [21]. The reduction in spontaneous curvature is rationalized as due to the relatively small headgroup of the DAG, relative to the acyl chains, which permits the headgroups of adjacent phospholipids to crowd closer together and adopt a negative curvature.

DAG is particularly interesting in that it is produced *in vivo* by phospholipase C action on phospholipids as a second messenger. It has long been speculated that DAG produced in this manner may be the mechanism by which fusion of secretory vesicles with cellular plasma membranes (exocytosis) is coupled to stimulus. There is an extensive literature showing that phospholipase C action *in vitro* induces membrane fusion in a fashion that correlates with the amount of DAG produced with increasing time. This connection has recently been reviewed by two of the principal authors of much of this work [66, 67]. In the course of these studies, they also investigated the effects of other lipid additives on the lamellar/inverted phase transitions, as well as liposome leakage and fusion.

In a series of elegant studies, Goñi, Alonso, and co-workers studied the effects of DAG produced *in situ* by the enzyme on liposomal fusion and leakage rates while monitoring the concentration of DAG produced by the enzyme as a function of time. They also studied the phase behavior of the enzyme-free lipid composition corresponding to different time points in the incubations of liposomes with enzyme, at different temperatures. They were able to show that maximal fusion activity occurred when the liposomal system was in a part of the pseudophase diagram where Q_{II} precursors (indicated by isotropic ^{31}P-NMR resonances) coexisted with the lamellar phase, and where H_{II} phase was not the predominant phase [68]. Figure 8.3 is a reproduction of this pseudophase diagram. Under conditions in which the H_{II} phase predominated (corresponding to longer times for enzyme action and/or higher temperatures in the liposome/enzyme system), more leakage than fusion was observed. The authors interpreted their results in terms of formation of Q_{II} phases of bicontinuous vs. micellar structure under the two different circumstances, although it should be noted that another explanation is possible. At higher temperatures or higher DAG contents, intermediate formation may be so fast that H_{II} phase formation and liposomal disruption are simply too fast for liposomes to endure long enough to register fusion signals (see Section 8.2.3

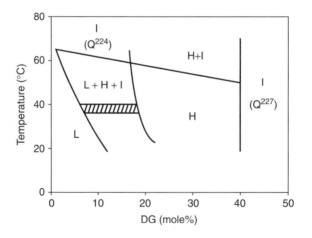

FIGURE 8.3 Pseudophase diagram of egg PC/egg PE/cholesterol/diacylglycerol in excess water constructed from ^{31}P NMR data. L, H, and I indicate lamellar, inverted hexagonal, and isotropic phases, respectively. *Isotropic phase* refers to Q_{II} phase precursors or Q_{II} phases. The nature of the Q_{II} phases, as determined from X-ray diffraction data at 50% w/w lipid/water, is indicated in parentheses. The shaded area corresponds to the region of temperature and composition at which one observes optimum liposome fusion induced by phospholipase C. The ratio of egg PC/egg PE/cholesterol is fixed at 2/1/1 mol/mol/mol. (Reprinted from [68], with permission.)

and [29, 69, 70]). Indeed, formation of well-defined unit cells of Q_{II} phase can only form after unilamellar liposomes have rearranged into multilamellar stacks, releasing their contents.

It is very interesting that the enzyme was found to play a catalytic role in more than one sense in these systems. The heat- or inhibitor-inactivated enzyme did not induce DAG production, fusion, or leakage [71]. However, a key result reported in [72] is that liposomes of egg PC/egg PE/cholesterol 2/1/1 mol/mol/mol were stable if up to 10 mol% of DAG was added in the absence of the phospholipase C. Thus, enzyme activity resulted in a faster fusion rate than observed for the same DAG concentration in the absence of enzyme. As summarized in [66, 73, 74], this is consistent with the results in [68] if two things are assumed:

First, the formation of fusion intermediates and Q_{II} phases may be very slow at this lipid composition, which would explain the apparent metastability of the liposomes in the absence of enzyme. This is compatible with the notoriously slow lamellar/Q_{II} transition kinetics in phospholipid systems far below T_H [40–43].

Second, it must be assumed that the localized production of high concentrations of DAG in the outer leaflet of the liposomal bilayers catalyzed the fusion reaction by encouraging formation of lamellar/inverted phase transition intermediates. Because the same authors found that fusion occurs more and more rapidly as the DAG concentration is raised, this seems reasonable.

In this sense, the enzyme is in a kinetic race to create high local concentrations of DAG and initiate fusion before the DAG diffuses away or equilibrates across the bilayer. This may be a close race indeed, because DAG has a transbilayer migration half-time that may be as short as 15 sec [73]. A further complication is that lipids that favor inverted phase formation also increase the activity of phospholipase C (see [66] for a review). It should be noted that the catalytic effects of DAG are noted at ca. 5 mol% and above. By comparison, the resting concentrations of DAG in cultured rat kidney cells is ca. 0.5 to 1 mol% of the total whole-cell phospholipid content [75]. It is conceivable that a subset of the subcellular membranes could transiently achieve higher concentrations.

Sphingomyelinase action on sphingomyelin produces ceramide, a lipid that also lowers lamellar/inverted phase transition temperatures [76, 77]. Ceramides are similar in structure to DAGs, and one might expect them to have similar effects on fusion. However, ceramides induced leakage rather than fusion, at least in the phospholipid systems investigated by Goñi et al. [77, 78]. However, it is interesting to note that joint action of phospholipase C and sphingomyelinase was synergistic, producing more fusion than in the presence of either enzyme alone [77, 77a].

Fatty acids are another class of biologically relevant lipids that can stabilize inverted phases: fatty acids, and especially unsaturated-chain fatty acids, have been shown to lower T_H [79]. It is inferred that their effects are also due to a reduction in spontaneous curvature (the headgroup cross-sectional area is small compared to the effective cross-sectional area of the unsaturated acyl chains). Basáñez et al. [80], working with the phospholipase C/liposome system, showed that addition of arachidonic acid increased the rate of fusion in the liposome/phospholipase C system by much more than its effect on the enzyme activity. This is consistent with a negative curvature effect on inverted phase formation and fusion intermediate formation. As discussed in Section 8.2.3, this effect has a sidedness with respect to the membranes that is mechanistically revealing.

Conversely, certain lipid additives can increase the spontaneous curvature (i.e., increase the curvature toward zero or positive values and stabilize lamellar phases at the expense of inverted phases). In lipid systems that are close to lamellar/inverted phase boundaries, addition of such lipids is observed to stabilize lamellar phases and decrease the rate of fusion. A principal example is lysophosphatidylcholine (LPC), which has a comparatively large headgroup relative to the single acyl chain, and which forms normal micelles in dilute solution. LPC increases (makes more positive) the spontaneous curvature of phospholipids [81], raises T_H of PE in a concentration-dependent fashion [64], and raises the lamellar/inverted phase transition temperature in N-methylated-DOPE

(DOPE-Me) [81a]. Yeagle et al. [82] showed that LPC inhibited fusion and leakage of N-methylated-DOPE LUVs under circumstances where it also inhibited formation of inverted phase precursors. LPC has also been observed to inhibit outer monolayer lipid mixing and subsequent fusion in PS/PE mixtures, as induced by Ca^{2+} [83]. In the phospholipase C/liposome system, LPC reduced the fusion rate by much more than its effect on the enzyme activity [80], consistent with this increase in spontaneous curvature. Similar observations were made for palmitoylcarnitine [80]. Epand et al. [84] showed that a variety of single-chain amphiphiles raise T_H of PE, although effects on fusion were not investigated. In addition, studies with the fusion of planar membranes and liposome/planar membrane fusion [85–87] also show inhibition of lipid mixing and fusion by incorporation of LPC. Interestingly, there is also extensive data showing that fusion in several biomembrane systems is reversibly inhibited by introduction of LPC into the membranes (e.g., [88–96]; see [7, 97] for older reviews.) At least in some systems, the inhibition is general for LPC and other C_0-increasing surfactants of different chemical structure [88]. Inhibition can generally be reversed by washing the surfactants out of the membrane and occurs at surfactant concentrations well below the levels that cause membrane lysis (e.g., [88]). All of these observations suggest that the effect of LPC and similar surfactants is, in fact, an effect on the physical properties of the lipids in the membranes, which is compatible with inhibition of stalk formation through an effect on C_0.

Some caution is necessary in interpreting the studies of LPC effects, however, especially in biomembrane studies. The LPC or other exogenous surfactant is often introduced by equilibrating the membrane preparations with aqueous surfactant solutions. The amount of surfactant incorporated into the membranes will depend on the surfactant concentration, the volume of solution used, and the mass of membrane lipid in the preparation. Quite large mole fractions of lipid can be introduced in this manner; perhaps enough to perturb (or lyse) the membranes very substantially. Moreover, LPC that is added to membranes in one step of the experiment can reequilibrate with suspending solutions and leave the membranes in subsequent steps. The study by Chernomordik et al. [98] was especially careful and well designed, and showed that 10 to 13 mole% of LPC and arachidonic acid were introduced by equilibration with measured amounts of host lipid. Such levels substantially perturb other properties of the lipids than solely C_0, a number of which are discussed in [97]. For example, such surfactant levels could increase the short-range repulsive forces between membranes [99], and LPC is observed to inhibit vesicle aggregation in at least one system [83]. LPC could also alter surface hydrophobicity and the susceptibility of opposed membranes to defect formation (e.g., during initiation of stalk formation), as do short-chain alcohols [100]. LPC has also been reported to alter peptide binding to lipid/water interfaces (e.g., by the so-called fusion peptides of fusion-mediating proteins in biomembranes [101,102]). LPC has also been reported to inhibit binding of virions to cell surface receptors [103, 104], although this occurs at higher LPC concentrations than inhibition of fusion, and subsequent work on arrested fusion intermediates indicates that this is not the primary effect of LPC [92]. In addition, LPC seems to act on the membrane lipids and not on the fusion protein itself [105], and the LPC works downstream of fusion peptide binding in at least one case, with no effect on virion binding [94]. As we will see in Section 8.2.3.1, mechanistically significant differences in the effects of LPC can occur when the distribution of LPC is asymmetric across the membrane (i.e., enriched in the facing or distal monolayers of interacting membranes). Such asymmetry in monolayer leaflet composition is not always easy to maintain. For instance, as pointed out by Chernomordik et al. [89], there is evidence that fatty acids can redistribute across the bilayer in as little as seconds. Moreover, Gaudin [94] inferred that lauroyl- and myristoyl-LPC in the outer monolayers of liposomes could cross to the inner monolayers under some circumstances, and concentration-dependent reversal of short-chain LPC inhibition of fusion was observed in [96].

It should be noted that lipid additives can, in theory, change the curvature energy of a lipid composition by affecting the monolayer bending modulus (k_c; Equation 8.1), as well as the spontaneous curvature. However, at least in the case of DAG, mole fractions of ca. 0.15 or more of the additive are necessary to have a substantial effect on k_c [62a]. As long as the mole fractions of additive are small and the bilayer lipid remains in the L_α phase (as opposed to the L_β or liquid-ordered phase), effects on k_c are likely to be small.

8.2.1.2 Chain-Packing Energy

In lipid systems with negative spontaneous curvature, inverted phase formation and membrane fusion are both stimulated by factors that lower the spontaneous curvature. A similar parallel response is seen when substances are added that affect the chain-packing energy in inverted phases.

8.2.1.2.1 Chain-Packing Energy in H_{II} Phases

The chain-packing energy is an unfavorable contribution to the total free energy of lipid in an inverted phase. This unfavorable free energy tends to oppose inverted phase formation from the lamellar phase, while inverted phase formation is driven by the reduction in curvature energy of going from flat bilayers to H_{II} tubes with curvature essentially equal to C_0 [69, 70, 62a, 106]. The chain-packing energy arises from an entropically driven need to pack the hydrophobic moieties of the lipids as uniformly as possible within the structure of the phases (for reviews see Chapter 5, [16, 17]). In the H_{II} phase, there are hydrophobic interstices between the tubes of the monolayer (indicated in black in Figure 8.1D, lower right). Similar interstitial spaces probably exist within intermediates in membrane fusion: in the process of fusion, the original lamellar structure must be disrupted, so that in some regions the monolayers bend away from each other to form intermembrane connections. Examples are shown as stippled areas in Figure 8.1B and Figure 8.1C. Factors that stabilize interstitial spaces in H_{II} phases should stabilize fusion intermediates, as well.

According to conventional models of nonlamellar phase stability (e.g., [53, 62a, 69, 70]), in H_{II} phases of a single pure lipid component, the interstices are stabilized in two ways. Acyl chains of the molecules in the curved monolayers lining the interstices stretch to fill them, which decreases the entropy and raises the free energy. Alternatively, the circular cylinders of monolayers may distort to a more hexagonal cross section to decrease the size of the interstices, which increases the curvature free energy. The unfavorable chain-packing free energy tends to oppose inverted phase formation from the lamellar phase, whereas inverted phase formation is driven by minimization of curvature energy (in the H_{II} phase, lipid monolayers have negative curvature that is essentially the same as their spontaneous curvature) [62a, 69, 70]. More recent theories indicate that tilt of lipid molecules in lipid monolayers, and gradients in tilt along the surface of lipid monolayers, may also play a role in stabilizing these interstices [107, 108]. This new approach will be described in Section 8.2.3.1.3.

8.2.1.2.2 Changes in Lipid Composition that Reduce the Chain-Packing Energy

Two sorts of changes in lipid composition can lower the chain-packing energy. The first is addition of long-chain alkanes or nonpolar oils. Addition of long-chain alkanes like tetradecane or hexadecane to PE, PE-PC mixtures of DOPE-Me drastically reduces T_H [64, 65, 109, 110]. The long-chain alkanes are concentrated in the interstitial regions in such mixtures [65, 111]: by filling the voids, they remove the necessity for chain-stretching or monolayer deformation and lower the free energy of the H_{II} phase. The alkanes have fairly minor effects on the monolayer curvature of the H_{II} phase at the concentrations that are found to drastically lower T_H [48, 65, 109, 110, 112], so it is fairly clear that the chief effect of the additives is on the chain-packing energy. Other nonpolar oils, like squalene [113], squalane [65], triglycerides [114], dolichol [115, 116], and retinal [117], have similar effects. These substances are very effective: for example, in pure PE 1 mole% of triglycerides lowers T_H by about 8°C [64]. Addition of nonpolar oils also increases the fusion rate at a given temperature and lowers the temperature at which a given fusion rate is achieved in pure lipid systems [117] and in the phospholipase C-induced fusion system [80]. Interestingly, in the phospholipase C system, hydrocarbon oils have a bigger effect on the contents-mixing (fusion) rate than the lipid-mixing rate [80], suggesting that the oils promote formation of fusion pores from the initial structures that form between the liposomes. The effectiveness of oils in promoting fusion indicates that small amounts of triglycerides, dolichol, or similar lipids in biomembranes may play an important role in fusion. Interestingly, nonpolar oils also accelerate Ca^{2+}-induced lipid mixing in bovine brain phosphatidylserine (PS) [118], a system that does not form inverted phases in excess water and neutral pH. This suggests that stabilizing interstices may also accelerate fusion in systems that fuse by different mechanisms, and that different mechanisms have some features in common (Section 8.2.3).

The effect of nonpolar oils on fusion raises the question of how common such oils are in biomembranes. Examples of nonpolar oils (neutral lipids) found in biomembranes are triglycerides [119, 120], dolichol and its derivatives [121], and fatty acid esters of cholesterol. The neutral lipid content of biomembranes is still fairly poorly known. Only a few weight percent of neutral lipids in membranes substantially affects their fusion behavior. It is clear that the nonpolar oil content of biomembranes can be this large, so that these lipids are available to fill interstices in fusion defects, at least to some extent. However, the neutral lipid levels are also quite variable, depending on the tissue and cellular membrane fraction. Triacylglycerols are approximately 5 mole% of unstimulated neutrophil plasma membrane lipids [120], but represent 1 wt% or less of the membrane lipids of most mammalian cell membranes (e.g., < 0.1 wt% of human red blood cell membranes [119]). Moreover, not all the neutral lipid in a membrane is disposed within the hydrophobic core of the membrane, where it would be immediately available for filling voids during intermediate formation. For example, the glycerol backbones of "neutral" lipids like triglycerides orient at the lipid/water interface. Egg PC membranes can accommodate up to 3 mol% of triacylglycerol, with this region aligned at the lipid/water interface [122–126] before excess triacylglycerol phase separates into lenses within the hydrophobic core of the bilayers. The presence of 30 mol% cholesterol in the membranes reduces this limit to 0.7 mol% [127]. Fatty acid esters of cholesterol behave similarly [123, 125, 128]. Better information is required on the levels of neutral (nonpolar) lipids in biomembranes before we can assume that there is enough to eliminate chain-packing energies in fusion defect formation.

The second sort of change in lipid composition that can lower the chain-packing energy in H_{II} phases is addition of a minor fraction of polar lipids (e.g., phospholipids) that have acyl chains substantially longer than the average length. Tate and Gruner demonstrated this by adding PC with a longer acyl chain to DOPE [110]. However, mole for mole, this effect is much smaller than the effect of nonpolar oils (e.g., a 4°C drop in T_H accompanies exchange of 4 mole% di-22:0-PC for 4 mole% di-18:1-PC [110]). The present author is not aware of studies on the effects of small mole fractions of long-chain polar lipids on fusion or lipid-mixing rates.

8.2.1.2.3 Chain-Packing Energy in Q_{II} Phases and Possible Effects of Bilayer Thickness Gradients

The chain-packing energy in bicontinuous Q_{II} phases has a more subtle origin than in the H_{II} phase. In these phases, the bilayer midplanes conform to one of a family of infinite periodic minimal surfaces [129], which have zero geometric curvature. However, the monolayers lie on surfaces that are displaced from the bilayer midplanes, and these surfaces are not surfaces of constant curvature. The tendency of the monolayers to reduce the curvature energy by complying with the constant curvature surfaces would require variation in bilayer thickness across the unit cell, which would result in an unfavorable chain-packing energy [130, 131]. The two requirements cannot be fulfilled simultaneously, which is referred to as *frustration* and is analogous to the balance of curvature and chain-packing energy encountered in H_{II} phase formation. For Q_{II} phases forming in phospholipids, the free energy penalty for bilayer thickness variation is estimated to be comparatively large, and bilayer thickness variation is estimated to be very small [130, 131]. In the case of the Q_{II} phase, addition of nonpolar oils does not directly affect the chain-packing energy. (In fact, addition of one or two mole percent of hexadecane to DOPE-Me induces H_{II} phases to form instead of Q_{II} phase [Siegel, D.P. and Banschbach, J.L., unpublished observations]). The chain-packing energy of the

Q_{II} phase might be reduced by additives that stabilize gradients in bilayer thickness within the phase. To the knowledge of the present author, this effect has not been demonstrated. Duesing et al. published a simplified model of chain-packing energies in inverted phases that qualitatively reproduces some aspects of observed inverted phase behavior as a function of C_0 and lipid monolayer thickness [132].

Our understanding of the principles determining the relative stability of lamellar and Q_{II} phases is still incomplete, and this is largely due to our poor understanding of chain-packing energies in this phase. A chief example is confusion over how long, transmembrane peptides stabilize Q_{II} phases. Transmembrane peptides of at least two compositions reduce both T_H and the temperature at which Q_{II} phase precursors form (as detected by ^{31}P NMR) [132–136] and can even promote formation of inverted phases in PC [137, 138], which does not normally form them in excess water. In one case, the effect of transmembrane peptides on T_Q, the lamellar/Q_{II} phase transition temperature, was measured directly via synchrotron-source, time-resolved X-ray diffraction experiments [136]. Depending on their length and sequence, and on the host lipid composition, transmembrane peptides can lower T_Q by as much as 18°C at concentrations of only 0.5 mole% [136], and 0.1 mole% lowers the onset of isotopic resonance formation by more than 10°C in dielaidoyl-PE [133]. These levels are well within expected physiological concentrations of transmembrane domains: in influenza virions, for example, the concentration of trimers of hemagglutinin transmembrane domains is approximately 0.5 mole% on a phospholipid basis (using data from [139]).

It is not easy to rationalize the length-dependent effects of these peptides on formation of Q_{II} phase and Q_{II} precursors in terms of peptide effects on curvature [133, 136]. An effect on monolayer curvature (due to bilayer mismatch between the peptides and the host lipid bilayers) [140] and effects on chain-packing in the H_{II} phase have been invoked to explain the effect of the shorter peptides in promoting H_{II} phase formation. This interpretation has been disputed [135]. However, an effect on curvature cannot explain all of the observations, especially the effects on Q_{II} phase formation and the effects of peptides that are longer than the thickness of the host lipid bilayer [133, 135, 136]. By inference, the peptides seem to affect the chain-packing energy of the Q_{II} phase, and it is not clear how this occurs. Resolution of this question may require more detailed knowledge of the conformation and orientation of the peptides in the host lipid bilayers, and a more detailed experimental understanding of chain-packing energies in Q_{II} phases (e.g., by studies of the effects of small mole fractions of longer-chain lipids).

8.2.2 New Experimental Methods

One difficulty in studies of fusion in membrane dispersions has been that of separating the kinetics of liposome aggregation from subsequent destabilization steps [141–143]. In order to get fusion rates fast enough to study conveniently, researchers have often chosen lipid compositions and conditions where the final phase of the membranes after fusion is either an inverted phase (Section 8.2.1) or a cochleate phase (Section 8.2.3.2). In such systems, it is difficult to stop membrane interactions at different stages of the fusion process. Another difficulty is that when one studies fusion in dispersions, one is observing the superposition of many simultaneous events at different stages of evolution. It is not easy to get direct evidence of the different stages in the overall fusion process in this manner.

Two general means of circumventing these difficulties have been used, which have yielded important results concerning the sequence of events in liposomal fusion. The first is to observe fusion in systems where the membranes are stable in the lamellar phase but are induced to fuse by enforced close opposition and sometimes by additional membrane curvature stress, by using small unilamellar vesicles (SUVs). This is most usually done via mixing of the SUVs with solutions of poly(ethylene glycol), or PEG. PEG-induced fusion has recently been reviewed by one of the principal practitioners [144]. This approach has the advantage that one can force membranes of refractory composition to fuse under hydration stress. One can then study fusion in systems with a broader range of lipid properties (e.g., C_0). Another advantage of this approach is that the membrane–membrane interaction process can be stopped by dilution of the suspending medium. This may permit "trapping" of intermediates in some cases (e.g., [145]), if the membrane composition reacts slowly enough to permit dilution on relevant time scales.

The second approach is to observe fusion of single liposomes at fast time resolution, and try to observe the sequence of events directly. Principal recent examples of this second approach are studies of giant unilamellar vesicle (GUV)/GUV fusion in systems where the liposomes bear large and opposite-sign charge densities [146, 147]; combined electrical conductivity and fluorescence study of GUV/planar bilayer fusion [148]; and the use of time-resolved cryoelectron microscopy to observe fusion or lipid phase transition intermediates. The use of these techniques will not be comprehensively reviewed here, but some important results relevant to determination of fusion mechanisms will be described.

8.2.2.1 PEG-Induced Fusion

Studies with PEG-induced fusion have produced important results. Work with this technique has implicated lipid packing defects (relative lipid depletion) in outer monolayer packing in initiating the first steps in fusion. It has also provided additional detailed evidence that lipid mixing and contents mixing represent different steps in the overall fusion process and can sometimes proceed independently, and that leakage can occur in parallel with fusion. The PEG-fusion technique also permits one to study the composition dependence of fusion in lipid systems that are normally too refractory for study via more conventional means. Finally, this work has identified a sequence of intermediate structures in the overall fusion process that is generally compatible with the stalk mechanism (Section 8.2.3.1).

8.2.2.1.1 Method

Close opposition is easily achieved by incubation in PEG solutions. High concentrations of PEG in the suspending medium have the effect of driving liposomes together, because the large hydrated PEG molecules are excluded from the water layers immediately surrounding liposomes. This raises the chemical potential of the water immediately adjacent to the liposomes and causes the system to minimize the total free energy by closely opposing as much membrane/membrane interface as possible. Enforced close opposition of the interfaces overcomes one of the major energetic barriers to membrane fusion.

There are some complications in relating results obtained in the PEG fusion system to fusion in other model membrane systems and in biomembrane systems. One complication is that, for lipid compositions that do not contain high mol fractions of PE, one often has to apply high bilayer curvature stress in order to get vesicle interaction rates that are experimentally convenient. The amount of this stress (i.e., the change in lipid free energy) is only indirectly quantifiable, which can make it hard to translate the findings into other systems. The bilayer curvature stress is applied by using SUVs (e.g., [149, 150]), where the bilayers are bent into such small radii of curvature (diameter ca. 20 nm) that the structures are at a higher lipid chemical potential than in the lamellar phase. LUVs (large unilamellar vesicles, ca. 100 nm in diameter) have less bilayer curvature stress and are more refractory to fusion than SUVs (i.e., LUVs require higher PEG concentrations to achieve the same rates of lipid mixing or fusion (e.g., [144, 149–155]). Also, there is a correlation between increasing SUV curvature and increasing susceptibility to fusion in PC SUVs [149] (which could be due in part to the decrease in membrane rupture tension with increasing unsaturation in PC [156].) Moreover, there is some evidence that the activation energy for lipid mixing between SUVs of 20 nm diameter is less than half that for 45 nm-diameter SUVs [157], implying that bilayer curvature stress is a big factor in liposome interactions.

Another complication is that the small size of SUVs may physically restrict the size of fusion intermediates (e.g., TMCs, HDs) that can form compared to larger structures, or make corresponding changes in the intermediate free energies. For example, the energy of a stalk can increase rapidly if one restricts the width of its base below 20 nm [158], which is approximately the diameter of an SUV. There are also large stresses in SUVs that could alter the nature or dynamics of the intermediates compared to behavior in LUV and biomembrane fusion. For example, because of the high bilayer curvature in SUVs, there are many more lipids in the outer monolayer than in the inner monolayer of the membrane. As one increases the radius of an SUV, the bilayer asymmetry decreases rapidly. When two SUVs fuse, the resulting structure will tend to regain a spherical shape to reduce the curvature energy relative to the two original SUVs (a major driving force for SUV fusion). At equilibrium, the resulting liposome will have a greater radius than the original SUVs, and thus a smaller area asymmetry. For example, it can be shown that the fraction of lipid on the inner monolayers of 20 nm-diameter SUVs must increase by about 22% in the first fusion event. The existence of this asymmetry may affect the evolution of fusion intermediates. This is especially true for expansion of the initial fusion pore, which makes the shape of the initial fusion product progressively more like that of a single, quasi-spherical liposome. At some point in the overall fusion process, there will be a flux of lipid from the outer monolayer to the inner monolayer. Using a clever assay, Lentz et al. observed such a flux [154, 157]. The induced bilayer asymmetry is much smaller in biomembrane fusion, where the membrane compartments are typically on the order of 100 nm or larger (40 nm for synaptic vesicles).

There are still some questions about the role of osmotic gradients in PEG-induced fusion. PEG was originally thought to induce a large osmotic stress in liposomes, tending to shrink them and create stacks of flattened LUVs with high-curvature edges [159]. These high-curvature edges would be under curvature stress and would be likely spots for membrane–membrane interactions to occur. However, Malinin et al. [153] suggested that this was not the case and used cryoelectron microscopy to show that LUVs in 10 wt% PEG were fairly round. This is consistent with the low osmotic gradient generated by 10% PEG (100 mOsm [153]). However, 10% is a modest PEG concentration (10% PEG causes only a very small amount of contents mixing in LUVs of the same composition, for example [153]). Higher concentrations could not be investigated due to interference of the PEG with contrast in cryoelectron microscopy [160]. However, the extent of PEG-induced fusion and leakage is larger for larger osmotic gradients (tending to shrink the liposomes) than used for the cryo-TEM experiments [153], and the extents of fusion and leakage decreased when the gradient decreased below this value and changed sign. Therefore, the osmotic gradient is important for reasons that are not clear [153].

8.2.2.1.2   Findings

Work in PEG-induced fusion systems shows that faults in outer monolayer leaflet packing (in this case, caused by removal of a small percentage of the total lipid) promote membrane fusion [161], at least in the case of comparatively refractory compositions like PC. LPC in the outer monolayer of LUVs tends to inhibit PEG-induced fusion [151, 161, 162]. If LPC is initially equally distributed between the inner and outer leaflets of the LUV bilayers, removal of the LPC from the outer leaflet by incubation with bovine serum albumin (BSA) potentiates fusion (lowers the PEG concentration needed to induce fusion [162]). If LPC and fatty acid are created in the external leaflet of LUV by phospholipase A_2, incubation with BSA to remove the hydrolysis products induces fusion in liposomes that were otherwise refractory [151, 161].

PEG-induced fusion shows similar C_0 dependence to fusion studied in other systems (Section 8.2.1). In both SUVs and LUVs, the extent of fusion increases with increasing mol fraction of PE in PC, with more highly unsaturated PEs being most effective [150]. The more highly unsaturated the PE, the lower T_H and (presumably) the lower C_0, when compared at constant temperature, so this finding mirrors findings obtained with different PEs in the absence of PEG [26]. The study by Yang et al. [150] is also notable in that they compared the effects of lyso-PE substitution for PE, to show that the effect of PE in fusion is not through a change in headgroup composition of the bilayer (and a concomitant reduction in repulsive hydration force). The same principle was underscored by Haque et al. [155], who showed that the lamellar repeat distance of a phospholipid/sphingomyelin/cholesterol 65/15/20 wt/wt/wt mixture was the same whether or not the phospholipid was pure PC or an almost equimolar mixture of PE and PC, despite the fact that the PC mixture was much less susceptible to fusion. Both these results show that it is unlikely that the effects of PE are due only to a reduction in repulsive forces between interfaces. In other recent PEG-induced fusion studies, PE/PC/cholesterol mixtures yielded more extensive fusion than mixtures of PCs [153], and addition of Ca^{2+} to a cardiolipin/DOPC 1/10 mixture induced fusion and lowered the PEG concentration threshold for lipid mixing [161]. Although the latter effect was interpreted in terms of defect creation in the outer monolayer [161], both these effects may also be due to a reduction in C_0 (Ca^{2+} induces inverted phases [57–59] and fusion [60] in cardiolipin). Chain-packing energy considerations also play a role in the liposome interactions induced by PEG, because addition of 5 mol% hexadecane increased the extent of fusion (liposomal contents mixing) of SUVs [152, 153, 128]. Hexadecane was also said to increase the extent of contact-mediated leakage, but not of lipid mixing, in [153], whereas there was no effect on leakage in [152, 168]. The authors suggested [152, 168] that hexadecane acts on the TMC or HD intermediate, lowering the activation energy for pore formation. It is not clear whether this is the explanation: hexadecane should greatly reduce the chain-packing stress, promote radial expansion of stalks, stabilize TMCs and hemifusion diaphragms, and reduce the driving force for pore formation [70, 163]. It could be argued that an increase in bilayer septum area makes fusion more probable. However, the only systems used for the hexadecane experiments in [152, 153] were SUVs (25 nm diameter in [152]), and it can be questioned how big a septum can form in such systems. An alternative suggestion [168] is that the hexadecane stabilizes a transient pore seen earlier in the fusion process than final pore formation [145].

A combination of data from inner and outer monolayer lipid-mixing assays, contents-mixing assays (including formation of Tb^{3+}/DPA fluorescent complex), and bilayer asymmetry assays led Lentz and co-workers to propose a sequence of events during SUV/SUV interactions [145, 157, 164], based on the rather slow interactions of PC SUVs. This scheme is depicted in Figure 8.4. The first step is lipid mixing between outer monolayers of the SUVs, which can occur at low PEG concentrations in the absence of content mixing or leakage [157]. This occurs within seconds after mixing with PEG. (Lipid mixing is often observed at extents substantially different than for contents mixing, in PEG-induced interaction of both SUVs and LUVs, e.g., [145, 151–153, 168], or in the complete absence of contents mixing, e.g., [165]). During this stage, I_1 (Figure 8.4), transient pore formation can occur that permits the transfer of protons between aqueous compartments, but not complete mixing of compartment contents [145]. It is possible that the I_1 stage is the "flickering" fusion pores observed in liposome/planar bilayer fusion [148]. If SUV dispersions in PEG are diluted at this stage, liposomes with the same size as the initial liposomes are recovered [145], implying that formation of the I_1 stage is reversible. The I_1 stage slowly converts to the I_2 stage, in which inner monolayer lipid mixing occurs. If reacting SUV dispersions are diluted at this point, much larger lipid aggregates are observed, indicating that formation of I_2 is not reversible. Moreover, in samples diluted at the I_2 stage, the fusion process proceeds in the absence of PEG, indicating that I_2 is a fusion-committed intermediate [145]. In the presence or absence of PEG, contents mixing and complete inner monolayer mixing proceed on about the same time scale from this stage [156]. Because of the outer monolayer lipid mixing observed at low PEG concentrations and early times at higher concentrations, the I_1 stage is assumed [152, 153] to represent a stalklike intermediate with a transient pore-forming activity. I_2 is assumed to be a stable bilayer septum between SUVs, similar to the TMC or HD in the stalk mechanism (Section 8.2.3.1), which subsequently forms a stable fusion pore. The outer-to-inner monolayer lipid flux in SUV fusion [154] occurs roughly simultaneously with fusion pore formation and contents mixing [157].

Activation energies for the steps in PC SUV/SUV interactions have been estimated [157, 164]. The activation energies and rate constants for formation of the intermediates in 45 nm-diameter SUVs of 85/15 mol/mol DOPC/1,2-dilinolenoyl-PC at 35°C in 17.5% PEG [164] are:

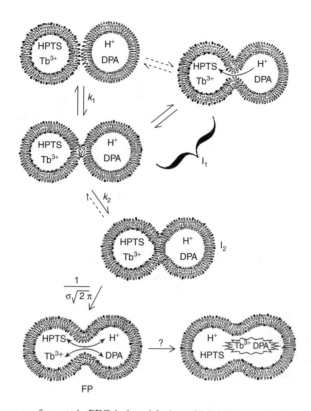

FIGURE 8.4 The sequence of events in PEG-induced fusion of PC SUVs, as deduced from fluorescent assays of lipid and contents mixing. See description in text. (Reprinted from [164], with permission.)

Intermediate	Activation Energy	Rate Constant
I_1	35–37 kcal/mol (ca. 60 k_BT)	2.2–2.4 sec^{-1}
I_2 from I_1	27–33 kcal/mol (ca. 50 k_BT)	0.031–0.044 sec^{-1}
Fusion pore from I_2	21–22 kcal/mol (ca. 36 k_BT)	0.017–0.020 sec^{-1}

Much lower activation energies (12 to 15 kcal/mol or 20 to 25 k_BT) were subsequently found for initial lipid mixing at 5 to 10% PEG for smaller SUVs of the same composition [157], and it was found that interactions could no longer be resolved into distinct steps. The activation energies may not represent the real activation energies of elemental processes at work. They were determined on the assumptions that formation of each of the intermediates is irreversible and occurs in a single step. As pointed out in [164], formation of I_1 is reversible, and activation energies estimated in this manner for a complex and reversible process could be misleading. It was noted in [164] that the observed rates and activation energies of SUV fusion were comparable to observed values for fusion in some biomembrane systems. This encourages us to believe that lipid fusion mechanisms are relevant to biomembrane fusion, even if we must remember that, in this case, fusion starts out in a highly stressed membrane system.

PEG also induces considerable leakage of interacting liposomes, depending on the lipid composition, as in studies of lipid system fusion by other techniques. Haque et al. [155] showed that the amount of leakage with respect to fusion could be minimized by using a certain ratio of PE, PC, cholesterol, and sphingomyelin, which approximated the composition of synaptic vesicle membranes. The role of the cholesterol and sphingomyelin was to reduce leakage without significantly reducing the extent of fusion. The authors inferred that these two lipids (especially sphingomyelin) might toughen the bilayer, reducing its susceptibility to rupture before the fusion intermediate could form [155]. Adhesion energy-induced deformation of aggregated liposomes induces tension in the membranes that can produce rupture and leakage [166, 167].

Researchers on PEG-induced fusion have also developed an assay designed to report the presence of hydrophobic voids in fusion intermediates [165]. A fluorescent probe (cholesteryl 1-pyrenebutyrate, or Py-CH) was identified that had a particularly large excimer/monomer fluorescence ratio change during cycling between the L_α and H_{II} phases in DiPoPE. This is thought to be due to concentration or more rapid diffusion of the fluorescent moieties in the hydrophobic interstices of the H_{II} phase [165], which represent a maximum of 9.3% of the volume of the phase if the H_{II} tubes are perfect circular cylinders (not 20%, as cited in [165]). Py-CH was incorporated into SUVs, and the excimer/monomer fluorescence ratio was monitored during fusion. This ratio increased by a few percent (vs. 150% for the same level of probe in the H_{II} phase) during lipid mixing and contents, and subsequently decayed back to initial levels as the reaction moved toward completion. This signal was observed in two PE-containing lipid systems: one produced only lipid mixing (hemifusion; DOPC/DOPE = 3/1 mol/mol) and the other produced contents mixing (DOPC/DOPE/cholesterol = 2/1/1 mol/mol/mol). No signal was observed in SOPC, which produced no lipid mixing or contents mixing under these conditions. These results suggest that the probe is reporting the formation of hydrophobic voids like those in the H_{II} phase during fusion, consistent with many observations of alkane effects on fusion rates (Section 8.2.1.2) and with theory of fusion intermediate structure (Section 8.2.3.1).

However, one aspect of the Py-CH results is not easy to understand. In the system showing only hemifusion, the signal attributed to hydrophobic void creation is maximal almost immediately (a few seconds) after mixing, while the lipid mixing is negligible in this period. Moreover, the Py-CH signal begins to decay well before most of the lipid mixing takes place (over 300 sec). It was suggested [165] that the intermediates might be flickering in and out of existence, existing for only a fraction of the time, and hence might be restricting bulk lipid mixing. However, for both contents mixing and lipid mixing, diffusion between two SUVs connected by a stable pore or continuous monolayer should occur on the millisecond time scale. Also, SUVs are small enough that the number of lipids in an intermediate is a significant fraction of the number of lipids in the entire structure. For example, the "waist" of a stalk subtends an area that is 1% of the outer monolayer area of a 20 nm-diameter SUV. Lipid molecules at the "waist" of a stalk could presumably be traded between SUVs with a single diffusional "hop" of nanosecond duration, in either direction. Thus, each "flicker" should result in the transfer of on the order of 1% of the lipid per SUV/SUV contact. It is questionable whether intermediates can be transient enough to allow for the time-averaged increase in Py-CH to appear in 1 second or less, while restricting lipid mixing to a half-time of 100 sec. It also may be that the Py-CH signal reports formation of some structure that does not permit lipid mixing between SUVs, but it is not obvious what that would be. It may be that the intermembrane intermediates restrict exchange of the fluorescent lipid probe much more than exchange of the bulk of the membrane lipids, however. A small fraction of fusion events in a GUV/planar bilayer fusion system [148] showed apparent restriction of lipid flow after initial fusion pore formation.

Lentz and coworkers also studied the influence of viral fusion peptides [151, 152] and transmembrane domains (TMDs) of a viral fusion protein [168] on PE/PC/cholesterol and PC liposomes. In general, the fusion peptides can enhance the extent of lipid mixing and content mixing of stressed liposomes at low peptide concentrations (e.g., below a peptide/lipid ration of 1/200 mol/mol) but can inhibit contents mixing at higher concentrations [151, 152]. Stress was applied either by partial depletion of lipid by phospholipase and subsequent incubation with BSA, followed by mixing with PEG [151], or merely by incubation with PEG [152]. Hexadecane increased the extent of contents mixing between SUVs more than did the fusion peptide at all peptide concentrations tested [152], and the effects were not synergistic when used together. The authors suggested [127] that the fusion peptide acts on the TMC or HD intermediate, lowering the activation energy for pore formation, as they inferred for hexadecane. The TMD study [168] used a wild-type and a mutant TMD of a viral fusion protein that had been shown to be fusion inactive at concentrations up to 0.2 mol%. Both the wild-type and mutant TMDs increased the rate of contents mixing and lipid mixing. The wild-type TMD did not affect the extent of contents mixing, whereas the mutant TMD decreased it, and neither TMD affected the extent of lipid mixing. The authors interpreted the results in terms of a TMD effect on both the formation rate of initial intermediates (lipid mixing) and the stability of transient fusion pores formed early in the SUV/SUV fusion process [145]. That is, the authors [168] speculated that the peptides do not change the number of final intervesicular contacts that are possible, but change the nature and evolution of those contacts. The fact that hexadecane also increased the rate of contents mixing, but did not affect the rate of lipid mixing, suggested to the authors [168] that the TMDs destabilize the bilayers of the liposomes, lowering the activation energy for formation of the intermediates, whereas hexadecane has much less effect. A similar speculation has been made for the effect of model TMDs on Q_{II} phase formation [136].

8.2.2.2 GUV/GUV Fusion

MacDonald et al. [146, 147, 169] studied the fusion of GUVs labeled with appropriate fluorescent dyes to determine the sequence of events and outcomes in isolated, pairwise vesicle–vesicle interactions. These studies provide a striking demonstration that hemifusion is a process that can either precede fusion or exist as an end state in itself. The method is also appropriate for studies of the role of adhesion-induced membrane tension in fusion intermediate evolution. Relation of results obtained in this system to other model membrane and biomembrane systems is hampered by the use of oppositely charged liposomes at high charge densities. There are few biologically relevant cationic membrane lipids. Divalent cations can induce adhesion between anionic lipids and generate similar adhesion-induced membrane tensions (Section 8.2.3.2), but the mol fraction of anionic lipid in biomembranes is usually low (ca. 10 mol%).

Only a comparatively small number of events can be observed with this technique, so one must use GUVs with compositions that will make them adhere to one another and react within a convenient time. Therefore, interactions between GUVs bearing opposite charges were studied in [146, 147, 169]. The anionic GUVs were made of different mixtures of dioleoylphosphatidyl-glycerol (DOPG) with DOPC, DOPE, or DOPC and cholesterol. The cationic GUVs were made with EDOPC (1,2-dioleoyl-sn-glycero-3-ethylphosphocholine) mixed with the same lipids. EDOPC is a DOPC derivative in which the phosphate oxygen is substituted with an ethyl moiety, leaving the molecule with a net positive charge [170].

The outcome of GUV/GUV encounters depended on the mol fraction of charged lipid in the membranes [146, 147, 169]. When the membranes were rich in charged lipids, the GUVs flattened to maximize the extent of intermembrane contact. This generates large tensions in the liposome membranes [166, 167]. Because of this, at mol fractions of about 70% and above, a little less than half of the events resulted in rupture of one or both of the GUVs. The primary outcome of GUV/GUV interactions was fusion (mixing of both membranes and contents). Hemifusion (exchange of lipid dye without contents mixing) was observed in a minority of events. As the mol fraction of charged lipid decreased to 40% and below, hemifusion became the primary outcome (64% of encounters at 20 mol%), whereas the fraction of events ending in fusion decreased more slowly and the fraction of rupture events dropped rapidly to essentially zero at 20% charged lipid [146]. Hemifusion could be a stable end state without subsequent fusion (minutes [147]), especially at charged lipid fractions ≤ 40%. Substitution of cholesterol for PC increased hemifusion at the expense of fusion [147], and substitution of PE for PC both increased hemifusion at the expense of fusion and made hemifusion immediate upon contact.

MacDonald and coworkers [146, 147, 169] interpreted these findings in terms of an adhesion/condensation mechanism of fusion [171, 173], which is related to the mechanism thought to underlie Ca^{2+}/PS fusion (Section 8.2.3.2). In this view, local charge neutralization of the oppositely charged lipid interfaces (or cation binding to anionic lipid interfaces) drives adhesion of the membranes. Neutralization of the charge density leads to contraction of the facing monolayers (the lipid headgroups no longer repel each other so strongly). This local contraction generates tension in, and hence rupture of, the contacting monolayers to form a bilayer septum (hemifusion). Rupture of the septum is presumed to occur due to the adhesion-induced tension in the bilayers. This picture is consistent with the increasing stability of the septum (hemifusion) and decreasing frequency of bilayer rupture (leakage) with decreasing charge density. The nature of the first intermembrane structure is unknown, but there are two pieces of evidence implying that it is related to the intermediates that form in the course of lamellar/inverted phase transitions (i.e., stalks). First, PE and cholesterol, which would reduce C_0, increase the rate and extent of hemifusion. Second, it was noted that EDOPC/DOPG mixtures (i.e., the products of GUV interactions) form inverted phases [146], as do EDOPC/cardiolipin mixtures [172]. Moreover, the lipid of ruptured GUVs and excess membrane of fused GUVs were observed to form small, compact lipid aggregates consistent with bulk inverted phase formation.

Lei and MacDonald recently coupled the technique with high-speed microfluorescence spectroscopy [169]. This permits observation of fluorescent label migration on the millisecond time scale. Work with this system may provide an estimate for the time scale of hemifusion and hemifusion/fusion intermediate conversion. Preliminary results [169] indicate that, at high charge density, substantial lipid mixing occurs within 5 msec and may precede contents mixing, implying the existence of transient hemifusion intermediates.

8.2.2.3 GUV/Planar Bilayer Fusion Observed via Simultaneous Conductance and Fluorescent Assays

Chanturiya et al. [148] monitored fusion of GUVs with planar membranes by simultaneous measurements of fusion pore conductance, lipid mixing, and contents mixing by fluorescent assays. This permitted detailed observation of the sequence of events, in particular GUV/planar membrane interactions. It was found that GUVs formed reversible hemifusion intermediates with planar lipid membranes (lipid dilution from the GUV could occur in stages), and that osmotic stress (swelling the GUV) increased the frequency of fusion events. It was also found that the initial fusion pores exhibited the "flickering" behavior (transient pore formation) that is inferred to occur in PEG-induced fusion [145] and that is also observed in many biomembrane systems. The distribution of pore conductance in the lipid system in [148] is comparable to the distribution of initial conductance in biological fusion pores. These data provide strong support for the contention that the fusion pores in biomembrane systems are lipidic and are not protein-lined channels.

8.2.2.4 Time-Resolved Cryoelectron Microscopy

Cryoelectron microscopy is an elaboration of transmission electron microscopy (TEM). A great advantage of the technique over conventional TEM is that no stains, cryoprotectants, drying, or solvent substitution procedures are applied to the specimens. A sample of the dispersion of interest is blotted to form a thin film (a fraction of a micron thick) on a special support, which is then vitrified by plunging into a cryogen (usually ethane at its melting point). This transforms aqueous suspensions to vitreous ice within an interval estimated to be less than ca. 0.1 msec. The specimens are then maintained under liquid nitrogen, inserted into a liquid nitrogen-cooled sample holder, and examined in a transmission electron microscope. The vitrification procedure effectively halts dynamic processes on this rapid time scale. Because of the nonperturbative and "stop-action" effect of vitrification, the technique is appropriate for studying membrane dynamics and membrane–membrane interactions (e.g., [22, 27–31, 33, 78, 153, 174, 175]), reconstitution of membrane proteins into liposomes, and macromolecular/supramolecular structures. It has been widely reviewed (e.g., [160, 176, 177]).

Time-resolved specimen preparation techniques can be applied to start dynamic processes in membrane dispersion on cryoelectron microscopy specimens at defined times before vitrification.

These intervals can range from minutes to milliseconds [28, 178]. Results from cryoelectron microscopy studies of fusion and phase transition intermediates in systems with lamellar/inverted phase transitions will be discussed in more detail in the context of the stalk model (Section 8.2.3.1). Cryoelectron microscopy in general is currently contrast limited. Contrast generation in TEM is a complex process, but generally membranes do not generate much contrast in TEM, and they generate most contrast when viewed in the plane of the membrane (i.e., when viewed edge-on, as at the periphery of a spherical vesicle). As demonstrated in [28], the contrast decreases as the length of membrane in the beam direction decreases, so that contrast decreases as the radii of membrane curvature decreases. Small unilamellar vesicles are much harder to "see" than LUVs. Objects the size of stalks would be extremely difficult or impossible to detect [28]. However, cryo-TEM is well suited to detecting the subsequent evolution of fusion intermediates into TMCs and/or hemifusion diaphragms in LUV or oligolamellar liposome dispersions.

An example of time-resolved cryoelectron microscopy data is given in Figure 8.5, reprinted from [29]. The left-hand image is a micrograph of a dispersion of dipalmitoleoyl-PE (DiPoPE), a

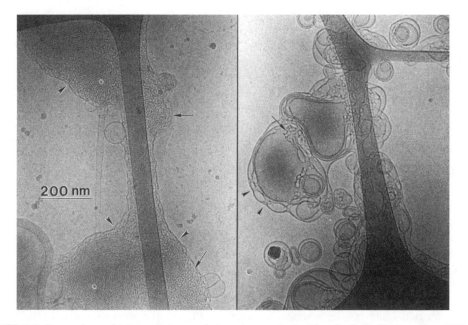

FIGURE 8.5 Comparison of intermediate morphology in a system with a facile L_α/H_{II} transition (DiPoPE; left) and a system that forms both Q_{II} and H_{II} phases (DOPE-Me; right). Both micrographs are shown at the same magnification, and the scale bar is 200 nm. In both cases, the transitions started in dispersions of LUVs near room temperature. For DiPoPE, the transition was started by rapid adjustment of the pH from 9.9 to 4.5 a few seconds before vitrification. For DOPE-Me, the liposomes were originally at pH 7.4, and the transition was triggered by a flashtube-induced temperature jump to about 90°C about 10 msec before vitrification. Note the appearance of many disordered intermembrane connections in the DiPoPE micrograph (left), indicated by arrows. The structure of these individual connections cannot be definitely assigned due to superposition effects, but they have the rough shapes of TMCs or ILAs, in views perpendicular to the axis of the structures (parallel to the surfaces of the planar membranes: arrowheads) and views looking down the axes of the structures (arrows). The dimensions (ca. 10 nm in diameter and height normal to the bilayer stack) are in the range predicted for TMCs [69, 70]. Asterisks mark areas that resemble projections through H_{II} phase aggregates. In DOPE-Me (right), the system rapidly forms well-defined ILAs (fusion pores), which are indicated by arrowheads when seen from the side and by arrows when viewed down the pore axis. These are larger, more well-defined structures than in DiPoPE, and they are often larger in both diameter and height in the direction normal to the membranes. What appear to be holes at the edges of folds of membrane elsewhere in this image are projections of ILAs (see Figure 1 of [22]). Adapted from [29].

lipid with a thermotropic L_α/H_{II} phase transition. The DiPoPE was prepared as a dispersion of LUVs at pH 9.9, a pH where the lipid is anionic and the phase transition cannot occur. A drop of this dispersion on the TEM grid was mixed with the same volume of a pH buffer that made the final pH permissive for the transition, and the specimen was blotted and vitrified within 6 to 10 sec. The micrograph shows that, in this time, the LUVs have aggregated to form much larger structures and have also formed extensive inverted phase structures. The arrows indicate areas where there are numerous disordered intermembrane connections. The structure of these individual connections cannot be definitely assigned due to superposition effects, but they have the rough shapes of TMCs or ILAs, in views perpendicular to the axis of the structures (parallel to the surfaces of the planar membranes: arrowheads) and views looking down the axes of the structures (arrows). The dimensions (ca. 10 nm in diameter and height normal to the bilayer stack) are in the range predicted for TMCs [69, 70]. There are also structures that look like projections through H_{II} phase aggregates (bottom center and top left, asterisks). At right is a similar micrograph obtained of a dispersion of LUVs of dioleoyl mono-methylated PE (DOPE-Me), a lipid that forms both Q_{II} and H_{II} phases when heated to about 60°C [28, 42, 43]. This specimen was subjected to a temperature jump from room temperature to about 90°C at a time about 10 msec before vitrification. Even within this short time, the LUVs have extensively aggregated and fused into oligolamellar structures, with numerous ILAs (fusion pores). The ILAs are larger and less numerous than the structures in DiPoPE and have well-defined pore structure, as viewed both from the side (arrowheads) and down the axes of the pore (arrows). Domains of H_{II} phase are also rarer in these samples. What appear to be holes in the bilayers are actually ILAs seen in projection at the edges of membrane folds (see Figure 1 of [22]). These data support the existence of stalklike structures during lamellar/inverted phase transitions and also make the important point that stalks must be able to form fusion pores (ILAs) without the existence of obvious hemifusion diaphragms within milliseconds after liposome interactions begin. This may help us distinguish between the predictions of different variants of the stalk mechanism (Section 8.2.3.1).

Some care is necessary in interpretation of intermediate morphology in cryo-TEM, because membranes and monolayers generate by far the most contrast when seen edge-on, and this can produce projection images that resemble holes or pits in membranes when in fact the structures are simple, continuous-membrane catenoids [22]. In addition, the low contrast of high-curvature segments of membranes or monolayers [7] makes some parts of larger structures almost disappear in micrographs, which can also give a misleading impression of the overall geometry. Whenever possible, images of the same area should be obtained at more than one defocus setting, although the susceptibility of specimens to beam damage can make this difficult. In addition, Siegel has pointed out [70] that the substantial bilayer/bilayer adhesion energies in PE lamellar phases might have the effect of lowering the temperature at which H_{II} phases can form from LUV dispersions by as much as 10 K with respect to T_H as measured in multilamellar dispersions. Siegel and Epand [29] observed transition intermediates and nonlamellar morphology at temperatures 20 to 36 K below T_H when liposome interactions were triggered in LUV dispersions of DiPoPE, so it is unlikely that the adhesion energy effect accounted for most of these observations, but the effect must be borne in mind. Several time-resolved cryo-TEM studies [22, 28, 29, 31] have made use of PE LUVs prepared at high pH to stabilize them in the L_α phase, so that the transition can be conveniently triggered by a drop in pH. These LUVs should be made immediately before TEM sample preparation and maintained on ice before use, because PE is subject to hydrolysis at high pH, which gradually lowers T_H [31].

8.2.2.5 Inner Monolayer Lipid Mixing and Transbilayer Lipid Migration Assays

Reliable assays for the mixing of inner monolayer lipid between liposomes have been developed (e.g., [83, 154, 179, 180, 180a]). These assays permit resolution of events that mix only the outer monolayers of liposomes (such as hemifusion) from processes that make both the inner and outer monolayers of the liposomes continuous (fusion). They can also be used to monitor transbilayer migration of lipid in individual liposomes (e.g., [154]). This is important in the study of hemifusion, which otherwise can only be inferred by lipid mixing in the absence of aqueous contents mixing. Use of contents-mixing assays is sometimes experimentally inconvenient. Moreover, when total membrane lipid mixing is observed in the absence of contents mixing, and there is also some liposome leakage, the lipid mixing might arise from rupture and resealing of membranes from different liposomes late in the evolution of lipid aggregates. Hence, these assays permit resolution of an important ambiguity and can represent a convenient fusion assay.

Malinin et al. [181] have also developed an improved fluorescence resonant energy transfer (FRET) assay for lipid mixing in general, based on a new pair of lipid probes.

8.2.3 FUSION INTERMEDIATES AND FUSION MECHANISMS

The last decade has seen major advances in our understanding of the mechanism of lipid membrane fusion. This has been due to three factors:

- First, a better understanding of the principles determining the stability of lipidic structures with respect to the lamellar phase was available, obtained through studies of nonbilayer phase stability.
- Second, better theoretical models were proposed, incorporating this knowledge.
- Third, new techniques were employed, which allowed successful tests of some of the main predictions of the new models. These new techniques included variations on the planar bilayer fusion system, time-resolved cryoelectron microscopy, and observation of the effects of exogenous lipids on fusion in cellular and viral biomembrane systems.

A recurring theme in these developments has been the need for detailed studies of lipid physical chemistry and phase behavior in order to achieve a mechanistic understanding of fusion.

Most progress has been made in understanding fusion of lipids where the lipids are initially in the liquid–crystalline (L_α) phase and where anionic lipids are not the major component. These systems are now thought to fuse via the stalk mechanism. This mechanism will be discussed in detail (Section 8.2.3.1). There is a very substantial literature regarding divalent cation-induced fusion of anionic lipids like PS, which has been expertly reviewed (e.g., [10–13]). This subset of anionic lipids does not form inverted phases at neutral pH or in the presence of the divalent cations. It is not clear what the intermediates in this process are, although a stalk mechanism is not out of the question [182]. These systems will be briefly discussed in Section 8.2.3.2.

First, some general comments about modeling fusion mechanisms are appropriate. In order for fusion to occur at appreciable rates, the activation energy for the process must be no larger than a few tens of $k_B T$, where k_B is Boltzmann's constant and T is the absolute temperature. In general, it will be almost impossible to know the detailed sequence of steps involved in fusion, because of the small length scale (< 10 nm) and fast time scale (from electrophysiological studies, < 0.1 msec) of the events. However, as argued in [183], we can generate testable models of the fusion process by trying to infer what configurations the lipid molecules adopt during different stages in the process and by estimating the relative free energy of each according to principles determined from studies of inverted phase stability. Fusion necessarily involves energetically intensive processes. These include bringing two opposed membranes into contact against strong repulsive (hydration) forces, disruption of the interactions between adjacent lipids in a monolayer (with transient exposure of acyl chains to water), bending monolayers into small radii of curvature, stabilizing interstitial spaces, and forming a pore in a bilayer. All these contributions to the free energy of activation must be minimized in a hypothetical mechanism. Generally, this argues for action of these forces over as small an area as possible, hence for as small an intermediate structure as possible. One of the biggest energetic barriers to fusion is rupture of a bilayer to compete the fusion process. In the absence of external energy inputs, like a bulk tension in one or more of the fusing membranes, the energy to perform this final rupture must be assembled by concentration of as much as possible of the stress in an intermediate at the point at which rupture must occur in order to produce fusion.

Hence, the object of fusion mechanism modeling is to create intermediates of minimum size, with minimal curvature and chain-packing energies, which nevertheless bring a maximum of these stresses to bear on the minimum area of monolayer whose rupture will produce a fusion pore. The amount of energy necessary to start the pore formation process is the most significant barrier to the overall process. As we will see, this quantity also plays a major role in the kinetics and hysteresis of lamellar/inverted phase transitions.

8.2.3.1 Stalk Mechanism

The "stalk" is postulated to be the first intermediate structure to form between two interacting membranes in both membrane fusion [29, 69, 70, 85, 98, 158, 163, 182, 184–190] and lamellar/inverted phase transitions [29, 69, 70, 163]. A "generic" cross section of this structure is depicted in Figure 8.1B: it is a cylindrically symmetrical, hourglass-shaped connection between apposed membranes, composed of lipids from the facing (*cis*) leaflets of the two membranes. Gingell and Ginsberg were apparently the first to suggest the existence of such an intermediate structure [190a]. Hui et al. proposed a stalk structure on the basis of morphological evidence [191]. The stalk fulfills several of the constraints on hypothetical fusion mechanisms. The intermediate is of the minimum size necessary to connect two apposed membranes. Thus, the two original membranes must be pushed close together and their interfaces disrupted over only a minimal area, requiring a minimal activation energy for stalk formation. The compound curvature of the stalk outer surface (one principal radius of curvature is positive, the other negative) also leads to a low curvature energy (see Equation 8.1). Thus, this model is attractive on several grounds, even before a detailed analysis of the energy required to form the intermediates.

Stalks have so far evaded direct observation in fusing systems. Stalks are small structures, and most models of stalk energetics predict that they are transient structures when they form in lamellar phases in excess water or in dispersions of fusing liposomes. Because of their small size and presumed short lifetimes, they seem to be undetectable via current electron microscopic techniques [28]. However, recently Yang and Huang described a rhombohedral phase forming in phospholipids at low hydration, which consists of stalk structures within a stack of planar membranes [192–194]. The stalk-based structure of these phases has been confirmed in diphytanoyl PC [192, 193] and in mixtures of DOPE and DOPC [194]. The stalk-based phases form only at low water activity and not in pure DOPE. These observations are extremely important and lend strong experimental support to the role of stalks as intermediates in membrane interactions. The fact that stalks are thermodynamically stable in phospholipids at room temperature, at least under some circumstances, strongly suggests that they are low enough in energy to form transiently as fusion intermediates between membranes of physiological compositions. This is in accord with predictions of more recent theoretical studies on the stalk mechanism [158, 188–190]. An electron density reconstruction of the rhombohedral phase is shown in Figure 8.6 (from [193]).

The first theoretical description of the stalk mechanism was given by Kozlov and Markin [184]. The original model has undergone two successive generations of refinements.

8.2.3.1.1 Stalk-Pore Model

Figure 8.7 shows the steps in the so-called stalk-pore models. In the first generation of stalk models [85, 98, 182, 184–187], the free energy of the structures with respect to the lamellar phase was assumed to arise solely from differences in curvature elastic energy, following the treatment of Helfrich [46]. The first treatments [85, 184, 185] showed that, if the spontaneous curvature (C_0) of the lipid in the *cis* monolayers is sufficiently negative, formation of stalks (Figure 8.7C and Figure 8.7F) was followed by radial expansion of the stalk in the plane of the opposed membranes (Figure

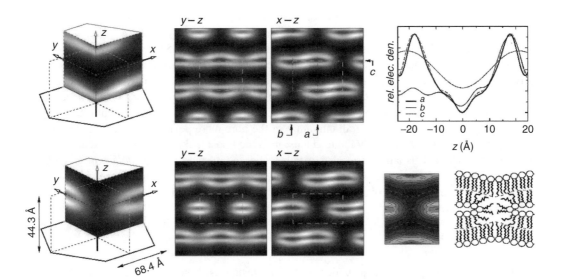

FIGURE 8.6 Electron density reconstruction of the unit cell of the rhombohedral phase lattice in diphytanoyl-PC. (High density is white, low density is black.) *Top row:* The dashed rectangles within each figure are the boundaries of the unit cell in each map. On the *x-z* map, the electron density profiles were taken along the lines a, b, and c, and are shown in the graph at top right. Also shown in the graph is a transbilayer electron density profile from the lamellar phase of the same lipid (heavy black line). *Bottom row:* The unit cell (left) in this set of views was obtained by translation from the previous one. A stalk structure appears in the center of the cell. The maps in the *y-z* and *x-z* planes are shown in the center. On the right, the electron density of the central part of the *y-z* map is plotted with density contours. At the extreme right is a diagram of the stalk structure imaged in the contour plot. (Adapted from [193], and used with permission.)

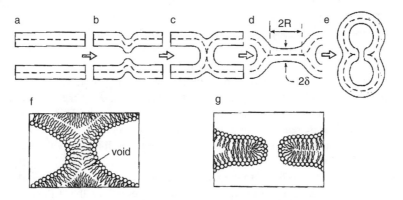

FIGURE 8.7 Stalk-pore fusion model. (A) Two opposed membranes. (B) Fluctuation resulting in a local close contact between membranes. (C) Stalk formation. (D) Radial expansion of a stalk resulting in formation of a contact bilayer (or hemifusion diaphragm). (E) Formation of a pore in the contact bilayer completing the fusion process. (F) Enlarged diagram of a pore, indicating one of the two hydrophobic voids in the stalk. (G) Enlarged diagram of the fusion pore. Dashed lines show the boundaries of the hydrophobic surfaces of two lipid monolayers. 2δ and $2R$ are the thickness of the lipid bilayer and the diameter of the contact bilayer, respectively. (Reprinted from [7], with permission.)

8.7D). This forms a trilamellar structure, or hemifusion diaphragm, which is a single-bilayer diaphragm between the two original liposomes. This stage is termed *hemifusion*, to emphasize that only the facing monolayers of the two opposed membranes have become connected and that lipid mixing can occur between these leaflets of the two membranes. In subsequent treatments [86, 182,

185a, 186], it was shown that radial expansion of the stalk creates an area asymmetry between the inner and outer monolayers of the liposomes. This area asymmetry produces a tension in the hemifusion diaphragm as the diaphragm continues to expand [182, 185a]. For sufficiently negative C_0 of the lipids in the outer monolayers and sufficiently small pore edge tension, fusion of the two original liposomes occurs (Figure 8.7E and Figure 8.7G). Pore formation would be favored by larger (more positive) values of C_0 for the distal monolayer lipid composition. This is because the edges of the pore would have one large, constant, positive principal curvature (Figure 8.7G). This would make the area-averaged overall curvature of the pore more positive than for the stalk for all diameters of the pore. Osmotically induced tension in the liposomal membranes could also encourage pore formation. This latter form of the stalk model is referred to as the *stalk-pore model*.

Formation of a fusion pore, ILA, (Figure 8.1D, upper left) from the pore in the single-bilayer diaphragm was not explicitly discussed, although this was discussed in a subsequent treatment by Chizmadzhev et al. [195].

The predictions of the stalk-pore model [86, 182, 185a, 186] are in accord with many observations of fusion behavior:

First, the model predicts that, under some circumstances, lipid mixing can occur before fusion or perhaps even without fusion. Stalks and radially expanded stalks connect the facing monolayers of two membranes (hemifusion) before pore formation (fusion) occurs. It is also possible that pore formation will not occur in all cases or that other processes (e.g., leakage via inverted phase formation in aggregates of liposomes) will occur before fusion can occur. Lipid mixing is often observed as a stage distinct from fusion, or instead of fusion, in liposomal dispersions of nonbilayer lipids (reviewed in Section 8.2.1). It is also observed in giant liposomes of lipids that do not form inverted phases [146, 147], in Ca^{2+}-induced interactions of PE/PS liposomes [83], and in both small and large unilamellar liposomes induced to fuse by high concentrations of polyethylene glycol [144, 145, 151–153, 157, 165, 168, 165] (see Section 8.2.2). Studies of fusion between planar lipid membranes [85, 86] clearly resolve lipid mixing (trilamellar structure formation) from fusion (diaphragm rupture), due to the formation of large single-bilayer diaphragms in these systems. Resolution of the two steps is also observed in giant liposome/planar bilayer interactions [87, 148]. Interestingly, pore formation was found to be reversible ("flickering pores") in the early stages of fusion in one GUV/planar bilayer study [148].

Second, the original model of stalks predicts that reductions in C_0 of the lipids in the facing monolayers of the membranes should lower the energy of the stalk. This agrees with many observations linking increases in lipid mixing and fusion rates with increases in the mol fraction of inverted phase-forming lipids, or of additives like DAG that lower C_0, as described in Section 8.2.1. Alternatively, lipids that raise the C_0 of all the lipids in the membranes, or of only the lipids in the facing (opposed) monolayers, should decrease the rate of stalk formation. This rationalized extensive observations that LPC and other surfactants that raise C_0 inhibit the rate of lipid mixing and fusion between model membranes [85–87] and biomembranes (e.g., [88, 98]; see [7, 97] for reviews).

Third, the model explains an interesting sidedness effect of LPC and similar additives on lipid mixing and fusion. The stalk model [85, 98, 182, 184–187] predicts that lipids that make C_0 more positive (like LPC) should increase the energy of stalks when added to the facing monolayers, inhibiting lipid mixing and fusion. However, adding LPC to the distal monolayers should increase the likelihood of pore formation in the bilayer diaphragms of radially expanded stalks, increasing the rate of fusion. The opposite effects for each monolayer are expected for additives like arachidonic acid (AA), which lowers C_0. The sidedness effect of LPC addition was observed in the fusion of macroscopic planar membranes [85, 86] and in the fusion of liposomes with planar membranes [98]. In the fusion of two planar membranes, LPC added between the membranes slows the rate of hemifusion diaphragm (trilamellar structure) formation [85]. In contrast, the presence of LPC in both monolayers of a membrane accelerates pore formation (rupture) of the bilayer diaphragm [86]. The edge tension of pores in single bilayers can be measured indirectly in these systems, and LPC is observed to lower the edge tension [86], consistent with its role in pore formation. When liposomes are brought into contact with a single planar membrane, addition of LPC to the side of the membrane with the liposomes inhibits fusion, whereas addition to the opposite side accelerates fusion [98]. The opposite corresponding effects of arachidonic acid were observed in [98]: decreased fusion when added to the distal monolayers of the planar membrane, although this was not observed with "solvent-free" (vs. decane-containing) planar membranes. Arachidonic acid had essentially no effect on fusion when added to the facing monolayers in [98], in contrast to the predictions of the theory, but arachidonic acid is not especially powerful, mole for mole, in lowering C_0, as judged by its small effect on T_H [61, 79]. There is also evidence for a distal monolayer effect of positive C_0 surfactants in some experiments on HA-induced cell fusion (e.g., [94, 196]). In these systems, however, the effects of changes in distal monolayer curvature on fusion pore enlargement, as opposed to initial pore formation, are more complex. They seem to reflect effects on curvature energy of a fusion pore (ILA), and not of a pore in a single bilayer [91].

However, there are some observations that are *not* readily explained by the stalk-pore model in its original form [86, 182, 185, 186]:

First is the effect of chain-packing energy (Section 8.2.1.2). Addition of long-chain alkanes has been observed to substantially increase the rate of fusion in liposomal systems [80, 118]. The stalk-pore model does not take chain-packing energies into account and cannot account for this. In addition, because the stalk structure is presumably the lowest energy intermediate that can form between L_α phase bilayers, the structure should be capable of forming the H_{II} phase (although this was not addressed in the stalk-pore model). T_H is quite sensitive to the presence of nonpolar oils, too (Section 8.2.1.2), and hence the stalk free energy should reflect this sensitivity.

Second, the stalk-pore model predicted that, for sufficiently negative values of C_0, the stalk structure had a lower free energy than lipid in the L_α phase. For membranes separated by a water layer thickness of 3 nm (i.e., full hydration of the L_α phase of lipids like PE and DOPE-Me), the stalk should be more stable than the L_α phase for $C_0 <$ ca. -0.24 nm^{-1}, and any stalk that forms should radially expand for $C_0 < -0.14$ nm^{-1} [182]. In contrast, H_{II} phases form in PE and DOPE-Me when C_0 is ca. -0.35 nm^{-1}, and only the L_α phase and Q_{II} phases are observed at higher (less negative) curvatures, which corresponds to a range of temperatures well below T_H [70]. That is, the stalk-pore model predicts that a phase of stalks should be observed in a temperature range below T_H in the presence of excess water, and no such phase is observed.

Third, there is evidence from time-resolved cryoelectron microscopy studies that is not readily compatible with the stalk-pore theory. There is no evidence for the formation of large, single-bilayer diaphragms between interacting liposomes from such studies, at least when the lipids are in the temperature range starting some 20°C below T_H [22, 27–31]. Extensive diaphragms *do* form in the fusion of planar bilayers [85, 86], GUV/planar bilayer fusion [87, 148], and GUVs [146, 147]. However, in the first two cases nonpolar oils are present to reduce the energy barrier to large hemifusion diaphragm formation. In the third, the adhesion energies between the interfaces are much larger than in typical phospholipid systems, which generates large membrane tensions, which in turn are a powerful driving force for hemifusion diaphragm formation. In addition, cryo-TEM and freeze-fracture TEM have established the existence of ILAs (Figure 8.1D, upper left), which are bilayer-lined channels connecting two opposed membranes (e.g., [22, 27–29, 31, 33]). The diameters of the "waists" of these structures are only ca. 12 to 15 nm in pure lipids like DOPE and DOPE-Me [22, 28], with aqueous pores 4 to 9 nm in diameter. It is not clear how these would form from stalks according to the stalk-pore model, given the predicted tendency of the stalks to radially expand into large bilayer diaphragms for lipids with these values of C_0 (approximately −0.3 nm^{-1}). However, it is possible that the expanding stalks transformed directly into ILAs consistently at a very early stage in the expansion. One could also argue that, following pore formation in the hemifusion diaphragm, the rim of the hemifusion structure rapidly contracts in radius to form ILAs of the observed dimensions, on the time scale of cryofixation (ca. 0.1 msec). However, it is not obvious that there is a strong enough driving force for this process. The radial expansion is driven by a reduction in curvature energy of the rim (in this and later versions of the stalk theory), so contraction is disfavored. The ILA is a lower-energy structure than the flat diaphragm, but, depending on the radius of the original hemifusion diaphragm, it can also be much smaller in total interfacial area, and hence higher in total energy.

These difficulties arise in part because early treatments [85, 98, 82, 84–187] did not calculate the chain-packing energy of the interstices that are associated with stalk formation and evolution (stippled areas in Figure 8.1B and Figure 8.1C). They also neglected explicit treatment of the curvature energy of the distal (*trans*) monolayers during radial expansion of the stalk.

8.2.3.1.2 The Modified Stalk Model

The modified stalk model [69, 70] accounted for both the chain-packing and the distal monolayer curvature energies. This model was used to show that the stalk mechanism was more probable than a mechanism based on inverted micellar intermediates [69].* These changes have several effects on the overall predictions of the model. The chain-packing energy restricts radial expansion of the stalk beyond formation of a structure known as a *trans* monolayer contact (TMC) (Figure 8.1C). The TMC is formed by radial expansion of the stalk and simultaneous pinching together of the two distal monolayers to form a local contact in the center of the structure. In pure lipid membranes, the interstices in the TMC restrict further radial expansion, because radial expansion would lengthen the perimeter of the interstitial spaces, increasing the chain-packing energy. TMCs are predicted to show only a slight tendency to expand radially by a few nm when C_0 approaches values found in H_{II} phase lipids [70]. However, TMCs should radially expand and form large bilayer diaphragms if the membranes contain nonpolar oils. Only a few weight percent of nonpolar oil in the lipids is sufficient.

In the modified model, fusion pore formation occurs within the small area of opposed distal monolayers in the TMC. A substantial fraction of the unfavorable chain-packing and monolayer curvature energy is concentrated in this "dimple" of the TMC [70]. This puts these monolayers under a stress that is comparable to the tensions observed to rupture bilayers [156, 197–200]. Pore formation is proposed [70] to occur under influence of this stress in oil-free systems. Rupture of the dimple of the TMC forms a structure with the dimensions of the fusion pore (ILA) directly (Figure 8.1C and Figure 8.1D, upper left), rather than by growth of a hole in a larger single-bilayer diaphragm and subsequent contraction of the radius of the walls of the radially expanded stalk.

The modified stalk mechanism, therefore, removed some of the difficulties with the first model.

First, the modified theory accounts for the effects of alkanes on the rate of fusion and lamellar/inverted phase transitions. Inclusion of chain-packing energy (void stabilization) in the model makes formation of stalks and subsequent intermediates sensitive to the

---

* Equation A1 in [69] contains a typographical error, which is corrected in the Appendix to [70].

presence of nonpolar oils like alkanes, as required by observations on fusion [80, 117] and lamellar/inverted phase transitions (Section 8.2.1.2).

Second, the modified theory predicts higher energies of the stalks and expanded stalks as a function of C_0 than the original model, and no longer predicts the existence of unobserved inverted phases. Addition of chain-packing energy and distal monolayer curvature energy to the model in [69, 70] were not elective options. Both those contributions to the total energy were necessary to give an accurate account of the total energy of the intermediates. In an important contribution, Kozlov et al. [53] showed that a combination of the curvature and chain-packing energies, with a hydration force component, was sufficient to explain the major features of the complicated low-water portion of the DOPE-water phase diagram. At the time, this gave reasonable confidence that the model used for calculating intermediate energies in [69, 70] would give a good estimate of the free energy of a fusion intermediate of given geometry.

Third, the TMC and ILA intermediates of the modified model are more compatible with cryoelectron microscopy [22, 27–31, 33] and freeze-fracture electron microscopy [22] data on LUV systems undergoing lamellar/inverted phase transitions. Well-defined ILAs and TMC-like structures are seen in large numbers, which contrasts with the stalk-pore model's prediction of rapid radial expansion of stalks past these stages of intermediate evolution (for lipids with $C_0 < 0.14$ nm^{-1} [182]).

Fourth, a mechanism involving TMCs can mediate transitions between the L_α, H_{II}, and Q_{II} phases, and is compatible with almost all observations about the relative stability of these phases, as well as with the relative rates of transitions between them [29, 70]. The Q_{II} phase forms from ILAs (Figure 8.1) by a mechanism described in [11]. The relevant intermediate ILA array was observed in [20, 22]. The H_{II} phase is proposed [29, 70] to form via aggregation of TMCs into arrays, which are nearly identical in cross section to a domain of H_{II} phase, and from which domains of H_{II} phase can grow directly by lipid diffusion from adjacent bilayers (see Figure 8.8). There is some cryoelectron microscopy evidence for such a mechanism [29, 31]: large numbers of TMC-like structures are observed to form in dispersions of aggregated LUVs, and aggregates resembling H_{II} phases in some projections were observed at temperatures as much as 20°C below T_H, implying that they are not equilibrium H_{II} phase domains. Recent work also suggests that the modified stalk mechanism rationalizes the apparent kinetic competition between H_{II} and Q_{II} phase formation in a lipid with hysteretic Q_{II} phase formation [43] as a competition of two parallel mechanisms for the intermediate TMCs.

However, although the modified theory solves some problems with the stalk-pore model, it still makes predictions that are in conflict with observations. One of these conflicts is general and obvious, and was highlighted as an apparent paradox in [70] as a challenge for further work. The other conflict is more subtle.

First, the predicted energies for stalks and TMC-like intermediates are very high (on the order of 100 $k_B T$ or more) in the modified model [70]. As noted in [70], these high energies present a paradox. Fusion intermediates must have low energies (tens of $k_B T$) in order to form at observable rates. The stalk intermediates have the lowest free energy of any intermediates yet proposed and seem, in principle, to be the lowest-energy intermembrane structures that could form, due to their compound curvature. How can such large energies be consistent with rapid intermediate formation? This paradox created an "energy crisis" in membrane fusion theory. The formation energy of stalks must be lower than predicted by the modified model, because stalks were recently found to be structural elements of phases that form spontaneously in phospholipids around room temperature, albeit only at low water activity and in a modest range of spontaneous curvature [192–194].

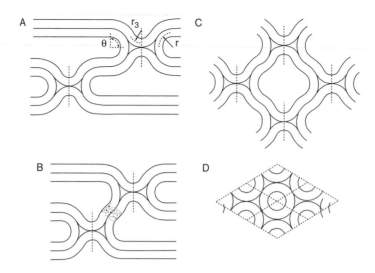

FIGURE 8.8 TMC aggregation to form H_{II} phase precursors. (A) A cross section through a pair of isolated TMCs in a stack of planar bilayers. The dashed vertical lines are the axes of the TMCs. The marginal radius r and marginal angle θ are indicated. (B) Sideways aggregation of the two TMCs reduces θ and increases r where they contact (stippled area of bilayer), which reduces the curvature free energy of the system [70], especially by the reduction in θ. The values of r and θ change in complementary fashion in the vicinity of a contact, so that the local spacing between bilayers in the vicinity of the TMCs does not change. (C) The aggregation process proceeds to form a TMC aggregate with body-centered or primitive tetragonal symmetry. A cross section in the 110 plane of this structure is shown in C. This cross section consists of closed cylinders of monolayer packed in a quasi-hexagonal fashion (i.e., each cylinder has six nearest neighbors). At temperatures $> T_H$, this structure can accumulate lipid molecules by diffusion from the adjoining bilayers and extend directly into a domain of H_{II} phase (out of the plane of the paper in C). (D) Cross section of a bundle of H_{II} tubes, to illustrate the similarity of the geometry in C. (Reprinted from [70], with permission.)

Siegel suggested [70] that the modified model might overestimate the energy in two ways. Both the original and modified stalk theories made an assumption about intermediate geometry that is probably not valid for intermediates of minimal energy. The outer mono-layers of stalks, TMCs, and ILAs were all assumed to be circular semitoroids of revolution. This assumption simplifies the mathematics. However, as noted in [70], an infinite periodic minimal surface exists that resembles the shape of the exterior monolayer surface of stalks, TMCs, and ILAs. Such a surface has zero mean curvature energy and should have a lower curvature energy than sections of circular toroids. More generally, the calculations in [69, 70] showed that the details of the intermediate geometry can have substantial effects on the computed energies. For example, the energy of stalks is sensitive to the dihedral angle ε between the edges of the stalks and TMCs and the membranes around the intermediate's periphery [69] (i.e., the energy depends on how much the membranes were bulged toward one another around the edges of the intermediates; see Figure 8.9). In addition, Siegel suggested that the estimates of chain-packing energy in [70] might yield overestimates of the energy, especially for stalks. The chain-packing energy in [69, 70] was estimated by geometrical analogy to the interstitial spaces in H_{II} phases, but the geometry of the interstitial spaces in TMCs is somewhat different. The geometry of the interstitial spaces in stalks is even more different. As we will see, subsequent work implies that the chain-packing energy is probably lower because of an additional degree of freedom available to lipids in monolayers.

Despite the high predicted energies for stalks and TMCs, the sequences of steps necessary to form the H_{II} and Q_{II} phases via these intermediates were nonetheless explored

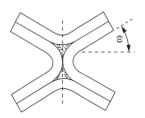

FIGURE 8.9 A stalk formed between two bilayers that are bulged toward each other. The marginal angle ε is the angle between the tangent and the stalk surface in the plane of the stalk axis, at the point where the compound-curvature surface of the stalk meets the conical surface of the bilayer. The energy of a stalk or TMC relative to the original bilayers decreases rapidly with increasing ε. Thus, the additional energy required to form a stalk is sensitive to the geometry of the interacting bilayers. (Reprinted from [70], with permission.)

in [70] in hopes that the principles discussed previously would lower the free energies of these two intermediates by comparable amounts. As we will see, the refinements in subsequent theories (especially [158, 163]) suggest that this is true.

Second, the modified stalk model makes predictions concerning the effects of distal mono-layer C_0 that are in apparent disagreement with some experimental results. In the modified model, increasing the distal monolayer C_0 generally increases the energy of TMCs, and so should decrease the rate of fusion [70]. (This is also predicted by two later stalk models [158, 163, 189] discussed in Section 8.2.3.1.3, and is predicted for most cases by a third, as well [190].) This is because of the curvature of the distal monolayers of the TMC. The stalk-pore model [86] predicts that LPC in the distal monolayers increases the rate of fusion by stimulating pore formation in the hemifusion diaphragm. There is evidence showing that distal monolayer LPC stimulates fusion (Section 8.2.3.1.1).

However, it is not clear whether the observed effect of LPC is due to an effect on the energy of the stalk or the radially expanded stalk as a whole. This observed effect could be due to an effect on the kinetics for pore formation within the TMC or hemifusion diaphragm. For example, LPC in the hemifusion diaphragm lowers the edge tension of nascent pores [86]. It could be that local fluctuations in concentration of LPC can increase the rate or decrease the activation energy for pore formation. Hence, the models could be predicting the correct energies for stalks, TMCs, and HDs, but missing this additional effect on the overall kinetics. Moreover, the acceleration of fusion by distal monolayer LPC is observed in systems in which the chain-packing energy term is essentially zero. As pointed out by Chanturiya et al. [100], the observations were made in systems using planar bilayer membranes made with nonpolar oil solvents [85–87, 98] or in biomembranes [196]. In biomembranes, there are small amounts of triglycerides, diglycerides, or terpenoid lipids like squalene (e.g., [119, 120]). In both types of system, the nonpolar oils would allow radial expansion of the TMC into an extensive hemifusion diaphragm, reduce the total energy of fusion intermediates, and reduce the chain-packing energy stress on the hemifusion diaphragm that drives pore formation. Therefore, there would be more hemifusion structures, and each structure would be more susceptible to rupture in the presence of LPC than in its absence, as observed in solvent-containing membranes [86]. (Chanturiya et al. found [100] that incorporation of tetradecane in liposomes interacting with a planar lipid membrane does not affect the rate of hemifusion. However, this is consistent with residual solvent already present in the planar membrane minimizing the chain-packing energy of the intermediates, so that further addition of nonpolar oil in the liposomes had no effect [100].) In contrast, the rapid formation of ILAs in studies of lamellar/nonlamellar phase transitions (e.g., via electron microscopy and ^{31}P NMR) implies that radial expansion of stalks past the TMC stage does not occur in those systems. Hence, LPC may have less effect in these systems. It would be very interesting to check the effect of distal monolayer effect in systems like these, by somehow introducing distal monolayer LPC into LUVs or GUVs consisting primarily of PE in the absence of nonpolar oils, and observing whether or not this increases the rate of fusion.

Another alternative explanation for the effect of distal monolayer LPC is that it affects the free energy of the initial fusion pore (ILA) that forms. Even with the lower curvature energy ILA shapes in [189], LPC in the distal monolayers substantially reduces the energy of ILAs with small pore diameters. Thus, the LPC could increase the driving force for fusion pore formation, rather than the energy of the intermediates themselves.

A related, underlying question is: How far does radial expansion of stalks proceed in model membrane fusion in the absence of nonpolar oils? To this reviewer's knowledge, the answer is currently unknown. The answer has substantial impact on our theoretical picture of fusion and phase transition intermediates. Neither is the answer obvious in some biomembrane systems. The fusion pore may form from a TMC-like structure (without formation of a large hemifusion diaphragm) in at least some circumstances. An extensive study of cell–cell fusion mediated by influenza hemagglutinin (HA) [201] found no extensive hemifusion diaphragms, only contact sites of the size observed and predicted for ILAs [70]. Expansion of initial fusion intermediates may be restricted by protein structures (e.g., [92]). This contrasts with the large hemifusion diaphragms observed in hemifusion of cells with planar bilayers made with squalene, mediated by an analog of HA that lacks a transmembrane domain. These diaphragms are about as large as the cells themselves [202]. However, in wild-type HA-mediated fusion of cells with planar bilayers, there is indirect evidence for rapid conversion of a single-bilayer pore (whose formation is LPC-sensitive) to a three-dimensional fusion pore (ILA) [91]. LPC and other C_0-altering additives appear to act in different ways at different stages of lipid intermediate evolution in biological systems (pore formation vs. pore expansion) [91].

In summary, the modified stalk theory accounts for more observations than does the original model and is qualitatively compatible with most data on membrane fusion, with the possible exception of the distal monolayer effect of LPC. However, the modified stalk theory has the important deficiency of predicting an unrealistically high energy for the intermediates. The modified stalk theory also raises unanswered questions about the evolution of fusion intermediates, especially the extent of hemifusion diaphragm formation in the absence of nonpolar oils.

8.2.3.1.3   Third-Generation Stalk Models

The "energy crisis" identified by the modified stalk model inspired several authors to look for ways in which the modified theory had overestimated the intermediate energy [158, 163, 188–190]. This set of models corrected the modified theory by introducing two new elements.

The first element is reduction of the bending elastic energy of the "neck" of the stalk. This is the region of the stalk nearest the axis of cylindrical symmetry, and it is the region where most of the unfavorable bending elastic energy is concentrated. Markin and Albanesi [189] recognized that the stalk would adopt a shape with lower bending elastic energy than the circular toroidal surfaces assumed in earlier versions of the stalk theory. They showed that the monolayers in the neck region can adopt a shape with zero curvature energy for every choice of C_0, which they refer to as a "stress-free" shape. This substantially reduces the curvature energies of stalks and intermediates that evolve from them early in the process of radial expansion. Kozlovsky and Kozlov [158] achieved this through a more subtle approach described later.

The second element is the introduction of gradients in chain tilt [158, 163, 188–190]. These models allow lipid molecules to tilt with respect to the local surface normal, and for gradients in chain tilt to exist along the monolayers. Monolayer domains with opposing tilts can be fitted together at the edges of domains with opposing directions of tilt so as to form nonsmooth lipid/water interfaces. This permits construction of stalklike intermediates with lower energies. In the case of Kozlovsky and Kozlov [158], a stalk is constructed that essentially lacks the high-energy neck region of earlier models, where the requirement for smooth interfaces (no tilt gradients or discontinuities) produced high-curvature stalk necks. In addition, introduction of tilt and tilt gradients

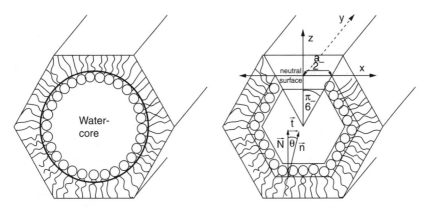

FIGURE 8.10 *Left:* Conventional structure of a unit cell of the H_{II} phase, assuming a homogeneously curved lipid/water interface and packing of the hydrophobic interstices (the corners of the hexagonal cross section) by local stretching of the lipid acyl chains. *Right:* Structure of the H_{II} phase when gradients in chain tilt exist in the lipid monolayers. The unit cell consists of six identical flat fragments of monolayer. (Reprinted from [107], with permission.)

provides an additional degree of freedom for the system to minimize deformation energy of curved monolayers [158, 163, 190].

The effect of chain tilt gradients is treated in most detail by Kozlovsky and Kozlov [158, 163]. Kozlov and Hamm previously showed [107, 108] that inclusion of tilt deformations with conventional bending elasticity is a consistent way to deal with free energy of lipid monolayers in inverted phases, and is consistent with much of what we know about phospholipid phase behavior. In this formalism, hydrophobic interstices like those in the H_{II} phase do not exist. Instead, a combination of gradients in chain tilt and bending deformations in the monolayers results in monolayer junctures that remove those interstices. For example, the H_{II} phase is viewed as consisting of hexagonal prisms of monolayer, instead of tubes lined with linear interstices [107] (see Figure 8.10). The unfavorable free energy required to stabilize (partially fill) the interstices via chain stretching in the previous simple bending model (Section 8.2.1) is instead expended in producing gradients in chain tilt along a monolayer. It is probably more apt to use the more general term of frustration energy [16] for this energy, because it arises due to the need to change the connectivity of the bilayers. Just as in the conventional view of inverted phase stability (e.g., [53]), the presence of a nonpolar oil would reduce the frustration energy by removing the necessity to produce the gradients in chain tilt. In this case, presumably the H_{II} phase would revert to a circular cylindrical geometry, with only simple bending elastic energy (Figure 8.10b) as a free energy with respect to the L_α phase [107]. Nonpolar oils would reduce the free energy of fusion intermediates for similar reasons.

Whether or not chain tilt gradients act to stabilize hydrophobic interstices in H_{II} phases as proposed in [107, 108] may be difficult to determine experimentally. The X-ray diffraction reconstructions of the H_{II} phase unit cell that are currently available [203, 204] do not have sufficient resolution to answer this question, although they do suggest that the monolayer tubes have at least a slight hexagonal distortion. (Parenthetically, an L_α/H_{II} phase transition mechanism is suggested in [203] that is not based on stalks. However, the inverted micellar intermediates involved would have much higher curvature energies than stalks [69], which are observed to form spontaneously under similar circumstances [192–194]. As noted in [29,70], only a very small fraction (ca. 10^{-3}) of the lipid in the system would have to exist as stalk or TMC structures at any instant to mediate a rapid phase transition. It is doubtful whether the contribution of such a small number of transient structures would have been registered in the diffraction patterns in [204].)

Tilt gradients and bending both give rise to chain splay, and the splay elastic energy can be calculated using the same measured elastic coefficients as simple bending [107, 108]. The analog to the spontaneous curvature, C_0, is the spontaneous splay, $\tilde{J}_s$, which is identical in value to C_0 as measured in H_{II} phases [48, 49]. The tilt elastic modulus can be calculated using dimensions of the H_{II} phase lattice at T_H and the curvature elastic modulus [107].

On surfaces with compound curvature like stalks, the interplay of splay arising from tilt gradients and simple bending can result in lower curvature energies than estimated on the basis of simple bending alone [158]. This is elegantly shown in Figure 8.11 (reproduced from [158]) and in Appendix A of [158]. Chain tilt was discussed in [189] as an alternative to the void energy estimation procedure used by Siegel [70]. A similar concept was also introduced in [188].

The introduction of chain tilt gradients allows the presence of nonsmooth interfaces in the intermediates [158, 163, 188, 190]. This allows segments of monolayer to come together at nonzero angles, eliminating hydrophobic interstices and also reducing the area of an intermediate (which can often reduce the curvature energy, especially of the distal monolayers). This effect makes substantial differences in the intermediate structure relative to the previous stalk models, as shown by the stalk structure in Figure 8.12 from Kozlovsky and Kozlov [158]. Note that the high-curvature, high-energy neck of the stalk is absent in this structure (compare with Figure 8.1B). Note also that the distal monolayers are already in single-point contact in this structure. In contrast, in the modified stalk theory [69, 170] the stalk must radially expand into a TMC to achieve any contact between these two monolayers (Figure 8.1). This raises the possibility of direct stalk-to-fusion pore conversion. However, the probability of rupture of the contact is proportional to its area, and the small area of the distal monolayer contacts in Figure 8.11 makes direct stalk/fusion pore conversion unlikely. Besides chain tilt gradients, May [190] adds acyl chain stretching to his model. His results are interesting because they show that inclusion of a narrow tether (cylindrical section) of lipid at the thinnest portion of the stalk can also reduce the chain-packing energy. This seems to improve the efficiency of chain packing with minimal chain stretching.

The use of stress-free surfaces [189] and nonsmooth interfaces [158, 163, 190] has two substantial influences on the energy of stalks and subsequent intermediates:

The first is that the new elements drastically reduce the intermediate energies. In the modified stalk model [70], stalks with $J_s = -0.1$ nm^{-1} had energies ranging between 40 and 200 k$_B$T, depending on the size of r, whereas the energy of stalks in [158] is about 45 k$_B$T for all choices of the separation. The energy decreases rapidly with decreasing J_s.

The second is that the dependence of the energy of stalks and radially expanded stalks on the local separation between the bilayers is more complicated than in previous models. For example, in [158] the energy of the intermediates is not very sensitive to the local interbilayer separation, as long as the radius of the stalk at the base is allowed to vary. In [189], the explicit dependence of the stalk energy on the interbilayer separation was not given, but there is a specific separation that minimizes the stalk energy for each value of C_0. In contrast, in the modified stalk theory [70] the energies depend strongly on the local bilayer separation through the size of the radius in the plane of the stalk axis, r. In [69, 70], the energy of stalks increases monotonically with increasing r, and there is an optimal r value for TMCs at each value of C_0.

There are significant differences among the third-generation models concerning radial expansion of stalks, which have substantial effects on fusion and phase transition behavior. Some [188, 189] propose the same sequence of events as Siegel [69, 70], as depicted in Figure 8.1, whereas May [190] only discusses stalks. Markin and Albanesi [189] attempted to plot the energy of a stalk intermediate as it radially expanded, in order to plot the energy vs. intermediate size and shape for the fusion process [189], but they did not seem to deal explicitly with the change in geometry between the stalk with two voids and the TMC with one ring-shaped void. They also showed that formation of TMCs by radial expansion of stalks is unlikely unless the C_0 value is substantially less than zero.

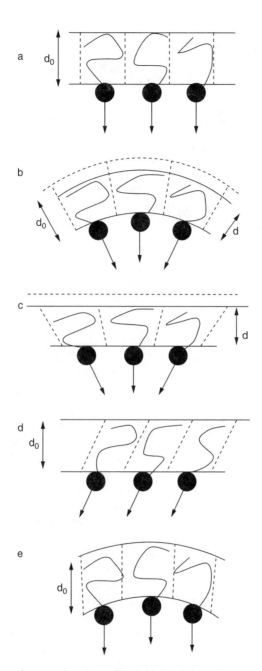

FIGURE 8.11 Deformation of a monolayer. (A) The initial undeformed state. (B) Bending with vanishing tilt. (C) Splay of the chains resulting from changing tilt (a tilt gradient) in a flat monolayer. (D) Homogeneous tilting of chains. (E) Mutual compensation of bending and tilt resulting in a bent monolayer with vanishing total splay of the chains. (Reprinted from [158], with permission.)

Kozlovsky et al. [163] make detailed predictions about the tendency of the stalk to expand as a function of the spontaneous splay and local membrane separation. Their work predicts that stalks have an increasing tendency to expand with decreasing $\tilde{J}_s$. For a particular range of $\tilde{J}_s$, a hemifusion diaphragm (HD) (Figure 8.7D) of restricted diameter forms, with unlimited expansion of the diameter occurring for lower values of $\tilde{J}_s$. (In practice, the extent of HD expansion will be limited by growing transbilayer area asymmetry or liposome size.) The range of $\tilde{J}_s$ for restricted

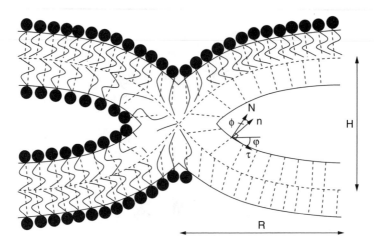

FIGURE 8.12 Stalk incorporating chain tilt gradients. There are no extensive hydrophobic interstices present (compare to Figure 8.1), and the distal monolayers are in contact at a single point in the center of the structure. (Reprinted from [158], with permission.)

HD diameter depends on the local bilayer separation, but it corresponds roughly to -0.19 to -0.3 nm^{-1}, with unlimited expansion for smaller values. Pore formation within the bilayer diaphragm is driven by curvature and chain-packing stress inherent in the intermediate structure, similar to the proposition of the modified theory [69, 70]. A detailed calculation of this lateral stress is done in [163], which shows that the stress is similar to the tensions observed to rupture macroscopic bilayers, and that pore formation should occur near the three-way bilayer junctures at the edge of the hemifusion diaphragm. As in [70], the tension in the diaphragm walls decreases as $\tilde{J}_s$ decreases, making rupture to a fusion pore likely for modest curvatures, but less likely for systems near the L_α/H_{II} phase boundary (which is at $\tilde{J}_s \approx -0.34$ nm^{-1} for lipids like DOPE). The spontaneous curvature decreases with increasing temperatures or increasing inverted phase lipid composition, so that one can approach the L_α/H_{II} phase boundary by increasing T and/or increasing the mole fraction of lipids like PE. The combination of lowered lateral tension in the HD walls and increasing tendency for HD expansion with decreasing $\tilde{J}_s$ suggests that extended hemifusion intermediates in pure lipids should only be observed for systems near the L_α/H_{II} phase boundary.

The prediction of unlimited stalk expansion for $\tilde{J}_s$ values < -0.3 nm^{-1} also has important implications for our understanding of the L_α/H_{II} and the L_α/Q_{II} phase transition mechanisms. Predicting the detailed mechanism of these phase transitions was not an aim of [158, 163], but the predicted evolution of intermediates past the stalk stage is relevant to these mechanisms. It is not obvious that unlimited radial expansion of stalks would form well ordered domains of H_{II} phase on the observed time scale of this transition (milliseconds, when subjected to large temperature jumps [28]). The rims of the expanded stalks are circular, whereas the H_{II} phase consists of linear bundles of tubes, and it is not clear how the expanding rims on alternate sides of a given bilayer could rapidly be brought into register, as must happen to produce H_{II} unit cells. In contrast, the modified stalk theory mechanism [29,70] (Figure 8.8) produces H_{II} domains directly from an aggregate of TMCs. The third-generation models [163, 189] predict that structures like TMCs are much lower in energy than in the modified stalk model [70], which makes the TMC aggregate transition mechanism much more viable, provided that the TMCs do not immediately expand into very large hemifusion structures. As mentioned in Section 8.2.3.1.1, an additional difficulty with rapid radial expansion of stalks past the TMC stage is that it is then not clear how ILAs form in large numbers in systems below T_H. It is also not easy to explain how Q_{II} phases form in the context of this model. Clearly, more information is required on the evolution of intermediate structures when lipids are near T_H.

8.2.3.1.4   Stalk Models: Summary

The predictions of the third-generation models [158, 163, 189, 190] are compatible with most observations of fusion behavior. They also predict stalk and intermediate energies that are low enough to be compatible with observed transition and fusion rates, while incorporating most of the major energetic terms in their equations. This makes them a substantial advance over the modified stalk theory [69, 70]. However, there are still four apparent inconsistencies between these theories and observations of fusion and inverted phase behavior. Further experimental work on the basis of these inconsistencies will improve our understanding of lipid fusion intermediates and how to use them to study protein-induced fusion.

First, two of the third-generation models [158, 163, 189] now predict unobserved phases of stalks (or radially expanded stalks) in excess water at $\tilde{J}_s$ values that are intermediate between L_α and H_{II} phases (in [158], < -0.26 nm^{-1}). This is qualitatively similar to predictions of the first versions of stalk theory (e.g., [182]). This prediction probably arises because some unknown component of the free energy is still missing from our models. The missing term may be the saddle splay (Gaussian curvature) energy. A very recent analysis suggests that the saddle splay energy contribution to the total stalk energy could be as large as the contribution of splay energy, and would tend to inhibit stalk formation [Kozlovsky, Siegel, and Kozlov, manuscript in preparation]. The model of May [190] predicts somewhat higher energies for stalks as a function of C_0 throughout the range of -0.3 to $+0.3$ nm^{-1} (for stalks that form between flat bilayers), at least for one particular set of materials constants.

Second, as discussed in Section 8.2.3.1.3, unlimited expansion of hemifusion diaphragms is predicted to occur for $\tilde{J}_s <$ ca. -0.3 nm^{-1} in pure lipid systems by one version of this theory [163], and it is not clear that these form in the absence of nonpolar oils. Neither is this prediction readily compatible with rapid formation of ILAs and Q_{II} phases in many lipid systems (see [70]). The reasons for this apparent discrepancy are not clear. It is unlikely that saddle splay (Gaussian) curvature plays a role in this, because this curvature energy depends only on the topology of the surfaces in the structure [129], which does not change during radial expansion of a stalk. In contrast, chain-packing energy is very important in restricting radial expansion, and it is possible that our models of this energy are still too approximate.

Third is the challenge posed by the demonstration of rhombohedral phases consisting of arrays of stalks in stacks of flat bilayers [192–194]. Yang et al. recently published a DOPC/DOPE/water phase diagram of the low water–content region containing this phase [194]. These phases do not form in excess water or in pure DOPE, in contrast to predictions of the third-generation models [158, 163, 189]. Preliminary work by Kozlovsky et al. (manuscript in preparation) indicates that significant features of this DOPC/DOPE/water phase diagram can be reproduced by accounting for the decrease in hydration repulsion between lamellae that accompany stalk formation, and by including a saddle splay curvature term in the expression for the stalk free energy.

Fourth, at least one third-generation model [189] now predicts that fusion pores (ILAs) are thermodynamically stable in systems in which they are not observed: lipid systems with curvatures as large as -0.05 nm^{-1} should form ILAs [189]. In contrast, DOPC, which has a lower C_0 of -0.115 nm^{-1} near room temperature [170], is not observed to form ILAs. The reason for this may be a combination of neglect of Gaussian curvature and the neglect of chain-packing energies in bilayer labyrinths [130, 131]. It could be that a combination of studies of the rhombohedral phase and ILA stability for DOPE/DOPC systems will set limits on the value of the Gaussian curvature elastic modulus for these two important model lipid systems.

A thread running through all of these remaining difficulties is the lack of good models for estimating Gaussian curvature and chain-packing energies [130, 131]. Currently, there is no practical way to measure the Gaussian curvature elastic modulus. More experimental and theoretical work on the relative stability of lamellar and inverted phases is necessary for us to progress in this area. Our current models of the curvature elastic energy are based on an approach that is formally applicable only to surfaces where the radii of curvature are much larger than lipid molecules [46]. For small structures like stalks, we may have reached the limit of applicability of such treatments.

Some other interesting observations of fusion behavior are not readily explained by the stalk theory. It is not immediately obvious why the fusion pores should flicker in the manner observed in GUV/planar bilayer fusion [148] and PEG-induced SUV fusion [145], as fusion pores are generally predicted to be considerably more stable than the intermediates from which they form and should continuously reduce their free energy by expanding (e.g., [70, 189]). Tension seems to be necessary to ensure that nascent fusion pores open (e.g., [148]).

8.2.3.2 Cation/Anionic Lipid Fusion Mechanisms

In several anionic lipid systems, divalent cations like Ca^{2+} and Mg^{2+} induce very rapid and extensive liposome lipid mixing, fusion, and aggregation-mediated leakage at neutral pH. The principal examples are phosphatidylserine (PS), phosphatidylglycerol (PG), and phosphatidic acid (PA). The very extensive studies on these systems have been reviewed by skilled practitioners (e.g., [10–13]) and will not be reviewed again here. Only some general principles will be described. Rapid fusion and the other aggregation-mediated phenomena in these systems occur even though there is no inverted phase in the phase diagram of the relevant lipids at neutral pH. In pure lipid systems, one would expect simple stalk-mediated processes to be very slow far from a lamellar/inverted phase boundary. Hence a fusion mechanism other than a simple stalk mechanism is probably at work in these cases. Even in the case of CL, where the cations drive formation of inverted phases, there is some evidence that another mechanism is also at work. For Sr^{2+} and Ba^{2+}, the aggregation-mediated events are correlated with T_H of the cation-lipid complex, consistent with a stalk-mediated model [60]. Presumably, the binding of these ions changes the C_0 of the lipid system, as well as removing the electrostatic repulsion between bilayers. However, for Ca^{2+} and Mg^{2+} the fusion mechanism appears to be different from a simple stalk model. As discussed in Section 8.2.1.1.1, Ca^{2+} and Mg^{2+} induce a much higher extent of fusion than do Sr^{2+} and Ba^{2+}, the fusion rates are not well correlated with the lamellar/inverted phase transition temperatures, and the lamellar-to-inverted phase transition kinetics are slower for Ca^{2+} and Mg^{2+} [60]. Hence, Ca^{2+} and Mg^{2+} may be working by another mechanism, instead of or in parallel with a stalk-type mechanism.

In the case of Ca^{2+}-mediated fusion of PS, fusion occurs via formation of a dehydrated Ca^{2+}-PS complex between the opposed liposomes. Prolonged incubation of the liposomes leads to formation of a cation-PS cochleate phase (which is a dehydrated multilamellar gel phase with frozen acyl chains [206, 207]). The Ca^{2+} is bound quite tightly in the dehydrated equilibrium phase, with a melting point in excess of 100°C [208–211]. Once formed, this phase is in equilibrium with submicromolar concentrations of the cation [212, 213], but millimolar-range Ca^{2+} concentrations are necessary to nucleate it between PS LUVs. However, formation of precursors to a cochleate phase is not required for fusion of PS as induced by other cations. PS fusion also occurs in the presence of Sr^{2+} and Ba^{2+} at temperatures either above or below the melting point of the cation-PS complexes [214, 215]. Furthermore, Mg^{2+} is not capable of fusing LUVs [216]. Clearly, the specific nature of the cation–headgroup interactions is important in this mechanism. Interestingly, when a 1:1 mixture of egg PE and PS was used, Mg^{2+} was able to induce fusion, fusion rates increased sharply with temperature around T_H, and more leakage occurred earlier than in comparable PS LUVs [214]. This suggests a changeover to an inverted phase-sensitive (stalk-based) mechanism. In the presence of Ca^{2+}, PS fusion always eventually results in leakage of the liposomal contents, because of eventual transformation of the LUVs into the cochleate phase. However, the first few rounds of fusion are almost completely nonleaky (reviewed in [13]). There is especially little leakage in the first several Ca^{2+}-induced fusion of DOPA/DOPC liposomes [217] and in La^{3+}-induced fusion of PS [218]. Oddly, in these two systems, addition of EDTA at late times in LUV/LUV interactions results in nearly complete lysis of the vesicles fused until that point.

Fusion in systems like PS, and in fusion events where the liposomes bear opposite charges (e.g., [146, 147]), has often been suggested to occur via a mechanism like the adhesion–condensation mechanism of Kozlov and Markin [171, 173]. The free energy of adhesion of the PS bilayers due to divalent cation binding or to the adsorption of oppositely charged membranes on one another, is estimated to be many erg/cm² (e.g., [167]), in contrast to the much lower adhesion energies found for uncharged PE membranes (ca. 0.15 erg/cm² [197]). These large adhesion energies result in extensive deformation of the vesicles in order to form as much membrane–membrane contact as possible. This generates large tensions in the membranes, which can be large enough to cause membrane rupture and contents mixing between the two compartments [166, 167]. (Formation of the Ca^{2+}-PS complex also reduces the area/lipid molecule at the area of contact between the liposomes, which creates dilatational stress in the outer monolayers.) The structure of the first intermembrane contacts is unclear. If the facing monolayers rupture, lipids from the distal monolayers might merge to form a single-bilayer diaphragm. This diaphragm may subsequently rupture because of the tensions in the two original membranes, resulting in aqueous compartment mixing. It is also possible that rupture of the facing and distal monolayers is a concerted event: rupture of the opposed monolayers is likely to produce a defect at the same site in the distal monolayers, due to coupling between monolayers of the same bilayer. Evidence for monolayer coupling comes from observations of the melting behavior of PS bilayers that are exposed to Mg^{2+} on only one side vs. both sides. Rather than observing independent melting of the lipid/cation complex in each monolayer, when Mg^{2+} is on only one side of the membrane the observed chain melting temperature is intermediate between the melting temperatures for bilayers exposed to Mg^{2+} and Mg^{2+}-free media on both sides [219]. This suggests that rupture of the facing monolayers in the contact area between two PS LUVs will have an immediate and local effect on the distal monolayers, so that pore formation may occur rapidly thereafter, at the same site.

The relevance of the PS membrane fusion (adhesion–condensation) mechanism to biomembrane fusion is unclear. It has often been cited as a possible mechanism of stimulus–secretion coupling. Stimulus of many cells triggers a rise in intracellular Ca^{2+} concentration, which is followed by fusion of secretory granules with the cell plasma membrane, and it has been postulated that the Ca^{2+} acts by inducing fusion between patches of lipid in the secretory granule and cell plasma membranes that are rich in acidic lipids like PS and PA. PS typically makes up only 5 to 9 weight% of the lipids of mammalian cell membranes, with almost all of the PS present in the cytoplasmic leaflet under normal circumstances [220]. The comparatively low concentration of PS is a serious constraint, because lipids like PC and sphingomyelin tend to inhibit PS/PC LUV fusion (e.g., [221]). If there is a preponderance of PC or sphingomyelin present, fusion will be unlikely to occur via the mechanism observed in pure PS systems. If a preponderance of PE is present, fusion will likely occur via stalk-type mechanism (if sufficient ion binding occurs to neutralize the negative charges on the membranes and allow close opposition). It is unlikely that Ca^{2+}-driven lateral phase separation of patches of nearly pure PS will occur under biological conditions. Physiological concentrations of Ca^{2+} generally do not produce rapid phase separation of unsaturated chain PS in mixtures with other unsaturated lipids in isolated vesicles (especially PC or sphingomyelin; see references in [13]), although rapid lateral phase separations can occur in PA/PC liposomes [217]. Moreover, the Ca^{2+} concentration necessary to induce fusion of PS liposomes is generally in the millimolar range, which is much higher than the intracellular concentration (micromolar range). It is conceivable that the actions of proteins could remove some of these constraints. For example, the threshold concentration for fusion of PS liposomes can be reduced to about 10 micromolar by action of aggregation-inducing proteins like synexin [222–224]. It is also possible that localized enzyme action with restrictions on diffusion of the products could produce locally high concentrations of acidic lipids. In most biomembrane systems, however, the lipid composition seems more conducive to a stalk-based mechanism, although divalent cations may certainly play a role in permitting close opposition of the membranes.

8.3 PROSPECTS

8.3.1 WHAT MUST WE DO TO BETTER UNDERSTAND LIPID FUSION MECHANISMS?

We can only trust our models of fusion intermediates if they are consistent with observed lipid phase behavior: our understanding of intermediate energies comes from studies of the relative stability of lamellar and inverted phases. As discussed in Section 8.2.3.1, our models are not fully consistent with some aspects of this behavior (relative stability of ILAs, Q_{II} phases, rhombohedral phases, and H_{II} phases). Most of the inconsistency is probably due to our incomplete understanding of Gaussian curvature elastic modulus and chain-packing energies. Information on these is likely to come from carefully designed studies of lamellar/nonlamellar phase stability. The recent discovery that the rhombohedral phase is a stalk-based structure [192–194] is a critical development. Experimentally, this phase could be used as a sort of fusion intermediate laboratory, to explore the effects of different additives (e.g., peptides of different types, nonpolar oils, polyvalent cations) on the stability of stalks. This should be very helpful in understanding the role of different parts of fusion proteins and of fusion cofactors in catalyzing fusion.

There is currently disagreement about the nature of stalk evolution: one version of the third-generation stalk models [158, 163] predicts unlimited expansion of hemifusion diaphragms when the spontaneous splay is < ca. −0.3 nm^{-1}, whereas the modified stalk model [69, 70] suggests that radial expansion does not proceed much further than the TMC, even at similar values of C_0. There are data supporting this latter view in some fusing systems (e.g., [148]) and systems that form ILAs and Q_{II} phases (reviewed in [70]). A careful study of intermediate evolution in pure lipid systems with and without nonpolar oil (e.g., a few mol% of hexadecane) might establish the circumstances in which stable hemifusion intermediates are likely to form.

Time-resolved cryoelectron microscopy (e.g., [22, 27–31, 33]) can be a very powerful tool for mechanistic studies for systems in excess water and might be a useful technique for such studies.

Our understanding of the factors controlling defect formation (interfacial disruption at the start of stalk formation), pore formation, and rupture [224a, 224b] in bilayers is also fragmentary. Studies of pore formation in isolated bilayers under different types of stress are needed. For example, Melikov et al. [225] recently found evidence for a population of "pre-pore" defects in planar membranes stressed by transbilayer electric fields. The nature of these defects is unknown, and they could have a strong influence on fusion through destabilization of hemifusion intermediates. It would also be interesting to know how peptides or other cofactors affect their stability and tendency to convert to membrane pores.

8.3.2 SIMULATION TECHNIQUES

In recent years, there has been considerable progress in molecular modeling and increasing interest in modeling membrane fusion. This is very exciting. In principle, molecular modeling techniques can yield information on processes that are unlikely to be directly observable, due to the short (nsec) half-lives and small sizes of the intermediates. This can provide useful checks on theoretical models and generate new experimental designs. However, it is important to consider how the assumptions and restrictions of the calculations affect interpretation of the results. Not all hypotheses about fusion are testable with current modeling capabilities.

Recently, Marrink and Tieleman published an atomistic molecular dynamics simulation of an inverted cubic phase in glycerol mono-oleate [226]. The simulation revealed the transition of the Q_{II} phase to the H_{II} phase by way of a number of intermediates with different topologies. The interconversion of intermediates involved the formation of TMCs and bilayer rupture events, which produce changes in the topology of the water channels in the structures. The apparent mechanism of the transition was similar to the one predicted by the modified stalk theory [70], in the sense that TMCs are involved and that a cubic array of water channels should exist at one point in the transition, as was observed in [226]. It is encouraging that the simulation shows formation of stalk and TMC-like structures, indicating that these structures violate no basic principles of intermolecular interactions in relevant lipid systems. The fact that the simulation by Marrink and Tieleman can generate a Q_{II}-to-H_{II} phase transition is a computational triumph.

Some limitations of the current methods must be noted, however. Atomistic molecular dynamics simulations like the one in [226] are computationally limited: the calculations are extraordinarily complex, and only a small unit cell can be simulated. This can restrict the range of structures that can appear. For example, the periodic boundary conditions used in simulations like the one in [226] can be a problem. Marrink and Tieleman noted that the periodic boundary conditions applied to the simulation tend to amplify fluctuations and may not do a good job of simulating cooperative phenomenon: "Phase transitions are cooperative phenomena over significant length scales, but because we simulate only one unit cell, cooperativity is likely to be enhanced, making the events fast in comparison with truly macroscopic systems" [226]. Moreover, if the fusion intermediate is larger than the size of the simulated cell, the periodic boundary conditions (which surround a cell only with copies of itself) may not permit formation of structures with larger unit cells. For instance, Marrink and Tieleman [226] used a cell with an edge length of 7.4 nm and noted that the TMC array proposed [29,70] as an intermediate in transitions to the H_{II} phase could not be observed in their simulation because of the size limitations. Finally, it was noted in [226] that the simulation method did not permit the original Q_{II} phase to change its surface area through changes in the unit cell dimensions. This is clearly an important degree of freedom available in real systems.

Moreover, in order to simulate accurately the process of membrane fusion, the simulation techniques must be very accurate in modeling the energies of large assemblies of molecules. Although great progress has been made, more may be necessary. In [226], for example, the simulated Q_{II} phase was unstable, although that phase is observed to be stable under the simulated conditions [227]. Marrink and Tieleman [226] stated that their technique was presently not accurate enough to simulate the details of the phase transition, noting that the free energy differences between the different Q_{II} phases and the H_{II} phase are small. The *enthalpy* of the Q_{II}/H_{II} transition that occurs at a higher temperature is only 1 kJ/mol, or < 0.4 $k_B T$ per molecule, so the *free energy* difference at temperatures near the transition is smaller than this figure. This degree of accuracy is problematic for the study of fusion intermediates. Fusion intermediates contain hundreds of lipid molecules (e.g., a patch of two opposed bilayers that is 5 nm in diameter contains about 500 phospholipids). An inaccuracy of 0.1 $k_B T$ per molecule in free energy of a lipid in an intermediate can lead to inaccuracies of 50 $k_B T$ in the energy of the entire fusion intermediate. This is larger than the energy difference between the lamellar phase and stalks for many lipid compositions according to the most recent models [158, 163, 189, 190].

Coarse-grained Monte Carlo simulations like those of Müller, Katsov, and Shick [228, 229] and Marrink and Mark [229a] can simulate larger structures. In such studies, the complexity of the calculations is reduced by modeling each of the molecules as a small number of beadlike subunits. Müller et al. [228, 229] have suggested an interesting fusion mechanism on the basis of a simulation designed to reproduce the elements of interactions in lamellar phases composed of either block copolymers or lipid–water moistures. Their simulations showed that fusion could occur by two mechanisms. One looked very much like a stalk-pore mechanism. The other involved formation of pores in individual bilayers. Müller et al. suggested that stalk formation weakens the bilayer so that pores form in the individual membranes adjacent to the stalk. The stalk structure subsequently extends into a more linear structure in the plane of the membrane, forming a curtain that will surround the single-bilayer pore, creating an ILA-like structure.

The coarse-grained simulation by Marrink and Mark [229a] indicates that fusion of two small vesicles proceeds by the stalk mechanism, with formation of stalks with no apparent hydrophobic voids and formation of a hemifusion diaphragm that subsequently forms a fusion pore, essentially as predicted in [158, 163]. The properties of the vesicles were changed to approximate different lipid compositions (pure POPE, pure DPPC, or mixtures of DPPC with 25% DPPE or lyso-PC). Interestingly, for pure POPE, the stalk and HD formed rapidly, but the HD was stable for a prolonged period (100 nsec), whereas the HD ruptured more quickly for the PC/PE mixture. Pure DPPC vesicles formed only stalks, and the vesicles had to be more closely opposed to do so. Lyso PC on the interior monolayers accelerated rupture of the HD. These last results are all compatible with the overall spontaneous curvature dependence of fusion observed (e.g., [7]) and predicted by the stalk theory [85]. The simulation by Marrink and Mark [229a] also shows that a transient pore can form in one vesicle after stalk formation, although in this case the pore is highly transient (5 nsec) and does not participate in the fusion event. After the pore joining one vesicle interior to the surrounding water closes, the stalk proceeds to form an HD and a fusion pore, as in the stalk theory. At least a few lipid molecules on the outer monolayer of one vesicle can enter the inner monolayer of the same vesicle when the transient pore forms. This seems compatible with observation of an outer-to-inner monolayer lipid flux in PEG-induced fusion of PC vesicles [154, 157].

The most notable aspect of the results in [228–229a] is the transient formation of a pore in the wall of one or both vesicles (leakage) prior to fusion, although this did not always happen and was not absolutely required for fusion to occur in either simulation. It is possible that transient formation of pores in the individual bilayers precedes fusion in real systems or forms in the stalk itself, compatible with the simulation results. Leakage is often observed in parallel with fusion in lipid systems, because the factors that drive fusion also induce considerable tension in the bilayers. Two principal examples are

- Osmotic and adhesion stress in PEG-induced fusion [153]
- Adhesion energy-induced stress in the fusion of oppositely charged bilayers [146–148], divalent cation-induced fusion of anionic lipids [10–13], and inverted phase lipids [197]

There are also indications that viral fusion protein-mediated events are in fact leakier than previously thought (e.g., [230]). However, it should also be noted that nonleaky fusion is not essential to viral propagation in many cases (influenza merely needs to enter the cytoplasm from endosomes), and the proteins may not be "designed" to achieve it. Moreover, nonleaky fusion is observed in some model membrane systems, at least in the first several rounds of fusion (e.g., systems cited in [10]). The occurrence of some degree of leakage during many fusion events cannot be ruled out on the basis of present data. It is always possible that the lifetime of the leaky intermediate is shorter (e.g., submicrosecond) than can be detected with present technology. The predicted pore lifetime in [229a] is only 5 nsec.

It must be noted that coarse-grained simulation methods also have limitations. The coarse-graining procedure involves parameterization of the interactions of the groups making up the molecules, in order to reflect faithfully the behavior of real systems. One example of parameterization is the selection of the size of the contact energy between hydrophilic and hydrophobic parts of the lipid molecules. The choice of this value seems subject to inexactly formulated criteria in [228, 229] but has large effects on the interfacial width and local structure of the bilayers, as well as (presumably) on the rupture tension of the bilayers. Another is choice of the relative size of the hydrophilic and hydrophobic groups of the lipids, which can generate very different sorts of apparent phase behavior (i.e., the types of structures that form after extended times [228, 229]). This raises concerns about how well the simulations reflect actual lipid membrane behavior, especially when fusion intermediates are increasingly found to be close to the lamellar phase in free energy for a wide variety of system compositions [158, 163, 188–190].

In addition, to observe fusion in a practically computable length of simulated time, researchers are often forced to stress the initial membranes so as to destabilize them. One must be sure that the magnitude of the applied stress is relevant to real systems, or one may observe behavior that is not typical of real membranes. The bilayers in [229] were dilated to an area 19% larger than the stress-free area (corresponding to action of a very large lateral tension). By comparison, most macroscopic phospholipid bilayers rupture if dilated by 3% or less [197–200]. Indeed, the bilayers simulated in [229] frequently underwent spontaneous rupture in the absence of fusion. Similarly, the original vesicles in [229a] had diameters of only about 15 nm. As the authors noted, this is smaller than the smallest SUV (20 nm). This suggests that the bilayers were under great curvature stress. One has to wonder if the transient formation of pores (leakage) in the simulations [228–229a] is due to extreme stresses applied to the original membranes. The pores might otherwise be too high in energy to form in real membranes.

Advances in modeling techniques will be followed eagerly by researchers in the fusion area.

8.3.3 THE "JOB DESCRIPTION" OF FUSION-CATALYZING PROTEINS

The rate of fusion in lipid systems that are not close to a lamellar/nonlamellar phase transition (e.g., phosphatidylcholine), and that are not mostly anionic lipid in the presence of millimolar concentrations of divalent cations, is generally very small. However, proteins seem able to mediate fusion even of fairly refractory lipid compositions. For example, influenza virus is able to fuse fairly rapidly with liposomes composed of DOPC and a small amount of ganglioside [231] and even (more slowly) with pure saturated-chain phosphatidylcholine [232]. Fusion-mediating proteins such as influenza hemagglutinin must be able to help the lipids of the system overcome several energetic and kinetic barriers to fusion. The weight of evidence is that, at least in some systems, protein-mediated fusion proceeds by formation of lipidic structures, rather than by initial formation of a protein-lined fusion pore (e.g., [3, 5, 6, 92, 233]). The fusion-mediating proteins must be able to help the lipid systems overcome energetic and kinetic barriers to fusion.

The barriers to fusion and some possible roles of proteins in overcoming them are described as follows (see also [9a]):

1. Close opposition of the membrane interfaces against strong repulsive hydration forces and against the steric hindrance imposed by the presence of large macromolecules on biomembrane surfaces, so that intermembrane structures can form. This may involve deformation of the membranes, which requires an input of curvature energy. It has been proposed that proteins around the fusion site act collectively to change the membrane curvature [234, 235], forming "dimples" or "hats" that locally deform one of the opposed membranes toward the other, reducing the local intermembrane separation. Interestingly, most of the curvature energy of the structure is concentrated in a small-radius spherical "cap" at the tip of the protrusion, which would be especially fusogenic and in closest proximity to the target membrane. The necessary energy comes from energy stored in fusion protein conformation, which is released by the conditions that trigger fusion (in the case of influenza hemagglutinin, by a drop in pH).

2. Production of faults in the membrane/water interfaces at which intermediate formation can begin (this may be related to the first step [186]). Insertion of hydrophobic peptides into the membrane interfaces may nucleate such defects, especially the so-called viral fusion peptides [2, 236].

3. Creation of the stalk intermediate. For lipid compositions that are not at or near the lamellar/inverted phase boundary (probably true for most biomembrane compositions, and certainly true for PC), this requires an input of curvature and chain-packing energy. The third-generation stalk models [158, 163, 188–190] predict that the curvature and chain-packing energy inputs for this step will be tens of k_BT. The modes of coupling protein conformational free energy into membrane deformation suggested by Kozlov and Chernomordik [234, 235] both create close opposition of the relevant membranes (step 1) and promote stalk formation. Intriguingly, these models concentrate most of the

curvature energy input into the membrane at a small, high-curvature cap that is brought into close proximity to the target membrane. This is just what is required to make the regions of membrane that contact each other most susceptible to fusion. Moreover, they appear to raise the energy of these small areas of bilayer by approximately the amount necessary to form stalks according to the third-generation stalk models. Both the collective action models in [234, 235] require formation of large (tens or even hundreds of nm) structures involving large numbers of fusion proteins. These are ideal targets for time-resolved cryoelectron microscopy, and a careful search for such morphology under fusion-triggering conditions seems in order.

It has been suggested that viral fusion peptides decrease the spontaneous curvature of membranes into which they are inserted, favoring fusion intermediate formation (see [237] for a review). However, the situation is probably more complicated: an influenza hemagglutinin fusion peptide either had no effect or slightly *increased* the spontaneous curvature of the host lipid [31]. Bentz [238] has suggested an interesting fusion mechanism in which energy stored in the fusion protein conformation is coupled directly into producing a large (nm-range size) hydrophobic defect in the viral membrane, which presumably recruits lipids from the target membrane to form an intermembrane intermediate, subsequently evolving in a fashion similar to a stalk.

4. Evolution of the stalk into a TMC or HD. This can be spontaneous, depending on the lipid composition, and is otherwise driven by tension in the membranes.
5. Formation of a pore in the TMC or HD. This membrane rupture step may arise from some molecular-scale defect and would be accelerated by tension in the membranes. One role of fusion-mediating proteins may be to produce such defects either by introduction of inclusions into the membrane or by application of a membrane tension. It is interesting that hydrophobic peptides of very mundane structure [197, 198] and a viral fusion peptide [199] have been found to reduce drastically the tension that must be applied to membranes in order to produce rupture. However, it is not clear how such peptides on the fusion-mediating proteins could reach the diaphragm of an HD or TMC, unless they were on the cytoplasmic domains of the proteins or acted in the apposed monolayers to induce rupture at the junction of the diaphragm with the "wall" of the TMC or HD.
6. Expansion of the pore into a fusion pore (ILA-like structure). Depending on the spontaneous curvature of the lipids and initial dimensions of the pore, this may require application of membrane tension (e.g., [195]). (The opening of nascent fusion pores in viral protein-mediated fusion seems to be determined in part by the C_0 of the lipid monolayers [91].) However, the free energy of pores as a function of C_0, pore dimensions, and applied tension needs to be revisited, using the stress-free geometry [189] as a basis for calculations. Tension in the membranes could be imposed by osmotic stresses or by differences in the gross lipid composition between the two original membranes, generating a flux of lipid into the other membrane, which would tend to increase pore diameter.

Further study of the physical properties of lipid membranes and development of our models of fusion intermediates will be critical to our understanding the mechanism of protein-induced fusion. Such studies will provide insights into the structure–function relationships for fusion-mediating proteins and the mechanism of fusion in different biomembranes.

NOTE ADDED IN PROOF

Very recent work has established that the Gaussian curvature modulus of phospholipids can be measured at the L_α/Q_{II} transition temperature (or when ILAs exist in large numbers), and the influence of Gaussian curvature energy on the energy of fusion intermediates is substantial: D.P. Siegel and M.M. Kozlov, The Gaussian curvature elastic modulus of N-monomethylated dioleoylphosphatidylethanolamine: relevance to membrane fusion and lipid phase behavior, *Biophysical J.* (in press, 2004); and Y. Kozlovsky, D.P. Siegel, and M.M. Kozlov, Stalk phase formation: effects of dehydration and saddle splay modulus, *Biophysical J.* (submitted).

ACKNOWLEDGMENTS

The author is grateful to M.M. Kozlov, L.V. Chernomordik, F. Goñi, and A. Alonso for helpful comments on the manuscript.

REFERENCES

1. Jahn, R., Lang, T., and Sudhof, T.C., Membrane fusion. *Cell*, 112, 519, 2003.
2. Chapter 16 in this work; J. Bentz, Viral Fusion Mechanisms.
3. Blumenthal, R. et al., Membrane fusion, *Chem. Rev.*, 103, 53, 2003.
4. Basáñez, G., Membrane fusion: the process and its energy suppliers, *Cell Mol. Life Sci.*, 59, 1478, 2002.
5. Stegmann, T., Membrane fusion mechanisms: the influenza hemagglutinin paradigm and its implications for intracellular fusion, *Traffic,* 1, 598, 2000.
6. Lentz, B.R. et al., Protein machines and lipid assemblies: current views of cell membrane fusion, *Curr. Opin. Struct. Biol.*, 10, 607, 2000.
7. Chernomordik, L.C., Kozlov, M.M., and Zimmerberg, J., Lipids in biological membrane fusion, *J. Membr. Biol.*, 146, 1, 1995.
8. Chernomordik, L.V. and Zimmerberg, J., Bending membranes to the task: structural intermediates in bilayer fusion, *Curr. Opin. Struct. Biol.*, 5, 541, 1995.
9. Burger, K.N.J., Greasing membrane fusion and fission machines, *Traffic,* 1, 605, 2000.
9a. Chernomordik, L.V. and Kozlov, M.M., Protein-lipid interplay in fusion and fission of biological membranes, *Annu. Rev. Biochem.*, 72, 175, 2003.
10. Bentz, J. and Ellens, H., Membrane fusion: kinetics and mechanism, *Colloids Surf.*, 30, 65, 1988.
11. Papahadjopoulos, D., Nir, S., and Dzgnes, N., Molecular mechanisms of calcium-induced membrane fusion, *J. Bioenerg. Biomembr.*, 22, 157, 1990.
12. Burger, K.N.J. and Verkleij, A.J., Membrane fusion, *Experientia*, 46, 631, 1990.
13. Wilschut, J., Membrane fusion in lipid vesicles systems: an overview, in *Membrane Fusion*, Wilschut, J. and Hoekstra, D., eds., Marcel Dekker, New York, 1990, pp. 89–126.
14. Lewis, R.N.A.H., Mannock, D.A., and McElhaney, R.N., Membrane lipid molecular structure and polymorphism, in *Lipid Polymorphism and Membrane Properties*, Epand, R.M., ed., Current Topics in Membranes, Vol. 44, Academic Press, New York, 1997, Chap. 2.
15. Lindblom, G. and Rilfors, L., Cubic phases and isotropic structures formed by membrane lipids — possible biological relevance, *Biochim. Biophys. Acta*, 988, 221, 1989.
16. Seddon, J.M., Structure of the inverted hexagonal (H_{II}) phase, and non-lamellar phase transitions in lipids, *Biochim. Biophys. Acta*, 1031, 1, 1990.
17. Tate, M.W. et al., Nonbilayer phases of membrane lipids, *Chem. Phys. Lipids*, 57, 147, 1991.
18. Cribier, S. et al., Cubic phases of lipid-containing systems: a translational diffusion study by fluorescence recovery after photobleaching, *J. Mol. Biol.* 229, 517, 1993.
18a. Mariani, P., Luzzati, V., and Delacroix, H., Cubic phases of lipid-containing systems. Structure analysis and biological implications, *J. Mol. Biol.*, 204, 165, 1988.
19. Ellens, H., Bentz, J., and Szoka, F.C., Fusion of phosphatidylethanolamine-containing liposomes and mechanism of the L_α/H_{II} phase transition, *Biochemistry*, 25, 4141, 1986.
20. Ellens, H. et al., Membrane fusion and inverted phases. *Biochemistry*, 28, 3692, 1989.
21. Siegel, D.P. et al., Physiological levels of diacylglycerols in phospholipid membranes induce membrane fusion and stabilize inverted phases, *Biochemistry*, 28, 3703, 1989.
22. Siegel, D.P. et al., Intermediates in membrane fusion and bilayer/non-bilayer phase transitions imaged by time-resolved cryo-transmission electron microscopy. *Biophys. J.*, 56, 161, 1989.

23. Ellens, H., Bentyz, J., and Szoka, F.C., Destabilization of phosphatidylethanolamine liposomes at the hexagonal phase transition temperature, *Biochemistry*, 25, 285, 1986.

24. Bentz, J. et al., On the correlation between H$_{II}$ phase and the contact-induced destabilization of phosphatidylethanolamine-containing membranes, *Proc. Natl. Acad. Sci. USA*, 82, 5742, 1985.

25. Ellens, H., Bentz, J., and Szoka, F.C., pH-induced destabilization of phosphatidylethanolamine-containing liposomes: role of bilayer contact, *Biochemistry*, 23, 1532, 1984.

26. Bentz, J., Ellens, H., and Szoka, F.C. Destabilization of phosphatidylethanolamine-containing liposomes: hexagonal phase and asymmetric membranes, *Biochemistry*, 26, 2105, 1987.

27. Frederik, P.M. et al., Lipid polymorphism as observed by cryo-electron microscopy. *Biochim. Biophys. Acta*, 1062, 133, 1991.

28. Siegel, D.P., Green, W.J., and Talmon, Y., The mechanism of lamellar-to-inverted hexagonal phase transitions: a study using temperature-jump cryo-electron microscopy. *Biophys. J.*, 66, 402, 1994.

29. Siegel, D.P. and Epand, R.M., The mechanism of lamellar-to-inverted hexagonal phase transitions in phosphatidylethanolamine: implications for membrane fusion mechanisms. *Biophys. J.*, 73, 3098, 1997.

30. Verkleij, A.J., Lipidic intramembranous particles, *Biochim. Biophys. Acta*, 779, 43, 1984.

31. Siegel, D.P. and Epand, R.M., Effect of influenza hemagglutinin fusion peptide on lamellar/inverted phase transitions in dipalmitoleoylphosphatidylethanolamine: implications for membrane fusion mechanisms. *Biochim. Biophys. Acta*, 1468, 87, 2000.

32. Hui, S.W. and Stewart, T.P., Lipidic particles are intermembrane attachment sites. *Nature* (London), 287, 166, 1981.

33. Johnsson, M. and Edwards, K., Phase behavior and aggregate structure in mixtures of dioleoylphosphatidylethanolamine and ploy(ethylene glycol)-lipids, *Biophys. J.*, 80, 313, 2001.

34. Siegel, D.P., Inverted micellar intermediates and the transitions between lamellar, cubic, and inverted hexagonal amphiphile phases. III. Isotropic and inverted cubic state formation via intermediates in transitions between L$_\alpha$ and H$_{II}$ phases. *Chem. Phys. Lipids*, 42, 279, 1986.

35. Hui, S.W., Stewart, T.P., and Boni, L.T., The nature of lipidic particles and their roles in polymorphic transitions. *Chem. Phys. Lipids*, 33, 113, 1983.

36. Shyamsunder, E. et al., Observation of inverted cubic phase in hydrated dioleoylphosphatidylethanolamine membranes, *Biochemistry*, 27, 2332, 1988.

37. Veiro, J.A., Khalifah, R.G., and Rowe, E.S., P-31 nuclear magnetic resonance studies of the appearance of an isotropic component in dielaidoylphosphatidylethanolamine, *Biophys. J.*, 57, 637, 1990.

38. Erbes, J. et al., On the existence of bicontinuous cubic phases in dioleoylphosphatidylethanolamine, *Ber. Bunsenges., Phys. Chem.*, 98, 1287, 1994.

39. Tenchov, B., Koynova, R., and Rapp, G., Accelerated formation of cubic phases in phosphatidylethanolamine dispersions. *Biophys. J.*, 75, 853, 1998.

40. Gagné, J. et al., Physical properties and surface interactions of bilayer membranes containing N-methylated phosphatidylethanolamines, *Biochemistry*, 24, 4400, 1985.

41. Gruner, S.M. et al., X-ray diffraction study of the polymorphic phase behavior of N-methylated dioleoylphosphatidylethanolamine, *Biochemistry*, 27, 2853, 1988.

42. Siegel, D.P. and Banschbach, J.L., Lamellar/inverted cubic phase transition in N-methylated dioleoylphosphatidylethanolamine. *Biochemistry*, 29, 5975, 1990.

43. Cherezov, V. et al., The kinetics of non-lamellar phase transitions in DOPE-Me: relevance to biomembrane fusion (submitted for publication).

44. Cullis, P.R., Hope, J., and Tilcock, C.P.S., Lipid polymorphism and the roles of lipids in membranes, *Chem. Phys. Lipids*, 40, 127, 1986.

45. Tilcock, C.P.S., Lipid polymorphism, *Chem. Phys. Lipids*, 40, 109m, 1986.

46. Helfrich, W., Elastic properties of lipid bilayers: theory and possible experiments, *Z. Naturforsch.*, 28C, 693, 1973.

47. Kozlov, M.M. and Winterhalter, M., Elastic moduli for strongly curved monolayers. Position of the neutral surface, *J. Phys. II France*, 1, 1077, 1991.

48. Rand, R.P. et al., Membrane curvature, lipid segregation, and structural transitions for phospholipids under dual-solvent stress, *Biochemistry*, 29, 76, 1990.

49. Rand, R.P. and Parsegian, V.A., Hydration, curvature, and bending elasticity of phospholipid monolayers, in *Lipid Polymorphism and Membrane Properties*, Epand, R.M., ed., Current Topics in Membranes, Vol. 44, Academic Press, New York, 1997, Chap. 4.

50. Keller, S.L. et al., Probability of alamethicin conductance states varies with nonlamellar tendency of bilayer phospholipids, *Biophys. J.*, 65, 23, 1993.

50a. Szule, J.A., Fuller, N.L., and Rand, R.P., The effects of acyl chain length and saturation of diacylglycerols and phosphatidylcholines on membrane monolayer curvature, *Biophys. J.*, 83, 977–984, 2002.

51. Lafleur, M. et al., Correlation between lipid plane curvature and lipid chain order, *Biophys. J.* 70, 2747, 1996.

52. Tate, M.W. and Gruner, S.M., Temperature dependence of the structural dimensions of the inverted hexagonal (H_{II}) phase of phosphatidylethanolamine-containing membranes, *Biochemistry*, 28, 4245, 1989.

53. Kozlov, M.M., Leikin, S., and Rand, R.P., Bending, hydration and interstitial energies quantitatively account for the hexagonal-lamellar-hexagonal reentrant phase transition in dioleoylphosphatidylethanolamine, *Biophys. J.*, 67, 1603, 1994.

54. Lai, M.-Z., Vail, W.J., and Szoka, F.C., Acid- and calcium-stabilized structural changes in phosphatidylethanolamine membranes stabilized by cholesteryl hemisuccinate, *Biochemistry*, 24, 1654, 1985.

55. Dzgne, N. et al., Proton-induced fusion of oleic acid-phosphatidylethanolamine liposomes, *Biochemistry*, 24, 3091, 1985.

56. Nayar, R. et al., N-succinyldioleoylphosphatidylethanolamine: structural preferences in pure and mixed model membranes, *Biochim. Biophys. Acta*, 937, 31, 1988.

57. Rand, R.P. and Sengupta, S., Cardiolipin forms hexagonal structures with divalent cations, *Biochim. Biophys. Acta*, 255, 484, 1972.

58. Cullis, P.R., Verkleij, A.J., and Ververgaert, P.H., Polymorphic phase behavior of cardiolipin as detected by ^{31}P NMR and freeze-fracture techniques. Effects of calcium, dibucaine, and chloropromazine, *Biochim. Biophys. Acta*, 513, 11, 1978.

59. De Kruijff, B. et al., Further aspects of the Ca^{2+}-dependent polymorphism of bovine heart cardiolipin, *Biochim. Biophys. Acta*, 693, 1, 1982.

60. Ortiz, A. et al., Membrane fusion and the lamellar-to-inverted hexagonal phase transition in cardiolipin vesicle systems induced by divalent cations, *Biophys. J.*, 77, 2003, 1999.

61. Janes, N., Curvature stress and polymorphism in membranes, *Chem. Phys. Lipids*, 81, 133, 1996.

62. Small, D.M., The physical chemistry of lipids, in *The Handbook of Lipid Research*, Vol. 4, Hanahan, J., series editor, Plenum Press, New York, 1986.

62a. Leikin, S. et al., Measured effects of diacylglycerol on structural and elastic properties of phospholipid membranes, *Biophys. J.*, 71, 2623, 1996.

63. Basáñez, G. et al., Diacylglycerol and the promotion of lamellar-hexagonal and lamellar-isotropic phase transitions in lipids: implications for membrane fusion, *Biophys. J.*, 70, 2299, 1996.

64. Epand, R.M., Diacylglycerols, lysolecithin, or hydrocarbons markedly alter the bilayer to hexagonal phase transition temperature of phosphatidylethanolamines, *Biochemistry*, 24, 7092, 1985.

65. Siegel, D.P., Banschbach, J.L., and Yeagle, P.L., Stabilization of H_{II} phase by low levels of diglycerides and alkanes: an NMR, calorimetric, and x-ray diffraction study, *Biochemistry*, 28, 5010, 1989.

66. Goñi, F.M. and Alonso, A., Membrane fusion induced by phospholipase C and sphingomyelinases, *Biosci. Rep.*, 20, 443, 2000.

67. Goñi, F.M. and Alonso, A., Structure and functional properties of diacylglycerols in membranes, *Progr. Lipid Res.*, 38, 1, 1999.

68. Nieva, J.L. et al., Topological properties of two cubic phases of a phospholipid:cholesterol:diacylglycerol aqueous system and their possible implications in the phospholipase C-induced liposome fusion, *FEBS Lett.*, 368, 143, 1995.

69. Siegel, D.P., Energetics of intermediates in membrane fusion: comparison of stalk and inverted micellar intermediate mechanisms, *Biophys. J.*, 65, 2124, 1993.

70. Siegel, D.P., The modified stalk theory of lamellar/inverted phase transitions and its implications for membrane fusion, *Biophys. J.*, 76, 291, 1999.

71. Nieva, J.-L., Goñi, F.M., and Alonso, A., Liposome fusion catalytically induced by phospholipase C, *Biochemistry*, 28, 7364, 1989,

72. Nieva, J.-L., Goñi, F.M., and Alonso, A., Phospholipase C-promoted membrane fusion. Retroinhibition by the end-product diacylglycerol, *Biochemistry*, 32, 1054, 1993.

73. Ganong, B.R. and Bell, R.M., Transmembrane movement of phosphatidylglycerol and diacylglycerol sulfhydryl analogues, *Biochemistry*, 23, 4977, 1984.

74. Goñi, F.M. et al., Interfacial enzyme activation, non-lamellar phase formation and membrane fusion. Is there a conducting thread? *Faraday Discuss.*, 111, 55, 1998.

75. Preiss, J. et al., Quantitative measurement of *sn*-1,2-diacylglycerols present in platelets, hepatocytes, and *ras*- and *sis*-transformed normal rat kidney cells, *J. Biol. Chem.*, 261, 8597, 1986.

76. Veiga, M.P. et al., Ceramides in phospholipid membranes: effects on bilayer stability and transition to nonlamellar phases, *Biophys. J.*, 76, 342, 1999.

77. Ruiz-Argüello, M.B. et al., Different effects of enzyme-generated ceramides and diacylglycerols in phospholipid membrane fusion and leakage, *J. Biol. Chem.*, 271, 26616, 1996.

77a. Ruiz-Argüello, M.B., Goñi, F.M., and Alonso, A., Vesicle membrane fusion induced by the concerted activities of sphingolmyelinase and phospholipase C, *J. Biol. Chem.*, 273, 22977, 1998.

78. Basáñez, G. et al., Morphological changes induced by phospholipase C and by sphingomyelinases on large unilamellar vesicles: a cryo-transmission electron microscopy study of liposome fusion, *Biophys. J.*, 72, 2630, 1997.

79. Epand, R.M. et al., Promotion of hexagonal phase formation and lipid mixing by fatty acids with varying degrees of unsaturation, *Chem. Phys. Lipids*, 57, 75, 1991.

80. Basáñez, G., Goñi, F.M., and Alonso, A., Effect of single chain lipids on phospholipase C-promoted vesicle fusion. A test for the stalk hypothesis of membrane fusion, *Biochemistry*, 37, 3901, 1998.

81. Fuller, N. and Rand, R.P., The influence of lysolipids on the spontaneous curvature and bending elasticity of phospholipid membranes, *Biophys. J.*, 81, 243, 2001.

81a. Darkes, J.M. et al., The effect of fusion inhibitors on the phase behavior of N-methylated dioleoylphosphatidylethanolamine, *Biochim. Biophys. Acta*, 1561, 119, 2002.

82. Yeagle, P.L. et al., Inhibition of membrane fusion by lysophosphatidylcholine, *Biochemistry*, 33, 1820, 1994.

83. Meers, P. et al., Novel inner monolayer fusion assays reveal differential monolayer mixing associated with cation-dependent membrane fusion, *Biochim. Biophys. Acta*, 1467, 227, 2000.

84. Epand, R.M. et al., Dependence of the bilayer to hexagonal phase transition on amphiphile chain length, *Biochemistry*, 28, 93998, 1989.

85. Chernomordik, L.V. et al., The shape of lipid molecules and monolayer membrane fusion, *Biochim. Biophys. Acta*, 812, 643, 1985.

86. Chernomordik, L.V., Melikyan, G.B., and Chizmadzhev, Y.A., Biomembrane fusion: a new concept derived from model studies using two interacting planar lipid bilayers, *Biochim. Biophys. Acta*, 906, 309, 1987.

87. Chernomordik, L. et al., The hemifusion intermediate and its conversion to complete fusion: regulation by membrane composition, *Biophys. J.*, 69, 922, 1995.

88. Chernomordik, L.V. et al., Lysolipids reversibly inhibit Ca^{2+}-, GTP- and pH-dependent fusion of biological membranes, *FEBS Lett.*, 318, 71, 1993.

89. Chernomordik, L.V. et al., The pathway of membrane fusion catalyzed by influenza hemagglutinin: restriction of lipids, hemifusion, and lipidic pore formation, *J. Cell Biol.*, 140, 1369, 1998.

90. Chernomordik, L.V. et al., An early stage of membrane fusion mediated by the low pH conformation of influenza hemagglutinin depends on membrane lipids, *J. Cell Biol*, 136, 81, 1997.

91. Razinkov, V.I. et al., Effects of spontaneous bilayer curvature on influenza virus-mediated fusion pores, *J. Gen. Physiol.*, 112, 409, 1998.

92. Chernomordik, L.V. et al., Structural intermediates in influenza hemagglutinin-mediated fusion, *Mol. Membr. Biol.*, 16, 33, 1999.

93. Melikyan, G.B. et al., Evidence that the transition of HIV-1 gp41 into a six-helix bundle, not the bundle configuration, induces membrane fusion, *J. Cell Biol.*, 151, 413, 2000.

94. Gaudin, Y., Rabies virus-induced membrane fusion pathway, *J. Cell Biol.*, 150, 601, 2000.

95. Leikina, E. et al., The 1-127 HA2 construct of influenza virus hemagglutinin induces cell-cell hemifusion, *Biochemistry*, 40, 8378, 2001.

96. Grote, E. et al., Geranylgeranylated SNAREs are dominant inhibitors of membrane fusion, *J. Cell Biol.*, 151, 453, 2000.

97. Chernomordik, L., Non-bilayer lipids and biological fusion intermediates, *Chem. Phys. Lipids*, 81, 203, 1996.

98. Chernomordik, L. et al., The hemifusion intermediate and its conversion to complete fusion: regulation by membrane composition, *Biophys. J.*, 69, 922, 1995.

99. McIntosh, T.J. et al., Experimental tests for protrusion and undulation pressures in phospholipid bilayers, *Biochemistry*, 34, 8520, 1995.

100. Chanturiya, A. et al., Short-chain alcohols promote an early stage of membrane hemifusion, *Biophys. J.*, 77, 2035, 1999.

101. Martin, I. et al., Lysophosphatidylcholine mediates the mode of insertion of the NH2-terminal SIV fusion peptides into the lipid bilayer, *FEBS Lett.*, 333, 325, 1993.

102. Martin, I. and Ruysschaert, J.-M., Lysophosphatidylcholine inhibits vesicle fusion induced by the NH2-terminal extremity of SIV/HIV fusogenic proteins, *Biochim. Biophys. Acta*, 1240, 95, 1995.

103. Guenther-Ausborn, S., Praetor, A., and Stegmann, T., Inhibition of influenza-induced membrane fusion by lysophosphatidylcholine, *J. Biol. Chem.*, 270, 29279, 1995.

104. Gunther-Ausborn, S. and Stegmann, T., How lysophosphatidylcholine inhibits cell-cell fusion mediated by the envelope glycoprotein of human immunodeficiency virus, *Virology*, 235, 201, 1997.

105. Baljinnyam, B. et al., Lysolipids do not inhibit influenza virus fusion by interaction with hemagglutinin, *J. Biol. Chem.*, 277, 20461, 2002.

106. Gruner, S.M., Parsegian, V.A., and Rand, R.P., Directly measured deformation energy of phospholipid H_{II} phases, *Faraday Disc. Chem. Soc.*, 81, 29, 1986.

107. Hamm, M. and Kozlov, M.M., Tilt model of inverted amphiphilic mesophases, *Eur. Phys. J. B.*, 6, 519, 1998.

108. Hamm, M.M. and Kozlov, M.M., Elastic energy of tilt and bending of fluid membranes, *Eur. Phys. J. E.*, 3, 323, 2000.

109. Kirk, G.L. and Gruner, S.M., Lyotropic effects of alkanes and headgroup composition on the L_α/H_{II} lipid liquid crystal phase transition: hydrocarbon packing *versus* intrinsic curvature, *J. Physique*, 46, 761, 1985.

110. Tate, M.W. and Gruner, S.M., Lipid polymorphism of mixtures of dioleoylphosphatidylethanolamine and saturated and monounsaturated phosphatidylcholines of various chain lengths, *Biochemistry*, 26, 231, 1987.

111. Turner, D.C., Gruner, S.M., and Huang, J.S., Distribution of decane within the unit cell of the inverted hexagonal (H_{II}) phase of lipid-water-decane systems determined by neutron diffraction, *Biochemistry*, 31, 1356, 1992.

112. Chen, Z. and Rand, R.P., Comparative study of the effects of several *n*-alkanes on phospholipid hexagonal phases, *Biophys. J.*, 74, 944, 1998.

113. Lohner, K. et al., Squalene promotes the formation of non-bilayer structures in phospholipid model membranes, *Biochim. Biophys. Acta*, 1152, 69, 1993.

114. Epand, R.M., Epand, R.F., and Lancaster, C.R.D., Modulation of the bilayer to hexagonal phase transition of phosphatidylethanolamines by acylglycerols, *Biochim. Biophys. Acta*, 945, 161, 1988.

115. Valtersson, C. et al., The influence of dolichol, dolichol esters, and dolichyl phosphate on phospholipid polymorphism and fluidity in model membranes, *J. Biol. Chem.*, 260, 2742, 1985.

116. Knudsen, M.J. and Troy, F.A., Nuclear magnetic resonance studies of polyisoprenols in model membranes, *Chem. Phys. Lipids*, 51, 205, 1989.

117. Boesze-Battaglia, K. et al., Retinal and retinol promote membrane fusion, *Biochim. Biophys. Acta*, 1111, 256, 1992.

118. Walter, A., Yeagle, P.L., and Siegel, D.P., Diacylglycerol and hexadecane increase divalent cation-induced lipid mixing rates between phosphatidylserine large unilamellar vesicles, *Biophys. J.*, 66, 366, 1994.

119. Lerique, B. et al., Triacylglycerol in biomembranes, *Life Sci.*, 54, 831, 1994.

120. May, G.L., Increased saturated triacylglycerol levels in plasma membranes of human neutrophils stimulated by lipopolysaccharide, *J. Lipid Res.*, 38, 1562, 1997.

121. Schutzbach, J.S. et al., Membrane structure and mannosyltransferase activities: the effects of dolichols on membranes, *Chemica Scripta*, 27, 109, 1987.

122. Hamilton, J.A. and Small, D.M., Solubilization and localization of triolein in phosphatidylcholine bilayers: a ^{13}C NMR study, *Proc. Natl. Acad. Sci. USA*, 78, 6878, 1981.

123. Hamilton, J.A., Miller, K.W., and Small, D.M., Solubilization of triolein and cholesteryl oleate in egg phosphatidylcholine vesicles, *J. Biol. Chem.*, 258, 12821, 1983.

124. Hamilton, J.A., Interactions of triglycerides with phospholipids: incorporation into the bilayer structure and formation of emulsions, *Biochemistry*, 28, 2514, 1989.

125. Hamilton, J.A., Fujito, D.T., and Hammer, C.F., Solubilization and localization of weakly polar lipids in unsonicated egg phosphatidylcholine: a ^{13}C MAS NMR study, *Biochemistry*, 30, 2894, 1991.

126. Li, R. et al., Solubilization of acyl heterogeneous triacylglycerol in phosphatidylcholine vesicles, *J. Agric. Food Chem.*, 51, 477, 2003.

127. Gorrissen, H., Tulloch, A.P., and Cushley, R.J., Deuterium magnetic resonance of triacylglycerols in phospholipid bilayers, *Chem. Phys. Lipids*, 31, 245, 1982.

128. Salmon, A. and Hamilton, J.A., Magic-angle spinning and solution ^{13}C NMR studies of medium- and long-chain cholesteryl esters in model bilayers, *Biochemistry*, 49, 16065, 1995.

129. Andersson, S. et al., Minimal surfaces and structures: from inorganic and metal crystals to cell membranes and biopolymers, *Chem. Rev.*, 88, 221, 1988.

130. Anderson, D.M., Gruner, S.M., and Leibler, S., Geometrical aspects of the frustration in the cubic phases of Lyotropic liquid crystals, *Proc. Natl. Acad. Sci. USA*, 85, 5364, 1988.

131. Schwartz, U.S. and Gompper, G., Bending frustration of lipid-water mesophases based on cubic minimal surfaces, *Langmuir*, 17, 2084, 2001.

132. Duesing, P.M., Templer, R.H., and Seddon, J.M., Quantifying packing frustration energy in inverse Lyotropic mesophases, *Langmuir*, 13, 351, 1997.

133. Morein, S.E. et al., The effect of peptide/lipid hydrophobic mismatch on the phase behavior of model membranes mimicking the lipid composition of *Escherichia coli* membranes, *Biophys. J.*, 78, 2475, 2000.

134. van der Wel, P.C.A. et al., Tryptophan-anchored transmembrane peptides promote formation of nonlamellar phase in phosphatidylethanolamine model membranes in a mismatch-dependent manner. *Biochemistry*, 39, 3124, 2000.

135. Liu, F. et al., A differential scanning calorimetric and ^{31}P NMR spectroscopic study of the effect of transmembrane peptides on the lamellar-reversed hexagonal phase transition of phosphatidylethanolamine model membranes, *Biochemistry*, 40, 760, 2001.

136. Siegel, D.P. et al., WALP peptides stabilize inverted cubic phases: implications for membrane fusion, (in preparation).

137. Killian, J.A. et al., Induction of nonbilayer structures in diacylphosphatidylcholine model membranes by transmembrane α-helical peptides: importance of hydrophobic mismatch and proposed role of tryptophans, *Biochemistry*, 35, 1037, 1996.

138. Morein, S.E. et al., Influence of membrane-spanning α-helical peptides on the phase behavior of the dioleoylphosphatidylcholine/water system, *Biophys. J.*, 73, 3078, 1997.

139. Lamb, R.A. and Krug, R.M., Orthomyxoviridae: the viruses and their replication, in *Virology*, Vol. 1, Fields, B.N., Knirpe, D.M., and Howley, P.M., eds., Lippincott-Raven, Philadelphia, 1996, pp. 1353–1395.

140. Killian, J.A., Hydrophobic mismatch between proteins and lipids in membranes, *Biochim. Biophys. Acta*, 1376, 401, 1998.

141. Bentz, J., Nir, S., and Wilschut, W., Mass action kinetics of vesicle aggregation and fusion, *Colloids Surfaces*, 6, 333, 1983.

142. Dzgne, N. and Bentz, J., Fluorescence assays for membrane fusion, in *Spectroscopic Membrane Probes*, Vol. 1, Lowe, L. ed., CRC Press, Boca Raton, FL, 1988, Chap. 6.

143. Nir, S., Modeling aggregation and fusion of phospholipid vesicles, in *Membrane Fusion*, Wilschut, J. and Hoekstra, D., eds., Marcel Dekker, Inc., New York, 1991, p. 127.

144. Lentz, B.R. and Lee, J., Poly(ethylene glycol) (PEG)-mediated fusion between pure lipid bilayers: a mechanism in common with viral fusion and secretory vesicle release? *Mol. Membr. Biol.*, 16, 279, 1999.

145. Lee, J. and Lentz, B.R., Evolution of lipidic structures during model membrane fusion and the relation of this process to cell membrane fusion, *Biochemistry*, 36, 6251, 1997.

146. Pantazatos, D.P. and MacDonald, R.C., Directly observed membrane fusion between oppositely charged phospholipid bilayers, *J. Membr. Biol.*, 170, 27, 1999.

147. Garcia, R.A. et al., Cholesterol stabilizes hemifused phospholipid bilayer vesicles, *Biochim. Biophys. Acta*, 1511, 264, 2001.

148. Chanturiya, A., Chernomordik, L.V., and Zimmerberg, J., Flickering fusion pores comparable with initial exocytotic pores occur in protein-free phospholipid bilayers, *Proc. Natl. Acad. Sci. USA*, 94, 14423, 1997.

149. Talbot, W.A., Zheng, L.X., and Lentz, B.R., Acyl chain unsaturation and vesicle curvature alter outer leaflet packing and promote poly(ethylene glycol)-mediated membrane fusion, *Biochemistry*, 36, 5827, 1997.

150. Yang, Q. et al., Effects of lipid headgroup and packing stress on poly(ethylene glycol)-induced phospholipid vesicle aggregation and fusion, *Biophys. J.*, 73, 277, 1997.

151. Haque, M.E. et al., Effects of hemagglutinin fusion peptide on poly(ethylene glycol-)-mediated fusion of phosphatidylcholine, vesicles, *Biochemistry*, 40, 14243, 2001.

152. Haque, M.E., and Lentz, B.R., Influence of gp41 fusion peptide on the kinetics of poly(ethylene glycol)-mediated model membrane fusion, *Biochemistry*, 41, 10866, 2002.

153. Malinin, V.S., Frederik, P., and Lentz, B.R., Osmotic and curvature stress affect PEG-induced fusion of lipid vesicles but not mixing of their lipids, *Biophys. J.*, 82, 2090, 2002.

154. Lentz, B.R. et al., Transbilayer lipid redistribution accompanies poly(ethylene glycol) treatment of model membranes but is not induced by fusion, *Biochemistry*, 36, 2076, 1997.

155. Haque, M.E., McIntosh, T.J., and Lentz, B.R., Influence of lipid composition on physical properties and PEG-mediated fusion of curved and uncurved model membrane vesicles: "nature's own" fusogenic bilayer, *Biochemistry*, 40, 4340, 2001.

156. Olbrich, K., Needham, D., and Evans, E., Water permeability and mechanical strength of polyunsaturated lipid bilayers, *Biophys. J.*, 79, 321, 2000.

157. Evans, K.O. and Lentz, B.R., Kinetics of lipid rearrangements during poly(ethylene glycol)-mediated fusion of highly curved unilamellar vesicles, *Biochemistry*, 41, 1241, 2002.

158. Kozlovsky, Y. and Kozlov, M.M., Stalk model of membrane fusion: solution of energy crisis, *Biophys. J.*, 82, 882, 2002.

159. Burgess, S.W., McIntosh, T.J., and Lentz, B.R., Modulation of poly(ethylene glycol)-induced lipid fusion by membrane hydration: importance of interbilayer separation, *Biochemistry*, 31, 2653, 1992.

160. Dubochet, J. et al., Cryo-electron microscopy of vitrified specimens, *Q. Rev. Biophys.*, 21, 129, 1988.

161. Lee, J.K. and Lentz, B.R., Outer leaflet-packing defects promote poly(ethylene glycol)-mediated fusion of large unilamellar vesicles, *Biochemistry*, 36, 421, 1997.

162. Wu, X., Zheng, L.W., and Lentz, B.R., A slight asymmetry in the transbilayer distribution of lyso-phosphatidylcholine alters the surface properties and poly(ethylene glycol)-mediated fusion of dipalmitoylphosphatidylcholine large unilamellar vesicles, *Biochemistry*, 35, 12602, 1996.

163. Kozlovsky, Y., Chernomordik, L.V., and Kozlov, M.M., Lipid intermediates in membrane fusion: formation, structure, and decay of the hemifusion diaphragm, *Biophys. J.*, 83, 2634, 2002.

164. Lee, J.K., and Lentz, B.R., Secretory and viral fusion may share mechanistic events with fusion between curved lipid bilayers, *Proc. Natl. Acad. Sci. USA*, 95, 9274, 1998.

165. Malinin, V.S. and Lentz, B.R., Pyrene cholesterol reports the transient appearance of nonlamellar intermediate structures during fusion of model membranes, *Biochemistry*, 41, 5913, 2002.

166. Evans, E.A. and Parsegian, V.A., Energetics of membrane deformation and adhesion in cell and vesicle aggregation, *Ann. N. Y. Acad. Sci.*, 416, 13, 1983.

167. Parsegian, V.A. and Rand, R.P., Membrane interaction and deformation, *Ann. N. Y. Acad. Sci.*, 416, 1, 1983.

168. Dennison, S.M. et al., VSV transmembrane domain (TMD) peptide promotes PEG-mediated fusion of liposomes in a conformationally sensitive fashion, *Biochemistry*, 41, 14925, 2002.

169. Lei, G. and MacDonald, R.C., Lipid bilayer vesicle fusion: intermediates captured by high-speed microfluorescence spectroscopy, *Biophys. J.* (September 2003).

170. MacDonald, R.C. et al., Physical and biological properties of cationic trimesters of phosphatidylcholine, *Biophys. J.*, 77, 2612, 1999.

171. Kozlov, M.M. and Markin, V.S., On the theory of membrane fusion. The adhesion-condensation mechanism, *Gen. Physiol. Biophys.*, 5, 379, 1984.

172. Tarahovsky, Y.S. et al., Electrostatic control of phospholipid polymorphism, *Biophys. J.*, 79, 3193, 2000.

173. Markin, V.S. and Kozlov, M.M., Theory of membrane fusion. Adhesion-condensation mechanism, *Biophysics*, 29, 265, 1984.

174. Pereira, F.B. et al., Interbilayer lipid mixing induced by the human immunodeficiency virus type-1 fusion peptide on large unilamellar vesicles: the nature of the nonlamellar intermediates, *Chem. Phys. Lipids*, 103, 11, 1999.

175. Agirre, A. et al., Interactions of the HIV-1 fusion peptide with large unilamellar vesicles and monolayers. A cryo-TEM and spectroscopic study, *Biochim. Biophys. Acta*, 1467, 153, 2000.

176. Unger, V.M., Electron cryomicrosocpy methods, *Curr. Opin. Struct. Biol.*, 11, 548, 2001.

177. Frank, J., Cryo-electron microscopy as an investigative tool: the ribosome as an example, *Bioessays*, 23, 725, 2001.

178. Talmon, Y. et al., Time-resolved cryotransmission electron microscopy, *J. Electr. Micros. Tech.*, 14, 6, 1990.

179. McIntyre, J.C. and Sleight, R.G., Fluorescence assay for phospholipid membrane asymmetry, *Biochemistry*, 30, 11819, 1991.

180. Meers, P. et al., Annexin I-mediated vesicular aggregation: mechanism and role in human neutrophils, *Biochemistry*, 31, 6372, 1992 (erratum in *Biochemistry* 32, 1390, 1993).

180a. Villar, A.V., Alonso, A., and Goñi, F.M., Leaky vesicle fusion induced by phosphatidylinositol-specific phospholipase C: observation of mixing of vesicular inner monolayers, *Biochemistry*, 39, 14012, 2000.

181. Malinin, V.S., Haque, M.E., and Lentz, B.R., The rate of lipid transfer during fusion depends on the structure of fluorescent lipid probes: a new chain-labeled lipid transfer probe pair, *Biochemistry*, 40, 8292, 2001.

182. Kozlov, M.M. et al., Stalk mechanism of vesicle fusion: intermixing of aqueous contents, *Eur. J. Biophys.*, 17,121, 1989.

183. Siegel, D.P., Modeling protein-induced fusion mechanisms: insights from the relative stability of lipidic structures, in *Viral Fusion Mechanisms*, Bentz, J., ed., CRC Press, Boca Raton, FL, 1993.

184. Kozlov, M.M. and Markin, V.S. Possible mechanism of membrane fusion, *Biofizika*, 28, 255, 1983.

185. Markin, V.S., Kozlov, M.M., and Borovjagin, V.L., On the theory of membrane fusion: the stalk mechanism, *Gen. Physiol. Biophys.*, 3, 361, 1984.

185a. Chernomordik, L.V. et al., Membrane fusion: local interactions and structural reorganizations, *Dokl. Ackad. Nauk SSRR*, 28, 1009, 1986 (in Russian).

186. Leikin, S.L. et al., Membrane fusion: overcoming of the hydration barrier and local restructuring, *J. Theor. Biol.*, 129, 411, 1987.

187. Nanavati, C.V. et al., The exocytotic fusion pore modeled as a lipidic pore, *Biophys. J.*, 63, 1118, 1992.

188. Kuzmin, P.I. et al., A quantitative model for membrane fusion based on low-energy intermediates, *Proc. Natl. Acad. Sci. USA*, 98, 7235, 2001.

189. Markin, V.S. and Albanesi, J.P., Membrane fusion: stalk model revisited, *Biophys. J.*, 82, 693, 2002.

190. May, S., Structure and energy of fusion stalks: the role of membrane edges, *Biophys. J.*, 83, 2969, 2002.

190a. Gingell, D. and Ginsberg, L., Problems in the physical interpretation of membrane interaction and fusion, in *Membrane Fusion*, Poste, G. and Nicolson, G.L., eds., Elsevier North Holland Biomedical Press, New York, 1978, pp. 369–385.

191. Hui, S.W., Stewart, T.P., and Boni, L.T., Membrane fusion through point defects in bilayers, *Science* (Washington, DC), 212, 921, 1982.

192. Yang, L. and Huang, H.W., Observation of a membrane fusion intermediate structure, *Science*, 297, 1877, 2002.

193. Yang, L. and Huang, H.W., A rhombohedral phase of lipid containing a membrane fusion intermediate structure, *Biophys. J.*, 84, 1808, 2003.

194. Yang, L., Ding, L., and Huang, H.W., New phases of phospholipids and implications to the membrane fusion problem, *Biochemistry*, 42, 6631, 2003.

195. Chizmadzhev, Y.A. et al., Membrane mechanics can account for fusion pore dilation in stages, *Biophys. J.*, 69, 2489, 1995.

196. Melikyan, G.B. et al., Inner but not outer monolayer leaflets control the transition from glycosylphosphatidylinositol-anchored influenza hemagglutinin-induced hemifusion to full fusion, *J. Cell Biol.*, 136, 995, 1997.

197. Evans, E. and Needham, D., Giant vesicle bilayers composed of lipids, cholesterol and peptides, *Faraday Disc. Chem. Soc.*, 81, 267, 1986.

198. Evans, E. and Needham, D., Physical properties of surfactant bilayer membranes: thermal transitions, elasticity, rigidity, cohesion, and colloidal interactions, *J. Phys. Chem.*, 91, 4219, 19897.

199. Longo, M.L., Waring, A.J., and Hammer, D.A., Interaction of influenza hemagglutinin fusion peptide with lipid bilayers: area expansion and permeation, *Biophys. J.*, 73, 1430, 1997.

200. Needham, D. and Hochmuth, R.M., Electro-mechanical permeabilization of lipid vesicles. Role of membrane tension and compressibility, *Biophys. J.*, 55, 1001, 1989.

201. Frolov, V.A. et al., Multiple local contact sites are induced by GPI-linked influenza hemagglutinin during hemifusion and flickering pore formation, *Traffic*, 1, 622, 2000.

202. Melikyan, G.B., White, J.M., and Chen, F.S., GPI-anchored influenza hemagglutinin induces hemi-fusion to both red blood cell and planar bilayer membranes, *J. Cell Biol.*, 131, 679, 1995.

203. Turner, D.C. and Gruner, S.M., X-ray diffraction reconstruction of the inverted hexagonal (H_{II}) phase in lipid-water systems, *Biochemistry*, 31, 1340, 1992.

204. Rappolt, M. et al., Mechansim of the lamellar/inverse hexagonal phase transition examined by high resolution x-ray diffraction, *Biophys. J.*, 84, 3111, 2003.

205. Chen, Z. and Rand, R.P., The influence of cholesterol on phospholipid membrane curvature and bending elasticity, *Biophys. J.*, 73, 267, 1997.

206. Papahadjopoulos, D. et al., Cochleate lipid cylinders: formation by fusion of unilamellar vesicles, *Biochim. Biophys. Acta*, 394, 483, 1975.

207. Papahadjopoulos, D. et al., Studies on membrane fusion. III. The role of Ca^{2+}-induced phase changes, *Biochim. Biophys. Acta*, 465, 579, 1977.

208. Newton, C. et al., Specificity of Ca^{2+} and Mg^{2+} binding to phosphatidylserine vesicles and resultant phase changes of bilayer membrane structure, *Biochim Biophys Acta*, 506, 281, 1978.

209. Portis, A. et al., Studies on the mechanism of membrane fusion: evidence for an intermembrane Ca^{2+}-phospholipid complex, synergism with Mg^{2+} and inhibition by spectrin, *Biochemistry*, 18, 780, 1979.

210. Ekerdt, R. and Paphadjopoulos, D., Intermembrane contact affects calcium binding to phospholipid vesicles, *Proc. Natl. Acad. Sci. USA*, 79, 2273, 1982.

211. Mattai, J. et al., Interactions of metal ions with phosphatidylserine bilayer membranes: effect of hydrocarbon chain unsaturation, *Biochemistry*, 28, 2322, 1989.

212. Feigenson, G.W., On the nature of calcium binding between phosphatidylserine lamellae, *Biochemistry*, 25, 5819, 1986.

213. Feigenson, G.W., Calcium ion binding between lipid bilayers: the four-component system of phos-phatidylserine, phosphatidylcholine, calcium chloride and water, *Biochemistry*, 28, 1270, 1989.

214. Dzgne, N. et al., Modulation of membrane fusion by ionotropic and thermotropic phase transitions, *Biochemistry*, 23, 3486, 1984.

215. Bentz, J. and Dzgne, N., Fusogenic capacities of divalent cations and the effect of liposome size, *Biochemistry*, 24, 5436, 1985.

216. Wilschut, J., Dzgne, N., and Paphadjopoulos, D., Calcium/magnesium specificity in membrane fusion: kinetics of aggregation and fusion of phosphatidylserine vesicles and the role of bilayer curvature, *Biochemistry*, 20, 3126, 1981.

217. Leventis, R., Divalent cation induced fusion and lipid lateral segregation in phosphatidylserine-phosphatidic acid vesicles, *Biochemistry*, 25, 6978, 1986.

218. Bentz, J. et al., La^{3+}-induced fusion of phosphatidylserine liposomes, *Biophys. J.*, 53, 593, 1988.

219. Dzgne, N. et al., Monolayer coupling in phosphatidylserine bilayers: distinct phase transitions induced by magnesium interacting with one or both monolayers, *Biochim. Biophys. Acta*, 944, 391, 1988.

220. Cullis, P.R. and Hope, M.J., Physical properties and functional roles of lipids in membranes, in *Biochemistry of Lipids and Membranes*, Vance, D.E., and Vance, J.E., eds., Benjamin Cummings, Menlo Park, CA, 1985, Chap. 2.

221. Hoekstra, D., Role of lipid phase separation and membrane hydration in phospholipid vesicle fusion, *Biochemistry*, 21, 2833–2840, 1982.

222. Hong, K., Dzgne, N., and Papahadjopoulos, D., Modulation of membrane fusion by calcium-binding proteins, *Biophys. J.*, 37, 297, 1982.

223. Hong, K. et al., Synexin facilitates fusion of specific phospholipid vesicles at divalent cation concen-trations found intracellularly, *Proc. Natl. Acad. Sci. USA*, 79, 4942, 1982.

224. Meers, P. et al., Synexin enhances the aggregation rate but not the fusion rate of liposomes, *Biochem-istry*, 26, 4430, 1988.

224a. Needham, D., Cohesion and permeability of lipid bilayer vesicles, in *Permeability and Stability of Lipid Bilayers*, Disalvo, E.A. and Simon, S.A., eds., CRC Press, Boca Raton, FL, 1995, Chap. 3.

224b. Needham, D. and Zhelev, D., Use of micropipet manipulation techniques to measure the properties of giant lipid vesicles, in *Giant Vesicles*, Luisi, P.L. and Walde, P., eds., John Wiley & Sons, New York, 2000, Chap. 9.

225. Melikov, K.C. et al., Voltage-induced nonconductive pre-pores and metastable single pores in unmod-ified planar lipid bilayer, *Biophys. J.*, 80, 1829, 2001.

226. Marrink, S.-J., and Tieleman, D.P., Molecular dynamics simulation of spontaneous membrane fusion during a cubic-hexagonal phase transition, *Biophys. J.*, 83, 2382, 2002.
227. Briggs, J., Chung, H., and Caffrey, M. The temperature-composition phase diagram and mesophases structure characterization of the monoolein/water system, *J. Phys. II*, 6, 723, 1996.
228. Müller, M., Katsov, K., and Shick, M., New mechanism of membrane fusion, *J. Chem. Phys.*, 116, 2342, 2002.
229. Müller, M., Katsov, K., and Shick, M., A new mechanism of membrane fusion determined from Monte Carlo simulation, *Biophys. J.* (September 2003).
229a. Marrink, S.J. and Mark, A.E., The mechanism of vesicle fusion as revealed by molecular dynamics simulations, *J. Am. Chem. Soc.* (September 2003).
230. Frolov, V.A. et al., Membrane permeability changes at early stages of influenza hemagglutinin-mediated fusion, *Biophys. J.* (September 2003).
231. Alford, D., Ellens, H., and Bentz, J., Fusion of influenza virus with sialic acid-bearing target membranes, *Biochemistry*, 33, 1977, 1994.
232. Stegmann, T., Influenza hemagglutinin-mediated membrane fusion does not involve inverted phase intermediates, *J. Biol. Chem.*, 268, 1716, 1993.
233. Razinkov, V.I., Melikyan, G., and Cohen, F.S., Hemifusion between cells expressing hemagglutinin of influenza virus and planar membranes can precede formation of fusion pores that subsequently fully enlarge, *Biophys. J.*, 77, 3144, 1999.
234. Kozlov, M.M. and Chernomordik, L.V., A mechanism of protein-mediated fusion: coupling between refolding of the influenza hemagglutinin and lipid rearrangements, *Biophys. J.*, 75, 1384, 1998.
235. Kozlov, M.M. and Chernomordik, L.V., The protein coat in membrane fusion: lessons from fission, *Traffic*, 3, 256, 2002.
236. Tamm, L.K. and Han, X., Viral fusion peptides: a tool set to disrupt and connect biological membranes, *Biosci. Rep.*, 20, 501, 2000.
237. Epand, R.M., Modulation of lipid polymorphism by peptides, in *Lipid Polymorphism and Membrane Properties*, Current Topics in Membranes, Vol. 44, Epand, R.M., ed., Academic Press, San Diego, CA, 1997, Chap. 6.
238. Bentz, J., Membrane fusion mediated by coiled coils: a hypothesis, *Biophys. J.*, 78, 886, 2000.

9 Membrane Rafts

Kathleen Boesze-Battaglia

CONTENTS

9.1   INTRODUCTION

The physical relationship between hydrophilic aqueous medium and hydrophobic fatlike molecules has intrigued scientists since the 1770s, with Benjamin Franklin's observation that any oily substance covered clearly half the surface area when compared to an equal volume of aqueous solution, implying a bilayer structure. As early as 1925, Gorter and Grendel [2], proposed the now-classic deduction that membrane lipids are arranged in a bilayer configuration, in which parallel sheets of phospholipids have polar or charged headgroups oriented toward the aqueous environment and acyl chains interacting within the hydrophobic membrane core. In the 1950s, it was with remarkable foresight that Davson and Danielli [1] captured the importance of biological cell membranes in their observation, "it can truly be said of living cells, that by their membranes ye shall know them." In 1972, Singer and Nicolson provided a model that took into consideration the dynamic nature of lipid–protein interactions, providing a matrix in which proteins have a degree of motion that, in turn, can have a dramatic impact on activity. Thus the fluid-mosaic model [3] became the framework and benchmark for our current understanding of membrane bilayers and their physiological function. The assumed homogeneous nature of membrane bilayers proposed in this model was called into question in the 1970s, with the observations that membranes contain a unique composition of lipid and protein components that are specific to cell type and subcellular localization. A heterogeneous distribution of lipids and proteins is even observed within spatially separated regions of the same membrane of Golgi [4] or apical and basolateral plasma membranes of polarized cells [5]. This model of the membrane bilayer was further redefined and refined to include fast-emerging hypotheses regarding the association of the cytoskeletal elements with membrane components [6]. Moreover, during this period, considerable effort was spent characterizing the lipids that surround an integral membrane protein. These annular, or boundary, lipids have a distinguishable mobility and phase transition enthalpy when compared to the bulk membrane lipids [7–10].

Within the past decade, a unifying theme describing the organization of lipid and membrane proteins has focused on localized regions within the membrane, known collectively as *membrane microdomains*. The last five years have seen an emergence of interest in a specific type of microdomain, known colloquially as a *membrane raft*. More precisely, these regions are globally defined as sphingolipid-cholesterol rich domains in the liquid ordered phase. These microdomains are proposed to be involved in a wide variety of cellular processes, including protein sorting [11]; signal transduction [12]; calcium homeostasis [13]; transcytosis [14]; potocytosis [15]; alternative routes of endocytosis [16]; internalization of toxins, bacteria, and viruses [17–19]; HIV-1 assembly and release [20]; and cholesterol transport [21, 22].

Previous chapters have described in detail the biophysical principles and molecular determinants of membrane bilayer structure that served as the theoretical and experimental foundation for the current model of lipid rafts. The raft hypothesis has described these microdomains and states that "separation of discrete liquid-ordered and liquid-disordered phase domains occurs in membranes containing sufficient amounts of sphingolipid and sterol" [23]. This chapter provides a biochemical and biophysical characterization of membrane rafts and defines the criteria used in identifying microdomains as rafts in model membranes and cell membranes. It discusses the microscopic and biochemical approaches used to identify rafts and the shortcomings associated with each. The later part of the chapter focuses on recent, novel applications of the membrane raft hypothesis in two unique cells: platelets and retinal photoreceptors. The well-characterized application of the membrane raft hypothesis in the sorting of proteins in MDCK cells [24] and toxin binding [25] has been reviewed extensively.

9.2 THEORETICAL AND STRUCTURAL CONSIDERATIONS

A *domain* is defined as a region that is distinctively marked by physical features that distinguish it from the surrounding landscape. Within membrane bilayers, short-range order of lipid and protein components is a hallmark of domains. Membrane rafts, cholesterol- and sphingolipid-enriched regions within a heterogeneous membrane bilayer, are specialized membrane domains or, more accurately, microdomains based on their size and unique biochemical properties. Molecules associated with a membrane domain most often contain a specific molecular address for that domain. This address is usually an inherent structural or chemical feature of the molecule and can be thought of as loosely analogous to a protein signal sequence.

Inherent in the structural features of cholesterol and sphingomyelin is part of the explanation for the behavior of these lipids in a complex membrane. Sphingomyelin, a prototypical membrane raft lipid, contains a ceramide backbone to which a phosphorylcholine headgroup is attached. Due to the free amide and free hydroxyl group on sphingomyelin, this lipid has both hydrogen bond–accepting and hydrogen bond–donating groups. In addition, the amide-linked acyl chains of sphingomyelin are most often saturated. Collectively, these properties allow this lipid to facilitate the formation of extensive hydrogen-bonded networks. In contrast, the predominant glycerophospholipids consist of a glycerol base with ester-linked fatty acids in the *sn*-1 and *sn*-2 positions. Phospholipid acyl chains are usually 16 to 18 carbons in length, most often unsaturated at the *sn*-2 position, and have only hydrogen bond–accepting capacity. Thus, they are organized within the membranes in a manner distinct from the sphingolipids. The most abundant headgroup moieties are phospho-ester linked, most commonly serine, choline, and ethanolamine.

The difference in organization of the phospholipids and the glycerosphingolipids in membrane bilayers is clearly seen when one compares the gel to liquid–crystal phase transition temperatures of these lipid species [26]. The phase behavior of lipid bilayers, as reviewed previously, depends on the packing, degree of order, and mobility of the constituent lipids. The two most divergent phases are the gel phase and liquid–crystalline phase. In a homogeneous lipid mix, the melting temperature of long-chain sphingomyelin is 40°C, in comparison to −3°C for 1-palmitoyl-2-oleoylphosphatidylcholine (POPC). The kinked structure of the unsaturated acyl chain impedes

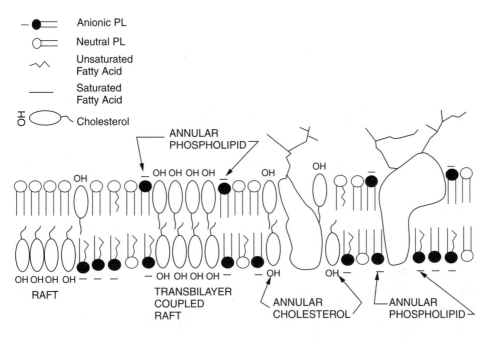

FIGURE 9.1 Schematic representation of plasma membrane lipid domains. The individual lipid components are shown in the figure legend at the top left. Annular phospholipid refers to the first molecular layer of phospholipid surrounding a cholesterol-rich domain or protein. Annular cholesterol refers to the first molecular layer of cholesterol surrounding a protein. Reprinted with permission from Dr. F. Schroeder (2001) *Proc. Soc. Exp. Biol. Med.* 213, p. 876.

straightening and tight packing of this lipid. Thus, the phase transition temperature of phosphatidylcholine (PC) decreases dramatically from −3°C to 49°C when the acyl chain monounsaturated fatty acid, 18:1, is replaced by 18:0.

The phospholipids and glycosphingolipids within cell membranes exist in an environment containing varying levels of a structurally unique lipid, cholesterol (as shown schematically in Figure 9.1). Cholesterol is intercalated between adjacent lipid acyl chains. The four fused rings of cholesterol allow for little conformational flexibility and are not easily accommodated within the acyl matrix. The 3-hydroxyl group on cholesterol allows the molecule to orient parallel to the lipid hydrocarbon chains, with the OH at the lipid/aqueous interface. These properties of cholesterol result in a decrease in the gel to liquid–crystalline phase transition of both sphingolipid and phospholipid bilayers [27–29].

The differential packing of sphingolipids (specifically sphingomyelin) and phospholipids is proposed to lead to membrane phase separations. Most likely, the sphingolipid-rich membrane regions coexist with phospholipid-enriched domains. In a region of the membrane, therefore, that is enriched in sphingomyelin and cholesterol, one may expect a third phase, a liquid-ordered phase (Io). The acyl chains of lipids in this phase are extended and tightly packed, as in the gel phase, but have a higher degree of lateral mobility [30]. Membrane rafts probably exist in the Io phase or in a state with very similar properties, given the large saturated acyl chain of sphingomyelin. The naturally occurring phospholipids, containing unsaturated acyl chains at the *sn*-2 position, tend to have a much lower affinity for the Io phase and are thus largely excluded from these domains [31]. A biochemical property of lipids in the Io phase is their insolubility in various detergents [32]. Thus Triton X-100, and to a lesser extent Brij, have been used as a tool to isolate membrane rafts consisting of the Io phase.

Because cholesterol can promote phase separation in mixtures of phospholipids and sphingolipids [33], it is hypothesized that rafts can still form in cell membranes with relatively low levels of cholesterol. Moreover, it also explains why depletion of membrane cholesterol collapses rafts and raft function. It is important to consider that lipids can be associated with any one of three states: solid gel, liquid ordered, or liquid disordered. Thus, in the plasma membranes of cells in which these phases all coexist, a specific type of lipid is found in more than one domain. Therefore, all of the membrane cholesterol and all of the glycosphingolipid is not always raft associated.

9.2.1 EARLY EXPERIMENTAL EVIDENCE IN SUPPORT OF RAFTS

A series of early, seminal observations led investigators to propose the existence of membrane microdomains, or rafts, in both model and biological cell membranes. Lipid domains were first detected in fibroblasts as large glycoprotein matrices that were detergent insoluble [34]. These complexes were enriched in glycosphingolipids [35] and supported a hypothesis that described the formation of cholesterol and glycosphingolipid microdomains that serve as platforms for protein and lipid transport from Golgi to plasma membrane [36]. Early experiments documented the colocalization of glycosphingolipids and GPI- (glycosylphosphoinositol) anchored protein to the apical membranes of MDCK cells [4, 36, 37]. It appeared that these proteins did not contain a cytoplasmic targeting signal and GPI anchoring was sufficient for targeting [37]. Subsequent extraction of the polarized MDCK cells with Triton X-100 yielded a detergent-resistant membrane (DRM) complex enriched in GPI-anchored proteins, cholesterol, and sphingomyelin. Thus, these investigators showed that the rafts observed using microscopic techniques appeared to be isolated as detergent-insoluble DRM complexes.

Surprising at the time was the association of a subset of signaling proteins found to be associated with the DRM; these signaling proteins included the src family of kinases. Scheek et al. [38] confirmed the coupling of cholesterol and sphingomyelin within membrane microdomains. In these studies, they showed that a decrease in fibroblast cell surface sphingomyelin affected the biosynthesis, uptake, and subsequent localization of cholesterol. The effect of sphingomyelin was to inhibit the cleavage of the membrane-bound transcription factor SREBP-2, which is involved in the regulation of these processes. These observations, coupled with earlier studies, argued that a functional interaction between cholesterol and sphingomyelin is necessary for normal intracellular transport processes [39, 40]. Collectively, these observations provided the biological evidence for the existence of membrane microdomains, which have been termed rafts.

This early work on biological membranes, coupled with model membrane studies (described later), provided substantial experimental evidence in support of membrane rafts. In model membrane systems, lipids in the Io phase were shown to be resistant to extraction by Triton X-100 [30]. The lipid composition of the Triton X-100–insoluble complexes extracted from whole cells suggested that Triton X-100–resistant membranes are most likely in the Io state [41]. The lipids contained the typical signature of Io phases: a high degree of acyl chain order but a substantial rotational and lateral mobility. Fluorescence quenching assays also showed that the Io phase occurs in membranes with a lipid composition similar to that of plasma membrane. In those studies, using diphenylhexatriene (DPH) and a PC-linked nitroxide quencher, Ahmed et al. [42] showed that cholesterol at 33 mole% induced the formation of sphingomyelin-enriched liquid ordered phase at 37°C in mixed vesicles of sphingomyelin/PC. Similar to plasma membranes, the sphingomyelin concentration of the vesicles was 10 to 13%. Moreover, these investigators confirmed that the Io phase correlates with detergent insolubility. These model membrane studies provided a strong correlation between lipids in the Io phase and those isolated as Triton X-100–resistant fractions. However, they provide little evidence to support the association of raft-specific proteins with these lipids.

Additional evidence linking the membrane rafts identified microscopically with biochemically isolated detergent-resistant fractions was, and still is, sketchy. It had been widely known that the crosslinking of GPI-anchored surface glycoproteins triggered the activation of specific signal transduction pathways in a number of cells, lymphocytes included [43]. Later, these GPI-anchored proteins were shown to be present in detergent-insoluble complexes with tyrosine kinases of the src family [44–47]. A more detailed characterization of these complexes showed DRMs to be enriched in GPI-anchored proteins associated with specific glycolipids in large structures that measured approximately 100 nm in diameter [44, 45, 48]. Association with the DRMs was dependent on the GPI anchor, because mutants with no GPI-linked anchor were not associated with the DRMs and were unable to associate with the src kinases [47].

At the time of these observations, membrane microdomains were shown to contain VIP21-caveolin [49], and a large number of subsequent studies led to the assumption that all or most membrane microdomains contain plasma membrane invaginations containing caveolae [50, 51]. One of the first studies to refute this generalization showed that the glycolipid-anchored protein Thy1 and the glycosphingolipid GM1 were present in DRMs isolated from lymphocytes. These DRMs were totally devoid of caveolin at both the mRNA and the protein levels [52]. These results led to the suggestion that, in lymphocytes, detergent insolubility does not correlate with the presence of caveolin and, further, that caveolae in these cells are not necessarily involved in signal transduction in lymphocytes. Thus, more recent studies have suggested that different types of rafts exist based on the presence of specific marker proteins and ultrastructural data [53]. The differences encompass membranes enriched in caveolin with a characteristic conelike appearance, in contrast to those devoid of caveolin with no clear morphological characteristics.

9.2.2 MEMBRANE RAFT ORGANIZATION

Membrane fractions that are insoluble in nonionic detergents have been referred to as DRMs (detergent-resistant membranes), DIGs (detergent-insoluble glycolipid-enriched membranes), GEMs (glycolipid-enriched membranes), and TIFFs (Triton-insoluble floating fractions). The nomenclature presently used in publications is highly variable and not standardized, although a concerted effort has been made to use the term DRMs. The term *lipid raft* includes caveolae, structures generally defined by both morphology and enrichment in caveolin, and noncaveolae liquid-ordered membrane microdomains. Regardless of the nomenclature used, the basic organization of the membrane microdomains described has some common characteristics. These characteristics are illustrated schematically in Figure 9.2. Most commonly, membrane rafts include enrich-

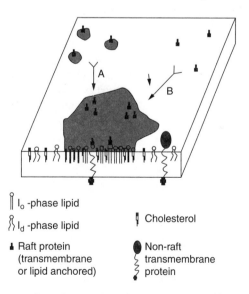

FIGURE 9.2 Schematic representation of membrane raft organization. Raft lipids in the Io phase are shown surrounded by nonraft lipids in the Id phase. It is proposed that clustering of raft proteins may cause small rafts to coalesce (A) or may increase the affinity of proteins for rafts (B). Reprinted with permission from Dr. D. Brown (2000) *J. Biol. Chem.* 275, p.1722.

ment in cholesterol, the tight packing of sphingolipids in the Io phase, and an increased affinity for gangliosides [54]. In addition, a variety of proteins are targeted to DRMs. These proteins, which include caveolin-1 caveolin-2, caveolin-3, hemagglutinin, and GPI-anchored proteins, are used extensively as membrane raft markers [for review, see 24, 55]. Although a growing number of proteins have been used as membrane raft markers, the identification of a detergent-insoluble or partially soluble fraction as a membrane raft based only on protein composition is not advised. A number of these proteins are not just raft-associated, and membrane rafts clearly have a unique lipid composition that should be characterized to avoid any artifacts due to detergent partitioning effects [29, 56, 57].

The general biochemical features of raft-marker proteins include modification with saturated chain lipids or GPI anchors. It is important to note that not all acyl chain modified proteins are in DRMs and not all DRM proteins are modified. In contrast, proteins that are prenylated are largely not raft associated. Although earlier work suggested that transmembrane proteins would be difficult to accommodate in a highly ordered raft domain, recent evidence refutes this predication. Members of the transmembrane 4 superfamily of membrane proteins are often localized to membrane rafts, as is the prototypical G-protein coupled receptor, rhodopsin [58]. Interestingly, the transmembrane 4 superfamily, also called the tetraspanins, comprises a large group of ubiquitously expressed proteins [59] that show various degrees of detergent insolubility. These proteins function in diverse processes, including membrane fusion, platelet aggregation, proliferation, and B- and T-cell activation [61, 62]. They function as molecular adaptors involved in the assembly of protein complexes. Tetraspanin protein associations are divided into three general categories based on the strength of the association. The most robust type of association is resistant to Triton X-100 and comprises tetraspanins that are raft associated, such as the CD151-containing complexes [63, 64]. The unifying mechanism by which this unique group of raft proteins exerts a biological effect remains under investigation. However, it does raise the questions of whether there are indeed compositionally distinct rafts within the same plasma membrane.

The definition of membrane rafts has been expanded recently to distinguish between *membrane caves* and *membrane barges*. Membrane caves refer to caveolae-enriched membrane microdomains. The cytoplasmic face of caveolae are covered with caveolins, a family of proteins between 21 and 25 kDa that form a membrane coat [65–68]. The caveolae function in endocytosis and signal transduction and contain receptor tyrosine kinases and GPI-anchored proteins. Membrane barges refer to microdomains with higher densities than rafts. These higher densities are proposed to be due to the presence of a high percentage of oligomerized complex assemblies, as seen with Gag proteins of HIV [69]. Lastly, a population of low buoyant density fractions has been identified as *lipid shells* [70]; these lipidic structures are so named because they contain a shell of cholesterol and sphingolipid. The predicted size of a lipid shell containing the GPI-anchored protein Thy and 80 molecules of cholesterol-sphingolipid is approximately 7 nm. It is proposed that these lipid shells are similar to hydration shells that surround ions or proteins in aqueous solutions. The role of such shells is to aid in the transition from the solute to the bulk phase. Lipid shells, by analogy, exist as mobile units in the plane of the membrane. Anderson and Jacobson [70] hypothesize that these shells have an affinity for preexisting membrane rafts and caves.

9.2.3 SIZE OF MEMBRANE RAFTS

Experimental evidence suggests that rafts vary in size from a few tenths of a nanometer to hundredths of a nanometer or even domains as large as microns [71–74]. As may be expected, the method by which the size of these domains is measured is crucial. Because GPI-anchored proteins and gangliosides are both enriched in rafts, it was not until the two were crosslinked or clustered with antibodies that they appeared to colocalize using microscopic techniques. Single-particle tracking experiments suggest that gangliosides and GPI-anchored proteins are confined to domains of 200 to 300 nm [72]. In contrast, by tracking the thermal positional fluctuation of HA and GPI-linked proteins in cholesterol enriched rafts, raft size was predicted to be 50 nm [74]. Of course, the size of the identifiable domains is dependent on the affinity of the fluorescent markers for the raft marker proteins. Thus, if these markers are not highly concentrated or if the affinity of the fluorescence marker is low, detection of rafts in biological membrane becomes difficult, if not impossible. Several approaches, however, do suggest that membrane rafts exist in biological membranes; these include fluorescence–resonance energy transfer between GPI-linked proteins [71], single-particle tracking [72], and single molecule microscopy [73]. Using pairs of raft and nonraft marker proteins and lipids, Harder et al. [75] identified raftlike lipid domains in cell plasma membranes. The raft-associated markers were found to form membrane patches, in contrast to the nonraft makers, which exhibited a more homogeneous distribution.

Recent work by Dietrich et al. [76, 77] has gotten around the dilemma of correlating membrane microdomains observed using microscopy techniques with the actual biochemical properties of such domains as predicted from Triton X-100 extraction studies. Using giant unilamellar vesicles, these authors showed that lipid extracts from brush border membranes (BBMs) contain membrane rafts and that the GPI-anchored protein Thy-1 partitions preferentially into these rafts in BBM lipids. The biophysical characteristics of these domains correspond reassuringly well to the properties of rafts described earlier, using other techniques. The 10 μm–sized domains had a degree of order expected of the Io phase, and the domains in the two leaflets were found to be coincident. It is expected that further experiments using monolayers and giant unilammellar vesicles will provide additional support for the formation of rafts in cell membranes.

9.3 WHEN IS A DRM A RAFT?

A challenge for the future becomes designing experiments that directly link the detergent-insoluble species and their localization to specific regions of the plasma membrane. It will become important to take into consideration that such domains are most likely transient, as the membrane bilayer is a dynamic structure. Experimentally, membrane rafts can be isolated using Triton X-100 at 4°C, because the sphingolipid-cholesterol-rich membrane microdomains are insoluble under these conditions. These lipid-rich complexes can be isolated at a low density using sucrose gradient centrifugation. Confirmation of a low buoyant density membrane fraction as a raft is obtained upon a thorough analysis of the lipid and protein composition. In addition, partial detergent solubilization can be controlled with the use of a detergent, such as OG, that will solubilize membrane rafts. Recent studies have suggested that detergent insolubility can result in the underestimation of domain association. However, the creation of artifactual domains from previously homogeneous bilayers and recruitment of unassociated proteins into rafts does not seem to occur [78]. Moreover, detergent-free techniques have been developed and used to isolate membrane fractions with characteristics similar to membrane rafts [79, 80]. Although the major limitation of either technique is that the original subcellular localization of the isolated rafts cannot be determined, it can be predicted that a raft is analogous to an isolated DRM fraction.

9.4 FUNCTIONS OF MEMBRANE RAFTS

The association of membrane lipids in a cholesterol-rich, tightly packed membrane domain allows for the formation of compartmentalized signaling complexes. Membrane rafts contain a high concentration of signaling molecules and have been classified as signaling centers. The mechanism by which these signaling centers function is not clearly understood; however, several intriguing hypotheses have been put forth [12, and references therein]. A number of protein and lipid signaling molecules have been localized to membrane rafts; these include the transmembrane protein, EGF receptor [81], bradykinin B2 receptor [82], TCR [83], and BCR [84]. The lipid signaling molecules (sphingomyelin [85], ceramide [86], phosphoinositides [87], and diacylglycerol), complementary signaling effectors (including the $G\alpha$ family of proteins [88], src-family kinases [81, 89, 90], eNOS [91], phospholipase C γ [92], and adenylate cyclase [93], to name a few) are all raft associated. An elegant and extensive review of this topic can be found in Zajchowski and Robbins [12].

The association of the components of signal transduction processes with membrane rafts has important functional and regulatory consequences. In the simplest case, the receptor may be a constituent resident of a lipid raft, ready to initiate signaling within this site; signaling through the src family of kinases most likely occurs by this means. In a more complex version of this mechanism, it has been proposed that the receptor ligand complex translocates out of the lipid raft, resulting in a down-regulation of the signal. Alternatively, the receptor–ligand complex may, upon translocation out of a raft, associate with other signaling complexes to initiate a response. IL-2R signaling is postulated to follow this mechanism [94]. A receptor may translocate to a lipid raft upon ligand binding, thus providing for some regulation of the signaling process based on raft association. The tetraspanin CD20 protein of B cells is proposed to follow such a mechanism [95]. Or, lastly, the receptor may be localized outside of a raft, but upon ligand binding initiate a signal within the raft domain leading to compartmentalization of the signal. In all of these mechanisms, the lipid raft plays an essential role in controlling and regulating signaling complexes. It remains to be determined whether the unique lipid composition of different types of rafts is capable of fine-tuning the response of signaling complexes. Evidence suggest that, in the case of the shaker potassium channel, the lipid composition of the raft alters the behavior of the signaling proteins [96]. Polyunsaturated fatty acids (PUFAs) inhibit T lymphocyte activation by displacing signaling proteins from membrane rafts [97].

9.4.1 MEMBRANE RAFTS IN PLATELETS

Platelets present a somewhat unique scenario. These small cells, 2 to 4 μm in diameter, undergo a stimulus-dependent translocation of phospholipids. Upon stimulation with thrombin, collagen, and ionomycin, PS and PE are postulated to translocate from the inner to the outer membrane monolayer. This translocation is necessary in the hemostatic process to mediate the formation of a prothrombinase complex [98]. A correlation between platelet responsiveness to stimulants (such as ADP, epinephrine [99], thrombin, [100, 101], thrombaxane A_2, [102], and collagen [103]) and membrane cholesterol content has been observed for the past three decades. Moreover, a stimulus-dependent translocation of cholesterol from the outer to the inner membrane monolayer has been observed [104]. The molecular basis of the mechanism responsible for these observations is largely unknown, but recent evidence suggests that cholesterol-enriched membrane microdomains (i.e., rafts) play an essential role.

In an early study, Triton X-100–resistant membranes (DRMs) were isolated from platelet vesicles [105]. These DRMs were enriched in cholesterol, with 41% of the total cholesterol present in these low buoyant density fractions. The population of DRMs was heterogeneous, consisting of vesicles ranging in size from 20 to 1000 nm in diameter, based on transmission scanning electron microscopy. The DRMs were specifically enriched in the membrane glycoprotein CD36 (a member of the tetraspanin protein family) and the surface protein GPIb, as well as Src p60 [c-src] and the src-related kinases lyn, p53/56 [lyn]. The association of CD36 with this membrane fraction was somewhat surprising, because CD36 is one of several transmembrane glycoproteins found in caveolae and no caveolae were detected in the platelet DRMs.

This initial work suggested that platelet membrane DRMs, in a manner analogous to other DRMs, were involved in platelet signal transduction. Recent work has provided more compelling data to support this hypothesis and, moreover, has met essential criteria in the characterization of platelet DRMs as membrane rafts: specifically, a detailed description of the lipid composition of the platelet DRMs [106]. In these studies, both resting platelets and platelets stimulated with thrombin were treated with Triton X-100. The resting platelets showed a heterogeneous distribution of vesicles within the DRM preparation, whereas the thrombin-stimulated platelet DRMs showed a modified curvature and a tendency to aggregate. Although significant changes in the lipid composition of the two different DRMs could explain the morphological differences, in fact, no substantial changes in lipid composition were detected. The cholesterol-to-phospholipid ratio in the DRMs was found to be 1.2, in contrast to 0.5 in whole platelets. The DRMs contained approximately 10% of the total phospholipid, with sphingomyelin accounting for almost 60% of the total phospholipids in the rafts. The fatty acid composition of the sphingomyelin in the DRMs was not very different from that found in whole platelets. In contrast, although the rafts contained small amounts of PC, PS+ PI, and PE, their fatty acid composition was distinct. In general, these phospholipids showed an increase in saturated fatty acids, including palmitic, stearic, and arachidic, consistent with a similar distribution seen in RBL-2H3 mast cells [107]. Whether this is a common feature of membrane rafts remains to be determined. This increase in saturated fatty acid content was paralleled by a decrease in the content of the unsaturated fatty acids oleate and arachidonate.

The phosphoinositol (PI) pool of phospholipids was the most significantly affected in DRMs isolated from thrombin-stimulated platelets. The change in DRM morphology in the thrombin-stimulated platelets correlated with an increase in phosphatidic acid and products of the phospho-inositide-3 kinase reaction. The increase in PtdIns $(3,4,5)$ P_3 concentration in the platelet rafts suggests that a family of tyrosine kinases important in platelet signaling may be recruited by the phosphoinositide metabolite. The recruitment of these proteins is postulated to involve a direct interaction between PI 3 kinase products and pleckstrin homology domains. Cholesterol depletion with methyl-β-cyclodextran dramatically decreased the thrombin-dependent production of these signaling molecules in the platelet DRMs.

Decreased platelet membrane cholesterol induced by β-methyl–cyclodextran also resulted in decreased levels of phosphatidylserine (PS) within the outer platelet membrane monolayer. The redistribution of PS from the inner to the outer membrane surface is a necessary step in the formation of the prothrombinase complex. The redistribution of PS was shown to be dependent on extracellular signal regulated kinase, the MEK kinase cascade found to be localized to platelet membrane rafts. These studies present the first clear link between phospholipid redistribution in platelet membranes and cholesterol localized to membrane rafts [108]. Collectively, this very new and growing body of work, coupled with the stimulus-dependent redistribution of PS, PE, and cholesterol, provides an ideal model system in which to study membrane raft formation and dissolution in response to stimuli, because platelets have both weak and strong stimulators. An added benefit is the strong correlation between the cholesterol content of platelets and dyslipoproteinemias leading to increased risk of thrombosis.

In a series of studies, Gousset et al. [109] were able to visualize membrane microdomains proposed to be rafts in platelets. Using fluorescence microscopy techniques, these authors documented that, upon exposure to cold or platelet agonists, the lipid-sensitive dye partitions into gellike lipid domains. This segregation was disrupted upon depletion of platelet membrane cholesterol. Separation of lipids into distinct phases was also followed using Fourier transform infrared spectroscopy (FTIR). In these studies, two phase transitions were observed, one at 15°C and a second at 30°C. The magnitude of the second phase separation was found to decrease upon cholesterol depletion, again suggesting multiple membrane domains within stimulated platelets. Moreover, analogous to other studies, the DRMs isolated from the platelets were enriched in CD36 but contained no detectable levels of CD55 or the major platelet integrin protein α IIbβ3A. These authors have suggested that raft formation is a dynamic, reversible event triggered by activation. Little evidence is provided to confirm the reversible nature of this domain formation. It has been well documented that some platelet agonists mediate reversible platelet aggregation, whereas stronger agonists mediate irreversible aggregation and rapid phospholipid translocation. It will be intriguing to determine how membrane raft formation is correlated with platelet phospholipid translocation and the reversible vs. irreversible processes.

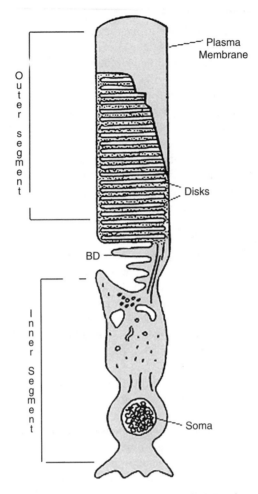

FIGURE 9.3 Schematic representation of photoreceptor rod cell. BD refers to newly forming basal disks.

9.4.2 PHOTORECEPTOR ROD CELLS

Photoreceptor rod cells are a unique, postmitotic cell type in which membrane rafts have just recently been isolated. To fully appreciate the significance and functional role of rafts in these cells, we must first consider the structure of the rod cell and its location in the retina. At the basal end, the synapse of the rod cell interacts with a complex array of retinal neural cells. These cells ultimately are responsible for transmitting the visual signal to the brain visual cortex via the optic nerve. The apical end of the outer segment is surrounded by the pigment epithelia. As illustrated in Figure 9.3, the rod cell consists of an inner segment and an outer segment. The inner segment contains nuclei and other organelles required for protein synthesis and the majority of the biochemical machinery for normal metabolic processes. The outer segment is composed of a stack of flattened membrane vesicles, the disks that are surrounded by the plasma membrane. It is the outer segment disk membranes that contain the requisite photopigment, rhodopsin, that carries out the visual transduction processes.

Although the disk membranes and the plasma membrane are distinct membranes with independent functions, rhodopsin is the only membrane protein within the ROS that is ubiquitously distributed. This prototypical GPCR [110] is the dominant protein in both the plasma membrane (60% total protein [111]) and, in slightly higher abundance, in disk membranes (85% total protein [112–114]). In contrast, the ion transport proteins responsible for maintaining normal retinal function, i.e., Na^{+1}-Ca^{+2} exchanger [115] and the cGMP-gated channel, α and β subunits [116], are found exclusively in the plasma membrane. Although the cell does not divide, the outer segment is constantly undergoing renewal as new disks are formed at the base, originally arising from evaginations of the plasma membrane [117], and are progressively displaced toward the apical tip. Old disks at the apical tip of the rod are shed and then phagocytosed by the overlying pigmented epithelium. Thus, the outer segment is in a constant state of degradation and renewal [118, 119]. In vertebrates, the transit of disks from the base to the tip of the outer segment requires approximately 10 days [120]. Within an individual rod outer segment, the disk membranes range from newly synthesized at the base to 10 days old at the apical tip.

The distribution of proteins within the various membranes of the ROS is mirrored by a heterogeneous distribution of lipid components. Certain lipids exhibit a preferential localization to the plasma membrane of the ROS compared to the disk membranes. For example, the plasma membrane is enriched in cholesterol [121], (fourfold compared to phospholipid), in unsaturated fatty acids species, in phosphatidylcholine (PC) relative to phosphatidylethanolamine (PE) [122], and in squalene, a precursor of both isoprenyl groups and cholesterol [123] relative to the majority of mature disk membranes. The disk membranes also exhibit a heterogeneous lipid distribution: newly formed disks have sixfold more cholesterol than disks at the apical tip of the ROS [124, 125], although there is little change in the phospholipid species. The predominant change occurs in the relative distribution of fatty acids within the various classes of phospholipids. Within the PC class, 16:0 dramatically decreases with disk age, whereas the 22:6 increases with disk age. Although the PE class exhibits some fatty acid changes, they are small. The PS class exhibits no significant changes in fatty acid composition. The PI class, which constitutes less than 2% of the total phospholipid, exhibits age-related changes in each of the fatty acids measured. Most notable of these is an increase in 20:4 as the disks are apically displaced.

The rhodopsin originally inserted into the newly forming disk membrane remains with the disk as it transits the length of the outer segment [126, 127]. In contrast, the lipid components of the disk bilayer exchange between disk membranes and undergo metabolic turnover [128, 129]. As the disks age, their lipid composition is altered. The remodeling of disk membranes may be related to the phototransduction process or to preparation for eventual disk phagocytosis. The functional impact of changes that occur in disk membranes as they transit from the base of the outer segment to the apical tip is of great interest. Disks must have the necessary properties to undergo morphogenesis at the base of the disk. They must then be competent to carry out visual transduction. Finally, they must undergo fusion with the plasma membrane to form packets that are then phagocytosed. Therefore, functional impact could be manifest in the ability of disks to carry out visual transduction or to maintain dynamic renewal of the outer segment. The modification of disk membranes as they are apically displaced in the outer segment places the modulation of rhodopsin function and structure in a dynamic context. It is likely that the disk membrane lipid composition facilitates the conformational changes of rhodopsin required for phototransduction.

9.4.2.1 Photoreceptor Membrane Microdomains

In a manner analogous to other biological membranes containing lipid rafts, early biophysical studies of rod outer segment membranes have implied a unique organization of lipids within lateral domains of the disks. More recently, lipid microdomains have been identified in disk membranes, with the observation of a cholesterol-dependent recruitment of di22:6-PC by rhodopsin into lateral domains [130]. The cluster of rhodopsin and lipid that formed the domain exceeded two adjacent lipid layers. This implies interactions between rhodopsin and the lipid environment that extend beyond direct protein–lipid interactions. The interplay of rhodopsin and the bilayer lipids is thus not simply one of providing a hydrophobic environment or of providing an environment of the appropriate "fluidity." Rather, it involves a process by which rhodopsin and the lipids are organized to encase the protein in an environment that permits the reversible changes of the visual cycle while simultaneously stabilizing the protein with respect to the kinetically driven denaturation [131]. Furthermore, it has recently been shown that the phospholipids of the rod outer segment reorganize in response to light [132].

Therefore, rhodopsin-lipid interactions can be viewed at several independent levels:

- Direct protein–lipid interactions, which likely involve a small number of lipids
- The effect of the lipids on protein stability
- The ability of the lipids and rhodopsin to organize into complex lateral domains

To begin to understand the interrelationship of rhodopsin structure, stability, and function in its membrane environment requires an appreciation of the rhodopsin–lipid complex as an integrated functional unit.

With these considerations in mind, the concept of rafts in the rod outer segment disks becomes an extremely interesting problem. DRMs have been isolated from rod outer segments [58]. In these initial studies, very little transducin or cGMP phosphodiesterase (PDE) was found to be DRM associated in dark-adapted ROS membranes. Upon exposure of the ROS to light, the isolated DRMs were shown to contain a larger percentage of transducin. Moreover, the addition of the nonhydrolyzable GTP analog GTPγS resulted in a decrease in raft-associated transducin in the light-stimulated membranes. Most interesting is the observation that, upon light stimulation and the addition of GTPγS, the PDE became raft associated. The depletion of membrane cholesterol with methyl-β-cyclodextran had no effect on transducin association with membrane rafts but decreased PDE association, suggesting that these two phototransduction proteins are associated with the raft through different mechanisms.

Although these studies provide an intriguing picture of the assembly of phototransduction molecules, they did not include an analysis of the lipid composition. Therefore, the similarity in membrane composition of these DRMs to those isolated from other cells or to later, ROS-specific DRMs was not established. Rhodopsin was found associated with both DRMs and soluble fractions, whereas guanylate cyclase was localized almost exclusively to the DRMs [133]. These studies document a light-dependent localization of transducin, RGS9-1-G 5L, and the p44 isoform of arrestin to DRMs.

More recently, Boesze-Battaglia et al. [134] have shown that Triton X-100–resistant disk membrane microdomains, high in cholesterol (0.24 cholesterol to phospholipid mole:mole ratio) and in sphingomyelin (0.11 sphingomyelin to phospholipid mole:mole ratio) contain caveolin and the tetraspanin rim–specific protein rom-1. When the Triton X-100–resistant membranes were treated with β-methyl-cyclodextran to deplete membrane cholesterol, the resultant membrane contained slightly lower levels of rom-1, specifically in the dimeric form. Cholesterol depletion also resulted in the collapse of the large caveolin complex to a monomeric caveolae.

The interpretation of these data derived from DRMs isolated from disk membranes is challenging because native disk membranes are known to have a nonuniform cholesterol composition [123, 124]. Cholesterol is high in the disk membranes of basal disks and low in the membranes of apical disks [124]. The fatty acyl composition of phospholipids is also modified as the disks are displaced, increasing the level of docosohexaenoic (DHA) 22:6 of apical disks.

It is likely that the disk membrane lipid composition optimizes the ability of rhodopsin to undergo conformational changes required for function. Therefore, microdomains of specific lipids are certainly relevant to visual transduction. It is important to understand the implications of the changes in disk membrane lipid composition. The isolation of cholesterol-enriched membrane rafts from ROS membranes complements earlier studies, in which the photolytic properties of rhodopsin were shown to be sensitive to the composition of the surrounding bilayer. The influence of lipid dynamics on rhodopsin photolysis has been investigated in rhodopsin–lipid reconstituted bilayers. These studies demonstrated that systematic changes in the cholesterol composition of reconstituted phosphatidylcholine–rhodopsin vesicles have profound changes on the activated form of rhodopsin, Metarhodopsin II. These changes corresponded to changes in the dynamics of the bilayer hydrocarbon region [135–139]. These studies indicate that cholesterol can affect rhodopsin function by reducing the partial free volume in the hydrocarbon core of the bilayer. The transition of Meta I to Meta II involves an expansion of the protein in the plane of the bilayer. At high levels of cholesterol, this volume is then unavailable to rhodopsin and the transition is inhibited.

As may be expected, cholesterol is not alone in its ability to alter the Meta I to Meta II transition. The phospholipid hydrocarbon chain composition in reconstituted bilayers has been shown to regulate the Meta I/Meta II equilibrium. In particular, docosahexaneoic acid (DHA) strongly promotes the formation of Meta II [140]. This fatty acid is conserved in the retina, accounting for over 50% of the phospholipid hydrocarbon chains in the disk membranes. Not only does this highly unsaturated lipid modulate rhodopsin function in the membrane, but deprivation of this essential fatty acid degrades visual acuity as measured by electroretinograms [141–144]. Thus, the affects seen on the molecular level have predicted consequences on retinal physiology. This suggests that the lipids are responsible for modulating the dynamic environment of the membrane, which in turn can affect the properties of the protein. It has been shown that the role of the unsaturated acyl chains in the sn-2 position of phospholipids is distinct from that of cholesterol in modulating the MetaI/MetaII equilibrium [143]. It clear that docosahexaenoic acid plays an important role in normal rod function, but the mechanism is not fully understood [142]. It is intriguing to consider that recent studies have shown that DHA contains a PE phase separate from membrane rafts, whereas the monounsaturated oleic acid containing PE does not [144]. Thus, the different subpopulations of PE in ROS may impart different characteristics to the ROS-specific membrane rafts.

9.5 SUMMARY

As newly developing technology allows us both to image and to quantitatively analyze membrane raft lipids [145, 146], we have the tools to correlate membrane raft localization in the plasma membrane with biochemical characterization of these domains. Analysis of the literature should proceed with caution, because a large number of studies do not provide even a cursory evaluation of the lipid composition of detergent-resistant fractions or correlate *in vitro* biochemical observations with images of microdomains within a given cell.

REFERENCES

1. Davson, H. and Danielli, J.F., *The Permeability of Natural Membranes*, Cambridge University Press, Cambridge, U.K., 1952, p. 365.
2. Gorter, E. and Grendel, F., On biomolecular layers of lipids on the chromatocytes of the blood, *J. Exp. Med.* 41, 439–443, 1925.
3. Singer, S.J. and Nicolson, G.L., The fluid mosaic model of the structure of cell membranes, *Science* 175, 720–731, 1972.
4. Simons, K. and van Meer, G., Lipid sorting in epithelial cells, *Biochemistry* 27, 6197–6202, 1988.
5. Rodriguez-Boulan, E. and Nelson, W.J., Morphogenesis of the polarized epithelial cell phenotype, *Science* 245, 718–725, 1989.
6. Repasky, E.A. and Gregorio, C.C., Plasma membrane skeletons, in *The Structure of Biological Membranes*, Yeagle, P., CRC Press, Boca Raton, FL, 1992, pp. 449–507.
7. Jost, P.C., Nadakavukaren, K.K., and Griffith, O.H., Phosphatidylcholine exchange between the boundary lipid and bilayer domains in cytochrome oxidase containing membranes, *Biochemistry* 16, 3110–3114, 1977.
8. Boggs, J.M., Clement, I.R., and Moscarello, M.A., Similar effects of proteolipid apoproteins from human myelin (lipophilin) and bovine white matter on the lipid phase transition, *Biochim. Biophys. Acta* 601, 134–151, 1980.

9. Lentz, B.R. et al., Ordered and disordered phospholipid domains coexist in membranes containing the calcium pump protein of sarcoplasmic reticulum, *Proc. Natl. Acad. Sci. USA* 80, 2917–2921, 1983.

10. Yeagle, P.L. and Kelsey, D., Phosphorus nuclear magnetic resonance studies of lipid-protein interactions: human erythrocyte glycophorin and phospholipids, *Biochemistry* 28, 2210–2215, 1989.

11. Simons, K. and Ikonen, E., Functional rafts in cell membranes, *Nature* 387, 569–572, 1997.

12. Zajchowski, L.D. and Robbins, S.M., Lipid rafts and little caves. Compartmentalized signalling in membrane microdomains, *Eur. J. Biochem.* 269, 737–752, 2002.

13. Isshiki, M. and Anderson, R.G.W., Calcium signal transduction from caveolae, *Cell Calcium* 26, 201–208, 1999.

14. Simionescu, N., Cellular aspects of transcapillary exchange, *Physiol. Rev.* 63, 1536–1560, 1983.

15. Anderson, R.G.W. et al., Potocytosis: sequestration and transport of small molecules by caveolae, *Science* 255, 410–411, 1992.

16. Smart, E.J. et al., Caveolins, liquid-ordered domains, and signal transduction, *Mol. Cell. Biol.* 19, 7289–7304, 1999.

17. Parton, R.G., Joggerst, B., and Simons, K., Regulated internalization of caveolae, *J. Cell Biol.* 127, 1199–1215, 1994.

18. Fivaz, M., Abrami, L., and van der Goot, F.G., Landing on lipid rafts, *Trends Cell. Biol.* 9, 212–213, 1999.

19. Shin, J.-S., Gao, Z., and Abraham, S.N., Involvement of cellular caveolae in bacterial entry into mast cells, *Science* 289, 785–788, 2000.

20. Ono, A. and Freed, E.O., Plasma membrane rafts play a critical role in HIV-1 assembly and release, *Proc. Natl. Acad. Sci. USA,* 98, 13925–13930, 2001.

21. Oram, J.F. and Yokoyama, S., Apolipoprotein-mediated removal of cellular cholesterol and phospholipids, *J. Lipid Res.* 37, 2473–2491, 1996.

22. Smart, E.J. et al., A role for caveolin in transport of cholesterol from endoplasmic reticulum to plasma membrane, *J. Biol. Chem.* 271, 29427–29435, 1996.

23. Brown, D.A., Seeing is believing: visualization of rafts in model membranes, *PNAS* 98, 10517–10518, 2001.

24. Prydz, K. and Simons, K., Cholesterol depletion reduces apical transport capacity in epithelial Madin-Darby canine kidney cells, *Biochem. J.* 357, 11–15, 2001.

25. van der Goot, F.G. and Harder, T., Raft membrane domains: from a liquid-ordered membrane phase to a site of pathogen attack, *Semin. Immunol.* 13, 89–97, 2001.

26. Brown, R.E., Sphingolipid organization in biomembranes: what physical studies of model membranes reveal., *J. Cell Sci.* 111, 1–9, 1998.

27. De Kruyjff, B. et al., The effect of the polar headgroup on the lipid-cholesterol interaction: a monolayer and differential scanning calorimetry study, *Biochim. Biophys. Acta* 307, 1–19, 1973.

28. Ipsen, J.H. et al., Phase equilibria in the phosphatidylcholine-cholesterol system, *Biochim. Biophys. Acta* 905, 162–172, 1987.

29. Maulik, P.R. and Shipley, G.G., Interactions of N-stearoyl sphingomyelin with cholesterol and dipalmitoylphosphatidylcholine in bilayer membranes, *Biophys. J.* 70, 2256–2265, 1996.

30. Brown, D.A. and London, E., Structure and origin of ordered lipid domains in biological membranes, *J. Membr. Biol.* 164, 103–114, 1998.

31. White, D.A., The phospholipid composition of mammalian tissues, in *Progress in Biophysics and Molecular Biology,* Anseel, G.B., Hawthorne, J.D., and Dawson, R.M.C., Eds., Elsevier Scientific Publication Company, New York, 1973, pp. 441–482.

32. Brown, D.A. and London, E., Structure of detergent-resistant membrane domains: does phase separation occur in biological membranes?, *Biochem. Biophys. Res. Commun.* 240, 1–7, 1997.

33. Sankaram, M.B. and Thompson, T.E., Cholesterol-induced fluid-phase immiscibility in membranes, *Proc. Natl. Acad. Sci. USA,* 88, 8686–8690, 1991.

34. Carter, W.G. and Hakomori, S., A new cell surface, detergent-insoluble glycoprotein matrix of human and hamster fibroblasts. The role of disulfide bonds in stabilization of the matrix, *J. Biol. Chem.* 256, 6953–6960, 1981.

35. Okada, Y. et al., Glycosphingolipids in detergent-insoluble substrate attachment matrix (DISAM) prepared from substrate attachment material (SAM). Their possible role in regulating cell adhesion, *Exp. Cell Res.* 155, 448–456, 1984.

36. Lisanti, M.P. et al., Polarized apical distribution of glycosyl-phosphatidylinositol-anchored proteins in a renal epithelial cell line, *Proc. Natl. Acad. Sci. USA* 85, 9557–9561, 1988.

37. Brown, D.A., Crise, B., and Rose, J.K., Mechanism of membrane anchoring affects polarized expression of two proteins in MDCK cells, *Science* 245, 1499–1501, 1989.

38. Scheek, S., Brown, M.S., and Goldstein, J.L., Sphingomyelin depletion in cultured cells blocks proteolysis of sterol regulatory element binding proteins at site 1, *Proc. Natl. Acad. Sci. USA,* 94, 11179–11183, 1997.

39. Slotte, J.P. and Bierman, E.L., Depletion of plasma-membrane sphingomyelin rapidly alters the distribution of cholesterol between plasma membranes and intracellular cholesterol pools in cultured fibroblasts, *Biochem. J.* 250, 653–658, 1988.

40. Lange, Y. and Steck, T.L., Quantitation of the pool of cholesterol associated with acyl-CoA:cholesterol acyltransferase in human fibroblasts, *J. Biol. Chem.* 272, 13103–13108, 1997.

41. Ge, M. et al., Electron spin resonance characterization of liquid ordered phase of detergent-resistant membranes from RBL-2H3 cells, *Biophys J.* 77, 925–933, 1999.

42. Ahmed, S.N., Brown, D.A., and London, E., On the origin of sphingolipid/cholesterol-rich detergent-insoluble cell membranes: physiological concentrations of cholesterol and sphingolipid induce formation of a detergent-insoluble, liquid-ordered lipid phase in model membranes, *Biochemistry* 36, 10944–10953, 1997.

43. Brown, D.A., The tyrosine kinase connection: how GPI-anchored proteins activate T cells, *Curr. Opin. Imunol.* 5, 349–354, 1993.

44. Stefanova, I. et al., GPI-anchored cell-surface molecules complexed to protein tyrosine kinases, *Science* 254, 1016–1019, 1991.

45. Cinek, T. and Horejsi, V., The nature of large noncovalent complexes containing glycosyl-phosphatidylinositol-anchored membrane glycoproteins and protein tyrosine kinases, *J. Immunol.* 149, 2262–2270, 1992.

46. Thomas, P.M. and Samelson, L.E., The glycophosphatidylinositol-anchored Thy-1 molecule interacts with the p60fyn protein tyrosine kinase in T cells, *J. Biol. Chem.* 267, 12317–12322, 1992.

47. Shenoy, S.A. et al., Signal transduction through decay-accelerating factor. Interaction of glycosyl-phosphatidylinositol anchor and protein tyrosine kinases p56lck and p59fyn 1, *J. Immunol.* 149, 3535–3541, 1992.

48. Brown, D.A. and Rose, J.K., Sorting of GPI-anchored proteins to glycolipid-enriched membrane subdomains during transport to the apical cell surface, *Cell* 68, 533–544, 1992.

49. Kurzchalia, T.V. et al., VIP21, a 21-kD membrane protein is an integral component of trans-Golgi-network-derived transport vesicles, *J. Cell Biol.* 118, 1003–1014, 1992.

50. Dupree, P. et al., Caveolae and sorting in the trans-Golgi network of epithelial cells, *EMBO J.* 12, 1597–1605, 1993.

51. Anderson, R.G.W., Caveolae: where incoming and outgoing messengers meet, *Proc. Natl. Acad. Sci. USA* 90, 10909–10913, 1993.

52. Fra, A.M. et al., Detergent insoluble glycolipid microdomains in lymphocytes in the absence of caveolae, *J. Biol. Chem.* 269, 30745–30748, 1994.

53. Volonte, D. et al., Flotillins/cavatellins are differentially expressed in cells and tissues and form a hetero-oligomeric complex with caveolins *in vivo*. Characterization and epitope-mapping of a novel flotillin-1 monoclonal antibody probe, *J. Biol. Chem.* 274, 12702–12709, 1999.

54. Hagmann, J. and Fishman, P.H., Detergent extraction of cholera toxin and gangliosides from cultured cells and isolated membranes, *Biochem. Biophys. Acta* 720, 181–187, 1982.

55. Anderson, R.G.W. and Jacobson, K., A role for lipid shells in targeting proteins to caveolae, rafts, and other lipid domains, *Science* 296, 1821–1825, 2002.

56. Mayor, S. and Maxfield, F.R., Insolubility and redistribution of GPI-anchored proteins at the cell surface after detergent treatment, *Mol. Biol. Cell* 6, 929–944, 1995.

57. Mayor, S., Rothberg, K.G., and Maxfield, F.R., Sequestration of GPI-anchored proteins in caveolea triggered by cross-linking, *Science* 264, 1948–1951, 1994.

58. Seno, K. et al., Light- and guanosine 5'-3-O-(thio)triphosphate-sensitive localization of a G protein and its effect on detergent-resistant membrane rafts in rod photoreceptor outer segments, *J. Biol. Chem.* 276, 20813–20816, 2001.

59. Claas, C., Stipp, C.S., and Hemler, M.E., Evaluation of prototype transmembrane 4 superfamily protein complexes and their relation to lipid rafts, *J. Biol. Chem.* 276, 7974–7984, 2001.

60. Wright, D.M. and Thomlinson, M.G., The ins and outs of the transmembrane 4 superfamily, *Immunol. Today* 15, 588–594, 1994.

61. Maecker, H.T., Todd, S.C., and Levy, S., The tetraspanin superfamily: molecular facilitators, *FASEB J.* 11, 428–442, 1997.

62. Hemler, M.E., Mannion, B.A., and Berditchevski, F., Association of TM4SF proteins with integrins: relevance to cancer, *Biochim. Biophys. Acta* 1287, 67–71, 1996.

63. Yauch, R.L. et al., Highly stoichiometric, stable, and specific association of integrin alpha3beta1 with CD151 provides a major link to phosphatidylinositol 4-kinase, and may regulate cell migration, *Mol. Biol. Cell* 9, 2751–2765, 1998.

64. Yauch, R.L. et al., Direct extracellular contact between integrin alpha(3)beta(1) and TM4SF protein CD151, *J. Biol. Chem.* 275, 9230–9238, 2000.

65. Glenney, J.R., Jr., The sequence of human caveolin reveals identity with VIP21, a component of transport vesicles, *FEBS Lett.* 314, 45–48, 1992.

66. Rothberg, K.G. et al., Caveolin, a protein component of cell membrane coats, *Cell* 68, 673–682, 1992.

67. Scherer, P.E. et al., Caveolin isoforms differ in their N-terminal protein sequence and subcellular distribution. Identification and epitope mapping of an isoform-specific monoclonal antibody probe, *J. Biol. Chem.* 270, 16395–16401, 1995.

68. Tang, Z. et al., Molecular cloning of caveolin-3, a novel member of the caveolin gene family expressed predominantly in muscle, *J. Biol. Chem.* 271, 2255–2261, 1996.

69. Lindwasser, O.W. and Resh, M.D., Multimerization of human immunodeficiency virus type 1 Gag promotes its localization to barges, raft-like membrane microdomains, *J. Virol.* 75, 7913–7924, 2001.

70. Anderson, R.G. and Jacobson, K., A role for lipid shells in targeting proteins to caveolae, rafts, and other lipid domains, *Science* 296, 1821–1825, 2002.

71. Varma, R. and Mayor, S., GPI-anchored proteins are organized in submicron domains at the cell surface, *Nature* 394, 798–801, 1998.

72. Sheets, E.D. et al., Transient confinement of a glycosylphosphatidylinositol-anchored protein in the plasma membrane, *Biochemistry* 36, 12449–12458, 1997.

73. Schutz, G.J. et al., Properties of lipid microdomains in a muscle cell membrane visualized by single molecule microscopy, *EMBO J.* 19, 892–901, 2000.

74. Pralle, A. et al., Sphingolipid-cholesterol rafts diffuse as small entities in the plasma membrane of mammalian cells, *J. Cell Biol.* 148, 997–1007, 2000.

75. Harder, T. et al., Lipid domain structure of the plasma membrane revealed by patching of membrane components, *J. Cell Biol.* 141, 929–942, 1998.

76. Dietrich, C. et al., Partitioning of Thy-1, GM1, and cross-linked phospholipid analogs into lipid rafts reconstituted in supported model membrane monolayers, *PNAS* 98, 10642–10647, 2001.

77. Dietrich, C. et al., Relationship of lipid rafts to transient confinement zones detected by single particle tracking, *Biophys. J.* 82, 274–284, 2002.

78. Ostermeyer, A.G. et al., Glycosphingolipids are not essential for formation of detergent-resistant membrane rafts in melanoma cells. Methyl-beta-cyclodextrin does not affect cell surface transport of a GPI-anchored protein, *J. Biol. Chem.* 274, 34459–34466, 1999.

79. Wu, C. et al., Tyrosine kinase receptors concentrated in caveolae-like domains from neuronal plasma membrane, *J. Biol. Chem.* 272, 3554–3559, 1997.

80. Shu, W. et al., Helical interactions in the HIV-1 gp41 core reveal structural basis for the inhibitory activity of gp41 peptides, *Biochemistry* 39, 1634–1642, 2000.

81. Furuchi, T. and Anderson, R.G., Cholesterol depletion of caveolae causes hyperactivation of extra-cellular signal-related kinase (ERK), *J. Biol. Chem.* 273, 21099–21104, 1998.

82. Haasemann, M. et al., Agonist-induced redistribution of bradykinin B2 receptor in caveolae, *J. Cell Sci.* 111, 917–928, 1998.

83. Xavier, R. et al., Membrane compartmentation is required for efficient T cell activation, *Immunity* 8, 723–732, 1998.

84. Cheng, P.C. et al., A role for lipid rafts in B cell antigen receptor signaling and antigen targeting, *J. Exp. Med.* 190, 1549–1560, 1999.

85. Fujimoto, T., GPI-anchored proteins, glycosphingolipids, and sphingomyelin are sequestered to caveolea only after cross-linking., *J. Histochem. Cytochem.* 44, 929–941, 1996.

86. Liu, P. and Anderson, R.G., Compartmentalized production of ceramide at the cell surface, *J. Biol. Chem.* 270, 27179–27185, 1995.

87. Hope, H.R. and Pike, L.J., Phosphoinositides and phosphoinositol utilizing enzymes in detergent-insoluble lipid domains, *Mol. Biol. Cell* 7, 843–851, 1996.

88. Stan, R.V. et al., Immunoisolation and partial characterization of endothelial plasmalemmal vesicles (caveolae), *Mol. Biol. Cell* 8, 595–605, 1997.

89. Robbins, S.M., Quintrell, N.A., and Bishop, J.M., Myristoylation and differential palmitoylation of the HCK protein-tyrosine kinases govern their attachment to membranes and association with caveolae, *Mol. Cell. Biol.* 15, 3507–3515, 1995.

90. Davy, A., Feuerstein, C., and Robbins, S.M., Signaling within a caveolae-like membrane microdomain in human neuroblastoma cells in response to fibroblast growth factor, *J. Neurochem.* 74, 676–683, 2000.

91. Oh, P. and Schnitze, R.J.E., Immunoisolation of caveolae with high affinity antibody binding to the oligomeric caveolin cage. Toward understanding the basis of purification, *J. Biol. Chem.* 274, 23144–23154, 1999.

92. Liu, J. et al., Organized endothelial cell surface signal transduction in caveolae distinct from glycosylphosphatidylinositol-anchored protein microdomains, *J. Biol. Chem.* 272, 7211–7222, 1997.

93. Schwencke, C. et al., Compartmentation of cyclic adenosine 3',5'-monophosphate signaling in caveolae, *Mol. Endo.* 13, 1061–1070, 1999.

94. Marmor, M.D. and Julius, M.J.A.B., Role for lipid rafts in regulating interleukin-2 receptor signaling, *Blood* 98, 1489–1497, 2001.

95. Deans, J.P. et al., Rapid redistribution of CD20 to a low density detergent-insoluble membrane compartment, *J. Biol. Chem.* 273, 344–348, 1998.

96. Martens, J.R. et al., Differential targeting of shaker-like potassium channels to lipid rafts, *J. Biol. Chem.* 275, 7443–7446, 2000.

97. Stulnig, T.M. et al., Polyunsaturated eicosapentaenoic acid displaces proteins from membrane rafts by altering raft lipid composition, *J. Biol. Chem.* 276 (40), 37335–37340, 2001.

98. Bevers, E.M. et al., Generation of prothrombin-converting activity and the exposure of phosphatidylserine at the outer surface of platelets, *Eur. J. Biochem.* 122, 429–436, 1982.

99. Shattil, S.J. et al., Platelet hypersensitivity induced by cholesterol incorporation, *J. Clin. Invest.* 55, 636–643, 1975.

100. Tandon, N. et al., Thrombin receptors define responsiveness of cholesterol-modified platelets, *J. Biol. Chem.* 258, 11840–11845, 1983.

101. Kramer, R.M. et al., Effect of membrane cholesterol on phospholipid metabolism in thrombin-stimulated platelets. Enhanced activation of platelet phospholipase(s) for liberation of arachidonic acid, *J. Biol. Chem.* 257, 6844–6849, 1982.

102. Tomizuka, T. et al., Hypersensitivity to thromboxane A2 in cholesterol-rich human platelets, *Thrombosis Haemostasis* 64, 594–599, 1990.

103. Schimmel, R.J. et al., Cholesterol enhances the adhesion of human platelets to fibrinogen: studies using a novel fluorescence based assay, *Platelets* 8, 261–267, 1997.

104. Boesze-Battaglia, K., Clayton, S.T., and Schimmel, R.J., Cholesterol redistribution within human platelet plasma membrane: evidence for a stimulus-dependent event, *Biochemistry* 35, 6664–6673, 1996.

105. Dorahy, D.J. et al., Biochemical isolation of a membrane microdomain from resting platelets highly enriched in the plasma membrane glycoprotein CD36, *Biochem. J.* 319, 67–72, 1996.

106. Bodin, S. et al., Production of phosphatidylinositol 3,4,5-triphosphate and phosphatidic acid in platelet rafts: evidence for a critical role of cholesterol-enriched domains in human platelet activation, *Biochemistry* 40(50), 15290–15299, 2001.

107. Fridriksson, E.K. et al., Quantitative analysis of phospholipids in functionally important membrane domains from RBL-2H3 mast cells using tandem high-resolution mass spectrometry, *Biochemistry* 38, 8056–8063, 1999.

108. Kunzelmann-Marche, C., Freyssine, T.J.M. and Martinez, M.C., Loss of plasma membrane phospholipid asymmetry requires raft integrity. Role of transient receptor potential channels and ERK pathway, *J. Biol. Chem.* 277, 19876–19881, 2002.

109. Gousset, K. et al., Evidence for a physiological role for membrane rafts in human platelets, *J. Cell. Phys.* 190, 117–128, 2002.

110. Applebury, M.L., Relationships of G-protein-coupled receptors. A survey with the photoreceptor opsin subfamily, *Soc. Gen. Phys. Ser.* 49, 235–248, 1994.

111. Molday, R.S., Structural properties of the peripherin/rds-rom-1 complex of disk membranes. [ARVO abstract], *Invest. Ophthalmol. Vis. Sci.* 39, 1998.

112. Nir, I. and Papermaster, D.S., Differential distribution of opsin in the plasma membrane of frog photoreceptors: an immunocytochemical study, *Invest. Ophthalmol. Vis. Sci.* 24, 868–878, 1983.

113. Hicks, D. and Molday, R.S., Differential immunogold-dextran labeling of bovine and frog rod and cone cells using monoclonal antibodies against bovine rhodopsin, *Exp. Eye Res.* 42, 55–71, 1986.

114. Polaws, A.S., Altman, L.G., and Papermaster, D.S., Immunocytochemical binding of anti-opsin N-terminal specific antibodies to the extracellular surface of rod outer segment plasma membranes, *J. Histochem. Cytochem.* 34, 659–664, 1986.

115. Reid, D.M. et al., Identification of the sodium-calcium exchanger as the major ricin-binding glyco-protein of bovine rod outer segments and its localization to the plasma membrane, *Biochemistry* 29, 1601–1607, 1990.

116. Cook, N.J. et al., The c-GMP-gated channel of bovine rod photoreceptors is localized exclusively in the plasma membrane, *J. Biol. Chem.* 264, 6996–6999, 1989.

117. Anderson, D.H., Fisher, S.K., and Steinberg, R.H., Mammalian cones: disc shedding, phagocytosis and renewal, *Invest. Ophthalmol* 17, 117–133, 1978.

118. Hall, M.O., Bok, D., and Bacharach, A.D., Visual pigment renewal in the mature frog retina, *Science* 161, 787–789, 1968.

119. Young, R.W., The renewal of photoreceptor cell outer segments, *J. Cell Biol.* 33, 61–72, 1967.

120. Young, R.W., Shedding of discs from rod outer segments in the rhesus monkey, *J. Ultrastruct. Res.* 34, 190–203, 1971.

121. Boesze-Battaglia, K. and Albert, A.D., Phospholipid distribution among bovine rod outer segment plasma membrane and disk membranes, *Exp. Eye Res.* 54, 821–823, 1992.

122. Fliesler, S.J. et al., Squalene is localized to the plasma membrane in bovine retinal rod outer segments, *Exp. Eye Res.* 64, 279–282, 1997.

123. Boesze-Battaglia, K., Hennessey, T., and Albert, A.D., Cholesterol heterogeneity in bovine retinal rod outer segment disk membranes, *J. Biol. Chem.* 264, 8151–8155, 1989.

124. Boesze-Battaglia, K., Fliesler, S.J., and Albert, A.D., Relationship of cholesterol content to spatial distribution and age of disc membranes in retinal rod outer segments, *J. Biol. Chem.* 265, 18867–18870, 1990.

125. Albert, A.D., Young, J.E., and Paw, Z., Phospholipid fatty acyl spatial distribution in bovine rod outer segment disk membranes, *Biochem. Biophys. Acta* 1368, 52–60, 1998.

126. Bibb, C. and Young, R.W., Renewal of fatty acids in the membranes of visual cells outer segments, *J. Cell Biol.* 61, 327–343, 1974.

127. Bibb, C. and Young, R.W., The renewal of glycerol in the visual cells and pigment epithelium of the frog retina, *J. Cell Biol.* 62, 378, 1974.

128. Anderson, R.E. et al., Synthesis and turnover of lipid and protein components of frog retinal rod outer segments, *Neurochemistry* 1, 29–42, 1980.

129. Anderson, R.E., Kelleher, P.A., and Maude, M.B., Metabolism of phosphatidylethanolamine in the frog retina, *Biochim. Biophys. Acta* 620, 227–235, 1980.

130. Polozova, A. and Litman, B.J., Cholesterol dependent recruitment of di22:6-PC by a G protein-coupled receptor into lateral domains, *Biophys J.* 79, 2632–2643, 2000.

131. Landin, J.S., Katragadda, M., and Albert, A.D., Thermal destabilization of rhodopsin and opsin by proteolytic cleavage in bovine rod outer segment disk membranes, *Biochemistry* 40, 11176–11183, 2001.

132. Hessel, E. et al., Light-induced reorganization of phospholipids in rod disc membranes, *J. Biol. Chem.* 276, 2538–2543, 2001.

133. Nair, K.S., Balasubramanian, N., and Slepak, V.Z., Signal-dependent translocation of transducin, RGS9-1-Gbeta5L complex, and arrestin to detergent-resistant membrane rafts in photoreceptors, *Curr. Biol.* 12, 421–425, 2002.

134. Boesze-Battaglia, K., Dispoto, J., and Kahoe, M.A., Association of a photoreceptor-specific tetraspanin protein, ROM-1, with Triton X-100-resistant membrane rafts from rod outer segment disk membranes, *J. Biol. Chem.* 277, 41843–41849, 2002.

135. Straume, M. and Litman, B.J., Influence of cholesterol on equilibrium and dynamic bilayer structure of unsaturated acyl chain phosphatidylcholine vesicles as determined from higher order analysis of fluorescence anisotropy decay, *Biochemistry* 26, 5121–5126, 1987.

136. Straume, M. and Litman, B.J., Equilibrium and dynamic bilayer structural properties of unsaturated acyl chain phosphatidylcholine-cholesterol-rhodopsin recombinant vesicles and rod outer segment disk membranes as determined from higher order analysis of fluorescence anisotropy decay, *Biochemistry* 27, 7723–7733, 1988.

137. Straume, M. et al., Interconversion of metarhodopsins I and II: a branched photointermediate decay model, *Biochemistry* 29, 9135–9142, 1990.

138. Straume, M. and Litman, B.J., Equilibrium and dynamic structure of large, unilamellar, unsaturated acyl chain phosphatidylcholine vesicles. Higher order analysis of 1,6-diphenyl-1,3,5-hexatriene and 1-[4-(trimethylammonio)phenyl]-6-phenyl-1,3,5-hexatriene anisotropy decay, *Biochemistry* 26, 5113–5120, 1987.

139. Mitchell, D.C. et al., Modulation of metarhodopsin formation by cholesterol-induced ordering of bilayer lipids, *Biochemistry* 29, 9143–9149, 1990.

140. Mitchell, D.C., Straume, M., and Litman, B.J., Role of sn-1-saturated,sn-2-polyunsaturated phospholipids in control of membrane receptor conformational equilibrium: effects of cholesterol and acyl chain unsaturation on the metarhodopsin I in equilibrium with metarhodopsin II equilibrium, *Biochemistry* 31, 662–670, 1992.

141. Hamosh, M. and Salem, N.J., Long-chain polyunsaturated fatty acids, *Biol. Neonate* 74, 106–120, 1998.

142. Mitchell, D.C. and Litman, B.J., Molecular order and dynamics in bilayers consisting of highly polyunsaturated phospholipids, *Biophys. J.* 74, 879–891, 1998.

143. Birch, E.E. et al., Visual acuity and the essentiality of docosahexaenoic acid and arachidonic acid in the diet of term infants, *Pediatric Res.* 44, 201–209, 1998.

144. Shaikh, S.R. et al., Monounsaturated PE does not phase-separate from the lipid raft molecules sphingomyelin and cholesterol: role for polyunsaturation?, *Biochemistry* 41, 10593–10602, 2002.

145. Ekroos, K. et al., Quantitative profiling of phospholipids by multiple precursor ion scanning on a hybrid quadrupole time-of-flight mass spectrometer, *Analytical Chem.* 74, 941–949, 2002.

146. Keller, P. et al., Multicolour imaging of post-Golgi sorting and trafficking in live cells, *Nat. Cell Biol.* 3, 140–149, 2001.

10 Passive and Facilitated Transport

Shinpei Ohki and Robert A. Spangler

CONTENTS

10.1 PASSIVE AND CARRIER-MEDIATED TRANSPORT

The passive transport of materials across homogeneous membrane systems will be the main focus of this chapter. Strictly speaking, passive translocation of a substance occurs as a result of a gradient in the electrochemical potential of the transported species, and always occurs in the downhill direction of that gradient. In contrast, the coupling of transmembrane movement of a solute with a biochemical reaction, such that its flow may be counter to its electrochemical potential gradient, is known as active transport; this phenomenon is described elsewhere (e.g., Stein, 1986; Lingrel and Kuntzweiler, 1994).

10.1.1 NONELECTROLYTE TRANSPORT

10.1.1.1 Equilibrium Relationships

At equilibrium, there is no net substance transport across the membrane. Equilibrium requires that the electrochemical potentials of any permeable species, including the solvent, be equal on the two sides of the membrane. In circumstances where the membrane is not permeable to at least one species (a semipermeable membrane), however, attainment of equilibrium in the solvent distribution may require the development of a hydrostatic pressure difference across the membrane. This manifestation of osmotic pressure and the physicochemical factors underlying it, are important considerations in biological membrane systems. Using a simple case as a model, we will briefly explore its nature.

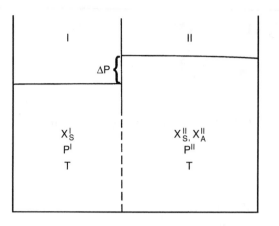

FIGURE 10.1 Schematic diagram for osmotic pressure across a semipermeable membrane.

Consider a system comprising two subsystems (denoted I and II), separated by a rigid membrane, which is ideally semipermeable to the solvent, s, alone. Dissolved in the fluid on side II is a quantity of solute A. As a result of the presence of A, the chemical potential of the solvent on side II will be lower than that on side I, with a consequent flow of solvent from I to II. This process will continue until a hydrostatic pressure difference develops that is sufficient to restore equilibrium (see Figure 10.1).

Equilibrium is achieved when the chemical potential, μ_s, is equal on the two sides of the membrane:

$$\mu_s^I = \mu_s^{II} \tag{10.1}$$

$$\mu_s^o (P^{II},T) + RT \ln\ X_s^{II} = \mu_s^o (P^I,T) \tag{10.2}$$

in which X_s^{II} is the mole fraction of solvent on side II ($X_s = n_s/(n_s + n_A)$ and n_j is the number of moles of the jth component). The dependence of chemical potential on pressure is given by:

$$\frac{\partial \mu_s}{\partial P} = \bar{v}_s \tag{10.3}$$

where $\bar{v}_s$ is the partial molar volume. This equation is derived from the differential form for the Gibbs free energy:

$$dG = VdP - SdT + \Sigma \mu_i dn_i \tag{10.4}$$

from which we have the partial derivatives

$$V = \frac{\partial G}{\partial P} \qquad \mu_i = \frac{\partial G}{\partial n_i} \tag{10.5}$$

and

$$\frac{\partial \mu_i}{\partial P} = \frac{\partial^2 G}{\partial P \partial n_i} = \frac{\partial^2 G}{\partial n_i \partial P} = \frac{\partial V}{\partial n_i} \equiv \bar{v}_s \tag{10.6}$$

Assuming constancy of the partial molar volume, that is, incompressibility of the solvent, the chemical potential is expressed as

$$\mu_s = \mu_s^o (P_o, T) + RT \ln X_s + \overline{v}_s (P - P_o) \tag{10.7}$$

From Equation 10.2 we have

$$RT \ln X_s^{II} = \overline{v}_s (P^I - P^{II}) \tag{10.8}$$

Because the mole fractions of solvent and solute are not independent, such that $X_s + X_A = 1$, the mole fraction of solvent can conveniently be expressed in terms of the solute. Moreover, if the solution is sufficiently dilute

$$P^I - P^{II} = \frac{RT}{\overline{v}_s} \ln(1 - X_A) \cong -\frac{RTX_A}{\overline{v}_s} \cong -RTC_A \tag{10.9}$$

where C_A is the concentration of solute on side II. Generalizing to the instance of finite solute concentration on both sides of the membrane, we have for the equilibrium hydrostatic pressure difference

$$P^{II} - P^I = RT \left(C_A^{II} - C_A^I \right) \tag{10.10}$$

Independent of the existence of equilibrium, however, the depression in the chemical potential of water resulting from dissolved solute(s), expressed in units of pressure, is denoted by π, the osmotic pressure of the solution. The osmotic pressure thus is equivalent to the hydrostatic pressure necessary to return the chemical potential of the water solvent to that of pure water at the reference pressure. For an arbitrary solution in phase I, then, we have

$$\pi^I = RT \sum_A C_A^I \tag{10.11}$$

and with the establishment of equilibrium between phases I and II,

$$P^{II} - P^I = \pi^{II} - \pi^I = RT \sum_A \left(C_A^{II} - C_A^I \right) \tag{10.12}$$

This relationship is commonly known as the van't Hoff equation. Its importance lies in the link it draws between hydrostatic pressure on the one hand and solute concentration on the other hand, as a driving force for the movement of water across biological membranes and other phase boundaries.

By measuring the effect of osmotic pressure on cells, for example, one can estimate the permeability of water through the plasma membrane. If the osmolarity of the solution external to the cell is changed away from its equilibrium value, net water flow across the membrane will take place, thus altering the cell size. The water permeability coefficient k_w can then be estimated from the following equation:

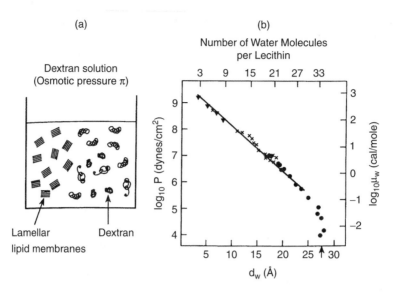

FIGURE 10.2 (a) By varying dextran concentration in the solution, different pressures can be applied to the lamellar lipid membranes normal to their surfaces, $P = \mu_w / \overline{V} = \pi$. (b) Experimentally measured repulsive pressure P between egg phosphatidylcholine bilayers as their separation varies. The arrow indicates the maximum separation attained in water where $P \to 0$. μ_ω gives the chemical potential of the interbilayers water relative to bulk water. The line represents a best-fit exponential to the experimental points where repulsion dominates attraction. ● = osmotic; × = hydraulic; ▼ = vapor pressure methods. At the point corresponding to P = 0, repulsive and attractive forces become equal. Therefore, if the attractive forces are known for the interbilayer interaction of the lipid lamellar system, one could obtain the magnitude of the repulsive force from this measurement. (From Rand RP, *Annu. Rev. Biophys. Bioeng.* 1981, 10:277.)

$$k = \frac{\partial V}{\partial t} \frac{1}{A(\pi_i - \pi_o)} = \frac{1}{\pi_i - \pi_o} \frac{\partial r}{dt} \qquad (10.13)$$

where V is the volume of the cell, $\pi_i - \pi_o$ is the osmotic pressure difference inside and outside the cell, A is the area of the cell, and r is the radius of an equivalent spherical cell. In this case, clearly, the system is not in equilibrium. A practical difficulty associated with this experiment arises from any permeability of molecular species other than water; the cell's rate of volume change may reflect the transport of these other species as well.

Another example of application of the osmotic pressure effect is found in the study of the forces of intermolecular interaction. Cell membranes usually have strongly hydrophilic surfaces. Consequently, water molecules adsorbed on the membrane surface exert a strong repulsive force against a closely apposed membrane. Introduction of a high molecular weight polymer, such as dextran, to a cell suspension often induces close adhesion of cell membranes. Because the large polymers cannot penetrate between closely adherent cells, the system behaves much like a sieve semipermeable to these polymers. Therefore, the osmotic pressure of the external solution is equivalent to a hydrostatic pressure acting to force the water out of the restricted space between cell membranes. By measuring the interspace distance of the adhered membranes (by X-ray diffraction techniques, for example) and knowing the osmotic pressure of the solution, one can estimate interaction energy between two membranes (see Figure 10.2, and LeNeveu et al., 1976).

10.1.1.2 Nonequilibrium Relationships

In the case of a nonequilibrium spatial distribution of a certain species, characterized by a nonuniform distribution of its chemical potential, a translational flow, or diffusion, of this species will occur in accordance with the gradient of its chemical potential. Before considering this process in a membrane system, some of the basic properties of matter flow in homogeneous systems will be described.

Two types of molecular transfer mechanisms can be identified: bulk flow and diffusion. The former is the result of *movement as a whole of a volume* of solution due to external forces, such as hydrostatic pressure gradients or gravitational force. In this case, the solvent and solute move together. The volume flow (J_v) is defined by the volume of solution moving across a surface of unit area, A, per unit time:

$$J_v = \frac{1}{A} \frac{\partial V}{\partial t} \tag{10.14}$$

where V is the volume of solution. Therefore, the flow or transfer of jth solute J_j across the same unit area per unit time will be

$$J_j = C_j J_v = \frac{C_j}{A} \frac{\partial V}{\partial t} = \frac{1}{A} \frac{\partial n_j}{\partial t} \tag{10.15}$$

where C_j is the concentration of solute and n is the molar quantity of the solute.

The second mechanism of molecular transfer is diffusion. For simplicity, only the case of one-dimensional diffusion will be discussed. If a gradient in the chemical potential of the jth species exists, a statistical force will be exerted on the jth particle distribution:

$$\text{Force} = -\frac{\partial \mu_j}{\partial x} \neq 0 \tag{10.16}$$

where

$$\mu_j = \mu_j^o \ (P,T) + RT \ln X_j$$

$$\cong \mu_j^{o'} \ (P,T) + RT \ln C_j, \quad (X_j \ll 1) \tag{10.17}$$

The average velocity achieved by the jth particle will be given by the product of its mobility and the force acting upon it:

$$\text{Velocity} = \text{force} \times \text{mobility} = -\frac{\partial \mu_j}{\partial x} U_j \tag{10.18}$$

where U_j is the mobility of the jth particle in a solution. The flux J_j of species j, the number of molecules crossing a surface of unit area per unit time, is

$$\text{Flux} = \text{velocity} \times \text{concentration} = -U_j \frac{\partial \mu_j}{\partial x} C_j \tag{10.19}$$

or

$$J_j = \frac{1}{A} \frac{dn_j}{dt} = -U_j C_j \frac{\partial \mu_j}{\partial x} = -U_j RT \frac{\partial C_j}{\partial x} \tag{10.20}$$

According to Fick's (first) law of diffusion, the flux J_j is given by

$$J_j = -D_j \frac{\partial C_j}{\partial x} \tag{10.21}$$

where D is the diffusion constant, a coefficient that is usually fairly constant. Comparing Equation 10.20 and Equation 10.21, the following relationship between the diffusion constant and mobility emerges:

$$D_j = U_j RT$$
$$= \frac{RT}{Nf_j} = \frac{kT}{f_j} \tag{10.22}$$

where f_j is the frictional coefficient of the particle, N is Avogadro's number and k is the Boltzmann constant. This relation was first derived by Einstein (1908). The diffusion coefficient, D, for solute particles that are large in comparison with the solvent molecule and are spherical in shape, was presented by Einstein and Stokes:

$$D = \frac{RT}{N} \frac{1}{6\pi\eta r} \tag{10.23}$$

in which the frictional coefficient, f, for a particle has been expressed in terms of the fluid viscosity, η, and the radius of the molecule r.

The molecular mobility, in another limiting case in the relative sizes of solute and solvent, has been resolved theoretically by Eyring (1936), from the point of view of absolute reaction rate theory. According to this theory, the diffusion of a particle involves a series of jumps between holes in the liquid lattice. The distance between two successive energy minima separated by an energy barrier is assumed to be λ. If the concentration varies over the x-direction only, its value at the first minimum, located at x_o, will be $C(x_o)$, while at the next, it will be

$$C(x_o) + \lambda \frac{\partial C}{\partial x}\Big|_{x_o}$$

Then the flux in the positive direction will be $\lambda\, k_1\, C(x_o)$; in the opposite direction, it will be

$$\lambda\, k_1 \left(C(x_o) + \lambda \frac{\partial C}{\partial x}\Big|_{x_o} \right)$$

where k_1 is the transition rate of a particle moving across the potential barrier between two adjacent holes within the liquid lattice (see Figure 10.3) (Glasstone et al., 1941):

$$k_1 = \frac{kT}{h} \exp\left(-\frac{\Delta G}{kT} \right) \tag{10.24}$$

where h is Planck's constant, and ΔG is the free energy height of the barrier. The net flux will be

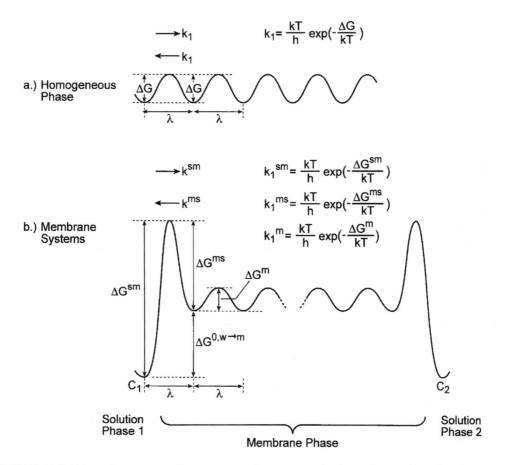

FIGURE 10.3 Schematic energy profile across a membrane:energy barrier at the membrane interface.

$$J = - \lambda^2 k_1 \frac{\partial C}{\partial x} \qquad (10.25)$$

From Equation 10.25 and Equation 10.21, the diffusion constant may be expressed

$$D = \lambda^2 k_1 \qquad (10.26)$$

The absolute rate for the transition k_1 may be obtained from the viscosity. An approximate expression for viscosity follows from the theory of absolute reaction rates (Eyring, 1936):

$$\eta = \lambda_1 \frac{RT}{N} (k_1 \lambda^2 \lambda_2 \lambda_3)^{-1} \qquad (10.27)$$

where λ_1 is the perpendicular distance between two neighboring layers of molecules sliding past each other, λ_2 is the distance between neighboring molecules in the direction of motion, and λ_3 the analogous distance in the direction normal to the other two. With Equation 10.26 and Equation 10.27 we have

$$D = \frac{RT}{N} \frac{\lambda_1}{\lambda_2 \lambda_3 \eta} \qquad (10.28)$$

This equation has been very successfully applied to the analysis of the diffusion of D_2O in H_2O, an approximation of molecular self-diffusion. Finally, we have two expressions for the diffusion constant, D:

$$D = \frac{RT}{N} \frac{1}{6\pi\eta r} \quad \text{(Einstein–Stokes)} \tag{10.29}$$

for a large solute, and

$$D = \frac{RT}{N} \frac{\lambda_1}{\lambda_2 \lambda_3 \eta} \quad \text{(Eyring)} \tag{10.30}$$

for a solute of a size comparable to that of the solvent. According to Einstein's relation (Equation 10.29) the diffusion constant D is inversely proportional to molecular radius, or $M^{1/3}$ (where M is the mass of a particle):

$$DM^{1/3} = \text{constant} \tag{10.31}$$

On the other hand, for a small molecule that is comparable in size to the lattice dimension of the solvent, the diffusion coefficient becomes inversely proportional to $M^{1/2}$ since the average thermal velocity of a particle is proportional to $(kT/M)^{1/2}$:

$$DM^{1/2} = \text{constant} \tag{10.32}$$

The measured results of the diffusion constant for various molecules appear to fit the above relations (see Figure 10.4, Table 10.1).

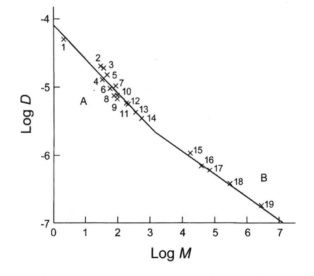

FIGURE 10.4 The diffusion coefficient D as a function of molecular weight M. Diffusing molecules as follows: 1, hydrogen; 2, nitrogen; 3, oxygen; 4, methanol; 5, carbon dioxide; 6, acetamide; 7, urea; 8, *n*-butanol; 9, *n*-amyl alcohol; 10, glycerol; 11, chloral hydrate; 12, glucose; 13, lactose; 14, raffinose; 15, myoglobin; 16, lactoglobulin; 17, hemoglobin; 18, edestin; 19, erythrocruorin. All are at or near 20°C. A is the relation $DM^{1/2}$ constant, B the relation $DM^{1/3}$ constant. (Adapted from Stein WD: in *Comprehensive Biochemistry*. Vol. 2. Elsevier, Amsterdam, 1962, Chap. 3. With permission.)

TABLE 10.1
Diffusion Constants for Various Substances in H$_2$O (25°C)

	D (10^{-5} cm²/s)	MW	DM$^{1/2}$ (10^{-5} cm²/s)
H$_2$O (self-diffusion)	2.5	18	10.6
Ethanol	1.25	46	8.4
Urea	1.38	60.1	10.7
Glycine	1.06	75.1	9.2
Sucrose	0.5	342	9.2
Insulin	0.15	5,100	10.7
Ribonuclease	10.7×10^{-2}	13,700	11.9
Human hemoglobin	6.8×10^{-2}	64,500	17.3

10.1.1.2.1 Diffusion in Membrane Systems

In the transport of a substance across a membrane, it must move through phase boundaries, such as liquid/membrane/liquid. In addition, the substance may have different affinities for partitioning in each phase encountered. In most cases, the diffusion of the substance is much slower within the membrane than it is in the solutions bathing the membrane. Diffusion of substance within a membrane is given by

$$\frac{1}{A}\frac{dn}{dt} = -D_m \frac{\partial C^m}{\partial x} \tag{10.33}$$

where D_m and C^m are the diffusion constant and concentration of a substance in the membrane, respectively. Because the concentration profile of the permeant substance within the membrane is not usually known, a constant concentration gradient is often assumed (see Henderson's approximation, Equation 10.74):

$$\frac{\partial C^m}{\partial x} = \frac{C_i^m - C_o^m}{\delta} \tag{10.34}$$

where C_i^m and C_o^m are the concentrations of the substance at the inside and outside boundaries within the membrane and δ is the thickness of the membrane. Equation 10.34 is then

$$\frac{1}{A}\frac{dn}{dt} = J = -\frac{D_m}{\delta}(C_i^m - C_o^m) \tag{10.35}$$

Experimentally, the concentrations of the permeant substance in the outer, C_o, and inner, C_i, bathing solution are known. When the diffusion of the substance is not too fast, relative to its penetration of the phase boundary (it is usually slow in the membrane), the following partition equilibrium for the permeant species is established at the membrane boundaries:

$$\frac{C_o^m}{C_o} = \frac{C_i^m}{C_i} = \text{constant} = K \tag{10.36}$$

where C^m and C are the concentrations of the permeant molecule just inside and outside the membrane, and K is the partition coefficient. Equation 10.36 is

TABLE 10.2
Permeability Constants for Various Cells in cm/s
(Multiplied by 10^5)

	Ox Erythrocyte	Arbacia	Chara	Beggiatoa	B[a]
Trimethylcitrate	—	—	6.7	—	0.047
Propionamide	—	2.3	3.6	—	0.0036
Acetamide	—	1.0	1.5	—	0.00083
Glycol	0.21	0.73	1.2	1.39	0.00049
Urea	7.8	—	0.11	1.58	0.00015
Malonamide	—	—	0.0039	—	0.00008
Diethylene glycol	0.075	0.43	—	—	0.005
Glycerol	0.0017	0.005	0.021	1.06	0.00007
Erythritol	—	—	0.0013	0.84	0.00003
Sucrose	—	—	0.0008	0.11	0.00003

[a] B is olive oil/water partition coefficient.

Adapted from Davson H and Danielli JF: *The Permeability of Natural Membranes.*
2nd ed. Cambridge University Press, Cambridge, 1952.

$$C_o^m = KC_o \text{ and } C_i^m = KC_i \qquad (10.37)$$

Because the permeant flux is continuous throughout the membrane

$$J = - D_m \frac{C_i^m - C_o^m}{\delta} = - \frac{D_m K}{\delta} (C_1 - C_o)$$

$$= - \frac{D_m K}{\delta} \Delta C = - P\Delta C \qquad (10.38)$$

where

$$P = \frac{D_m K}{\delta} \qquad (10.39)$$

P is called the *permeability constant.* The permeability constant is proportional to its diffusion constant in the membrane and the partition coefficient between the membrane and aqueous phases, while being inversely proportional to the thickness of the membrane. P has dimensions of $[P] = cm/s$, while the dimensions of D are $[D] = cm^2/s$. Some typical values of permeability constants are listed in Table 10.2.

Experimentally, the permeability of a solute through a membrane is determined from Equation 10.38 by the measurement of its flux developed with the membrane having a known concentration difference between the two sides of the membrane.

10.1.1.2.2   Partition between the Aqueous and Membrane Phases
The permeability of substances through cell membranes has been of great interest and importance for elucidation of many biological cell functions. Each cell membrane has inherently its own specific permeability for various permeants. Although most metabolically important substances are transported across the membrane by active or specific transport, a general transport property

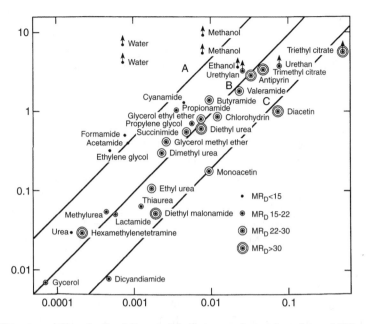

FIGURE 10.5 The permeability of cells of *C. ceratophylla* to organic nonelectrolytes of different oil solubility and different molecular sizes. Ordinate: $PM^{1/2}$ (P in cm h^{-1}); abscissa: olive oil–water partition coefficients. MR_D is the molar refraction of the molecules depicted, a parameter proportional to the molecular volume. (From Collander R: *Trans Faraday Soc.* 1937, 33:985. With permission.)

for many permeants has been observed for a variety of membranes. At the beginning of the last century, Overton (1902) and others observed that the permeability of substances across cell membranes is proportional to the relative partition constant (or coefficient) of the permeant substance between oil and water phases. This view was supported later by Ferguson et al. from a thermodynamic viewpoint. In later years, Collander (1937), Danielli (1952), and Stein (1967) have done further experiments and found that the product of permeability P and the square root of molecular weight of the permeant has a better correlation with the partition coefficient ($PM^{1/2} \propto K$), than does the permeability alone. Figure 10.5 shows such a relationship between the $PM^{1/2}$ product and the partition coefficient K for various membranes and permeating species; both quantities are plotted on a logarithmic scale. The relatively good correlation so observed has led Collander to view penetration as being governed not only by the lipid solubility of the permeable substances, but also by a sievelike character of the membrane, because small molecules (methanol, water, formamide, and ethylene glycol) penetrate faster than would be predicted from their oil–water partition coefficients.

For larger molecules, however, one would expect the product of permeability with the cube root, rather than the square root of the molecular weight, to be most closely correlated with the partition coefficient, in accord with Equation 10.31.

In addition to the size of the permeant substance, its partition coefficient with the membrane plays a major role for most transport processes across the membrane. The energy necessary for a permeant to enter from the aqueous phase to the membrane phase is the most significant factor for the membrane transport.

The partition K of a substance between two phases (e.g., aqueous and membrane) may be expressed by the standard Gibbs free energy difference between those in each phase of the substance:

$$\frac{[A]^m}{[A]} = K = \exp\left[\frac{\mu_A^o(w) - \mu_A^o(m)}{RT}\right] = \exp\left[\frac{-\Delta G_A^{o,w \to m}}{RT}\right] \qquad (10.40)$$

where $[A]^m$ and $[A]$ are the concentrations of the substance A in the membrane and in the aqueous phases, respectively, and $\Delta G_A^{o, w \rightarrow m}$ is the difference of the standard Gibbs free energies of solvation of the substance in the two phases (water and membrane).

This relation can be derived from the equilibrium condition for the substance to be distributed into two phases:

$$\mu_A \text{ (water)} = \mu_A \text{ (membrane)} \tag{10.41}$$

using the chemical potential expression (Equation 10.7) for the substance in each phase. Major energy differences of the permeant between the two phases come from hydrogen bonding energy of the substance or hydration energy. In addition, nonpolar molecular interaction and entropy terms also contribute to the energy differences. If one of the media is water, substances with $\Delta G^{o,w \rightarrow m} > 0$ may be characterized as hydrophilic, and those with $\Delta G^{o,w \rightarrow m} < 0$ as hydrophobic. Partition coefficients for some substances between the water–oil phases are listed in Table 10.3.

The plot of the logarithm of $PM^{1/2}$ vs. the number of hydrogen bonds for permeant molecules gives a straight line relation, lending partial support to the previous statement (Figure 10.6). Therefore, the partition of a permeant at the water/membrane phases is greatly dependent on the degree of the polar nature of the permeant. The more polar the substance is, the higher the free energy is for the substance to enter in the membrane phase. For example, an increase in the number of OH-groups (an OH group forms two hydrogen bonds, and the free energy to break a mole of hydrogen bonds is about 1.6 kcal [Stein]) results in a smaller partition of the permeant in the membrane and, consequently, its lower permeability through the membrane. The permeability and partition in the membrane decrease in an approximately exponential manner with the number of OH groups of the permeating molecule.

TABLE 10.3
Partition Coefficients of Various Substances

	H⁺	2.3×10^{-6}
	Li⁺	2.1×10^{-7}
	Na⁺	6.8×10^{-7}
	K⁺	4.0×10^{-5}
	Ca²⁺	3.9×10^{-13}
	Mg²⁺	1.4×10^{-12}
	Choline⁺	1.1×10^{-2}
	Cl⁻	5×10^{-6}
Hydrophilic substances	CNS⁻	2.0×10^{-3}
	Lauryl sulfate	2.0×10^{-1}
	Sucrose*	3.0×10^{-5}
	Glycerol*	7.0×10^{-5}
	Urea*	1.5×10^{-4}
	Glycol*	4.9×10^{-4}
	Propionapride*	3.6×10^{-3}
	Diethylene glycol*	5.0×10^{-3}
	Trimethyl citrate*	4.7×10^{-2}
	Picrate⁻	6.3
	Et HN⁺	10.2
Hydrophobic substances	Procaine (neutral)*	45
	Tetracaine (neutral)*	274
	Tetraphenylborate⁻	1.7×10^6

* Oil (olive oil)/water. Others: nitrobenzin/water.

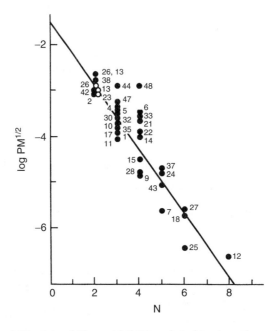

FIGURE 10.6 The permeability data of Figure 10.5 (●) replotted to show the variation of log PM $^{1/2}$ (P in cm s^{-1}) on the ordinate with the number N of hydrogen-bond-forming groups in the permeant on the abscissa. (○): Data of Dainty and Ginzburg (1964) on the related alga *C. australis*. Permeants have been assigned the following numbers: 1, acetamide; 2, antipyrin; 3, butanol; 4, butyramide; 5, cyanamide; 6, diacetin; 7, dicyandiamide; 8, diethylene glycol; 9, diethyl malonamide; 10, diethyl urea; 11, dimethyl urea; 12, erythrithol; 13, ethanol; 14, ethylene glycol; 15, ethyl urea; 16, ethyl urethane; 17, formamide; 18, glycerol; 19, glycerol + CO$_2$; 20, glycerol + Cu^{2+}; 21, glycerol ethyl ether; 22, glycerol methyl ether; 23, isopropanol; 24, lactamide; 25, malonamide; 26, methanol; 27, methylol urea; 28, methyl urea 29, monoacetin; 30, monochlorohydrin; 31, propanol; 32, propionamide; 33, (α, β)–propylene glycol; 34, (α, γ)-propylene glycol; 35, succinimide; 36, tetraethylene glycol; 37, thiourea; 38, triethyl citrate; 39, triethylene glycol; 40, trihydroxybutane; 41, trihydroxybutate + Cu^{2+}; 42, trimethyl citrate; 43, urea; 44, urethane; 45, urethylan, 46, urotropin; 47, valeramide; 48, water; 49, 2,3-butylene glycol. (From Stein WD: *The Movement of Molecules across Cell Membranes*. Academic Press, New York, 1967. Copyright Czechoslovak Academy of Sciences.)

10.1.1.2.3   Effects of Unstirred Layers

Unstirred layers are regions of slow laminar flow parallel to the membrane. The effective thickness of these unstirred layers may be as much as 500 μm and cannot be reduced below some 20 μm even by vigorous mechanical stirring. Typical diffusion coefficients for small solute particles in water are about 10^{-5} cm^2/s. One can express the permeability coefficient of an unstirred layer of 100 μm thick as 10^{-5} cm^2/s/10^{-2} cm = 10^{-3} cm/s. If the permeability coefficient of a membrane under study is of the order of 10^{-3} cm/s, the unstirred layer would become a significant factor in determining the transmembrane flux. Therefore, under these conditions, a correction for this effect is necessary. Suppose that the measured flux is J, and the concentration difference is ΔC; the experimental permeability is then

$$P_{measured} = J/\Delta C \tag{10.42}$$

However, the true permeability corrected for the unstirred layer effect is

$$P_{true} = J/[\Delta C - (\Delta C)_1 - (\Delta C)_2] \tag{10.43}$$

Each of the two concentration differences developed across the unstirred layers is related to the layers' thickness h and to the approximate diffusion coefficient D_1 or D_2:

$$(\Delta C)_1 = \frac{Jh_1}{D_1}, \quad (\Delta C)_2 = \frac{Jh_2}{D_2} \tag{10.44}$$

Therefore,

$$P_{true} = \left(\frac{1}{P_{measured}} - \frac{h_1}{D_1} - \frac{h_2}{D_2} \right)^{-1} \tag{10.45}$$

10.1.2 ELECTROLYTE TRANSPORT

Another kind of transport becomes possible when transported particles are electrically charged, because the movement of such particles is subject not only to the gradient of concentration but also to the force an electric field exerts on a charged particle. Biological fluids contain various charged particles formed by the dissociation of salts, acids, bases, and other molecules. These molecules, which possess electrical charges in an aqueous solution, are called electrolytes. An electric field, E, is frequently written in terms of the gradient of an electrical potential φ:

$$E = - \text{grad } \varphi \tag{10.46}$$

The flux, J_j^{elec}, of the jth charged particles due to an electric field is expressed as

$$J_j^{elec} = C_jU_jz_jFE = - C_jU_jz_jF \text{ grad}\varphi \tag{10.47}$$

where C_j is the concentration of the jth charged particle, U_j, its mobility, z_j the valency of the jth charged particle, and F the Faraday constant. The flux of the particles in the direction of the x-axis can be written as a sum of terms due to its concentration gradient (Equation 10.20) and the electric field:

$$J_j = - U_jRT \frac{\partial C_j}{\partial x} - z_jFC_jU_j \frac{\partial \varphi}{\partial x} \tag{10.48}$$

As a consequence, the chemical potential of a charged particle in an electric field can be written conveniently as

$$\tilde{\mu}_j = \mu_j^\circ + RT \ln C_j + z_jF\varphi \tag{10.49}$$

Expression 49 is called the *electrochemical potential* (Guggenheim, 1929).

10.1.2.1 Equilibrium Relationships

When an electrolyte solution is separated by a boundary (e.g., a membrane) and the electrolytes are in equilibrium across the boundary, the following relation holds for the electrochemical potential of each permeable ion:

$$\tilde{\mu}_{jo} = \tilde{\mu}_{ji} \tag{10.50}$$

where o and i refer to the two spaces separated by a boundary, and j refers to the jth ionic species. In the case of nonpermeable ionic species across the membrane, using Equation 10.49 and Equation 10.50, the potential difference E_m across the membrane at equilibrium is

$$E_m = \varphi_i - \varphi_o = \frac{RT}{z_j F} \ln \frac{C_{jo}}{C_{ji}} \tag{10.51}$$

where C_{jo} and C_{ji} are the concentrations of the permeable species in the two spaces separated by a membrane. Such a potential is called the *Nernst equilibrium potential.*

10.1.2.1.1 Example 1
Suppose a membrane that is permeable to positive ions, but not to negative ions, separates a salt (e.g., KCl) solution into two compartments at different concentrations. Such a system can attain an equilibrium distribution with respect to the permeable ionic species:

$$\tilde{\mu}_{K,o} = \tilde{\mu}_{K,i} \tag{10.52}$$

Then, from Equation 10.49 and Equation 10.50, the potential difference across the membrane, $E_m = \varphi_i - \varphi_o$ is

$$E_m = \frac{RT}{zF} \ln \frac{C_o}{C_i} = \frac{RT}{F} \ln \frac{[K^+]_o}{[K^+]_i} \tag{10.53}$$

10.1.2.1.2 Example 2
When a membrane is permeable to small ions but not to a large macromolecular ion, such as protein, at equilibrium, the potential difference across the membrane is obtained from Equation 10.49 and Equation 10.50.

$$E_m = \frac{RT}{F} \ln \frac{\left(C_{jo}\right)^{1/z_j}}{\left(C_{ji}\right)^{1/z_j}} \tag{10.54}$$

where z_j is the valency of any j^{th} ion, and the following relationship holds for all permeable species:

$$\frac{\left(C_{jo}\right)^{1/z_j}}{\left(C_{ji}\right)^{1/z_j}} = \gamma \tag{10.55}$$

where γ is called the *Donnan ratio.* If the inner space is surrounded by a rigid membrane, as shown in Figure 10.7, so that its volume is fixed and the concentration $[R^{n-}]$ of a nonpermeating polyvalent anion only in the inner space is constant, but the membrane is permeable to both potassium cations and chloride anions, then the equilibrium membrane potential will be established as

$$E_m = \frac{RT}{F} \ln \frac{[K^+]_o}{[K^+]_i} = \frac{RT}{F} \ln \frac{[Cl^-]_i}{[Cl^-]_o} \tag{10.56}$$

where the concentrations of ions are interrelated due to electroneutrality:

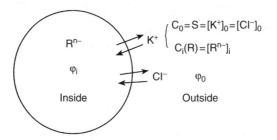

FIGURE 10.7 An example of membrane equilibrium. For explanation, see text.

$$\begin{cases} \left[K^+\right]_o = \left[Cl^-\right]_o = S \\ n\left[R^{n-}\right]_i + \left[Cl^-\right]_i = \left[K^+\right]_i \end{cases} \tag{10.57}$$

Solving Equation 10.56 and Equation 10.57, one obtains

$$\left[Cl^-\right]_i = -\frac{n\left[R^{n-}\right]}{2} + \sqrt{\left(\frac{n\left[R^{n-}\right]}{2}\right)^2 + S^2} \tag{10.58}$$

and

$$\left[K^+\right]_i = \frac{n\left[R^{n-}\right]}{2} + \sqrt{\left(\frac{n\left[R^{n-}\right]}{2}\right)^2 + S^2}$$

From Equation 10.56, then, the potential difference across the membrane is

$$E_m = -\frac{RT}{F} \ln\left(\frac{n\left[R^{n-}\right]}{2S} + \sqrt{\left(\frac{n\left[R^{n-}\right]}{2S}\right)^2 + 1}\right) \tag{10.59}$$

10.1.2.2   Nonequilibrium Case

A nonequilibrium situation for an ion occurs when there is a gradient in its electrochemical potential with respect to the space coordinates. In such a case, an ionic particle will experience a statistical force expressed in terms of its electrochemical potential

$$\text{Force} = -\text{grad }\tilde{\mu}$$

$$= -\frac{RT}{C}\frac{\partial C}{\partial x} - zF\frac{\partial \varphi}{\partial x} \quad \text{(one dimensional case)} \tag{10.60}$$

Therefore, the flux of the jth ionic species is:

$$J_j = -C_j U_j \frac{\partial \tilde{\mu}_j}{\partial x} = -RT\, U_j \frac{\partial C_j}{\partial x} - C_j U_j z_j F \frac{\partial \varphi}{\partial x} \tag{10.61}$$

where U_j is the mobility of the jth species ($u = FU$ is often called *electrical mobility*). The flux equation (Equation 10.61) is called the *Nernst–Planck diffusion equation*. The most rigorous solution of this equation was obtained by Planck (Kotyk and Janacek, 1970).

A number of attempts have been made to obtain the transmembrane potential using the Nernst–Planck equation. However, because the exact profiles of both the concentrations of ions, as well as the electrical potential across a membrane, are usually not known, the exact solution for the membrane potential equation is difficult to obtain. However, under either of the following simplifying assumptions, a constant gradient in concentration or in the electrical potential, the equation can be solved easily. Using the continuity condition for an ionic flux (no diffusing ions are destroyed or newly created)

$$\frac{\partial C}{\partial t} = - \text{div } J \tag{10.62}$$

which, for the one-dimensional case, becomes

$$\frac{\partial C}{\partial t} = - \frac{\partial J}{\partial x} \tag{10.63}$$

Then the flux equation (Equation 10.61) can be rewritten as

$$\frac{\partial C_j}{\partial t} = U_j \left[RT \frac{\partial^2 C_j}{\partial x^2} + z_i F \frac{\partial}{\partial x} \left(C_j \frac{\partial \varphi}{\partial x} \right) \right] \tag{10.64}$$

Considering for the moment a solution of only univalent ions ($z_j = \pm 1$), then because electroneutrality conditions prevail in most cases, at any point,

$$\sum_j C_j^+ = \sum_j C_j^- \tag{10.65}$$

and therefore

$$\sum_j \frac{\partial C_j^+}{\partial t} - \sum_j \frac{\partial C_j^-}{\partial t} = 0 \tag{10.66}$$

Making use of the electroneutrality condition, Equation 10.65, Equation 10.64 can be rewritten as

$$\sum_j U_j^+ RT \frac{\partial^2 C_j^+}{\partial X^2} + \sum_j U_j^+ F \frac{\partial}{\partial x} \left(C_j^+ \frac{\partial \varphi}{\partial x} \right) - \sum_j U_j^- RT \frac{\partial^2 C_j^-}{\partial x^2}$$

$$+ \sum_j U_j^- F \frac{\partial}{\partial x} \left(C_j^- \frac{\partial \varphi}{\partial x} \right) = 0 \tag{10.67}$$

By setting

$$\sum_j U_j^+ C_j^+ \equiv U \text{ and } \sum_j U_j^- C_j^- \equiv V \qquad (10.68)$$

Equation 10.67 may be solved for

$$\frac{\partial \varphi}{\partial x} = \frac{RT \dfrac{\partial (U - V)}{\partial X}}{F(U + V)} \qquad (10.69)$$

This equation cannot be easily solved for the electrical potential profile because the dependence of the concentration on x is not known. However, the equation is useful in discussing liquid junction phenomena: when two electrolyte solutions form an interface, an electric potential difference will usually develop across the boundary. For example, consider two electrolyte solutions consisting of the same uni-univalent salt (e.g., NaCl), but having two different concentrations, in contact at an interface. The resultant potential difference can be obtained by integrating Equation 10.69 across the liquid interface.

$$\varphi_i - \varphi_o = \int_{out}^{in} \frac{\partial \varphi}{\partial x} dx = -\int_{out}^{in} \frac{RTd(U - V)}{F(U + V)} \qquad (10.70)$$

which for a uni-univalent salt becomes

$$\varphi_i - \varphi_o = \frac{RT}{F} \frac{U^+ - U^-}{U^+ + U^-} \ln \frac{C_o}{C_i} \qquad (10.71)$$

A potential difference due to the difference in mobility of two ions ($U_{Na} > U_{Cl}$) is produced. Such a potential is called *liquid junction potential* or *concentration potential* (Nernst, 1888):

$$E_m = \varphi_i - \varphi_o = \frac{RT}{F} \frac{U^+ - U^-}{U^+ + U^-} \ln \frac{C_o}{C_i} \qquad (10.72)$$

By introducing the transference numbers for cation and anion, defined as $t^+ = U^+/(U^+ + U^-)$ for cations and $t^- = U^-/(U^+ + U^-)$ for anions, respectively, Equation 10.72 is rewritten as

$$E_m = \frac{RT}{F} (2t^+ - 1)\ln \frac{C_o}{C_i} \qquad (10.73)$$

For example, the selectivity of ions across a membrane can be obtained by measuring the concentration potential for a membrane separating the same kind of electrolyte, but having different concentrations on the two sides of the membrane (Figure 10.8). By performing such experiments, the ratio of U^+/U^- can be measured (Hopfer et al., 1970).

Returning to the Nernst–Planck diffusion equation, other approximate solutions of the equation, with respect to the potential difference, can be obtained in two extreme cases:

1. Constant concentration gradient: $\partial c/\partial x = $ constant
2. Constant field: $\partial \varphi/\partial x = $ constant

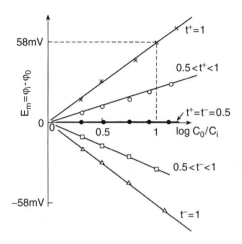

FIGURE 10.8 Membrane potential as a function of the ratio of the outside of uni-univalent electrolyte solutions to the inside concentration; t: ion transference number $t^+ = U^+/(U^+ + U^-)$, $t^- = U^-/(U^+ + U^-)$.

For the case of a constant concentration gradient, by expressing the concentration of the jth ionic species in the membrane by

$$C_j(x) = C_{jo} + \frac{C_{ji} - C_{jo}}{\delta} x \tag{10.74}$$

where suffixes i and o refer to the inside and the outside of the membrane, x is the position within the membrane relative to the outside surface, and δ is the membrane thickness. Equation 10.48 becomes

$$J_j = - RT\, U_j \frac{C_{ji} - C_{jo}}{\delta} - z_j\, FU_j \left(C_{jo} + \frac{C_{ji} - C_{jo}}{\delta} x \right) \frac{\partial \varphi}{\partial x} \tag{10.75}$$

By assuming the total current is zero, $\Sigma_j z_j F\, J_j = 0$, Equation 10.75 can be expressed as

$$\frac{\partial \varphi}{\partial x} = - \frac{RT}{F\delta} \frac{\displaystyle\sum_j z_j U_j \left(C_{ji} - C_{jo} \right)}{\displaystyle\sum_j z_j^2 U_j C_{jo} + \frac{x}{\delta} \sum_j z_j^2 U_j \left(C_{ji} - C_{jo} \right)} \tag{10.76}$$

This equation can then be integrated from the outside to inside surfaces of the membrane, yielding

$$E_m = \frac{RT \displaystyle\sum_j z_j U_j \left(C_{ji} - C_{jo} \right)}{F \displaystyle\sum_j z_j^2 U_j \left(C_{ji} - C_{jo} \right)} \ln \frac{\displaystyle\sum_j z_j^2 U_j C_{jo}}{\displaystyle\sum_j z_j^2 U_j C_{ji}} \tag{10.77}$$

This equation for the membrane potential is called the *Henderson equation* (Henderson, 1908). It is an expression for the potential often used in analysis of the membrane potential observed for polymer resin systems.

Another approximation formula for the membrane potential is obtained under the second assumption, that of a constant electric field at a steady state of transmembrane flux.

$$\frac{\partial \varphi}{\partial x} = \frac{\varphi_i - \varphi_o}{\delta} = \text{constant} \tag{10.78}$$

Then, the flux equation (Equation 10.61) for the jth ionic species is

$$J_j = - RT\, U_j \frac{\partial C_j}{\partial x} - z_j F C_j U_j \frac{\varphi_i - \varphi_o}{\delta} \tag{10.79}$$

Because the steady state of flux, $J_j = $ constant, is assumed for the jth species, we have

$$dx = - \frac{RT U_j dC_j}{J_j + z_j F C_j U_j \dfrac{\varphi_i - \varphi_o}{\delta}} \tag{10.80}$$

Equation 10.80 can be integrated easily from the outside to inside surfaces of the membrane, resulting in

$$J_j = z_j F U_j \frac{\varphi_i - \varphi_o}{\delta} \frac{C_{ji} - C_{jo} e^{-z_j F(\varphi_i - \varphi_o)/RT}}{e^{-z_j F(\varphi_i - \varphi_o)/RT} - 1} \tag{10.81}$$

Assuming that the net current across the membrane is zero

$$\sum_j z_j F J_j = 0 \tag{10.82}$$

we obtain the following equation from Equation 10.81 and Equation 10.82 for univalent electrolyte systems:

$$\varphi_i - \varphi_o = E_m = \frac{RT}{F} \ln \frac{\sum U_j^+ C_{jo}^+ + \sum U_j^- C_{ji}^-}{\sum U_j^+ C_{ji}^+ + \sum U_j^- C_{jo}^-} \tag{10.83}$$

This equation is often called *Goldman's constant field equation* for the membrane potential (Goldman, 1943). The flux equation (Equation 10.81) does not include the special effect of the membrane boundary, such as ion partition difference between the two phases (aqueous and membrane). Nor does the Goldman equation. Hodgkin and Katz took this effect into consideration in extending the Goldman equation, as follows. The flux equation (Equation 10.81) can be rewritten by using the partition coefficient $K = C^m/C$ where C^m and C are the ionic concentrations in the membrane and the bulk solution, respectively.

$$J_j = z_j F U_j \frac{\varphi_i - \varphi_o}{\delta} K \frac{C_{ji} - C_{jo}\, e^{-z_j F(\varphi_i - \varphi_o)/RT}}{e^{-z_j F(\varphi_i - \varphi_o)/RT} - 1} \tag{10.84}$$

Introducing permeability coefficient P as

$$P \equiv \frac{RTUK}{\delta}, \tag{10.85}$$

the flux equation (Equation 10.83) is expressed using the concentration of ions in the bulk solution:

$$J_j = z_j F P_j \frac{\varphi_i - \varphi_o}{RT} \frac{C_{ji} - C_{jo}\, e^{-z_j F(\varphi_i - \varphi_o)/RT}}{e^{-z_j F(\varphi_i - \varphi_o)/RT} - 1} \tag{10.86}$$

With Equation 10.82 and Equation 10.86, Hodgkin and Katz rewrite the membrane potential equation in the following form:

$$E_m = \frac{RT}{F} \ln \frac{\sum P_j^+ C_{jo}^+ + \sum P_j^- C_{ji}^-}{\sum P_j^+ C_{ji}^+ + \sum P_j^- C_{jo}^-} \tag{10.87}$$

This equation is called the *Goldman-Hodgkin-Katz equation* (G-H-K equation), which has been widely and successfully used to explain observed biological membrane potentials. This membrane potential arises from the movements of different ions (*diffusion potential*) across the membrane. It is distinct from the equilibrium potential described earlier (Table 10.4).

TABLE 10.4
Ion Distributions, Relative Ionic Permeabilities, and Resting Potential of Various Cells

Specimen	Extracellular (Intracellular) Ion Concentration (mM)			Relative Permeability	Resting Potential $F = \varphi_i - \varphi_o$ (mV) at Normal Physiological pH
	K+ (K+)	Na+ (Na+)	Cl- (Cl-)		
Loligo nerve axon	10 (400)	440 (50)	560 (40–150)	K+ > Cl- > Na+	−60
Frog sartorius muscle	2 (125)	110 (15)	77 (1.2)	K+ > Cl- > Na+	−95
Dog skeletal muscle	4 (140)	150 (12)	120 (−)	K+ > Cl- > Na+	
Visceral smooth muscle					−50 to 54
Red cells					
Man	5.0 (.36)	155 (19)	112 (78)	Cl- > K+ > Na+	−15
Rabbit	5.5 (142)	150 (22)	110 (80)	Cl- > K+ > Na+	−15
Dog	4.8 (10)	153 (135)	112 (87)	Cl- > K+	
L-cell	5.6 (171)	136 (8.6)	147 (70)	Cl- > K+ > Na+	
Mouse fibroblasts L-cell					−10
Cortical cell (guinea pig fetus)					0 to 7.6
Fibroblast line (hamster kidney)	6.3 (−)	148.7 (−)		K+ > Na+	−55
HeLa cell					−17

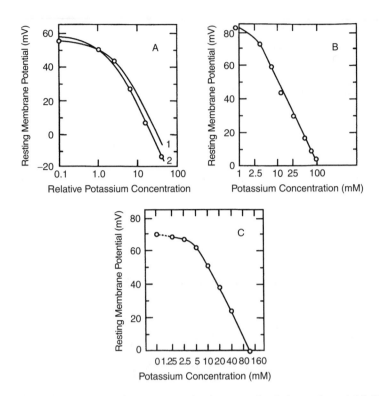

FIGURE 10.9 Relation between potassium concentration in external solution and potential difference across resting membrane. (A) Squid giant axon (data of Curtis and Cole). (B) Frog sartorius muscle (data of Ling and Gerald). (C) Frog myelinated nerve (data of Huxley and Stumpfli). Abscissa: potassium concentration on logarithmic scale, (B) and (C) in mM; (A) as multiple of concentration in standard solution (13 mM. Ordinate: potential difference across resting membrane (outside potential minus inside potential). The curves in (A) give the calculated changes in resting potential, drawn according to the G-H-K equation, with $P_K:P_{Na}:P_{Cl} = 1:0.04:0.45$ (curve 1) and $P_K:P_{Na}:P_{Cl} = 1:0.025:0.3$ (curve 2). (For details, see Hodgkin and Katz, 1949). The membrane potentials in (A) and (B) are the observed values and have not been corrected for the junction potentials between axoplasm and myoplasm and the electric field. In (C), the potentials are given with reference to a node that has been depolarized with isotonic potassium chloride (117 mM). (From Hodgkin AL, *Biol. Rev.* 1951, 26:360.)

The G-H-K equation can be verified by inserting the experimental results of membrane potential measurements, ionic concentrations, and their permeability constants (or relative permeability). One method of obtaining the ionic permeability of membranes is to use the flux equation (Equation 10.86) based on the constant field assumption. The permeability constant P_j of the jth ionic species, can be obtained by measuring its flux J_j, using a radioisotope tracer method, and by knowing the transmembrane potential ($\varphi_i - \varphi_o$) and all ionic concentrations in both compartments. The latter quantities are also measurable. Hodgkin and Katz and others have found that, for certain muscle and nerve cells, the potential equation (Equation 10.87) holds quite well in explaining the measured membrane potential (see Figure 10.9).

The net flux (Equation 10.86) can be rewritten in terms of unidirectional fluxes $\vec{J}_j$ and $\bar{J}_j$

$$J_j = \frac{-z_j P_j FE_m}{RT} \frac{C_{jo} - C_{ji} e^{z_j FE_m/RT}}{1 - e^{z_j FE_m/RT}} \qquad (10.88)$$

$$= \vec{J}_j - \bar{J}_j \qquad (10.89)$$

TABLE 10.5
Ionic Flux across Squid Axon Membrane

Ions	Concentration (mM)		Flux in pmol/cm/s		
	Extracellular	Intracellular	Inward	Outward	Inward/Outward
K$^+$	97	246 (300)	17 (11)	58 (34)	0.3 (0.32)
Na$^+$	458	110	61	31	2

Adapted from Keynes, RD: *J. Physiol.* 1951, 114:119.

where $\vec{J}_j$ and $\bar{J}_j$ are defined as ionic influx and efflux, respectively:

$$\vec{J}_j = -z_j P_j \frac{FE_m}{RT} \frac{C_{jo}}{1 - e^{z_j FE_m / RT}} \tag{10.90}$$

$$\bar{J}_j = -z_j P_j \frac{FE_m}{RT} \frac{C_{ji}\, e^{z_j FE_m / RT}}{1 - e^{z_j FE_m / RT}} \tag{10.91}$$

Therefore, the ratio of $\vec{J}_j$ to $\bar{J}_j$ is

$$\frac{\vec{J}_j}{\bar{J}_j} = \frac{C_{jo}}{C_{ji}} \exp\left(-\frac{z_j FE_m}{RT}\right) = \exp\left[z_j\left(E_j - E_m\right)F / RT\right] \tag{10.92}$$

where E_j is the equilibrium potential of the jth ion:

$$E_j = \frac{RT}{z_j F} \ln \frac{C_{jo}}{C_{ji}} \tag{10.93}$$

Keynes (1951) first measured the ionic fluxes across the squid axon by means of a radioisotope method. See Table 10.5.

According to Equation 10.90 and Equation 10.91, and the experimental data of Table 10.5, the flux ratio gives

$$\vec{J}_K / \bar{J}_K = \exp\left[-(E_m - E_k)\, z_k F/RT\right] \cong 0.34$$

$$\vec{J}_{Na} / \bar{J}_{Na} = \exp\left[-E_m - E_{Na}\right) z_{Na} F/RT\right] \cong 50 \sim 100$$

for K$^+$ and Na$^+$, respectively. The flux ratio for K$^+$ calculated from theoretical estimates and experimental values agrees well. Such is not the case with Na$^+$. However, taking into account Na$^+$ active transport (the Na$^+$ efflux was reduced by a factor of 20 upon application of ouabain), both theoretical and experimental values are comparable. For example, the experimental value of the outward flux ratio of K$^+$ to Na$^+$ is

$$\bar{J}_K / \bar{J}_{Na} = \frac{58}{31} = 1.87$$

Because

$$[K^+]_i/[Na^+]_i \cong 3 \sim 4$$

the permeability ratio under these conditions yields

$$\frac{P_{Na}}{P_K} \cong 1.6$$

However, using the value for Na efflux under the application of ouabain

$$\bar{J}_K / \bar{J}_{Na} = \frac{58}{1.5} = 38$$

Therefore, the permeability ratio of P_K/P_{Na} increases to 13, which is a reasonable value for the ratio of passive K^+ to Na^+ permeability across squid membranes (see Figure 10.9).

10.1.2.2.1 Surface Charge Effect

The treatment presented thus far for ion transport across membranes does not include the effect of any surface charge on the membrane. Because most biological membranes possess fixed charges on their surface, its effect on ion transport will be discussed here (Jones, 1975). Although the surface charges are distributed in a discrete manner on the membrane, for simplicity, the case of uniform surface charge distribution will be described. A discussion of the discrete case can be found in Ohki and Ohshima (1995).

The assumption of uniform charge distribution does not give results much different from those of the discrete charge approach, unless one is concerned with the behavior of ions quite near the fixed charge sites or with nonuniform membrane systems, such as ionic channels or surface protein aggregates. When a membrane having a uniform charge density σ is in an electrolyte solution, according to the treatment of Gouy-Chapman (Davies and Rideal, 1963), a diffuse ionic layer forms near the surface, thus modifying the electrical potential it generates. The magnitude of this effect depends on the surface charge density, the kinds of electrolytes, and their concentrations. The surface potential $\psi(0)$ at the membrane surface is given by

$$\sigma = \left(\frac{\varepsilon RT}{2000\pi}\right)^{1/2}\left[\sum_j C_j\left(e^{-z_j F\psi(o)/RT} - 1\right)\right]^{1/2} \qquad (10.94)$$

where σ is the electronic charges per cm^2, C_j is the molar concentration of the jth species in the bulk solution, and ε is the dielectric constant of the solution.

For complex electrolyte systems including polyvalent ions, there is no explicit analytical form for $\psi(0)$, and only numerical methods can be used to estimate the surface potential. However, for a simple uni-univalent electrolyte, e.g., NaCl solution (C:concentration), the Gouy-Chapman diffused layer potential can be expressed (Rice and Nagasawa, 1961) as follows (see Figure 10.10):

$$\psi(x) = \frac{2RT}{F}\ln\frac{1+\alpha\exp(-\kappa x)}{1-\alpha\exp(-\kappa x)} \qquad (10.95)$$

where

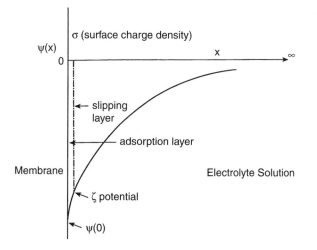

FIGURE 10.10 A schematic diagram of the Gouy-Chapman diffused electrical potential $\psi(x)$. Membrane surface set at $x = 0$.

$$\alpha = \frac{\exp\left(F\psi(0) / 2\,RT\right) - 1}{\exp\left(F\psi(0) / 2\,RT\right) + 1}$$

$$\kappa = \left(\frac{8\pi z^2 F^2 C}{1000\varepsilon RT}\right)^{1/2} \tag{10.96}$$

and the surface potential $\psi(0)$ is

$$\psi(0) = \frac{2RT}{F}\sinh^{-1}\left[\sigma\left(\frac{500\pi}{\varepsilon RTC}\right)^{1/2}\right] \tag{10.97}$$

The parameter kappa in these equations is known as the Debye constant; it has the units of reciprocal length. Expressions 10.95 and 10.97 can be derived by combining and solving the following two equations with appropriate boundary conditions.

$$\frac{\partial^2 \psi}{\partial x^2} = -\frac{4\pi}{\varepsilon}\rho \quad \text{(Poissen equation)} \tag{10.98}$$

$$\rho = \sum \rho_1 = \sum_1 \frac{Fz_i C_i(\infty)}{1000}\exp\left(\frac{-Fz_i\,\psi(x)}{RT}\right) \quad \text{(Boltzmann equation)} \tag{10.99}$$

where ρ is the total charge per unit volume and $C_i(\infty)$ is the molar concentration of the ith ionic species in the bulk phase. The boundary conditions are

$$\psi(x)\big|_{x=0} = \psi(0)$$

$$\frac{\partial \psi}{\partial x}\bigg|_{x=0} = -\frac{4\pi\sigma}{\varepsilon}$$

$$\psi(x)\big|_{x=\infty} = 0 \qquad (10.100)$$

and

$$\frac{\partial\psi}{\partial x}\bigg|_{x=\infty} = 0$$

Here, the membrane surface is set at x = 0, and the x-coordinate, which is normal to the membrane surface, extends in the electrolyte solution to ∞. In the case of a negative surface charged membrane, the surface potential is negative and therefore, cation concentrations near the membrane surface are greatly enhanced over those in the bulk by the surface potential.

$$C_i^{Surface} = C_i(\infty)\exp\left(-\frac{z_i F\psi(0)}{RT}\right) \qquad (10.101)$$

A simple example is shown in Table 10.6.

The surface potential of charged membranes could be obtained by measuring the electrical conductance of the membrane. In the presence of a neutral carrier, such as valinomycin or monactin (polyene antibiotics), the conductance, G, of a charged lipid membrane has been shown to be equal to the conductance of a neutral lipid membrane multiplied by an exponential function of the surface potential $\psi(0)$:

$$(G^+)^{charged} = (G^+)^{uncharged} \cdot \exp\left(-\frac{F\psi(0)}{RT}\right) \qquad (10.102)$$

A similar expression is given for a negatively charged permeant species. Such an equation is valid only if the product of the mobility and partition coefficient of the permanent species are the same for the charged and neutral membranes. Applied to lipid bilayer membranes, at the limit of zero applied voltage, the treatment is found to be valid. Therefore, by knowing the conductances for the charged and uncharged membranes, one can estimate the surface potential of the membranes. The experimental result to demonstrate this is shown in Figure 10.11.

When a membrane possesses a surface potential on each side of the membrane surface, the unidirectional ionic fluxes (Equation 10.90 and 10.91) should be modified in the following manner:

$$\vec{J}_j = -z_j P_j \frac{E_m^* F}{RT} \frac{C_{jo}\, e^{-z_j\psi_o F/RT}}{1-e^{z_j E_m^* F/RT}} \qquad (10.103)$$

$$\overleftarrow{J}_j = -z_j P_j \frac{E_m^* F}{RT} \frac{C_{ji}\, e^{z_j E_m^* F/RT}\, e^{-z_j\psi_i F/RT}}{1-e^{z_j E_m^* F/RT}} \qquad (10.104)$$

$$E_m^* = E_m + \psi_i - \psi_o \qquad (10.105)$$

E_m is the transmembrane potential between the two bulk solutions, E_m^* is the membrane potential difference across the membrane proper, and ψ_i and ψ_o are the surface potentials at the inner and outer membrane surfaces, respectively.

TABLE 10.6
Enhancement of Ion Concentration at Charged Membrane Surfaces (25°C)

C_o	H+ $C = C_o \exp(-e\psi_o/kT)$			M+ $C = C_o \exp(-e\psi_o/kT)$		M- $C = C_o \exp(+e\psi_o/kT)$		M2+ $C = C_o \exp(-2e\psi_o/kT)$	
	$\psi_o = 0$	$\psi_o = -20$ mV	$\psi_o = -60$ mV	$\psi_o = -20$ mV	$\psi_o = -60$ mV	$\psi_o = -20$ mV	$\psi_o = -60$ mV	$\psi_o = -20$ mV	$\psi_o = -60$ mV
10^{-6} (M)	6 (pH)	5.66 (pH)	4.96 (pH)	2.2×10^{-6} (M)	1.1×10^{-5} (M)	0.45×10^{-6} (M)	0.9×10^{-8} (M)	4.9×10^{-6} (M)	1.2×10^{-4} (M)
10^{-5}	5	4.66	3.96	2.2×10^{-5}	1.1×10^{-4}	0.45×10^{-5}	0.9×10^{-7}	4.9×10^{-5}	1.2×10^{-3}
10^{-4}	4	3.66	2.96	2.2×10^{-4}	1.1×10^{-3}	0.45×10^{-4}	0.9×10^{-6}	4.9×10^{-4}	1.2×10^{-2}
10^{-3}	3	2.66	1.96	2.2×10^{-3}	1.1×10^{-2}	0.45×10^{-3}	0.9×10^{-5}	4.9×10^{-3}	1.2×10^{-1}
10^{-2}	2	1.66	0.96	2.2×10^{-2}	1.1×10^{-1}	0.45×10^{-2}	0.9×10^{-4}	4.9×10^{-2}	1.2
10^{-1}	1	0.66		2.2×10^{-1}	1.1	0.45×10^{-1}	0.9×10^{-3}	4.9×10^{-1}	12.0

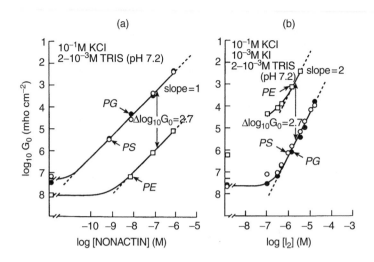

FIGURE 10.11 (a) The effect of the antibiotic nonactin on the conductance of a neutral (phosphatidyletha-nolamine, PE) and two negatively charged (phosphatidylserine, PS and phosphatidylglycerol, PG) bilayer membranes. The permeant species is the positively charged nonactin-K complex. (b) The effect of the neutral iodine molecule on the conductance of a bilayer formed from the same lipids. The permeant species is the negatively charged complex. The change of 2.7 log units in the conductance implies the surface potential at the surface of the negatively charged membrane to be −158 mV. (From McLaughlin SGA et al.: *J. Gen. Physiol.* 1971, 58:667. With permission.)

$$\vec{J}_j = -z_j \vec{P}_j \frac{FE_m^*}{RT} \frac{C_{jo}}{1 - e^{z_j FE_m^*/RT}} \tag{10.106}$$

$$\vec{J}_j = -z_j \vec{P}_j \frac{FE_m^*}{RT} \frac{C_{ji} e^{z_j FE_m^*/RT}}{1 - e^{z_j FE_m^*/RT}} \tag{10.107}$$

where

$$\vec{P}_j = P_j e^{-z_j \psi_o F/RT}$$

and

$$\vec{P}_j = P_j e^{-z_j \psi_i F/RT} \tag{10.108}$$

The surface potentials can be included in the expression for the membrane potential by the following:

$$E_m = \frac{RT}{F} \ln \frac{\sum P_j^+ C_{jo}^+ + \sum P_j^- C_{ji}^- e^{F(\psi_o + \psi_i)/RT}}{\sum P_j^+ C_{ji}^+ + \sum P_j^- C_{jo}^- e^{F(\psi_o + \psi_i)/RT}} \tag{10.109}$$

where $E_m = \psi_o - \psi_i + E_m^*$. Here, when $\psi_o = \psi_i = 0$, the potential equation is reduced to the G-H-K equation (Equation 10.87). However, frequently biological membranes possess surface charges of the asymmetric on the two sides of the membrane. Therefore, $\psi_o = \psi_i = 0$ is not expected. In the case of the squid axon membrane, the internal surface potential is low because of low surface charge density, whereas the external surface potential is also low because of adsorption of divalent cations on the outer membrane. In this instance, the G-H-K equation generally holds well.

In an extreme experimental condition, on the other hand, as in solutions of nonphysiological ionic strength or with a low divalent cation content, the G-H-K equation must be modified to include the effect of surface charges and surface potential, accordingly (Ohki, 1985).

An alternative to the Nernst–Planck equation (Equation 10.48 or Equation 10.61), describing the passive flux of an ion, uses finite differences of the electrochemical potentials across the membrane:

$$J_j = -L_j \left(\tilde{\mu}_{ji} - \tilde{\mu}_{jo} \right) \tag{10.110}$$

The electrical current carried across the membrane by the jth ionic species may be expressed as follows:

$$I_j = z_j F J_j = - z_j F L_j \left(RT \ln C_{ji} + z_j F \varphi_i - RT \ln C_{jo} - z_j F \varphi_o \right)$$

$$= z_j^2 F^2 L_j \left[\frac{RT}{z_j F} \ln \frac{C_{jo}}{C_{ji}} - E_m \right] = G_j \left(E_j - E_m \right) \tag{10.111}$$

where G_j is the conductivity of the membrane for the jth ion ($G_j = z_j^2 F^2 L_j$), E_j is its equilibrium potential (Equation 10.56), and E_m is the membrane potential.

Under the condition of zero total electrical current,

$$\sum_j I_j = 0 \tag{10.112}$$

(assuming that ion fluxes due to active transport processes etc. are either absent or electrically silent), the steady-state open circuit membrane potential can then be expressed as

$$E_m = \frac{\sum G_j E_j}{\sum G_j} = \sum t_j E_j \tag{10.113}$$

where t_j is the transference number of the jth ionic species:

$$t_j = \frac{G_j}{\sum G_j}$$

Equation 10.113 is known as the *Hodgkin–Horowitz equation*, and it appears to describe most cell membrane potentials better than the corresponding Goldman equation. This indicates that the approach of relating fluxes linearly to finite differences of electrochemical potentials across the membrane is preferable to the integration of the differential electrodiffusion equation. The complication of the membrane boundary effects can be avoided, because the conductivity of the membrane is usually determined experimentally, and these boundary effects are included in the empirical parameter.

In certain instances, the membrane conductance of specific ions may vary with the transmembrane potential difference, and exhibit complex time dependence with changing membrane potentials.

Such behavior can generate non-linear effects which have vital biological functions, and underlie the nerve impulse and action potentials in muscle, for example. In the case of the propagated nerve impulse, Hodgkin and Huxley (1952) demonstrated quantitatively through voltage clamp studies of the squid giant axon, that the time–voltage dependence of Na^+ and K^+ conductances can account for the propagated nerve spike, or action potential. The total membrane current is expressed as

$$I = C_m \frac{dV}{dt} + G_{Na}(V - E_{Na}) + G_k(V - E_k) + G_L(V - E_L) \tag{10.114}$$

in which C_m is the membrance capacitance, V is the membrane potential, and E_j is the equilibrium potential for the particular ion. The subscript L refers to a composite "leakage" conductance comprising minor ionic currents other than that of sodium and potassium. G_{Na} and G_K are the time- and voltage-dependent conductivities elucidated by the studies of Hodgkin and Huxley. Note that the sign of this expression differs from that used above since the neurophysological convention is to assign outward flowing currents as positive.

In the resting state, the membrane potential is about –60 mV (inside of the nerve axon relative to external solution) which is close to the equilibrium potential for potassium. Na^+, however, is far from its equilibrium state, which corresponds to a membrane potential of about +40 mV. The resting potential results from the relatively higher K^+ conductivity than that for sodium ion, although both conductances are low at the resting membrane potential. If the membrane potential is perturbed in a more positive direction, however, both ionic conductances increase. Given a sufficiently large perturbation, a self-regenerating effect occurs, leading to the nerve impulse. This effect occurs because the response time of increased sodium conductivity to the more positive membrane potential is more rapid than that of potassium, allowing sodium ions to enter the cell and drive the membrane potential even more positive. Thus, the membrane potential rises rapidly (within a fraction of a millisecond) toward the equilibrium potential for sodium. This process is eventually interrupted by the slower turn-on of potassium, and a comparably slow turn-off mechanism in the sodium perme-ation system. Thus, the membrane potential returns once again to its resting state. Propagation of the action potential along the nerve axon occurs because the presence of an action potential at one location of the nerve raises the membrane potential at an adjacent segment beyond its threshold value.

Based upon their measurements of the voltage and time dependence of the Na^+ and K^+ conductivities, Hodgkin and Huxley calculated using Equation (10.114) the form of the nerve impulse, as well as the velocity of propagation of the action potential. Both theoretical predictions agreed quite well with experimental results. More recently (Patlak, 1991; Bezanilla and Stefani, 1994), the voltage-dependent sodium and potassium conductances observed by Hodgkin and Huxley have been shown to correspond to the ensemble average over a population of stochastic individual Na^+ and K^+ channels. It should be noted that the nerve impulse and other similar biological action potentials arise from the purely passive movements of ions across the cell membrane, although an independent active transport process is required to maintain the non-equilibrium distribution of ionic species across the membrane.

10.1.2.2.2 Electrogenic Pump Potential

Ions transported by the active transport process contribute an additional potential to those arising from passive ion transport processes. According to the principles of thermodynamics of nonequi-librium processes (see Section 10.1.3), the flux of a jth ion across a membrane is expressed as

$$J_j = -L_j(\Delta \tilde{\mu}_j) + L_{jr} A_r \tag{10.115}$$

where A_r is the affinity of the reaction driving the electrogenic ion pump. The corresponding current will be

$$I_j = z_j F J_j = L_j F^2 z^2_j (E_j - E_m) + L_{jr} F^2 z_j A_r / F$$

$$= G_j (E_j - E_m) + G^{Act}_j E^{Act}_j \qquad (10.116)$$

where G^{Act}_j stands for $L_{jr} F^2 z_j$, and E^{Act}_j stands for A_r / F.

The steady-state open-circuit membrane potential will be

$$E_m = \frac{\sum_j G_j E_j + G^{Act}_i E^{Act}_i}{\sum_j G_j} = \frac{\sum G_j E_j}{\sum G_j} + \Delta \qquad (10.117)$$

where

$$\Delta = \frac{G^{Act}_i E^{Act}_i}{\sum_j G_j} = \frac{L_{ir} F A_r}{\sum_j G_j}$$

represents the contribution of the electrogenic ion pump to the membrane potential. It has been reported that not all actively transported ions contribute an electrogenic term to the membrane potential. An active ion pump is often partially electrogenic, with a portion that is electrically silent. In the silent transport process, the actively transported ionic species (e.g., Na^+) travels together with a counter ion (e.g., Cl^-), or alternatively is coupled with the exchange transport of a co-ion (e.g., K^+) in the opposite direction. The active transport can be split into two parts, corresponding to the portion of the process involving net charge transfer:

$$L_{jr} A_r = \left(1 - \frac{1}{\alpha}\right) L_{jr} A_r + \left(\frac{1}{\alpha}\right) L_{jr} A_r \qquad (10.118)$$

where α is the number of actively transported ion per co-ion transported in the opposite direction. Then, the current carried by the jth species then will be

$$I_j = G_j (E_j - E_m) + \left(1 - \frac{1}{\alpha}\right) L_{jr} A_r \qquad (10.119)$$

In this case, we define $G^{Act}_j = (1 - 1/\alpha) L_{jr} z_j F^2$. Some electrogenic potentials measured in various tissues are shown in Table 10.7. In the squid giant axon, for example, two potassium ions are exchanged for every three Na^+ extruded. Thus, $\alpha = 3/2$ in this case.

10.1.2.2.3 Ion Selectivity

Membranes generally have specific selectivities or relative permeabilities for various ions. There are several factors responsible for these selectivities, including ionic size, the degree of hydration of ions, electrostatic interactions (including polar interaction), image charge forces, ionic coordination number, entropic contributions, and membrane pore size. For ion transport across membranes, a three-step process may be involved: first, ion adsorption or binding to the membrane surface or sites; second, the entry of ions into the membrane phase; and third, ion diffusion across membrane proper. The factors mentioned previously may contribute to these latter two processes. Relative to the second process, the partition of an uncharged particle between aqueous and membrane phases has been discussed earlier. In the case of ions, a dominant factor is the hydration shell

TABLE 10.7
Electrogenic Potentials of Various Cells at a Steady State

Cells	Electrogenic Potential (mV)
Helix	+2–3
Crayfish stretch receptor neurons	+10
Barnacle muscle	+6
Squid axon	+1.4
Limulus photoreceptor	+5
Aplysia neurons	+22
Leech sensory neuron	+5

around an ion due to the interaction between the charge (ion) and the electric dipole of water molecules. Born proposed the solvation energy of ion entering into water from a vacuum as

$$\Delta E^{vac \to w} = \frac{z^2 e^2}{8\pi \varepsilon_o a} \left(\frac{1}{\varepsilon_w} - \frac{1}{\varepsilon_{vac}} \right) \tag{10.120}$$

where a is the radius of the ion, ε_o is the permittivity of vacuum, and ε_w and ε_{vac} are the dielectric constants of water and vacuum, respectively. This energy is considered as hydration energy. Thus, when an ion enters from an aqueous phase into a membrane phase, the hydration energy becomes an energy barrier for the ion. Such energy, $\Delta G^{w \to m}$, can be calculated as a difference between the solvation energies (ΔE) of an ion in water and oil phases as computed according to Born:

$$\Delta G^{w \to m} = \Delta E^{vac \to m} - \Delta E^{vac \to w} = \frac{z^2 e^2}{8\pi \varepsilon_o} \left(\frac{1}{\varepsilon_m} - \frac{1}{\varepsilon_w} \right) \tag{10.121}$$

The hydration energies calculated using Equation 10.120 for various monovalent ions are listed in Table 10.8. Latimer et al. modified this approach, taking into account the effective dielectric constant of the media around the ions, with results as shown in the table. In addition, the energy change, $\Delta G^{w \to m}$ as computed according to Equation 10.121, is listed in Table 10.8. The permeability of monovalent ions through a homogeneous oil membrane generally follows the order of these hydration energies; the greater the ion hydration energy, the less permeable the ion is to the membrane. Most biological membranes, however, are charged membranes. Therefore, for the biological membranes, the first factor mentioned above bearing upon membrane selectivity and ion permeability also becomes important.

Eisenman (1961) examined ionic selectivities for various glass membranes and found that there are restricted numbers of selectivity series for various ions. For five alkaline cations (Na^+, Li^+, K^+, Rb^+, and Cs^+), the sequences of selectivities observed comprise only 11 of the large number possible. These are listed in Table 10.9. Sequence 1 is the lyotropic series, with ions arranged in order of increasing hydrated ion size, whereas sequence 11 is arranged in order of increasing nonhydrated ion size. The other sequences are seen in various types of glass membranes and various biological membranes (Eisenman, 1965).

Eisenman proposed a unified theory accounting for these observed selectivities in which differences in anionic site strength in the membrane determine the sequences. According to his theory, in order for a cation in solution I^+ to bind to a membrane site X^-, the ion must detach itself from the associated water molecules W:

TABLE 10.8
Hydration Energy of Various Monovalent Ions

Ions	Ionic Radius[a] (Å)	Born	Latimer[b] (Kcal/mole)	Experimental Data	Hydration Number[c]	$\Delta G^{w \rightarrow m}$ [d] (Kcal/mole)
Li^+	0.78	239	136	131	4	120 (68)
Na^+	0.98	166	114	116	3	83 (57)
K^+	1.33	123	94	92	2	62 (47)
Rb^+	1.49	109	87	87	1	55 (43)
Cs^+	1.65	99.5	80	79	—	50 (40)
F^-	1.33	122	97	94	3	61 (58)
Cl^-	1.81	90	65	67	2	45 (32.5)
Br^-	1.96		57	63	—	— (28)
I^-	2.20		47	49	0.7	— (23)

[a] Goldschmit radii.

[b] Latimer et al. (1939) used the effective ionic radii in their calculation.

[c] Hydration numbers are according to Bocris.

[d] $\Delta G^{w \rightarrow m}$ was calculated using the dielectric constant of membrane of 2.0, and the values in parenthesis is using the solvation energy obtained by Latimer et al.

TABLE 10.9
Observed Sequences of Alkali-Cation Selectivities

1. $Cs^+ > Rb^+ > K^+ > Na^+ > Li^+$
2. $Rb^+ > Cs^+ > K^+ > Na^+ > Li^+$
3. $Rb^+ > K^+ > Cs^+ > Na^+ > Li^+$
4. $K^+ > Rb^+ > Cs^+ > Na^+ > Li^+$
5. $K^+ > Rb^+ > Na^+ > Cs^+ > Li^+$
6. $K^+ > Na^+ > Rb^+ > Cs^+ > Li^+$
7. $Na^+ > K^+ > Rb^+ > Cs^+ > Li^+$
8. $Na^+ > K^+ > Rb^+ > Li^+ > Cs^+$
9. $Na^+ > K^+ > Li^+ > Rb^+ > Cs^+$
10. $Na^+ > Li^+ > K^+ > Rb^+ > Cs^+$
11. $Li^+ > Na^+ > > K^+ > Rb^+ > Cs^+$

From Eisenman G: *Symposium on Membrane Transport and Metabolism.* Eds. Kleinzeller A, Kotyk A, Academic Press, New York, 1961, 163–179. Copyright Czechoslovak Academy of Sciences.

$$I^+W + X^- \leftrightharpoons I^+X^- + W \qquad (10.122)$$

The standard free energy change for this reaction is

$$\Delta G^{o, w \rightarrow x} = \Delta G^o_{ion-site} - \Delta G^o_{ion-water} \qquad (10.123a)$$

It is assumed that the significant differences between $\Delta G^o_{ion-site}$ and $\Delta G^o_{ion-water}$ are due to the differences in electrostatic energy, U, for ion-site and ion-water bindings:

$$\Delta G^{o, w \rightarrow x} \cong \Delta U = \Delta U^{elec}_{ion-site} - \Delta U_{ion-water} \qquad (10.123b)$$

TABLE 10.10
Hydration Energies and Goldschmidt
Crystal Radii for the Alkali Cations

	r+, Å	$\Delta U_{ion\text{-}water}$, Kcal/mol
Li$^+$	0.78	−24.2
Na$^+$	0.98	−20.5
K$^+$	1.33	−15.8
Rb$^+$	1.49	−14.1
Cs+	1.65	−12.7

Note: Hydration energies based on coulombic forces between single Rawlinson-type molecules and each cation (see Eisenman, 1961).

The first of these electrostatic contributions is

$$\Delta U^{elec}_{ion-site} = \frac{332 q^+ q^-}{r^+ + r^-}$$ (10.124)

where

$\Delta U^{elec}_{ion-site}$ = ion-site binding energy.
q^+ = charge of the site, electronic charge unit
q^- = charge of the site, electronic charge unit
r^+ = crystal radius of the cation, Å
r^- = crystal radius of the anionic site, Å

In relation to the standard free energy of an ion in water, the hydration energy of an ion was calculated from the electrostatic interaction between a cation and a multipolar water molecule (after Rawlinson), making use of the crystal radius for various ions (Goldschmidt, 1926; Table 10.10). Although the calculation is based upon a physical model different from that used by Born and others mentioned earlier, the results obtained by the two different methods are qualitatively similar. The difference in the standard free energy between ion-site and ion-water can be calculated with the use of Equation 10.124 and the hydration energy of the ion as a function of r^- and q^-, again making use of the crystal radius of each ion (Goldschmidt, 1926). The results are shown in Figure 10.12. It is understood that the lower ΔU is for an ion, the more readily it penetrates a membrane. As seen in Figure 10.12, the order of ion selectivity depends on the anionic site radius r^-.

A similar analysis may also be carried out, but varying q^- instead of r^-. Membrane selectivity studies for the halides F$^-$, Cl$^-$, I$^-$, and Br$^-$, and for the divalent cations have been calculated according to Eisenman's theory (Diamond and Wright, 1969). This approach to the analysis of ionic selectivity and membrane permeability can be extended to include several other factors, such as pore size and the degree of hydration. Some effort along these lines to explain the relative permeability across membranes through ionic channels has been carried out (Hille, 1984; Miller, 1986). Experimental results indicate that the permeability of ions through Na$^+$ channels, depending upon species, falls in either the ion selectivity series 11 or 10. K$^+$ channels correspond to the series 5, while the end plate channel (acetylcholine (ACh) receptor) belong to sequence 1 (Hille, 1984).

In recent years, the structures and functional properties of various ionic channels have become evident. There are at least three major excitable ionic channel families: (1) voltage gated channels (Na$^+$, K$^+$, Ca^{2+}); (2) ligand gated channels (ACh, Glycine, GABA, Glutamate, and possibly ATP,

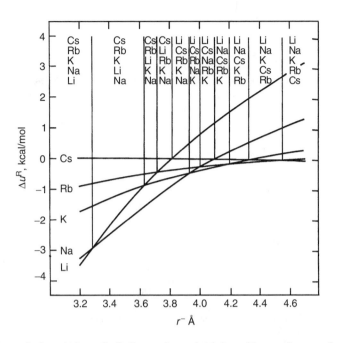

FIGURE 10.12 Theoretical variation of alkaline-cation selectivity with membrane anionic site radius r. $\Delta U_{ion\text{-}site} - \Delta U^R$ = difference in energies of site binding and dehydration relative to that for cesium (Adapted from Eisenman G: *Symposium on Membrane Transport and Metabolism.* Academic Press, New York, 1961, pp. 163–179.)

among others); and (3) mechano-sensitive activated channels. Besides these, a number of non-excitable ionic channels or large membrane pores have been found. Some of the ionic channels are categorized in Table 10-11. The ionic channels are proteins, consist of either single polypeptides or oligomeric assemblies, and contain transmembrane subunits. For example, the main component of the Na^+ channel is a single polypeptide consisting of four repeating subunits, each having six transmembrane segments. The four subunits form an ionic channel across the membrane. A voltage sensitive segment, S-4, contains many positively charged amino-acids, while an ion selective site is at the loop between segments S5 and S6 (see Figure 10.13). The channel can be opened by applying an electric field of proper orientation across the membrane. The gating of the channel is thus dependent upon the membrane voltage (Hodgkin and Huxley, 1952).

Although several kinds of K^+ channels exist, the main morphology of the various channels is similar, consisting of an oligomer of four independent α-subunits, which form a channel. The ion selectivity site is also located at the loop segment between segments 5 and 6, which is located on the extracellular side (as is the case with the Na^+ channel). This channel also has voltage dependent characteristics for ionic conductance gating as does the sodium channel.

Ca^{2+} channels consist of an oligomer of several different subunits $(\alpha,\beta,\gamma,\delta)$. The α_1 subunit forms a core of the ionic channel, similar to the Na^+ channel, but possessing a more complex structure with other subunit forms. Different types of Ca^{2+} channels exist which exhibit different kinetics in ionic conductances upon excitation (L, P, and N). Ligand gated ionic channels are activated by various ligands binding at regions on the extracellular side. The ion selective region seems to be intracellular, in contrast to voltage gated channels. Upon binding of the ligand (e.g., acetylcholine) to two or more sites, a conformational change occurs and the channel opens for a short period. It is thought that the pore opening depends upon the kinetics of at least two different processes; ligand binding and the protein conformation change.

Mechano-sensitive channels are the ion channels, the opening probability of which is altered by mechanical stress applied to the cell membrane. The activity of sense organs such as Pacinian

TABLE 10.11
Ion Channels

Channels	Oligomers	Subunits	Ligands	Ions	Types	Involved In:
Voltage-Gated Channels						
Na^+		$\alpha\beta$ $(\alpha\beta^2)$		Li^+ and Na^+	Unique	Action potential in nerves and muscles
K^+		$\alpha^4\beta$		K^+	Variety	Action potential
Ca^{2+}		$\alpha^2\beta\gamma\delta$		Ca^{2+}	L, P, N	Cardiac AP
						Calcium transport in cells
Ryanodine receptor	Tetramer	α^4		Ca^{2+}		Calcium release from SR in muscles
Ligand-Gated Channels						
Acetylcoline	Pentamer	$\alpha^2\beta\gamma\delta(\alpha^2\beta^3)$	ACh	M^+ (cations)		Synaptic transmission (EPP, EPSP)
GABA	Pentamer	$\alpha^2\beta\gamma\delta$	GABA	Cl^- (anions)		Synaptic transmission (IPSP)
Glycine	Pentamer	$\alpha^3\beta^2$	Glycine	Cl^- (anions)		Synaptic transmission (IPSP)
Glutamate receptor	Tetramer		Glutamate	M^+ Ca^{2+}, M^+	AMPA-kainate NMDA	Synaptic EPSP and plasticity
Cyclic nucleotide-gated channels	Tetramer		cAMP, cGMP			Visual and olfactory transduction
Ca^{2+}-activated K^+ channels			Ca^{2+}	K^+		
IP_3 receptor	Tetramer	α^4	ATP and IP_3	Ca^{2+}		Ca^{2+} release from SR in cells
P_{2X} receptors	Trimer		ATP	M^+, Ca^{2+}		Nociception (sensing of pain)
Mechano-Sensitive Channels						
				M^+, M^-, Ca^{2+}		Muscle cells, auditory cells
Other Non-Excitable Channels						
Chloride channels	Tetramer?			Cl^-, M^- (anions)		CT and anions transport in cells
Gap junction channels	Hexamer		Ca^{2+}, H^+, voltage	Ions and small molecules		Epithel and cardiac cells, synapses

corpuscles and hair cells is related to these channel activities. Muscle stretch receptors have long been studied. More recently, the patch clamp technique has demonstrated the presence of the mechano-sensitive channels in a large number of non-sensory cells.

Many channels populating the biological membrane operate in a stochastic discrete fashion, fluctuating between open and closed states. The activating agent then modifies the dwell time in either or both states, or in other words, modulates the probability of being open at any particular instant. The apparent ionic conductivity, however, averaged over a sufficiently long period in the case of a single channel, or averaged over a population of channels, can appear as a continuous graded response to the magnitude of the controlling factor, thus disguising the underlying stochastic mechanisms. Nonetheless, despite the molecular and kinetic complexity, one would expect the

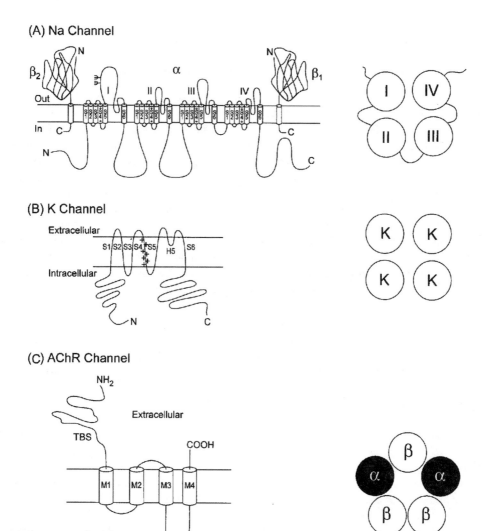

(A) Na Channel

(B) K Channel

(C) AChR Channel

FIGURE 10.13 (a) Proposed transmembrane topology (left) and model top view (right) of voltage dependent Na$^+$ channels. It consists of one α-subunit and two β-subunits. The α-subunit contains four domains (I, II, III, and IV) in series, each of which has six transmembrane segments (S1, S2, S3, S4, S5, and S6) shown in the figure. These four repeating units form a core of the Na$^+$ channel. The S4 segment contains basic amino acid residues and would serve a voltage sensor; however, the ion selectivity filter is located at the loop segment between S5 and S6. (b) A typical voltage dependent K$^+$ channel is considered to consist of four separated subunits forming an oligomer (right) and, in addition, four β-subunits (not shown in the figure) similar to those of Na$^+$ channels. Each subunit (left) spans six times through the membrane like in the Na$^+$ channel. The S4 is a voltage sensor and the loop between S5 and S6 seems to pertain to the selectivity of ions. The N-terminus of the α subunit corresponds to the cytoplasmic inactivation gate (the ball), which interacts with the S4-S5 loop and occludes the ionic pore. (c) Current view of the ligand-gated ion channel subunit topology (left). Two transmitter-binding sites (TBS) per pentamer are formed in the extracellular N-terminal domain. The channel pore is lined, for the most part, by the five M2 segments.

membrane permeation processes in such channels to conform to the basic physico-chemical principles illustrated in this chapter.

10.1.3 Interaction among Fluxes (Nonequilibrium Thermodynamics)

Thus far, molecular diffusion across membranes has been described under circumstances in which no interaction between permeant species occurs, except for the solvent. In this formulation, the only thermodynamic force affecting the molecular diffusion of a species arises from its own electrochemical potential gradient. From a general point of view, however, there is no justification for excluding the possibility of kinetic interaction between different permeant species, such that the flow of one is influenced by gradients of the electrochemical potentials of others. The principles and formalism of nonequilibrium thermodynamics (NET) provide a convenient framework for the treatment of interactions of this sort. Although devoid of mechanistic information (as is classical thermodynamics), the developments of NET have been surprisingly utilitarian in defining constraints on coupled processes and the relationships between kinetic parameters describing these processes.

Thermodynamics of irreversible processes is founded upon the axiom that any naturally occurring process must be accompanied by a net production of entropy. The change of entropy of a system in contact with its environment can be expressed by

$$dS = d_eS + d_iS \qquad (10.125)$$

in which d_eS represents the change resulting from the flow of entropy across the boundary of the system, and d_iS is the entropy produced by the occurrence of a nonequilibrium process within the system.

Thus, for the rate of entropy production, we have the requirement

$$d_iS/dt \geq 0 \qquad (10.126)$$

with this rate vanishing only at equilibrium.

Should the system under consideration be isolated or taken to be the entire universe, the entropy exchange term vanishes, and $dS = d_iS$. For simplicity, the subscript i will be dropped in the following discussion. If the entropy production of the system is now expanded in a series about a reference equilibrium state, the first-order terms vanish because entropy is at an extremum at equilibrium. The quadratic terms of the series take the form of a sum over products of flows (J) and forces (X)

$$dS/dt = \sum_j X_j J_j \qquad (10.127)$$

in which the index j ranges over every independent degree of freedom in defining the state of the system. The paired forces and fluxes in this summation are termed *conjugate pairs*. Both forces and fluxes vanish at equilibrium. The set of forces X_j, moreover, in specifying the perturbations away from equilibrium, is sufficient to define completely the state of the system, including the magnitudes of the resultant fluxes. Thus, one can expand each flux as a series in the forces, yielding

$$J_k = \sum_j (\partial J_k / \partial X_j) X_j + \text{higher order terms} \qquad (10.128)$$

Retaining only the first order terms, and writing L for the partial derivatives, the linear phenomenological law results:

$$J_k = \sum_j L_{kj} X_j \qquad (10.129)$$

Arising as it does as the leading term in an expansion about an equilibrium state, the validity of the linear relation (Equation 10.129) is restricted to a "sufficiently small" region around equilibrium. Just how large the region of linearity is depends on the specific processes involved; in many instances, the extent of linear behavior proves to be surprisingly large.

The interesting feature of this phenomenological law is the introduction of coupling between a particular flux and forces nonconjugate to it. Thus, the translational flux of one species can be directly influenced by the electrochemical potential gradient of another, provided the corresponding coefficient L_{ji} is nonzero.

Important properties of the L matrix have been established, which are key to the utility of the nonequilibrium thermodynamic formulation. Writing once again the entropy production

$$dS/dt = \sum_i X_i J_i = \sum_{ij} X_i L_{ij} X_j \qquad (10.130)$$

Onsager (1931) showed that with a complete set of force and fluxes satisfying the entropy production equation, then

$$L_{ij} = L_{ji} \qquad \text{any } i, j \qquad (10.131)$$

Known as *Onsager's reciprocal relations,* the equality asserted by Equation 10.131 is important in reducing the number of independent kinetic parameters necessary to characterize the system fully. Furthermore, the requirement that entropy production (Equation 10.130) be positive definite places constraints on the value that the cross-coupling coefficient L_{ij} can assume:

$$(L_{ij})^2 < L_{ii} L_{jj} \qquad \text{any } i, j \qquad (10.132)$$

Derivations of this constraint, as well as Onsager's relations, may be found in the monographs by Denbigh (1958), Katchalsky and Curren (1965), and DeGroot (1963).

The treatment of transmembrane water and uncharged solute fluxes elaborated by Kedem and Katchalski (1958) provides an illuminating example of the application of nonequilibrium thermodynamics. Of particular interest are the transformations in the forces and fluxes possible within the formalism. For the flux of water and solute i, in an isothermal system, we have

$$J_w = L_w \Delta\mu_w + \sum_j L_{wj} \Delta\mu_j$$

$$J_i = L_{iw} \Delta\mu_w + \sum_j L_{ij} \Delta\mu_j \qquad (10.133)$$

in which $\Delta\mu = \mu$ (side 1) $- \mu$ (side 2), with the direction of positive flux being from side 1 to side 2. For clarity, water is denoted by the explicit subscript w. The index j, then, is taken over all nonwater constituents of the solutions on either side of the membrane, even though the membrane may be completely impermeable to certain of these species. The reason for their inclusion will become apparent shortly. Now, for the entropy production function, we have

$$T \, dS/dt = J_w \Delta\mu_w + \sum_i J_i \Delta\mu_i \tag{10.134}$$

Note that these expressions are valid only after a steady state has been achieved within the membrane. One can expect the phenomenological coefficients of Equation 10.133 to satisfy Onsager's reciprocal relations.

Maintaining the invariance of the entropy production, Equation 10.134 offers a convenient recipe for the transformation of the description of the system into an alternative, and perhaps more convenient set of forces and fluxes. For example, expanding the force terms into pressure and purely composition-dependent terms, as denoted by the superscripted c, yields

$$\Delta\mu_i = \bar{v}_i \Delta P + \Delta^c\mu_i \tag{10.135}$$

The $\bar{v}_i$ are partial molar volumes. With these substitutions, the entropy production then becomes

$$T \frac{dS}{dt} = J_v \Delta P + J_w \Delta^c\mu_w + \sum_i J_i \Delta^c\mu_i \tag{10.136}$$

where J_v is defined to be the volume flow, given by

$$J_v = \bar{v}_w J_w + \sum_i \bar{v}_i J_i \tag{10.137}$$

Expression 10.136 is flawed, however, in that it contains an additional force and flux; thus they cannot all be independent. This problem is resolved by eliminating the explicit water force term through the substitution

$$\Delta^c\mu_w = -\sum_i \frac{\tilde{C}_i \Delta^c\mu_i}{C_w} \tag{10.138}$$

This expression is obtained from the Gibbs–Duhem relation, written in the form

$$\frac{S}{V} dT - dP + C_w d\mu_w + \sum_i C_i d\mu_i = 0 \tag{10.139}$$

$$\frac{S}{V} dT - dP\left(1 - C_w \bar{v}_w - \sum_i C_i \bar{v}_i\right) + C_w d^c\mu_w + \sum_i C_i d^c\mu_i = 0$$

Because $C_w \bar{v}_w + \Sigma C_i \bar{v}_i = 1$, for an isothermal (dT = 0) system we obtain

$$C_w d^c\mu_w = -\sum_i C_i d^c\mu_i \tag{10.140}$$

This expression can then be integrated through a continuum of states (which need not physically exist within the membrane) from the composition on one side of the membrane to that on the other, arriving at relation Equation 10.138 with the definition

$$\tilde{C}_i \equiv \frac{\int_1^2 C_i d^c \mu_i}{\Delta^c \mu_i} \tag{10.141}$$

and assuming the concentration of water, being the major constituent, to be essentially constant.

If the composition dependence of the ith species chemical potential is assumed to be logarithmic in its concentration, Equation 10.141 becomes

$$\tilde{C}_i = \frac{\Delta C_i}{\Delta \ln C_i} \tag{10.142}$$

the average concentration term first introduced in the work of Kedem and Katchalsky (1958). In any case, $\tilde{C}_i$ represents some sort of averaged concentration. Making use of the approximation $J_w/C_w \cong J_v$, valid for dilute solutions, Equation 10.137 finally becomes

$$T \, dS/dt = J_v \Delta P + \sum_i \left(J_i - \tilde{C}_i J_v \right) \Delta^c \mu_i = J_v \Delta P + \sum_i J_i' \Delta^c \mu_i \tag{10.143}$$

in which the entropy production is expressed in terms of the volume flow with the pressure difference as its conjugate force, and the newly defined fluxes, J_i'. From the form of these fluxes, they may be readily interpreted as the flow of species i relative to the flow of the solution as a whole. Hence, they represent the purely diffusive flux, separated from the contribution of convection in the flow of that species.

Having retained a valid expression for the entropy production through the transformation in forces and fluxes, we can confidently write a new set of phenomenological equations and expect reciprocity to be retained:

$$J_v = L_p \Delta P + \sum_j L_{pj} \Delta^c \mu_j \tag{10.144a}$$

$$J_i' = L_{ip} \Delta P + \sum_j L_{ij} \Delta^c \mu_j \tag{10.144b}$$

with $L_{pj} = L_{jp}$ and $L_{ij} = L_{ji}$.

L_p is known as the *hydraulic coefficient* and, as the name implies, relates the bulk flow of solution across the membrane to the difference in pressure in the absence of any solute concentration differences. Considering the case of a single solute s, should the membrane structure be open and exhibit no discrimination between water and solute, one would expect a concentration difference in s across the membrane to have no influence on the hydrodynamic flow. Thus, under these circumstances, $L_{ps} = 0$. At the other extreme, should the membrane be completely impermeable to s, one would expect the hydrostatic pressure difference and the osmotic pressure difference to be equivalent in their effects on the flow of volume, now consisting entirely of the water flux. Under these circumstances, from Equation 10.144, one must have a relation of the form

$$L_{ps}\Delta^c\mu_s = -L_p\Delta\pi_s \qquad (10.145)$$

in which $\Delta\pi_s$ is the osmotic pressure of s, as developed earlier in this chapter.

These ideas can be expressed more concretely through the use of the reflection coefficient σ, introduced by Staverman. For this purpose, we rearrange Equation 10.144b and eliminate ΔP from the result using Equation 10.144a:

$$J_i = \tilde{C}_i J_v\left(1 + \frac{L_{pi}}{\tilde{C}_i L_p}\right) + \sum_j \left(L_{ij} - \frac{L_{pi}L_{pj}}{L_p}\right)\Delta^c\mu_j \qquad (10.146)$$

Upon introducing the definition of reflection coefficients as

$$\sigma_i = -\frac{L_{pi}}{\tilde{C}_i L_p} \qquad (10.147)$$

Equation 10.144 and Equation 10.146 can be rewritten:

$$J_v = L_p\left(\Delta P - \sum_i \sigma_i \tilde{C}_i \Delta^c\mu_i\right)$$

$$J_i = \tilde{C}_i J_v(1 - \sigma_i) + \sum_j L'_{ij}\Delta^c\mu_j$$

$$L'_{ij} \equiv L_{ij} - \frac{L_{pi}L_{pj}}{L_p} \qquad (10.148)$$

Finally, making the approximation of Equation 10.142 for $\tilde{C}_i$ leads to

$$J_v = L_p\left(\Delta P - \sum_i \sigma_i \Delta\pi_i\right)$$

$$J_i = \tilde{C}_i J_v(1 - \sigma_i) + \sum_j L'_{ij}\Delta^c\mu_j \qquad (10.149)$$

These are the well known "practical" equations of Kedem and Katchalsky, so called because they express system dynamics in conveniently measured quantities. The nature of the reflection coefficient σ is evident from this equation. Should this parameter be zero, the solute flux (in the absence of any concentration gradients) is simply that carried through by bulk flow of the solution. That is, there is no discrimination (reflection) of the solute. On the other hand, should σ be unity, there is complete reflection of the solute from the bulk flow stream, corresponding to the circumstances identified in Equation 10.145. The primary contribution of NET to this analysis is that of establishing the connection of the reflection coefficient, as a parameter in the flow of solute resulting from solvent drag, to its role as an effectiveness coefficient in the osmotic pressure difference of that solute species.

10.1.3.1 Electro-Osmosis

Electro-osmosis and the related phenomenon of a streaming potential are examples of coupled transport processes amenable to treatment by nonequilibrium thermodynamics. Electro-osmosis refers to the flow of volume across a membrane, in the absence of a hydrostatic pressure difference, as the result of electrical current flow through the membrane. Streaming potentials, on the other hand, are electrical potential differences that develop across a membrane as fluid is forced through it by means of a hydrostatic pressure difference. The appearance of these phenomena requires the presence of ionic solutes in the bathing solutions and a differential permeability of the membrane to cations and anions. Charged groups affixed to the membrane structure, at the surface, or lining pores through the membrane provide a likely mechanism by which anion–cation differential permeability can arise.

The entropy production function, given in Equation 10.143, can be readily generalized to the case of ionic solutes:

$$T \frac{dS}{dt} = J_v \Delta P + \sum_i \left(J_i - \tilde{C}_i J_v\right) \Delta^c \mu_i + \sum z_i F J_i \Delta \varphi \tag{10.150}$$

in which $\Delta \varphi$ is the electrical potential difference across the membrane.

For the sake of simplicity, we can consider the case of a uni-univalent ionic solution on both sides of the membrane, such that

$$\tilde{C}_a = \tilde{C}_c = \tilde{C}_s, \ \Delta^c \mu_s = \Delta^c \mu_a + \Delta^c \mu_c \text{ and } \Delta^c \mu_a = \Delta^c \mu_c \tag{10.151}$$

where the subscripts a, c, and s refer to anion, cation, and neutral salt, respectively. Making use of the definitions $J_s = (J_c + J_a)/2$ and $I = F(J_c - J_a)$, where F is the Faraday, the entropy production can be rearranged into

$$T \, dS/dt = J_v \Delta P + I \Delta \varphi + (J_s - \tilde{C}_s J_v) \, \Delta^c \mu_s \tag{10.152}$$

The newly defined fluxes I and J_s represent the electrical current flow and the flow of neutral salt, respectively. Phenomenological relations describing the system are then

$$J_v = L_p \Delta P + L_{pE} \Delta \varphi + L_{ps} \Delta^c \mu_s$$

$$I = L_{Ep} \Delta P + L_E \Delta \varphi + L_{Es} \Delta^c \mu_s$$

$$J_s = L_{sp} \Delta P + L_{sE} \Delta \varphi + L_s \Delta^c \mu_s \tag{10.153}$$

in which L_p is again the hydraulic conductivity and L_E is the electrical conductivity of the membrane.

Invoking Onsager's reciprocity, we have

$$L_{pE} = L_{Ep}, \ L_{sp} = L_{ps}, \text{ and } L_{Es} = L_{sE} \tag{10.154}$$

These cross-coupling coefficients express the interactions between the three independent permeation processes occurring within the system, interactions that underlie electro-osmosis and the streaming potential. In considering the electrokinetic properties of the membrane system, we will assume the neutral salt driving force $\Delta^c\mu_s$ to be zero.

Electro-osmosis is measured as the ratio of volume flow, at zero transmembrane pressure difference, to electric current through the membrane. From Equation 10.153 we obtain

$$J_v/I = L_{pE}/L_E \quad (\Delta P, \Delta^c\mu_s = 0) \qquad (10.155)$$

The streaming potential, on the other hand, is the transmembrane potential difference developed, with zero current flow, as the result of a pressure difference across the membrane:

$$\Delta\varphi/\Delta P = -\frac{L_{pE}}{L_E} \quad (I, \Delta^c\mu_s = 0) \qquad (10.156)$$

Thus, based on the principle of reciprocity, the coefficients of these two electrokinetic phenomena are numerically equivalent, apart from their sign.

A rather simple mechanistic model can be proposed to illustrate the plausibility of these relationships. Consider, for example, the membrane to be penetrated by a number of pores, each lined by fixed anionic charges. As a result, the solution within the pore will contain an excess of cations, to balance the fixed charge density. If now an electrical potential difference is impressed across the membrane, the force exerted by this field on the ions within the pore is transmitted to the fluid matrix by frictional interaction. Because these ions are predominantly positively charged, there will be a net force exerted on the fluid, resulting in a flow of volume. On the other hand, in the absence of an electric field, any flow of volume would sweep an excess of the predominant free ion, the cations, through the membrane, resulting in a current flow. In order for the current flow to vanish, a potential difference sufficient to counterbalance the current of convective origin must appear. This compensating field is the streaming potential. Because both phenomena arise in proportion to the net free charge density within the pore, the numeric equivalence of the magnitudes of these effects is not surprising. Nonequilibrium thermodynamics places these qualitative relationships on a firm theoretical foundation.

10.1.3.2 Active Transport

The transmembrane transport of many physiologically important ions and metabolites proceeds by means of a mechanism in which the translational flow is coupled with an exergonic process. Because such transport can take place counter to the electrochemical potential gradient of the transported species, drawing the required energy from the coupled process, it is known as *active transport*. In most organisms, the transport of hydrogen ions, sodium, and calcium are coupled with biochemical reactions such as the hydrolysis of ATP. Translocation processes of this sort are denoted as *primary active transport*. A number of anions, on the other hand, as well as most neutral metabolites, are transported at the expense of a difference of the electrochemical potential of either H or Na, by processes known as *secondary active transport*. Examples of transport of this sort are the movements of monosaccharides and amino acids coupled with Na^+ at intestinal epithelium and kidney tubules, and H-driven citrate transport in mitochondria. Nonequilibrium thermodynamics is particularly useful for delineating primary and secondary active transport, and for quantitating the strength of the coupling between the source reaction and the driven flux. A criterion of active transport, based on the phenomenological coefficients, has been offered by Kedem (1958).

The usual phenomenological equations can be extended to include a scalar reaction process, yielding

$$J_i = \sum_j L_{ij} \Delta\mu_j + L_{ir} A_r$$

$$J_r = \sum_j L_{rj} \Delta\mu_j + L_r A_r \qquad (10.157)$$

where J_r is the rate of the reaction, and A_r its affinity, the negative of the Gibb's free energy change for the reaction. In this expression, the reaction flux J_r is coupled to a translational motion of species i, through the coefficient L_{ir}. Since the flux of i is directional, whereas the reaction flux is not, the directional property must be inherent in the L coefficient, reflecting the fundamental anisotropy of the membrane in which the coupling occurs. The existence of active transport, in these terms, hinges on a nonzero value of the L_{ir} for a transported species i.

Kedem's criterion, however, is based on a transformation of the relations of Equation 10.157 into a form involving resistive, rather than conductive, coefficients. In this form, readily measured experimental variables are more conveniently interpreted within the theoretical framework. Performing such a transformation yields

$$\Delta\mu_i = \sum_j R_{ij} J_j + R_{ir} J_r$$

$$A_r = \sum_j R_{rj} J_j + R_r J_r \qquad (10.158)$$

in which the Rs are algebraic functions of the L-coefficients.

The coefficients R have the character of a resistance. These relations can be further rearranged into the form

$$J_i = \frac{1}{R_i}\left(\Delta\mu_i - \sum_{j \neq i} R_{ij} J_j - R_{ir} J_r \right)$$

$$J_r = \frac{1}{R_r}\left(A_r - \sum_j R_{rj} J_j \right) \qquad (10.159)$$

expressing each individual flow as a function of its conjugate force, and all other fluxes within the membrane. Clearly, the primary active transport of species i requires R_{ir} to be nonzero. If such is the case, J_i may proceed counter to its conjugate force, as a result of coupling to the reaction process. At zero flow of species i, its conjugate force can be nonzero, again depending on the reaction flux. This latter condition is denoted as a *static head* and is often chosen as an experimental condition to test the existence of an active transport mechanism.

In secondary active transport, the translational flow of species i can be coupled to the trans-membrane movement of another species j, which is in turn coupled to a scalar reaction process. Secondary active transport thus requires a nonzero value for R_{ij}. In secondary active transport, it is generally assumed that the carrier can transiently bind both the transported solute (often a nonelectrolyte) and the driving ion (either H^+ or Na^+). The energy used for the transport is available from a gradient of the driving ion oriented oppositely to that formed by the solute during transport.

TABLE 10.12
Ionic Selectivity of Lipid Bilayers: Modified by Some
Carrier Compounds

	Charge (pH 7)	Ionic Selectivity
Valinomycin	0	$H^+ > Rb^+ > K^+ > Cs^+ >> Na^+ > Li^+$
Monactin	0	$NH_4^+ > K^+ > Rb^+ > Cs^+ > Na^+ >> Li^+$
Enniatin A	0	$Rb^+ > K^+ > Cs^+ > Na^+ > Li^+$
Nigericin	—	$K^+ > Rb^+ > Na^+$
2,4-Dinitrophenol		H^+
Dicoumarol		H^+
FCCP		H^+
Channel-forming compounds		
Gramicidin A	0	$H^+ > NH_4^+ > Cs^+ > Rb^+ > K^+ > Na^+ > Li^+$
Alamethicin	+	Cation
Nystatin	–	Anion
Ampoterscin B	–	Anion

Ultimately, the energy required for secondary active transport is derived from the reaction process powering the primary active transport of the driving species.

10.1.4 FACILITATED DIFFUSION

The transport of most solutes across biological membranes involves a process other than simple diffusion. Thus, many substances show a much higher rate of passage than would be predicted from a consideration of their concentration gradient, oil-water partition coefficient, and electric potential gradient. Furthermore, the activation energy is often found to be considerably lower than that expected from simple diffusion through membranes.

In addition to membrane bulk transport (cytotic transport), transport across biological membranes may be categorized into three groups according to the mechanisms involved:

- Molecular diffusion across membranes, including diffusion through small membrane pores
- Facilitated (or carrier-mediated) diffusion
- Active pump (carrier-mediated uphill) transport

The first two groups are classes of passive transport. In this section, the second type of transport, passive facilitated transport, will be considered briefly.

In an earlier view, a carrier molecule (often called a *mobile carrier*) is envisioned as a vehicle translationally moving along and across the membrane while being able to bind and carry along the solute from one side of the membrane to the other. A typical example for carrier substances is represented by various cyclic polypeptide antibiotics (see Table 10.12). For example, valinomycin can bind K^+ much more readily than most other monovalent cations. It can then selectively carry K^+ through the membrane in a manner shown in Figure 10.15c. Experimental results demonstrating such membrane transport are shown in Figure 10.14a. In biological membrane systems, however, such mobile carriers are rarely found. Binding sites within integral membrane proteins that do not migrate across the membrane have the property of being alternately accessible from both sides of the membrane (Widdas, 1952), which is a more common mechanism for the facilitated transport. There are two schemes for such transport:

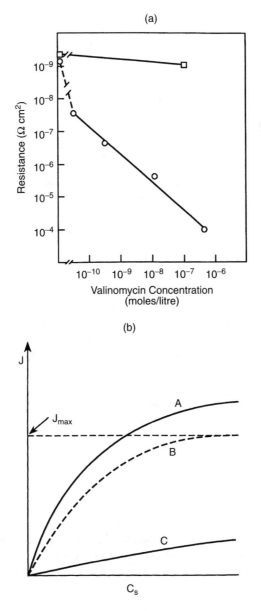

FIGURE 10.14 (a) Resistance as a function of valinomycin concentration in 0.1 M NaCl (□) and 0.1 KCl (○) (Andreoli et al., 1967). (b) Substrate flux across a membrane in the presence (A) and absence (C) of a carrier in the membrane with respect to the substrate concentration C_s. (B) Net substrate flux due to the carrier transport.

1. Protein transconformational changes
2. The successive sequence of binding through a protein channel formed across the membrane

The protein conformational change mechanism is modeled as follows: after binding with a specific substrate (ion), the protein is susceptible to a change in its conformation. As it flips from a left-facing orientation of the binding site to one facing to the right, the bound substrate can cross the membrane (Lieb and Stein, 1970). A schematic diagram for such a transport system is given in Figure 10.15. Examples of such biological carriers are ion transport protein (band 3 protein) and

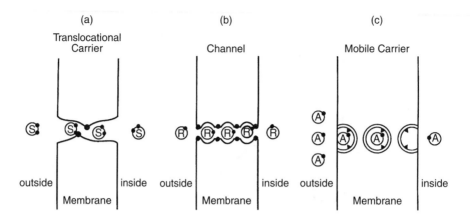

FIGURE 10.15 Schematic models for different types of carriers in membranes.

glucose carrier protein (e.g., band 4.5) in erythrocyte membranes. The former molecule facilitates anion transport across the membrane, while the latter facilitates D-glucose transport across erythrocyte membranes preferentially over L-glucose (LeFevre, 1948; Widdas, 1952) (see Figure 10.14b).

Another mechanism of nonmobile facilitation is the channel (pore) transporter. These pores are protein lined and are permanent structures (see Figure 10.15b). Specificity for the channel transport arises from pore diameter, charge distribution within the pore, and the presence of specific binding sites at the pore entrance, core, and exit (Figure 10.15b) (Hille, 1984; Miller, 1986).

For both cases of carrier transport (mobile carrier and nonmobile transporter), kinetic description for carrier-mediated transported may be expressed in similar equations. One of the simplest models of carrier transport can be formulated as follows: the substrate S has a certain binding affinity with carrier X at side 1 of the membrane

$$S_1 + X \underset{k_{-1}}{\overset{k_1}{\rightleftharpoons}} SX \tag{10.160}$$

and a substrate-carrier complex SX is formed. Upon rotation or conformational change, it can dissociate, releasing the solute on the opposite side of the membrane. The carrier may show substrate specificity through its affinity for the substrate (see Table 10.13), and may have a different affinity for the substrate on the two surfaces of the membrane (see Figure 10.16).

$$SX \overset{k_2}{\rightarrow} S_2 + X \tag{10.161}$$

For simplicity, it is assumed that the substrate is transported in only one direction. That is, the substrate concentration on side 2 is assumed to be neglibible. The total number of carrier molecules or binding sites is constant:

$$X^T = X + XS = \text{constant} \tag{10.162}$$

Because by assumption there is no reverse flux at the inner surface of the membrane, the flux J is

$$J = \frac{dS_2}{dt} = k_2 C_{XS} \tag{10.163}$$

TABLE 10.13
Affinities of Different Sugars for the Carrier
Mechanism in the Human Erythrocyte

Sugar	(K_m, M)
2-deoxy-D-glucose	0.005
D-glucose	0.007
D-mannose	0.02
D-galactose	0.03
D-xylose	0.006
L-arabinose	0.01
D-ribose	0.2
D-lyxose	0.3
D-arabinose	1.5
L-fucose	2.0
L-rhamnose	ca. 3.0
L-glucose	3.0
L-galactose	3.0
L-xylose	3.0

Adapted from LeFevre PG and Marshall JK: *J. Biol. Chem.*
1959, 234:3022.

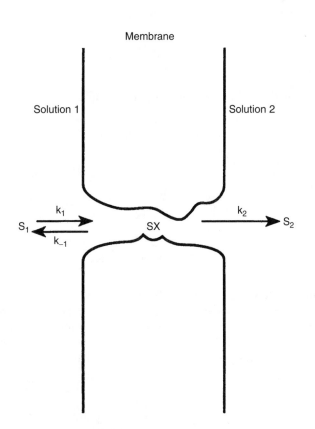

FIGURE 10.16 Simplest scheme for either translocational carrier or channel carrier kinetics.

The change of the XS complex with respect to time is

$$\frac{dC_{xs}}{dt} = k_1 C_s C_X - (k_{-1} + k_2)\, C_{XS} \tag{10.164}$$

where k_1, k_{-1}, and k_2 are the rate constants of the forward and reverse reactions on side 1, and that of the forward reaction on side 2. C refers to concentration. If the concentration of S is sufficiently large, the rate of transport of S is constant, with the flux reaching a steady state. Therefore, $dC_{xs}/dt = 0$. Then, Equation 10.164 becomes

$$k_1 C_s C_X = (k_{-1} + k_2)\, C_{XS} \tag{10.165}$$

Because the total number of carriers is constant (Equation 10.162), in terms of the concentrations

$$C_X^T = C_X + C_{XS} \tag{10.166}$$

With Equation 10.165 and Equation 10.166, C_{XS} is

$$C_{XS} = \frac{C_X^T C_S}{\dfrac{k_{-1} + k_2}{k_1} + C_S} = \frac{C_X^T C_S}{K_m + C_S} \tag{10.167}$$

where $K_m = (k_{-1} + k_2)/k_1$. K_m is called the *steady state dissociation constant* (or the *Michaelis constant*, by analogy with enzyme kinetics). Therefore, the flux J can be expressed as

$$J = \frac{k_2 C_X^T C_S}{K_m + C_S} \tag{10.168}$$

With $C_s \gg K_m$, J approaches the maximum achievable flux:

$$J \cong k_2\, C_X^T \equiv J_{max} \tag{10.169}$$

Equation 10.168 is rewritten as

$$J = \frac{J_{max} C_S}{K_m + C_S}$$

$$\Rightarrow \frac{J_{max}}{2} \quad \text{when } C_S = K_m \tag{10.170}$$

A relationship of flux vs. substrate is shown in Figure 10.17.

From Equation 10.170, it can be seen that the substrate concentration at which $J = J_{max}/2$ is the steady-state dissociation constant K_m. A characteristic feature of all carrier-mediated transport is a saturation kinetics at high substrate concentration. At low substrate concentration, the flux is linearly related to the substrate concentration C_s. Taking the inverse of Equation 10.170 yields

$$\frac{1}{J} = \frac{K_m}{J_{max}} \frac{1}{C_S} + \frac{1}{J_{max}} \tag{10.171}$$

This relation between $1/J$ and $1/C_S$ is linear and is shown in Figure 10.18.

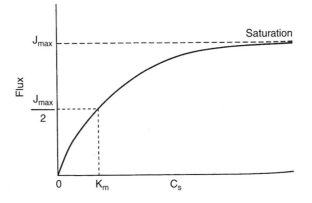

FIGURE 10.17 Saturation kinetics due to carrier transport.

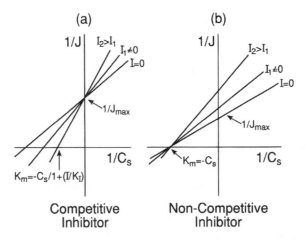

FIGURE 10.18 Flux J with respect to substrate concentration C_s for carrier-mediated transport in the absence (I-0) and the presence of inhibitors.

In Figure 10.18, the value of $1/J$, at $1/C_S = 0$, is $1/J_{max}$. The intercept point at $1/C_S$ axis satisfies $K_m = -C_S$.

10.1.4.1 Competitive Inhibition

When substrate and competitive inhibitors coexist in the same kinetic system as developed previously, the kinetic equations are

$$S + X \underset{k_{-1}}{\overset{k_1}{\rightleftharpoons}} XS \overset{k_2}{\rightarrow} X + S$$

$$X + I \underset{k'_{-1}}{\overset{k'_1}{\rightleftharpoons}} XI \tag{10.172}$$

yielding at the steady state the relations

$$XS = \frac{X \cdot S}{K_m}$$

$$XI = \frac{X \cdot I}{K_I} \tag{10.173}$$

in which K_m is the Michaelis constant for the substrate-carrier reaction, and $K_I = k'_{-1}/k'_1$ is the dissociation constant for the inhibitor-carrier binding. Conservation requires that the total carrier present in all forms be constant:

$$X^T = X + SX + IX \tag{10.174}$$

With use of Equation 10.172 and Equation 10.173, Equation 10.174 becomes

$$C_X^T = C_X\left(1 + \frac{C_S}{K_m} + \frac{C_I}{K_I}\right) \tag{10.175}$$

Then

$$XS = \frac{X \cdot S}{K_m} = \frac{X^t \cdot S}{K_m\left(1 + \frac{S}{K_m} + \frac{I}{K_I}\right)} \tag{10.176}$$

The flux of the substrate crossing the membrane is expressed by using Equation 10.176 and the rate of dissociation of the complex, k_2.

$$J = k_2 XS = \frac{J_{max} \cdot S}{K_m\left(1 + \frac{S}{K_m} + \frac{I}{K_I}\right)} \tag{10.177}$$

The inverse of Equation 10.175 is

$$\frac{1}{J} = \frac{1}{J_{max}} + \frac{K_m\left(1 + \frac{I}{K_I}\right)}{J_{max}}\frac{1}{C_s} \tag{10.178}$$

A plot of $1/J$ vs. $1/C_S$ with the concentration C_I as a parameter, is shown in Figure 10.18a. In the presence of a competitive inhibitor, J_{max} remains unchanged; the effective value of K_m, however, depends on the concentration of inhibitor.

On the other hand, if the inhibitor is noncompetitive with the substrate (the inhibitor binds the carrier independently of the substrate), the kinetics display different features. The reaction equations for this case are then

$$X + S \overset{k_1}{\underset{k_{-1}}{\rightleftharpoons}} XS \overset{k_2}{\rightarrow} X + S$$

$$X + I \underset{k'_{-1}}{\overset{k'_1}{\rightleftharpoons}} XI$$

$$XS = I \underset{k'_{-1}}{\overset{k'_1}{\rightleftharpoons}} XSI \tag{10.179}$$

When steady state is achieved, the reactant concentrations satisfy the relations

$$XS = \frac{X \cdot S}{K_m}$$

$$XI = \frac{X \cdot I}{K_I}$$

$$XSI = \frac{XS \cdot I}{K_I} = \frac{X \cdot S \cdot I}{K_I K_m} \tag{10.180}$$

Again, the total carrier is a conserved quantity, such that

$$X^T = X + SX + XI + SXI = \text{constant} \tag{10.181}$$

which, with the use of Equation 10.180, becomes

$$C_X^T = C_X \left(1 + \frac{C_S}{K_S} + \frac{C_I}{K_I} + \frac{C_S \cdot C_I}{K_m K_I} \right) \tag{10.182}$$

With rearrangement

$$C_X = \frac{C_X^T}{1 + \dfrac{C_I}{K_I}} \frac{1}{1 + \dfrac{C_S}{K_m}} \equiv \frac{C_X^{T*}}{1 + \dfrac{C_S}{K_m}} \tag{10.183}$$

The transport flux then becomes

$$J = k_2 XS = \frac{k_s X^{t*} S}{K_m + S} \equiv \frac{J_{max}^* S}{K_m + S} \tag{10.184}$$

in which

$$C_X^{T*} = \frac{C_X^T}{1 + C_I / K_i} \qquad J_{max}^* = \frac{J_{max}}{1 + C_I / K_I} \tag{10.185}$$

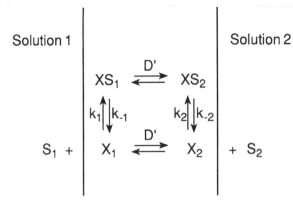

FIGURE 10.19 Simplest kinetics for mobile carriers.

Inversion of Equation 10.184 gives

$$\frac{1}{J} = \frac{1}{J_{max}^{*}} + \frac{K_m}{J_{max}^{*}} \cdot \frac{1}{C_S} \tag{10.186}$$

The plot of $1/J$ vs. $1/C_S$ is linear (Figure 10.18b), as is the case with competitive inhibition. In contrast to the case of competitive inhibition, however, the effective value of K_m is unchanged while J_{max} is reduced by the presence of a non-competitive inhibitor.

These examples are for rather extreme cases. The actual experimental data often suggest a mixture of these modes of inhibition, or an even more complicated mechanism. The kinetic analysis for such experimental results can be pursued in a manner similar to those presented previously, but in more complex forms (Kotyk et al., 1988).

The expressions for transmembrane substrate flux become somewhat more complex, and more interesting, with the presence of a finite substrate concentration on both sides of the membrane. A simple kinetic scheme, again applicable to both mobile and conformational carriers, is as follows (Widdas, 1952): Assume that (1) only one substrate molecule is bound to each carrier molecule, (2) the reaction between the carrier and its substrate proceeds at a much greater rate than the actual movement of carrier and carrier-substrate complex, so that an equilibrium exists at both sides of the membrane with respect to substrate binding to the carrier, and (3) the permeation kinetic coefficient, D', of the free carrier and of the carrier-substrate complex are equal (Figure 10.19).

The rate constants of reaction at both membrane surfaces (1, 2) are designated k_1, k_{-1}, k_2, and k_{-2}, respectively; hence, $k_{-1}/k_1 = K_1$, and $k_{-2}/k_2 = K_2$ are dissociation constants of the carrier-substrate complex. We also assume that the system is symmetric, so that $K_1 = K_2 = K_{SX}$. In the case of a mobile carrier, the permeation coefficient incorporates both the diffusion constant and membrane thickness δ, such that $D' = D/\delta$. The total carrier in all forms located at side I is given by $X^t_1 = X_1 + XS_1 = X_1 (1 + S_1/K_{xs})$. The net flux of S transported through the membrane is then

$$J_{s1} = X^t_1 D' \frac{S_1}{K_{xs} + S_1} = J'_{max} \frac{S_1}{K_{xs} + S_1} \tag{10.187}$$

At a steady state in the carrier distribution in the membrane, there is no net flux of the carrier (in all forms) across the membrane; consequently,

$$J_{s2} = X^t_2 D' \frac{S_2}{K_{xs} + S_2} = -J''_{max} \frac{S_2}{K_{xs} + S_2} \tag{10.188}$$

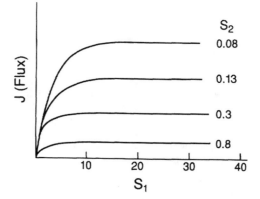

FIGURE 10.20 Schematic plot of E-kinetics of transport. The carrier movement is rate limiting; saturation concentrations of substrate are used.

Then from Equation (10.187), the substrate flux becomes

$$J_{1\to2} = X^t \cdot D' \left(\frac{S_1}{S_1 + K_{xs}} - \frac{S_2}{S_2 + K_{xs}} \right)$$

$$= J_{max} \left(\frac{S_1}{S_1 + K_{xs}} - \frac{S_2}{S_2 + K_{xs}} \right) \tag{10.189}$$

(a) When the carrier is far from saturation, e.g., $S \ll K_{sx}$, Equation 10.189 becomes

$$J_{1\to2} = \frac{J_{max}}{K_{xs}} \left(S_1 - S_2 \right) \tag{10.190}$$

in which the flux is first order in substrate concentration.

(b) In another extreme case, in which both substrate concentrations are much greater than K_{XS} (saturation),

$$J_s = J_{max} K_{xs} \frac{S_1 - S_2}{\left(S_1 + K_{xs} \right)\left(S_2 + K_{xs} \right)}$$

$$\to J_{max} K_{xs} \left(\frac{1}{S_2} - \frac{1}{S_1} \right) \quad (S \gg K_{XS}) \tag{10.191}$$

The rate is markedly affected by the concentration of substrate on side II (see Figure 10.20).

In the presence of two substrates R and S which compete for the single carrier binding site, the permeation rates for the two become interdependent. In such circumstances, the total carrier is given by

$$X^t = X + XS + XR = X + \frac{S \cdot X}{K_{XS}} + S \frac{R \cdot X}{K_{XR}} \quad (i = 1, 2) \tag{10.192}$$

The rate of transport R is then

$$J_R = J_{max} \left(\frac{R}{R_1 + K_{XR}\left(1 + \dfrac{S_1}{K_{XS}}\right)} - \frac{R_2}{R_2 + K_{XR}\left(1 + \dfrac{S_2}{K_{XS}}\right)} \right) \tag{10.193}$$

The form of this equation is identical to that derived for competitive inhibition. By setting $S' = S/K_{XS}$ and $R' = R/K_{XR}$, Equation 10.193 can be rewritten:

$$J_R = J_{max} \left(\frac{R_1'}{R_1' + S_1' + 1} - \frac{R_2'}{R_2' + S_2' + 1} \right) \tag{10.194}$$

By using Equation 10.194, a countertransport phenomenon can be explained: suppose that the cells are preincubated with labeled solute R, and then nonlabeled solute S (R and S are chemically the same) is added. Between A and B, the labeled solute R moves out of the cell against its concentration gradient (see Figure 10.21). When the system is initially equilibrated with labeled solute R, the flux Equation 10.194 becomes

$$J_{R'} = J_{max} \left(\frac{R'}{R' + 1} - \frac{R'}{R' + 1} \right) = 0 \tag{10.195}$$

After equilibration, the addition of nonlabeled solute S on side 1 (extracellular) generates a net flux of R from side 2 to side 1:

$$J_{R'} = J_{max} \left(\frac{R'}{R' + S' + 1} - \frac{R'}{R' + 1} \right) \tag{10.196}$$

Thus $J_{R-} < 0$, indicating that R will move out of the cell against its concentration gradient. This is called *countertransport*. This movement, however, is not independent but is rather closely

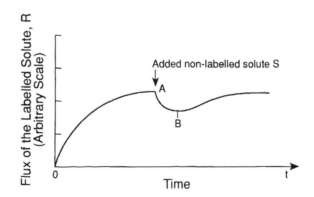

FIGURE 10.21 Countertransport of solute R. The cells are preincubated with labeled solute R (preincubated states at t = 0), and the nonlabeled solute S (R and S are chemically the same) is added to the extracellular compartment at time A. Between A and B, the labeled solute R moves out of the cells against its concentration gradient.

associated with the simultaneous movement of S, which, immediately after its addition to the system, proceeds according to

$$J_{S'} \propto \frac{S'}{S' + R' + 1} \tag{10.197}$$

The outward movement or R persists until the condition is reached when

$$\frac{R_1}{R_2} = \frac{S_1' + 1}{S_2' + 1} \tag{10.198}$$

From there on, substrate R begins to move back into the cell until equilibrium is attained with $R_1 = R_2$ and $S_1 = S_2$.

10.1.4.2 Kinetics of a Two-Site Carrier

Multisite carriers are known to exist, and it is generally observed that the binding sites exhibit kinetic cooperativity. Using the simplest carrier equation for a single substrate, one can develop carrier kinetic equations for two binding sites.

In an experimental procedure similar to that for countertransport described previously, the kinetics of a two-site carrier predict a movement of a preequilibrated substrate R upon addition of a second substrate S. Chemically, R and S are the same. The direction of this movement, however, depends on the concentrations of the substrates:

- $J_R < 0$ for relatively high concentrations of R and S.
- $J_R > 0$ for very low concentrations of R and S.

The former case corresponds to countertransport, whereas the latter is an instance of co-transport. However, this transport phenomenon is different from that of solute flow stoichiometrically coupled with the transport of H^+ or Na^+, which in turn is driven by biochemical energy input (active transport). This latter form of transport is called *symport* and is an example of secondary active transport. The most well-known symport systems are monosaccharide transport coupled with active sodium transport in the intestine, and amino acid transport in many cells.

Another secondary active transport mechanism is *antiport,* in which a solute flux is coupled with active ion transport such that its flow is in the direction opposite to that of the primary actively transported ion. Although the counter transport discussed earlier within the context of passive transport may correspond to antiport, the thermodynamic basis is different in these two transport processes (see Table 10.14).

TABLE 10.14
Classification of Membrane Transport[a]

Passive transport:

Molecular diffusion — { Through membrane core
Through aqueous pore (ionic channels and large pores)

Carrier mediated diffusion:
    Mobile carrier — {Valinomycin, monactin, dinitrophenol, etc.

    Channels — { Gramicidin, alamethicin, etc.
    Some ionic channels in cells

    Translocation of carriers — { Monosaccharides transport in erythrocytes (uniport)
    ADP/ATP exchanges in mitochondria,
    Na^+/Ca^{2+} exchanger (antiports), etc.

Active transport:

    Primary active — { Na^+/K^{+b}, Ca^{2+} through ATPase with ATP-hydrolysis
    H^+ with use of oxidation free energy, H^+/K^+ with ATPase

    Secondary active — Complex transport with the primary active:

Group translocation — Glucose becomes glucose-6p while crossing the membrane (bacteria)   { Lactose in bacteria with the primary H^+ active (symport)
Glycine in tumors with the primary Na^+ active (symport)

Membrane bulk transport: Amino acids in cells with the primary Na^+ active (symport)
    Exocytosis
    Endocytosis
    Without vesiculation

[a] Stein, 1986.
[b] Lingrel and Kuntzweiler, 1994.

REFERENCES

Aidley DJ and Stanfield PR: *Ion Channels,* Cambridge University Press, Cambridge, U.K., 1996.
Andreoli TE, Tieffenberg M, and Tosteson DC: *J. Gen. Physiol.* 1967, 50:2527.
Bezanilla F and Stefani E: *Ann. Rev. Biophysics.* 1994, 23:819.
Bockris JOM: *Q. Rev. London* 1949, 3:173.
Born M: *Z. Physik* 1920, 1:45.
Collander R: *Trans. Faraday Soc.* 1937, 33:985.
Diamond JM and Wright EM: *Annu. Rev. Physiol.* 1969, 31:581.
Dainty J and Ginzburg BZ: *Biochem. Biophys. Acta* 1964, 79:122.
Danielli JF: *The Permeability of Natural Membranes.* Eds. Dawson H and Danielli JF, Cambridge University Press, Cambridge, U.K., 1952.
Davson H and Danielli JF: *The Permeability of Natural Membranes.* 2nd ed. Cambridge University Press, Cambridge, U.K., 1952.
Davies JT and Rideal EK: *Interfacial Phenomena.* Academic Press, New York, 1963.
De Groot SR: *Thermodynamics of Irreversible Processes.* North-Holland, Amsterdam, 1963.
Dembigh KG: *The Thermodynamics of the Steady State.* Methuen and Co., London, and J. Wiley & Sons, New York, 1958.
Donnan FG: *Z. Electrochem.* 1911, 17:572.
Einstein A: *Z. Electrochem.* 1908, 14:235.
Eisenman G: *Symposium on Membrane Transport and Metabolism.* Eds. Kleinzeller A and Kotyk A, Academic Press, New York, 1961, pp. 163–179.
Eisenman G: *Proc. 23 Int. Congr. Physiol. Soc.* (Tokyo) Excerpta Medica, Amsterdam, 1965, pp. 489–506.

Eyring H: *J. Chem. Phys.* 1936, 4:283.

Ferguson J: *Proc. R. Soc. London, Ser. B.* 1939, 127:387.

Glasstone S, Laidler KJ, and Eyring H: In *The Theory of Rate Processes.* McGraw-Hill, New York, 1941.

Goldman DE: *J. Gen. Physiol.* 1943, 27:37.

Goldschmidt VM: *Skrifter det Norske Vindenskaps — Akad. Oslo I, Matem Naturvid,* 1926.

Guggenheim EA: *J. Phys. Chem.* 1929, 33:842.

Henderson P: *Z Phys. Chem.* 1908, 63:325.

Hille B: *Ionic Channels of Excitable Membranes.* Sinauer Associates, Sunderland, MA, 1984.

Hodgkin AL: *Biol. Rev.* 1951, 26:360.

Hodgkin AL and Horowitz P: *J. Physiol.* 1959, 145:405.

Hodgkin AL and Huxley AF: *J. Physiol.* 1952, 117:500.

Hodgkin AL and Katz B: *J. Physiol.* 1949, 108:37

Hopfer V, Lehninger AL, and Lennard WJ: *J. Membr. Biol.* 1970, 2:41.

Jain MK: *The Biomolecular Lipid Membranes.* van Nostrand-Reinhold, New Jersey, 1972, Chap. 6.

Jones MN: *Biological Interfaces.* Elsevier, Amsterdam, 1975.

Katchalski A and Curran PE: *Non Equilibrium Thermodynamics in Biophysics.* Harvard University Press, Cambridge, MA, 1965.

Kedem O and Katchalski A: *Biochim. Biophys. Acta* 1958, 27:229.

Keynes RD: *J. Physiol.* 1951, 114:119.

Kotyk A and Janacek K: *Cell Membrane Transport.* Plenum Press, New York, 1970.

Kotyk A, Janacek K, and Koryta J: *Biophysical Chemistry of Membrane Functions.* John Wiley & Sons, New York, 1988.

Latimer WM, Pitzer KS, and Slansky CM: *J. Chem. Phys.* 1939, 7:108.

LeFevre PG: *J. Gen. Physiol.* 1948, 31:505.

LeFevre PG and Marshall JK: *J. Biol. Chem.* 1959, 234:3022.

LeNeveu DM, Rand RP, and Parsegian VA: *Nature (London)* 1976, 259:601.

Lieb WR and Stein WD: *Biophys. J.* 1970, 10:585.

Lingrel JS and Kuntzweiler T: *J. Biol. Chem.* 1994, 289:19659.

McLaughlin SGA, Szabo G, and Eisenman G: *J. Gen. Physiol.* 1971, 58:667.

Miller C: (Ed.) *Ionic Channels Reconstitution.* Plenum Press, New York, 1986.

Nernst W: *Z. Phys. Chem.* 1888, 2:613 and 634.

Ohki S: *Comprehensive Treatise of Electrochemistry.* Vol. 10. eds. Srinivasan S, Chizmadzhev YA, Bockris JO'M, Conway BE, and Yeager E, Plenum Press, New York, 1985, p. 1–130.

Ohki S and Ohshima H: Electrochemistry in colloidal systems: Double layer phenomena, in *Bioelectrochemistry: Principles and Practice,* vol. 1, Caplan SR, Miller IR, and Milazzo G, Eds., Birkhauser Verlag, Basel, Switzerland, 1995, pp. 211–287.

Onsager L: *Phys. Rev.* 1931, 37:405.

Overton E: *Pfluegers Arch. Gesamte Physiol.* 1902, 92:346.

Patlak J: *Physiol. Rev.* 1991, 71:1047.

Rand RP: *Annu. Rev. Biophys. Bioeng.* 1981, 10:277.

Rice SA and Nagasawa M: *Polyelectrolyte Solutions.* Academic Press, New York, 1961, p. 118.

Planck M: *Ann. Physik.* 1890, 54:561.

Stein WD: *Diffusion and Osmosis in Comprehensive Biochemistry.* Vol. 2. Eds. Florkin M and Stotz EM, Elsevier, Amsterdam, 1962, Chap. 3.

Stein WD: *The Movement of Molecules across Cell Membranes.* Academic Press, New York, 1967.

Stein WD: *Transport and Diffusion Across Cell Membranes,* Academic Press, New York, 1986.

Widdas WF: *J. Physiol.* 1952, 118:23.

Wilbrandt W and Rosenberg Th: *Helv. Physiol. Acta* 1956, 8:C82.

11 Membrane Protein Dynamics: Rotational Dynamics

Richard J. Cherry

CONTENTS

11.1 INTRODUCTION

Measurements of rotational and lateral diffusion of membrane proteins performed during the early 1970s were of considerable importance in both generating and establishing the familiar fluid-mosaic model of membrane structure (Singer and Nicholson, 1972). Interest in investigating the extent to which membrane proteins freely diffuse in the fluid lipid bilayer led to the development of a variety of powerful techniques for measuring protein mobility. Application of these techniques has continued unabated long after the requirement to demonstrate protein diffusion in membranes *per se* had been met. Two principal reasons can be identified for this continuing interest. First, there has been a growing awareness that the ability of proteins to diffuse in the membrane and to form dynamic associations through random collision is likely to be of functional significance in a number of situations. Second, the motion of an individual protein may be considerably influenced by its association with other proteins. Thus, mobility measurements provide a powerful approach to the study of protein–protein interactions and hence to elucidation of a more detailed understanding of membrane structure. This is particularly true of rotational diffusing particles.

0-8493-1403-8/05/$0.00+$1.50
© 2005 by CRC Press LLC

In this review, the theoretical framework of rotational diffusion of membrane proteins is outlined and the principal methods of measurement are described. The considerable body of data obtained from both model membrane systems and cell membranes is surveyed and evaluated. The potentially important, though so far less well developed area of detecting segmental motion in membrane proteins is also discussed. Previous reviews of rotational diffusion of membrane proteins have been published by Thomas (1985, 1986), Jovin and Vaz (1989), and Cherry (1979).

11.2 THEORY

Molecules in fluid solution undergo a series of small amplitude transverse and angular random displacements known as Brownian motion. On a macroscopic scale, the behavior of an ensemble of molecules is described by a diffusion equation analogous to Fick's second law for diffusion down a concentration gradient, i.e., uniaxial rotation:

$$\frac{dC_{(\Theta,t)}}{dt} = D_R \frac{d^2 C_{(\Theta,t)}}{dt^2} \tag{11.1}$$

where $C_{\Theta,t}$ is the concentration of molecules at time t oriented at an angle to a defined axis, and D_R is the rotational diffusion coefficient. Equation 11.1 describes the rate at which an initially oriented population of molecules approaches equilibrium. Because current methods of measuring rotational diffusion of membrane proteins require observation of a molecular ensemble, analysis of the data is based on solution of diffusion equations with appropriate boundary conditions. However, conceptually it is helpful to consider diffusion at a microscopic level, where Einstein showed that individual molecules make a mean square rotation $<\Delta\Theta^2>$ in time Δt given (for uniaxial rotation) by

$$<\Delta\Theta^2> = 2D_R\Delta t \tag{11.2}$$

The diffusion coefficient D_R in Equation 11.1 and Equation 11.2 is the same, provided the molecules are noninteracting. Einstein also showed that

$$D_R = \frac{kT}{F} \tag{11.3}$$

where F is the friction coefficient describing the retardation of motion by the surrounding medium. Although Stokes had earlier derived expressions for the friction coefficients of particles in solution, the method of derivation breaks down for the two-dimensional case of membranes. Saffman and Delbrück (1975) provided approximate solutions to this problem for a model in which the protein is represented by a cylinder of radius a traversing the membrane of thickness h, which, in combination with Equation 11.3, yields

$$D_R = \frac{kT}{4\pi a^2 h\eta} \tag{11.4}$$

where the surrounding membrane is treated as a medium of viscosity (see also Saffman, 1975). In a further theoretical treatment, Hughes et al. (1981) showed that Equation 11.4 is valid provided $\varepsilon < 0.1$, where ε is given by

$$\varepsilon = \left(\frac{a}{h}\right)\left(\frac{\eta_1 + \eta_2}{\eta}\right) \tag{11.5}$$

and η_1, η_2 are the viscosities of the solutions adjacent to the two sides of the membrane. Because the viscosity of lipid bilayers is about 100 times that of water, the condition $\varepsilon < 0.1$ is normally fulfilled. Equation 11.4 also neglects the effects of any viscous drag by the aqueous medium on portions of the protein protruding from the membrane. Again, this is justified in most situations because of the relatively high viscosity of the lipid bilayer.

The importance of Equation 11.4 is that it relates diffusion measurements to molecular parameters. In particular, it shows that rotational diffusion is strongly dependent on the radius of the rotating particle. This is the basis of one of the major applications of rotational diffusion measurements, namely, the investigation of protein associations in membranes.

11.3 MEASUREMENT TECHNIQUES

A variety of experimental methods can, in principle, be used to measure molecular rotation. In practice, however, only two types of method, based, respectively, on optical spectroscopy and electron paramagnetic resonance (EPR), have been used to any appreciable extent in the study of membrane proteins. Description of methodology will thus be confined to these two approaches.

11.3.1 OPTICAL ANISOTROPY DECAY

Optical spectroscopic methods of measuring rotational motion were first introduced by Perrin (1936) and later developed by Weber (1953) for application to proteins in aqueous solution. These methods depend on creating an anisotropic molecular orientation by the process of photoselection. When a randomly oriented population of chromophores is excited by linearly polarized light, those chromophores, whose transition dipole moment for the selected absorption lies in or near to the direction of polarization, are selectively excited (see Figure 11.1). Thus, an anisotropic distribution of molecules in the excited state is photoselected from the initial random distribution. If the excitation is by a brief pulse of light, then the initial anisotropy created by the flash decays at a rate determined

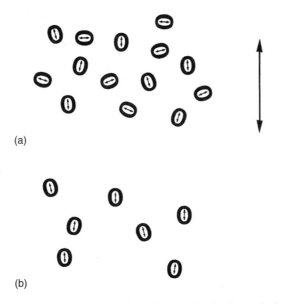

(a)

(b)

FIGURE 11.1 Photoselection. An oriented population of molecules in the excited state (b) is obtained by absorption of linearly polarized light by a randomly oriented population (a). The small arrows represent the transition dipole moment of the electronic transition, and the large arrow indicates the direction of polarization of the incident radiation.

by the rotational diffusion coefficient D_R. By continuously monitoring the anisotropy, the value of D_R can, in principle, be determined.

Variations of the method arise through the detection of different signals in order to measure the anisotropy. Most commonly, the anisotropy is measured by detecting polarization of fluorescent emission from the molecules in the excited state. However, rotational diffusion can only be measured if a significant rotation occurs during the lifetime of the detected signal. This restricts the classical fluorescence method to studying rotations in the nanosecond time range (or faster), which is appropriate for proteins in aqueous solution but not for proteins in membranes where rotations occur in the microsecond or even millisecond time range.

Two related ways of extending the optical anisotropy decay method to longer time ranges were developed in early 1970s. The first measurements of rotational diffusion of a membrane protein were made by Cone (1972) with rhodopsin in rod outer segment disc membranes. The measurement exploited the long-lived changes in the absorption spectrum, which occur when the retinal chromophore of rhodopsin is excited. Anisotropy decay can thus be measured by following the dichroism of the absorbance change at a given wavelength. A few other membrane proteins, notably bacteriorhodopsin and some cytochromes, also possess intrinsic chromophores that permit rotational diffusion to be similarly measured.

The second method, put forward by Razi Naqvi et al. (1973), was designed to be generally applicable to any membrane protein. The idea was to utilize the triplet state of a suitable probe molecule bound to the protein in order to monitor slow rotations. Triplet states can have millisecond lifetimes at room temperature and hence provide suitably long-lived signals for measuring rotation of membrane proteins. Since excitation to the triplet state changes the molecule's absorption spectrum, anisotropy decay can be monitored by observing dichroism of the absorbance changes, just as in the case of intrinsic chromophores. Following the development of a suitable triplet probe, the method was first applied to band 3 proteins in the human erythrocyte membrane (Cherry et al., 1976).

The previously described methods, which involve absorption detection, are generally known as *transient dichroism.* Suitable instrumentation for performing transient dichroism measurements has been described in some detail (Cherry, 1978; Thomas, 1986; Kinosita, Jr., and Ikegami, 1988). The technique works well for proteins that are relatively abundant in the membrane. However, sensitivity is inherently limited because absorption methods require detection of a change in the intensity of transmitted light. Emission methods, in which the signal is detected against a zero background, offer the possibility of greater sensitivity. Emission methods of detecting the triplet state include phosphorescence and delayed fluorescence. Methods based on the polarization of phosphorescence have been developed and extensively used (Austin et al., 1979; Moore et al., 1979; Garland and Moore, 1979; Restall et al., 1984; Thomas, 1986; Jovin and Vaz, 1989), whereas the use of delayed fluorescence has only occasionally been reported (Razi Naqvi and Wild, 1975; Greinert et al., 1979; Jovin and Vaz, 1989). Virtually all measurements with the previously described methods employ pulse excitation, although the use of steady-state phosphorescence has also been discussed (Strambini and Galley, 1976; Murray et al., 1983). Different methods of triplet state detection are summarized in Figure 11.2.

A particularly promising variation, known as *fluorescence depletion,* was introduced by Johnson and Garland (1981). When molecules are excited into the triplet state, there is a corresponding fall in the concentration of molecules in the ground state. In the transient dichroism method, this is detected by a decrease in the intensity of the lowest singlet–singlet absorption band. Johnson and Garland showed that a corresponding decrease in fluorescence could be measured with a spectacular increase in sensitivity. The gain in sensitivity is such that microscopic measurements on single cells are feasible. Further developments of the fluorescence depletion technique have been reported (Yoshida and Barisas, 1986; Corin et al., 1987), including the proposal that still higher sensitivities may be obtained by employing phase modulation methods (Garland, 1986; Yoshida et al., 1988). A variation of fluorescence depletion in which the probe is permanently photobleached by polarized light may be employed for the measurement of very slow rotational motion (Smith et al., 1981;

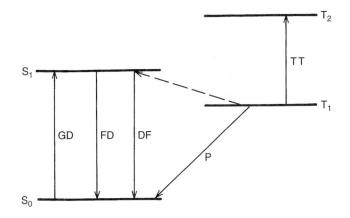

FIGURE 11.2 Electronic energy level diagram illustrating different methods of detecting molecules in the triplet state T_1. Phosphorescence (P) is emission accompanying a radiative transition back to the ground state S_0 via the excited singlet state S_1. Population of T_1 depletes the ground state, resulting in loss of intensity of singlet–singlet absorption (GD = ground state depletion) and prompt fluorescence (FD = fluorescence depletion) under continuous illumination. Finally, molecules in the triplet state undergo triplet–triplet absorption (TT) when illuminated with light of appropriate wavelength. T_1 is populated by intersystem crossing from S_1 after exciting the S_0-S_1 transition. For simplicity, vibrational levels are omitted.

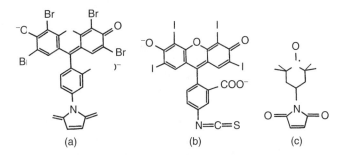

FIGURE 11.3 Commonly used probes for rotational diffusion measurements: (a) eosin-5-maleimide and (b) erythrosine-5-isothiocyanate are employed for optical methods, and the spin label (c) 6-MSL (N-[1-oxyl-2,2,6,6-tetramethyl-4-piperidinyl]maleimide) is employed for ST-EPR.

Velez and Axelrod, 1988). Descriptions of the instrumentation for the various emission techniques can be found in the references cited previously.

The most commonly used triplet probes are halogenated derivatives of fluorescein. Eosin (2,4,5,7-tetrabromofluorescein) has a quantum yield for triplet state formation of about 70% and is particularly versatile. For phosphorescence, erythrosin (2,4,5,7-tetraiodofluorescein) is often preferred because of its even higher triplet state quantum yield (98%) and low fluorescence. A disadvantage of erythrosin is its relatively short triplet lifetime, about ten times less than that of eosin. In the case of fluorescence depletion, probes with a high quantum efficiency for fluorescence are required, such as fluorescein or rhodamine. A number of reactive derivatives to these probes are available to enable them to be coupled covalently to proteins of interest (see Figure 11.3). Isothiocyanate and maleimide derivatives are most commonly used, permitting coupling to amino and thiol groups, respectively.

11.3.2 SATURATION TRANSFER–EPR

In an EPR spectrum of a spin label, the position of the EPR absorption line depends on the orientation of the label relative to the applied magnetic field. For nitroxide spin labels, hyperfine interaction of the unpaired electron with the nitrogen nucleus results in three lines in the EPR spectrum. The g-value (position) and A-value (hyperfine splitting) both depend on the orientation of the principal axis of the nitroxide with respect to the field. The "powder" spectrum obtained for an immobile, randomly oriented population of spin labels corresponds to the sum of the spectra over all possible orientations. If the spin labels in a randomly oriented sample undergo rotational motion such that $\tau_T < T_2$, where the rotational transfer time τ_T is the average time for rotation through about 4° and T_2 is the transverse relaxation time (typically 24 nsec for nitroxides), the spectrum is narrowed by motional averaging. Thus, lineshape analysis of conventional EPR spectra provides information on submicrosecond rotational motions.

The introduction of saturation transfer–EPR (ST-EPR) permitted extension of rotational diffusion measurements to the microsecond time domain, which is applicable for many membrane proteins (Thomas et al., 1976). In conventional EPR, the exciting microwave radiation is of sufficiently low intensity not to significantly perturb the spin system from equilibrium. In ST-EPR, equilibrium is perturbed by application of an intense microwave field. The amount of saturation depends on competition between the rates of excitation and relaxation back to the ground state (determined by T_1, the longitudinal relaxation time) and on rotational motion. T_1 is about 7 μsec for slowly rotating nitroxides. If $\tau_T < T_1$, rotational diffusion transfers probes into and out of the saturated angular range during the excited state lifetime and thus decreases the saturation at resonance. This, in turn, affects the net steady-state absorption and hence the lineshapes. For nitroxides, saturation transfer is detectable for correlation times less than about 1 msec with optimal sensitivity in the range of 10 to 100 μsec.

The spin labels used for ST-EPR experiments and their covalent attachment to proteins are the same as those that have been used for conventional EPR and extensively described (Gaffney, 1976; Morrisett, 1976; Keana, 1979). The maleimide derivative MSL (*N*-[1-oxyl-2,2,6,6-tetramethyl-r-piperidinyl]maleimide) has proved particularly popular as a protein thiol reagent (Figure 11.3). This probe may be immobilized in its binding site by hydrogen bonding to its nitroxide group, which is helpful in eliminating independent probe motion that complicates the spectra.

The conventional (V_1) EPR spectrum shows only small changes in intensity and lineshape in response to saturation and is thus not used in ST-EPR experiments. Sensitivity is greatly increased by rapid field modulation and out-of-phase detection. The optimal signal for many applications is the V_2' spectrum, i.e., out-of-phase second harmonic detection of absorption. This signal is found to have good sensitivity to saturation in terms of both intensity and lineshape, as illustrated in Figure 11.4.

Steady-state ST-EPR has been used for virtually all applications to membrane proteins to date. Instrumentation for time-resolved EPR has also been developed, which potentially can give more direct information about rates and amplitudes of rotational motion (Hyde, 1979; Forrer et al., 1980; Kusumi et al., 1982; Fajer et al., 1986). In saturation recovery experiments, the instrument measures the transient EPR absorption following a saturating pulse of radiation. Spin-echo techniques can also be used to detect time-resolved transfer of spins due to rotational motion. ELDOR (electron–electron double resonance), in which the relative positions of excitation and detection are varied, maximizes the information that can be obtained from spin-echo experiments. Pulsed EPR is technically demanding, and applications to biological problems have so far been limited. Further details of these techniques and discussion of instrumental aspects of ST-EPR can be found in an excellent review by Thomas (1986).

11.4 DATA ANALYSIS

11.4.1 Optical Anisotropy Decay

The time-dependent anisotropy r(t) is defined by

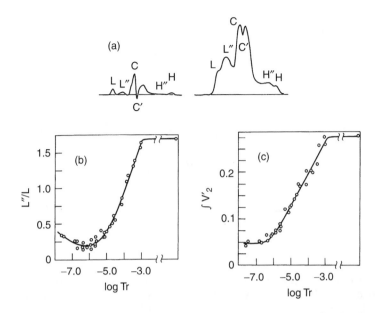

FIGURE 11.4 (a) ST-EPR spectrum of MSL-hemoglobin in 50% glycerol at –12°C showing the parameters used to quantify rotational diffusion. (b) Plot of L"/L as defined in (a) and obtained from ST-EPR spectra of MSL-hemoglobin in aqueous glycerol solutions of varying viscosity. The correlation times τ_r were calculated according to Equation 11.11. (c) Dependence of the normalized integral of the V_2' spectrum on the correlation time. Other details as in (b). (From Thomas DD: in *Techniques for the Analysis of Membrane Proteins*. Eds. Ragan CI and Cherry RJ, Chapman and Hall, London, 1986. pp. 337–431. With permission.)

$$r(t) = \frac{A_{II}(t) - A_\perp(t)}{A_{II}(t) + 2A_\perp(t)} \qquad (11.6)$$

where t is the time after excitation. For transient dichroism, $A_{II}(t)$, $A_\perp(t)$ are the absorbance changes for light polarized parallel and perpendicular, respectively, to the polarization of the exciting pulse. The same equation, with absorbances replaced by intensities, applies to phosphorescence and delayed fluorescence measurements. The anisotropy decay is normally determined only by rotational motion, for complications can occur when the excited state decay is multiexponential.

Analysis of the anisotropy decay usually involves fitting r(t) to a sum of exponential terms by a nonlinear least squares procedure, i.e.,

$$r(t) = \sum_{i=1}^{n} r_i \exp(-t/\phi_i) + r_\infty \qquad (11.7)$$

where ϕi are the rotational correlation times and r_∞ is the residual anisotropy. The initial anisotropy at t = o is given by

$$r_o = \sum_{i=1}^{n} r_i + r_\infty \qquad (11.8)$$

In most cases, the fit of Equation 11.7 to the experimental data is not improved by including terms beyond n = 2.

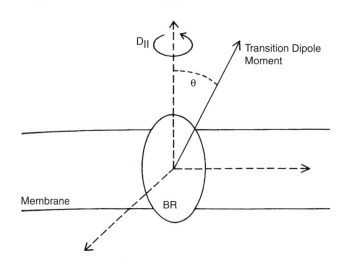

FIGURE 11.5 Uniaxial rotation model used to analyze anisotropy decays for bacteriorhodopsin (BR) in model membranes. (From Cherry RJ and Godfrey RE: *Biophys. J.* 1981, 36:257. With permission.)

The relationship between correlation times and rotational diffusion coefficients (and hence molecular parameters) depends on an assumed model for molecular rotation (Cherry, 1979; Lipari and Szabo, 1980; Kawato and Kinosita, 1981). For membrane proteins, the simplest model is that of uniaxial rotation, in which the protein can rotate only about an axis normal to the plane of the membrane with rotational diffusion coefficient D_{11} (Figure 11.5). In this case

$$r(t)/r_o = A_1 \exp(-t/\phi_1) + A_2 \exp(-t/\phi_2) + A_3 \qquad (11.9)$$

where

$$\phi_1 = 1/D_{11}, \phi_2 = 1/4D_{11}$$

The coefficients A_1, A_2, and A_3 are functions of the angle ϕ between the probe's transition dipole moment and the membrane normal. This model has been shown to be in good accord with experimental data in the case of bacteriorhodopsin reconstituted into lipid vesicles (Cherry and Godfrey, 1981).

The existence of two correlation times in even the simple uniaxial rotation model causes considerable difficulty in quantitative analysis. If more than one rotating species is present, it is not usually possible to resolve their individual contributions to r(t). Where analysis by Equation 11.9 is possible, it is helpful to define a rotational relaxation time $\phi_{11} = 1/D_{11}$ to facilitate comparison with other measurements. The previous equations and discussion apply to a random membrane suspension; considerable simplification occurs for oriented membranes (Cone, 1972; Johnson and Garland, 1981), although this fact has been rather little exploited.

An alternative model is that of restricted rotation, or "wobbling in a cone." Such a model would apply to a rigid protein wobbling about the membrane normal or the segmental motion within a protein that results in a similar motion. For the case in which wobbling occurs freely within a cone of half-angle θ_c, it has been shown that (Kinosita et al., 1977; Kawato and Kinosita, 1981)

$$r(t)/r_o = A \exp(-t/\phi_w) + r_\infty/r_o \qquad (11.10)$$

where

$$r_\infty/r_0 = [(1/2)\cos\theta_c(1 + \cos\theta_c)]^2$$

The expression $(r_\infty / r_0)^{1/2}$ is equal to the familiar order parameter S. The rather limited evidence obtained so far indicates that the uniaxial rotation model is the more appropriate for whole-body rotation of proteins with a substantial membrane-embedded domain. However, for most cases, the experimental data are not of sufficient precision to distinguish between the two models.

11.4.2 ST-EPR

Steady-state ST-EPR data are analyzed by comparison with reference spectra obtained by either computer simulation or experiments with model systems. Hemoglobin labeled with MSL is most commonly employed as a model system (Thomas et al., 1976; Squier and Thomas, 1986). ST-EPR spectra of MSL-hemoglobin are obtained as a function of viscosity in glycerol–water solutions. Because hemoglobin behaves like a rigid sphere, rotational correlation times can be calculated from the Stokes–Einstein equation

$$\tau_r = \frac{V\eta_w}{kT} \tag{11.11}$$

where V is the molecular volume of the hemoglobin and η_w is the viscosity of the solution.

Comparison of spectra is most simply achieved by defining spectral parameters that are particularly sensitive to rotational motion. For this purpose, it is usual to calculate the line height ratios L"/L, C"/C, and H"/H from the low, center, and high field regions of the spectrum, respectively. Figure 11.4 defines these parameters and shows a plot of the most commonly used parameter L"/L against correlation time. Absolute intensities of the V_2' spectrum are also sensitive to rotational motion (Squier and Thomas, 1986), and the normalized integral of the spectrum has been used to characterize rotational motion (Evans et al., 1981; Horváth and Marsh, 1983).

Theoretical simulations of the reference spectra have been reasonably successful for model systems, such as MSL-hemoglobin, in which rotational motion is isotropic (Thomas et al., 1976). However, the rotation of membrane proteins is highly anisotropic; thus it is not surprising that the lineshapes for ST-EPR spectra measured for proteins in membranes do not match spectra corresponding to isotropic motion. Because a simple model system analogous to hemoglobin is not available for anisotropic rotation, the interpretation of these spectra must rely on computer simulation. Theoretical simulations of uniaxial rotation have been achieved (Robinson and Dalton, 1980, 1981; Marsh, 1980) and compared with experimental ST-EPR spectra for a number of membrane proteins (Thomas, 1986; Esmann et al., 1987; Fajer et al., 1989; Horváth et al., 1989). It is found that uniaxial rotation tends to affect the center of the spectrum more than it affects the wings, particularly if the axis of rotation coincides with the nitroxide's z-axis (Marsh, 1980). In general, simulations indicate that the effective correlation times deduced from isotropic reference spectra are most likely to be greater than or equal to the actual correlation times for an anisotropically rotating protein. This is also true if the amplitude of the angular motion is restricted, as in the "wobbling in a cone" model (Lindahl and Thomas, 1982; Thomas, 1985).

11.5 SURVEY OF MEASURED CORRELATION TIMES

11.5.1 PROTEINS RECONSTITUTED INTO LIPID BILAYERS

Numerous measurements of rotational diffusion have been performed with membrane proteins that have been purified and reconstituted into lipid bilayer vesicles. Such vesicles provide a relatively simple system for evaluating methodology and investigating effects of lipid composition, lipid phase transitions, and lipid:protein mole ratios (L:P).

TABLE 11.1
Correlation Times of Proteins in Reconstituted Lipid Vesicles

Protein	Lipid	°C	Correlation Time (μsec)	Method	Reference
Band 3	Egg-PC	20	8*	TD	Morrison et al. (1986); Mühlebach and Cherry (1985)
Band 3	DMPC	37	73	ST	Sakaki et al. (1982)
Bacteriorhodopsin	DMPC	30	4*	TD	Godfrey and Cherry (1981)
Cytochrome P-450	PC:PE:PS	20	24	TD	Kawato et al. (1982)
ADP/ATP translocator	PC:PE:CL	20	60	TD	Müller et al. (1984)
ADP/ATP translocator	PC:PE:CL	30	1–5	ST	Horváth et al. (1989)
Cytochrome b_5	DMPC	35	0.4	TD	Vaz et al. (1979)
Rhodopsin	DPPC	44	10	ST	Kusumi and Hyde (1982)
Lactose permease	DMPC	28	6*	P	Dornmair et al. (1985)
Glycophorin	DMPC	30	<2	P	Van Hoogevest et al. (1985)
Ca^{2+}-ATPase	DPPC	45	~30	TD	Hoffman et al. (1980)
Ca^{2+}-ATPase	DOPC	37	25	ST	Napier et al. (1987)
Cytochrome reductase	Asolectin	4	70	ST	Quintanilha et al. (1982)
Cytochrome oxidase	PC:PE:CL	22	130	TD	Kawato et al. (1981)
Cytochrome oxidase	Asolectin	4	40	ST	Swanson et al. (1980)
Cytochrome oxidase	Asolectin	4	34	ST	Ariano and Azzi (1980)
Cytochrome oxidase	DMPC	30	25	ST	Fajer et al. (1989)
H^+-ATPase	Asolectin	4	100–180	P	Musser-Forsyth and Hammes (1990)

Note: PC = phosphatidylcholine, PE = phosphatidylethanolamine, PS = phophatidylserine, CL = cardiolipin, DMPC = dimyristoylphosphatidylcholine, DOPC = dioleylphosphatidylcholine, TD = transient dichroism, ST = saturation transfer–EPR, P = phosphorescence anisotropy. Where multiple correlation times are detected, only the fastest is given in the table. Data marked by * were fitted to Equation 11.9, i.e., fastest correlation time is $1/4D_{11}$.

Correlation times of membrane proteins measured in reconstituted vesicles are collected in Table 11.1. Data are included only for temperatures at which the lipids are in a liquid–crystalline phase. In the gel phase, proteins become immobile, as might be expected, although the temperatures at which all measurable mobility is lost may be several degrees below the phase transition temperature of the pure lipid. ST-EPR measurements normally yield only a single correlation time. For optical measurements, the situation is more complicated. Data may be fitted to Equation 11.7 to yield up to three correlation times. Occasionally, it has proved possible to fit the data by Equation 11.9 to obtain D_{11}. For simplicity, where more than one correlation time is obtained, only the fastest is listed in Table 11.1. Where D_{11} has been determined, then the value is again the faster correlation time, i.e., $1/4D_{11}$.

Clearly, any comparison of the correlation times listed in Table 11.1 must be made with caution. The number obtained is dependent on the measurement technique and the way the data are analyzed. Nevertheless, it appears that correlation times for medium-sized membrane proteins rotating freely in fluid lipid bilayers are typically in the order of 5 to 100 μsec. As would be expected from Equation 11.4, faster correlation times are detected for glycophorin and cytochrome b_5 (Vaz et al., 1979), which are anchored in the membrane by relatively small hydrophobic segments (also for the polypeptide gramicidin A, where the correlation time was determined by NMR [Macdonald and Seelig, 1988]).

The most detailed quantitative analysis of rotational diffusion of a membrane protein has been performed with bacteriorhodopsin, in part because measurements employing its intrinsic retinal chromophore obviate some of the problems encountered with probes. Optical anisotropy decays

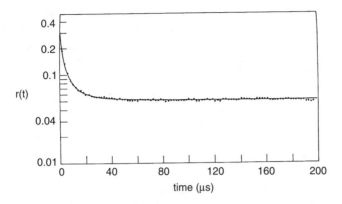

FIGURE 11.6 Anisotropy decay curve for bacteriorhodopsin in dimyristoylphosphatidylcholine vesicle (L/P = 220) at 28°C. The solid line is the best fit of Equation 11.9 to the experimental points with the parameters $r_o = 0.28$, $\phi = 12.6$ µsec, and $\Theta = 77$°C.

measured for bacteriorhodopsin in fluid DMPC bilayer have been successfully fitted by Equation 11.9 with only r_o, D_{11}, and θ as adjustable parameters, as illustrated in Figure 11.6 (Cherry and Godfrey, 1981). The value of θ obtained is in good agreement with independent determinations. Moreover, combination of D_{11} with measurements of the lateral diffusion coefficient permitted a determination of the diameter of bacteriorhodopsin that was in good accord with structural studies (Peters and Cherry, 1982). There are thus grounds for confidence in both the uniaxial rotation model and the Saffmann–Delbrück equations for membrane diffusion.

The rotational diffusion coefficient D_{11} for bacteriorhodopsin in fluid DMPC bilayers obtained in the previous studies is 5×10^4 s^{-1}($\phi_{11} = 20$ µsec). This value may be used as a benchmark for other membrane proteins, especially as bacteriorhodopsin is well characterized. Thus, Dornmair et al. (1985) determined $D_{11} = 4 \times 10^4$ s^{-1} for eosin-labeled lactose permease in the same lipid system. By comparison with bacteriorhodopsin, they calculated the protein radius in the plane of the membrane to be 2 ± 0.2 nm and hence deduced the protein to be monomeric, both in the absence and the presence of a membrane potential.

Evaluation of molecular size from rotational diffusion data, as described earlier, requires the protein to be present in the membrane as a single rotating species (monomer, dimer, etc.). When multiple species are present, it is very difficult to separate individual contributions with any confidence. In ST-ESR, where correlation times are determined from lineshapes, it is normally impossible to tell whether the correlation time can be attributed to a single species or is the average of multiple species. This problem may eventually be overcome by pulsed ESR methods. For optical anisotropy decays, even a single rotating species gives two correlation times, differing by a factor of 4 (Equation 11.9). The detection of fast correlation times differing by much more than a factor of 4 is indicative of heterogeneity arising from protein association. Because correlation times depend on the square of the particle radius in the plane of the membrane, anisotropy decays are particularly sensitive to such association. Aggregates of sufficient size may not have any significant rotation on the time scale of the experiment, in which case they cause an increase in the residual anisotropy r_∞.

Protein association is usually detected at sufficiently high protein concentration in the membrane. Thus, in the bacteriorhodopsin measurements described previously, it was necessary that L:P > 100 for the protein to be in a fully monomeric state (Cherry et al., 1978; Cherry and Godfrey, 1981). The monomeric state of the protein could be confirmed in this case by circular dichroism measurements, as well as by analysis of the anisotropy decays. In the case of another membrane protein, band 3, it was found that a small amount of irreversible self-aggregation occurred during the reconstitution procedure (Mühlebach and Cherry, 1985). Thus, even at very high L:P it was not possible to obtain a single rotating species and hence make a precise analysis of individual anisotropy decays. A solution to this difficulty was found by making the reasonable assumption that in each reconstituted sample, the major component was the lowest oligomeric form of band 3 but each sample additionally contained a variable aggregated component (Morrison et al., 1986). By applying the technique of global analysis (Knutson et al., 1983) to data sets obtained from different reconstitutions, it was possible to deduce a value of $D_{11} = 3 \times 10^4$ s^{-1} ($\phi_{11} = 33 \pm 3$ μsec). This number indicates a protein radius in the plane of the membrane of about 3.8 nm, which is compatible with either dimers or tetramers of band 3.

A further useful feature of analysis of anisotropy decays by Equation 11.9 is that it yields information on chromophore orientation. θ may be obtained from the best fit of the coefficients A_1, A_2, and A_3 or calculated from A_3 ($\equiv r_\infty/r_o$), provided there are no immobile species present. Of course, the analysis is only valid if there is a single binding site for the chromophore on the protein. For this reason, most determinations of chromophore orientation have been performed with intrinsic retinal (Heyn et al., 1977) or heme chromophores (Kawato et al., 1981, 1982), although θ has also been determined for eosin probes attached to lactose permease (Dornmair et al., 1985) and to band 3 (Morrison et al., 1986), where the criterion of a single binding site appears to be fulfilled.

11.5.2 PROTEINS IN CELL MEMBRANES

In order to study rotational motion of proteins in complex cell membranes, it is necessary to detect signals selectively from the proteins of interest. This may be achieved by several different means. The simplest is to utilize intrinsic probes, such as retinal of rhodopsin and bacteriorhodopsin and the heme-CO complex of cytochrome oxidase and cytochrome P-450. In these cases, there is no problem with unwanted signals from other membrane components. Unfortunately, such intrinsic chromophores are not widespread and are not applicable to ST-EPR measurements. If intrinsic triplet probes or spin labels are employed, then the question of selective labeling must be considered. In some instances, the problem can be sidestepped by studying membranes in which most of the protein is a single species, e.g., sarcoplasmic reticulum. In more complex membranes, selective labeling of a major component can sometimes be achieved with a probe of apparent low selectivity. Thus, reaction of eosin-5-maleimide with intact red blood cells results in highly selective labeling of band 3, even though this constitutes only 25% of the membrane protein (Nigg and Cherry, 1979). This is because the probe is membrane impermeable and only a limited number of proteins are exposed on the outer surface of the cell. Eosin also probably has an affinity for the anion binding site on band 3, because it is an inhibitor of anion transport (Nigg et al., 1979). Similar selective labeling using eosin-5-maleimide has been achieved for the ADP/ATP transporter of the mitochondrial inner membrane (Mueller et al., 1982, 1984). Finally, membrane proteins may be labeled using molecules with specific binding affinity. Examples include growth factor receptors (Zidovetzki et al., 1986a), F_c receptors, (Zidovetzki et al., 1986b), and proteins labeled with specific antibodies (Damjanovich et al., 1983).

Using the previously described methods, it has proved possible to measure correlation times for a wide range of membrane proteins (Table 11.2). As with reconstituted systems, where more than one correlation time is detected, only the fastest is given in the table. By and large, these correlation times are comparable to those obtained in simple reconstituted systems. Probably they correspond to proteins freely rotating in the membrane. However, additional slower or immobile components are frequently observed in cell membranes. In part, this may reflect the relatively high protein concentration in most cell membranes. Direct evidence that this is the case has been obtained for cytochrome oxidase in mitochondrial membranes by manipulating the lipid:protein ratio (Kawato et al., 1982b). In reconstituted systems, similar slow or immobile components are also detected at sufficiently high protein concentration due to protein association. Such associations could be quite short-lived; their lifetimes only need to be in the order of a few milliseconds to produce the observed effects on the anisotropy decays or ST-EPR spectra (Napier et al., 1987). Additional restraints may be imposed by interactions with peripheral proteins or cytoskeletal

TABLE 11.2
Correlation Times of Proteins in Cell Membranes

Protein	Membrane	°C	Correlation Time (μsec)	Method	Reference
Rhodopsin	Rod outer segment	20	5*	TD	Cone (1972)
Rhodopsin	Rod outer segment	20	20	ST	Kusumi et al. (1978); Baroin et al. (1977)
Rhodopsin	Rod outer segment	25	28	P	Coke et al. (1986)
Rhodopsin	Anthropod microvilli	22	Immobile	TD	Goldsmith and Wehner (1977)
Ca^{2+}-ATPase	SR	37	40	TD	Bürkli and Cherry (1981)
Ca^{2+}-ATPase	SR	4	30–60	ST	Thomas and Hidalgo (1978); Squier and Thomas (1985)
Ca^{2+}-ATPase	SR	25	100	P	Restall et al. (1979)
Ca^{2+}-ATPase	SR	20	5	P	Birmachu and Thomas (1990)
Band 3	Human erythrocyte ghosts	37	150	TD	Nigg and Cherry (1979)
Band 3	Human erythrocyte ghosts	38	40	P	Austin et al. (1979)
Band 3	Human erythrocyte ghosts	40	8–80	ST	Beth et al. (1986)
Bacteriorhodopsin	Purple membrane	20	Immobile	TD	Razi Naqvi et al. (1973)
Cytochrome oxidase	Bovine heart mitochondria	22	90	TD	Kawato et al. (1980)
Cytochrome P-450	Rat liver microsomes	37	15	TD	Kawato et al. (1982)
Cytochrome P-450	Rat liver microsomes	20	480	ST	Schwarz et al. (1982)
Cytochrome P-450	Adrenocortical mitochondria	20	80	TD	Kawato et al. (1988)
ADP/ATP translocator	Bovine heart mitochondria	37	60	TD	Müeller et al. (1984)
ADP/ATP translocator	Bovine heart mitochondria	0	2–20	ST	Horváth et al. (1989)
Spike glycoproteins	Sendai virus	37	100–200	TD	Lee et al. (1983)
Hemagglutinin	Influenza virus	36	30	TD	Junankar et al. (1986)
Con A receptors	Friend erythroleukemia cells	37	Immobile	P	Austin et al. (1979)
CFI	Thylakoid	6	1700	TD	Wagner and Junge (1980)
Acetylcholine receptor	Torpedo electric organ	20	Immobile	P, ST	Lo et al. (1980); Bartholdi et al. (1981); Rousselet et al. (1981)
Na^+/K^+-ATPase	Squalus acanthias rectal gland	20	50	ST	Esmann et al. (1987)
EGF-receptors	A431 cells	4	16–20	P	Zidovetzki et al. (1986a)
IgE-receptors	Rat basophilic leukemia cells	38	23	P	Zidovetzki et al. (1986b)
H-2K antigens	T41 cells	25	10–20	P	Damjanoviatch et al. (1983)

Note: SR = sarcoplasmic reticulum, other abbreviations as in Table 11.1. Where multiple correlation times are detected, only the fastest is given in the table. Correlation time marked by * corresponds to $1/4D_{11}$.

structures. Elucidation of such constraints is of considerable interest and is considered more fully in a subsequent section.

11.6 SEGMENTAL MOTIONS OF PROTEINS

So far, this review has been concerned with global motion of proteins in membranes. Using bacteriorhodopsin as a benchmark, it can reasonably be deduced that submicrosecond correlation times are unlikely to reflect global motion, at least for proteins with comparable or larger membrane domains. The presence of submicrosecond motion has, in some cases, been deduced from unusually low values of the initial anisotropy in optical measurements. Low anisotropies can, in principle, result from either independent probe motion or segmental motion (flexibility) of the protein. These may be distinguished by crosslinking agents, such as glutaraldehyde, which are expected to have a much greater effect on segmental motion than on independent probe motion. A better method is to perform time-resolved fluorescence polarization measurements, which permit direct detection of subnanosecond motion.

Perhaps the best studied example of segmental motion is that of Ca^{2+}-ATPase of sarcoplasmic reticulum. Bürkli and Cherry (1981) originally deduced the existence of segmental motion from transient dichroism studies of eosin-labeled membranes. Further investigations by phosphorescence depolarization (Speirs et al., 1983; Restall et al., 1985) were in agreement with this conclusion. More recently, Suzuki et al. (1989) have observed segmental motion directly by time-resolved fluorescence polarization of site-specific probes. Correlation times of 69 nsec and 190 nsec were measured for two different sites on the ATPase. Interestingly, the 69 nsec correlation time was selectively changed by Ca^{2+} binding and solubilization in deoxycholate. Other examples of implied segmental motion have been reported for the spike proteins of influenza virus (Junankar et al., 1986) and Sendai virus (Lee et al., 1983) and for the oligosaccharide chains of glycophorin (Cherry et al., 1980b). Using ST-EPR, Esmann et al. (1987) deduced the presence of segmental motion in Na^+/K^+-ATPase from discrepancies between rotational correlation times obtained from line-height ratios and those obtained from spectral integrals.

11.7 APPLICATIONS OF ROTATIONAL DIFFUSION MEASUREMENTS TO UNDERSTANDING MEMBRANE STRUCTURE AND FUNCTION

Tables 11.1 and 11.2 demonstrate that protein rotational diffusion has been investigated for a substantial number of different systems; hence, it is not feasible to discuss each of these in detail. In this section, selected examples are used to illustrate the principal applications of rotation measurements. The power of the approach arises largely from the sensitivity of the rotational diffusion coefficient to the size of the rotating particle (Equation 11.4). Thus, measurement of correlation times can readily detect protein association phenomena, such as self-association and complex formation between different proteins.

11.7.1 ASSOCIATIONS OF INTEGRAL MEMBRANE PROTEINS

Many integral membrane proteins are believed to exist in an oligomeric state in cell membranes. In principle, an accurate measurement of rotational diffusion coefficient permits calculation of the diameter of the rotating particle, as discussed in an earlier section. Knowledge of the diameter should readily distinguish monomers from oligomeric forms but may fail to resolve dimers, trimers, and tetramers. One of the most extensively investigated proteins is the Ca-ATPase of sarcoplasmic reticulum, whose rotational diffusion has been measured by transient dichroism (Bürkli and Cherry, 1981; Hoffman et al., 1979, 1980), phosphoresence anisotropy (Restall et al., 1984; Birmachu and Thomas, 1990), and ST-EPR (Thomas and Hidalgo, 1978; Kirino et al., 1978; Bigelow et al., 1986; Bigelow and Thomas, 1987; Napier et al., 1987; Squier and Thomas, 1988; Squier et al., 1988a,b). There is general agreement that Ca-ATPase exists in different aggregation states in the SR membrane. The sensitivity of rotational diffusion of the Ca-ATPase to protein–protein interactions was tested by Squier et al. by measuring ST-EPR spectra as a function of chemical crosslinking. They found that the effective correlation time was proportional to the mean molecular weight of the crosslinked aggregate (see Figure 11.7). The most detailed attempt to distinguish different oligomeric species of Ca-ATPase in SR was reported by Birmachu and Thomas (1990) using phosphorescence anisotropy. They analyzed their anisotropy decay curves to give three correlation times, of which the lowest component dominated at 4°C. Upon increasing the temperature, the faster

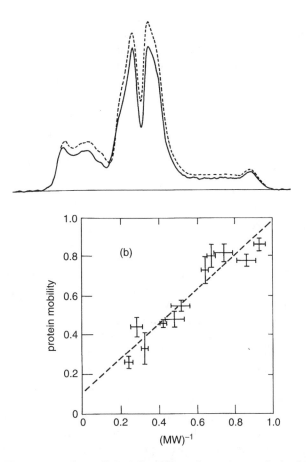

FIGURE 11.7 (a) ST-EPR spectra of crosslinked Ca-ATPase. Spectra were obtained at 4°C for the control (solid line) and a crosslinked sample (dashed line; 5 mM dithiobissuccinimidyl propionate, 5 min). The effective correlation times (τ_{ro} and τ_t) for the control and crosslinked sample are 44 and 166 μsec, respectively. The baseline is 100 G wide. (b) Dependence of rotational mobility of crosslinked Ca-ATPase on $\langle M_w \rangle^{-1}$. Protein mobility is measured by ST-EPR as in (a) and plotted as (τ_{ro}/τ_t), i.e., relative to control. $\langle M_w \rangle$ is the mean molecular weight relative to the monomer. (From Squier TC, Hughest SE, and Thomas DD: *J. Biol. Chem.* 1988, 263:9162–9170. With permission.)

components increased in intensity, which was interpreted as reflecting disaggregation of the protein. This agrees with the transient dichroism results of Bürkli and Cherry (1981) but not those of Hoffmann et al. (1979), who claimed that the Ca-ATPase becomes more associated at higher temperature. Above 20°C, Birmachu and Thomas found that the fastest component became increasingly intense. The correlation time of this component (5 μsec at 20°C) is most consistent with the monomeric form of the protein. On the basis of this and previous ST-EPR studies (Bigelow and Thomas, 1987; Squier and Thomas, 1988; Squier et al., 1988a), it is suggested that decreased protein–protein interaction is associated with an increase in enzyme activity.

An alternative method for investigating the oligomeric state of membrane proteins is to combine rotational diffusion measurements with specific crosslinking reactions. Thus, band 3 in human erythrocyte membranes can be crosslinked into covalent dimers by Cu-phenanthroline catalyzed disulfide bridge formation. This crosslinking has no effect on rotational mobility of band 3, indicating that the dimers are present prior to the crosslinking reaction (Nigg and Cherry, 1980).

Protein association is of particular interest in relation to receptor biochemistry. In many instances, ligand binding may promote clustering of receptors from a randomly dispersed state.

The Structure of Biological Membranes, Second Edition

Rotational diffusion measurements have been used to investigate clustering of epidermal growth factor (EGF) receptors on human epidermal carcinoma A431 cells and on membrane vesicles prepared from these cells (Zidovetzki et al., 1981, 1986a). Time-resolved phosphorescence polarization measurements of erythrosin-labeled DGF bound to receptors were performed as a function of temperature. At 4°C, the correlation times were consistent with rotation of single hormone-receptor complexes or small microclusters (two to three receptors). Incubation at 37°C resulted in significantly longer correlation times, indicating microaggregation of the receptors. The same group has also investigated rotational diffusion of F_c receptors in rat basophilic leukemia cells by measuring phosphorescence depolarization of erythrosin-labeled immunoglobin E bound to the receptor (Zidovetzki et al., 1986b). Rotational correlation times varied from 65 to 23 µsec over the temperature range 5.5 to 38°C and were consistent with dispersed receptors freely rotating in the membrane. Thus receptor aggregation does not occur prior to crosslinking with multivalent antigens or anti-IgE antibodies. Tilley et al. (1988) have labeled low-density lipoproteins with an eosinyl fatty acid probe. From phosphorescence depolarization measurements, they find that global rotation of the particle is abolished upon binding to membranes isolated from adrenal cortex.

The ability of rotational diffusion measurements to detect protein associations has been further exploited to study protein aggregation phenomena. For instance, a variety of agents, including melittin, divalent cations, polylysine, and other cationic polypeptides, have been shown to aggregate band 3 proteins in erythrocyte membranes (Dufton et al., 1984a,b; Clague and Cherry, 1986, 1988, 1989; Hui et al., 1990). The effect of melittin on the anisotropy decay of eosin-labeled band 3 is illustrated in Figure 11.8. Hydrophilic agents such as polylysine appear to aggregate band 3 by electrostatic interactions, which can be reversed at high ionic strength. For amphiphilic agents such as melittin, the situation is more complex. The positively charged residues of melittin are involved in protein aggregation, because their modification reduces or eliminates the effect. However, protein aggregation occurs at both high and low ionic strength. Melittin and related peptides are anchored in the membrane by a hydrophobic moiety. As well as facilitating hydrophobic interactions with membrane proteins, this could restrict electrostatic interactions to ionic groups close to the lipid bilayer surface (Clague and Cherry, 1989). Rotational diffusion measurements have also been used to demonstrate that lipid peroxidation promotes aggregation of cytochrome P-450 in rat liver microsomes (Gut et al., 1985) and that the presence of cholesterol causes aggregation of bacteriorhodopsin in lipid vesicles (Cherry et al., 1980). In erythrocyte membranes, band 3 microaggregation also appears to be sensitive to cholesterol content of the membrane (Mühlebach and Cherry,

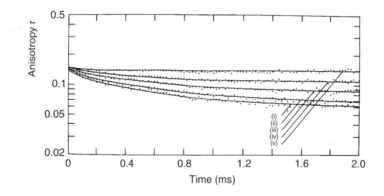

FIGURE 11.8 Anisotropy decay curves for eosin-labeled band 3 in erythrocyte membranes at different concentrations of native melittin; r(t) was measured by transient dichroism at 37°C with membranes suspended in 5 mM sodium phosphate buffer, pH 7.4. Concentrations of melittin are expressed as moles per milligram of ghost protein. (i) Without melittin. (ii) 1.8×10^{-8} mol/mg. (iii) 3.5×10^{-8} mol/mg. (iv) 4.9×10^{-8} mol.mg. (v) 1.1×10^{-7} mol/mg. (From Dufton MJ, Hider RC, and Cherry RJ: *Eur. Biophys. J.* 1984(a), 11:17. With permission.)

1982). Ohta et al. (1990) showed that conversion of cholesterol to pregnenolone in adrenocortical mitochondria increased the fraction of mobile cytochrome P-450 molecules in the membrane. Rhodopsin in rod outer segment disc membranes has been shown by ST-EPR to become aggregated upon delipidation or prolonged illumination (Baroin et al., 1979).

The formation of complexes between two different membrane proteins can also be investigated by measuring rotational diffusion. One technique is to crosslink specific proteins in the membrane with antibodies and test for the effect on rotational mobility of a second protein. If the two proteins exist as a complex in the membrane, the second protein should be immobilized, whereas there should be no effect if the two proteins are independent. This method has been used to demonstrate complex formation between glycophorin A and band 3 in the human erythrocyte membrane (Nigg et al., 1980) and between cytochrome P-450 and NADPH-cytochrome P-450 reductase in reconstituted vesicles (Gut et al., 1983). Virus particles have also been used to crosslink glycophorin and have similar effects to antibodies on band 3 mobility (Nigg et al., 1980b). In a somewhat different experiment, it was found that the rotational mobility of cytochrome oxidase reconstituted into lipid vesicles was unaffected by the coreconstitution of cytochrome bc_1 complex, suggesting an absence of a specific association (Kawato et al., 1981). ST-EPR has also been used to study interactions between components of the mitochondrial electron transport chain (Poore et al., 1982). The existence or otherwise of complexes in these systems is, of course, of considerable interest in relation to the mechanisms of electron transport.

11.7.2 INTERACTIONS OF INTEGRAL MEMBRANE PROTEINS WITH CYTOSKELETAL STRUCTURES AND PERIPHERAL PROTEINS

Integral membrane proteins that provide attachment sites for cytoskeletal structures are expected to have restricted mobility. Disruption of these attachments should, in turn, lead to increased mobility. Thus mobility measurements, in principle, offer a powerful method of investigating membrane–cytoskeletal interactions. The most detailed application of this approach has been to the human erythrocyte membrane. The erythrocyte cytoskeleton consists of a network of peripheral proteins that control cell shape and provide mechanical strength (Bennett, 1989). Band 3 proteins are believed to provide major points of attachment of this cytoskeleton to the membrane.

Early studies of the rotational mobility of band 3 by transient dichroism revealed at least two components of the anisotropy decay measured at 37°C (Nigg and Cherry, 1979). The faster component is thought to correspond to freely mobile band 3, with the slower component corresponding to a population of band 3 with restricted motion. It has been shown that this restriction is in part removed by release of spectrin, actin, ankyrin, and band 4.1 from the membrane. Furthermore, tryptic cleavage of a 43 kDa cytoplasmic fragment of band 3 produces a similar, though larger, effect, as illustrated in Figure 11.9 (Nigg and Cherry, 1980). These results were interpreted as indicating restricted motion of a population of band 3, resulting from a linkage via ankyrin or band 4.1 of the cytoplasmic domain of band 3 to the spectrin-actin network. This conclusion was in accord with studies of the binding properties of isolated proteins.

More recent work suggests, however, that the situation is more complex than previously envisaged (Clague et al., 1989). It was found that *in situ* proteolytic cleavage of ankyrin had no effect on band 3 rotational mobility. There was also no effect of heating membranes to a temperature at which ankyrin is expected to be denatured. Finally, in confirmation of earlier work (Cherry et al., 1976), removal of most spectrin and actin from the membrane failed to significantly alter band 3 rotation. More recent work on the flexibility of spectrin helps to explain these observations (Learmonth et al., 1989; Clague et al., 1990). Spectrin is found to undergo rapid flexing motions, even when forming part of the erythrocyte cytoskeleton. Thus, the cytoskeleton is probably too flexible to restrict angular displacement of band 3 over $\pm\pi/2$, which is what determines the anisotropy decay. On this interpretation, the observed restriction to band 3 mobility is likely to result from further interactions involving peripheral proteins other than spectrin.

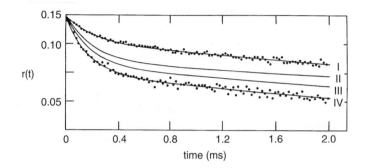

FIGURE 11.9 Effect of trypsin treatment on anisotropy decay of eosin-labeled band 3 in human erythrocyte membranes, measured by transient dichroism at 37°C. Ghosts were treated with trypsin at concentration of I, 0; II, 0.125; III, 10.25; IV, 0.5 mg/ml. For clarity, experimental points are omitted from curves II and III. (From Nigg EA and Cherry RJ: *Proc. Natl. Acad. Sci. U.S.A.* 1980, 77:4702. With permission.)

The rotational mobility of acetylcholine receptors in *Torpedo* electric organ has been studied by both phosphorescence depolarization (Lo et al., 1980; Bartholdi et al., 1981) and ST-EPR (Rousselet et al., 1981). For these measurements, the probe was either eosin- or erythrosin-labeled α-bungarotoxin or MSL. The receptor was found to be immobile but could be mobilized by alkali removal of a 43 kDa peripheral protein or by reduction with dithiothreitol. It thus appears likely that the 43 kDa protein is involved in stabilizing aggregation or oligomeric association of the receptor.

Wagner et al. (1982) investigated the rotational diffusion of ferredoxin-NADP$^+$ reductase rebound to thylakoid membranes. Using transient dichroism, they found the correlation time was considerably increased by the presence of ferredoxin, suggesting that ferredoxin mediates the formation of a ternary complex with photosystem I. Proteins present in the aqueous phase show a large increase in correlation time on binding to membranes. Thus, ST-EPR was employed by Cassoly (1982) to study hemoglobin binding to erythrocyte membranes.

11.8 SUMMARY

A range of measurement techniques employing either optical anisotropy decay or ST-EPR is available for measuring rotational diffusion of proteins in membranes. Because anisotropy decays are analyzed by fitting the data to between one and three correlation times and ST-EPR spectra are generally analyzed by comparison of lineshapes with reference spectra for isotropic rotation, a strict comparison between different measurements is not possible. Nevertheless, within this limitation there appears to be reasonable agreement between the different methods in most cases. Only in a few instances in reconstituted systems has it proved possible to obtain a value for the rotational diffusion coefficient D_{II} and hence to determine the radius of the rotating particle. From the available data, it appears likely that most polytopic transmembrane proteins freely diffusing in a fluid lipid bilayer have values of D_{II} between 10^3 and 10^4 s^{-1}, with the precise value depending on the size of the protein and the viscosity of the lipid bilayer according to Equation 11.4.

Rotational diffusion is particularly sensitive to size, and hence its measurement provides a powerful way of studying protein associations in membranes. The oligomeric state of membrane proteins, reversible and irreversible clustering of proteins, and complex formation between different proteins have all been investigated by measuring rotational diffusion. In cell membranes, many proteins appear to be rotating freely but additional, more slowly rotating components are frequently observed. Elucidation of the associations responsible for slow rotation can provide considerable information on membrane structure. Of particular interest are restrictions on free rotation imposed by interactions of integral proteins with cytoskeletal structures or peripheral proteins. Such interactions also restrict lateral diffusion, and the combination of rotational and lateral diffusion measurements constitute a valuable approach to their study.

Biological membranes are complex structures, and unraveling the details of their molecular organization is a slow and difficult task. Diffusion measurements have already contributed much to our understanding of membrane structure and dynamics and are likely to continue to play an important role in the years to come.

REFERENCES

Ariano BH and Azzi A: *Biochem. Biophys. Res. Commun.* 1980, 93:478–485.

Austin RH, Chan SS, and Jovin TM: *Proc. Natl. Acad. Sci. U.S.A.* 1979, 76:5650–5654.

Baroin A, Thomas DD, Osborne B, and Devaux PF: *Biochem. Biophys. Res. Commun.* 1977, 78:442–447.

Baroin A, Bienvenue A, and Devaux PF: *Biochemistry* 1979, 18:1151–1155.

Bartholdi M, Barrantes FJ, and Jovin TM: *Eur. J. Biochem.* 1981, 120:389–397.

Bennett V: *Biochim. Biophys. Acta* 988:107–121.

Beth AH, Conturo TE, Venkataramu SD, and Staros JV: *Biochemistry* 1986, 25:3824–3832.

Bigelow DJ and Thomas DD: *J. Biol. Chem.* 1987, 262:13449–13456.

Bigelow DJ, Squier TC, and Thomas DD: *Biochemistry* 1986, 25:194–202.

Birmachu W and Thomas DD: *Biochemistry* 1990, 29:3904–3914.

Brown IM: in *Time Domain Electron Spin Resonance.* Eds. Kevan L and Schwartz RN, John Wiley & Sons, New York, 1979, 196–229.

Bürkli A and Cherry RJ: *Biochemistry* 1981, 20:138–145.

Cassoly R: *Biochim. Biophys. Acta* 1982, 689:203–209.

Cherry RJ, Müller U, Henderson R, and Heyn MP: *J. Mol. Biol.* 1978, 121:283–298.

Cherry RJ: *Biochim. Biophys. Acta* 1979, 559:289–327.

Cherry RJ: *Methods Enzymol.* 1978, 54:47–61.

Cherry RJ and Godfrey RE: *Biophys. J.* 1981, 36:257–276.

Cherry RJ, Bürkli A, Busslinger M, Schneider G, and Parish GR: *Nature* 1976, 263:389–393.

Cherry RJ, Müller U, Holenstein C, and Heyn MP: *Biochim. Biophys. Acta* 1980, 596:145–151.

Cherry RJ, Nigg EA, and Beddard GS: *Proc. Natl. Acad. Sci. U.S.A.* 1980(b), 77:5899–5903.

Clague MJ and Cherry RJ: *Biochem. J.* 1988, 252:791–794.

Clague MJ and Cherry RJ: *Biochim. Biophys. Acta* 1989, 980:93–99.

Clague MJ, Harrison JP, and Cherry RJ: *Biochim. Biophys. Acta* 1989, 981:45–50.

Clague MJ and Cherry RJ: *Biochem. Soc. Trans.* 1986, 14:883–884.

Clague MJ, Harrison JP, Morrison IEG, Wyatt K, and Cherry RJ: *Biochemistry* 1990, 29:3898–3904.

Coke M, Restall CJ, and Chapman D: *Biochemistry* 1986, 25:513–518.

Cone RA: *Nature New Biol.* 1972, 236:39–43.

Corin AF, Blatt E, and Jovin TM: *Biochemistry* 1987, 26:2207–2217.

Damjanovich S, Tron L, Szollosi J, Zidovetzki R, Vaz WLC, Regateiro F, Arndt-Jovin D, and Jovin TM: *Proc. Natl. Acad. Sci. U.S.A.* 1983, 80:5985–5989.

Dornmair K, Corin AF, Wright JK, and Jähnig F: *EMBO J.* 1985, 4:3633–3638.

Dufton MJ, Cherry RJ, Coleman JW, and Stanworth DR: *Biochem. J.* 1984(b), 223:67–71.

Dufton MJ, Hider RC, and Cherry RJ: *Eur. Biophys. J.* 1984(a), 11:17–24.

Esmann M, Horváth LI, and Marsh D: *Biochemistry* 1987, 26:8675–8683.

Evans CA et al.: *J. Magn. Res.* 1981, 44:109–116.

Fajer P, Thomas DD, Feix JB, and Hyde JS: *Biophys. J.* 1986, 50:1195–1202.

Fajer P, Knowles PF, and March D: *Biochemistry* 1989, 28:5634–5643.

Forrer JE, Wubben RC, and Hyde JS: *Bull. Magn. Reson.* 1980, 2:441–446.

Gaffney FJ: in *Spin Labeling.* Ed. Berliner L, Academic Press, New York, 1976, pp. 183–238.

Garland PB and Birmingham JJ: *Biochem. Soc. Trans.* 1986, 14:383–839.

Garland PB and Moore CH: *Biochem. J.* 1979, 183:561–572.

Goldsmith TH and Wehner R: *J. Gen. Physiol.* 70:453–490.

Greinert R, Stärk H, Stier A, and Weller A: *J. Biochem. Biophys. Methods* 1979, 1:77–83.
Gut J, Kawato S, Cherry RJ, Winterhalter KH, and Richter C: *Biochim. Biophys. Acta* 1985, 817:217–228.
Gut J, Richter C, Cherry RJ, Winterhalter KH, and Kawato S: *J. Biol. Chem.* 1983, 258:8588–8594.
Heyn MP, Cherry RJ, and Mueller U: *J. Mol. Biol.* 1977, 117:607–620.
Hoffman W, Sarzala MG, and Chapman D: *Proc. Natl. Acad. Sci. U.S.A.* 1979, 76:3860–3864.
Hoffman W, Sarzala MG, Gomez-Fernandez JC, Goni FM, Restall CJ, and Chapman D: *J. Mol. Biol.* 1980, 141:119–132.
Horváth LI and Marsh D: *J. Magn. Res.* 1983, 54:363–373.
Horváth LI, Munding A, Beyer K, Klingenberg M, and Marsh D: *Biochemistry* 1989, 28:407–414.
Hughes BD, Pailthorpe BA, and White LR: *J. Fluid Mech.* 1981, 110:349–372.
Hui SW, Stewart CM, and Cherry RJ: *Biochim. Biophys. Acta* 1990, 1023:335–340.
Hyde JS: in *Time Domain Electron Spin Resonance.* Eds. Kevan L and Schwartz RN, John Wiley & Sons, New York, 1979, pp. 1–30.
Hyde JS, Froncisz W, and Mottley C: *Chem. Phys. Lett.* 1984, 110:621–625.
Johnson P and Garland PB: *FEBS Lett.* 1981, 132:252–256.
Jovin TM and Vaz WLC: *Methods Enzymol.* 1989, 172:471–573.
Junankar P and Cherry RJ: *Biochim. Biophys. Acta* 1986, 854:198–206.
Kawato S and Kinosita K Jr.: *Biophys. J.* 1981, 36:277–296.
Kawato S, Sigel E, Carafoli E, and Cherry RJ: *J. Biol. Chem.* 1980, 255:5508–5510.
Kawato S, Gut J, Cherry RJ, Winterhalter KH, and Richter C: *J. Biol. Chem.* 1982, 257:7023–7029.
Kawato S, Lehner C, Mueller M, and Cherry RJ: *J. Biol. Chem.* 1982(b), 257:6470–6476.
Kawato S, Sigel E, Carafoli E, and Cherry RJ: *J. Biol. Chem.* 1981, 256:7518–7527.
Kawato S, Mitani F, Iizuka T, and Ishimura Y: *J. Biochem.* 1988, 104:188–191.
Keana JFW: in *Spin Labeling.* Ed. Berliner L, Academic Press, New York, 1979, pp. 115–172.
Kinosita K Jr. and Ikegami A: *Subcell. Biochem.* 1988, 13:55–86.
Knutson JR, Beecham JN, and Brand L: *Chem. Phys. Lett.* 1983, 102:501–507.
Kusumi A and Hyde JS: *Biochemistry* 1982, 21:5978–5983.
Kusumi A, Ohnishi S, Ito T, and Yoshizawa T: *Biochim. Biophys. Acta* 1978, 507:539–543.
Kusumi A, Subczynski T, and Hyde JS: *Proc. Natl. Acad. Sci. U.S.A.* 1982, 79:1854–1858.
Learmonth RP, Woodhouse AG, and Sawyer WH: *Biochim. Biophys. Acta* 1989, 987:124–128.
Lee PM, Cherry RJ, and Bächi T: *Virology* 1983, 128:65–76.
Lindahl KM and Thomas DD: *Biophys. J.* 1982, 37:31a.
Lipari G and Szabo A: *Biophys. J.* 1980, 30:489–506.
Lo MMS, Garland PB, Lamprecht J, and Barnard EA: *FEBS Lett.* 1980, 111:407–412.
Macdonald PM and Seelig J: *Biochemistry* 1988, 27:2770–2779.
Marsh D: *Biochemistry* 1980, 19:1632–1637.
Millhauser GL and Freed JH: *Chem. Phys. Lett.* 1984, 81:37–48.
Moore CH, Boxer DH, and Garland PB: *FEBS Lett.* 1979, 108:161–166.
Morrisett JD: in *Spin Labeling.* Ed. Berliner L, Academic Press, New York, 1976, pp. 273–338.
Morrison IEG, Mühlebach T, and Cherry RJ: *Biochem. Soc. Trans.* 1986, 14:885–886.
Mühlebach T and Cherry RJ: *Biochemistry* 1982, 21:4225–4228.
Mühlebach T and Cherry RJ: *Biochemistry* 1985, 24:975–983.
Müller M, Krebs JJR, Cherry RJ, and Kawato S: *J. Biol. Chem.* 1984, 259:3037–3043.
Müller M, Krebs JJR, Cherry RJ, and Kawato S: *J. Biol. Chem.* 1982, 257:1117–1120.
Murray EK, Restall CJ, and Chapman D: *Biochim. Biophys. Acta* 1983, 732:347–351.
Musier-Forsyth K and Hammes GG: *Biochemistry* 1990, 29:3236–3241.
Napier RM, East JM, and Lee AG: *Biochim. Biophys. Acta* 1987, 903:365–373.
Nigg E and Cherry RJ: *Nature* 1979, 277:493–494.
Nigg E, Kessler M, and Cherry RJ: *Biochim. Biophys. Acta* 1979, 550:328–340.
Nigg EA and Cherry RJ: *Proc. Natl. Acad. Sci. U.S.A.* 1980, 77:4702–4706.
Nigg EA, Bron C, Girardet M, and Cherry RJ: *Biochemistry* 1980, 19:1887–1893.
Nigg EA, Cherry RJ, and Bächi T: *Virology* 1980(b), 107:552–556.
Nigg EA and Cherry RJ: *Biochemistry* 1979, 18:3457–3465.
Ohta Y, Mitani F, Ishimura Y, Yanagibashi K, Kawamura M, and Kawato S: *J. Biochem.* 1990, 107:97–104.
Perrin F: *J. Phys. Radium* 1936, 7:1–11.

Peters R and Cherry RJ: *Proc. Natl. Acad. Sci. U.S.A.* 1982, 79:4317–4321.

Poore VM, Fitzsimons JTR, and Ragan CI: *Biochim. Biophys. Acta* 1982, 693:113–124.

Quintanilha AT, Thomas DD, and Swanson M: *Biophys. J.* 1982, 37:68–69.

Razi Naqvi K and Wild UP: *Chem. Phys. Lett.* 1975, 36:222–224.

Razi Naqvi K, Gonzalex-Rodrigüz J, Cherry RJ, and Chapman D: *Nature New Biol.* 1973, 245:249–251.

Restall CJ, Dale RE, Murray EK, Gilbert CW, and Chapman D: *Biochemistry* 1984, 23:6765–6776.

Robinson BR and Dalton LR: *Chem. Phys.* 1981, 54:253–259.

Robinson BR and Dalton LR: *J. Chem. Phys.* 1980, 72:1312–1324.

Rousselet A, Cartaud J, and Devaux PF: *Biochim. Biophys. Acta* 1981, 648:169–185.

Saffman PG: *J. Fluid Mech.* 1976, 75:593–602.

Saffman PG and Delbrück M: *Proc. Natl. Acad. Sci. U.S.A.* 1975, 72:3111–3113.

Sakaki T, Tsuji A, Chang C-H, and Ohnishi S: *Biochemistry* 1982, 21:2366–2372.

Schwarz D, Pirrwitz J, and Ruckpaul K: *Arch. Biochem. Biophys.* 1982, 216:322–328.

Singer SJ and Nicolson GL: *Science* 1972, 175:720–731.

Smith LM, Weis Rm, and McConnell HM: *Biophys. J.* 1981, 36:73–91.

Spiers A, Moore CH, Boxer DH, and Garland PB: *Biochem. J* 1983, 213:67–74.

Squier TC, Bigelow DJ, and Thomas DD: *J. Biol. Chem.* 1988(b), 263:9178–9186.

Squier TC, Hughes SE, and Thomas DD: *J. Biol. Chem.* 1988(a), 263:9162–9170.

Squier TC and Thomas DD: *Biophys. J.* 1986, 49:921–935.

Squier TC and Thomas DD: *J. Biol. Chem.* 1988, 263:9171–9177.

Strambini GB and Galley WC: *Nature* 1976, 260:554–556.

Suzuki S, Kawato S, Kouyama T, Kinosita K, Jr., Ikegami A, and Kawakita M: *Biochemistry* 1989, 28:7734–7740.

Swanson MS, Quintanilha AT, and Thomas DD: *J. Biol. Chem.* 1980, 25:7494–7502.

Thomas DD: in *The Enzymes of Biological Membranes*. Vol. 1. Ed. Martonosi AN, Plenum Press, New York, 1985, pp. 287–312.

Thomas DD: in *Techniques for the Analysis of Membrane Proteins*. Eds. Ragan CI and Cherry RJ, Chapman and Hall, London, 1986, pp. 377–431.

Thomas DD and Hidalgo C: *Proc. Natl. Acad. Sci. U.S.A.* 1978, 75:5488–5492.

Thomas DD, Dalton LR, and Hyde JS: *J. Chem. Phys.* 1976, 65:3006–3024.

Tilley L, Sawyer WH, Morrison JR, and Fidge NH: *J. Biol. Chem.* 1988, 263:17541–17547.

Van Hoogevest P, de Kruiff B, and Garland PB: *Biochim. Biophys. Acta* 1985, 813:1–9.

Vaz WLC, Austin RH, and Vogel H: *Biophys. J.* 1979, 26:415–426.

Velez M and Axelrod D: *Biophys. J.* 1988, 53:575–591.

Wagner R and Junge W: *FEBS Lett.* 1980, 114:327–333.

Wagner R, Carrillo N, Junge W, and Vallejos RH: *Biochim. Biophys. Acta* 1982, 680:317–330.

Weber G: *Adv. Protein Chem.* 1953, 8:415–519.

Yoshida TM, Zarrin F, and Barisas BG: *Biophys. J.* 1988, 54:277–288.

Yoshida TM and Barisas BG: *Biophys. J.* 1986, 50:41–53.

Zidovetzki R, Yarden Y, Schlessinger J, and Jovin TM: *Proc. Natl. Acad. Sci. U.S.A.* 1981, 78:1337–1341.

Zidovetzki R, Yarden Y, Schlessinger J, and Jovin TM: *EMBO J.* 1986(a), 5:247–250.

Zidovetzki R, Bartholdi M, Arndt-Jovin D, and Jovin TM: *Biochemistry* 1986(b), 25:4397–4401.

12 Translational Diffusion of Membrane Proteins

Michael Edidin

CONTENTS

12.1 INTRODUCTION

The translational diffusion of membrane proteins was one of the key observations used to establish the current fluid-mosaic model of cell membranes (Singer and Nicolson, 1972). Measurements of translational diffusion of membrane proteins are interesting in two respects. First, such measurements address some physical problems about diffusion in the protein-rich, two-dimensional fluids that constitute cell membranes, in particular, problems about the forces retarding diffusion. Second, studies of translational dynamics of membrane proteins address questions of membrane organization and function. Measurements of translational diffusion are used to analyze the fine structure of cell membrane organization and to detect associations among membrane proteins and between these proteins and the cell cytoplasm. Such associations are biologically important in forming functioning receptors and in signaling by such receptors.

This review covers fluorescence and other optical methods for measuring translational diffusion of membrane proteins. This restriction is not as stringent as it might first seem. Though translational diffusion of membrane proteins has been measured using magnetic resonance techniques, the relatively low sensitivity of these methods compared to fluorescence has limited their application to biological membranes (see Edidin, 1987).

12.2 EARLY MEASUREMENTS OF THE LATERAL MOBILITY OF MEMBRANE PROTEINS

Diffusion of proteins in the plane of a cell surface membrane was first demonstrated qualitatively in heterokaryons, cells formed by fusing cultured fibroblasts of two different species (Frye and Edidin, 1970). At first, fluorescent labels for mouse or human antigens were segregated in separate regions of the heterokaryon surface, but within 5 min of fusion the membranes of some cells contained large regions that were doubly labeled, indicating that the surface antigens of mouse and human membranes had intermixed in this short time. Controls and later experiments on intermixing of surface antigens in fixed cells (Edidin and Wei, 1982) showed that this intermixing was not metabolically driven but was due to diffusion in the plane of the heterokaryon membrane.

The experiment described provides a vivid qualitative picture of lateral diffusion and has been used to show lateral diffusion in erythrocyte (Fowler and Branton, 1977; Koppel and Sheetz, 1981) and plant protoplast membranes (Mastrangelo and Mitra, 1981). It allows a rough estimate of D, the translational diffusion coefficient, in the population of fused cells. This estimate proves to be in good agreement with D measured using other techniques (Edidin and Wei, 1982). However, the method is cumbersome and slow. It also describes cell populations, not single cells.

The first measurements of D of a single species of protein were made by two groups (Poo and Cone, 1974; Liebman and Entine, 1974). Each group used a microspectrophotometer to quantitate the visual pigment rhodopsin in the disk membranes of rod outer segments in vertebrate retinas. Changes in light absorption at a wavelength characteristic of unbleached rhodopsin monitored the return of unbleached molecules to an area that had been bleached by a brief, intense light pulse. This return was rapid, requiring only tens of seconds, and there was a corresponding loss of unbleached rhodopsin molecules from the area of the disks that had not been exposed to the intense light pulse. Diffusion coefficients for rhodopsin calculated from these data are 3.5 to 5×10^{-9} cm^2 s^{-1}. This is about D expected for a protein of about 50 kDa diffusing an oil of viscosity ~1 to 6 pose (100 to 600 × the viscosity of water). This viscosity had previously been estimated for disc membranes from measurements of rotational diffusion (Cone, 1972).

Both the results and the technique of these experiments are extremely important for all subsequent work in translational diffusion of membrane proteins and lipids. The result, that the measured D is close to that calculated for unrestricted diffusion in an oil, is the benchmark for all other measurements of D, many of which, as we will see, are orders of magnitude smaller.

The technique of using light to create an inhomogeneity in a uniformly labeled membrane and to follow the diffusion of particles into a marked region is the basis for the most commonly used present-day method for measuring D, fluorescence photobleaching and recovery (FPR, also known as fluorescence recovery after photobleaching [FRAP]). Here, a gradient of fluorescence is created by bleaching; dissipation of the gradient is measured in terms of recovery of fluorescence in the bleached spot. This greatly increases the sensitivity of the method and also allows the tracking of as many different membrane proteins as can be labeled by fluorophores.

The fluorescence photobleaching method was first attempted by Peters and coworkers (Peters et al., 1974) in an experiment to measure D of fluorescein-labeled erythrocyte membrane proteins. No recovery of fluorescence was detected after bleaching half of a fluorescent erythrocyte ghost using a mercury arc lamp, but the approach pointed the way to a powerful and general method for measuring translational diffusion. We will discuss this method in more detail in later sections.

12.3 CURRENT METHODS FOR MEASURING TRANSLATIONAL DIFFUSION IN MEMBRANES

12.3.1 RELAXATION AFTER ELECTROPHORESIS

The surface distribution of membrane proteins is perturbed when cells are held in an electric field with a potential drop of 1 to 10 V/cm (Jaffe, 1977; Poo and Robinson, 1977). The 2 to 3 mV potential across a cell of 20 μm diameter is sufficient to cause redistribution of membrane proteins toward one or the other electrodes in the system. This redistribution is due mainly to electrophoresis of charged species, balanced by diffusion and in some cases by electro-osmotic water flow (review, Poo, 1981). In principle, the diffusion coefficient of the redistributing species can be inferred from its equilibrium distribution. In practice, D is most reliably estimated from the relaxation time of the field-induced asymmetric distribution of membrane proteins. This postelectrophoresis relaxation time (PER time) may be roughly quantitated by measuring the fraction of a cell population maintaining an asymmetry at particular times after the electric field is removed (Poo et al., 1978). A more accurate estimate of relaxation time is made by measuring the fluorescence on opposite (cathodal and anodal) sides of the cell surface and calculating an asymmetry index (Poo et al., 1979). The method is only applicable if proteins are not irreversibly aggregated by the electrophoresis step. Higher voltages or prolonged exposure to the field result inhibit the back-diffusion of concanavalin A binding sites (Poo, 1981) and another protein, the acetylcholine receptor of *Xenopus* myocytes, is irreversibly aggregated by the field; its diffusion is undetectable.

The proteins whose diffusion is estimated by PER need not be labeled until they must be visualized in samples of the cell population taken during the relaxation phase of the PER experiment. PER then is the only quantitative method available that estimates D of unlabeled, native, membrane proteins. All other methods for measuring translational diffusion require labeling of the molecules of interest that could affect D. On these grounds alone, it is expected that D from PER measurements will be larger than D from other sorts of measurements, such as FPR, that require large fluorescent or particle labels. This expectation is also raised by theoretical work comparing self-diffusion, the random walks of single labeled molecules measured by FPR, with mutual diffusion, the process involved in the relaxation of a protein concentration gradient and measured by PER. This work predicts that, even for equally labeled proteins, $D_{PER} > D_{FPR}$. This is because protein–protein interactions increase D in the case of mutual diffusion, whereas they tend to reduce D in the case of self-diffusion (Scalettar et al., 1988; Abney et al., 1989a,b; Ryan et al., 1988). Some calculations suggest that D_{PER}/D_{FPR} could be as large as ~50 (Scalettar et al., 1988).

Table 12.1 summarizes D estimated from PER results and gives some indications of the magnitude of these relative to estimates from FPR or other methods. The PER results for ConA receptors are not strictly comparable with other D for ConA, since the latter were not measured on amphibian myocytes. The great difference between D by PER and D by FPR also strikingly illustrates perturbation by the large, multivalent ConA label. The other PER results are compared with FPR results on the same types of cells or membranes. There is no consistent relationship between D_{PER} and D_{FPR}, suggesting that postfield relaxation involves interactions of the diffusing species with other molecules.

12.3.2 RANDOM WALK OF MEMBRANE-BOUND PARTICLES

Recent advances in microscopy and in computer-based image processing have allowed the ultimate measurement of self-diffusion: description of the random walk of a single particle on a cell surface, with a resolution of nanometers. This method is still in its infancy, but it promises to teach us a good deal about cell surface dynamics. The experimental data are images of the cell surface showing particle position with time. Such images contain information about the following:

TABLE 12.1
Translational Diffusion of Membrane Proteins Estimated for Postelectrophoresis Relaxation (PER)

Cells	Protein	DPER*	DPER Relative to DFPR	References
Spindle-shaped *Xenopus* myocytes	Unlabeled concanavalin A "receptors"	4–7	>	Poo et al., 1978
Round *Xenopus* myocytes	Unlabeled concanavalin A "receptors"	34–85 (22°C)	>>	Poo et al., 1979
		~8 (10°C)	>>	Stollberg and Fraser, 1988
		~1 (0°C)		
		25		
NS-20 mammalian cells	Unlabeled concanavalin A "receptors"	1–3 (19°C)	≈	Zagyansky and Jard, 1979
Human fibroblasts	LDL receptor	20 (37°C)		Tank et al., 1985
		11 (22°C)	>>	
		3.5 (8–10°C)	>>	
Rat basophilic leukemia	Fc epsilon receptor	2–4	=	McCloskey et al., 1984
Rat liver mitochondria inner membrane	Intramembrane particles	8	≈	Sowers and Hackenbrock, 1981

* D is in units of 10^{-10} cm^2 sec^{-1}. All D are for receptors labeled after relaxation.

- D
- Anisotropy of D
- Membrane flow
- Internalization of particles
- Molecular clustering
- Membrane compartmentation

This information is available on both nanometer and micrometer scales, and hence resolves some interactions that are averaged into D measured by FPR, a micrometer-scale method, and some that are missed by FCS, a nanometer-scale method.

To date, only two laboratories have reported on this method. One uses an intensely fluorescent derivative of a large ligand, low-density lipoprotein (LDL) (Barak et al., 1981) as a label for specific receptors of human fibroblasts. A preliminary report indicates that D calculated from particle tracking (Gross and Webb, 1986) is ~4 × 10^{-11} cm^2 s^{-1} (Gross and Webb, 1988), about the same as that measured by FPR (Barak and Webb, 1982). Note that this is still much smaller than D estimated from PER experiments on the free receptor (Table 12.1).

Another approach to particle tracking uses nanometer-sized colloidal gold beads, coated with lectin, to mark cell surface proteins (Sheetz et al., 1989). The position of the particles can be determined to within a few nanometers using video-enhanced differential interference contrast microscopy (Gelles et al., 1988). D for concanavalin A–coated beads adhering to mouse macrophage membranes is estimated as ~4 × 10^{-11} cm^2 s^{-1}, at the low end of a range of D for fluorescent concanavalin A estimated by FPR (Table 12.2). This is not surprising because the coated beads are probably multivalent, so their random walk is that of a small aggregate of glycoproteins, rather than that of a single membrane protein. Some preliminary experiments with antibody-coated beads give D closer to that measured by FPR (M.P. Sheetz and M. Edidin, unpublished).

The great potential of both single-particle methods lies in their ability to record combinations of directed particle movement (due to membrane flow or to anchorage to contractile cytoskeleton) and free diffusion, and to resolve and differentiate diffusion within a bounded region, a membrane

TABLE 12.2
Lateral Diffusion Coefficients of Membrane Proteins Determined by FPR and Related Methods

Molecule Labeled	Cell/Membrane	D (10 e-10 cm^2/sec)	Mobile	References
Proteins whose mass is largely within the lipid bilayer				
Rhodopsin	Vertebrate rod outer segments	30–50	100%	Poo and Cone, 1984; Liebman and Entine, 1984; Wey et al., 1981
Acetylcholine receptor	Rat myotubes diffuse receptor	0.5–1	—	Axelrod et al., 1976; Stya and Axelrod, 1984
	Xenopus myocytes diffuse receptor	2.5–5	60–80	Kuromi and Kidokoro, 1984
	Chicken myotubes	6–8 (16 × or 40 × obj)	40	Dubinsky et al., 1989
		1 (100 × obj)		
Na$^+$/K$^+$ -Dog MDCK ATPase	Chicken photoreceptors	5	50	Jesatis and Yguerabide, 1986
		20 (undifferentiated)	65	Madreperla et al., 1989
		20 (differentiated)	40	
Anion channel (band 3)	Human erythrocytes	0.3–0.5	Low	Golan and Veatch, 1980; Sheetz et al., 1980; Change et al., 1981
Voltage-dependent Na$^+$ channels	Rat neuronal cells	10–20 (cell body)	80–90	Angelides et al., 1988
	Frog skeletal muscle	2 (axon hillock)	30–40	Weis et al., 1986
			Immobile	
Voltage-dependent K$^+$ channels	Frog skeletal muscle	0.5	25	Weis et al., 1986
Cytochrome b-c$_1$	Mouse inner mitochondrial membrane	5–7	85–90	Gupte et al., 1984; Chazotte and Hackenbrock, 1988
Proteins whose mass lies largely outside the bilayer				
Concanavalin-A "receptors"	Mouse 3T3/SV3T3	0.5–1	~40	Jacobson et al., 1976
	Rat myoblasts	0.3	~66	Schlessinger et al., 1976a
	Mouse 3T3	0.4	~45	Schlessinger et al., 1977
	Human HELA	2	—	Thomas and Webb, 1987
	Cleaving frog eggs	0.06	—	Ku et al., 1988
	Soybean protoplasts	1.3	90	Metcalf et al., 1986
	Mouse lymphocytes	0.7	80–85	Henis and Gutmann, 1983
		3	40%	

TABLE 12.2 (Continued)
Lateral Diffusion Coefficients of Membrane Proteins Determined by FPR and Related Methods

Molecule Labeled	Cell/Membrane	D (10 e-10 cm²/sec)	Mobile	References
Wheatgerm agglutinin "receptors"	Human granulocytes and HL60 cells	2	30–50	Johansson et al., 1987
	Mouse 3T3 wild-type	0.2 on substrate	50	Swaisgood and Schindler, 1989
		4 in suspension		
	Mutant	0.2 on substrate	50	
		20 in suspension		
	Frog skeletal muscle	0.6	50	Weis et al., 1986
	Nuclear membrane	4	–	Schindler et al., 1985
	Soybean protoplasts	1.5–4	85	Metcalf et al., 1986
Fibronectin receptor	Chick neural crest	2 motile cells	66	Duband et al., 1988
		2 at focal contacts	16	
PolyIg receptor	HT29 (human)	7	66–77	Gustafsson et al., 1988
MHC antigens mouse, H-2 class I				
H-2K^k	L cells	~8	30	Edidin and Wei, 1982;
	L cells	7	47	Wier and Edidin, 1986
	L cells clone 1	3	50	
	L cells clone 2	2	30	Wier and Edidin, 1988
	Spleen lymphocytes	7	73 (immobile on 50% of cells)	Henis and Gutman, 1983
		2–4	30–45	Henis, 1984
	Spleen B-cells	2–4	30–45	Mecheri et al., 1990
	T-lymphoma	5	15–40	Damjanovich et al., 1983
H-2L^d (transfect)	L cells	12	30–60	Edidin and Zuniga, 1984
	L cells clone 1	6–8	37	
	L cells clone 2	13	32	Wier and Edidin, 1988

H-2L^d (endogenous)	B-lymphoblasts	4-6	50-60	Wade et al., 1989
Rat class I (AgB)	Spleen lymphocytes	~5	83	Woda et al., 1970
Human class I (HLA)	Neutrophils	5	—	Petty et al., 1980
	Lymphocytes	7	(Immobile in ~50% of cells)	Bierer et al., 1987; Wier and Edidin, 1986
	B-lymphoblasts	15-20	79	Stolpen et al., 1988a
	Skin fibroblasts	8-10	60 (sparse)	
		10	—	Wier and Edidin, 1986
	Transformed fibroblasts	6	32	Wier and Edidin, 1986
	Endothelial cells + IFN	15-20		
		30		
	Endothelial cells + IFN and TNF	6		Stolpen et al., 1988b
Dog, DLA	Kidney-derived MDCK	10	65 (immobile in ~50% of cells)	Salas et al., 1988
Mouse MHC class II endogenous	B-lymphoblasts	2	60	Barisas et al., 1988; Mecheri et al., 1990
	Normal B-cells	6-9	30-60 (depends on genotype)	
Transfected	B-lymphoblasts	1	73	Wade et al., 1989; Griffith et al., 1988
		40	65	
EGF receptor	A431 cells High affinity	—	Immobile	
	A431 cells Low affinity	3	80	Rees et al., 1984
	A431 all	6	60	Hillman and Schlessinger, 1982
	Transfected 3T3 cells	7	20	Livneh et al., 1986
	Transfected COS cells	1	60-80	
Fc Epsilon receptor (IgE R)	Rat peritoneal mast cells	1-2	50-80	Schlessinger et al. 1976b
	Rat basophilic leukemia cells (2H3)	2-3	70-80	Wolf et al. 1980; Menon et al., 1985
		3	84	

domain, from unbounded diffusion. The shapes of the curves for mean square displacement vs. time differ for these three cases and are readily distinguished in analyses of high-quality video records (Sheetz et al., 1989; Haft and Edidin, 1989).

12.3.3 Fluorescence Correlation Spectroscopy (FCS)

The time-dependent fluctuation of fluorescence intensity within a small spot on a membrane measures the random walk of fluorophores into and out of the region of interest (reviews, Elson and Webb, 1975; Elson and Qian, 1989). FCS has two unique positive features. First, it probes the Å-nanometer scale, rather than a micrometer scale, and second, it observes a system in thermodynamic equilibrium throughout the measurement. No concentration gradient need be created initially. Rather, the fluorescence autocorrelation function gives information about the transport coefficients (and any interactions) of the labeled molecules. Unfortunately, these two advantages are more than offset by the need to collect data over long time periods: tens of minutes or more. This is both because individual fluctuations in fluorescence intensity are so small that they cannot be measured with high precision, and because the time course of any given fluctuation is only stoichiastically determined by diffusion coefficients, reaction rate constants, and so forth. The data are compromised by cell movement or mechanical drift of instruments, which are likely over the times needed for measurement, and hence FCS is limited to measurement of $D > 1 \times 10^{-9}$ cm^2 s^{-1}, a value only occasionally measured for cell membrane proteins. There is one published report using FCS to measure D for a lipid probe in a cell membrane (Elson et al., 1976) and another on the diffusion of lipid probe in artificial bilayers (Fahey et al., 1977).

If a constant translational diffusion is imposed on the system, a scanning FCS technique may be used to study the spatial, rather than the time-dependent, fluctuations of concentration, a measurement that detects molecular aggregates in membranes (Petersen, 1986). Results by this method on viral glycoproteins agreed with measurements by FPR but gave greater insight than FPR into the extent of aggregation of different glycoproteins (Petersen et al., 1986). We conclude this section by quoting a comment on the FCS technique from a review by two talented experimentalists: "FCS is an experimentally difficult method" (Jovin and Vaz, 1988).

12.3.4 Continuous Fluorescence Microphotolysis (CFM)

Fading of fluorescence under observation, a problem in most schemes for measuring translational diffusion in proteins, is here made an essential part of the measurement. A spot is irradiated at constant laser power, and the fading of fluorescence in the spot is followed with time. The fading rate is determined by the photochemical reactions leading to bleaching of fluorophores and by diffusion of unbleached fluorophores. The early, steep part of the fluorescence decay curve is dominated by the reaction rate for bleaching, whereas the later, shallow part is dominated by diffusion. Fluorescence decay curves can be fit to yield D and reaction rate constants for both first- and second-order reactions (Peters et al., 1981; Ferrieres et al., 1989). The method has a higher signal-to-noise ratio than most others used to measure D. The beam intensity for CFM is ~100 times greater than the intensity used for measuring fluorescence recovery in FPR. For best results, $tau_{photolysis} \sim tau_d = a^2/D$. This relationship can be adjusted by varying intensity and/or a^2, the area bleached. The computations required for fitting the bleaching curve are cumbersome, but new approaches may popularize this method (Ferrieres et al., 1989).

12.3.5 Fluorescence Photobleaching and Recovery (FPR/FRAP)

The background to the FPR method was given in the historical section of this review. The method has proved to be a versatile and sensitive technique for measuring translational diffusion on the micrometer scale in a wide range of native and artificial bilayer membranes. Its essence is the use of a focused laser beam to define a region in a fluorescently labeled membrane. The attenuated beam is used to monitor fluorescence in the region before and after a fraction of the fluorophores is bleached by a millisecond pulse of laser light at full intensity. The (asymptotic) extent of recovery of fluorescence after bleaching, $F_{infinity}$, is a measure of the mobile fraction of labeled molecules diffusing in the time of an FPR measurement (tens of seconds to a few minutes). The half-time to reach the maximum recovery is proportional to the area bleached and to the diffusion coefficient of the labeled species. D is calculated from

$$t_{1/2} = \frac{\omega^2}{4D} \gamma$$

where ω^2 is the $1/e^2$ radius of the focused laser spot and γ is a function of the fraction of unbleached molecules. For bleaches between 5 and 95% of total fluorescence, $1 < \gamma < 2$.

The method is useful for diffusion coefficients in the range 10^{-8} cm^2 s^{-1} to $\sim 10^{-12}$ cm^2 s^{-1}. FPR's origins, methodology, and results have been usefully summarized in sections of two reviews (Jovin and Vaz, 1988; Matko et al., 1988).

The FPR instrument couples a laser light source and a photomultiplier detector to a standard microscope. Most FPR machines use 1 W and larger water-cooled argon lasers, but useful FPR work has been done with low-power, air-cooled HeCd lasers (Woda et al., 1979; Zagyansky and Edidin, 1976). A simple beam-splitting device is used to produce coaxial measuring and bleaching beams. An auxiliary lens is positioned to focus the laser in a secondary image plane, so that the objective brings it to a sharp focus at the object plane. An aperture, limiting depth of field of collected light, is placed between the objective and the photomultiplier. This reduces the amount of fluorescence collected from the cell's interior. In fact, the FPR microscope is operated in a confocal mode. The photomultiplier tube (PMT) detector may vary greatly in quality, sensitivity, and noise level. Many FPR microscopes use cooled, gated, end-window tubes. We find that quite good measurements can be made with a simple and rugged (but selected) side-window PMT. PMT output is usually collected at a computer/controller, but in many early experiments data were taken as chart recorder tracings. A user-friendly and thorough discussion of the design and construction of an FPR instrument has been published (Wolf, 1989).

Analysis of FPR recovery curves is complicated by the Gaussian intensity profile of the laser beam, which bleaches a corresponding profile in the fluorescence in the spot of interest. The series solution to the diffusion equation for this boundary condition converges very slowly (Axelrod et al., 1976b). In practice, recovery curves have been fit by various approximations (Axelrod et al., 1976) particularly nonlinear least squares methods (Barisas and Leuther, 1977; Wolf, 1989). Though, in principle, they could be made for more than 1 D per curve, the fits all assume a single recovery. Some linear transformations are good approximations that allow ready dissection of recovery curves into two components (Yguerabide et al., 1982; Van Zoelen et al., 1983).

The shape of the recovery curve in a spot FPR measurement and the calculation of D from the curve depend critically on the membrane area defined by the focused laser beam. Measurement of the laser spot's size is critical to the evaluation of D. One convenient way to do this is to translate a small (submicron-size) fluorescent bead through the laser beam while recording the fluorescence at the pmt. The intensity profile should be Gaussian and the $1/e^2$ radius of the beam is measured directly from the scan. Unfortunately, the precision micrometer stage required, translating in 0.4 μm steps, is expensive. It is cheaper and easier to measure the beam radius reflected off a first-surface mirror into a conventional video camera, using some sort of commonly available image analyzer to quantitate the intensity profile. The image of the fluorescence excited from a layer of fluorescent dye may also be used for quantitation with a more sensitive video camera (for example, Johansson et al., 1987; Dubinsky et al., 1989). The spot size probably cannot be determined to better than 20%, implying about a 40% uncertainty in D. Hence, most measurements of D are better made as comparisons within an experiment, rather than as absolute values. Having said this,

I must also state that we have obtained measurements of D for a given cell type and label, made a year apart, that are within 10% of one another; we have also quantitatively reproduced literature values for D of particular membrane proteins.

12.3.6 Other Forms of the FPR Experiment

The spot FPR experiment provides information from a limited sample of a single cell surface, and the sample area in turn is largely defined by the intensity profile and focus of the laser beam. There are three drawbacks to this arrangement:

- Only a limited sample is taken of the surface of any given cell.
- The geometry of the experiment makes fitting of recovery curves somewhat clumsy.
- By integrating over the entire spot, a few square micrometers, the method loses information about the spatial distribution of fluorescence (note that such information was obtained in the pre-FPR experiments of Poo and Cone, 1974 and Liebman and Entine, 1974).

Several methods have been developed that overcome one or more of the drawbacks mentioned.

12.3.7 Pattern Photobleaching

In pattern photobleaching, a pattern of bright and dark stripes is imposed on the labeled surface, either by illuminating through a mask (Smith and McConnell, 1978) or by creating a pattern of interference fringes on the surface (Davoust et al., 1982). Recovery of fluorescence intensity I, after bleaching the bright regions of the pattern, is due to diffusion from the dark regions and is described by a simpler function than that required for spot bleaching:

$$I_t = \pi A - 2B_c^{-Da^2t} - 2/9\,Bc^{-9Da^2t} \ldots$$

where coefficients A and B are determined by the contrast of the projected pattern and the extent of bleaching a is the spatial frequency of the stripe pattern. The third term of the equation can be neglected by arranging $t > 0.1/Da^2$, so the recovery curve is fit by a single exponential.

The method of making a pattern with a mask can easily be used for large synthetic bilayers; a commercially available Ronchi ruling is used to make the pattern (Smith and McConnell, 1978). Creating a pattern on cells requires "contact printing" through masks with submicron rulings that may be hard to obtain. A greater problem is that due to the finite thickness of the mask, the stripe width is expanded at the cell surface from its width in the mask, so the correct value of a is hard to determine. The method also requires more intense laser light sources than spot photobleaching because the illuminating beam is expanded over many micrometers.

Stripe patterns made by interference fringes overcome the drawbacks of mask-determined patterns. The pattern is oscillated on the cell surface by mechanical vibration of a mirror. Therefore, the recovering fluorescence signal, seen as a decay of contrast, can be picked out of the noise using a lock-in amplifier set to the oscillation frequency.

The stripe patterns produced by either the mask or the fringe method also contain information about the spatial distribution of fluorophores, but this is neglected for the sake of a simple analysis of fluorescence recovery. FPR with significant spatial information can indicate whether recoveries are isotropic and whether the recoveries are in fact due to flow as well as Brownian motion. Isotropy of recovery cannot be estimated from spot photobleaching. Though, in principle, the shape of the recovery curve indicates flow, in practice a significant flow component cannot be separated from the diffusional recovery in spot FPR data.

12.3.8 SCANNING SPOT PHOTOBLEACHING

Both the problems raised in the preceding paragraph are addressed by a method of multipoint scanning, in which the cell surface is interrogated with a spot or a line at points to either side of the bleached region (Koppel, 1979; Koppel et al., 1980; Koppel, 1985). Systematic drift produces a phase shift in the pattern that is readily detectable. The size of the measuring and bleaching spot is defined by the scans, and so spot size does not have to be determined separately as in FPR. The multipoint scanning method has been used in several biological systems with notable results. The method is also the one used in the only commercially available FPR microscopy (ACAS, Meridien Instruments).

12.3.9 INTENSIFIED VIDEO IMAGING FPR

Some spatial information is lost even in multipoint scanning, because the beam only samples a small fixed number of points on the surface. A much higher spatial resolution can be obtained from video images of the labeled surface (Kapitza et al., 1985).

The richness of the images raises a problem in sampling cells of a population. Spot FPR allows measurements on hundreds of cells a day. Small changes in D and R, correlated with changes in cell physiology, can easily be detected in such large data sets. Video-imaging FPR will show up anisotropic recovery and other features of the surface of a single cell, but it has limited application in determining D and R when these vary over a wide range. A combination of conventional video and FPR may be useful in such cases (Duband et al., 1988).

12.4 FLUORESCENT LABELS FOR FPR

The range of practical fluorescent labels for FPR is enormous. It extends from small covalent labels (for example, fluorescein directly coupled to cell surfaces [Edidin et al., 1976] through antibodies and antibody fragments) to large markers, such as the fluorescent LDL cited earlier in this chapter (Barak and Webb, 1981, 1982). There are trade-offs to be made in preparing such labels for proteins, among parameters of small size, high signal, affinity, and specificity.

Direct coupling of fluorescein isothiocyanate and the relaxed (4,6-dichlorotriazinyl) amino fluorescein (DTAF) have been especially useful for work on erythrocytes where these derivatives, as well as other fluorescein derivatives (for example, erythrosin isothiocyanate), selectively label band 3 (Peters et al., 1974; Fowler and Branton, 1977; Golan and Veatch, 1980; Schindler et al., 1980). A sulfhydryl-reactive iodoacetamido derivative of tetramethylrhodamine labeled rhodopsin in disc membranes of vertebrate rod outer segments (Wey et al., 1981). The more polar iodoacetamido fluorescein labeled the disc membranes, but the label was not coupled to rhodopsin (Philips and Cone, 1986; R.A. Cone, personal communication). This indicates that not only the chemistry of covalent coupling, but also the charge and hydrophobicity of the dye used, will determine the proteins labeled. Of course, extremely hydrophobic dyes (for example, pyrene isothiocyanate) will permeate the surface and label proteins in the cell interior, as well as those on the surface.

Lectins and antibodies are readily available noncovalent labels for cell surfaces. The broad specificity of lectins makes them particularly tempting labels, but this temptation should, in most cases, be resisted. Lectins are difficult labels for FPR on two counts. First, most are multivalent and cannot readily be cleaved to monovalent forms. This means that they crosslink membrane proteins as they label, and aggregates of such crosslinked proteins are certainly liable to diffuse differently than the monomeric proteins. A second count against lectins is their broad specificity. FPR results using these labels are averages of the diffusion of a collection of membrane glycoproteins and glycolipids. Although D_{lat} is relatively insensitive to the size of the diffusing species, it is not reasonable to calculate D for an ensemble ranging over three orders of magnitude in MW.

Antibodies and antibody fragments, Fab, offer much greater scope for controlled and specific labeling than lectins. Our preferred label for most experiments is the monovalent Fab fragment of either a monoclonal antibody or a high-titered polyclonal antibody (IgG) mixture. Either Fab fragment may be labeled with three or four fluoresceins without affecting its specificity. If intact monoclonal IgG can be shown to bind monovalently, this can be used instead of Fab (Dower et al., 1984). Many mAb are IgM. These are too large and multivalent to be useful labels. Unfortunately, there is no consistent procedure for cleaving IgM to monovalent fragments, and, even if produced, the affinity of these fragments may be too low to be useful for cell labeling.

Fluorescent ligands for specific cell surface receptors have been prepared. Fluorescent epidermal growth factor (EGF), fluorescent toxins (labeling ion channels), and the light chain of class I MHC antigens (beta-2-microglobulin) have all been used for FPR studies of their receptors (Schlessinger et al., 1978; Angelides, 1989; Angelides et al., 1988; Salas et al., 1988). Fluorescent cytochrome c derivatives have also been used to investigate the dynamics of mitochondrial inner membranes by FPR (Hochman et al., 1982). Other ligands used for studies of receptor dynamics, for example, fluorescent insulin, show little specificity of binding (Shechter et al., 1978).

All labels should be regarded with suspicion until they have been thoroughly checked for specificity and affinity of binding. Binding of fluorescent derivatives should be blocked to a large extent by the unlabeled molecules from which they are made. Fluorescent agonists, such as EGF, should also be checked for biological potency. After showing that a fluorescent derivative binds specifically, its apparent affinity of binding should be considered. If this is sufficiently low, recovery of fluorescence after bleaching may be due to rebinding of fluorescent labels that have previously dissociated and diffused through the medium in the vicinity of the membrane before rebinding. A practical check for this possibility is to measure the average fluorescence intensity per cell before and after the FPR experiment. This will decline significantly with time if the label is dissociating from the cell surface. A recent theoretical and experimental treatment of this problem is a helpful guide (Goldstein et al., 1989).

Extended discussions of these points have been published elsewhere. (Angelides, 1989; Edidin, 1989; Maxfield, 1989).

12.4.1 Problems and Pitfalls in FPR Measurements

In an FPR experiment, high-intensity light is focused into a small area that has been sensitized (by dye labeling) to capture much of this light. Power densities during the bleaching phase of the experiment (tens of msec) may reach 1 MW/cm^2. This raises the possibility of local heating, and perhaps the more serious possibility of photoinduced crosslinking or other damage to the labeled proteins. Local heating is a problem in pigmented cells. Intact erythrocytes cannot be probed with the usual labels, such as fluorescein, because the cell hemoglobin absorbs so much energy that the cells burst (Bloom and Webb, 1984; Edidin, unpublished observation). However, if no pigment is present in or beneath the membrane, it can be shown that local heating due to the laser flash is well under 1°C (Axelrod, 1976). This is largely because of the high thermal diffusivity of water.

Photolysis of the label during bleaching could produce excited-state species of molecules with high probability of reaction with other excited-state molecules or with other proteins in the spot. Such crosslinked species would likely diffuse more slowly than the native proteins, so that FPR would report erroneously low D. We will discuss this issue at greater length in a later section. Here, we can only raise the question whether or not crosslinking occurs in an FPR experiment. Formerly, the analysis of the proteins in a single bleached spot — or even in 1000 such spots — was out of reach. Therefore, all of the experiments bearing on crosslinking were of two sorts. One sort of experiment, which showed that irradiation of labeled proteins produced crosslinked products, used suspensions of labeled cells or membranes irradiated for long times (seconds) at power densities much lower than obtained in an FPR experiment. The most careful of these experiments concluded that, for a constant total light flux, production of crosslinked proteins was inversely proportional to the intensity of irradiation (Sheetz and Koppel, 1979). Extrapolating the data of this experiment suggests that no significant crosslinking occurs under the conditions of an actual FPR experiment.

The second sort of experiment to control for photobleaching damage is one in which conditions of an FPR experiment are varied in ways that ought to enhance or reduce photodamage. Variation in D and R for different conditions of the experiment would be strong evidence for photodamage. In one such set of experiments, multiple bleaches of the same spot did not affect the measured D; R, the mobile fraction, systematically increased with successive bleaches, as expected if incomplete recovery is due to immobilization of a fraction of the labeled molecules (Jacobson et al., 1978). Other experiments in the same series found no effect of quenchers or traps for singlet oxygen or free radicals (suspected mediators of photodamage) on measured D and R. In another experiment, a surface protein, the high-affinity receptor for IgE, was labeled with a mixture of rhodamine and fluorescein on each receptor (Wolf et al., 1980). This was done by labeling the receptor with either R-IgE or Fl-IgE, and then labeling with either Fl-Fab anti-IgE or R-Fab anti-IgE. FPR measurements were then made on cells in which one label had been bleached by exposure to light from a mercury arc lamp or from an expanded laser beam before the FPR measurements. For example, all the fluorescein label on a cell was bleached to 50% of its initial intensity, and this was followed by an FPR measurement using laser excitation of the rhodamine label. It would be expected that any crosslinking or other photodamage from the first bleach of the fluorescein label would be reflected in the FPR measurement using the second label, rhodamine in our example. This was not seen, even though the first bleach required between 10 s (if the expanded laser beam was used) and 5 min (if the arc lamp was used). No evidence for photodamage or other artifacts was found, even in conditions that approached those used in the experiments on photodamage to cell suspensions.

The only experiments in which there is evidence for photodamage or photoinduced crosslinking are those on FPR of labeled microtubules or microfilaments in which high concentrations of labeled proteins are irradiated in solution for times ranging from milliseconds to minutes (Leslie et al., 1984; Simon et al., 1988). Here, irradiation results in crosslinking (tubulin) or breakage (actin). However, acceptable measurement of D can even be made in these systems if conditions are carefully controlled (see the treatment in McIntosh and Koonce, 1989, and in Simon et al., 1988).

Other, indirect evidence that D and R are not artifacts of the FPR experiment will be discussed when we compare data obtained from FPR with those obtained using other methods for measuring translational mobility. The issue is important not only for understanding the FPR experiment, but also because the data of FPR have, on occasion, been vigorously attacked by articulate but mistaken proponents of theories of cell locomotion and membrane flow, because the theories (Bretscher, 1976) will not accommodate diffusion coefficients as small as 10^{-10} cm^2 s^{-1}, values often measured for membrane proteins.

12.5 TRANSLATIONAL DIFFUSION OF MEMBRANE PROTEINS MEASURED BY FPR

Over 70 diffusion coefficients for different membrane proteins are tabulated in Table 12.2 and Table 12.3; these are only a sample of all values in the literature. There are data for three different types of membrane proteins:

- Proteins with a large part of their bulk lying in the bilayer, for example, rhodopsin and the acetylcholine receptor
- Proteins with large endo- and ectodomains, linked to the lipid bilayer by a single transmembrane sequence of 20 to 23 mainly hydrophobic amino acids
- Proteins linked to the bilayer through glycosylphosphatidyl inositol (GPI) links

For each group of proteins, we will consider first the range of diffusion coefficients measured and their reproducibility among laboratories. Next, we will comment on factors, such as cytosk-

TABLE 12.3
Lateral Diffusion of GPI-Linked Proteins

Molecule Labeled	Cell/Membrane	D (10 e-10 cm²/sec)	Mobile	References
Tryopanosomal		1	80 (Fab label)	Bulow et al., 1988
VSG in trypanosomes		0.2	33 (IgG label)	
Trypanosomal VSG in BHK (added from detergent soln)		0.7	56	
Thy-1	Mouse lymphocytes and fibroblasts	20–40	40–60	Ishihara et al., 1987
VSVG-Thy-1	COS cells	4		Zhang et al., 1989
PH-20	Guinea pig cauda sperm	2 (before acrosome reaction) 50 (after acrosome reaction)	73	Cowan et al., 1987
PH-20	Guinea pig testicular sperm	0.003–3		Phelps et al., 1988
Alkaline phosphate	Rat osteosarcomas			Noda et al., 1987
	RO 17/2.8	6	~70	
	UMR106	20	~80	
	Rat transfected fibroblasts	10	~90	
Decay accelerating factor (DAF)	HELA cells	16	<<100	Thomas and Webb, 1987

eleton or protein–protein interactions, that affect the observed mobilities. Finally, we will consider the correlations between changes in cell biology and changes in translational mobility of membrane proteins.

12.5.1 Translational Diffusion of Membrane Proteins That Are Largely Embedded in the Membrane Bilayer

We commented early in this review that the lateral diffusion of rhodopsin measured by a microspectrophotometric technique was about that calculated for a sphere freely diffusing in three dimensions in an oil of ~1 P, $D = 3$ to 5×10^{-9} cm^2 s^{-1}, 100% mobile. These values were also obtained when rhodopsin was labeled on –SH groups with tetramethylrhodamine and its diffusion measured by spot FPR.

D of Na$^+$/K$^+$-ATPase in chicken photoreceptors approaches that of rhodopsin, though the highest mobile fraction measured is significantly less than 100%. In contrast, D for the same type of protein in MDCK cells is about one-tenth that of rhodopsin, and again, only 50% of labeled molecules are mobile. D for the voltage-dependent Na$^+$ channel, labeled by a fluorescent toxin, also approaches that of the freely diffusing rhodopsin and ~100% of molecules are mobile, but this is only when diffusion of the channel is measured in the cell body of a cultured neuron. D is an order of magnitude smaller if channel mobility is measured at the axon hillock and R is also reduced.

Four other surface membrane channels are significantly less mobile under physiological conditions than the three discussed so far. Only 25% of K$^+$ channels and none of the Na$^+$ channels are mobile in frog skeletal muscle when these are probed by a variant of photobleaching in which native channels are destroyed by local UV irradiation (Roberts et al., 1986). Mobility of erythrocyte ghost band 3, the anion channel, is highly restricted; D is 100-fold smaller than that of rhodopsin and R may be as low as 10%. The widest range of mobilities of any of the proteins in this group is measured for acetylcholine receptors (AChR). All measurements were made on diffusely distributed receptors in cultured cells. D ranges from 1 to ~10 × 10^{-10} cm^2 s^{-1} and depends on the size of the area measured.

Our list also includes the cytochrome b-c$_1$ complex of the mitochondrial inner membrane. Almost all the complex is mobile, but D is about one-tenth that of rhodopsin.

With a few exceptions, the mobilities of this entire group of proteins are smaller than those predicted for free diffusion in a lipid bilayer. The constraints to translational diffusion of these molecules varies from cell type to cell type. Diffusion of the Na$^+$/K$^+$ ATPase appears to be restricted by association with elements of a spectrin-rich cytoskeleton (Rodriguez-Boulan and Nelson, 1989). A spectrin cytoskeleton is certainly important in the constraint of band 3, because the protein's diffusion increases by orders of magnitude in spectrin-deficient erythrocytes (Sheetz et al., 1980) or in ghosts that have been stripped of cytoskeleton (Golan and Veatch, 1980). Mobility of the Na$^+$ channel is also constrained by interaction with the spectrin cytoskeleton (Srinivasan et al., 1988; Wood and Angelides, 1988).

The theme of spectrin-dominated constraints to lateral diffusion is broken by observations on the cytochrome b-c$_1$ complex and by AChR. Mobility of both of these proteins is greatly affected by protein–protein interactions. AChR aggregate when they are concentrated, for example, by electrophoresis *in situ,* and this may be a major factor in maintaining patches of receptor. D of the b-c$_1$ complex increases 20-fold to D > 10^{-8} cm^2 s^{-1}, when the proteins of inner mitochondrial membrane are diluted by lipid (Chazotte and Hackenbrock, 1988) indicating that crowding, if not specific protein–protein interactions, largely determines D and R measured for native membranes.

AChR, the Na$^+$/K$^+$-ATPase and the voltage-dependent Na$^+$ channel are all immobilized when they are localized to particular regions of a cell surface. Diffuse AChR is mobile but immobilizes when patches form. Sixty-five percent of the Na$^+$/K$^+$-ATPase is mobile in the round chicken photoreceptor cells, but the mobile fraction is reduced by 50% in morphologically polarized cells. Voltage-dependent Na$^+$ channels are mobile when they are diffusely distributed but are largely immobilized when concentrated at the junction of cell body and axonal process. Though immobilization is not complete in any of these cases, it does appear that significant changes in mobility are associated with functional differentiation of cell surfaces. It has been shown that a voltage-dependent Ca^{++} channel also is immobilized as cells mature functionally and morphologically (Jones et al., 1989).

12.5.2 TRANSLATIONAL DIFFUSION OF PROTEINS WITH LARGE ENDO- AND ECTODOMAINS

Most FPR measurements have been made on proteins whose bulk lies outside the membrane bilayer. Some proteins of this group, for example, EGF receptors, have nearly equal extracellular (exo-) and cytoplasmic (endo-) domains. The bulk of others, for example, class I and class II MHC antigens, is mainly external to the bilayer. There is an additional group of lectin (concanavalin A and wheat germ agglutinin) "receptors" that is heterogeneous and may even include glycolipids.

The results on lectin "receptors" are included here mainly to show that D for these collections of molecules varies only about tenfold over a range of laboratories and experiments. There are only two measurements that lie outside this range: one in cleaving frog eggs where the label may be largely in extracellular coats, and another in mutant 3T3 cells, selected for a high D for WGA "receptors." The mobility measured for lectin labels, particularly of concanavalin A and its putatively monovalent derivative, sCon A, depends critically on the density of surface glycoproteins and glycolipids and on the concentration of ConA or sCon A used for labeling. In general, high concentrations (> 20 μg/ml) of lectin immobilize the "receptors." This may be due to crosslinking or to a general "anchorage modulation" of the translational diffusion of all membrane proteins (Edelman, 1976). This second possibility is raised in experiments in which binding conA platelets to a small fraction of cell surface immobilizes all molecules on that cell surface (Schlessinger et al., 1977).

Class I MHC antigens are expressed on the surfaces of almost all adult cells, with the exceptions of cells in the nervous system and cells of some tumors. The structure of one of these polymorphic molecules, HLA A2, has been determined to atomic resolution. The importance of MHC antigens in immune responses has resulted in the creation and characterization of many mAb to them, and of many mutant antigen molecules. This rich material has been the subject of numerous studies of lateral diffusion. The data shown in Table 12.2 summarize results from different laboratories, using different cells, that vary over an order of magnitude, but not much more. These data, presented without error estimates, mask the observation made in many of the papers that the variation in D and R within a cell population measured in a single laboratory may be as great as the variation in D and R between cell types and between laboratories. This is vexing if we consider the accurate determination of D and R as the main aim of FPR measurements. However, from a biologist's point of view, these wide ranges imply some interesting cell-to-cell biological variation that is reflected in D and R.

There are multiple constraints to diffusion in class I MHC antigens and other proteins, such as EGFR. Diffusion of mutants of either of these molecules, lacking almost all of the cytoplasmic domain, is no different from that of wild-type molecules (Edidin and Zuniga, 1984; Livneh et al., 1986; Scullion et al., 1987) and D is much less than that for rhodopsin. D for mutants lacking glycosylation sites is $\sim 2 \times 10^{-9}$ cm^2 s^{-1}, almost at the limit set by lipid viscosity, compared to 5 to 7×10^{-10} cm^2 s^{-1} of wild-type class I antigen (Wier and Edidin, 1988). This, like the results on tailless mutants, implies that diffusion is largely constrained by interactions between exodomains of class I antigens and other proteins on the cell surface. However, other data on mobility of human class I antigens in isolated membranes (Su et al., 1984) suggest that the cytoskeleton is an important constraint to translational diffusion of class I antigens. This suggestion is reinforced by observations that class I antigens are immobile in about half of all mouse and human lymphocytes (Petty et al., 1980; Henis and Gutman, 1983) and that their mobility changes when cells contact extracellular matrix (Wier and Edidin, 1986). Other evidence for cytoskeletal constraints to the diffusion of class I antigens is the observation that these molecules are largely immobilized when MDCK cells polarize and differentiate a spectrin cytoskeleton (Salas et al., 1988) and the observation that tumor necrosis factor (TNF) modulates D of HLA antigens on endothelial cells (Stolpen et al., 1988b).

These results do not need to be reconciled so much as synthesized. It is likely that both cytoplasmic elements and protein–protein interactions in the plane of the bilayer constrain lateral diffusion of membrane proteins of this type. In this regard, it is interesting to note that the basal rate of endocytosis of class I antigens in L-cells, the cells used to express the class I MHC antigen tailless mutants and the underglycosylated mutants mentioned earlier, is very much smaller than the rate of endocytosis of the same antigens in T-lymphoblasts (Capps et al., 1989). Thus, there is reduced functional coupling between surface and cytoplasm. This reduced functional coupling is reflected in the amount of class I MHC antigens that remains associated with cytoskeleton after detergent extraction. We predict that truncation of cytoplasmic tail will affect D of class I antigens if these are expressed in T-lymphoblasts.

The prediction just made is supported by observations on class II MHC antigens, homologus molecules, truncated in the cytoplasmic domain and expressed in cells in which the class II antigens function as signaling elements. In this instance, D increases and signaling efficiency falls as the cytoplasmic domain is truncated. D for the wild-type class II antigen is $\sim 1 \times 10^{-10}$ cm^2 s^{-1}, and that for tailless class I molecules is an order of magnitude higher but still about one-fifth of that for a molecule of this size in a pure lipid bilayer (Wade et al., 1989). It appears that, although cytoplasmic constraints dominate lateral mobility of wild-type antigens, lateral mobility of the tailless mutants is constrained by protein–protein interactions in the plane of the surface membrane.

The last few paragraphs have suggested something about the biological variation in translational diffusion of class I MHC antigens and similar transmembrane molecules. We also note the observation that the high-affinity EGF receptors are immobile (Rees et al., 1984) and that IgE receptors are immobilized soon after crosslinking by IgE and antigen (Menon et al., 1985).

12.5.3 LATERAL DIFFUSION OF GPI-LINKED PROTEINS

The range of D and R in this group of "lipid-linked" proteins is greater than that in any of the others that we have discussed. The first two measurements of GPI-linked proteins were made on a small antigen common to mouse T-cells and fibroblasts, Thy-1, and on decay-accelerating factor (DAF) (Ishihara et al., 1987; Thomas and Webb, 1987). D of each of these proteins was similar to that of a lipid probe measured in the same cells, 2 to 4×10^{-9} cm^2 s^{-1}. Surprisingly, only about half of labeled molecules were free to diffuse at all, implying that they were anchored in some way. D of a third GPI-linked protein, alkaline phosphatase, is somewhat smaller than for Thy-1 and DAF, and varies with cell type. Mobile fractions of alkaline phosphatase are higher than those for Thy-1 and DAF but are still less than 100%. Together, these results imply that D for conventional membrane glycoproteins is largely constrained by associations with the cytoplasm, because GPI-linked proteins do not have cytoplasmic domains. On the other hand, the low mobile fraction of these proteins suggest that somehow, perhaps through lateral associations with anchored glycoproteins, a fraction of the GPI-linked molecules are themselves anchored.

Results on two other proteins, the variable surface glycoprotein of trypanosomes and a sperm antigen, PH-20, suggest that even in GPI-linked proteins, D and R are constrained by interactions of the extracellular domains of proteins. The most striking result is that for PH-20. The translational mobility of this GPI-linked protein varies over four orders of magnitude, depending upon the stage of maturity of the sperm. D is ~10^{-13} cm^2 s^{-1} in one stage of testicular sperm at any early stage of maturation, and reaches 5×10^{-9} cm^2 s^{-1} in mature sperm that have undergone the acrosome reaction. These results, like the results with VSG, imply that mobility of GPI-linked proteins can be constrained by interactions of protein exodomains with other molecules in the plane of the membrane. Such an interpretation is reinforced by the data, quoted earlier, on diffusion of mutant class I MHC antigens and by a finding that the translational mobility of hybrid proteins expressing the exodomain of VSV G protein (D ~3×10^{-10} cm^2 s^{-1}) and the GPI-membrane anchor of Thy-1 (D ~3×10^{-9} cm^2 s^{-1}) is like that of the complete VSV G protein, and not like that of Thy-1. Exodomain interactions, not lipid viscosity, dominate diffusion in this case (Zhang et al., 1989).

12.6 TRANSLATIONAL MOBILITY OF MEMBRANE PROTEINS BY FPR COMPARED TO THAT DETERMINED BY OTHER METHODS

We have anticipated this discussion in the sections on PER and in our discussion of possible artifacts of FPR. Here, we reiterate the point that discrepancies between D, determined by different methods, particularly PER vs. FPR, seem mainly to turn on the comparison of diffusion of receptors bound with ligand bound with diffusion of free receptors, and the comparison of self-diffusion, in which marked particles are followed in an otherwise unperturbed surface, and mutual diffusion in which all proteins are nonuniformly distributed in a surface. Considering this, there is fairly good agreement between PER AND FPR results. There is certainly good agreement between measurements of random walks of single particles and FPR, that is, in a comparison of self-diffusion of particle/ligand complexes.

The single greatest discrepancy between FPR results and other measurements of translational diffusion is for D of diffusely distributed AChR measured by FPR vs. D measured by a chemical analog of the method in which a patch of receptors is inactivated by toxin, and then the patch is monitored for return of free AChR, detected as recovery of an electrical response to ACh (Poo, 1982; Young and Poo, 1983). By this method, D is ~2.5×10^{-9} cm^2 s^{-1}, whereas the FPR measurements on rat myocytes, estimated D at least an order of magnitude smaller (Axelrod et al., 1976;

Stya and Axelrod, 1984). Dubinsky and coworkers (1989) have measured D of AChR in chicken myotubes. They found that D is a function of the size of the area measured with larger areas estimating larger D (see Table 12.2). They estimate D ~6 to 8×10^{-10} cm^2 s^{-1}, still about three times smaller than the estimates by Poo, but much more nearly in line with his measurements, which themselves are based on some assumptions about the relationship between ACh sensitivity and AChR concentration.

The resolution of this single numerical discrepancy between D estimated by different methods removes the last rational basis for claiming that FPR gives fundamentally flawed results (Bretscher, 1980, 1984). The persistence of this claim in the face of the evidence has been due only to the influence of some theories of membrane lipid flow that cannot accommodate D < 10^{-9} cm^2 s^{-1} for membrane proteins. The FPR results, and data on random walks of particles, should have laid some theories to rest, though it seems that the ghosts will be with us for a while longer (Bretscher, 1989).

12.7 CONCLUDING REMARKS

The measurement of translational diffusion of membrane proteins has progressed from the crude demonstration of translational mobility, through reliable quantitation of this mobility, to the use of such measurements to investigate issues in cell biology. However, the field is still young and underdeveloped, and we are left with unanswered questions about all levels of the problem of translational diffusion.

Theories on translational diffusion, particularly on the interaction between molecules in the plane of the membrane, have begun to evolve (Saxton, 1982, 1987; Eisinger et al., 1986; Abney et al., 1989a,b). However, much more work is needed here, particularly comparing theory with results on model proteins in reconstituted bilayers that are less complex than a native cell surface or an organelle membrane.

The coupling between cell surface and cytoplasm detected by translational diffusion measurements is poorly understood, and we need to know much more about the importance of particular molecules of the cytoskeleton in anchoring membrane proteins in impeding their diffusion. The role, if any, of translational mobility in signaling, for example, through G-proteins, is still not understood (see Peters, 1988). Indeed, for most cells it is not clear whether translational diffusion on a small, nanometer scale involves the same processes and interactions as translational diffusion on the micrometer scale that is measured by FPR, PER, and the other techniques discussed here (see Yechiel and Edidin, 1987, for one aspect of this problem).

Recent results have emphasized striking changes in translational mobility of membrane proteins with changes in cell biology. Those results, discussed previously, as well as others on developmental changes in mobility (for example, Pollerberg et al., 1986) hold great promise for our understanding of how cell surface and endomembranes are organized functionally and how the organization of membranes changes with changing function.

REFERENCES

Abney JR, Scalettar BA, and Owicki JC: Self diffusion of interacting membrane proteins. *Biophys. J.* 1989(a), 55:817–833.

Abney JR, Scalettar BA, and Owicki JC: Mutual diffusion of interacting membrane proteins. *Biophys. J.* 1989(b), 56:315–326.

Angelides K: Fluorescent analogs of toxins. *Methods Cell Biol.* 1989, 29:29–58.

Angelides KJ, Elmer LW, Loftus D, and Elson E: Distribution and lateral mobility of voltage-dependent sodium channels in neurons. *J. Cell Biol.* 1988, 106:1911–1925.

Axelrod D: Cell surface heating during photobleaching recovery experiments. *Biophys. J.* 1976, 18:129–131.

Axelrod D, Ravdin P, Koppel DE, Schlessinger J, Elson EL, and Podleski T: Lateral motion of fluorescently labeled acetylcholine receptors in membranes of developing muscle fibers. *Proc. Natl. Acad. Sci. U.S.A.* 1976(a), 73:4594–4598.

Axelrod D, Koppel DE, Schlessinger J, Elson E, and Webb WW: Mobility measurements by analysis of fluorescence photobleaching recovery experiments. *Biophys. J.* 1976(b), 16:1055–1069.

Barak LS and Webb WW: Fluorescent low density lipoprotein for observation of dynamics of individual receptor complexes on cultured human fibroblast. *J. Cell Biol.* 1981, 90:595–604.

Barak LS and Webb WW: Diffusion of low density lipoprotein-receptor complex on human fibroblasts. *J. Cell Biol.* 1982, 95:846–852.

Barisas BG and Leuther MD: Fluorescence photobleaching recovery measurement of protein absolute diffusion constants. *Biophys. Chem.* 1979, 10:221–229.

Barisas BG, Roess DA, Grey HM, and Jovin TM: Rotational and lateral dynamics of lymphocyte surface proteins involved in antigen presentation. *J. Cell Biol.* 1988, 107:68a.

Bierer B, Herrmann SH, Brown CS, Burakoff SJ, and Golan D: Lateral mobility of class I histocompatibility antigens in B-lymphoblastoid cell membranes: modulation by cross-linking and effect of cell density. *J. Cell Biol.* 1987, 105:1147–1152.

Bloom JA and Webb WW: Photodamage to intact erythrocyte membranes at high laser intensities: methods of assay and suppression. *J. Histochem. Cytochem.* 1984, 32:608–616.

Bretscher M: Directed lipid flow in cell membranes. *Nature* 1976, 260:21–23.

Bretscher M: Lateral diffusion in eukaryotic cell membranes. *Trends Biochem. Sci.* 1980, 5:R6–R7.

Bretscher M: Endocytosis: relation to capping and cell locomotion. *Science* 1984, 224:681–686.

Bretscher M: Particle migration on cells. *Nature* 1989, 341:491–492.

Bulow R, Overath P, and Davoust J: Rapid lateral diffusion of the variant surface glycoprotein in the coat of *Trypanosoma brucei*. *Biochemistry* 1988, 27:2384–2388.

Capps GG, van Kampen M, Ward CL, and Zuniga MC: Endocytosis of the class I major histocompatibility antigen via a phorobol myristate acetate-inducible pathway is a cell-specific phenomenon and requires the cytoplasmic domain. *J. Cell Biol.* 1989, 108:1317–1329.

Chang C-H, Takeuchi H, Ito T, Machida K, and Ohnishi S-I: Lateral mobility of erythrocyte membrane proteins studies by the fluorescence photobleaching recovery technique. *J. Biochem.* 1981, 90:997–1004.

Chazotte B and Hackenbrock CR: The multicollisional, obstructed, long-range diffusional nature of mitochondrial electron transport. *J. Biol. Chem.* 1988, 263:14359–14367.

Cone RA: Rotational diffusion of rhodopsin in the visual receptor membrane. *Nature* 1972, 236:39–43.

Cowan AE, Myles DG, and Koppel DE: Lateral diffusion of the PH-20 protein on guinea pig sperm: evidence that barriers to diffusion maintain plasma membrane domains in mammalian sperm. *J. Cell Biol.* 1987, 104:917–923.

Damjanovich S, Tron L, Szollosi J, Zidovetski R, Vaz WLC, and Jovin TM: Distribution and mobility of murine histocompatibility H-2K^k antigen in the cytoplasmic membrane. *Proc. Natl. Acad. Sci. U.S.A.* 1983, 80:5985–5989.

Davoust J, Devaux PF, and Leger L: Fringe pattern photobleaching, a new method for the measurement of transport coefficients of biological molecules. *EMBO J.* 1982, 1:1233–1238.

Dower SK, Ozato K, and Segal DM: The interaction of monoclonal antibodies with MHC class I antigens on mouse spleen cells. I. Analysis of the mechanism of binding. *J. Immunol.* 1984, 132:751–758.

Duband JL, Nuckolls GH, Ishihara A, Hasegawa T, Yamada KM, Thiery JP, and Jacobson K: Fibronectin receptor exhibits high lateral mobility in embryonic muscle cells but is immobile in focal contacts and fibrillar streaks in stationary cells. *J. Cell Biol.* 1988, 107:1385–1396.

Dubinsky JM, Loftus DJ, Fischbach GD, and Elson EL: Formation of acetylcholine receptor clusters in chick myotubes: migration of new insertion? *J. Cell Biol.* 1989, 109:1733–1743.

Edelman GM: Surface modulation in cell recognition and cell growth. *Science* 1976, 192:218–226.

Edidin M: Rotational and lateral diffusion of membrane proteins and lipids: phenomena and function. *Curr. Topics Membr. Trans.* 1987, 29:91–127.

Edidin M: Fluorescent labeling of cell surfaces. *Methods Cell Biol.* 1989, 29:87–102.

Edidin M, Zagyansky Y, and Lardner TJ: Measurement of membrane protein lateral diffusion in single cells. *Science* 1976, 191:466–468.

Edidin M and Wei T: Lateral diffusion of H-2 antigens on mouse fibroblasts. *J. Cell Biol.* 1982, 95:458–462.

Edidin M and Zuniga M: Lateral diffusion of wild-type and mutant L^d antigens in L cells. *J. Cell Biol.* 1984, 99:2333–2335.

Eisinger J, Flores J, and Petersen WP: A milling crowd model of local and long-range obstructed diffusion. *Biophys. J.* 1986, 49:987–1001.

Elson EL and Webb WW: Concentration correlation spectroscopy: a new biophysical probe based on occupation number fluctuations. *Annu. Rev. Biophys. Bioeng.* 1975, 4:311–334.

Elson EL and Qian H: Interpretation of fluorescence correlation spectroscopy and photobleaching recovery in terms of molecular interactions. *Methods Cell Biol.* 1989, 30:307–332.

Elson EL, Schlessinger J, Koppel DE, Axelrod D, and Webb WW: Measurement of lateral transport on cell surfaces, in *Membranes and Neoplasia: New Approaches and Strategies.* Alan Liss, New York, 1976, pp. 137–147.

Fahey PF, Koppel DE, Barak LS, Wolf DE, Elson EL, and Webb WW: Lateral diffusion in planar lipid bilayers. *Science* 1977, 195:305–306.

Ferrieres X, Lopez A, Altibelli A, Dupou-Cezanne L, Lagounalle J-L, and Toconne J-F: Continuous fluorescence microphotolysis of anthracene-labeled phospholipids in membranes. Theoretical approach of the simultaneous determination of their photodimerization and lateral diffusion rates. *Biophys. J.* 1989, 55:1081–1091.

Fowler V and Branton D: Lateral mobility of human erythrocyte integral membrane proteins. *Nature* 1977, 268:23–26.

Frye LD and Edidin M: The rapid intermixing of cell surface antigens after formation of mouse-human heterokaryons. *J. Cell Sci.* 1970, 7:319–335.

Gelles J, Schnapp BJ, and Sheetz MP: Tracking kinesin-driven movements with nanometer-scale precision. *Nature* 1988, 331:450–453.

Golan DE and Veatch W: Lateral mobility of band 3 in the human erythrocyte membrane studies by fluorescence photobleaching recovery: evidence for control by cytoskeleton interactions. *Proc. Natl. Acad. Sci. U.S.A.* 1980, 77:2537–2541.

Goldstein B, Posner RG, Torney DC, Erickson J, Holowka D, and Baird B: Competition between solution and cell surface receptors for ligand. Dissociation of hapten bound to surface antibody in the presence of solution antibody. *Biophys. J.* 1989, 56:955–966.

Griffith IJ, Ghogawala Z, Nabavi N, Golan D, Myer A, McKean DJ, and Glimcher LH: Cytoplasmic domain affects membrane expression and function of an Ia molecule. *Proc. Natl. Acad. Sci. U.S.A.* 1988, 85:4847–4851.

Gross D and Webb WW: Molecular counting of low-density lipoprotein particles as individuals and small clusters on cell surfaces. *Biophys. J.* 1986, 49:901–911.

Gross DJ and Webb WW: Cell surface clustering and mobility of the liganded LDL receptor measured by digital video fluorescence microscopy, in *Spectroscopic Membrane Probes.* Ed. Loew LM, CRC Press, Boca Raton, FL, 1988, pp. 19–45.

Gupte S, Wu E-S, Hoechli L, Hoechli M, Jacobson K. Sowers AE, and Hackenbrock CR: Relationship between lateral diffusion, collision frequency and electron transfer of mitochondrial inner membrane oxidation-reduction components. *Proc. Natl. Acad. Sci. U.S.A.* 1984, 81:2606–2610.

Gustafsson M, Sundqvist T, and Magnusson KE: Lateral diffusion of the secretory component (SC) in the basolateral membrane of the human colon carcinoma cell line HT29 assessed with the fluorescence recovery after photobleaching method. *J. Cell Physiol.* 1988, 137:608–611.

Haft D and Edidin M: Modes of particle transport. *Nature* 1989, 340:262–263.

Henis YI and Gutman O: Lateral diffusion and patch formation of H-2K^k antigens on mouse spleen lymphocytes. *Biochim. Biophys. Acta* 1983, 762:281–288.

Henis Y: Mobility modulation by local concanavalin A binding. *J. Biol. Chem.* 1984, 259:1515–1519.

Hillman GM and Schlessinger J: Lateral diffusion of epidermal growth factor complexed to its surface receptors does not account for the thermal sensitivity of patch formation and endocytosis. *Biochemistry* 1982, 21:1667–1672.

Hochman JH, Schindler M, Lee JG, and Ferguson-Miller S: Lateral mobility of cytochrome c on intact mitochondrial membranes as determined by fluorescence redistribution after photobleaching. *Proc. Natl. Acad. Sci. U.S.A.* 1982, 79:6866–6870.

Ishihara A, Hou Y, and Jacobson K: The Thy-1 antigen exhibits rapid lateral diffusion in the plasma membrane of rodent lymphoid cells and fibroblasts. *Proc. Natl. Acad. Sci. U.S.A.* 1987, 84:1290–1293.

Jacobson K, Wu E, and Poste G: Measurement of the translational mobility of concanavalin A in glycerol-saline solutions and on the cell surface by fluorescence recovery after photobleaching. *Biochim. Biophys. Acta* 1976, 433:215–222.

Jacobson K, Hou Y, and Wojieszyn J: Evidence for lack of damage during photobleaching measurements of the lateral mobility of cell surface components. *Exp. Cell Res.* 1978, 116:178–189.

Jaffe LF: Electrophoresis along cell membranes. *Nature* 1977, 265:600–602.

Jesatis AJ and Yguerabide J: The lateral mobility of the (Na⁺,K⁺)-dependent ATPase in Madin-Darby canine kidney cells. *J. Cell Biol.* 1986, 102:1256–1263.

Johansson N, Sundqvist T, and Magnusson KE: Regulation of the lateral diffusion of WGA-labeled glyco-conjugates in human leukocytes. Comparison between adult granulocytes and differentiating promy-elocytic HL60 cells. *Cell Biophys.* 1987, 10:233–244.

Jones OT, Kinze DL, and Angelides KJ: Localization and mobility of omega-conotoxin-sensitive Ca^{2+} channels in hippocampal Ca1 neurons. *Science* 1989, 244:1189–1193.

Jovin TM and Vaz WLC: Rotational and translational diffusion in membranes measured by fluorescence and phosphorescence methods. *Methods Enzymol.* 1988, 172:471–513.

Kapitza HG, McGregor G, and Jacobson KA: Direct measurement of lateral transport in membranes by using time-resolved spatial photometry. *Proc. Natl. Acad. Sci. U.S.A.* 1985, 82:4122–4126.

Koppel DE: Redistribution after photobleaching. A new multipoint analysis of membrane translational dynam-ics. *Biophys. J.* 1979, 28:281–292.

Koppel DE: Normal-mode analysis of lateral diffusion on a bounded membrane surface. *Biophys. J.* 1985, 47:337–347.

Koppel DE and Sheetz MP: Fluorescence photobleaching does not alter the lateral mobility of erythrocyte membrane glycoproteins. *Nature* 1981, 293:159–161.

Koppel DE, Sheetz MP, and Schindler M: Lateral diffusion in biological membranes. A normal-mode analysis of diffusion on a spherical surface. *Biophys. J.* 1980, 30:187–192.

Kuromi H and Kidokoro Y: Nerve disperses preexisting acetylcholine receptor clusters prior to induction of receptor accumulation in *Xenopus* muscle cultures. *Dev. Biol.* 1984, 103:53–61.

Ku KY, Xu CT, and Zhang KH: Cleavage formation of Rana amurensis eggs observed with fluorescence pattern photobleaching. *Cell Biol. Int. Rep.* 1988, 12:175–187.

Leslie RJ, Saxton WM, Mitchison TJ, Neighbors B, Salmon ED, and McIntosh JR: Assembly properties of fluorescein-labeled tubulin *in vitro* before and after fluorescence bleaching. *J. Cell Biol.* 1984, 99:2146–2156.

Liebman PA and Entine G: Lateral diffusion of pigment in photoreceptor disk membranes. *Science* 1974, 185:457–459.

Livneh E, Benveniste M, Prywes R, Felder S, Kam Z, and Schlessinger J: Large deletions in the cytoplasmic kinase domain of the EGF-receptor do not affect its lateral mobility. *J. Cell Biol.* 1986, 103:327–331.

McIntosh JR and Koonce MP: Mitosis. *Science* 1989, 246:622–628.

Madreperla SA, Edidin M, and Adler R: Na⁺,K⁺-adenosine triphosphatase polarity in retinal photoreceptors: a role for cytoskeletal attachments. *J. Cell Biol.* 1989, 109:1483–1493.

Mastrangelo IA and Mitra J: Chinese hamster ovary chromosomes and antigens in tobacco/hamster heter-okaryons. *J. Hered.* 1981, 72:81–86.

Matko J, Szollosi J, Tron L, and Damjanovich S: Luminescence spectroscopic approaches in studying cell surface dynamics. *Q. Rev. Biophys.* 1988, 21:479–544.

Maxfield F: Fluorescent analogs of peptides and hormones. *Methods Cell Biol.* 1989, 29:13–28.

Mecheri S, Edidin M, Dannecker G, and Hoffman MK: Immunogenic Ia-binding peptides immobilize the Ia molecule and facilitate its aggregation on the B-cell membrane. Control by the *Mls-1* gene. *J. Immunol.* 1990, 144:1361–1368.

Menon AK, Holowka D, Webb WW, and Baird B: Clustering, mobility and triggering activity of small oligomers of immunoglobulin E on rat basophilic leukemia cells. *J. Cell Biol.* 1985, 102:534–540.

Metcalf TN, III, Wang JL, Schubert KR, and Schindler M: Lectin receptors on the plasma membrane of soybean cells. Binding and lateral diffusion of lectins. *Biochemistry* 1983, 22:3969–3975.

Noda M, Yoon K, Rodan G, and Koppel DE: High lateral mobility of endogenous and transfected alkaline phosphatase: a phosphatidylinositol-anchored membrane protein. *J. Cell Biol.* 1987, 105:1671–1677.

Peters R: Lateral mobility of proteins and lipids in the red cell membrane and the activation of adenylate cyclase by beta-adrenergic receptors. *FEBS Lett.* 1988, 234:1–7.

Peters R, Peters J, Tewes KH, and Bahr W: A microfluorimetric study of translational diffusion in erythrocyte membranes. *Biochim. Biophys. Acta* 1974, 367:282–294.

Petty HR, Smith L, Fearon DT, and McConnel HM: Lateral distribution and diffusion of the C3b receptor of complement, HLA antigens and lipid probes in peripheral blood leukocytes. *Proc. Natl. Acad. Sci. U.S.A.* 1980, 77:6587–6591.

Peters R, Brunger A, and Schulten K: Continuous fluorescence microphotolysis: a sensitive method for study of diffusion processes in single cells. *Proc. Natl. Acad. Sci. U.S.A.* 1981, 78:962–966.

Petersen NO: Scanning fluorescence correlation spectroscopy. I. Theory and simulation of aggregation measurements. *Biophys. J.* 1986, 49:809–815.

Peterson NO, Johnson DC, and Schlesinger MJ: Scanning fluorescence correlation spectroscopy. II. Application to virus glycoprotein aggregation. *Biophys. J.* 1986, 49:817–820.

Phelps BM, Primakoff P, Koppel D, Low GM, and Myles DG: Restricted lateral diffusion of PH-20, a PI-anchored sperm membrane protein. *Science* 1988, 240:1780–1782.

Philips E and Cone RA: Do diffusion rates in the cytoplasm or the membrane limit the response speed of the vertebrate rod? *Biophys. J.* 1986, 49:277a.

Pollerberg GE, Schachner M, and Davoust J: Differentiation sate-dependent surface mobilities of two forms of the neural cell adhesion molecule. *Nature* 1986, 324:462–465.

Poo M-M: *In situ* electrophoresis of membrane components. *Annu. Rev. Biophys. Bioeng.* 1981, 10:245–276.

Poo M-M: Rapid lateral diffusion of functional ACh receptors in embryonic muscle culture. *Nature* 1982, 295:332–334.

Poo M-M and Cone RA: Lateral diffusion of rhodopsin in the photoreceptor membrane. *Nature* 1974, 247:438–441.

Poo M-M and Robinson KR: Electrophoresis of concanavalin A receptors along embryonic muscle cell membrane. *Nature* 1977, 265:602–605.

Poo M-M, Poo W-JH, and Lamm JW: Lateral electrophoresis and diffusion of concanavalin A receptors in the membrane of embryonic muscle cell. *J. Cell Biol.* 1978, 76:483–501.

Poo M-M, Lamm JW, Orida N, and Chao AW: Electrophoresis and diffusion in the plane of the cell membrane. *Biophys. J.* 1979, 26:1–22.

Rees AR, Gregoriou M, Johnson P, and Garland PB: High affinity epidermal growth factor receptors on the surface of A431 cells have restricted lateral diffusion. *EMBO J.* 1984, 3:1843–1847.

Rodriguez-Boulan E and Nelson WJ: Morphogenesis of the polarized epithelial cell phenotype. *Science* 1989, 245:718–725.

Ryan TA, Meyers J, Holowka D, Baird B, and Webb WW: Molecular crowding at the cell surface. *Science* 1988, 239:61–64.

Salas PJI, Vega-Salas DE, Hochman J, Rodriguez-Bouland E, and Edidin M: Selective anchoring in the specific plasma membrane domain: a role in epithelial cell polarity. *J. Cell Biol.* 1988, 107:2363–3476.

Saxton M: Lateral diffusion in an archipelago: effects of impermeable patches on lateral diffusion in a cell membrane. *Biophys. J.* 1982, 39:165–173.

Saxton M: Lateral diffusion in an archipelago: the effect of mobile obstacles. *Biophys. J.* 1987, 52:989–997.

Scalettar BA, Abney JR, and Owicki JC: Theoretical comparison of the self diffusion and mutual diffusion of interacting membrane proteins. *Proc. Natl. Acad. Sci. U.S.A.* 1988, 85:6726–6730.

Schecter Y, Schlessinger J, Jacobs S, Chang K-J, and Cuatrecasas P: Fluorescent labeling of hormone receptors in viable cells: preparation and properties of highly fluorescent derivatives of epidermal growth factor and insulin. *Proc. Natl. Acad. Sci. U.S.A.* 1978, 75:2135–2139.

Schindler M, Holland JF, and Hogan M: Lateral diffusion in nuclear membranes. *J. Cell Biol.* 1985, 100:1408–1414.

Schindler M, Koppel DE, and Sheetz M: Modulation of membrane protein lateral mobility by polyphosphates and polyamines. *Proc. Natl. Acad. Sci. U.S.A.* 1980, 77:1457–1461.

Schlessinger J, Elson EL, Webb WW, Yahara I, Rutishauser U, and Edelman GM: Receptor diffusion on cell surfaces modulated by locally bound concanavalin A. *Proc. Natl. Acad. Sci. U.S.A.* 1977, 74:1110–1114.

Schlessinger J, Koppel DE, Axelrod D, Jacobson K, Webb W, Elson EL: Lateral transport on cell membranes: mobility of concanavalin A receptors on myoblasts. *Proc. Natl. Acad. Sci. U.S.A.* 1976(a), 73:2409–2413.

Schlessinger J, Webb WW, Elson EL, and Metzger H: Lateral motion and valence of Fc receptors on rat peritoneal mast cells. *Nature* 1976(b), 264:550–552.

Scullion BF, Hou Y, Puddington L, Rose JK, and Jacobson K: Effects of mutations in three domains of the vesicular stomatitis viral glycoprotein on its lateral diffusion in the plasma membrane. *J. Cell Biol.* 1987, 105:69–75.

Sheetz MP and Koppel DE: Membrane damage caused by irradiation of fluorescent concanavalin A. *Proc. Natl. Acad. Sci. U.S.A.* 1979, 76:3314–3317.

Sheetz M, Schindler M, and Koppel D: Lateral mobility of integral membrane proteins is increased on spherocytic erythrocytes. *Nature* 1980, 285:510–512.

Sheetz MP, Turney S, Qian H, and Elson EL: Nanometre-level analysis demonstrates that lipid flow does not drive membrane glycoprotein movements. *Nature* 1989, 340:284–288.

Simon JR, Gough A, Urbanik E, Wang F, Lanni F, Ware BR, and Taylor DL: Analysis of rhodamine and fluorescein-labeled F-actin diffusion *in vitro* by fluorescence photobleaching recovery. *Biophys. J.* 1988, 54:801–815.

Singer SJ and Nicolson G: The fluid mosaic model of the structure of cell membranes. *Science* 1972, 175:720–731.

Smith BA and McConnell HM: Determination of molecular motion in membranes using periodic pattern photobleaching. *Proc. Natl. Acad. Sci. U.S.A.* 1978, 75:2759–2763.

Sowers AE and Hackenbrock CR: Rate of lateral diffusion of intramembrane particles: measurement by electrophoretic displacement and randomization. *Proc. Natl. Acad. Sci. U.S.A.* 1981, 78:6246–6250.

Srinivasan Y, Elmer L, Davis J, Bennett V, and Angelides K: Ankyrin and spectrin associate with voltage-dependent sodium channels in brain. *Nature* 1988, 333:177–180.

Stollberg J and Fraser SE: Acetylcholine receptors and concanavalin A-binding sites on cultured *Xenopus* muscle cells: electrophoresis, diffusion and aggregation. *J. Cell Biol.* 1988, 107:1397–1408.

Stolpen AH, Pober JS, Brown CS, and Golan DE: Class I major histocompatibility complex proteins diffuse isotropically on immune interferon-activated endothelial cells despite anisotropic cell shape and cytoskeletal organization: application of fluorescence photobleaching recovery with an elliptical beam. *Proc. Natl. Acad. Sci. U.S.A.* 1988(a), 85:1844–1848.

Stolpen AH, Golan DE, and Pober JS: Tumor necrosis factor and immune interferon act in concert to slow the lateral diffusion of proteins and lipids in human endothelial cell membranes. *J. Cell Biol.* 1988(b), 107:781–789.

Stya M and Axelrod D: Mobility of extrajunctional acetylcholine receptors on denervated adult muscle fibers. *J. Neurosci.* 1984, 4:70–74.

Su Y-X, Lin S, and Edidin M: Lateral diffusion of human histocompatibility antigens in isolated plasma membranes. *Biochim. Biophys. Acta* 1984, 776:92–96.

Swaisgood M and Schindler M: Lateral diffusion of lectin receptors in fibroblast membranes as a function of cell shape. *Exp. Cell Res.* 1989, 180:515–528.

Tank D, Fredericks WJ, Barak LS, and Webb WW: Electric field-induced redistribution and postfield relaxation of low density lipoprotein receptors on cultured human fibroblasts. *J. Cell Biol.* 1985, 101:148–157.

Thomas J and Webb WW: Decay accelerating factor diffuses rapidly. *Biophys, J.* 1987, 51:522a.

Van Zoelen EJ, Tertoolen LGJ, and DeLaat SW: Simple computer method for evaluation of lateral diffusion coefficients from fluorescent photobleaching recovery kinetics. *Biophys. J.* 1983, 42:103–108.

Wade WF, Freed JH, and Edidin M: Translational diffusion of class II major histocompatibility complex antigens is constrained by their cytoplasmic domains. *J. Cell Biol.* 1989, 109:3325–3331.

Weis RE, Roberts WM, Stuhmer W, and Almers W: Mobility of voltage-dependent ion channels and lectin receptors in the sarcolemma of frog skeletal muscle. *J. Gen. Physiol.* 1986, 87:955–983.

Wey C-L, Edidin MA, and Cone RA: Lateral diffusion of rhodopsin in photoreceptor cells by fluorescence photobleaching and recovery (FPR). *Biophys. J.* 1981, 33:225–232.

Wier ML and Edidin M: Effects of cell density and extracellular matrix on the lateral diffusion of major histocompatibility antigens in cultured fibroblasts. *J. Cell Biol.* 1986, 103:215–222.

Wier ML and Edidin M: Constraint of the translational diffusion of a membrane glycoprotein by its external domains. *Science* 1988, 242:412–414.

Woda BA, Yguerabide J, and Feldman J: Mobility and density of AgB "Ia" and Fc receptors on the surface of lymphocytes from old and young rats. *J. Immunol.* 1979, 123:2161–2167.

Wolf DE: Designing, building and using a fluorescence recovery after photobleaching instrument. *Methods Cell Biol.* 1989, 30:271–306.

Wolf DE, Edidin M, and Dragsten PR: Effect of bleaching light on measurements of lateral diffusion in cell membranes by the fluorescence photobleaching recovery method. *Proc. Natl. Acad. Sci. U.S.A.* 1980, 77:2043–2045.

Wood J and Angelides KJ: Association of ankyrin and spectrin with sodium channels control sodium channel mobility in nerve. *J. Cell Biol.* 1988, 107:26a.

Yechiel E and Edidin M: Micrometer scale domains in fibroblast plasma membranes. *J. Cell Biol.* 1987, 105:755–760.

Yguerabide J, Schmidt JA, Yguerabide EE: Lateral mobility in membranes as detected by fluorescence recovery after photobleaching. *Biophys. J.* 1982, 40:69–75.

Young SH and Poo M-M: Rapid lateral diffusion of extrajunctional acetylcholine receptors in the developing muscle membrane of *Xenopus* tadpole. *J. Neurosci.* 1983, 3:225–231.

Zagyansky Y and Edidin M: Lateral diffusion of concanavalin A receptors in the plasma membrane of mouse fibroblasts. *Biochim. Biophys. Acta* 1976, 433:209–214.

Zagyansky Y and Jard S: Does lectin-receptor complex produce zones of restricted mobility within the membrane? *Nature* 1979, 280:591–593.

Zhang F, Crise B, Hou Y, Rose J, and Jacobson K: Lateral mobility of lipid-linked proteins expressed by transfection in COS-1 cells. *J. Cell Biol.* 1989, 109–133a.

13 Inorganic Anion Transporter AE1

Michael L. Jennings

CONTENTS

0-8493-1403-8/05/$0.00+$1.50

13.1 INTRODUCTION

In the first edition of this volume, the chapter on inorganic anion transport was focused on general principles of coupled anion exchange and cotransport. At that time, the red blood cell band 3 protein (AE1) and the related proteins AE2 and AE3 were the only coupled inorganic anion transporters whose sequences were known. In the past ten years, the sequences of entire new families of coupled anion exchangers and cotransporters have been determined, including Na^+-HCO_3^- cotransporters [1, 2], Cl^--coupled alkali cation cotransporters [3–6], the $SO_4^=$ transporter superfamily [7, 8], and Na^+/Cl^--coupled organic amine transporters [9]. In addition, the sequences of many Cl^- channels are now known [10], and high resolution structures of Cl^- channels have been determined [11, 12]. It is impossible in a single chapter to cover all these classes of anion transporter. Instead, this chapter will focus on a single transporter, the Cl^-/HCO_3^- exchanger AE1, which continues to be an important model system for mechanistic studies of coupled transport. Native erythrocyte AE1 (band 3 protein), because of its abundance, is well suited for biochemical, biophysical, and kinetic studies. Another reason that AE1/band 3 is an instructive system is that there are many known human mutations in the AE1 gene. Several of these mutations have provided valuable information on the molecular basis of both renal and erythrocyte disorders.

The goal of this chapter is to summarize the current state of understanding of the structure and function of AE1, derived from a wide variety of experimental approaches, including molecular biology, biochemistry, biophysics, immunology, cell biology, and physiology. An attempt is made to compare conclusions from heterologous expression/mutagenesis experiments with those obtained from biochemical or biophysical studies of the native protein. Emphasis is on advances made in the past ten years. Readers interested in more detailed discussion of previous work are referred to earlier reviews [13–17].

13.2 BICARBONATE TRANSPORTER FAMILY

AE1 is a member of the bicarbonate transporter family, also known as *solute carrier family 4* (SLC4). This family includes Na^+-independent anion exchangers AE1, AE2, AE3, and AE4 [18, 19]; Na^+-dependent anion exchangers [20, 21]; and electrogenic and electroneutral Na^+-HCO_3^- cotransporters [22, 23]. Figure 13.1 shows a radial tree dendrogram of the membrane domains of selected members of SLC4. The known transport functions of members of SLC4 do not assort strictly according to sequence relatedness. For example, two laboratories have found that AE4 is a Na^+-independent Cl^-/HCO_3^- exchanger [19, 24], even though the sequence of AE4 is more closely related to Na^+-HCO_3^- cotransporters than to AE1, AE2, and AE3.

There are no ORFs with significant homology to the bicarbonate transporter family in any of the known prokaryote genomes. There are, however, ORFs in many eukaryotic genomes (e.g., *C. elegans*, *A. thaliana,* and *S. cerevisiae*) that are clearly members of SLC4; the functions of most

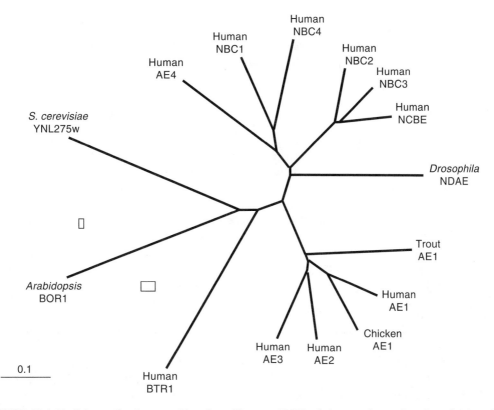

FIGURE 13.1 Radial tree dendrogram (Treeview, Glasgow, U.K.) of the membrane domains of selected members of SLC4. Sequences starting at the residue corresponding to human 381 (i.e., in the hydrophilic tether between cytoplasmic and membrane domains) were aligned using ClustalW (www.ebi.ac.uk/clustalw) using the default parameters. The Na+-independent anion exchangers AE1, AE2, and AE3 form one group of closely related sequences. The ORF designated NBC3 is now known to be a Na+-dependent Cl-/HCO3- exchanger (NCBE), and the sequence of the NBC3 membrane domain is closely related to another NCBE (see [33]). *Drosophila* NDAE is a Na+-dependent anion exchanger that does not require HCO3- [20]. The recently described human bicarbonate transporter related protein BTR1 [307] is not included in either of the major groups (Na+-dependent vs. Na+-independent). The anion exchanger AE4 is grouped with the Na+-HCO3- cotransporters NBC1 and NBC4; this is also true if the entire sequences, rather than the membrane domains, are compared. The plant and yeast homologs are in a distinct group, along with several other ORFs (not shown) of unknown function from *C. elegans* and *A. thaliana*.

of these gene products are unknown, with the following exception. A member of SLC4 from *A. thaliana* known as BOR1 was recently shown to have a role in boron metabolism in plants [25]. *A. thaliana* BOR1 protein is believed to act as a boric acid efflux transporter that facilitates xylem loading [25]. It is worth noting that boric acid (H_3BO_3) is known to diffuse relatively rapidly through unmodified lipid bilayers and biological membranes [26, 27]. Despite this rapid, unmediated diffusion, however, plants apparently use BOR1 as a facilitated transport system for boric acid.

The *Saccharomyces cerevisiae* homolog YNL275w (the sole SLC4 ORF in *S. cerevisiae*) appears also to act as a boron efflux transporter, because strains lacking this gene have a much higher boron content when preincubated in media containing 0.5 mM H_3BO_3 than do wild-type strains [25]. Although the evidence for boric acid transport by *A. thaliana* BOR1 is convincing [25], it would be premature to conclude that yeast YNL275wp functions solely or even primarily as a boric acid transporter. Zhao and Reithmeier [28] had previously shown that YNL275wp is a plasma membrane protein that exhibits stilbenedisulfonate binding that is inhibited by various anions, and it is possible that the protein functions as an anion transporter or sensor. The discovery of boric acid transport by *A. thaliana* BOR1p and *S. cerevisiae* YNL275wp adds to the considerable functional diversity of the bicarbonate transporter family. The study of structure and function of AE1 should eventually provide insights into the other members of SLC4, but, at present, relatively little is known about the molecular basis of the functional diversity of this family of proteins. Readers interested in more detailed discussion of other SLC4 family members are referred to recent reviews [29–33].

13.3 GENERAL PROPERTIES OF THE AE1 PROTEIN AND GENE

Band 3 protein (erythroid AE1) represents about 25% of the protein of the mature human erythrocyte membrane [34]. The polypeptide consists of 911 amino acid residues, a single N-linked carbohydrate chain at N642 [35], fatty acylation of cysteine at C843 [36], and several sites of phosphorylation [37–42]. The carbohydrate chain in erythroid AE1 is of variable length and results in a band that appears as a broad band of ~95 to >110 kDa on SDS polyacrylamide gels [13, 34].

Human erythroid AE1 protein consists of a large N-terminal hydrophilic cytoplasmic domain (cdb3, residues 1–360), a hydrophobic membrane domain (mdb3, residues 361–878), and a short hydrophilic C-terminal cytoplasmic domain (residues 879–911). (K360 is conveniently defined as the boundary between cdb3 and mdb3 because it is the site of trypsin cleavage of band 3 in unsealed membranes [43]; the sequence from G361 to D399 is a hydrophilic tether that connects the two domains but will be considered here as part of the membrane domain.) The three-domain arrangement of the AE1 polypeptide is observed in mammals [44–46], chicken [47], and trout [48]. The sequence of the membrane domain is somewhat better conserved among species than are either the N- or the C-terminal hydrophilic domains. Within the membrane domain, there are two subdomains (G361–Y553 and M559–F878), separated by an extracellular loop of variable sequence.

The human AE1 gene spans about 20 kb, and the erythroid transcript contains 20 exons [49, 50]. The gene is located on chromosome 17q21-q22 and is expressed mainly in erythrocytes and the α-intercalated cells of the renal collecting tubule (see later section). The renal form is expressed with an alternate promoter located in the third erythroid intron and therefore lacks the first three exons of the erythroid transcript. The erythroid promoter region, which is well conserved between human and mouse, contains no TATA or CAAT elements but has potential binding sites for several transcription factors associated with other erythroid genes [49–51].

13.4 PHYSIOLOGICAL FUNCTIONS OF ERYTHROCYTE BAND 3

13.4.1 ROLE OF ANION EXCHANGE IN CO_2 TRANSPORT

Erythrocyte AE1 (band 3) was shown to be responsible for inorganic anion transport on the basis of covalent labeling with H_2DIDS (4,4'-diisothiocyanatodihydrostilbene-2,2'-disulfonate) [52] and other transport inhibitors [16, 53]. The most important mode of transport physiologically is a tightly coupled 1:1 exchange of Cl^- for HCO_3^-, which increases the CO_2 carrying capacity of blood [54]. In tissue capillaries, CO_2 enters the red cell by diffusion through the lipid bilayer, through aquaporin [55], and possibly through proteins of the Rh complex [56]. After entering the cell, CO_2 is hydrated to $HCO_3^- + H^+$ by carbonic anhydrase, which is present in high concentrations in red cell cytoplasm and is also bound to the cytoplasmic face of band 3, as discussed later. The formation of HCO_3^- creates a driving force for outward transport of HCO_3^- in exchange for extracellular Cl^-. The anion exchange does not require ATP and is driven by the anion gradients until Donnan equilibrium is reached: $[Cl^-]_{in}/[Cl^-]_{out} = [HCO_3^-]_{in}/[HCO_3^-]_{out}$. In pulmonary capillaries, the entire process takes place in reverse (Figure 13.2).

13.4.2 OTHER FUNCTIONS

Many anions, both inorganic and organic, are transported by band 3 (see [17, 57]). Even though most of these are transported relatively slowly, band 3-mediated transport of ions other than Cl^-

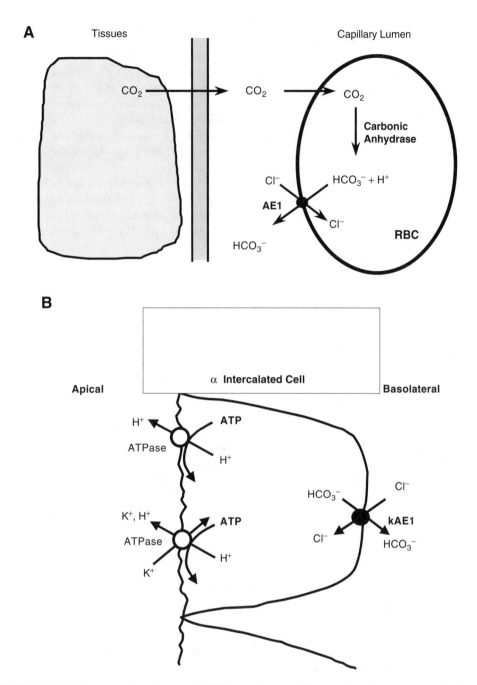

FIGURE 13.2 (A) Function of erythrocyte Cl^-/HCO_3^- exchange in CO_2 transport. In systemic capillaries, CO_2 diffuses into the capillary luman and then into the red blood cell, where it is hydrated by carbonic anhydrase to $HCO_3^- + H^+$. The incoming CO_2 causes an increase in the intracellular HCO_3^- concentration, which drives a net HCO_3^- efflux in exchange for extracellular Cl^- through the band 3 (AE1) protein. The sames steps happen in reverse in pulmonary capillaries. (B) Simplified diagram of a renal collecting duct α-intercalated cell, showing apical H^+ pumps (H^+-ATPase and K^+,H^+-ATPase) and basolateral Cl^-/HCO_3^- exchanger (kAE1), which mediates net HCO_3^- efflux to balance the H^+ pumped into the lumen. Much of the Cl^- that enters the cell in exchange for HCO_3^- is recycled across the basolateral membrane by a conductive channel not shown in the figure (see [308]).

and HCO_3^- (e.g., superoxide [58] or peroxynitrite [59]) could be of physiological or pathophysiological significance. An example of AE1-mediated anion transport in a pathophysiological context is the antidote for cyanide poisoning (nitrite and thiosulfate to generate cyanomethemoglobin and subsequently $S=C=N^-$, thereby allowing hemoglobin to be a sink for CN^-). This antidote is effective only because all these anions (nitrite, thiosulfate, and $S=C=N^-$) can enter and leave the erythrocyte rapidly via band 3. Band 3 also catalyzes the slow transport phosphoenolpyruvate (PEP), which makes it possible to use PEP as a metabolic supplement in stored blood [60, 61].

In addition to the transport of soluble anions, band 3 can act as a "flippase" for amphipathic molecules [62, 63]. There is also good evidence that trout [64] and skate [65] erythrocyte band 3 act as a swelling-activated osmolyte channel. The focus of this chapter is mammalian AE1, and the osmoregulatory function of other vertebrate AE1s will not be discussed further in this chapter.

Erythrocyte AE1 has important functions other than transport. As discussed later, the cytoplasmic domain is a major attachment site for the membrane skeleton via ankyrin [66] and also binds hemoglobin and glycolytic enzymes, and may be involved in the regulation of red cell metabolism [67]. Band 3 also has a role as a signal for the removal of senescent red cells (see [67, 68]) and in the alterations of the red cell surface associated with invasion by *Plasmodium falciparum* [69, 70].

13.5 FUNCTION OF RENAL AE1

Renal AE1 is expressed under the control of a promoter that is downstream from the erythroid promoter, and the renal AE1 polypeptide lacks N-terminal residues 1–65 (human) or 1–79 (mouse) of erythroid AE1 [71–73]. Kidney AE1 (kAE1) is expressed in the basolateral membrane of acid-secreting α-intercalated cells in the renal collecting duct [74, 75]. The function of kAE1 is as a Cl^-/HCO_3^- exchanger, which provides a HCO_3^- exit pathway across the basolateral membrane (Figure 13.2B) to balance the acid secreted across the apical membrane. Acid conditions cause up-regulation of basolateral AE1 in α-intercalated cells by increasing mRNA expression and trafficking of kAE1 protein from intracellular vesicles to the basolateral membrane [76–79].

A separate cell type in the renal cortical collecting tubule is the β-intercalated cell, which alkalinizes urine by secreting HCO_3^- via an apical Cl^-/HCO_3^- exchanger [80, 81]. It was proposed several years ago that, depending on the acid/base status of the animal, the α- and β-intercalated cells of the cortical collecting duct can change polarity [82]. Considerable evidence exists for this polarity switching, and an extracellular matrix protein known has hensin has been implicated as necessary for the process (see [83]). However, there are several lines of evidence against the idea that AE1 is the apical exchanger in β-intercalated cells. First, antibodies against AE1 consistently stain the basolateral membranes of α-intercalated cells but do not stain the apical membranes of β-intercalated cells [74, 75, 84]. In addition, the expression of AE1 mRNA is far higher in α-intercalated than in β-intercalated cells [77]. Finally, the functional properties of the apical anion exchanger in β-intercalated cells are different in many respects from those of the basolateral exchanger in α-intercalated cells [85]. Some of these differences could be caused by different lipid composition of the apical vs. basolateral membrane [86]. However, given the functional differences and the apparent lack of expression of either AE1 mRNA or protein in normal β-intercalated cells, it seems unlikely that AE1 is the apical anion exchanger in these cells.

The identity of the apical anion exchanger of β-intercalated cells has been difficult to define. In mouse, rat, and human kidneys, the SLC26 transporter Pendrin is localized to the apical membrane of a population of intercalated cells that is distinct from those staining for AE1 [87]. AE4, a member of the bicarbonate transporter family, carries out DIDS-insensitive Cl^-/HCO_3^- exchange and is the apical anion exchanger of rabbit β-intercalated cells [24]. However, Ko et al. [19] have shown that, in rat and mouse, AE4 is in the basolateral membrane of α-intercalated cells and carries out DIDS-sensitive anion exchange. In rabbit, AE4 may indeed be the apical exchanger in base-secreting cells, but in other species this function is probably carried out by Pendrin. The presence of a second anion exchanger, AE4, in the basolateral membrane of α-intercalated cells in some species raises the question of whether AE1 or AE4 is the more important protein functionally. The main information on this point at present is that, in humans with certain mutations in the AE1 gene (see later discussion), urinary acidification is compromised, indicating that, at least in humans, no other anion exchanger is able to compensate for deficient anion exchange by kAE1.

13.6 N-TERMINAL CYTOPLASMIC DOMAIN OF ERYTHROCYTE AE1

13.6.1 INTERACTION WITH OTHER PROTEINS

The 43 kDa N-terminal cytoplasmic domain of erythrocyte AE1 (cdb3) has no known enzymatic or transporter activity. Its function is to bind other proteins on the cytoplasmic surface of the cell, including ankyrin [88], protein 4.1 [89], protein 4.2 [90], several glycolytic enzymes [13], deoxy-hemoglobin [91], tyrosine kinase p72^{syk} [92], and tyrosine phosphatase SHP-2 [93]. Low and coworkers [94] have determined the crystal structure of cdb3 at 2.6 Å resolution. Cdb3 is a tightly associated dimer, each subunit of which consists of 11 β strands and 10 α-helical segments. The dimer is stabilized by eight backbone-to-backbone H-bonds, several of which are in an intersubunit antiparallel β sheet. The N-terminal ~54 residues are disordered in the crystal; this part of the protein binds hemoglobin as well as glycolytic enzymes and probably adopts different conformations when bound to different cytoplasmic proteins. The tether linking cdb3 with the membrane domain is also disordered in the crystal and is likely to be flexible in the intact protein, because it is very susceptible to proteolysis under mild conditions [95].

The crystal structure of cdb3 provides many other insights into the role of this part of band 3 as an organizational center for membrane proteins (see [94]). Chang and Low [96] have recently used the structure of cdb3 as a guide for mutagenesis studies that demonstrate that a β-hairpin loop (residues 175–185) constitutes a major ankyrin-binding region of band 3. Removal of the N-terminal 50 residues does not prevent ankyrin binding, but removal of the N-terminal 65 residues (i.e., the truncation in normal human renal AE1) prevents ankyrin binding, as was previously known [97]. The reason for the major difference in the effects of truncation of 50 vs. 65 residues is that the latter disrupts a β-sheet and changes the structure substantially. Although the N-terminal 50 residues do not physically bind ankyrin, they are close enough to the ankyrin binding site for antibodies bound to the N-terminus to block ankyrin binding to residues 175–185 (see [96].)

13.6.2 TARGETING SIGNALS IN THE N-TERMINAL CYTOPLASMIC DOMAIN

Cox and coworkers have shown that there are three splice variants of chicken kidney AE1 that differ in the N-terminal sequence [98] and that these differences affect the targeting of the protein [99]. When expressed in MDCK cells, the chicken AE1 variant that is missing the N-terminal 63 residues is targeted primarily to the apical membrane, whereas the variant that retains these residues is directed to the basolateral membrane [100]. In both MDCK cells and in embryonic chick erythroid cells, AE1 reaches its final form in the plasma membrane by a novel pathway, in which newly synthesized AE1 protein with core glycosylation traffics through the Golgi apparatus to the plasma membrane and is then recycled back to the Golgi, where the carbohydrate is processed into the mature form before final trafficking to the plasma membrane [100, 101]. Recycling of band 3 is not unique to avian cells; Hanspal et al. [102] have shown that in late erythroblasts of mouse, band 3 moves rapidly to the plasma membrane and then is recycled through a subcellular compartment before stable incorporation into the plasma membrane.

Mutagenesis studies using fusion proteins between N-terminal residues of chicken AE1 and mouse F_c receptor lacking its normal cytoplasmic tail indicate that a tyrosine residue in the conserved sequence YVEL (corresponding to Y58 of human AE1) has a critical role as a sorting signal [103]. The homologous tyrosine residue in skate AE1 is phosphorylated in response to osmotic stress

[104]; it is not clear whether this phosphorylation is related to a targeting role for this residue. Although the YVEL sequence in the chicken renal AE1 variant AE1-4 is clearly important in directing this protein to the basolateral membrane, it is not known whether there are N-terminal targeting signals in mammalian renal AE1, because the main renal transcript in mammals lacks the YVEL sequence. As discussed later in connection with renal tubular acidosis, a C-terminal targeting signal is believed to be of critical importance in mammalian AE1.

13.7 C-TERMINAL HYDROPHILIC DOMAIN — CARBONIC ANHYDRASE BINDING

The cytoplasmic C-terminal 30 residues of human AE1 are highly acidic and contain a binding site for carbonic anhydrase II (CAII) [105]. The CAII binding site has been localized to the sequence LDADD (residues 886–890) [106]. CAII bound to the cytoplasmic surface of AE1 facilitates the changes in cytoplasmic pH (detected by BCECF fluorescence) associated with Cl^-/HCO_3^- exchange mediated by AE1 expressed in HEK 293 cells [107]. Mutations that prevent CAII binding, or the coexpression of a catalytically inactive form of CAII, inhibit these pH changes. These data indicate that the complex of CAII and AE1 constitutes a transport metabolon, in which HCO_3^- entering the cells through AE1 is dehydrated preferentially by bound CAII. The process works in the efflux direction as well; HCO_3^- formed by bound CAII is preferentially used as a substrate for efflux through AE1. Thus, although CAII is not necessary for anion transport itself, the binding of CAII facilitates the overall process of anion exchange in series with CO_2 hydration or HCO_3^- dehydration. The presence of sequences similar to the LDADD CAII binding sequence of human AE1 in other HCO_3^- transport proteins (AE2, AE3, NBCs, Pendrin, DRA) raises the possibility that substrate channeling of HCO_3^- between carbonic anhydrase and transporters is a general phenomenon [107].

13.8 INTERACTION BETWEEN AE1 AND GLYCOPHORIN A

Functional expression of AE1 does not require any other heterologous gene expression in either *Xenopus* oocytes or mammalian cells [108–111]. However, the cell surface expression of AE1 is enhanced by coexpression of glycophorin A (GPA) in *Xenopus* oocytes, as well as in other heterologous systems, including yeast [112–115]. Mutations in glycophorin A that prevent GPA dimerization do not prevent GPA from facilitating surface expression of AE1 in *Xenopus* oocytes [116]. Mice with targeted deletion of AE1 have erythrocytes lacking glycophorin A [117], indicating that there is a mutual dependency between AE1 and GPA for erythroid expression. In mice with no expression of erythroid AE1, GPA mRNA is made, but the protein is rapidly degraded in the absence of AE1 expression. In transgenic mice expressing human GPA, the amount of erythrocyte AE1 is unchanged, but the amount of mouse GPA is decreased, indicating a tight coupling between the total amount of GPA and AE1 in the mature membrane [118].

The mutual interactions between GPA and AE1 during erythroid expression of these proteins appear to enhance the rate of folding and maturation of AE1 and stabilize both polypeptides, which makes it possible to express a large amount of protein in the membrane in a relatively limited time during erythroid maturation. The Wright b blood group antigen depends on sequences from both AE1 (E658) and GPA (R61) [119, 120], indicating that, at least in the presence of antibodies directed agains this antigen, the two proteins are in close contact. However, it is not clear how stable the AE1-GPA complex is in normal erythrocytes [120], and it would not be correct to consider GPA as a β subunit of band 3, because band 3 alone, without GPA, can transport anions, and because band 3 can be separated from glycophorin under relatively mild conditions [121]. Nonetheless, it is clear that the two proteins interact closely during biosynthesis and targeting.

13.9 OLIGOMERIC STRUCTURE

It is very clear that the band 3 polypeptide is a tightly associated dimer [122] with strong subunit interactions in both the membrane domain [123, 124] and the cytoplasmic domain [94, 125]. Adjacent cytoplasmic domains can be -SS- crosslinked with o-phenanthroline [126], and membrane domains can be crosslinked by treatment of intact cells with the homobifunctional reagent bis(sulfosuccinimidyl) suberate (BS^3) [127]. Double crosslinking by both methods produces only a small amount of covalent tetramer [128, 129], indicating that each dimer is tightly associated in both the membrane domains and the cytoplasmic domains and that the main covalent crosslinks are within a given dimer and not between two dimers.

Although it has been reported that monomeric band 3 can transport anions [130], there is abundant evidence that the band 3 dimer is very stable and that dissociation of the dimer requires unfolding of the protein [131, 132]. Numerous gel filtration studies have shown that dimeric band 3 is stable for the times necessary for chromatography, i.e., there is no detectable dissociation into monomers under nondenaturing conditions [129, 133, 134]. Additional evidence for the stability of the dimer is that the distribution of long polylactosaminyl and short complex carbohydrate chains in band 3 dimers is not random. Instead, one population of dimers appears to be processed to contain long polylactosaminyl chains, and another population of dimers is not processed in this way; the subunits of the two populations do not appear to interchange during the long life of the red cell [135]. After very long incubations *in vitro* (days at 37°C), the dimer dissociates into monomers [136], but only after unfolding of at least one of the subunits [134].

Many lines of evidence, including chemical crosslinking [126] and spectroscopic measurements of rotational mobility [137, 138], indicate that, in addition to dimers, band 3 can form tetramers. The isolated membrane domain does not form tetramers [133, 139], indicating that associations between adjacent dimers to form tetramers requires the cytoplasmic domain. The oligomeric form of band 3 that is associated with the membrane skeleton (via ankyrin) is a tetramer or higher oligomer, as indicated by the fact that the subset of the band 3 population that is extractable by nondenaturing detergents is dimeric [133]. The maximum stoichiometry of ankyrin binding to band 3 in intact membranes is 1:4, suggesting that the ankyrin-binding complex is the band 3 tetramer [140]. There are two different band 3 binding sites on each ankyrin molecule [141], and the most likely configuration of the complex is a pseudotetramer, in which a subunit of two different band 3 dimers binds to each of the two binding sites on a single ankyrin molecule, giving a 1:4 complex (see [142]). This idea is supported by the fact that membranes of ankyrin-deficient mice have very little, if any, tetrameric band 3 [143]. Further evidence of a 1:4 ankyrin:band 3 association is that a complex of one ankyrin and four band-3 polypeptides has been observed in analytical ultracentrifugation experiments [144].

In native membranes, the number of ankyrin molecules is only about 10 to 15% of the number of copies of band 3 [88]. If all copies of ankyrin are complexed with band 3 pseudotetramers, then about half the copies of band 3 should exist as dimers not associated with ankyrin. Lateral mobility measurements by fluorescence recovery after photobleaching are consistent with this idea: at near-physiological ionic strength and 37°C, about half the copies of band 3 have lateral mobility sufficiently high to measure by fluorescence recovery [145], consistent with the idea that these copies of band 3 are in tetramers associated with ankyrin.

It is not known to what extent, in the normal human erythrocyte membrane, authentic tetramers of band 3 exist (other than the pseudotetramers held together by ankyrin). Certainly there are many examples of preparations containing tetrameric band 3 observed experimentally in the absence of ankyrin. Salhany and coworkers [146] have found that crosslinking with BS^3 produces covalent dimers that can associate into noncovalent tetramers sufficiently stable to be observable on SDS gels of samples that are not heated in boiling water. Purified band 3 can form tetramers under conditions of analytical ultracentrifugation in nonionic detergent [147–149]. These preparations with detectable band 3 tetramers do not appear to require ankyrin, and it is not yet clear what relationship these tetramers have to the complex formed among ankyrin and two band-3 dimers. Under some conditions, tetramers of isolated band 3 may form as an artifact during chromatography in detergent solutions [129], and there is good evidence that oxidants in nonionic detergents can alter the oligomeric state of the protein, in this case by stabilizing band 3 dimers [147, 150].

In summary, a complete understanding of the oligomeric state of band 3 has been elusive, most likely because variations in experimental conditions (e.g., pH of extraction, detergent used, temperature, etc.) can strongly affect the behavior of isolated band 3 and because the protein has a strong tendency to form aggregates in isolation, even in the presence of detergent. It is clear that the band 3 dimer is the main oligomeric state of the protein and that two dimers can be held together by a single ankyrin molecule to form a 1:4 complex. Authentic tetramers of band 3 in the membrane, formed solely through interactions among band 3 subunits and not requiring ankyrin, may not exist in significant amounts.

13.10 ANION TRANSPORT MECHANISM OF THE AE1 MEMBRANE DOMAIN

13.10.1 TRANSPORT CATALYTIC CYCLE

Inorganic anion transport in erythrocytes has been reviewed in detail previously [14–17], and only a few aspects of the transport mechanism will be discussed further here. The most striking characteristic of red cell Cl^- transport is that it consists almost entirely of an obligatory 1:1 exchange [15]. For the monovalent anions Cl^- and HCO_3^-, the turnover number for the catalysis of anion exchange is quite high: ~50,000/s at 37°C [151]. The mechanism of obligatory exchange is believed to be of the ping-pong class [14, 15, 17, 152], in which there are distinct inward-facing and outward-facing forms of the transporter (Figure 13.3), and the translocation event is a conformational change between the inward-facing and outward-facing states of the protein with a bound substrate anion. There is a small conductive Cl^- flux mediated by band 3, but the conductive flux is roughly 10,000-fold smaller than the exchange flux [153]. The mechanism of the conductive flux relative to the catalytic cycle for exchange is not well understood but does not appear to be "slippage," i.e., the binding, outward translocation, and release of a Cl^- ion followed by the return of the empty site to the inward-facing state [154, 155].

13.10.2 TRANSPORT ASYMMETRY

The process of net Cl^-/HCO_3^- exchange in red cells operates nearly equally well under physiological conditions in each direction, i.e., the small outward HCO_3^- gradient in the systemic capillaries and the small inward HCO_3^- gradient in the pulmonary capillaries are dissipated at similar rates. This superficial appearance of transport symmetry does not imply that the actual, individual kinetic events in the catalytic cycle are symmetric. In fact, there are major mechanistic asymmetries in the binding and translocation of anions through AE1. The inward-facing empty transporter (E_{in}) and the inward-facing transporter with bound Cl^- (ECl_{in}) are each energetically more stable than, respectively, the outward-facing empty transporter (E_o) and the outward-facing Cl^- complex (ECl_o) [156, 157]. Because of this asymmetry, the inward translocation rate constant for Cl^- is much higher than the outward translocation rate constant (Figure 13.3).

Surprisingly, the translocation rate constants of the other main physiological substrate, HCO_3^-, have a completely different asymmetry from those of Cl^- [158]. The outward-facing HCO_3^- complex is considerably more stable than the inward-facing complex (Figure 13.3), and, therefore, the inward translocation rate constant for HCO_3^- is slower than the outward translocation rate for HCO_3^-. Despite the underlying major asymmetries in the mechanisms of Cl^- and HCO_3^- binding and/or translocation events, the net exchange of these two anions under physiological conditions is nearly the same in each direction, because the rate of net exchange is, in general, a function of the affinities

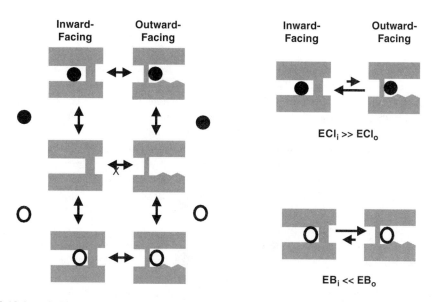

FIGURE 13.3 *Left*: Ping-pong catalytic mechanism of anion exchange. The transporter is depicted as having two possible conformations: inward-facing and outward-facing. The transition between these conformations is very slow in the absence of a bound transportable anion at the translocation site. The nature of the conformational transition is unknown but is the equivalent of a gate opening in front of the transported anion and another gate closing behind it, resulting in diffusional access to the opposite solution from where the anion originated. The rates of the translocation events depend on the particular substrate anion bound at the translocation site. *Right:* Asymmetry in the translocation events for Cl^- (above, filled circle) and HCO_3^- (below, open circle), as recently demonstrated by Knauf et al. [158]. The inward translocation rate for Cl^- is more rapid than the outward translocation rate; the opposite is true of HCO_3^-. This difference in translocation rates is a consequence of the fact that HCO_3^- has a higher true affinity for the outward-facing state than the inward-facing state; that is, the outward-facing protein with bound HCO_3^- (EB_o) is energetically more stable than the inward-facing state (EB_i). For Cl^-, the inward-facing state (ECl_i) is more stable than the outward-facing state (ECl_o).

and translocation rate constants for influx and efflux of both exchange partners (see [158]). Underlying asymmetries in the mechanism can be observed only when transport is measured in the presence of very asymmetric ion gradients.

13.10.3 INHIBITOR BINDING SITES

Although the basic features of the ping-pong mechanism are reasonably well established, it is clear that there are sites for transportable anions and transport inhibitors other than the translocation site. For example, it is well established that very high Cl^- concentrations inhibit Cl^-/Cl^- exchange [159]. The most plausible explanation of this inhibition is that there is an additional site, known as the *modifier site*, that causes transport inhibition when occupied by Cl^- or other anions [15, 16]. The self-inhibitory site is accessible from the extracellular medium [160].

Many arylsulfonates and other organic acids are inhibitors of band 3-mediated anion transport (see [57]). The kinetic patterns of inhibition are variable, with some compounds acting as competitive inhibitors and others noncompetitive. Some noncompetitive inhibitors bind to a site that is sensitive to the proportions of inward-facing and outward-facing states; therefore, the conformational transition between inward-facing and outward-facing sites affects the accessibility not only of the transport site itself but also of the site occupied by noncompetitive inhibitors such as flufenamic acid [161], eosin-5-maleimide [162], and WW781 [156].

The stilbenedisulfonate derivative H_2DIDS binds reversibly to AE1 before forming a covalent bond with a lysine residue [16]. The covalent reaction is sufficiently slow that the action of H_2DIDS

as a reversible inhibitor can be studied at low temperature. As an inhibitor of Cl$^-$/Cl$^-$ exchange at 0°C, H$_2$DIDS behaves as a competitive inhibitor [163]. The reversibly acting stilbenedisulfonate DNDS is also a competitive inhibitor [164]. However, the fact that stilbenedisulfonates appear to be competitive inhibitors of Cl$^-$ transport in a steady-state transport experiment does not imply that Cl$^-$ binding completely prevents stilbenedisulfonate binding. Salhany and coworkers have shown that Cl$^-$ lowers stilbenedisulfonate (DBDS) affinity by increasing the rate constant for release of the inhibitor, without causing a major change in the forward rate constant for binding [165]. This indicates that Cl$^-$, DBDS, and band 3 can form a ternary complex [166] in which DBDS is released more rapidly than it is in the absence of Cl$^-$. The relationship between the Cl$^-$ binding site in the DBDS complex and the Cl$^-$ transport site in normal band 3 is not yet known, but the relative apparent affinities for most inorganic anions for the two sites are similar [167]. Iodide, however, has a much lower than expected affinity for the band 3-DBDS complex, suggesting that DBDS binds to a site that is in an access channel leading to the transport site, which can be reached by many monovalent anions but not by iodide [167].

The high number of copies of erythrocyte band 3 per cell makes it possible to study Cl$^-$ binding to the protein by nuclear magnetic resonance (NMR). Early studies by Chan and coworkers [168–170] showed that ^{35}Cl NMR line broadening by erythrocyte membranes is blocked by DNDS in a manner consistent with the idea that the line broadening is a consequence of Cl$^-$ binding to the transport site. Falke et al. [171] also used ^{35}Cl$^-$ NMR to estimate the rate of binding and release of Cl$^-$ from the presumed substrate site; at 0°C, binding and release are considerably faster than translocation. More recently, Knauf and coworkers have used ^{35}Cl NMR to investigate the transport asymmetry discussed earlier [157], as well as the nature of inhibitor binding sites. Eosin-5-maleimide, which inhibits anion transport at a site that overlaps with but is not identical to the stilbenedisulfonate site, does not block rapid Cl$^-$ binding and release from band 3, as detected by ^{35}Cl line broadening, indicating that this agent does not inhibit transport by blocking an access channel to the transport site [172].

In summary, the previous spectroscopic and kinetic measurements of inhibitor binding all indicate that there are complex interactions among transport sites and inhibitor sites that can be reached from the extracellular surface of band 3. Eosin 5-maleimide interferes with stilbenedisulfonate binding, even though the two compounds bind to different sites (see later discussion). Cl$^-$ lowers DBDS affinity not by binding to the same site but by binding to a separate site and increasing the DBDS dissociation rate. The transport site conformation (inward vs. outward) affects the binding of several inhibitors to noncompetitive sites. A detailed understanding of these effects will require further structural information about the protein and its complexes with various inhibitors.

13.10.4 STOICHIOMETRY OF THE CATALYTIC CYCLE

The fact that band 3 is a tightly associated dimer raises the question of whether the catalysis of transport requires the concerted actions of both subunits. There is one H$_2$DIDS binding site per band 3 polypeptide [173], and there is a very good linear relationship between stilbenedisulfonate binding and transport inhibition [174–176]. These findings are consistent with the idea that each subunit of the dimer is capable of carrying out anion exchange, even if the opposite subunit is inhibited by H$_2$DIDS or DIDS.

The linear relationship between stilbenedisulfonate binding and transport inhibition does not rule out the possibility that the dimer is the functional unit of transport [177]. For example, transport catalysis could require both subunits of the dimer, but binding of stilbenedisulfonate to one subunit could lower the transport rate through the dimer by exactly 50% [165]. Another possibility is that the mechanism normally involves concerted interactions between subunits, but binding of stilbenedisulfonate disrupts these interactions. There are many kinetic features of anion transport and stilbenedisulfonate binding and release that indicate the presence of site/site interactions, some of which involve two different subunits. For example, the binding of DIDS to one subunit affects the thermal unfolding of the opposite subunit of the dimer [178]. As discussed earlier, DBDS release is accelerated by Cl⁻, and this acceleration could be mediated by interactions between subunits of the dimer [179].

Red cell anion transport kinetics themselves are complicated by self-inhibition, but there is little evidence of major positive or negative cooperativity in most of the data (e.g., [151, 180, 181]). The transport of some anions does exhibit negative cooperativity, which is relieved by binding of stilbenedisulfonate to one subunit of the dimer [182, 183]. This cooperativity may indicate the presence of subunit interactions in the catalytic cycle. It is not known with certainty, however, whether, in a complete catalytic cycle, the two exchanging anions cross the membrane by interacting solely with a single subunit, or whether the mechanism requires communication between the two subunits.

13.10.5 SINGLE TURNOVER MEASUREMENTS

Because of the abundance of erythrocyte band 3 (~15 nmol/ml cells), it is possible to measure the Cl⁻ flux associated with a single turnover of the transporter [184]. The stoichiometry of a single turnover is consistent with a catalytic cycle of one ion moving outward and one moving inward per band 3 subunit. It is also possible to detect a transient *uphill* efflux of Cl⁻ following addition of H_2DIDS, which binds preferentially to the outward-facing conformation [185]. In this experiment (Figure 13.4), intact red cells are initially at Donnan equilibrium in a CO_2-free Na-phosphate medium containing about 80 µM Cl⁻. Inorganic phosphate is transported about 10,000 times more slowly than Cl⁻ [16]. Therefore, even though phosphate is much more abundant than Cl⁻, the unidirectional flux of Cl⁻ is higher than that of phosphate. At the indicated time, Cl⁻-free H_2DIDS is added. The binding of H_2DIDS to outward-facing states upsets the equilibrium between inward-facing and outward-facing states, and the most rapid way for inward-facing states to become outward-facing is as the Cl⁻ complex. Therefore, addition of impermeant H_2DIDS causes a rapid transient uphill efflux of Cl⁻. The stoichiometry of the transient uphill efflux is indistinguishable from that predicted if each band 3 subunit in the dimer is a functional unit of transport [185]. This does not imply that the subunits can act completely independently, but it does argue that each subunit is capable of carrying out transport. The issue of subunit interactions in transport will be discussed further later in connection with genetic variants of AE1.

13.10.6 LACK OF MAJOR REGULATION OF AE1-MEDIATED ANION TRANSPORT

Red cell band 3 is constitutively active (i.e., functional, not active in the sense of an ATP-driven pump) as a transporter, as would be expected from its physiological function. In both the pulmonary and the systemic capillaries, net Cl⁻/HCO_3^- exchange mediated by band 3 is driven by the small HCO_3^- gradients that are generated by incoming or outgoing CO_2 and intracellular carbonic anhydrase. In order to react to these gradients during the short (1 sec or less) capillary transit time, the protein needs to be transport competent before the blood arrives at the capillary. During most of the time in the circulation, Cl⁻ and HCO_3^- are at Donnan equilibrium, and band 3 is carrying out the self-exchange of each of these anions with no net flux; this process has no energy cost and has no adverse consequences for the cell. There is little reason, then, to suspect that the anion transport function of erythrocyte AE1 is regulated.

Although band 3 is constitutively activated as a transporter, ATP depletion causes a moderate (~35 to 50%) decrement in anion transport [186–188]. The mechanism of this decrease is not clear. It has been reported that okadaic acid, which inhibits serine/threonine protein phosphatases types 1 and 2a [189], causes a significant increase in the rate constant for red cell oxalate transport [190], suggesting that a phosphorylation event activates band 3-mediated transport. However, Jennings and Adame [188] found no effect of okadaic acid on red cell oxalate transport. In the same cell preparation, okadaic acid inhibited K⁺/Cl⁻ cotransport, as expected from work with rabbit red cells [191]. Accordingly, although ATP depletion does have an effect on anion transport, there does not

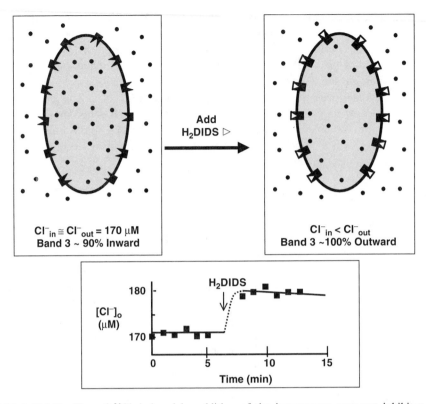

FIGURE 13.4 Uphill efflux of $^{36}Cl^-$ induced by addition of the impermeant transport inhibitor H_2DIDS. Initially, intact human erythrocytes are at Donnan equilibrium in a medium consisting of 150 mM Na-phosphate, pH 6.2, and 0.17 mM $Na^{36}Cl$, 23°C. The acid pH is needed to facilitate the complete removal of HCO_3^- by flushing the medium with N_2. Under these conditions, about 90% of the transporters face inward [185]. After cells and medium reach a stable Donnan equilibrium, H_2DIDS is added to a final concentration of 200 μM. H_2DIDS binds to outward-facing states, and, by mass action, causes initially inward-facing states to be recruited outward (mainly as the Cl^- complex). The final state (upper right) consists of nearly all the band 3 in the outward-facing state, bound with H_2DIDS. The H_2DIDS-induced outward recruitment of transporters is associated with a net, uphill loss of $^{36}Cl^-$ from the cells, as shown in the lower part of the figure (data from one of the experiments summarized in Table 1 of [185]).

appear to be major modulation of band 3-mediated transport by a phosphorylation/dephosphory-lation cycle.

There are several sites of tyrosine and serine/threonine phosphorylation on band 3, mainly on the cytoplasmic domain [37–42]. These sites, and the kinases and phosphatases that act on them, represent a continuing area of investigation, and the functions of these phosphorylation events are not yet clear. The other major posttranslational modifications of AE1 are palmitoylation of C843 and glycosylation of N642. Neither palmitoylation [192] nor glycosylation [193, 194] is necessary for anion transport through AE1.

13.11 AE1 MEMBRANE DOMAIN TOPOLOGY AND STRUCTURE

The membrane domain of AE1 (mdb3), which has identical amino acid sequence in erythroid and renal tissue, consists of 12 to 14 transmembrane helices and is responsible for anion transport. Complete removal of cdb3 by trypsin does not have a major effect on $SO_4^=$ transport [195, 196], and heterologous expression of the membrane domain without the cytoplasmic domain results in anion transport at rates comparable to those of full-length AE1 [113].

The membrane and cytoplasmic domains of human band 3 unfold independently of each other [197]. Moreover, proteolysis of unsealed membranes causes release of cdb3 without the need for extremes of temperature, salt, pH, divalent cations, or detergent, indicating that, in the absence of the covalent link between the two domains, cdb3 does not bind strongly to mdb3. *In situ*, however, the local cdb3 concentration is necessarily high, and the two domains may form a low-affinity, noncovalent association that is difficult to detect in an experiment using the separated domains. If such an association exists, it must be readily reversible.

13.11.1 2-D CRYSTALS OF AE1 MEMBRANE DOMAIN

The recent successes in determining high-resolution structure of several multispanning membrane proteins [11, 198, 199] provides encouragement for ongoing efforts to crystallize band 3 membrane domain. The crystals of mdb3 that have been prepared thus far, however, are not suitable for high-resolution X-ray diffraction [200]. Wang et al. [123, 124] have determined the structure of the membrane domain dimer (residues 361–911) at a resolution of 20 Å, using cryoelectron microscopy of two-dimensional crystals (sheets and tubes) (see also [201]). The dimer is tightly associated, and the electron density map shows a stain-filled canyon between the two subunits. Each subunit of the dimer consists of three apparent subdomains, two of which are in the plane of the membrane and may represent two different helical bundles. The third domain extends into the cytoplasm; this domain was in a different position relative to the other two domains in the sheet vs. tube preparations, indicating that its connection to the rest of mdb3 is flexible. It is not yet possible to make definitive assignments of electron densities to the specific parts of the sequence.

13.11.2 TOPOLOGY OF THE AE1 MEMBRANE DOMAIN

13.11.2.1 Biochemical Studies of the Native Erythrocyte Protein

The abundance of red cell band 3 has made it an attractive system for biochemical studies of the topology of the membrane domain [202–204]. Circular dichroism of the isolated membrane domain [205] showed that the membrane domain has a high α-helix content. NMR studies on the first and second TM segments also indicate a high α-helix content [206, 207]. Accordingly, the conceptual framework for building a model of the structure of mdb3 is based on transmembrane α-helices connected by hydrophilic loops.

From the known structures of multispanning membrane proteins (e.g., [11]), it is clear that intramembrane α-helices can be oriented at angles that are far from perpendicular to the plane of the membrane and that significant nonhelical structures are often found within the boundaries of the hydrophobic part of the bilayer. Therefore, working models consisting of TM helices and surface hydrophilic connector loops are almost certainly oversimplifications. Nonetheless, in the absence of information on tilt angles of individual α-helices in AE1, the following discussion will be framed in terms of TM helices and hydrophilic surface connector loops (denoted EC or IC, depending on sidedness).

On the basis of the polypeptide sequence [44], chemical labeling [16, 17], *in situ* proteolysis [202, 208, 209], and antibody binding [210, 211], a model of the membrane domain was developed in which there are 14 transmembrane helices (Figure 13.5). Some aspects of the topological model in Figure 13.5 are established very firmly. For example, both the N- and C-termini are definitely cytoplasmic [13, 210, 211]. It is also clear that certain residues, e.g., the site of N-glycosylation (N642 [35]), the reaction sites for impermeant chemical probes such as eosin 5-maleimide [212] and H_2DIDS [213], or sites of proteolytic cleavage of intact cells [202, 208] are accessible from the extracellular medium. Further markers for extracellular sites come from the study of variant blood group antigens, which are presumably extracellular. Jarolim et al. [214] have identified mutations associated with several low-incidence blood group antigens (R432W in EC1; Y555H, P561S, G565A, N569K in EC3; R656C and R656H in EC4). These regions of the sequence are

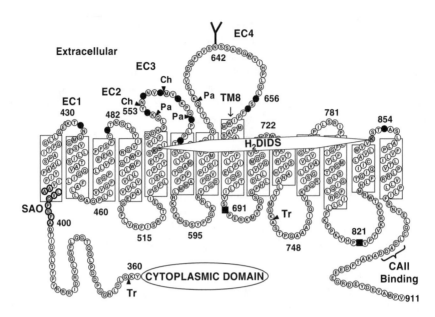

FIGURE 13.5 Model of the 2-dimensional arrangement of the polypeptide in the membrane domain of human AE1. Locations of the two lysine residues (K539 abd K851) that react covalently with the same molecule of the anion transport inhibitor H$_2$DIDS are indicated [213]. Sites of *in situ* proteolysis of the nondenatured protein by trypsin (Tr), chymotrypsin (Ch), and papain (Pa) [208, 209] are indicated. Residues in black with white letters are sites of point mutations resulting in variant blood group antigens: R432W, Y555H, P561S, G565A, N569K, R656C, and R656H [214]; E480K [215]; E658K [120]; and P854L [216]. The site of binding of carbonic anhydrase II (D887-D890) is indicated [106]. The nine residues (A400-A408) that are deleted in Southeast Asian ovalocytes [274–276] are shown as thick circles. Two residues (K691 and D821) identified by chemical labeling experiments as being cytoplasmic [309, 310] are indicated as black squares with white lettering. Extracellular loops 1–4 (EC1, etc.) and transmembrane segment 8 are labeled because they are discussed in more detail later in the text. No attempt has been made to incorporate into this model the NMR structures of synthetic peptides corresponding to the first two TM segments [206] and the loop immediately preceeding TM13 [311].

all known to be accessible to chemical labeling and/or proteolysis at the extracellular surface [16, 17, 212]. In addition, the variant E480K (in EC2) is associated with the Froese blood group [215]. Finally, one of the Diego variants is associated with a mutation of P854 [216], which is close to a lysine residue (K851) that reacts with the impermeant anion transport inhibitor H$_2$DIDS [213].

13.11.2.2   Heterologous Expression and Mutagenesis

In addition to using biochemical and immunological markers in native erythrocyte AE1, it is possible to study the topology in AE1 expressed in functional form in *Xenopus* oocytes or cultured mammalian cells. Expression in yeast (*S. cerevisiae*) is also possible, although most of the protein does not traffic to the plasma membrane [114, 217]. Topological studies of heterologously expressed AE1 have been based on site-directed mutagenesis to insert consensus sites for N-glycosylation [218, 219] or cysteine residues for chemical labeling [220, 221]. Much of the data using these approaches is in agreement with biochemical studies in red cells. For example, it is possible to remove the normal glycosylation site (N642) and insert consensus sites for N-glycosylation elsewhere in the sequence. As long as the extracellular loop is sufficiently long (~30 residues [222]), insertion of a consensus site for N-glycosylation in known extracellular loops gives rise to the expected glycosylated polypeptide in either a cell-free translation system [218] or intact COS or HEK293 cells [219]. However, substitution of asparagines for K743 (which gives a consensus N-glycosylation site without any other modifications in the protein, because serine is in position 745), results in a partially glycosylated product in the cell-free translation system [218]. Therefore, during synthesis/translocation, residue 743 is exposed to the lumen of the ER in a loop that is sufficiently long for N743 to receive a high-mannose carbohydrate chain in about half the copies of the protein. Glycosylation of the K743N mutant is surprising, because K743 had previously been thought to be cytoplasmic, on the basis of proteolysis experiments in intact cells and resealed ghosts [209].

In contrast to the *in vitro* translation experiments, expression of the K743N construct in COS cells does not result in a stable N-glycosylated product [219], indicating that copies of the polypeptide that receive N-glycosylation during synthesis are degraded before they reach the cell surface. The disparity between the results of cell-free translation and whole-cell expression of K743N mutants can be explained by a model in which residue 743 is transiently exposed to the lumen of the ER during cotranslational insertion. If the residue at this site is asparagine, at least some of the copies of the protein receive a high-mannose carbohydrate chain, which is detected in the cell-free translation experiments. However, in intact cells these copies of the protein are presumably recognized as misfolded and are degraded before reaching the plasma membrane.

There are other indicators that the topology of the protein in the vicinity of K743 is unusual. Fujinaga et al. [220] have expressed AE1 containing cysteine insertions at various position, in a Cys-less background protein, which is functional because no cysteine residue is essential for AE1-mediated anion transport [110]. Cys substitutions at positions 742, 745, or 751 give rise to protein that can carry out anion exchange, and each inserted Cys can be labeled by Lucifer yellow iodoacetamide, which appears to be impermeant. By this criterion, these residues are exposed on the extracellular surface. The cysteine mutagenesis experiments are clearly at odds with proteolysis experiments using intact cells, resealed ghosts, unsealed membranes, and inside-out vesicles [209, 223], all of which indicate that K743 is accessible on the cytoplasmic, but not the extracellular side of the membrane.

Heterologous expression with cysteine substitution mutagenesis has provided evidence that hydrophilic loops downstream from K743 are exposed to the extracellular medium [224]. Cysteine substitutions and chemical modification showed that cysteine residues inserted in the positions of S852-A855 are accessible to extracellular medium. These residues are expected to be extracellular on the basis of reactivity of K851 with H_2DIDS [213] and the blood group antigen associated with P854 [216]. However, the same cysteine scanning study also showed that residues in the sequence P815-K829 are extracellular. There was previous evidence that this sequence is extracellular in erythrocytes infected with *Plasmodium falciparum* [69], but antibody binding experiments with normal erythrocytes indicated that the sequence is intracellular [225]. A topological model of mdb3 beginning with TM9, based on the results of Zhu et al. [224], is shown in Figure 13.6, with further discussion of the topology of residues 800–840 in the legend.

It is not clear why there is such disagreement regarding the topology of the protein in the sequences surrounding K743 and P815-K829. The cysteine scanning experiments were well controlled, with transport assays to confirm that the mutated proteins are functional [220]. It is possible that the topology of the protein in these regions is labile and that a minor perturbation represented by a cysteine substitution may cause loops that are normally cytoplasmic to be folded into a conformation with extracellular exposure. If there were two populations of expressed protein, with two alternate conformations, the transport assay would detect the normally folded conformation, and the labeling assay could detect the abnormally folded conformation. Another potential explanation is that the sidedness of a loop in the mature protein can actually change with different protein conformations as part of the conformational changes associated with anion transport events. However, the turnover number for monovalent anion transport through band 3 is quite high (see earlier discussion), and it is difficult to envision a conformational change involving an entire loop of polypeptide (ten or more residues) completely changing sidedness so rapidly.

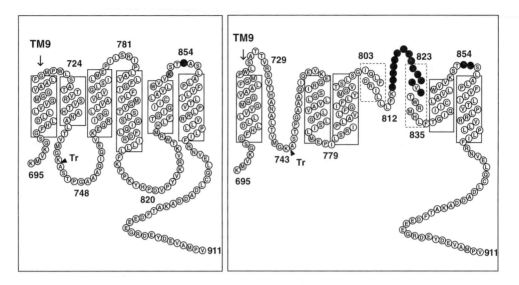

FIGURE 13.6 Alternative topological models of the membrane domain of human band 3, beginning after TM8. The model on the left is as in Figure 13.5 for comparison. The model on the right is derived from that of Zhu et al. [224] and is based on cysteine scanning mutagenesis/chemical modification experiments in AE1 expressed in HEK 293 cells [220, 224]. Residues depicted in black with white lettering are reactive with an impermeant cationic reagent (bromotrimethylammoniumbimane bromide) in intact HEK293 cells expressing the protein with cysteine inserted at this position. The extracellular localization of these residues is not in agreement with data from antibody binding [224], chemical labeling [310], or glycosylation scanning [218], indicating that this sequence is cytoplasmic in the native erythrocyte protein. On the other hand, a peptide from this sequence blocks cytoadherence of erythrocytes infected with *Plasmodium falciparum* [69], indicating that this sequence is extracellular, at least in infected erythrocytes. The structure of the synthetic peptide corresponding to residues 796–840 has been determined by NMR to have two helical stretches (one α helix and one 3_{10} helix, each of which is amphipathic) in 30% trifluoroethanol [311]. The positions of these helices (I803-L810 and Y824-L835) are indicated by the dashed boxes on the right side of the figure to indicate a possible topology for this sequence. It is not known whether the structure of this sequence in 30% trifluoroethanol is the same as in the intact protein.

13.11.3 TWO CATEGORIES OF TRANSMEMBRANE HELICES

Hamasaki and coworkers have pointed out that, in general, there should be two categories of TM helix in a multispanning protein [226]:

* Category 1 — Those that interact directly with the lipid bilayer
* Category 2 — Those that are surrounded entirely by other polypeptides or by an aqueous cavity

In mdb3, exposure to alkaline pH causes partial unfolding of the protein and exposes new sites of proteolytic cleavage; some peptides that are almost certainly in transmembrane helices can be released from the membrane by alkaline exposure followed by trypsin proteolysis [227, 228]. For example, the topology of the sequence comprising TM1 is reasonably well defined. The cytoplasmic end is near P403 [207], and the extracellular end must be fairly close to K430, the site of labeling by extracellular eosin 5-maleimide [212], and R432, the site of a mutation causing a rare cell surface antigen [214]. Alkaline exposure causes this TM segment to become susceptible at several positions to proteolytic enzymes [228]. These findings can be explained by a model in which TM1 is in Category 2, i.e., not stabilized by the lipid bilayer sufficiently to resist unfolding at alkaline pH. TM 2, 3, 5, 6, and 10 in the model in Figure 13.5 also behave as Category 2 TM helices [228].

If these TM segments are not in contact with the bilayer, many possible arrangements of the TM helices in the plane of the membrane can be ruled out.

13.11.4 TOPOGENESIS OF TM SEGMENTS

An important aspect of the study of the topology of a membrane protein is to understand how a given topology is generated during the synthesis/insertion of the protein. Because so many aspects of AE1 are relatively well characterized, AE1 is an attractive system for the study of topogenic signals of a complex (12 to 14 TM) multispanning membrane protein. Individual TM segments of AE1 have been tested for their properties as signal anchor I (SAI)/stop transfer function (resulting in a cytoplasmic C-terminus of the TM segment) or signal anchor II (SAII, resulting in a cytoplasmic N-terminus of the TM segment) [229]. Odd-numbered TM segments are expected to act as SAII sequences and direct the C-terminus of the segment to the lumen of the ER; interestingly, neither TM3 nor TM5 served this function in the translation assay used by Ota et al. [229]. However, TM4 and TM6, as expected for even-numbered segments, exhibit strong SAI activity. The most reasonable interpretation of this finding is that, in the synthesis and insertion of the whole protein, the strong SAI function of TM4 and TM6 makes up for the weak SAII activity of TM3 and TM5, respectively, resulting in the insertion of TM3-TM4 and TM5-TM6 as hairpins in the proper orientation (i.e., helix-loop-helix with extracellular loop).

The rules for AE1 insertion/topogenesis are complex, however. In addition to effects of sequences immediately upstream or downstream of a given TM segment, there are topogenic effects of sequences that are much more distant. Kanki et al. [230] recently showed that the sidedness of sequences near K743 depend on the rate of elongation of the polypeptide and on the presence or absence of sequences as distant as TM1-3 and 12-14. These data, as well as those discussed previously [218, 219], all indicate that K743 is transiently exposed to the lumen of the ER during synthesis/insertion and is then folded back into the membrane and is actually exposed on the cytoplasmic surface in the mature protein [223].

13.11.5 COEXPRESSION OF COMPLEMENTARY FRAGMENTS

Tanner and coworkers have performed a systematic study of the expression of combinations of fragments of AE1 in *Xenopus* oocytes to assess the ability of the resultant polypeptides to assemble, traffic to the surface membrane, and carry out anion transport. The first such studies used two pairs of fragments: TM1-8 + TM9-14, or TM1-12 + TM13-14 (TM segments defined as in Figure 13.5). In both cases, the resultant assembly exhibits functional anion transport, indicating that the cytoplasmic loops connecting TM8 to TM9 and TM12 to TM13 do not need to be intact in order for the protein to function [231].

Subsequent studies used complementary fragments that are believed normally to be connected by extracellular loops [232]. To ensure that the downstream fragment would be oriented with extracellular N-terminus, the fragment was fused with the glycophorin A leader sequence, which is cotranslationally cleaved (confirmed by Edman degradation), leaving the N-terminus of the AE1 fragment in the lumen of the ER. Pairs of fragments separated by EC3 or EC4 could form transport-competent assemblies. The EC3 result is expected from the lack of effect of extracellular chymotrypsin (which cleaves EC3) on anion transport in intact erythrocytes [16]. The EC4 result is reasonably consistent with earlier studies with intact red cells. Extracellular papain cleaves Q630, i.e., in EC4 near the end of TM7 [208], and has complex effects on anion transport, including inhibition of Cl⁻ efflux [16] and acceleration of $SO_4^=$ influx [233], suggesting that papain cleavage of EC4 stabilizes the inward-facing conformation. The fact that cleavage of EC4 by papain does not completely inhibit Cl⁻ transport in erythrocytes is consistent with the fact that complementary fragments separated by EC4 can form a functional assembly [232].

In contrast to EC3 and EC4, complementary fragments normally connected by EC1 or EC2 do not transport anions [232]. The EC1-separated fragments (i.e., TM1 and TM2-14) do not traffic to the plasma membrane, but the fragments on either side of EC2 (TM1-3 and TM4-14) form a stable assembly that does not transport anions, suggesting that the integrity of EC2 (which is rather short) is necessary for transport. In fact, none of the extracellular or intracellular loops in TM1-5 can be interrupted without a loss of transport function of the fragments expressed in *Xenopus* oocytes [234]. This finding, and the fact that EC3 (following TM1-5) is poorly conserved and may be cleaved without loss of function, indicates that TM1-5 is a tightly folded subdomain of mdb3. This idea is also consistent with the fact that no extracellular or intracellular loop within TM1-5 has been cleaved by *in situ* proteolysis except under denaturing conditions.

Tanner and coworkers have expressed complementary fragments in groups of three or four rather than two; they have also expressed noncomplementary fragments, in which one or more TM segment is missing or in which there is an extra TM segment [235]. Most of the constructs with missing segments do not support transport; surprisingly, however, the construct that lacks TM6-7 is able carry out stilbenedisulfonate-sensitive anion transport. This helix-loop-helix hairpin contains a cluster of six positive charges at or near the cytoplasmic surface of the membrane, one of which (K590) reacts covalently with phenylisothiocyanate under conditions in which this agent inhibits transport [236]. Despite this association with a transport inhibitor, TM6-7 do not appear to be absolutely essential for anion exchange.

13.11.6 ARRANGEMENT OF TM HELICES IN THE PLANE OF THE MEMBRANE

There have been several attempts to define amino acid residues associated with the subunit interface in the membrane domain dimer. Crosslinking studies with intact erythrocyte have provided evidence that the third extracellular loop (EC3), between TM5 and TM6 (Figure 13.5), contains sites that may be involved in subunit interactions. The homobifunctional amino reagent BS^3 (bis[sulfosuccinimidyl]suberate) was originally shown by Staros [127, 237] to form an intermolecular crosslink between two complementary chymotryptic fragments of erythrocyte AE1. Jennings and Nicknish [128] subsequently showed that at least one end of the BS^3 intermolecular crosslink involves a lysine residue (K551 or K562) close to the extracellular chymotryptic cleavage sites (Y553 and L558). Therefore, either K551 or K562 (or both) in EC3 is sufficiently close to a lysine residue in the adjacent subunit to make a crosslinked with BS^3 possible.

Casey and coworkers [238] have used cysteine mutagenesis to insert cysteine residues that were then crosslinked by homobifunctional sulfhydryl reagents or by oxidation with Cu^{2+}/o-phenanthroline. Using this approach, intermolecular -SS- crosslinks could be formed in AE1 with a cysteine residue inserted into position 431 (EC1), 456 (IC1), 486 (EC2), 555 (EC3), 565 (EC3), 656 (EC4), 731, and 751 (IC5 or EC5, depending on the model; see earlier discussion). Crosslinking did not change the hydrodynamic radius of the protein, indicating that, as expected, the protein is a stable, noncovalent dimer before crosslinking. The efficiency of crosslinking is variable (0.2 to 0.8), but the results clearly indicate that several different sites in the AE1 sequence are sufficiently close to the same residue in the neighboring subunit to form an -S-S- crosslink. As pointed out by Taylor et al. [238], even a "zero-length" -S-S- crosslink in an extramembranous loop still allows a significant distance between adjacent TM helices, especially if the crosslink site is several residues from the end of the TM helix. Accordingly, precise statements about proximity relations in the dimer cannot yet be made on the basis of these crosslinking experiments. It is clear, however, that several different parts of the sequence can be crosslinked in the band 3 dimer. The finding that sites in EC3 in adjacent subunits can be crosslinked is in agreement with earlier work on erythrocytes [128].

Groves and Tanner [235] have coexpressed fragments of AE1 with intact mdb3 (or whole band 3) and immunoprecipitated the complex to determine what portions of the polypeptide are associated with the adjacent subunit of the dimer. All fragments except the last two TM segments coprecipitated with mdb3. A working model of the subunit interface based on these and previous coexpression studies was proposed by Groves and Tanner [235]. Figure 13.8 depicts the main features of this model, along with indications of the approximate positions of intrasubunit (H_2DIDS [213]) and intersubunit [128, 238] crosslinks. The model depicts TM13-14 as a separate subdomain from the rest of the membrane domain, based on the finding that these two TM segments (plus the C-terminal cytoplasmic domain) can be inserted into ER in a stable manner even if the rest of the protein is absent [231], suggesting that it is a relatively self-contained domain. However, K851 (in the short EC loop between TM13 and TM14) can be crosslinked by H_2DIDS to K539 (near the C-terminal end of TM5) in the same subunit [213], indicating that TM13-14 are associated with the rest of the protein closely enough for this crosslinking to take place. The dimer model of Groves and Tanner does not depict EC3 or EC4 as being near the subunit interface, even though these loops can participate in intersubunit crosslinking [128, 238]. As mentioned previously, both EC3 and EC4 are long and presumably flexible enough to be crosslinked to the adjacent subunit, even if TM5-8 are not themselves at the subunit interface.

An additional set of potential constraints on the arrangement of helices in the plane of the membrane is that TM 1, 2, 3, 5, 6, and 10 behave as Category 2 TM helices and are probably not in close contact with lipid bilayer [228]. Even with these constraints, and those imposed by the lengths of each of the EC and IC connector loops, it is not possible to propose a unique arrangement of TM helices; with 14 (or even 12) helices, the number of possible combinations is enormous. As emphasized by Groves and Tanner [235], models such as that in Figure 13.8 are not intended to be a unique representation of the experimental facts but rather are presented as a guide for future experiments. It is clear that much further structural work is needed before definitive statements about the subunit interface can be made.

13.12 FUNCTIONAL ROLES OF SPECIFIC AMINO ACID RESIDUES IN TRANSPORT

13.12.1 LIMITATIONS OF CHEMICAL MODIFICATION OR MUTAGENESIS EXPERIMENTS

There have been many chemical modification and site-directed mutagenesis experiments on AE1; the general goal of these experiments is to identify amino acid residues that are important in anion transport. It is important to recognize the limitations of such experiments. The replacement of a particular residue in AE1 may alter anion transport by a wide variety of mechanisms. Even if the mutated protein is properly folded and traffics normally to the plasma membrane, it is exceedingly difficult to draw valid conclusions about the events in transport catalysis that are altered by mutagenesis. Suppose, for example, that a given amino acid substitution raises the concentration of extracellular Cl⁻ needed to half-saturate the ^{36}Cl⁻ influx. It would be tempting to conclude that this residue has a role in the binding of extracellular Cl⁻. Such a conclusion would not necessarily be correct, because the half-maximal substrate concentration depends not only on the true binding affinity but also on the unidirectional rate constants for both inward and outward translocation (e.g., [14, 158]). A mechanistic interpretation of site-directed mutagenesis experiments therefore requires a very thorough kinetic characterization of each mutant protein; this has not been possible for most of the amino acid residues discussed subsequently.

13.12.2 LYSINE RESIDUES

The membrane-impermeant bifunctional anion transport inhibitor H_2DIDS reacts with two different lysine residues within the same band-3 subunit to form an intramolecular crosslink between two different chymotryptic fragments [173]. The H_2DIDS-reactive lysine residues are K539 and K851 [213]; in the folded protein, the ε amino groups of these two residues must be close enough (~13 Å) to each other to be crosslinked by H_2DIDS. Neither of these lysine residues is necessary for anion transport in the *Xenopus* oocyte expression system [108, 109]. However, substitution of cysteine for K851 strongly inhibits anion exchange in AE1 expressed in HEK293 cells [224]. In the same study, substitution of cysteine for several other lysine residues (K814, K817, K826, K829) had relatively minor effects on transport [224].

Another lysine residue that has been studied by chemical modification is K430, which is near the extracellular end of TM1 and is the site of covalent reaction with eosin-5-maleimide site [212]. Eosin-5-maleimide inhibits anion transport by binding to an exofacial site that is distinct from the transport site and from the stilbenedisulfonate site, as discussed earlier. The possible role of K430 itself in transport is not clear. Another lysine residue that has been chemically modified is K590, the site of covalent reaction with transport inhibitor phenylisothiocyanate [236]. This residue is one of a cluster of six positive charges in a cytoplasmic loop between TM6 and TM7. The positive charges in this sequence are well conserved among SLC4 family members, but K590 itself is not; there is a glutamine in this position in both chicken and trout AE1. Also, as discussed previously, the entire TM6-7 hairpin can be eliminated without complete loss of transport [235]. It is unlikely, therefore, that K590 has a critical role in anion transport.

13.12.3 Arginine Residues

Band 3-mediated monovalent anion transport is only very slightly affected by either intracellular or extracellular pH in the physiological range (6.8 to 7.8) [180, 239]. (The lack of effect of pH in this range on AE1 contrasts with substantial effects on AE2 [240–242]. The activation of AE2 by alkaline cytoplasmic pH is necessary in order for AE2 to act as a base extrusion system to protect cells from alkalinization. Amino acid residues of importance in mediating pH effects on AE2 are beginning to be identified [243].) Although pH has little effect on AE1-mediated Cl^- transport in the physiological range, more extreme pH changes have strong effects. Very high pH inhibits Cl^- transport, consistent with a role for one or more arginine residues in anion binding and/or translocation [244–246].

Chemical modification with arginine-selective reagents, such as phenylglyoxal, cause irreversible inhibition of anion exchange [247–249]. Mutagenesis of mouse R509 (human R490) results in an unstable protein [250]; naturally occurring mutations at this site result in hereditary spherocytosis caused by band 3 instability [251, 252]. Site-directed mutation of R730 to either glutamine or lysine does not prevent stable expression of the protein in *Xenopus* oocytes but causes inhibition of transport [250], suggesting that this residue has an essential role in transport. R730 is conserved in all known AE1, AE2, and AE3 sequences, but a hydrophobic residue is in this position in most other members of SLC4, including the sodium bicarbonate cotransporters and Na^+-dependent Cl^- HCO_3^- exchangers. This residue may have an important role in the Na^+-independent anion exchangers, but other arginine residues may also be part of the transport pathway. Substitution of cysteine for R808 or R870 strongly inhibits anion exchange in the HEK293 expression system [224]. Mutation of other arginines in the C-terminal ~100 residues has either no effect (R832, R871) or causes slight inhibtion (R827, R879) [224]. In summary, it is clear that certain arginine residues are important for AE1 function, but exact roles for specific arginine residues are not yet known.

13.12.4 Role of Glutamate 681

Acid extracellular or intracellular pH inhibits Cl^- transport and at the same time accelerates the transport of monovalent anions such as $SO_4^=$ [16, 253]. The opposite effects of pH on monovalent and divalent anion transport led Gunn [254] to propose that protonation of a titratable group in the acid pH range (pK ~5.5) converts band 3 from a monovalent to a divalent anion transporter. In keeping with this model, the net exchange of Cl^- for $SO_4^=$ in red cells is accompanied by an obligatory, stoichiometric cotransport of H^+ with $SO_4^=$ [253, 255].

Chemical modification with Woodward's reagent K (WRK) and BH_4^- has demonstrated that a specific glutamate residue, E681, is the H^+-titratable group that converts band 3 from the monovalent to the divalent form [256, 257]. Removal of the negative charge on this residue by either WRK/BH_4^- modification [258] or site-directed mutagenesis [259] strongly accelerates AE1-mediated $SO_4^=$

transport and causes net $Cl^-/SO_4^=$ exchange to be an electrogenic 1:1 exchange rather than an electroneutral exchange of Cl^- for $SO_4^= + H^+$. These effects are exactly what would be predicted if E681 were the titratable residue associated with $SO_4^=/H^+$ cotransport. The precise role of E681 in monovalent anion exchange is not known, but this residue is clearly closely associated with the transport pathway. However, it is important to point out that modification of this residue with WRK does not prevent Cl^- and stilbenedisulfonate binding [260].

Glutamate 681 is located toward the cytoplasmic side of transmembrane helix 8 (TM8). E681 can be labeled from the extracellular surface of intact cells at 0°C by short exposures to WRK, a zwitterionic, hydrophilic reagent [257], indicating that the residue is outside the permeability barrier. Tang et al. [221, 261] have used heterologous expression, mutagenesis, and chemical modification to examine the sidedness of individual residues in TM8. Cysteine residues inserted into some sites in TM8 give rise to transport that is inhibited by SH reagents. Inserted cysteine residues on the N-terminal side of E681 can be reached from the extracellular surface by impermeant hydrophilic reagents, but residues on the C-terminal side of E681 cannot be labeled by these reagents [221]. These data strongly indicate that TM8 is an α-helix leading from the extracellular surface to a site, E681, which is critically important for transport and is very close to the barrier that separates intracellular from extracellular (Figure 13.7).

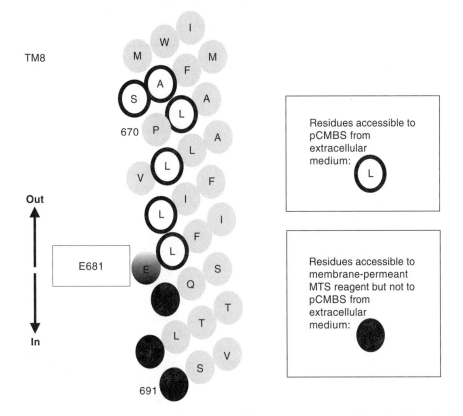

FIGURE 13.7 Amino acid residues in TM8, showing differing sidedness of residues on either side of E681, derived from heterologous expression and cysteine scanning mutagenesis of AE1 in HEK293 cells [221, 261]. Residues accessible to reaction with extracellular pCMBS are depicted as black circles with white interiors and black lettering. Residues accessible to a membrane-permeant MTS reagent but not extracellular pCMBS are depicted as solid black circles with white lettering. Residues that can be labeled are all on the same face of the helix as E681. Location of E681 at the permeability barrier agrees very well with chemical modification experiments in erythrocytes (see text). Further confirmation of the cytoplasmic orientation of K691, at the C-terminal end of TM8, comes from the chemical modification experiments of Erickson and Kyte [309].

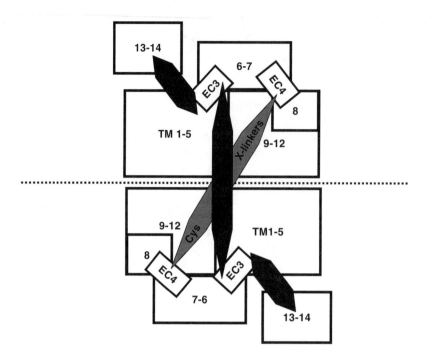

FIGURE 13.8 Model for the arrangement of 14 TM helices in the plane of the membrane, based on that of Groves and Tanner [235]. The position of the known H_2DIDS crosslink between K539 (C-terminal end of TM5) and K851 (C-terminal end of TM 13) within the same subunit is indicated [213]. Sites of intermolecular crosslinking with either BS^3 in EC3 [128] or crosslinkers reactive with inserted cysteine residues in EC3 or EC4 [238] are indicated. The TM helices believed to be surrounded by protein rather than lipid (Category 2) are TM 1, 2, 3, 5, 6, and 10 [226]. If this is correct, then only TM4 in the TM1-5 subdomain is in contact with the bilayer.

Salhany and coworkers [262] recently showed that modification of E681 with WRK/BH_4^- results in the appearance of a second binding site for Cl⁻, as detected by the kinetics of Cl⁻-stimulated DBDS release. The effects of WRK/BH_4^- modification on the kinetics of DBDS release are consistent with a model in which modification with WRK/BH_4^- of one subunit of the dimer exposes a cryptic Cl⁻ binding site; the binding of Cl⁻ to this site causes an increase in DBDS fluorescence [262]. There may be interpretations of the effects of WRK/BH_4^- on DBDS release kinetics that do not involve subunit interactions, but, in any case, the removal of the negative charge on E681 causes the apparent formation of a new Cl⁻ binding site. Interestingly, mutagenesis of a glutamate residue in the *E. coli* chloride channel causes an additional Cl⁻ binding site to appear [12]; in the native channel, the glutamate side chain occupies the site instead of Cl⁻. In the *E. coli* Cl⁻ channel, the glutamate side chain is involved with gating the channel. It is possible that, in the anion exchanger AE1, the E681 side chain acts in an analogous way by mimicking a Cl⁻ ion. Neutralization of this residue by low pH, chemical modification, or mutagenesis may make it possible for Cl⁻ to bind to the site that was previously occupied by the E681 side chain. Also, as pointed out by Salhany et al. [262], competition between Cl⁻ and the E681 side chain in normal band 3 could have a role in self-inhibition of Cl⁻ transport at high substrate concentrations.

13.12.5 HISTIDINE MODIFICATION AND MUTAGENESIS

Chemical modification studies with diethylpyrocarbonate (DEPC) have indicated that one or more histidine residues have a role in erythrocyte anion transport [263, 264]. Mutagenesis of the mouse residues corresponding to human H703, H819, and H834 to glutamine, or human H734 to serine, causes inhibition of $^{36}Cl^-/Cl^-$ exchange in the protein expressed in *Xenopus* oocytes [265]. The H703Q and H834Q mutants showed essentially no activity. In all cases, detectable polypeptide was expressed, but it is not clear what proportion of the protein reached the cell surface. Interestingly, introduction of a second mutation in the position of the H_2DIDS-reactive lysine residue K539 rescued transport in the H703Q, H819Q, and H834Q mutants. The mechanism of this rescue is not yet known. Diethylpyrocarbonate inhibits transport in oocytes expressing each of the histidine muta-tions, indicating that more than one histidine residue (or possibly reaction of DEPC with some other amino acid side chain) is responsible for DEPC inhibition of transport. Inhibition by DEPC was slowest in the H734S mutant, suggesting that this residue may be the primary target of DEPC [265]. In a separate study, Müller-Berger et al. [266] showed that mutagenesis of H743 to glutamine and of E681 to aspartate causes similar changes in the extracellular pH dependence of Cl^-/Cl^- exchange, suggesting that these two residues may act in concert to produce the native pH dependence.

13.13 GENETIC ABNORMALITIES IN AE1

13.13.1 EFFECTS OF DELETION OF AE1

The physiological consequences of the complete or near absence of AE1 have been studied in several settings. A strain of cattle lacking AE1 was discovered several years ago in Japan [267]; these animals are homozygous for a mutation that introduces a stop codon in the position corre-sponding to human R646. The resultant polypeptide is not stably incorporated into the membrane, and red cells from these animals have no detectable band 3. The cells also lack protein 4.2 (as is also true of the AE1 deletion mouse [268, 269]) and have significant decreases in the amounts of spectrin, actin, and G3PD. Analysis of arterial and venous blood gases demonstrated that arterial blood is somewhat acidotic (pH 7.30), with normal PCO_2. Venous pH is acid (pH 7.17), and venous PCO_2 is elevated. The blood gas values are consistent with the idea that, without Cl^-/HCO_3^- exchange, the tissue-to-lung PCO_2 gradient is larger than normal (see [270]).

Peters et al. [268] deleted the AE1 gene in mice and showed that, although AE1 -/- mice are severely anemic as a result of shortened red cell lifespan, the red cell membrane skeleton is essentially normal, despite the fact that AE1, the main attachment site for the skeleton, is absent. Southgate et al. [269] prepared mice with targeted deletion of erythroid but not renal AE1. In agreement with Peters et al. [268], red cells lacking AE1 can assemble a normal membrane skeleton. Both studies demonstrated that the membrane itself is quite unstable in the absence of AE1, with spontaneous shedding of lipid vesicles, indicating that band 3 normally stabilizes the membrane by interacting with membrane lipids. Stabilization of the membrane through lipid/band 3 interactions could help explain the hereditary spherocytosis associated with band 3 deficiency (see later section).

13.13.2 NATURALLY OCCURRING MUTATIONS IN HUMAN AE1

13.13.2.1 Memphis Mutation

The first human AE1 genetic variant was identified by Mueller and Morrison [271] on the basis of altered mobility of the 60 kDa N-terminal chymotryptic fragment on SDS polyacrylamide gels. This variant, designated the *Memphis mutation,* was subsequently shown to result from a substi-tution of glutamate for lysine in position 56 [272, 273]. The Memphis mutation is common and does not cause major abnormalities in either anion transport or the interactions of band 3 with the membrane skeleton.

13.13.2.2 Southeast Asian Ovalocytosis

An AE1 variant with deletion of residues 400–408 is relatively common in Malaysia and Papua New Guinea and causes a condition known as *Southeast Asian ovalocytosis* (SAO) [274-276]. The deletion is at the N-terminal end of the first transmembrane segment (Figure 13.5) and causes the TM helices to be abnormally organized but not unfolded [277]. The protein is stably incorporated into the membrane, but SAO band 3 does not exhibit detectable anion transport [278]. SAO red cells are abnormally rigid and are resistant to *Plasmodium falciparum*, thus providing an explanation of the prevalence of the gene in regions of the world where malaria is endemic. SAO red cells contain a higher proportion of band 3 tetramers (i.e., the fraction of band 3 with low lateral mobility) than normal red cells [279].

Red cells of heterozygotes for the SAO deletion have nearly equal amounts of normal and SAO polypeptide in the membrane [175]. Anion transport in red cells of SAO heterozygotes is about half the normal value (see later discussion). SAO band 3 polypeptide does not bind stilbenedisulfonates with high affinity [175]; this property makes it possible to test for the presence of heterodimers between normal and SAO polypeptide by forming an intramolecular crosslink with [³H] H_2DIDS between the two major chymotryptic fragments of the normal polypeptide [173, 213], followed by intermolecular crosslinking with BS³ [127]. The resultant product included [³H]H_2DIDS-labeled normal band 3 crosslinked to chymotryptic fragments of SAO band 3, indicating the presence of heterodimers [280]. The proportions of labeled crosslinked fragments were consistent with random association between normal and SAO subunits.

Jennings and Gosselink [280] reported that the $SO_4^=$ flux in SAO cells is 55% of normal under conditions of saturating $SO_4^=$ concentration. SAO cells have about 55% of the normal amount of wild-type band 3 [281], suggesting that the presence of SAO polypeptide in the heterodimer has no effect on $SO_4^=$ transport in the normal subunit of the dimer. However, these data were from a single donor, and the more extensive data from Tanner and coworkers [175, 281] indicate that both $SO_4^=$ and Cl⁻ transport in SAO cells (per DIDS binding) site is 75% of normal. Moreover, coexpression of SAO with normal AE1 (and GPA) in *Xenopus* oocytes causes 30% inhibition of Cl⁻ transport relative to normal AE1 and GPA alone [278]. Therefore, there is a detectable effect of the nontransporting SAO polypeptide on the normal subunit of the dimer. It is clear, however, that a dimer with one normal subunit and one SAO subunit can transport anions reasonably rapidly. Further evidence for functional effects of SAO on the neighboring subunit is that the rate of release of bound H_2DIDS from the SAO heterodimer is about fourfold slower than that from wild-type band 3 [282]. Also, as discussed later, there are individuals who are compound heterozygotes for the SAO mutation and a mutation associated with distal tubular acidosis. In these individuals, the SAO subunit causes a large decrease in transport mediated by the opposite subunit.

NMR studies with peptides encompassing the SAO deletion and TM1 indicate that the peptide from R389 to K430 consists of three helical regions (P391-A400, P403-A416, and a short stretch from I421-F423) [207]. There is a sharp bend at P403, which may represent the cytoplasm/membrane boundary of TM1. Solvent exchange indicates that the P403-A416 and I421-F423 helices are much more stable than the P391-A400 helix. Deletion of A400-A408 (i.e., the SAO deletion) results in the formation of a stable helix from P391-A416. However, the SAO deletion destabilizes TM1 by forcing hydrophilic residues into its N-terminal end. In keeping with the idea that the SAO deletion destabilizes TM1, a fragment consisting of cdb3 plus normal TM1 can insert normally into endoplasmic reticulum, but the corresponding fragment with the SAO deletion cannot [207].

Kanki et al. [283] have recently performed an extensive mutagenesis study of residues 361–408, i.e., starting at the trypsin cleavage site between the cytoplasmic and membrane domains and extending through the deletion associated with SAO. Deletion of the residue 400–406 does not interfere with proper insertion of TM1, but deletion of 400–408 (SAO) completely blocks insertion. Insertion of three leucine residues in the C-terminal end of SAO TM1 restores ability to insert, in agreement with the idea that SAO TM1 is unstable because the hydrophobic length of the helix is too short. Partial deletions of residues 400–408 were also tested for effects on anion transport. As expected, the deletions that inhibit insertion of TM1 exhibit very little transport. However, shorter deletions that support insertion of TM1 in the cell-free translation experiment do not give rise to a functioning transporter. Protein containing some of these shorter deletions reaches the surface membrane, indicating that proper insertion of the deletion mutants is not sufficient for transport.

Other mutagenesis experiments by Kanki et al. [283] showed that removal of all three arginine residues (alanine substitution) at positions 387–389 near the cytoplasmic surface does not interfere with the transport function of AE1. This cluster of positive charge is well conserved in SLC4 but is not necessary for anion transport in AE1 expressed in *Xenopus* oocytes. Expression of the membrane domain (361–911) or a slightly truncated membrane domain (381–911) gives rise to high levels of transport, but a further truncation (386–911) was expressed on the cell surface but not functional as a transporter. This indicates that the residues from G381 to D385 are of functional importance for the transporter.

13.13.2.3 Band 3 Coimbra

Rebeiro et al. [284] recently described an individual homozygous for the AE1 point mutation (V488M), known as band 3 Coimbra, which results in erythrocytes completely lacking band 3. Heterozygotes for this allele have relatively mild hereditary spherocytosis with about 20% reduction in band 3 [285]. The homozygous individual was delivered by emergency Caesarian section and at birth had intense pallor, edema, ascites, and hepatosplenomegaly [284]. Erythrocytes had abnormal morphology, including spikelike projections indicative of membrane instability and vesicle shedding. Electrophoresis and immunoblotting showed that the erythrocytes have no detectable band 3 or band 4.2. Band 6 (G3PDH) and glycophorin A were also markedly diminished, in qualitative agreement with the properties of the mice with targeted AE1 deletion [117, 269]. The individual also has distal renal tubular acidosis and does not produce urine with pH below 6.6, consistent with a lack of acid secretion in the collecting tubule. The last published report on this individual indicated that she was doing reasonably well at age three with frequent transfusions and bicarbonate ingestion to counter the renal tubular acidosis [284].

13.13.2.4 Mutations Causing Hereditary Spherocytosis

A variety of mutations in several different proteins result in a family of hemolytic anemias that are collectively grouped under the heading of hereditary spherocytosis (HS) [286]. The most common causes of HS are mutations in spectrin or ankyrin. However, a substantial number of cases of HS (23% of 166 kindreds examined by Jarolim et al. [287]) are the consequence of mutations in AE1. Many of these mutations cause premature termination of the polypeptide in heterozygotes, and in these cases the only AE1 mRNA present in reticulocytes is that of the normal allele. Several other mutations do not affect mRNA stability but instead lead to an unstable or mistargeted protein [288–291]. Most AE1 HS mutations result in a decreased total amount of erythrocyte band 3, which destabilizes the membrane and causes HS. This destabilization probably is a consequence of the fact that AE1 is a major component of normal erythrocyte membranes and that lack of a full complement of band 3 leads to blebbing of lipid vesicles and loss of surface area. Mutations that cause dominant HS do not have any measurable renal consequences, presumably because the lack of the unstable allele does not interfere with expression of the normal protein. The single copy of the normal gene is sufficient to produce adequate kidney AE1.

13.13.2.5 Band 3 Montefiore

A recessive AE1 mutation causing a condition similar to HS is a lysine substitution for E40 (band 3 Montefiore [292]). Individuals homozygous for this mutation have hemolytic anemia, osmotically fragile red cells, and a major deficiency (12% of normal) in protein 4.2 [292]. Rotational mobility measurements showed that band 3 from red cells of homozygotes for band 3 Montefiore or from cells deficient in protein 4.2 have increased mobility, indicating that interactions with protein 4.2 stabilize interaction of band 3 with the membrane skeleton [293].

13.13.2.6   Mutations Causing Distal Renal Tubular Acidosis

As discussed earlier, the AE1 gene is expressed in erythrocytes and in the acid-secreting intercalated cells of the renal collecting duct. In several families with dominant familial distal renal tubular acidosis (dRTA), mutations have been found in the AE1 gene [294–297]. The most frequent mutations are substitutions of histidine, serine, or cysteine for arginine in position 589. A *de novo* mutation at this site was recently identified, indicating that codon 589 is a mutational hotspot [298]. In erythrocytes from individuals heterozygous for R589 point mutants, the total number of band 3 molecules per cell is normal, and $SO_4^=$ transport is slightly reduced. In another mutation (S613F) associated with dominant dRTA, $SO_4^=$ transport in erythrocytes of heterozygotes is increased as a result of a higher apparent affinity for $SO_4^=$ [294]. Expression of either erythroid or renal AE1 R589H, R589C, or S613F mutations in *Xenopus* oocytes gives rise to Cl⁻ transport that is not dramatically different from that of normal AE1 [294]. When expressed in HEK293 cells, normal erythroid AE1, normal renal AE1, and erythroid AE1 R589H mutant protein were all targeted to the cell surface, but renal R589H was retained mainly in subcellular vesicles [299]. Coexpression of normal AE1 with the R589H mutant showed that R589H exerts a dominant negative effect, presumably from formation of heterodimers that do not traffic to the cell surface [299].

Another mutation causing dRTA, band 3[WALTON], is an insertion causing premature termination leading to an 11-amino acid deletion at the C-terminus of the protein [297, 300]. In nonpolarized renal cells (HEK293 or nonpolarized MDCK), both the erythroid and renal products of AE1[WALTON] exhibit impaired trafficking to the surface membrane [300, 301]. Interestingly, expression of (epitope-tagged) AE1[WALTON] in polarized MDCK cells resulted in cell surface expression on both the apical and basolateral membranes [302]. Fusion constructs between C-terminal AE1 sequences and CD8 (which is normally targeted to the apical membrane) demonstrated that there is a basolateral targeting signal in the C-terminal 11 residues of AE1; Y904 is a critical residue for this targeting signal [302]. Abnormal targeting of AE1 dRTA mutants to the apical as well as the basolateral membrane of α-intercalated cells would explain the dominant effect of these mutations.

Although most known AE1 mutations associated with dRTA are dominant, there is a known recessive mutation of AE1 causing dRTA. Individuals homozygous for the G701D mutation have dRTA with apparently normal erythrocyte anion transport [303, 304]. The dissociation between erythroid and renal phenotype of the G701D homozygotes is explained by the fact that the stable folding and targeting of G701D AE1 is absolutely dependent on coexpression of glycophorin A [303], which is expressed in erythroid but not renal cells. Therefore, G701D AE1 is present in normal amounts in erythrocytes but not in the kidney.

Bruce et al. [281] have described several families from Malaysia and Papua New Guinea with members who are compound heterozygotes with various combinations of band 3 mutations: G701D, A858D, ΔV850, and the SAO deletion. The normally recessive G701D and ΔV850 mutations were pseudodominant for dRTA in individuals who are compound heterozygotes for either of these mutations and the SAO deletion. In red cells from G701D/SAO compound heterozygotes, anion ($SO_4^=$) transport is half the normal value, indicating that the SAO polypeptides do not interfere with anion transport by the G701D polypeptides. The dRTA in these individuals is caused by the absolute GPA requirement for trafficking to the surface membrane, as discussed earlier.

Red cells from individuals heterozygous for the A858D mutation (A858D/N) have anion transport that is 65% of normal. Compound heterozygotes (A858D/SAO), however, have very low anion transport (3% of normal). Transport per DIDS binding site is only 17% of normal. *Xenopus* oocyte expression experiments showed that the A858D mutant protein has considerably less transport activity than normal and even less activity when coexpressed with SAO AE1. All these findings indicate that the A858D mutation by itself causes moderate impairment of transport, but that, in a heterodimer with SAO band 3, the A858D subunit has severely impaired transport. The ΔV850 mutation alone does not cause impaired anion transport, but, when present in a compound heterozygote with SAO, transport (per DIDS binding site) is only about 50% of normal. These experiments show that the abnormally folded SAO subunit can perturb the opposite subunit in the dimer more significantly in the presence of mutations at position 850 or 858.

13.13.2.7 Band 3 HT

There is one known mutation in AE1 that causes elevated anion transport. The mutation is P868L [305] and causes abnormal cell shape, as well as a decrease in the number of ankyrin binding sites and about a one-and-a-half- to twofold increase in the maximum $SO_4^=$ flux [306]. The number of H_2DIDS binding sites is normal, but band 3 HT is less readily labeled with [^{3}H]H_2DIDS [305]. The site of the P868L substitution is toward the cytoplasmic end of the last TM helix; the substitution very likely removes a proline kink and stabilizes this helix. The proline in this position is well conserved in AE1, AE2, and AE3 of all known species, but other members of SLC4 generally have an aliphatic hydrophobic residue in this position. The reasons for the abnormal cell shape and abnormally high transport rate in band 3 HT are unclear, but further study of this mutation should lead to new insights into band 3 structure–function relations.

13.14 CONCLUDING REMARKS

The AE1 protein is clearly among the most thoroughly studied transport proteins, and there has been continuing progress in understanding its structure and function. However, in the absence of a high-resolution structure of the membrane domain, it is not yet possible to make definitive statements about the molecular mechanism of anion exchange. It has proved to be very difficult to prepare crystals of the AE1 membrane domain suitable for X-ray diffraction. It is possible that a crystal structure of the membrane domain of some other member of SLC4 will be obtained; a structure of one member of the family would undoubtedly lead to a much improved structural model of AE1. However, it may turn out that no crystal structure of any member of this family will be prepared in the foreseeable future.

    If no crystal structure is available for this family of proteins, what will be the next steps toward understanding AE1? Magnetic resonance and optical spectroscopic techniques will undoubtedly continue to improve, and the abundance of erythrocyte band 3 will continue to make it a very attractive system for biophysical studies using various kinds of spectroscopy for direct studies of the kinetics of binding of substrates and inhibitors and of conformational changes associated with transport. The solution structures of synthetic peptides from parts of band 3 have been determined by NMR [206, 207, 311], and more such structures should provide important insights into the structure of the whole protein. The spectacular recent advancements in mass spectrometry of peptides have not yet been fully applied to band 3, and it is probable that the application of mass spectrometry to localizing sites of chemical modification, crosslinking, or proteolysis of the erythrocyte protein will yield useful information. As computational methods for modeling protein conformations continue to improve, it should become possible to construct testable structural models of band 3 subdomains, such as TM1-5 or perhaps the entire membrane domain. A high-resolution crystal structure would obviously be a major breakthrough in the study of this protein, but it is very likely that progress toward understanding coupled ion transport through AE1 will continue, even without a crystal structure.

ACKNOWLEDGMENT

The author gratefully acknowledges the support of NIH Grant R01 GM026861-22.

REFERENCES

1. Romero, M.F., Hediger, M.A., Boulpaep, E.L., and Boron, W.F., Expression cloning and characterization of a renal electrogenic Na$^+$/HCO$_3^-$ cotransporter, *Nature*, 387, 409, 1997.

2. Romero, M.F., The electrogenic Na^+/HCO_3^- cotransporter, NBC, *J. Pancr.*, 2, 182, 2002.

3. Gamba, G., Saltzberg, S.N., Lombardi, M., Miyanoshita, A., Lytton, J., Hediger, M.A., Brenner, B.M., and Hebert, S.C., Primary structure and functional expression of a cDNA encoding the thiazide-sensitive, electroneutral sodium-chloride cotransporter, *Proc. Natl. Acad. Sci. USA*, 90, 2749, 1993.

4. Payne, J.A., Xu, J.-C., Haas, M., Lytle, C.Y., Ward, D., and Forbush, B., III, Primary structure, functional expression, and chromosomal localization of the Bumetanide-sensitive Na-K-Cl cotransporter in human colon, *J. Biol. Chem.*, 27, 17977, 1995.

5. Xu, J.-C., Lytle, C., Zhu, T.T., Payne, J.A., Benz, E., Jr., and Forbush, B., III, Molecular cloning and functional expression of the bumetanide-sensitive Na-K-Cl cotransporter, *Proc. Natl. Acad. Sci. USA*, 91, 2201, 1994.

6. Gillen, C.M., Brill, S., Payne, J.A., and Forbush, B., III, Molecular cloning and functional expression of the K-Cl cotransporter from rabbit, rat, and human, *J. Biol. Chem.*, 271, 16237, 1996.

7. Bissig, M., Hagenbuch, B., Stieger, B., Koller, T., and Meier, P.J., Functional expression cloning of the canalicular sulfate transport system of rat hepatocytes, *J. Biol. Chem.*, 269, 3017, 1994.

8. Hastbacka, J., de la Chapelle, A., Mahtani, M.M., Clines, G., Reeve-Daly, M.P., Daly, M., Hamilton, B.A., Kusumi, K., Trivedi, B., Weaver, A., Coloma, A., Lovett, M., Buckler, A., Kaitila, I., and Lander, E.S., The diastrophic dysplasia gene encodes a novel sulfate transporter: positional cloning by fine-structure linkage disequilibrium mapping, *Cell*, 78, 1073, 1994.

9. Blakely, R.D., DeFelice, L.J., and Hartzell, H.C., Molecular physiology of norepinephrine and sero-tonin transporters, *J. Exp. Biol.*, 196, 263, 1994.

10. Jentsch, T.J., Friedrich, T., Schriever, A., and Yamada, H., The CLC chloride channel family, *Pflug. Arch.*, 437, 783, 1999.

11. Dutzler, R., Campbell, E.B., Cadene, M., Chait, B.T., and MacKinnon, R., X-ray structure of a ClC chloride channel at 3.0 Å reveals the molecular basis of anion selectivity, *Nature*, 415, 287, 2002.

12. Dutzler, R., Campbell, E.B., and MacKinnon, R., Gating the selectivity filter in ClC chloride channels, *Science*, 300, 108, 2003.

13. Steck, T.L., The band 3 protein of the human red cell membrane: a review, *J. Supramolec. Struct.*, 8, 311, 1978.

14. Fröhlich, O. and Gunn, R.B., Erythrocyte anion transport: the kinetics of a single-site obligatory exchange system, *Biochim. Biophys. Acta*, 864, 169, 1986.

15. Knauf, P.A., Erythrocyte anion exchange and the band 3 protein: transport kinetics and molecular structure, *Curr. Top. Membr. Trans.*, 12, 249, 1979.

16. Passow, H., Molecular aspects of band 3 protein-mediated anion transport across the red blood cell membrane, *Rev. Physiol. Biochem. Pharmacol.*, 103, 62, 1986.

17. Jennings, M.L., Structure and function of the red blood cell anion transport protein, *Annu. Rev. Biophys. Biophys. Chem.*, 18, 397, 1989.

18. Alper, S.L., The band 3-related anion exchanger (AE) gene family, *Annu. Rev. Physiol.*, 53, 549, 1991.

19. Ko, S.B.H., Luo, X., Hager, H., Rojek, A., Choi, J.Y., Licht, C., Suzuki, M., Muallem, S., Nielsen, S., and Ishibashi, K., AE4 is a DIDS-sensitive Cl^-/HCO_3^- exchanger in the basolateral membrane of the renal CCD and the SMG duct, *Am. J. Physiol. Cell*, 283, C1206, 2002.

20. Romero, M.F., Henry, D., Nelson, S., Harte, P.J., Dillon, A.K., and Sciortino, C.M., Cloning and characterization of a Na^+-driven anion exchanger (NDAE1), *J. Biol. Chem.*, 275, 24552, 2000.

21. Grichtchenko, I.I., Choi, I., Zhong, X., Bray-Ward, P., Russell, J.M., and Boron, W.F., Cloning, characterization, and chromosomal mapping of a human electroneutral Na^+-driven $Cl-HCO_3$ exchanger, *J. Biol. Chem.*, 276, 8358, 2001.

22. Romero, M.F., Fong, P., and Berger, U.V.H.M.A.B.W.F., Cloning and functional expression of rNBC, an electrogenic $Na^+-HCO_3^-$ cotransporter from rat kidney, *Am. J. Physiol. Renal*, 274, F425, 1998.

23. Choi, I., Aalkjaer, C., Boulpaep, E.L., and Boron, W.F., An electroneutral sodium/bicarbonate cotrans-porter NBCn1 and associated sodium channel, *Nature*, 405, 571, 2000.

24. Tsuganezawa, H., Kobayashi, K., Iyori, M., Araki, T., Koizumi, A., Watanabe, S., Kaneko, A., Fukao, T., Monkawa, T., Yoshida, T., Kim, D.K., Kanai, Y., Endou, H., Hayashi, M., and Saruta, T., A new member of the HCO_3^- transporter superfamily is an apical anion exchanger of beta intercalated cells in the kidney, *J. Biol. Chem.*, 276, 8180, 2001.

25. Takano, J., Noguchi, K., Yasumori, M., Kobayashi, M., Gajdos, Z., Miwa, K., Hayashi, H., Yoneyama, T., and Fujiwara, T., Arabidopsis boron transporter for xylem loading, *Nature*, 420, 337, 2002.

26. Dordas, C. and Brown, P.H., Permeaability and the mechanism of transport of boric acid across the plasma membrane of *Xenopus laevis* oocytes, *Biol. Trace Elem. Res.*, 81, 127, 2001.

27. Dordas, C. and Brown, P.H., Permeability of boric acid across lipid bilayers and factors affecting it, *J. Membr. Biol.*, 175, 95, 2000.

28. Zhao, R. and Reithmeier, R.A.F., Expression and characterization of the anion transporter homologue YNL275w in *Saccharomyces cerevisiae*, *Am. J. Physiol. Cell*, 281, C33, 2001.

29. Alper, S.L., Darman, R.B., Chernova, M.N., and Dahl, N.K., The AE gene family of Cl/HCO$_3$ exchangers, *J. Nephrol.*, 15, S41-S53, 2002.

30. Soleimani, M., Na$^+$:HCO$_3^-$ cotransporters (NBC): expression and regulation in the kidney, *J. Nephrol.*, 15, S32-S40, 2002.

31. Gross, E. and Kurtz, I., Structural determinants and significance of regulation of electrogenic Na$^+$-HCO$_3^-$ cotransporter stoichiometry, *Am. J. Physiol. Renal*, 283, F876, 2002.

32. Romero, M.F. and Boron, W.F., Electrogenic Na$^+$/HCO$^-_3$ cotransporters, NBC: cloning and physiology, *Annu. Rev. Physiol.*, 61, 1, 1999.

33. Soleiman, M. and Burnham, C.E., Na$^+$:HCO$_3^-$ cotransporters (NBC): cloning and characterization, *J. Membr. Biol.*, 183, 71, 2001.

34. Fairbanks, G., Steck, T.L., and Wallach, D.F.H., Electrophoretic analysis of the major polypeptides of the human erythrocyte membrane, *Biochemistry*, 10, 2606, 1971.

35. Jay, D.G., Glycosylation site of band 3, the human erythrocyte anion-exchange protein, *Biochemistry*, 25, 554, 1986.

36. Okubo, K., Hamasaki, N., Hara, K., and Kageura, M., Palmitoylation of cysteine 69 from the COOH-terminal of Band 3 protein in the human erythrocyte membrane, *J. Biol. Chem.*, 266, 16420, 1991.

37. Yannoukakos, D., Vasseur, C., Piau, J.-P., Wajcman, H., and Bursaux, E., Phosphorylation sites in human erythrocyte band 3 protein, *Biochim. Biophys. Acta*, 1061, 253, 1991.

38. Low, P.S., Allen, D.P., Zioncheck, T.F., Chari, P., Willardson, B.M., Geahlen, R.L., and Harrison, M.L., Tyrosine phosphorylation of band 3 inhibits peripheral protein binding, *J. Biol. Chem.*, 262, 4592, 1987.

39. Wang, C.C., Tao, M., Wei, T., and Low, P.S., Identification of the major casein kinase I phosphorylation sites on erythrocyte band 3, *Blood*, 89, 3019, 1997.

40. Brunati, A.M., Bordin, L., Clari, G., James, P., Quadroni, M., Baritono, E., Pinna, L.A., and Donella-Deana, A., Sequential phosphorylation of protein band 3 by Syk and Lyn tyrosine kinases in intact human erythrocytes: identification of primary and secondary phosphorylation sites, *Blood*, 96, 1550, 2000.

41. Barbul, A., Zipser, Y., Nachles, A., and Korenstein, R., Deoxygenation and elevation of intracellular magnesium induce tyrosine phosphorylation of band 3 in human erythrocytes, *FEBS Lett.*, 455, 87, 1999.

42. Dekowski, S.A., Rybicki, A., and Drickamer, K., A tyrosine kinase associated with the red cell membrane phosphorylates band 3, *J. Biol. Chem.*, 258, 2750, 1983.

43. Mawby, W.J. and Findlay, J.B.C., Characterization and partial sequence of di-iodosulphophenyl isothiocyanate-binding peptide from human erythrocyte anion-transport protein, *Biochem. J.*, 205, 465, 1982.

44. Kopito, R.R. and Lodish, H.F., Primary structure and transmembrane orientation of the murine anion exchange protein, *Nature*, 316, 234, 1985.

45. Lux, S.E., John, K.M., Kopito, R.R., and Lodish, H.F., Cloning and characterization of band 3, the human erythrocyte anion-exchange protein (AE1), *Proc. Natl. Acad. Sci. USA*, 86, 9089, 1989.

46. Tanner, M.J.A., Martin, P.G., and High, S., The complete amino acid sequence of the human erythrocyte membrane anion-transport protein deduced from the cDNA sequence, *Biochem. J.*, 256, 703, 1988.

47. Cox, J.V., Moon, R.T., and Lazarides, E., Anion transporter: highly cell-type-specific expression of distinct polypeptides and transcripts in erythroid and nonerythroid cells, *J. Cell Biol.*, 100, 1548, 1985.

48. Hubner, S., Michel, F., Rudloff, V., and Appelhans, H., Amino acid sequence of band-3 protein from rainbow trout erythrocytes derived from cDNA, *Biochem. J.*, 285, 17, 1992.

49. Sahr, K.E., Taylor, W.M., Daniels, B.P., Rubin, H.L., and Jarolim, P., The structure and organization of the human erythroid anion exchanger (AE1) gene, *Genomics*, 24, 491, 1994.

50. Schofield, A.E., Martin, P.G., Spillett, D., and Tanner, M.J.A., The structure of the human red blood cell anion exchanger (EPB3, AE1, band 3) gene, *Blood*, 84, 2000, 1994.

51. Kopito, R.R., Andersson, M., and Lodish, H.F., Structure and organization of the murine band 3 gene, *J. Biol. Chem.*, 262, 8035, 1987.

52. Cabantchik, Z.I. and Rothstein, A., Membrane proteins related to anion permeability of human red blood cells. I. Localization of disulfonic stilbene binding sites in proteins involved in permeation, *J. Membr. Biol.*, 15, 207, 1974.

53. Ho, M.K. and Guidotti, G., A membrane protein from human erythrocytes involved in anion exchange, *J. Biol. Chem.*, 250, 675, 1975.

54. Wieth, J.O., Anderson, O.S., Brahm, J., Bjerrum, P.J., and Borders, C.L., Jr., Chloride-bicarbonate exchange in red blood cells: physiology of transport and chemical modification of binding sites, *Phil. Trans. R. Soc. Lond. B*, 299, 383, 1982.

55. Cooper, G.J., Zhou, Y., Bouyer, P., Gritchenko, I.I., and Boron, W.F., Transport of volatile solutes through AQP1, *J. Physiol. (Lond.)*, 542.1, 17, 2002.

56. Bruce, L.J., Beckmann, R., Ribeiro, M.L., Peters, L.L., Chasis, J.A., Delaunay, J., Mohandas, N., Anstee, D.J., and Tanner, M.J.A., A band 3-based macrocomplex of integral and peripheral proteins in the red cell membrane, *Blood*, 101, 4180, 2003.

57. Cabantchik, Z.I. and Greger, R., Chemical probes for anion transporters of mammalian cell membranes, *Am. J. Physiol. Cell*, 262, C803, 1992.

58. Lynch, R.E. and Fridovich, I., Permeation of the erythrocyte stroma by superoxide radical, *J. Biol. Chem.*, 253, 4697, 1978.

59. Macfadyen, A.J., Reiter, C., Zhuan, Y., and Beckman, J.S., A novel superoxide dismutase-based trap for peroxynitrite used to detect entry of peroxynitrite into erythrocyte ghosts, *Chem. Res. Toxicol.*, 12, 223, 1999.

60. Hamasaki, N., Matsuyama, H., and Hirota-Chigita, C., Characterization of phosphoenolpyruvate transport across the erythrocyte membrane. Evidence for the involvement of band 3 in the transport system, *Eur. J. Biochem.*, 132, 531, 1983.

61. Hamasaki, N. and Yamamoto, M., Red blood cell function and blood storage, *Vox. Sang.*, 79, 191, 2000.

62. Ortwein, R., Oslender-Kohnen, A., and Deuticke, B., Band 3, the anion exchanger of the erythrocyte membrane, is also a flippase, *Biochim. Biophys. Acta*, 1191, 317, 1994.

63. Kleinhorst, A., Oslender, A., Haest, C.W.M., and Deuticke, B., Band 3-mediated flip-flop and phosphatase-catalyzed cleavage of a long-chain alkyl phosphate anion in the human erythrocyte membrane, *J. Membr. Biol.*, 165, 111, 1998.

64. Guizouarn, H., Gabillat, N., Motais, R., and Borgese, F., Multiple transport functions of red blood cell anion exchanger, tAE1: its role in cell volume regulation, *J. Physiol. (Lond.)*, 535, 497, 2001.

65. Musch, M.W., Leffingwell, T.R., and Goldstein, L., Band 3 modulation and hypotonic-stimulated taurine efflux in skate erythrocytes, *Am. J. Physiol.*, 266, R65, 1994.

66. Bennett, V. and Stenbuck, P.J., Association between ankyrin and the cytoplasmic domain of band 3 isoloated from the human erythrocyte membrane, *J. Biol. Chem.*, 255, 6424, 1980.

67. Low, P.S., Willardson, B.M., Thevenin, B.J.M., Kannan, R., Mehler, E., Geahlen, R.L., and Harrison, M.L. The other functions of erythrocyte membrane band 3. In Hamasaki, N. and Jennings, M.L., eds. *Anion Transport Protein of the Red Blood Cell Membrane*. Amsterdam, Elsevier. 1989, pp. 103–118.

68. Rettig, M.P., Low, P.S., Gimm, J.A., Mohandas, N., Wang, J., and Christian, J.A., Evaluation of biochemical changes during *in vivo* erythrocyte senescence in the dog, *Blood*, 93, 376, 1999.

69. Crandall, I., Collins, W.E., Gysin, J., and Sherman, I.W., Synthetic peptides based on motifs present in human band 3 protein inhibit cytoadherence/sequestration of the malaria parasite *Plasmodium falciparum*, *Proc. Natl. Acad. Sci. USA*, 90, 4703, 1993.

70. Oh, S.S., Chishti, A.H., Palek, J., and Liu, S.C., Erythrocyte membrane alterations in *Plasmodium falciparum* malaria sequestration, *Curr. Opin. Hematol.*, 4, 148, 1997.

71. Brosius, F.C., III, Alper, S.L., Garcia, A.M., and Lodish, H.F., The major kidney band 3 gene transcript predicts an amino-terminal truncated band 3 polypeptide, *J. Biol. Chem.*, 264, 7784, 1989.

72. Kollert-Jons, A., Wagner, S., Hubner, S., Appelhans, H., and Drenckhahn, D., Anion exchanger 1 in human kidney and oncocytoma differs from erythroid AE1 in its NH2 terminus, *Am. J. Physiol. Renal*, 265, F813, 1993.

73. Kudrycki, K.E. and Shull, G.E., Rat kidney band 3 Cl-/HCO-3 exchanger mRNA is transcribed from an alternative promoter, *Am. J. Physiol. Renal*, 264, F540, 1993.

74. Drenckhahn, D., Schluter, K., Allen, D.P., and Bennett, V., Colocalization of band 3 with ankyrin and spectrin at the basal membrane of intercalated cells in the rat kidney, *Science*, 230, 1287, 1985.

75. Schuster, V.L., Bonsib, S.M., and Jennings, M.L., Two types of collecting duct mitochondria-rich (intercalated) cells: lectin and band 3 cytochemistry, *Am. J. Physiol. Cell*, 251, C347, 1986.

76. Verlander, J.W., Madsen, K.M., Cannon, J.K., and Tisher, C.C., Activation of acid-secreting intercalated cells in rabbit collecting duct with ammonium chloride loading, *Am. J. Physiol. Renal*, 266, F633, 1994.

77. Fejes-Toth, G., Chen, W.-R., Rusvai, E., Moser, T., and Naray-Fejes-Toth, A., Differential expression of AE1 in renal HCO_3 -secreting and reabsorbing intercalated cells, *J. Biol. Chem.*, 269, 26717, 1994.

78. Sabolic, I., Brown, D., Gluck, S.L., and Alper, S.L., Regulation of AE1 anion exchanger and H(+)-ATPase in rat cortex by acute metabolic acidosis and alkalosis, *Kidney Int.*, 51, 125, 1997.

79. Huber, S., Asan, E., Jons, T., Kerscher, C., Puschel, B., and Drenckhahn, D., Expression of rat kidney anion exchanger 1 in type A intercalated cells in metabolic acidosis and alkalosis, *Am. J. Physiol. Renal*, 277, F841, 1999.

80. Schuster, V.L., Function and regulation of collecting duct intercalated cells, *Annu. Rev. Physiol.*, 55, 267, 1993.

81. Schuster, V.L. and Stokes, J.B., Chloride transport by the cortical and outer medullary collecting duct, *Am. J. Physiol. Renal*, 253, F203, 1987.

82. Schwartz, G.J., Barasch, J., and Al-Awqati, Q., Plasticity of functional epithelial polarity, *Nature Lond.*, 318, 368, 1985.

83. Schwartz, G.J., Tsuruoka, S., Vijayakumar, S., Petrovic, S., Mian, A., and Al-Awqati, Q., Acid incubation reverses the polarity of intercalated cell transporters, an effect mediated by hensin, *J. Clin. Invest.*, 109, 89, 2002.

84. Alper, S.L., Natale, J., Gluck, S., Lodish, H.F., and Brown, D., Subtypes of intercalated cells in rat kidney collecting duct defined by antibodies against erythroid band 3 and renal vacuolar H+-ATPase, *Proc. Natl. Acad. Sci. USA*, 86, 5429, 1989.

85. Emmons, C., Transport characteristics of the apical anion exchanger of rabbit cortical collecting duct β cells, *Am. J. Physiol. Renal*, 276, F635, 1999.

86. van't Hof, W., Malik, A., Vijayakumar, S., Qiao, J., van Adelsberg, J., and Al-Awqati, Q., The effect of apical and basolateral lipids on the function of the band 3 anion exchange protein, *J. Cell Biol.*, 139, 941, 1997.

87. Royaux, I.E., Wall, S.M., Karniski, L.P., Everett, L.A., Suzuki, K., Knepper, M.A., and Green, E.D., Pendrin, encoded by the Pendred syndrome gene, resides in the apical region of renal intercalated cells and mediates bicarbonate secretion, *Proc. Natl. Acad. Sci. USA*, 98, 4221, 2001.

88. Bennett, V. and Stenbuck, P.J., The membrane attachment protein for spectrin is associated with band 3 in human erythrocyte membrane, *Nature*, 280, 468, 1979.

89. Lombardo, C.R., Willardson, B.M., and Low, P.S., Localization of the protein 4.1-binding site on the cytoplasmic domain of erythrocyte membrane band 3, *J. Biol. Chem.*, 267, 9540, 1992.

90. Korsgren, C. and Cohen, C.M., Associations of human erythrocyte band 4.2. Binding to ankyrin and to the cytoplasmic domain of band 3, *J. Biol. Chem.*, 263, 10212, 1988.

91. Walder, J.A., Chatterjee, R., Steck, T.L., Low, P.S., Musso, G.F., Kaiser, E.T., Rogers, P.H., and Arnone, A., The interaction of hemoglobin with the cytoplasmic domain of band 3 of the human erythrocyte membrane, *J. Biol. Chem.*, 259, 10238, 1984.

92. Harrison, M.L., Isaacson, C.C., Burg, D.L., Geahlen, R.L., and Low, P.S., Phosphorylation of human erythrocyte band 3 by endogenous p72syk, *J. Biol. Chem.*, 269, 955, 1994.

93. Bordin, L., Brunati, A.M., Donella-Deana, A., Baggio, B., Toninello, A., and Clari, G., Band 3 is an anchor protein and a target for SHP-2 tyrosine phosphatase in human erythrocytes, *Blood*, 100, 276, 2002.

94. Zhang, D., Kiyatkin, A., Bolin, J.T., and Low, P.S., Crystallographic structure and functional interpretation of the cytoplasmic domain of erythrocyte membrane band 3, *Blood*, 96, 2925, 2000.

95. Steck, T.L., Koziarz, J.J., Singh, M.K., Reddy, G., and Kohler, H., Preparation and analysis of seven major, topographically defined fragments of band 3, the predominant transmembrane polypeptide of human erythrocyte membranes, *Biochemistry*, 17, 1216, 1978.

96. Chang, S.H. and Low, P.S., Identification of a critical ankyrin-binding loop on the cytoplasmic domain of erythrocyte membrane band 3 by crystal structure analysis and site-directed mutagenesis, *J. Biol. Chem.*, 278, 6879, 2003.

97. Ding, Y., Casey, J.R., and Kopito, R.R., The major kidney AE1 isoform does not bind ankyrin (ANK1) *in vitro, J. Biol. Chem.*, 269, 32201, 1994.

98. Cox, K.H., Adair-Kirk, T.L., and Cox, J.V., Variant chicken kidney AE1 anion exchanger transcripts are derived from a single promoter by alternative splicing, *Gene*, 173, 221, 1996.

99. Cox, K.H., Adair-Kirk, T.L., and Cox, J.V., Four variant chicken erythroid AE1 anion exchangers. Role of the alternative N-terminal sequences in intracellular targeting in transfected human erythroleukemia cells, *J. Biol. Chem.*, 270, 19752, 1995.

100. Adair-Kirk, T.L., Cox, K.H., and Cox, J.V., Intracellular trafficking of variant chicken kidney AE1 anion exchangers: role of alternative NH_2 termini in polarized sorting and Golgi recycling, *J. Cell Biol.*, 147, 1237, 1999.

101. Ghosh, S., Cox, K.H., and Cox, J.V., Chicken erythroid AE1 anion exchangers associate with the cytoskeleton during recycling to the Golgi, *Mol. Biol. Cell*, 10, 455, 1999.

102. Hanspal, M., Golan, D.E., Smockova, Y., Yi, S.J., Cho, M.R., Liu, S.-C., and Palek, J., Temporal synthesis of band 3 oligomers during terminal maturation of mouse erythroblasts, *Blood*, 92, 329, 1998.

103. Adair-Kirk, T.L., Dorsey, F.C., and Cox, J.V., Multiple cytoplasmic signals direct the intracellular trafficking of chicken kidney AE1 anion exchangers in MDCK cells, *J. Cell Sci.*, 116, 655, 2003.

104. Musch, M.W., Hubert, E.M., and Goldstein, L., Volume expansion stimulates $p72^{syk}$ and $p56^{lyn}$ in skate erythrocytes, *J. Biol. Chem.*, 274, 7923, 1999.

105. Vince, J.W. and Reithmeier, R.A.F., Carbonic anhydrase II binds to the carboxyl terminus of human band 3, the erythrocyte Cl^-/HCO_3^- exchanger, *J. Biol. Chem.*, 273, 28430, 1998.

106. Vince, J.W. and Reithmeier, R.A.F., Identification of the carbonic anhydrase II binding site in the Cl^-/HCO_3^- anion exchanger AE1, *Biochemistry*, 39, 5527, 2000.

107. Sterling, D., Reithmeier, R.A., and Casey, J.R., A transport metabolon. Functional interaction of carbonic anhydrase II and chloride/bicarbonate exchangers, *J. Biol. Chem.*, 276, 47886, 2001.

108. Bartel, D., Hans, H., and Passow, H., Identification by site-directed mutagenesis of Lys-558 as the covalent attachment site of dihydro DIDS in the mouse erythroid band 3 protein, *Biochim. Biophys. Acta*, 985, 355, 1989.

109. Bartel, D., Lepke, S., Layh-Schmitt, G., Legrum, B., and Passow, H., Anion transport in oocytes of *Xenopus laevis* induced by expression of mouse erythroid band 3 protein — encoding cRNA and of a cRNA derivative obtained by site-directed mutagenesis at the stilbene disulfonate binding site, *EMBO J.*, 8, 3601, 1989.

110. Casey, J.R., Ding, Y., and Kopito, R.R., The role of cysteine residues in the erythrocyte plasma membrane anion exchange protein, AE1, *J. Biol. Chem.*, 270, 8521, 1995.

111. Timmer, R.T. and Gunn, R.B., Inducible expression of erythrocyte band 3 protein, *Am. J. Physiol. Cell*, 276, C66, 1999.

112. Groves, J.D. and Tanner, M.J.A., The effects of glycophorin A on the expression of the human red cell anion transporter (band 3) in *Xenopus* oocytes, *J. Membr. Biol.*, 140, 81, 1994.

113. Groves, J.D. and Tanner, M.J.A., Glycophorin A facilitates the expression of human band 3-mediated anion transport in *Xenopus* oocytes, *J. Biol. Chem.*, 267, 22163, 1992.

114. Groves, J.D., Falson, P., LeMaire, M., and Tanner, M.J.A., Functional cell surface expression of the anion transport domain of human red cell band 3 (AE1) in the yeast *Saccharomyces cerevisiae*, *Proc. Natl. Acad. Sci. USA*, 93, 12245, 1996.

115. Groves, J.D., Parker, M.D., Askin, D., Falson, P., LeMaire, M., and Tanner, M.J., Heterologous expression of the red-cell anion exchanger (band 3; AE1), *Biochem. Soc. Trans.*, 27, 917, 1999.

116. Young, M.T., Beckmann, R., Toye, A.M., and Tanner, M.J.A., Red-cell glycophorin A-band 3 interactions associated with the movement of band 3 to the cell surface, *Biochem. J.*, 350, 53, 2000.

117. Hassoun, H., Hanada, T., Lutchman, M., Sahr, K.E., Palek, J., Hanspal, M., and Chishti, A.H., Complete deficiency of glycophorin A in red blood cells from mice with targeted inactivation of the band 3 (AE1) gene, *Blood*, 91, 2146, 1998.

118. Auffray, I., Marfatia, S., Jong, K., Lee, G., Huang, C.-H., Paszty, C., Tanner, J.A., Mohandas, N., and Chasis, J., Glycophorin A dimerization and band 3 interaction during erythroid membrane biogenesis: *in vivo* studies in human glycophorin A transgenic mice, *Blood*, 97, 2872, 2001.

119. Leddy, J.P., Wilkinson, S.L., Kissel, G.E., Passador, S.T., Falany, J.L., and Rosenfeld, S.I., Erythrocyte membrane proteins reactive with IgG (warm-reacting) anti-red blood cell autoantibodies: II. Antibodies coprecipitating band 3 and glycophorin A, *Blood*, 84, 650, 1994.

120. Bruce, L.J., Ring, S.M., Anstee, D.J., Reid, M.E., Wilkinson, S., and Tanner, M.J., Changes in the blood group Wright antigens are associated with a mutation at amino acid 658 in human erythrocyte band 3: a site of interaction between band 3 and glycophorin A under certain conditions, *Blood*, 85, 299, 1995.

121. Yu, J. and Steck, T.L., Isolation and chacterization of band 3, the predominant polypeptide of the human erythrocyte membrane, *J. Biol. Chem.*, 250, 9170, 1975.

122. Cuppoletti, J., Goldinger, J., Kang, B., Jo, I., Berenski, C., and Jung, C.Y., Anion carrier in the human erythrocyte exists as a dimer, *J. Biol. Chem.*, 260, 15714, 1985.

123. Wang, D.N., Kuhlbrandt, W., Sarabia, V.E., and Reithmeier, R.A.F., Two dimensional structure of the membrane domain of human band 3, the anion transport protein of the erythrocyte membrane, *EMBO J.*, 12, 2233, 1993.

124. Wang, D.N., Sarabia, V.E., Reithmeier, R.A.F., and Kuhlbrandt, W., Three-dimensional map of the dimeric membrane domain of the human erythrocyte anion exchanger, band 3, *EMBO J.*, 13, 3230, 1994.

125. Appell, K.C. and Low, P.S., Partial structural characterization of the cytoplasmic domain of the erythrocyte membrane protein, band 3, *J. Biol. Chem.*, 256, 11104, 1981.

126. Steck, T.L., Cross-linking the major proteins of the isolated erythrocyte membrane, *J. Mol. Biol.*, 66, 295, 1972.

127. Staros, J.V., N-hydroxysulfosuccinimide active esters: bis(N-hydroxysulfosuccinimide) esters of two dicarboxylic acids are hydrophilic, membrane-impermeant, protein cross-linkers, *Biochemistry*, 21, 3950, 1982.

128. Jennings, M.L. and Nicknish, J.S., Localization of a site of intermolecular cross-linking in human red blood cell band 3 protein, *J. Biol. Chem.*, 260, 5472, 1985.

129. Pinder, J.C., Pekrun, A., Maggs, A.M., Brain, A.P.R., and Gratzer, W.B., Association state of human red blood cell band 3 and its interaction with ankyrin, *Blood*, 85, 2951, 1995.

130. Lindenthal, S. and Schubert, D., Monomeric erythrocyte band 3 protein transports anions, *Proc. Natl. Acad. Sci. USA*, 88, 6540, 1991.

131. Boodhoo, A. and Reithmeier, R.A.F., Characterization of matrix-bound band 3, the anion transport protein from human erythrocyte membranes, *J. Biol. Chem.*, 259, 785, 1984.

132. Salhany, J.M., Cordes, K.A., and Sloan, R.L., Gel filtration chromatographic studies of the isolated membrane domain of band 3, *Mol. Memb. Biol.*, 14, 71, 1997.

133. Casey, J.R. and Reithmeier, R.A.F., Analysis of the oligomeric state of band 3, the anion transport protein of the human erythrocyte membrane, by size exclusion high performance liquid chromotography, *J. Biol. Chem.*, 266, 15726, 1991.

134. Salhany, J.M., Cordes, K.A., and Sloan, R.L., Mechanism of band 3 dimer dissociation during incubation of erythrocyte membranes at 37°C, *Biochem. J.*, 345, 33, 2000.

135. Landolt-Marticorena, C., Charuk, J.H.M., and Reithmeier, R.A., Two glycoprotein populations of band 3 dimers are present in human erythrocytes, *Mol. Membr. Biol.*, 15, 158, 1998.

136. Van Dort, H.M., Moriyama, R., and Low, P.S., Effect of band 3 subunit equilibrium on the kinetics and affinity of ankyrin binding to erythrocyte membrane vesicles, *J. Biol. Chem.*, 273, 14819, 1998.

137. Matayoshi, E.D. and Jovin, T.M., Rotational diffusion of band 3 in erythrocyte membranes. 1. Comparison of ghosts and intact cells, *Biochemistry*, 30, 3527, 1991.

138. Hustedt, E.J. and Beth, A.H., Analysis of saturation transfer electron paramagnetic resonance spectra of a spin-labeled integral membrane protein, band 3, in terms of the uniaxial rotational diffusion model, *Biophys. J.*, 69, 1409, 1995.

139. Colfen, H., Boulter, J.M., Harding, S.E., and Watts, A., Ultracentrifugation studies on the transmembrane domain of the human erythrocyte anion transporter band 3 in the detergent C12E8, *Eur. Biophys. J.*, 27, 651, 1998.

140. Thevenin, B.J.M. and Low, P.S., Kinetics and regulation of the ankyrin-band 3 interaction of the human red blood cell membrane, *J. Biol. Chem.*, 265, 16166, 1990.

141. Michaely, P. and Bennett, V., The ANK repeats of erythrocyte ankyrin form two distinct but cooperative binding sites for the erythrocyte anion exchanger, *J. Biol. Chem.*, 270, 22050, 1995.

142. Bennett, V. and Baines, A.J., Spectrin and ankyrin-based pathways: metazoan inventions for integrating cells into tissues, *Physiol. Rev.*, 81, 1353, 2001.

143. Yi, S.J., Liu, S.-C., Derick, L.H., Murray, J., Barker, J.E., Cho, M.R., Palek, J., and Golan, D.E., Red cell membranes of ankyrin-deficient nb/nb mice lack band 3 tetramers but contain normal membrane skeletons, *Biochemistry*, 36, 9596, 1997.

144. Mulzer, K., Kampmann, L., Petrasch, P., and Schubert, D., Complex associations between membrane proteins analyzed by analytical ultracentrifugation: studies on the erythrocyte membrane proteins band 3 and ankyrin, *Colloid Polym. Sci.*, 268, 60, 1990.

145. Golan, D.E. and Veatch, W., Lateral mobility of band 3 in the human erythrocyte membrane studied by fluorescence photobleaching recovery: evidence for control by cytoskeletal interactions, *Proc. Natl. Acad. Sci. USA*, 77, 2537, 1980.

146. Salhany, J.M., Sloan, R.L., and Cordes, K.A., *In situ* cross-linking of human erythrocyte band 3 by bis(sulfosuccinimidyl)suberate, *J. Biol. Chem.*, 265, 17688, 1990.

147. Pappert, G. and Schubert, D., The state of association of band 3 protein of the human erythrocyte membrane in solutions of nonionic detergents, *Biochim. Biophys. Acta*, 730, 32, 1983.

148. Schuck, P., Legrum, B., Passow, H., and Schubert, D., The influence of two anion-transport inhibitors, 4,4'-diisothiocyanatodihydrostilbene-2,2'-disulfonate, on the self-association of erythrocyte band 3 protein, *Eur. J. Biochem.*, 230, 806, 1995.

149. Taylor, A.M., Boulter, J., Harding, S.E., Colfen, H., and Watts, A., Hydrodynamic properties of human erythrocyte band 3 solubilized in reduced Triton X-100, *Biophys. J.*, 76, 2043, 1999.

150. Schubert, D., Boss, K., Dorst, H.-J., Flossdorf, J., and Pappert, G., The nature of the stable noncovalent dimers of band 3 protein from erythrocyte membranes in solutions of Triton X-100, *FEBS Lett.*, 163, 81, 1983.

151. Brahm, J., Temperature-dependent changes of chloride transport kinetics in human red cells, *J. Gen. Physiol.*, 70, 283, 1977.

152. Gunn, R.B. and Fröhlich, O., Asymmetry in the mechanism for anion exchange in human red blood cell membranes (evidence for reciprocating sites that react with one transported anion at a time), *J. Gen. Physiol.*, 74, 351, 1979.

153. Knauf, P.A., Fuhrmann, G.F., Rothstein, S., and Rothstein, A., The relationship between exchange and net anion flow across the human red blood cell membrane, *J. Gen. Physiol.*, 69, 363, 1977.

154. Fröhlich, O., Relative contributions of the slippage and tunneling mechanisms to anion net efflux from human erythrocytes, *J. Gen. Physiol.*, 84, 877, 1984.

155. Knauf, P.A., Law, F.-Y., and Marchant, P., Relationship of net chloride flow across the human erythrocyte membrane to the anion exchange mechanism, *J. Gen. Physiol.*, 81, 95, 1983.

156. Knauf, P.A., Raha, N.M., and Spinelli, L.J., The noncompetitive inhibitor WW781 senses changes in erythrocyte anion exchanger (AE1) transport site conformation and substrate binding, *J. Gen. Physiol.*, 115, 159, 2000.

157. Liu, D., Kennedy, S.D., and Knauf, P.A., Source of transport site asymmetry in the band 3 anion exchange protein determined by NMR measurements of external Cl affinity, *Biochemistry*, 35, 15228, 1996.

158. Knauf, P.A., Law, F.-Y., Leung, T., Gehret, A.U., and Perez, M.L., Substrate-dependent reversal of anion transport site orientation in the human red blood cell anion-exchange protein, AE1, *Proc. Natl. Acad. Sci. USA*, 99, 10861, 2002.

159. Dalmark, M., Effects of halides and bicarbonate on chloride transport in human red blood cells, *J. Gen. Physiol.*, 67, 223, 1976.

160. Knauf, P.A. and Mann, N.A., Location of the chloride self-inhibitory site of the human erythrocyte anion exchange system, *Am. J. Physiol. Cell*, 251, C1, 1986.

161. Knauf, P.A., Spinelli, L.J., and Mann, N.A., Flufenamic acid senses conformation and asymmetry of human erythrocyte band 3 anion transport protein, *Am. J. Physiol. Cell*, 257, C277, 1989.

162. Knauf, P.A., Strong, N.M., Penikas, J., Wheeler, R.B., Jr., and Liu, S.-Q.J., Eosin-5-maleimide inhibits red cell Cl exchange at a noncompetitive site that senses band 3 conformation, *Am. J. Physiol. Cell*, 264, C1144, 1993.

163. Shami, Y., Rothstein, A., and Knauf, P.A., Identification of the Cl- transport site of human red blood cells by a kinetic analysis of the inhibitory effects of a chemical probe, *Biochim. Biophys. Acta*, 508, 357, 1978.

164. Fröhlich, O., The external anion binding site of the human erythrocyte anion transporter: DNDS binding and competition with chloride, *J. Membr. Biol.*, 65, 111, 1982.

165. Salhany, J.M., Stilbenedisulfonate binding kinetics to band 3 (AE 1): relationship between transport and stilbenedisulfonate binding sites and role of subunit interactions in transport, *Blood Cells Mol. Dis.*, 27, 127, 2001.

166. Salhany, J.M., Sloan, R.L., Cordes, K.A., and Schopfer, L.M., Kinetic evidence for ternary complex formation and allosteric interactions in chloride and stilbenedisulfonate binding to band 3, *Biochemistry*, 33, 11909, 1994.

167. Salhany, J.M., Anion binding characteristics of the band 3/4,4-dibenzamidostilbene-2,2-disulfonate binary complex: evidence for both steric and allosteric interactions, *Biochem. Cell Biol.*, 77, 543, 1999.

168. Falke, J.J., Pace, R.J., and Chan, S.I., Chloride binding to the anion transport binding sites of band 3, *J. Biol. Chem.*, 259, 6472, 1984.

169. Falke, J.J., Pace, R.J., and Chan, S.I., Direct observation of the transmembrane recruitment of band 3 transport sites by competitive inhibitors: a ^{35}Cl NMR study, *J. Biol. Chem.*, 259, 6481, 1984.

170. Falke, J.J. and Chan, S.I., Evidence that anion transport by band 3 proceeds via a ping-pong mechanism involving a single transport site, *J. Biol. Chem.*, 260, 9537, 1985.

171. Falke, J.J., Kanes, K.J., and Chan, S.I., The kinetic equation for the chloride transport cycle of band 3, *J. Biol. Chem.*, 260, 9545, 1985.

172. Liu, D., Kennedy, S.D., and Knauf, P.A., ^{35}Cl nuclear magnetic resonance line broadening shows that eosin-5-maleimide does not block the external anion access channel of band 3, *Biophys. J.*, 69, 399, 1995.

173. Jennings, M.L. and Passow, H., Anion transport across the erythrocyte membrane, *in situ* proteolysis of band 3 protein, and cross-linking of proteolytic fragments by 4,4'-diisothiocyano-dihydrostilbene-2,2'-disulfonate, *Biochim. Biophys. Acta*, 554, 498, 1979.

174. Cabantchik, Z.I. and Rothstein, A., Membrane proteins related to anion permeability of human red blood cells as determined by studies with disulfonic stilbene derivatives, *J. Membr. Biol.*, 15, 207, 1974.

175. Schofield, A.E., Reardon, D.M., and Tanner, M.J.A., Defective anion transport activity of the abnormal band 3 in hereditary ovalocytic red blood cells, *Nature*, 355, 836, 1992.

176. Wieth, J.O., Bicarbonate exchange through the red cell membrane determined with (^{14}C)-HCO$_3^-$, *J. Physiol. (Lond.)*, 294, 521, 1979.

177. Salhany, J. M. *Erythrocyte Band 3 Protein.* CRC Press, Boca Raton, FL. 1990.

178. Van Dort, H.M., Low, P.S., Cordes, K.A., Schopfer, L.M., and Salhany, J.M., Calorimetric evidence for allosteric subunit interactions associated with inhibitor binding to band 3 transporter, *J. Biol. Chem.*, 269, 59, 1994.

179. Salhany, J.M., Allosteric effects in stilbenedisulfonate binding to band 3 protein (AE1), *Cell. Mol. Biol.*, 42, 1065, 1996.

180. Gunn, R.B., Dalmark, M., Tosteson, D.C., and Wieth, J.O., Characteristics of chloride transport in human red blood cells, *J. Gen. Physiol.*, 61, 185, 1973.

181. Gunn, R. and Frohlich, O., Asymmetry in the mechanism for anion exchange in human red blood cell membranes, *J. Gen. Physiol.*, 74, 351, 1979.

182. Salhany, J.M. and Swanson, J.C., Kinetics of passive anion transport across the human erythrocyte membrane, *Biochemistry*, 17, 3354, 1978.

183. Salhany, J.M. and Gaines, E.D., Steady state kinetics of erythrocyte anion exchange. Evidence for site-site interactions, *J. Biol. Chem.*, 256, 11080, 1981.

184. Jennings, M.L., Stoichiometry of a half-turnover of band 3, the chloride transport protein of human erythrocytes, *J. Gen. Physiol.*, 79, 169, 1982.

185. Jennings, M.L., Whitlock, J., and Shinde, A., Pre-steady state transport by erythrocyte band 3 protein: uphill countertransport induced by the impermeant inhibitor H$_2$DIDS, *Biochem. Cell Biol.*, 76, 807, 1998.

186. Motais, R., Baroin, A., and Baldy, S., Chloride permeability in human red cells: influence of membrane protein rearrangement resulting from ATP depletion and calcium accumulation, *J. Membr. Biol.*, 62, 195, 1981.

187. Bursaux, E., Hilly, M., Bluze, A., and Poyart, C., Organic phosphates modulate anion self-exchange across the human erythrocyte membrane, *Biochim. Biophys. Acta*, 777, 253, 1984.

188. Jennings, M.L. and Adame, M.F., Characterization of oxalate transport by the human erythrocyte band 3 protein, *J. Gen. Physiol.*, 107, 145, 1996.

189. Cohen, P., Holmes, C.F.B., and Tsukitani, Y., Okadaic acid: a new probe for the study of cellular regulation, *TIBS*, 15, 98, 1990.

190. Baggio, B., Bordin, L., Clari, G., Gambaro, G., and Moret, V., Functional correlation between the Ser/Thr-phosphorylation of band-3 and band-3-mediated transmembrane anion transport in human erythrocytes, *Biochim. Biophys. Acta*, 1148, 157, 1993.

191. Jennings, M.L. and Schulz, R.K., Okadaic acid inhibition of KCl cotransport. Evidence that protein dephosphorylation is necessary for activation of transport by either cell swelling or N-ethylmaleimide, *J. Gen. Physiol.*, 97, 799, 1991.

192. Kang, D., Karbach, D., and Passow, H., Anion transport function of mouse erythroid band 3 protein (AE1) does not require acylation of cysteine residue 861, *Biochim. Biophys. Acta*, 1194, 341, 1994.

193. Casey, J.R., Pirraglia, C.A., and Reithmeier, R.A.F., Enzymatic deglycosylation of human band 3, the anion transport protein of the erythrocyte membrane, *J. Biol. Chem.*, 267, 11940, 1992.

194. Groves, J.D. and Tanner, M.J.A., Role of N-glycosylation in the expression of human band 3 mediated anion transport, *Mol. Membr. Biol.*, 11, 31, 1994.

195. Lepke, S. and Passow, H., Effects of incorporated trypsin on anion exchange and membrane proteins in human red blood cell ghosts, *Biochim. Biophys. Acta*, 455, 353, 1976.

196. Grinstein, S., Ship, S., and Rothstein, A., Anion transport in relation to proteolytic dissection of band 3 protein, *Biochim. Biophys. Acta*, 507, 294, 1978.

197. Appell, K.C. and Low, P.S., Evaluation of structural interdependence of membrane-spanning and cytoplasmic domains of band 3, *Biochemistry*, 21, 2151, 1982.

198. Doyle, D.A., Morais Cabral, J., Pfuetzner, R.A., Kuo, A., Gulbis, J.M., Chait, B.T., and MacKinnon, R., The structure of the potassium channel: molecular basis of K^+ conduction and selectivity, *Science*, 280, 69, 1998.

199. Toyoshima, C., Nakasako, M., Nomura, H., and Ogawa, H., Crystal structure of the calcium pump of sarcoplasmic reticulum at 2.6 Å resolution, *Nature*, 405, 647, 2000.

200. Lemieux, M.J., Reithmeier, R.A., and Wang, D.N., Importance of detergent and phospholipid in the crystallization of the human erythrocyte anion-exchanger membrane domain, *J. Struct. Biol.*, 137, 322, 2002.

201. Vince, J.W. and Reithmeier, R.A.F., Structure of the band 3 transmembrane domain, *Cell. Mol. Biol.*, 42, 1041, 1996.

202. Steck, T.L., Ramos, B., and Strapazon, E., Proteolytic dissection of band 3, the predominant transmembrane polypeptide of the human erythrocyte memrbane, *Biochemistry*, 15, 1154, 1976.

203. Drickamer, L.K., Orientation of the band 3 polypeptide from human erythrocyte membranes, *J. Biol. Chem.*, 253, 7242, 1978.

204. Jenkins, R.E. and Tanner, M.J.A., The structure of the major protein of the human erythrocyte membrane, *Biochem. J.*, 161, 139, 1977.

205. Oikawa, K., Lieberman, D.M., and Reithmeier, R.A.F., Conformation and stability of the anion transport protein of human erythrocyte membranes, *Biochemistry*, 24, 2843, 1985.

206. Gargaro, A.R., Bloomberg, G.B., Dempsey, C.E., Murray, M., and Tanner, M.J.A., The solution structures of the first and second transmembrane-spanning segments of band 3, *Eur. J. Biochem.*, 221, 445, 1994.

207. Chambers, E.J., Bloomberg, G.B., Ring, S.M., and Tanner, M.J., Structural studies on the effects of the deletion in the red cell anion exchanger (band 3, AE1) associated with South East Asian ovalocytosis, *J. Mol. Biol.*, 285, 1289, 1999.

208. Jennings, M.L., Adams-Lackey, M., and Denney, G.H., Peptides of human erythrocyte band 3 protein produced by extracellular papain cleavage, *J. Biol. Chem.*, 259, 4652, 1984.

209. Jennings, M.L., Anderson, M.P., and Monaghan, R., Monoclonal antibodies against human erythrocyte band 3 protein. Localization of proteolytic cleavage sites and stilbenedisulfonate-binding lysine residues, *J. Biol. Chem.*, 261, 9002, 1986.

210. Lieberman, D.M. and Reithmeier, R.A.F., Localization of the carboxyl terminus of band 3 to the cytoplasmic side of the erythrocyte membrane using antibodies raised against a synthetic peptide, *J. Biol. Chem.*, 263, 10022, 1988.

211. Wainwright, S.D., Tanner, M.J.A., Martin, G.E.M., Yendle, J.E., and Holmes, C., Monoclonal antibodies to the membrane domain of the human erythrocyte anion transport protein, *Biochem. J.*, 258, 211, 1989.

212. Cobb, C.E. and Beth, A.H., Identification of the eosinyl-5-maleimide reaction site on the human erythrocyte anion-exchange protein: overlap with the reaction sites of other chemical probes, *Biochemistry*, 29, 8283, 1990.

213. Okubo, K., Kang, D., Hamasaki, N., and Jennings, M.L., Red blood cell band 3: lysine-539 and lysine 851 react with the same 4,4'-diisothiocyanodihydrostilbene-2,2'-disulfonate molecule, *J. Biol. Chem.*, 269, 1918, 1994.

214. Jarolim, P., Rubin, H.L., Zakova, D., Storry, J., and Reid, M.E., Characterization of seven low incidence blood group antigens carried by erythrocyte band 3 protein, *Blood*, 92, 4836, 1998.

215. McManus, K., Lupe, K., Coghlan, G., and Zelinski, T., An amino acid substitution in the putative second extracellular loop of RBC band 3 accounts for the Froese blood group polymorphism, *Transfusion*, 40, 1246, 2000.

216. Bruce, L.J., Anstee, D.J., Spring, F.A., and Tanner, M.J.A., Band 3 Memphis variant II. Altered stilbene disulfonate binding and the Diego (Di^a) blood group antigen are associated with the human erythrocyte band 3 mutation Pro^{854}->Leu, *J. Biol. Chem.*, 269, 16155, 1994.

217. Sekler, I., Kopito, R., and Casey, J.R., High level expression, partial purification, and functional reconstitution of the human AE1 anion exchanger in *Saccharomyces cerevisiae*, *J. Biol. Chem.*, 270, 21028, 1995.

218. Popov, M., Tam, L.Y., Li, J., and Reithmeier, R.A.F., Mapping the ends of transmembrane segments in a polytopic membrane protein, *J. Biol. Chem.*, 272, 18325, 1997.

219. Popov, M., Li, J., and Reithmeier, R.A.F., Transmembrane folding of the human erythrocyte anion exchanger (AE1, band 3) determined by scanning and insertional N-glycosylation mutagenesis, *Biochem. J.*, 339, 269, 1999.

220. Fujinaga, J., Tang, X., and Casey, J.R., Topology of the membrane domain of human erythrocyte anion exchange protein, AE1, *J. Biol. Chem.*, 274, 6626, 1999.

221. Tang, X., Fujinaga, J., Kopito, R., and Casey, J.R., Topology of the region surrounding Glu^{681} of human AE1 protein, the erythrocyte anion exchanger, *J. Biol. Chem.*, 273, 22545, 1998.

222. Nilsson, I. and von Heijne, G., Determination of the distance between the oligosaccharyltransferase active site and the endoplasmic reticulum membrane, *J. Biol. Chem.*, 268(8), 5798, 1993.

223. Kuma, H., Shinde, A.A., Howren, T., and Jennings, M.L., Topology of the anion exchange protein AE1: the controversial sidedness of lysine 743, *Biochemistry*, 41, 3380, 2002.

224. Zhu, Q., Lee, D.W.K., and Casey, J.R., Novel topology in C-terminal region of the human plasma membrane anion exchanger, AE1, *J. Biol. Chem.*, 278, 3112, 2003.

225. Wainwright, S.D., Mawby, W.J., and Tanner, M.J.A., The membrane domain of the human erythrocyte anion transport protein. Epitope mapping of a monoclonal antibody defines the location of a cytoplasmic loop near the C-terminus of the protein, *Biochem. J.*, 272, 265, 1990.

226. Hamasaki, N., Kuma, H., Ota, K., Sakaguchi, M., and Mihara, K., A new concept in polytopic membrane proteins following from the study of band 3 protein, *Biochem. Cell Biol.*, 76, 729, 1998.

227. Kang, D., Okubo, K., Hamasaki, N., Kuroda, N., and Shiraki, H., A structural study of the membrane domain of band 3 by tryptic digestion. Conformational change of band 3 *in situ* induced by alkali treatment, *J. Biol. Chem.*, 267, 19211, 1992.

228. Hamasaki, N., Okubo, K., Kuma, H., Kang, D., and Yae, Y., Proteolytic cleavage sites of band 3 protein in alkali-treated membranes: fidelity of hydropathy prediction for band 3 protein, *J. Biochem.*, 122, 577, 1997.

229. Ota, K., Sakaguchi, M., Hamasaki, N., and Mihara, K., Assessment of topogenic functions of anticipated transmembrane segments of human band 3, *J. Biol. Chem.*, 273, 28286, 1998.

230. Kanki, T., Sakaguchi, M., Kitamura, A., Sato, T., Mihara, K., and Hamasaki, N., The tenth membrane region of band 3 is initially exposed to the luminal side of the endoplasmic reticulum and then integrated into a partially folded band 3 intermediate, *Biochemistry*, 41, 13973, 2002.

231. Groves, J.D. and Tanner, M.J., Co-expressed complemetary fragments of the human red cell anion exchanger (Band 3, AE1) generate stilbene disulfonate-sensitive anion transport, *J. Biol. Chem.*, 270, 9097, 1995.

232. Wang, L., Groves, J.D., Mawby, W.J., and Tanner, M.J.A., Complementation studies with co-expressed fragments of the human red cell anion transporter (band 3; AE1): the role of some exofacial loops in anion transport, *J. Biol. Chem.*, 272, 10631, 1997.

233. Jennings, M.L. and Adams, M.F., Modification by papain of the structure and function of band 3, the erythrocyte anion transport protein, *Biochemistry*, 20, 7118, 1981.

234. Groves, J.D. and Tanner, M.J.A., Topology studies with biosynthetic fragments identify interacting transmembrane regions of the human red-cell anion exchanger (band 3; AE1), *Biochem. J.*, 344, 687, 1999.

235. Groves, J.D. and Tanner, M.J., Structural model for the organization of the transmembrane spans of the human red-cell anion exchanger (band3: AE1), *Biochem. J.*, 344, 699, 1999.

236. Brock, C.J., Tanner, M.J.A., and Kempf, C., The human erythrocyte anion-transport protein, *Biochem. J.*, 213, 577, 1983.

237. Staros, J.V. and Kakkad, B.P., Cross-linking and chymotryptic digestion of the extracytoplasmic domain of the anion exchange channel in intact human erythrocytes, *J. Membr. Biol.*, 74, 247, 1983.

238. Taylor, A.M., Zhu, Q., and Casey, J.R., Cysteine-directed cross-linking localizes regions of the human erythrocyte anion-exchange protein (AE1) relative to the dimeric interface, *Biochem. J.*, 359, 661, 2001.

239. Funder, J. and Wieth, J.O., Chloride transport in human erythrocyte and ghosts: a quantitative comparison, *J. Physiol. (Lond.)*, 262, 679, 1976.

240. Humphreys, B.D., Jiang, L., Chernova, M.N., and Alper, S.L., Functional characterization and regulation by pH of murine AE2 anion exchanger expressed in *Xenopus* oocytes, *Am. J. Physiol. Cell*, 267, C1295, 1994.

241. Jiang, L., Stuart-Tilley, A., Parkash, J., and Alper, S.L., pH and serum regulate AE2-mediated Cl-/HCO3-exchange in CHOP cells of defined transient transfection status, *Am. J. Physiol. Cell*, 267, C845, 1994.

242. Stewart, A.K., Chernova, M.N., Kunes, Y.Z., and Alper, S.L., Regulation of AE2 anion exchanger by intracellular pH: critical regions of the NH_2-terminal cytoplasmic domain, *Am. J. Physiol. Cell*, 281, C1344, 2001.

243. Alper, S.L., Chernova, M.N., and Stewart, A.K., How pH regulates a pH regulator: a regulatory hot spot in the N-terminal cytoplasmic domain of the AE2 anion exchanger, *Cell Biochem. Biophys.*, 36, 123, 2002.

244. Wieth, J.O. and Bjerrum, P.J., Titration of transport and modifier sites in the red cell anion transport system, *J. Gen. Physiol.*, 79, 253, 1982.

245. Liu, S.-Q.J., Ries, E., and Knauf, P.A., Effects of external pH on binding of external sulfate, 4,4'-dinitro-stilbene-2,2'-disulfonate (DNDS), and chloride to the band 3 anion exchange protein, *J. Gen. Physiol.*, 107, 293, 1996.

246. Liu, S.-Q.J., Law, F.-Y., and Knauf, P.A., Effects of external pH on substrate binding and on the inward chloride translocation rate constant of band 3, *J. Gen. Physiol.*, 107, 271, 1996.

247. Bjerrum, P.J., Wieth, J.O., and Borders, C.L., Jr., Selective phenylglyoxalation of functionally essential arginyl residues in the erythrocyte anion transport protein, *J. Gen. Physiol.*, 81, 453, 1983.

248. Zaki, L., Anion transport in red blood cells and arginine-specific reagents. The location of [^{14}C]phenylglyoxal binding sites in the anion transport protein in the membrane of human red cells, *FEBS Lett.*, 169, 234, 1984.

249. Bohm, R. and Zaki, L., Towards the localization of the essential arginine residues in the band 3 protein of human red blood cell membranes, *Biochim. Biophys. Acta*, 1280, 238, 1996.

250. Karbach, D., Staub, M., Wood, P.G., and Passow, H., Effect of site-directed mutagenesis of the arginine residues 509 and 748 on mouse band 3 protein-mediated anion transport, *Biochim. Biophys. Acta*, 1371, 114, 1998.

251. Lima, P.R., Sales, T.S., Costa, F.F., and Saad, S.T., Arginine 490 is a hot spot for mutation in the band 3 gene in hereditary spherocytosis, *Eur. J. Haematol.*, 63, 360, 1999.

252. Dhermy, D., Bournier, O., Bourgeois, M., and Grandchamp, B., The red blood cell band 3 variant (band 3 [Bicetrel]: R490C) associated with dominant hereditary spherocytosis causes defective membrane targeting of the molecule and a dominant negative effect, *Mol. Membr. Biol.*, 16, 305, 1999.

253. Milanick, M.A. and Gunn, R.B., Proton-sulfate cotransport: external proton activation of sulfate influx into human red blood cells, *Am. J. Physiol. Cell*, 247, C247, 1984.

254. Gunn, R.B., A titratable carrier model for both mono- and di-valent anion transport in human red blood cells. In Rørth, M. and Astrup, P., eds. *Oxygen Affinity of Hemoglobin and Red Cell Acid Base Status*. 1972, Munksgaard, Copenhagen, pp. 823–827.

255. Jennings, M.L., Proton fluxes associated with erythrocyte membrane anion exchange, *J. Membr. Biol.*, 28, 187, 1976.

256. Jennings, M.L. and Al-Rhaiyel, S., Modification of a carboxyl group that appears to cross the permeability barrier in the red blood cell anion transporter, *J. Gen. Physiol.*, 92, 161, 1988.

257. Jennings, M.L. and Smith, J.S., Anion-proton cotransport through the human red blood cell band 3 protein. Role of glutamate 681, *J. Biol. Chem.*, 267, 13964, 1992.

258. Jennings, M.L., Rapid electrogenic sulfate-chloride exchange mediated by chemically modified band 3 in human erythrocytes, *J. Gen. Physiol.*, 105, 21, 1995.

259. Chernova, M.N., Jiang, L., Crest, M., Hand, M., Vandorp, D.H., Strange, K., and Alper, S.L., Electrogenic sulfate/chloride exchange in *Xenopus* oocytes mediated by murine AE1 E699Q, *J. Gen. Physiol.*, 109, 345, 1997.

260. Bahar, S., Gunter, C.T., Wu, C., Kennedy, S.D., and Knauf, P.A., Persistence of external chloride and DIDS binding after chemical modification of Glu-681 in human band 3, *Am. J. Physiol.*, 277, C791–C799, 1999.

261. Tang, X., Kovacs, M., Sterling, D., and Casey, J.R., Identification of residues lining the translocation pore of human AE1, plasma membrane anion exchange protein, *J. Biol. Chem.*, 274, 3557, 1999.

262. Salhany, J.M., Sloan, R.L., and Cordes, K.S., The carboxyl side chain of glutamate 681 interacts with a chloride binding modifier site that allosterically modulates the dimeric conformational state of band 3 (AE1). Implications for the mechanism of anion/proton cotransport, *Biochemistry*, 42, 1589, 2003.

263. Hamasaki, N., Izuhara, K., Okubo, K., Kanazawa, Y., Omachi, A., and Kleps, R.A., Inhibition of chloride binding to the anion transport site by diethylpyrocarbonate modification of band 3, *J. Membr. Biol.*, 116, 87, 1990.

264. Izuhara, K., Okubo, K., and Hamasaki, N., Conformational change of band 3 protein induced by diethyl pyrocarbonate modification in human erythrocyte ghosts, *Biochemistry*, 28, 4725, 1989.

265. Muller-Berger, S., Karbach, D., Koenig, J., Lepke, S., Wood, P.G., Appelhans, H., and Passow, H., Inhibition of mouse erythroid band 3-mediated chloride transport by site-directed mutagenesis of histidine residues and its reversal by second site mutation of Lys558, the locus of covalent H_2DIDS binding, *Biochemistry*, 34, 9315, 1997.

266. Muller-Berger, S., Karbach, D., Kang, D., Aranibar, N., Wood, P.G., Rueterjans, H., and Passow, H., Roles of histidine 752 and glutamate 699 in the pH dependence of mouse band 3 protein-mediated anion transport, *Biochemistry*, 34, 9325, 1995.

267. Inaba, M., Yawata, A., Koshino, I., Sato, K., Takeuchi, M., Takakuwa, Y., Manno, S., Yawata, Y., Kanzaki, A., Sakai, J., Ban, A., Ono, K., and Maede, Y., Defective anion transport and marked spherocytosis with membrane instability caused by hereditary total deficiency of red cell band 3 in cattle due to a nonsense mutation, *J. Clin. Invest.*, 97, 1804, 1996.

268. Peters, L.L., Shivdasani, R.A., Liu, S.-C., Hanspal, M., John, K.M., Gonzalez, J.M., Brugnara, C., Gwynn, B., Mohandas, N., Alper, S.L., Okrin, S.H., and Lux, S.E., Anion exchanger 1 (band 3) is required to prevent erythrocyte membrane surface loss but not to form the membrane skeleton, *Cell*, 86, 917, 1996.

269. Southgate, C.D., Chishti, A.H., Mitchell, B., Yi, S.J., and Palek, J., Targeted disruption of the murine erythroid band 3 gene results in spherocytosis and severe haemolytic anaemia despite a normal membrane skeleton, *Nature Genetics*, 14, 227, 1996.

270. Wieth, J.O. and Brahm, J. Cellular anion transport. In Seldin, D.W. and Giebisch, G., eds. *The Kidney: Physiology and Pathophysiology.* New York, Raven Press. 1985, 49–89.

271. Mueller, T.J. and Morrison, M., Detection of a variant of protein 3, the major transmembrane protein of the human erythrocyte, *J. Biol. Chem.*, 252, 6573, 1977.

272. Yannoukakos, D., Vasseur, C., Draincourt, C., Blouquit, Y., Delaunay, J., Wajcman, H., and Bursaux, E., Human erythrocyte band 3 polymorphism (band 3 Memphis): characterization of the structural modification (Lys 56 - Glu) by protein chemistry methods, *Blood*, 78, 1117, 1991.

273. Jarolim, P., Rubin, H.L., Zhai, S., Sahr, K.E., Liu, S.-C., Mueller, T.J., and Palek, J., Band 3 Memphis: a widespread polymorphism with abnormal electrophoretic mobility of erythrocyte band 3 protein caused by substitution AAG->GAG (Lys->Glu) in Codon 56, *Blood*, 80, 1592, 1992.

274. Jarolim, P., Palek, J., Amato, D., Hassan, K., Sapak, P., Nurse, G.T., Rubin, H.L., Zhai, S., Sahr, K.E., and Liu, S.-C., Deletion in erythrocyte band 3 gene in malaria-resistant Southeast Asian ovalocytosis, *Proc. Natl. Acad. Sci. USA*, 88, 11022, 1991.

275. Liu, S.C., Zhai, S., Palek, J., Golan, D.E., Amato, D., Hassan, K., Nurse, G.T., Babona, D., Coetzer, T., Jarolim, P., Zaik, M., and Borwein, S., Molecular defect of the band 3 protein in southeast asian ovalocytosis, *New Engl. J. Med.*, 323, 1530, 1990.

276. Schofield, A.E., Tanner, M.J.A., Pinder, J.C., Clough, B., Bayley, P.M., Nash, G.B., Dluzewski, A.R., Reardon, D.M., Cox, T.M., Wilson, R.J.M., and Gratzer, W.B., Basis of unique red cell membrane properties in hereditary ovalocytosis, *J. Mol. Biol.*, 223, 949, 1992.

277. Moriyama, R., Ideguchi, H., Lombardo, C.R., Van Dort, H.M., and Low, P.S., Structural and functional characterization of band 3 from Southeast Asian ovalocytes, *J. Biol. Chem.*, 267, 25792, 1992.

278. Groves, J.D., Ring, S.M., Schofield, A.E., and Tanner, M.J.A., The expression of the abnormal human red cell anion transporter from south-east asian ovalocytes (band 3 SAO) in *Xenopus* oocytes, *FEBS Lett.*, 330, 186, 1993.

279. Liu, S.-C., Palek, J., Nichols, P.E., Derick, L.H., Chiou, S.-S., Amato, D., Corbett, J.D., Cho, M.R., and Golan, D.E., Molecular basis of altered red blood cell membrane properties in Southeast Asian ovalocytosis: role of the mutant band 3 protein in band 3 oligomerization and retention by the membrane skeleton, *Blood*, 86, 349, 1995.

280. Jennings, M.L. and Gosselink, P.G., Anion exchange in Southeast Asian ovalocytes: heterodimer formation between normal and variant subunits, *Biochemistry*, 34, 3588, 1995.

281. Bruce, L.J., Wrong, O., Toye, A., Young, M.T., Ogle, G., Ismail, Z., Sinha, A., McMaster, P., Hwaih-wanje, I., Nash, G., Hart, S., Lavus, E., Palmer, R., Othman, A., Unwin, R., and Tanner, J.A., Band 3 mutations, renal tubular acidosis and South-East Asian ovalocytosis in Malaysia and Papua New Guinea: loss of up to 95% band 3 transport in red cells, *Biochem. J.*, 350, 41, 2000.

282. Salhany, J.M. and Schopfer, L.M., Interactions between mutant and wild-type band 3 subunits in hereditary Southeast Asian ovalocytic red blood cell membranes, *Biochemistry*, 35, 251, 1996.

283. Kanki, T., Young, M.T., Sakaguchi, M., Hamasaki, N., and Tanner, M.J.A., The N-terminal region of the transmembrane domain of human erythrocyte band 3: residues critical for membrane insertion and transport activity, *J. Biol. Chem.*, 278, 5564, 2003.

284. Ribeiro, M., Alloisio, N., Almeida, H., Gomes, C., Texier, P., Lemos, C., Mimoso, G., Morle, L., Bey-Cabet, F., Rudigoz, R.C., Delaunay, J., and Tamagnini, G., Severe hereditary spherocytosis and distal renal tubular acidosis associated with the total absence of band 3, *Blood*, 96, 1602, 2000.

285. Alloisio, N., Texier, P., Vallier, A., Ribeiro, M.L., Morlé, L., Bozon, M., Bursaux, E., Maillet, P., Gonçalves, P., Tanner, M.J.A., Tamagnini, G., and Delaunay, J., Modulation of clinical expression and band 3 deficiancy in hereditary spherocytosis, *Blood*, 90, 414, 1997.

286. Delaunay, J., Molecular basis of red cell membrane disorders, *Acta Haematol.*, 108, 210, 2002.

287. Jarolim, P., Murray, J.L., Rubin, H.L., Taylor, W.M., Prchal, J.T., Ballas, S.K., Snyder, L.M., Chrobak, L., Melrose, W.D., Brabec, V., and Palek, J., Characterization of 13 novel band 3 gene defects in hereditary spherocytosis with band 3 deficiency, *Blood*, 88, 4366, 1996.

288. Quilty, J. and Reithmeier, R.A.F., Trafficking and folding defects in hereditary spherocytosis mutants of the human red cell anion exchanger, *Traffic*, 1, 987, 2000.

289. Jarolim, P., Palek, J., Rubin, H.L., Prchal, J.T., Korsgren, C., and Cohen, C.M., Band 3 Tuscaloosa: Pro^{327}»»Arg^{327} substitution in the cytoplasmic domain of erythrocyte band 3 protein associated with spherocytic hemolytic anemia and partial deficiency of protein 4.2, *Blood*, 80, 523, 1992.

290. Jarolim, P., Rubin, H.L., Brabec, V., Chrobak, L., Zolotarev, A.S., Alper, S.L., Brugnara, C., Wichterle, H., and Palek, J., Mutations of conserved arginines in the membrane domain of erythroid band 3 lead to a decrease in membrane-associated band 3 and to the phenotype of hereditary spherocytosis, *Blood*, 85, 634, 1995.

291. Jarolim, P., Rubin, H.L., Liu, S.-C., Cho, M.R., Brabec, V., Derick, L.H., Yi, S.J., Saad, S.T.O., Alper, S., Brugnara, C., Golan, D.E., and Palek, J., Duplication of 10 nucleotides in the erythroid band 3 (AE1) gene in a kindred with hereditary spherocytosis and band 3 protein deficiency (band 3 Prague), *J. Clin. Invest.*, 93, 121, 1994.

292. Rybicki, A.C., Qiu, J.J., Musto, S., Rosen, N.L., Nagel, R.L., and Schwartz, R.S., Human erythrocyte protein 4.2 deficiency associated with hemolytic anemia and a homozygous 40 glutamic acid to lysine substitution in the cytoplasmic domain of band 3 (band 3 Montefiore), *Blood*, 81, 2155, 1993.

293. Rybicki, A.C., Schwartz, R.S., Hustedt, E.J., and Cobb, C.E., Increased rotational mobility and extractability of band 3 from protein 4.2-deficient erythrocyte membranes: evidence for a role for protein 4.2 in strengthening the band 3-cytoskeleton linkage, *Blood*, 88, 2745, 1996.

294. Bruce, L.J., Cope, D.L., Jones, G.K., Schofield, A.E., Burley, M., Povey, S., Unwin, R.J., Wrong, O., and Tanner, M.J.A., Familial distal renal tubular acidosis is associated with mutations in the red cell anion exchanger (band 3, AE1) gene, *J. Clin. Invest.*, 100, 1693, 1997.

295. Jarolim, P., Shayakul, C., Prabakaran, D., Jiang, L., Stuart-Tilley, A., Rubin, H.L., Simova, S., Zavadil, J., Herrin, J.T., Brouillette, J., Somers, M.J.G., Seemanova, E., Brugnara, C., Guay-Woodford, L.M., and Alper, S.L., Autosomal dominant distal renal tubular acidosis is associated in three families with heterozygosity for the R589H mutation in the AE1 (band 3) Cl⁻/HCO⁻₃ exchanger, *J. Biol. Chem.*, 273, 6380, 1998.

296. Wrong, O., Bruce, L., Unwin, R.J., Toye, A.M., and Tanner, J.A., Band 3 mutations, distal renal tubular acidosis, and Southeast Asian ovalocytosis, *Kidney Int.*, 62, 10, 2002.

297. Karet, F.E., Gainza, F.J., Gyory, A.Z., Unwin, R.J., Wrong, O., Tanner, M.J.A., Nayir, A., Alpay, H., Santos, F., Hulton, S.A., Bakkaloglu, A., Ozen, S., Cunningham, M.J., Di Petro, A., Walker, W.G., and Lifton, R.P., Mutations in the chloride-bicarbonate exchanger gene AE1 cause autosomal dominant but not autosomal recessive distal renal tubular acidosis, *Proc. Natl. Acad. Sci. USA*, 95, 6337, 1998.

298. Sritippayawan, S., Kirdpon, S., Vasuvattakul, S., Wasanawatana, S., Susaengrat, W., Waiwawuth, W., Nimmannit, S., Malasit, P., and Yenchitsomanus, P.T., A *de novo* R589C mutation of anion exchanger 1 causing distal renal tubular acidosis, *Pediatr. Nephrol.*, in press, 2003.

299. Quilty, J.A., Li, J., and Reithmeier, R.A.F., Impaired trafficking of distal renal tubular acidosis mutants of the human kidney anion exchanger kAE1, *Am. J. Physiol. Renal*, 282, F810, 2002.

300. Toye, A., Bruce, L., Unwin, R., Wrong, O., and Tanner, M.J.A., Band 3 Walton, a C-terminal deletion associated with distal renal tubular acidosis, is expressed in the red cell membrane but retained internally in kidney cells, *Blood*, 99, 342, 2002.

301. Quilty, J.A., Cordat, E., and Reithmeier, R.A.F., Impaired trafficking of human kidney anion exchanger (kAE1) caused by hetero-oligomer formation with a truncated mutant associated with distal renal tubular acidosis, *Biochem. J.*, 368, 895, 2002.

302. Devonald, M.A.J., Smith, A.N., Poon, J.P., Ihrke, G., and Karet, F.E., Non-polarized targeting of AE1 causes autosomal dominant distal renal tubular acidosis, *Nature Genetics*, 33, 125, 2003.

303. Tanphaichitr, V.S., Sumboonnanonda, A., Ideguchi, H., Shayakul, C., Brugnara, C., Takao, N., Veerakul, G., and Alper, S.L., Novel AE1 mutations in recessive distal renal tubular acidosis, *J. Clin. Invest.*, 102, 2173, 1998.

304. Yenchitsomanus, P., Vasuvattakul, S., Kirdpon, S., Wasanawatana, S., Susaengrat, W., Sreethiphayawan, S., Chuawatana, D., Mingkum, S., Sawasdee, N., Thuwajit, P., Wilairat, P., and Malasit, P., Autosomal recessive distal renal tubular acidosis caused by G701D mutation of anion exchanger 1 gene, *Am. J. Kidney Dis.*, 40, 21, 2002.

305. Bruce, L.J., Kay, M.M.B., Lawrence, C., and Tanner, M.J.A., Band 3 HT, a human red-cell variant associated with acanthocytosis and increased anion transport, carries the mutation Pro-868 — >Leu in the membrane domain of band 3, *Biochem. J.*, 293, 317, 1993.

306. Kay, M.M.B., Bosman, G.J.C.G.M., and Lawrence, C., Functional topography of band 3: specific structural alteration linked to functional aberrations in human erythrocytes, *Proc. Natl. Acad. Sci. USA*, 85, 492, 1988.

307. Parker, M.D., Ourmozdi, E.P., and Tanner, M.J.A., Human BTR1, a new bicarbonate transporter superfamily member and human AE4 from kidney, *Biochem. Biophys. Res. Commun.*, 282, 1103, 2001.

308. Weinstein, A.M., A mathematical model of the outer medullary collecting duct of the rat, *Am. J. Physiol. Renal*, 279, F24, 2000.

309. Erickson, H.K. and Kyte, J., Lysine-691 of the anion exchanger from human erythrocytes is located on its cytoplasmic surface, *Biochem. J.*, 336, 443, 1998.

310. Erickson, H.K., Cytoplasmic disposition of aspartate 821 in the anion exchanger from human erythrocytes, *Biochemistry*, 33, 9958, 1997.

311. Askin, D., Bloomberg, G.B., Chambers, E.J., Tanner, M.J.A., NMR solution structure of a cytoplasmic surface loop of the human red cell anion transporter, band 3, *Biochemistry*, 37, 11670, 1998.

14 The Structures of G-Protein Coupled Receptors

Philip L. Yeagle

CONTENTS

14.1 INTRODUCTION

G-protein coupled receptors (GPCRs), which are involved in many kinds of signal transduction, represent a large family of integral membrane proteins located at the plasma membranes of cells. Estimates vary, but one organism may have hundreds of such receptors, regulating vision, olfaction, taste, and many specific cellular functions through hormones and other signals.

Because of this central role in biological regulation, GPCRs are of intense interest to the drug discovery community. It has been estimated that as much as 60% of the targets for drug discovery currently are GPCRs. Three-dimensional structural information is useful to drug discovery, yet until recently no three-dimensional structural information was available for any GPCR.

GPCRs are known to consist of a bundle of seven transmembrane helices. Considerable homology among GPCRs allowed the development of models (based ultimately on the sequence of rhodopsin, the first GPCR for which there was any structural information, including primary structure). Initially, sequence alignment was done against rhodopsin, and models of the helical bundle were proposed based on the sequence alignment and proposals for fundamental principles of packing for a helical bundle. Three-dimensional models were subsequently built using bacteriorhodopsin, the only membrane protein with seven transmembrane helices for which a three-dimensional structure was available at that time.

Because virtually all the structural information available for GPCRs is from rhodopsin, this chapter will emphasize rhodopsin. The homology in primary structure among GPCRs suggests that what we learn for rhodopsin will be useful for other GPCRs, as well. For these and other reasons, rhodopsin is often considered the prototype of all GPCRs.

Visual signal transduction is mediated by GPCRs. Two distinct kinds of cells provide the extant sensitivity to the spectrum of visible light of higher animals. Rod cells in the retina are exquisitely sensitive to low levels of light, acting effectively as single photon counters. These cells respond to light in black and white. The photopigment in these cells is rhodopsin. The other photoreceptor cell type in the retina is the cone cell. Cone cells are perhaps 100 times less sensitive to light than rod cells, but they provide color vision. In human retinas, three cone subtypes are known, each sensitive to different bandwidths of visible light. Each of these cells contains a photopigment that is related, but not identical, to rhodopsin. Only the structure of rhodopsin, however, has been studied in detail. Some of this material was recently reviewed (Albert and Yeagle 2002).

14.2 THE RETINAL ROD CELL

The retinal rod photoreceptor cells in vertebrate and invertebrate retinas are part of a complex array of neural and epithelial cells that are responsible for the capture of light and the transmission of the resulting nerve impulse to the brain. Rod cells make vision at low light levels possible, and the degeneration of these cells leads to night blindness. The rod cell is a highly polarized, postmitotic cell specializing in signal transduction.

Rod cells are divided morphologically into an inner segment and an outer segment. The outer segment of the rod cell encloses a stack of densely packed, closed, flattened membrane sacs referred to as disks, which are stacked along the long dimension of the outer segment. These disks turn over as an integral part of the rod cell physiology. The disks are formed from evaginations of the rod outer segment (ROS) plasma membrane at the base of the ROS. The disks thus form initially from the plasma membrane of the cell. With time, the disks move up the outer segment as new disks are formed at the base, lengthening the outer segment. Old disks at the apical tip of the rod are shed in a packet encompassing the terminal disks of the outer segment through a process that likely involves fusion of one disk with the plasma membrane (Boesze-Battaglia and Goldberg 2002). This packet of old disks is then phagocytosed by the overlying pigmented epithelium, shortening the outer segment. Thus, the outer segment is in a constant state of degradation and renewal (Bok 1986; Young 1967). In vertebrates, the transit of disks from the base to the tip of the outer segment requires approximately ten days.

14.2.1 ROS Disk Membranes

The ROS disk membranes contain lipids and proteins, in approximately equal amounts by weight (Daemen 1973; Fliesler and Anderson 1983). Phospholipids represent almost 90 mole% of the total ROS lipids, while cholesterol accounts for less than 10 mole% on average (Fliesler and Anderson 1983; Fliesler and Schroepfer 1982). Cholesterol is not uniformly distributed among the ROS disk membranes, however. Newly formed disks are high in cholesterol content, similar to the plasma membrane from which they form. As the disks age, the cholesterol content decreases, until it reaches 5 mole% or less in the old disks (Boesze-Battaglia and others 1990, 1989). The change in cholesterol content with age of the disk membrane is related to a modulation of activity of rhodopsin in that disk membrane. High cholesterol leads to inhibition of receptor function (Boesze-Battaglia and Albert 1990; Mitchell and others 1990). Glycerolipid constituents are not constrained to remain with the disks into which they were initially assembled; rather, individual lipid classes have distinct turnover rates that are considerably more rapid than those of the membrane proteins (Anderson and others 1980a, 1980b; Anderson and Maude 1972; Bibb and Young 1974a, 1974b; Masland and Mills 1979; Mercurio and Holtzman 1982). Phospholipid fatty acid composition among disks is also a function of disk age (Boesze-Battaglia and Albert 1992).

In striking contrast to the complexity of the ROS lipid molecular species composition (Boesze-Battaglia and Albert 1989, 1992), rhodopsin (the visual pigment) accounts for about 95% of the total ROS membrane protein (Krebs and Kuhn 1977; Papermaster and Dreyer 1974). This GPCR, once assembled into a disk, remains associated with that disk throughout its lifetime in the ROS. Thus, the turnover of rhodopsin as a receptor parallels the basal-to-apical transit time of the disks in the ROS, and the relative location of a disk along the length of the ROS reflects the age of its protein constituents (i.e., the basalmost disks contain the most recently synthesized proteins) (Hall and others 1969; Young and Droz 1968). The binding of transducin by rhodopsin is also a function of disk age (Young and Albert 2000).

It is interesting to note that rhodopsin is located both in the plasma membrane of the ROS and in the disk membranes. Normally one thinks of GPCRs as being located exclusively in the plasma membranes of cells. However, rhodopsin is a GPCR that functions primarily in intracellular organelle membranes, rather than in the plasma membrane. As a photopigment, rhodopsin is largely inactive in the plasma membrane because of the high cholesterol (Boesze-Battaglia and Albert 1990).

14.3 FUNCTION OF RHODOPSIN AS A GPCR

Rhodopsin, as with most GPCRs, responds to a bound ligand. In the case of rhodopsin, the bound ligand in the inactive state is 11-*cis* retinal, which is bound by a Schiff base to lys296. The retinal is photosensitive and confers on the receptor the sensitivity to light. When light strikes the ROS and is absorbed by the photopigment, rhodopsin goes through a series of spectrally defined intermediates. Early in the process is a rapid isomerization of the 11-*cis* retinal to all-*trans* retinal. This is followed by changes in protein conformation on a slower time scale. The transition of the intermediate Metarhodopsin I to Metarhodopsin II (Meta II, the active form of the receptor) stimulates the binding of the G protein, transducin, to rhodopsin. The all-*trans* retinal is then expelled from the receptor and is cycled, through the adjacent pigmented epithelium, to 11-*cis* retinal, ready to recombine with a bleached opsin (rhodopsin without retinal) to re-form rhodopsin. Binding of transducin to the activated receptor causes the activation of this G protein. Activation of transducin is accomplished by a conformational change in the protein concomitant with the exchange of GTP for GDP at a specific binding site on the G protein. The α subunit of transducin initiates the cGMP cascade (Bennett and others 1982) by dissociating from the heterotrimeric transducin and binding to the phosphodiesterase, which stimulates the hydrolysis of cGMP and thus a reduction in the concentration of this second messenger. Reduction in cGMP levels leads to closure of the plasma membrane Na+ channels, which results in a hyperpolarization of the plasma membrane that is transmitted to the synapse at the base of the rod cell.

Rhodopsin is a member of a large family of G-protein coupled receptors, and the mechanism of signaling by rhodopsin is analogous to the system used by other GPCRs. All of these receptors couple to heterotrimeric G proteins as the means to convert an extracellular signal into an intracellular signal. Many of these receptors bind ligands from the cell exterior, which induce a conformational change in the cytoplasmic face of the receptor, enabling binding of the G protein. When the G protein binds to an active receptor, an exchange of GTP for GDP on the α subunit occurs, and the α subunit separates from the $\beta\gamma$ subunit. In some systems, both the α subunit and the $\beta\gamma$ complex can function in signaling. Upon activation, the α subunit of the G protein can modulate the activity of a target protein, often an enzyme. The α subunit is itself a GTPase, and thus with time, the GTP bound to the α subunit is hydrolyzed. The GDP-bound form of the α subunit is inactive and can reassociate with the $\beta\gamma$ complex to recycle the G protein.

Rhodopsin is the most intensively studied member of the G-protein coupled receptor family because it is the only member that is naturally present in high abundance in biological tissues. Earlier reviews have been published (for example, see Hargrave and McDowell 1992). After methods were developed to isolate ROS disk membranes (Smith and others 1975), relatively large amounts of natural membrane containing predominantly one membrane protein could be obtained for study. Subsequently, the purification of rhodopsin on an affinity column in detergent was reported (Stubbs and others 1976), and reconstitution of purified rhodopsin into membranes of defined lipid content could be achieved (Jackson and Litman 1982; Rothschild and others 1980).

14.4 EARLY STRUCTURE RESULTS

Because of the availability of rhodopsin, as described previously, this was the first GPCR to be studied with techniques designed to reveal structure. Early data on the three-dimensional structure of bovine rhodopsin came from circular dichroism (CD) studies of sonicated disks (Litman 1972) and of the purified protein (Shichi 1971; Shichi and Shelton 1974) in detergents. CD is sensitive to the secondary structure of proteins, and the data analysis of the CD of rhodopsin was consistent with a structure containing a bundle of seven transmembrane helices (Albert and Litman 1978). Data also suggested that when light was absorbed by rhodopsin, movement occurred within the bundle of helices. Later FTIR experiments also provided data confirming the dominance of α-helices in the secondary structure of rhodopsin (Pistorius and deGrip 1994).

The next advance in understanding the three-dimensional structure of bovine rhodopsin was achieved with the completion of the primary sequence (Hargrave and others 1983; Ovchinnikov and others 1982), a significant accomplishment because the only means at the time to obtain the sequence was with chemical sequencing of the protein, a very difficult task with a protein as hydrophobic as rhodopsin. The primary sequence is represented in Figure 14.1. The hydropathy plot for this sequence suggested several hydrophobic segments in the protein. In particular, six hydrophobic segments are relatively obvious from such a plot (Albert and Yeagle 2002), consistent with six of the seven putative transmembrane helices. As is now known, the seventh helix is quite polar in the middle, with a lys-ser-thr sequence, and thus does not show well in the hydropathy analysis.

The prior development of the three-dimensional structure of bacteriorhodopsin heavily influenced thinking about rhodopsin. Like rhodopsin, bacteriorhodopsin hydropathy plots suggested the presence of seven transmembrane helices. Early structure results from two-dimensional crystals of bacteriorhodopsin showed a bundle of seven transmembrane helices, many running approximately perpendicular to the plane of the bilayer (Henderson and others 1990; Leifer and Henderson 1983). This result firmly established the concept of a bundle of helices connected by loops as the dominant model for the likely three-dimensional structure of rhodopsin.

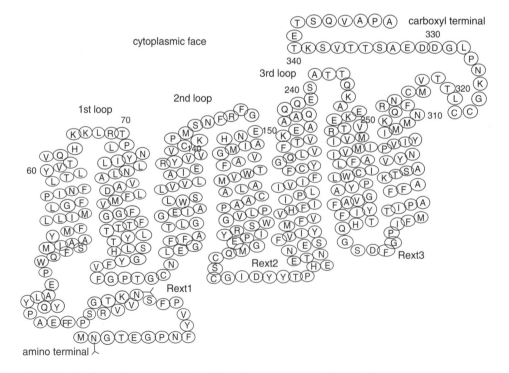

FIGURE 14.1 A schematic representation of the primary sequence of rhodopsin.

The next milestone in the development of structural information for rhodopsin came about a decade later, with the publication of a low-resolution structural analysis from two-dimensional crystals of bovine rhodopsin (Schertler and others 1993), using similar methods that had proved successful for bacteriorhodopsin. The electron density maps showed direct evidence for most of the putative transmembrane helices, and these helices were assigned using a model developed by Baldwin (Baldwin and others 1997). Baldwin performed an extensive analysis of the available sequences of GPCRs (with the advent of modern molecular biology, the number of GPCR sequences had increased from one in 1983 to several dozens in the early 1990s). She identified a number of conserved residues that were located in the putative transmembrane segments (as identified in the hydropathy plot) of rhodopsin. She developed a useful nomenclature for labeling these residues and some suggested rules for the likely disposition of such conserved residues within the structure. Among these principles were the suggestions that polar residues in putative transmembrane segments, as well as conserved residues in transmembrane segments, likely were in contact with other helices of the bundle, sequestered from the lipid bilayer.

This advance was followed by a low-resolution, three-dimensional structure for bovine rhodopsin in which, for the first time, one could see the bundle of transmembrane helices and trace most of them through the electron density. One of the first observations from such analysis was that, contrary to many expectations, the structure of the helical transmembrane bundle for bovine rhodopsin was not the same as for bacteriorhodopsin (Unger and others 1997). While in general the model of a bundle of seven transmembrane helices was valid, the packing of the helices differed between these two photosensitive proteins. This difference was confirmed by separate studies on two-dimensional crystals of frog rhodopsin (Schertler and Hargrave 1995) and squid rhodopsin (Davies and others 1996). These studies showed strong similarity of the helical packing in frog rhodopsin, squid rhodopsin, and bovine rhodopsin and significant differences from the helical packing observed in bacteriorhodopsin. These observations called into question some of the early models for bovine rhodopsin built on the template of bacteriorhodopsin. However, more recent models, using some limited experimental data and the principles enunciated by Baldwin, proved to be more accurate (Herzyk and Hubbard 1995).

As details emerged about the packing of the helices in rhodopsin, several distinctive features were identified. One feature is the tilt of helix 3. Helix 3 is severely tilted with respect to all the other helices of the bundle in the dark-adapted structure of rhodopsin. At the other extreme, helices 4, 6, and 7 were suggested to be close to perpendicular to the membrane surface. These features have survived into the most recent three-dimensional structure determinations for rhodopsin, as will be seen later.

14.5 SITE-TO-SITE DISTANCES WITHIN RHODOPSIN

For many years, three-dimensional structures were unobtainable for membrane proteins like rhodopsin, but a variety of approaches were developed in the 1990s to measure specific site-to-site distances within the intact membrane protein, or to probe close contacts within the membrane protein structure. Many of these studies employ one of two different methods. One method is site-directed spin labeling, developed as a collaboration between the laboratories of Hubbell and Khorana for use on bovine rhodopsin. The other, described in detail by Oprian and colleagues, engineers disulfide bonds to explore close contacts within the protein.

Site-directed spin labeling effectively provides long-range distances between sites in the intact protein (Hubbell and others 2000) labeled with a spin label. Using site-specific mutagenesis, one can construct a cysteineless protein and then reintroduce two cysteines at specific positions within the primary sequence. Cysteines can be chemically labeled with various reactive reagents. Labeling this protein with spin labels produces two discrete labeled sites. The dipolar interaction between the two spin labels at the two sites can be measured and, because the dipolar interaction is distance dependent, the magnitude of that dipolar interaction can be interpreted in terms of a distance between the labels. Distances between about 5 and 25 Å can be determined by this method. This technique has now been used on several membrane proteins.

The laboratories of Hubbell and of Khorana have used this technique to provide a wealth of data for rhodopsin (Altenbach and others 1999b, 1999c, 1996; Cai and others 1999a, 1999b, 1997; Farahbakhsh and others 1992, 1993; Farrens and others 1996; Klein-Seetharaman and others 1999; Langen and others 1999, 1998; Voss and others 1997; Wu and others 1996; Yang and others 1996b, 1996c) consisting of a set of distances between specific residues on the intact protein. These experiments have provided data on distances between specific sites in dark-adapted rhodopsin and have also provided data on interresidue distances in the transient state of Metarhodopsin II, the activated state of rhodopsin. From these data, researchers concluded that movements in helix 3 and helix 6 are among the conformational changes that occur when rhodopsin converts to Metarhodopsin II.

The spin label experiments also provide indirect information on secondary structure. Periodicities in the EPR spectra from labels at sequential positions in the sequence point to the occurrence of α helix, in some cases. With this kind of experiment, researchers were able to locate the carboxyl end of helix 5, for example (Altenbach and others 1996). These experiments also reveal information about the exposure of sites to the aqueous phase and about dynamics of the polypeptide chain at the spin label position.

The other method referred to earlier, engineered disulfide bonds, has provided considerable information about contacts between helices in the transmembrane helical bundle of rhodopsin. Much work in this area has come from Oprian and colleagues (Yu and others 1995). Putative contact points are probed by introducing cysteines at key positions in the protein and looking for the formation of disulfide bonds between those two cysteines. This approach has produced a number of contact points within rhodopsin (Struthers and Oprian 2000; Struthers and others 1999; Yu and Oprian 1999).

These experiments have yielded important information about the conformational change from dark-adapted rhodopsin to Metarhodopsin II. In some cases, the disulfide bonds inhibit the change in conformation, suggesting that those contact points may no longer be in contact in Metarhodopsin II. In other cases, the presence of the disulfide bonds does not inhibit the formation of Metarhodopsin II, thus indicating that those points of contact are present in both the dark-adapted state and the active state of the receptor (Struthers and others 2000; Yu and others 1999).

The success of these disulfide experiments depended on another important observation about the structure of rhodopsin. Oprian and colleagues were able to express rhodopsin in two pieces, each piece containing two or more transmembrane helices. For example, one of the two pieces contained four of the transmembrane helices of rhodopsin and the corresponding linking loops, and the other piece contained the remaining three of the transmembrane helices and the connecting links. These researchers noted that the separate coexpression of these pieces of rhodopsin led to the formation of pigment, that is, a protein that incorporated the ligand and retinal, even though the two pieces were not covalently linked. By some process after expression, the two pieces came together in the membrane in a nativelike conformation. This work suggests considerable stability within the helical bundle, which will be discussed later in this chapter.

If a metal ion binding site can be engineered between two or more helices, with residues from each of the helices contributing ligands to the metal ion, then one can identify a close approach of those helices. Such an experiment was performed in rhodopsin, which showed contact between helices 3 and 6. Furthermore, this contact was apparently broken upon activation of the receptor (Sheikh and others 1996).

14.6 THREE-DIMENSIONAL STRUCTURE OF RHODOPSIN: X-RAY CRYSTALLOGRAPHY

The X-ray crystal structure of rhodopsin was recently reported (Palczewski and others 2000). This is a milestone in the understanding of the structure and function of GPCRs. The importance of this

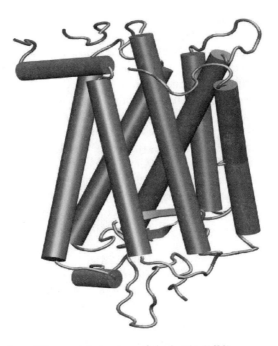

FIGURE 14.2 Representation of the crystal structure of rhodopsin (1f88).

contribution cannot be overemphasized. This is the first three-dimensional structure of any member of the GPCR family.

Crystals were obtained from purified bovine rhodopsin, isolated in the presence of Zn^{+2}. The crystals were twinned. Interestingly, it was not possible to use the known structure of bacteriorhodopsin in the molecular replacement protocol, consistent with the previous data (see earlier discussion) that suggested the structures of rhodopsin and bacteriorhodopsin were significantly different. Structure solution was achieved instead with MAS phasing. The crystals contained the dark-adapted form of rhodopsin. Exposure of the crystals to light led to loss of the diffraction pattern.

The crystal structure shows the major features of this receptor (for a recent review, see Stenkamp and others 2002). As expected, a bundle of seven transmembrane helices makes up the core of the protein (Figure 14.2). These helices are not packed in the same manner as in bacteriorhodopsin. Some of the helices are bent or broken at points, often at a proline. Helix 3 is the most seriously tilted from the remainder of the bundle.

A surprising discovery in the crystal structure is the presence of helix 8. This helix appears after helix 7 and is oriented approximately perpendicular to the bundle of transmembrane helices. The helix terminates in the palmitoylation sites (cys322, cys323), which are palmitoylated in this structure. It may be stabilized by an interface (see later) like the membrane surface.

The intradiskal surface (corresponding to the extracellular surface of other GPCRs) is well defined in the crystal structure. It consists of three loops, one of which (the loop connecting transmembrane helices 4 and 5) is extended across this face of the protein, covering the retinal binding pocket. The amino terminus joins with this loop to form a limited β-sheet, stabilizing the structure. Also seen in the crystal structure are the glycosylation sites on the amino terminus.

14.7 THREE-DIMENSIONAL STRUCTURE OF RHODOPSIN: NMR

Several years before the publication of this structure, another approach to the structure of rhodopsin was begun (Yeagle and others 1995). This approach, described in detail elsewhere (Albert and Yeagle 2002), ultimately led to the publication of a second structure of rhodopsin, in good agreement with the crystal structure but providing greater detail in the cytoplasmic face of the protein. A somewhat simpler version of this method has been used to determine, in part, the backbone fold for other proteins (Gross and others 1999; Perozo and others 1998; Poirer and others 1998). This approach circumvents the need to have crystals and is applicable to structure determination for other GPCRs and perhaps other membrane proteins, as well.

A growing body of data suggests that solution structures of peptides derived from some classes of proteins, including membrane proteins, retain the secondary structure of the parent protein because of the dominance in α-helices and turns of short-range interactions (Yang and others 1996a) that can be captured in peptides (Callihan and Logan 1999; Cox and others 1993; Fan and others 1998; Gao and others 1999; Gegg and others 1997; Hamada and others 1995; Hunt and others 1997; Jimenez and others 1999; Ramirez-Alvarado and others 1997; Wilce and others 1999) (Adler and others 1995; Blanco and Serrano 1994; Blumenstein and others 1992; Campbell and others 1995; Chandrasekhar and others 1991; Cox and others 1993; Ghiara and others 1994; Goudreau and others 1994; Wilce and others 1999) (Abdulaev and others 2000; Arshava and others 1998; Askin and others 1998; Barsukov and others 1992; Berlose and others 1994; Chopra and others 2000; Franzoni and others 1999; Haris 1988; Hunt and others 1997; Katragadda and others 2000; Lemmon and others 1992; Lomize and others 1992; Mierke and others 1996; Pervushin and others 1994; Popot and Engelman 2000; Xie and others 2000; Yeagle and others 1997, 2000a, 2000b). In some cases, the entire sequence of a helical bundle protein has been incorporated in a series of peptides spanning that sequence, and the individual peptides have reported the secondary structure of much of the native protein with fidelity (Behrends and others 1997; Blanco and Serrano 1994; Dyson and others 1992; Padmanabhan and others 1999; Reymond and others 1997). Glycophorin is a membrane protein that can be cleaved into fragments that retain the secondary structure of the native protein after separation (Schulte and Marchesi 1979). Both rhodopsin (Albert and Litman 1978) and the Na+K+ATPase (Esmann and others 1994) are membrane proteins for which proteolytic fragments retain structure characteristic of the native protein. Oprian has shown that rhodopsin can be expressed as fragments that spontaneously assemble after expression into a functional unit in the membrane (Yu and others 1995). Similar studies have been done on the α-factor receptor (Martin and others 1999). The cytoplasmic loops of rhodopsin have a function separate from the remainder of the protein (Abdulaev and others 2000; Konig and others 1989; Takemoto and others 1985) and have structure, as well (Yeagle and others 1995, 1997).

It can therefore be hypothesized that the intrinsic structures of peptides containing the amino acid sequences for turns or for transmembrane helices of membrane proteins built of helical bundles will be the same that those sequences adopt in the native protein. This hypothesis was tested on the membrane protein bacteriorhodopsin, a protein whose structure was already known. The structure of bacteriorhodopsin consists of a bundle of seven transmembrane helices connected by turns. Engelman and coworkers have found that the transmembrane helices of this protein are independently stable folding units (Hunt and others 1997) and thus can be considered protein domains (Popot and Engelman 2000). Several X-ray crystal structures are available for this membrane protein (Gouaux 1998; Grigorieff and others 1996; Luecke and others 1999; Pebay-Peyroula and others 1997). A series of peptides was designed for bacteriorhodopsin (Katragadda and others 2001a). The structures of these peptides in solution were determined by two-dimensional homonuclear ^{1}H NMR, as described in detail previously (Katragadda and others 2000, 2001a). Peptides corresponding to helices A, B, C, D, E, and F of bacteriorhodopsin formed helices in solution that agreed well with the crystal structure (Katragadda and others 2001a). The peptides corresponding to all six turns from bacteriorhodopsin form turns with the same residues in both the crystal structure and the peptide (Katragadda and others 2001a). Exploiting the overlap of adjacent peptides, a continuous construct of all the peptides can be made by superimposing the backbone atoms of the overlapping regions. All available experimental distance constraints were then written on this construct. Simulated annealing was used to optimize the conformation of the protein with respect

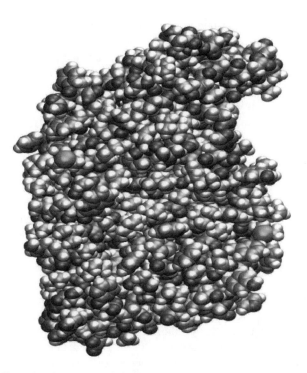

FIGURE 14.3 Three-dimensional structure of rhodopsin obtained from NMR data (1jfp).

to all the experimental constraints simultaneously. The result is shown in Figure 14.2 (Katragadda and others 2001a). Good agreement was observed between this structure and the structure determined by diffraction techniques.

The successful test of this novel approach to structure on bacteriorhodopsin encouraged the application of the approach to bovine rhodopsin. A series of overlapping peptides spanning the rhodopsin sequence was synthesized. Each peptide was designed to represent either a transmembrane helix of the protein or a turn. Solution structures for all the peptides were determined by two-dimensional homonuclear ^{1}H NMR as described in part previously (Yeagle and others 1995, 1996, 1997). Peptides from helices showed helices in agreement with the X-ray crystal structure of rhodopsin (Chopra and others 2000; Katragadda and others 2001b; Yeagle and others 2000a). Experimental distance constraints were written into the mol2 file for this construct in SYBYL (Tripos). The 11-*cis* retinal was constrained by the solid-state NMR data of Watts et al. (Grobner and others 2000). The construct with the distance constraints was subjected to simulated annealing (1000 fs at 1000 K followed by 1500 fs cooling to 200 K).

The result is a compact structure, strictly from experimental data (no modeling), showing a bundle of seven helices connected by six turns (Figure 14.3) (Yeagle and others 2001). Superposition of this structure on the previously published crystal structure of rhodopsin (Palczewski and others 2000) shows good agreement with the crystal structure in the transmembrane region.

This structure is consistent with information from other experiments for dark-adapted rhodopsin (which were not used in the structure determination). CD data suggested that 60% of the sequence was helical, in good agreement with 64% in this structure (Albert and Litman 1978). Indirect measurements with spin labels on the intact protein predicted the termination of helix 5 within one residue of that found in this structure (Altenbach and others 1996). The inferred termination of the seventh transmembrane helix from similar measurements (Altenbach and others 1999a) agrees within one residue with this structure. The short, antiparallel β-strands in the carboxyl terminus of this structure agree with FTIR data on the intact protein (Pistorius and deGrip 1994). Specific interactions between residues 338 and 242 indicate defined structure in the carboxyl terminus (Cai and others 1997). The third cytoplasmic loop projects toward the carboxyl terminus in agreement with AFM data (Heymann and others 2000).

One significant difference between the crystal structure and the structure derived from the NMR data is found in the carboxyl terminus. In the former structure, the region of residues 311–321 is in an α-helix called helix 8 (Palczewski and others 2000). In the latter structure, this region is a loop, indicating that there is little intrinsic tendency (from the primary sequence) to form a helix in this region. Therefore, in rhodopsin the formation of the helix may well be dependent on an interface, that is, the lipid bilayer. These observations could be important to structure and function in this protein.

14.8 STRUCTURE OF METARHODOPSIN II

The structure just described is the structure of this receptor in the dark-adapted state. To understand how this receptor system operates, information on the structure of the active state is required. Of the several intermediates in the photocycle, Metarhodopsin II (Meta II) is the intermediate that is responsible for binding and activating the G protein, transducin. Therefore, the structure of this intermediate is critical to understanding the mechanism of signal transduction with this receptor system.

Meta II is a transient species. However, it is possible to trap this species by lowering the temperature. With this approach, Hubbell and Khorana were able to use site-directed spin labeling to obtain distance information between specific sites in the protein while it was in the Meta II state (Altenbach and others 2001; Hubbell and others 2000). These data are analogous to the data used (see earlier discussion) to build the tertiary structure of dark-adapted rhodopsin. Oprian and colleagues have found that some of their engineered disulfide bonds do not inhibit the transition from dark-adapted rhodopsin to Meta II, suggesting that the close contacts implied by the disulfide bond formation are preserved in the active state (Yu and others 1999). These data provide additional distance constraints for the active state. These long-range distance constraints for the activated state of this receptor include a contact between the β-ionone ring of the retinal and A169 of helix 4 (Borhan and others 2000).

The structure of the dark-adapted state of rhodopsin was used as the starting point for this structure determination of Meta II. All the long-range interactions specific for the dark-adapted structure were removed, and the long-range interactions specific for Meta II were added. Simulated annealing was once again used to fold a structure consistent with the new data set of experimental long-range distance constraints. The resulting medium-resolution structure of Meta II is shown in Figure 14.4.

Significant structural changes occur upon conversion of dark-adapted rhodopsin to Meta II. The second and third cytoplasmic loops move apart and change conformation. As a result, a basic groove opens in the cytoplasmic face of the receptor when it is activated, leading to an exposure of a portion of the cytoplasmic face that is occluded in the dark-adapted state. Exposure of this new basic surface is likely the signal for G-protein binding, because it represents a contact surface on the receptor for the G protein and the face of transducin that binds to rhodopsin is acidic (Lambright and others 1996). That contact surface for the binding of transducin can be mapped on the cytoplasmic face of Meta II using sequences of peptides that inhibit the interaction between the receptor and transducin (Konig and others 1989; Marin and others 2000) (Ernst and others 2000). This mapping points to binding of the G protein in the basic groove on the cytoplasmic surface of the receptor that is exposed upon the transition to Meta II. Chemical crosslinking studies on transducin binding show contact between sites on the receptor and sites on transducin (Cai and others 2001; Itoh and others 2001). Based on these studies, the N and C termini of transducin α are near the third cytoplasmic loop of Meta II in the complex. These crosslinking data and the structure of Meta II suggest that the acidic portion of the amino terminal helix of transducin α (see GRASP picture of transducin, [Lambright and others 1996]) may lie in the basic groove on the receptor when the G protein is bound to Meta II.

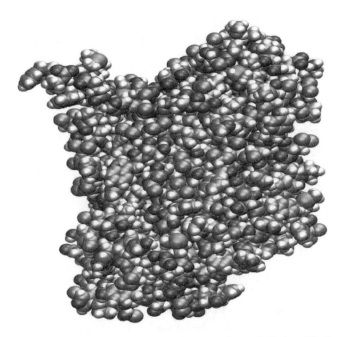

FIGURE 14.4 Three-dimensional structure of Metarhodopsin II from NMR data (1ln6).

Based on the structure of Meta II, several previously published experiments with respect to the cytoplasmic face of this receptor can be better understood. First, the conformational change in cytoplasmic loops two and three may have been detected previously. These loops contain β-turns (Yeagle and others 1997), and previous FTIR data suggest a change in β-turns in the conformation change from rhodopsin to Metarhodopsin II (DeGrip 1988). Second, both the crystal structure and the NMR structure of rhodopsin (Palczewski and others 2000; Yeagle and others 2001) show interactions between R135, E134, and E247. The structure of Meta II shows that, upon activation of rhodopsin, the interaction between R135 and E247 is broken due to the conformational change. This disruption of interactions between R135 and E247 was predicted previously for the corresponding residues in other receptors in the same GPCR family (Ballesteros and others 2001; Shapiro and others 2002).

The transmembrane region is defined by fewer experimental long-range distance constraints than are available for the cytoplasmic face. Available data (Struthers and others 2000) indicate that the arrangement of helices 1, 2, 5, and 7 is not changed upon conversion of rhodopsin to Meta II. However, helices 3, 4, and 6 move relative to their positions in the rhodopsin structure and alter in particular the positions of the helices in the cytoplasmic face of the receptor. Specifically, during the photoactivation and the *cis-trans* isomerization of retinal chromophore, helix 3 is forced outward, probably as a result of the steric interaction between C9 methyl group of the chromophore and residues on helix 3. Within the constraints imposed by the work of Oprian (Struthers and others 2000), an inward movement of helix 4 is necessary to counteract the outward movement of helix 3. During this cooperative rearrangement, helix 4 is rotated, with A169 coming in contact with the retinal β-ionone ring. The model of Baldwin predicts that helix 4 would be capable of rotation because it does not have a side restricted to interaction with other members of the helical bundle by the polarity (Baldwin 1993).

14.9 RHODOPSIN STABILITY

Integrated with the three-dimensional structure of membrane proteins are the questions of folding and stability. Folding is a very difficult subject with membrane proteins, because they cannot be studied in solution. No protein folding studies (analogous to the large amount of data available for many soluble proteins) are available for rhodopsin. Few studies are available on the stability of membrane proteins. The structures of integral membrane proteins must accommodate interactions with two different environments: the hydrocarbon core of the lipid bilayer in which they are imbedded and the aqueous phase into which they protrude. From available reports, the protein regions that interact with the aqueous phase seem to behave in a manner similar to soluble proteins, whereas those within the hydrophobic core are more structurally stable (Haltia and Freire 1995) (White and Wimley 1999). For example, the two transitions in the circular dichroism spectra of cytochrome b_5, taken as a function of denaturant concentration, were interpreted to represent the sequential denaturation of aqueous (at lower denaturant concentrations) and transmembrane (at higher denaturant concentrations) domains (Tajima and others 1976). These experiments were possible because the hydrophobic portion of this protein is small and the protein can be studied in solution (probably as protein micelles).

For membrane proteins that traverse the bilayer multiple times, the relative importance of the extramembraneous regions and intramembraneous regions to overall protein stability poses an interesting problem. The complexity of this was demonstrated for bacteriorhodopsin. As stated earlier, some of the helices of bacteriorhodopsin will associate properly without covalently linked loop regions (Kahn and others 1992; Marti 1998). This suggests that the loops may not be important in the stability. However, based on thermal denaturation studies, it was calculated that loss of all loops would lead to instability of bacteriorhodopsin at room temperature (Kahn and others 1992).

The thermal denaturation of rhodopsin is likely to involve multiple steps, including isomerization of 11-*cis* retinal, unfolding of regions exposed to the aqueous media, and/or reorientation of the helices. Differential scanning calorimetry (DSC) has proved a useful tool for studying the thermal stability of rhodopsin in native disk membranes (Khan and others 1991; Landin and others 2001; Miljanich and others 1985; Shnyrov and Berman 1988) and in disk membranes with altered lipid compositions (Albert and others 1996a; Polozova and Litman 2001). Interpretation of these studies is complicated by the irreversible nature of the denaturation. However, these studies showed that both the chromophore and the lipid bilayer affect the transition temperature of rhodopsin denaturation.

Investigations of the scan rate dependence of the calorimetric transition temperature indicated that at the transition temperature there is significant formation of the final, irreversibly denatured species (Khan and others 1991; Landin and others 2001). Thus, the formation of irreversibly denatured rhodopsin is rapid in comparison to reversible processes. The observed asymmetry of the calorimetric transition is also consistent with this interpretation. The denaturation process of rhodopsin is therefore subject to kinetic constraints. These constraints may also be important factors in the stability of cytochrome c oxidase (Morin and others 1990), bacteriorhodopsin (Galisteo and Sanchez-Ruiz 1993), and the GLUT 1 receptor (Epand and others 1999). These studies suggest that rhodopsin may have a finite, kinetically determined lifetime as a native protein in the membrane. Spontaneous denaturation occurs even at physiological temperature. This observation may be important in relation to the transit time of disks in the ROS from the point of biogenesis at the base of the outer segment to the apical tip of the outer segment. If in disease states the denaturation process is accelerated, denaturation of the receptor could occur before the disk completes its transit of the outer segment. Such a process would be expected to result in disk degeneration (see next section).

14.10 LIPID–RHODOPSIN INTERACTIONS IN THE DISK MEMBRANE

Rhodopsin is an integral membrane protein, with much of its mass buried in the lipid bilayer. Therefore, the interactions between the lipid and the protein may be important to protein structure and function. This subject requires considerably more investigation. Nevertheless, some structural studies have been reported on the interaction of the lipid component with rhodopsin in the disk membrane. ^{31}P NMR experiments have revealed that in the disk membrane about 15 to 20 of the phospholipids are interacting significantly with the protein, perhaps binding to sites on the protein (Albert and Yeagle 1983). Chemical labeling experiments have shown that some of the phosphatidylserine in the membrane is protected by protein from labeling (Crain and others 1978). Some of the phospholipid in the membrane is protected from phospholipase action; that protected phospholipid component is enriched in phosphatidylserine (Breugel and others 1978). These studies collectively suggest that a subset of the phospholipids of the membrane may be binding to rhodopsin and perhaps modulating its structure and function.

Rhodopsin is an important influence on the stability of the lipid bilayer of the disk. The lipids in the disk are unstable in a lipid bilayer in the absence of rhodopsin (Albert and others 1984). Nonlamellar structures are formed by the pure disk membrane lipids. This has important ramifications to the studies described previously on rhodopsin denaturation. Denaturation of rhodopsin in the disk membrane corresponds, at least in part, to an aggregation of the protein. Aggregation of the protein would leave areas of the membrane depleted of its major integral membrane protein. According to the previously mentioned studies, these lipids should not be stable in a bilayer but would spontaneously form nonlamellar structures. Such behavior corresponds to a loss of the integrity of the membrane bilayer and thus a degeneration of the membranes of the outer segment.

Biochemical studies have examined in some detail the influence of the lipid environment on rhodopsin function. Membrane cholesterol significantly inhibits rhodopsin function (Boesze-Battaglia and Albert 1990; Mitchell and others 1990). Highly unsaturated phospholipids enhance the activation of rhodopsin to Metarhodopsin II (Mitchell and others 1992). It has now been clearly shown that this modulation is propagated through the structure and dynamics of the lipid bilayer. Recent studies suggest that polyunsaturated phospholipids partition directly into the lipid/protein interface, preferentially to other phospholipid species in the disk membrane (Polozova and Litman 2001). Other studies have suggested that one molecule of cholesterol may also bind to the surface of rhodopsin within the membrane at a site with structural specificity for cholesterol (Albert and others 1996b).

REFERENCES

Abdulaev NG, Ngo T, Chen R, Lu Z, Ridge KD. 2000. Functionally discrete mimics of light-activated rhodopsin identified through expression of soluble cytoplasmic domains. *J. Biol. Chem.* 275:39354–39363.

Adler M, Sato MH, Nitecki DE, Lin J-H, Light DR, Morser J. 1995. The structure of a 19 residue fragment from the C loop of the fourth epidermal growth factor-like domain of thrombomodulin. *J. Biol. Chem.* 270:23366–23372.

Albert AD, Boesze-Battaglia K, Paw Z, Watts A, Epand RM. 1996a. Effect of cholesterol on rhodopsin stability in disk membranes. *Biochim. Biophys. Acta* 1297:77–82.

Albert AD, Litman BJ. 1978. Independent structural domains in the membrane protein bovine rhodopsin. *Biochemistry* 17(19):3893–3900.

Albert AD, Sen A, Yeagle PL. 1984. The effect of calcium on the bilayer stability of lipids from bovine rod outer segment disk membranes. *Biochim. Biophys. Acta* 771:28–34.

Albert AD, Yeagle PL. 1983. Phospholipid domains in bovine retinal rod outer segment disk membranes. *Proc. Natl. Acad. Sci. U.S.A.* 80:7188–7191.

Albert AD, Yeagle PL. 2002. Structural studies on rhodopsin. *Biochim. Biophys. Acta* 1565(2):183–195.

Albert AD, Young JE, Yeagle PL. 1996b. Rhodopsin-cholesterol interactions in bovine rod outer segment disk membranes. *Biochim. Biophys. Acta* 1285:47–55.

Altenbach C, Cai K, Khorana HG, Hubbell WL. 1999a. Structural features and light-dependent changes in the sequence 306–322 extending from helix VII to the palmitoylation sites in rhodopsin: a site-directed spin-labeling study. *Biochemistry* 38:7931–7937.

Altenbach C, Cai K, Khorana HG, Hubbell WL. 1999b. Structural features and light-dependent changes in the sequence 306–322 extending from helix VII to the palmitoylation sites in rhodopsin: a site-directed spin-labeling study. *Biochemistry* 38(25):7931–7937.

Altenbach C, Cai K, Klein-Seetharaman J, Khorana HG, Hubbell WL. 2001. Structure and function in rhodopsin: mapping light-dependent changes in distance between residue 65 in helix TM1 and residues in the sequence 306–319 at the cytoplasmic end of helix TM7 and in helix H8. *Biochemistry* 40(51):15483–15492.

Altenbach C, Klein-Seetharaman J, Hwa J, Khorana HG, Hubbell WL. 1999c. Structural features and light-dependent changes in the sequence 59–75 connecting helices I and II in rhodopsin: a site-directed spin-labeling study. *Biochemistry* 38(25):7945–7949.

Altenbach C, Yang K, Farrens DL, Farahbakhsh ZT, Khorana HG, Hubbell WL. 1996. Structural features and light dependent changes in the cytoplasmic interhelical E-F loop region of rhodopsin: a site-directed spin-labeling study. *Biochemistry* 35:12470–12478.

Anderson RE, Kelleher PA, Maude MB. 1980a. Metabolism of phosphatidylethanolamine in the frog retina. *Biochim. Biophys. Acta* 620:227–235.

Anderson RE, Kelleher PA, Maude MB, Maida TM. 1980b. Synthesis and turnover of lipid and protein components of frog retinal rod outer segments. *Neurochemistry* 1:29–42.

Anderson RE, Maude M. 1972. The effects of essential fatty acid deficiency on the phospholipids of the photoreceptor membranes of rat retinas. *Arch. Biochim. Biophys.* 151:270–276.

Arshava B, Liu SF, Jiang H, Breslav M, Becker JM, Naider F. 1998. Structure of segments of a G protein-coupled receptor: CD and NMR analysis of the *Saccharomyces cerevisiae* tridecapeptide pheromone receptor. *Biopolymers* 46:343–357.

Askin D, Bloomberg GB, Chambers EJ, Tanner MJA. 1998. NMR solution structure of a cytoplasmic surface loop of the human red cell anion transporter, band 3. *Biochemistry* 37:11670–11678.

Baldwin JM. 1993. The probable arrangement of the helices in G protein-coupled receptors. *EMBO J.* 12(4):1693–1703.

Baldwin JM, Schertler GFX, Unger VM. 1997. An alpha-carbon template for the transmembrane helices in the rhodopsin family of G-protein-coupled receptors. *J. Mol. Biol.* 272:144–164.

Ballesteros JA, Jensen AD, Liapakis G, Rasmussen SGF, Shi L, Gether U, Javitch JA. 2001. Activation of the β2-adrenergic receptor involves disruption of an ionic lock between the cytoplasmic ends of transmembrane segments 3 and 6. *J. Biol. Chem.* 276:29171–29177.

Barsukov IL, Nolde DE, Lomize AL, Arseniev AS. 1992. Three-dimensional structure of proteolytic fragment 163–231 of bacterioopsin determined from nuclear magnetic resonance data in solution. *Eur. J. Biochem.* 206(3):665–672.

Behrends HW, Folkers G, Beck-Sickinger AG. 1997. A new approach to secondary structure evaluation: secondary structure prediction of porcine adenylate kinase and yeast guanylate kinase by CD spectroscopy of overlapping synthetic peptide segments. *Biopolymers* 41(2):213–231.

Bennett N, Michel-Villay M, Kühn H. 1982. Light-induced interaction between rhodopsin and the GTP-binding protein. Metarhodopsin II is the major photoproduct involved. *Eur. J. Biochem.* 127:97–103.

Berlose J, Convert O, Brunissen A, Chassaing G, Lavielle S. 1994. Three dimensional structure of the highly conserved seventh transmembrane domain of G-protein-coupled receptors. *FEBS Lett.* 225:827–843.

Bibb C, Young RW. 1974a. Renewal of fatty acids in membranes of visual cell outer segments. *J. Cell Biol.* 61:327–343.

Bibb C, Young RW. 1974b. Renewal of glycerol in visual cells and pigment epithelium of the frog retina. *J. Cell Biol.* 62:378–389.

Blanco FJ, Serrano L. 1994. Folding of protein G B1 domain studied by the conformational characterization of fragments comprising its secondary structure elements. *Eur. J. Biochem.* 230:634–649.

Blumenstein M, Matsueda GR, Timmons S, Hawiger J. 1992. A β turn is present in the 392–411 segment of the human fibrinogen γ chain. Effects of structural changes in this segment on affinity to antibody 4A5. *Biochemistry* 31:10692–10698.

Boesze-Battaglia K, Albert A. 1990. Cholesterol modulation of photoreceptor function in bovine rod outer segments. *J. Biol. Chem.* 265:20727–20730.

Boesze-Battaglia K, Albert AD. 1989. Fatty acid composition of bovine rod outer segment plasma membrane. *Exp. Eye Res.* 49:699–701.

Boesze-Battaglia K, Albert AD. 1992. Phospholipid distribution in bovine rod outer segment membranes. *Exp. Eye Res.* 54:821–823.

Boesze-Battaglia K, Fliesler SJ, Albert AD. 1990. Relationship of cholesterol content to spatial distribution and age of disk membranes in retinal rod outer segments. *J. Biol. Chem.* 265:18867–18870.

Boesze-Battaglia K, Goldberg AF. 2002. Photoreceptor renewal: a role for peripherin/rds. *Int. Rev. Cytol.* 217:183–225.

Boesze-Battaglia K, Hennessey T, Albert AD. 1989. Cholesterol heterogeneity in bovine rod outer segment disk membranes. *J. Biol. Chem.* 264:8151–8155.

Bok D. 1986. *Invest. Opthamol. Visual Sci.* 26:1659–1694.

Borhan B, Souto ML, Imai H, Shichida Y, Nakanishi K. 2000. Movement of retinal along the visual transduction path. *Science* 288:2209–2212.

Breugel PJGMv, Geurts PHM, Daemen FJM, Bonting SL. 1978. Biochemical aspects of the visual process. XXXVIII. Effects of lateral aggregation on rhodopsin in phospholipase C-treated photoreceptor membranes. *Biochim. Biophys. Acta* 509:136–147.

Cai K, Itoh Y, Khorana HB. 2001. Mapping of contact sites in complex formation between light-activated rhodopsin and transducin by covalent crosslinking: use of a photoactivatable reagent. *Proc. Natl. Acad. Sci. U.S.A.* 98:4877–4882.

Cai K, Klein-Seetharaman J, Hwa J, Hubbell WL, Khorana HG. 1999a. Structure and function in rhodopsin: effects of disulfide cross-links in the cytoplasmic face of rhodopsin on transducin activation and phosphorylation by rhodopsin kinase. *Biochemistry* 38:12893–12898.

Cai K, Klein-Seetharaman J, Farrens D, Zhang C, Altenbach C, Hubbell WL, Khorana HG. 1999b. Single-cysteine substitution mutants at amino acid positions 306-321 in rhodopsin, the sequence between the cytoplasmic end of helix VII and the palmitoylation sites: sulfhydryl reactivity and transducin activation reveal a tertiary structure. *Biochemistry* 38(25):7925–-7930.

Cai K, Langen R, Hubbell WL, Khorana HG. 1997. Structure and function in rhodopsin: topology of the C-terminal polypeptide chain in relation to the cytoplasmic loops. *Proc. Natl. Acad. Sci. U.S.A.* 94:14267–14272.

Callihan DE, Logan TM. 1999. Conformations of peptide fragments from the FK506 binding protein: comparison with the native and urea-unfolded states. *J. Mol. Biol.* 285:2161–2175.

Campbell AP, McInnes C, Hodges RS, Sykes BD. 1995. Comparison of NMR structures of the receptor binding domains of *Pseudomonas aeruginosa* Pili strains PAO, KB7, and PAK: implications for receptor binding and synthetic vaccine design. *Biochemistry* 34:16255–16268.

Chandrasekhar K, Profy AT, Dyson HJ. 1991. Solution conformation preferences of immunogenic peptides derived from the principal neutralizing determinant of the HIV-1 envelope glycoprotein gp120. *Biochemistry* 30:9187–9194.

Chopra A, Yeagle PL, Alderfer JA, Albert A. 2000. Solution structure of the sixth transmembrane helix of the G-protein coupled receptor, rhodopsin. *Biochim. Biophys. Acta* 1463:1–5.

Cox JPL, Evans PA, Packman LC, Williams DH, Woolfson DN. 1993. Dissecting the structure of a partially folded protein. *J. Mol. Biol.* 234:483–492.

Crain RC, Marinetti GV, O'Brien DF. 1978. Topology of amino phospholipids in bovine retinal rod outer segment disk membranes. *Biochemistry* 17:4186–4192.

Daemen. FJM. 1973. Vertebrate rod outer segment membranes. *Biochim. Biophys. Acta* 300:255–288.

Davies A, Schertler GF, Gowen BE, Saibil HR. 1996. Projection structure of an invertebrate rhodopsin. *J. Struct. Biol.* 117:36–44.

DeGrip WJ, Gray D, Gillespie J, Bovee PHM, Van den Berg LJ, Rothschild KJ. 1988. Photoexcitation of rhodopsin: conformation changes in the chromophore, protein and associated lipids as determined by FTIR difference spectroscopy. *Photochem. Photobiol.* 48:497–504.

Dyson HJ, Merutka G, Waltho JP, Lerner RA, Wright PE. 1992. Folding of peptide fragments comprising the complete sequence of proteins. *J. Mol. Biol.* 226:795–817.

Epand RF, Epand RM, Jung CY. 1999. Glucose-induced thermal stabilization of the native conformation of GLUT 1. *Biochemistry* 38:454–458.

Ernst OP, Meyer CK, Marin EP, Henklein P, Fu W-Y, Sakmar TP, Hofmann KP. 2000. Mutation of the fourth cytoplasmic loop of rhodopsin affects the binding of transducin and peptides derived from the carboxyl-terminal sequences of transducin α and γ subunits. *J. Biol. Chem.* 275:1937–1943.

Esmann M, Karlish SJD, Sottrup-Jensen L, Marsh D. 1994. Structural integrity of the membrane domains in extensively trypsinized NaKATPase from shark rectal glands. *Bioichemistry* 33:8044–8050.

Fan JS, Cheng HC, Zhang M. 1998. A peptide corresponding to residues asp177 to asn208 of human cyclin a forms an alpha helix. *Biochem. Biophys. Res. Commun.* 253:621–627.

Farahbakhsh ZT, Altenbach C, Hubbell WL. 1992. Spin labeled cysteines as sensors for protein-lipid interaction and conformation in rhodopsin. *Photochem. Photobiol.* 56(6):1019–1033.

Farahbakhsh ZT, Hideg K, Hubbell WL. 1993. Photoactivated conformation changes in rhodopsin: a time-resolved spin label study. *Science* 262:1416–1419.

Farrens DL, Altenbach C, Yang K, Hubbell WL, Khorana HG. 1996. Requirement of rigid-body motion of transmembrane helices for light activation of rhodopsin. *Science* 274:768–770.

Fliesler SJ, Anderson RE. 1983. Chemistry and metabolism of lipids in the verbrate retina. *Prog. Lipid Res.* 22:79–131.

Fliesler SJ, Schroepfer GJ, Jr. 1982. Sterol composition of bovine retinal rod outer segment membranes and whole retinas. *Biochim. Biophys. Acta* 711:138–148.

Franzoni L, Nicastro G, Pertinhez TA, Oliveira E, Nakaie CR, Paiva AC, Schreier S, Spisni A. 1999. Structure of two fragments of the third cytoplasmic loop of the rat angiotensin II AT1A receptor. Implications with respect to receptor activation and G-protein selection and coupling. *J. Biol. Chem.* 274(1):227–35.

Galisteo ML, Sanchez-Ruiz JM. 1993. Kinetic study into the irreversible thermal denaturation of bacterio-rhodopsin. *Eur. Biophys. J.* 22:25–30.

Gao J, Li Y, Yan H. 1999. NMR solution structure of domain 1 of human annexin I shows an autonomous folding unit. *J. Biol. Chem.* 274:2971–2977.

Gegg CV, Bowers KE, Matthews CR. 1997. Probing minimal independent folding units in dihydrofolate reductase by molecular disection. *Prot. Sci.* 6:1885–1892.

Ghiara JB, Stura EA, Stanfield RL, Profy AT, Wilson IA. 1994. Crystal structure of the principal neutralization site of HIV-1. *Science* 264:82–85.

Gouaux E. 1998. It's not just a phase: crystallization and x-ray structure determination of bacteriorhodopsin in lipidic cubic phases. *Structure* 6:5–10.

Goudreau N, Cornille F, Duchesne M, Parker F, Tocqué B, Garbay C, Roques BP. 1994. NMR structure of the N-terminal SH3 domain of GRB2 and its complex with a proline-rich peptide from Sos. *Nature Struct. Biol.* 1:898–907.

Grigorieff N, Ceska TA, Downing KH, Baldwin JM, Henderson R. 1996. Electron-crystallographic refinement of the structure of bacteriorhodopsin. *J. Mol. Biol.* 259:393–421.

Grobner G, Burnett IJ, Glaubitz C, Choi G, Mason AJ, Watts A. 2000. Observations of light-induced structural changes of retinal within rhodopsin. *Nature* 405:810–813.

Gross A, Columbus L, Hideg K, Altenbach C, Hubbell WL. 1999. Structure of the KcsA potassium channel from *Streptomyces lividans*: a site directed spin labeling study of the second transmembrane segment. *Biochemistry* 38:10324–10335.

Hall MO, Bok D, Bacharach ADE. 1969. *J. Mol. Biol.* 45:397–406.

Haltia T, Freire E. 1995. Forces and factors that contribute to the structural stability of membrane proteins. *Biochim. Biophys. Acta* 1228:1–27.

Hamada D, Kuroda Y, Tanaka T, Goto Y. 1995. High helical propensity of the peptide fragments derived from β-lactoglobulin, a predominantly β-sheet protein. *J. Mol. Biol.* 254:737–746.

Hargrave PA, McDowell JH. 1992. Rhodopsin and phototransduction: a model system for G protein-linked receptors. *FASEB J.* 6:2323–2331.

Hargrave PA, McDowell JH, Curtis DR, Wang JK, Juszczak E, Fong SL, Rao JKM, Argos P. 1983. The structure of bovine rhodopsin. *Biophys. Struct. Mech.* 9:235–244.

Haris PI. 1988. Synthetic peptide fragments as probes for structure determination of potassium ion-channel proteins. *Biosci. Rep.* 18:299–312.

Henderson R, Baldwin JM, Ceska TA, Zemlin F, Beckmann E, Downing KH. 1990. Model for the structure of bacteriorhodopsin based on high-resolution electron cryo-microscopy. *J. Mol. Biol.* 213(4):899–929.

Herzyk P, Hubbard RE. 1995. Automated method for modeling seven-helix transmembrane receptors from experimental data. *Biophys. J.* 69:2419–2442.

Heymann JB, Pfeiffer M, Hildebrandt V, Kaback HR, Fotiadis D, Groot BD, Engel A, Oestserhelt D, Miller DJ. 2000. Conformations of the rhodopsin third cytoplasmic loop grafted onto bacteriorhodopsin. *Structure* 8:643–653.

Hubbell WL, Cafiso DS, Altenbach C. 2000. Identifying conformational changes with site-directed spin labeling. *Nature Struct. Biol.* 7:735–739.

Hunt JF, Earnest TN, Bousche O, Kalghatgi K, Reilly K, Horvath C, Rothschild KJ, Engelman DM. 1997. A biophysical study of integral membrane protein folding. *Biochemistry* 36:15156–15176.

Itoh Y, Cai K, Khorana HB. 2001. Mapping of contact sites in complex formation between light-activated rhodopsin and transducin by covalent crosslinking: use of a chemically preactivated reagent. *Proc. Natl. Acad. Sci. U.S.A.* 98:4883–4887.

Jackson M, Litman BJ. 1982. Rhodopsin-phospholipid reconstitution by dialysis removal of octyl-glucoside. *Biochemistry* 21:5601–5607.

Jimenez MA, Evangelio JA, Aranda C, Lopez-Brauet A, Andreu D, Rico M, Lagos R, Andreu JM. 1999. Helicity of alpha(404-451) and beta(394-445) tubulin C-terminal recombinant peptides. *Prot. Sci.* 8:788–799.

Kahn TW, Sturtevant JM, Engelman DM. 1992. Thermodynamic measurements of contributions of helix-connecting loops and of retinal to the stability of bacteriorhodopsin. *Biochemistry* 31:8829–8839.

Katragadda M, Alderfer JL, Yeagle PL. 2000. Solution structure of the loops of bacteriorhodopsin closely resemble the crystal structure. *Biochim. Biophys. Acta* 1466:1–6.

Katragadda M, Alderfer JL, Yeagle PL. 2001a. Assembly of a polytopic membrane protein structure from the solution structures of overlapping peptide fragments of bacteriorhodopsin. *Biophys. J.* 81:1029–1036.

Katragadda M, Chopra A, Bennett M, Alderfer JL, Yeagle PL, Albert AD. 2001b. Structures of the transmembrane helices of the G-protein coupled receptor, rhodopsin. *J. Peptide Res.* 58:79–89.

Khan SMA, Bolen W, Hargrave PA, Santoro MM, McDowell JH. 1991. Differential scanning calorimetry of bovine rhodopsin in rod-outer-segment disk membranes. *Eur. J. Biochem.* 200:53–59.

Klein-Seetharaman J, Hwa J, Cai K, Altenbach C, Hubbell WL, Khorana HG. 1999. Single-cysteine substitution mutants at amino acid positions 55–75, the sequence connecting the cytoplasmic ends of helices I and II in rhodopsin: reactivity of the sulfhydryl groups and their derivatives identifies a tertiary structure that changes upon light-activation. *Biochemistry* 38(25):7938–7944.

Konig B, Arendt A, McDowell JH, Kahlert M, Hargrave PA, Hofmann KP. 1989. Three cytoplasmic loops of rhodopsin interact with transducin. *Proc. Natl. Acad. Sci. U.S.A.* 86:6878–6882.

Krebs W, Kuhn H. 1977. *Exp. Eye Res.* 25:511–526.

Lambright DG, Sonked J, Bohm A, Skiba NP, Hamm HE, Sigler PB. 1996. The 2.0 Å crystal structure of a heterotrimeric G protein. *Nature* 379:311–319.

Landin JS, Katragadda M, Albert AD. 2001. Thermal destabilization of rhodopsin and opsin by proteolytic cleavage in bovine rod outer segment disk membranes. *Biochemistry* 40:11176–11183.

Langen R, Cai K, Altenbach C, Khorana HG, Hubbell WL. 1999. Structural features of the C-terminal domain of bovine rhodopsin: a site-directed spin-labeling study. *Biochemistry* 38(25):7918–7924.

Langen R, Kai K, Khorana HG, Hubbell WL. 1998. Structure and dynamics of the C-terminal domain in rhodopsin probed by site-directed spin labeling and disulfide cross-linking. *Biophys. J.* 74:A290.

Leifer D, Henderson R. 1983. Three dimensional structure of orthorhombic purple membrane at 6.5 Å resolution. *J. Mol. Biol.* 163:451–466.

Lemmon MA, Flanagan JM, Hunt JF, Adair BD, Bormann B-J, Dempsey CE, Engelman DM. 1992. Glyco-phorin A dimerization is driven by specific interactions between transmembrane α-helices. *J. Biol. Chem.* 267:7683–7689.

Litman BJ. 1972. Effect of light scattering on the circular dichroism of biological membranes. *Biochemistry* 11:3243.

Lomize AL, Pervushin KV, Arseniev AS. 1992. Spatial structure of (34-65)bacterioopsin polypeptide in SDS micelles determined from nuclear magnetic resonance data. *J. Biomol. NMR* 2(4):361–372.

Luecke H, Schobert B, Richter H-T, Cartailler J-P, J.K. Lanyi J. 1999. Structure of bacteriorhodopsin at 1.55 angstrom resolution. *J. Mol. Biol.* 291:899.

Marin EP, Krishna AG, Zvyaga TA, Isele J, Siebert F, Sakmar TP. 2000. The amino terminus of the fourth cytoplasmic loop of rhodopsin modulates rhodopsin-transducin interaction. *J. Biol. Chem.* 275:1930–1936.

Marti T. 1998. Refolding of bacteriorhodopsin from expressed polypeptide fragments. *J. Biol. Chem.* 273:9312–9322.

Martin NP, Leavitt LM, Sommers CM, Dumont ME. 1999. *Biochemistry* 38:682–695.

Masland RH, Mills JM. 1979. *J. Cell Biol.* 83:159–178.

Mercurio AM, Holtzman E. 1982. *J. Neurocytol.* 11:263–293.

Mierke DF, Royo M, Pelligrini M, Sun H, Chorev M. 1996. Third cytoplasmic loop of the PTH/PTHrP receptor. *J. Am. Chem. Soc.* 118:8998–9004.

Miljanich GP, Brown MF, Mabrey-Gaud S, Dratz EA. 1985. Thermotropic behavior of retinal rod membranes and dispersions of extracted phospholipids. *J. Membrane Biol.* 85:79–86.

Mitchell D, Straume M, Miller J, Litman BJ. 1990. Modulation of Metarhodopsin formation by cholesterol-induced ordering of bilayers. *Biochemistry* 29:9143–9149.

Mitchell DC, Straume M, Litman BJ. 1992. Role of sn-1-saturated, sn-2-polyunsaturated phospholipids in control of membrane receptor conformational equilibrium: effects of cholesterol and acyl chain unsaturation on the Metarhodopsin I-Metarhodopsin II equilibrium. *Biochemistry* 31:662–670.

Morin PE, Diggs D, Freire E. 1990. Thermal stability of membrane-reconstituted yeast cytochrome c oxidase. *Biochemistry* 29:781–788.

Ovchinnikov YA, Abdulaev NG, Feigina MY, Artamonov ID, Zolotarev AS, Kostina MB, Bogachuk AS, Miroshnkov AI, Martinov VI, Kudelin AB. 1982. The complete amino acid sequence of visual rhodopsin. *Bioorg. Khim.* 8:1011–1014.

Padmanabhan S, Jimenez MA, Rico M. 1999. Folding propensities of synthetic peptide fragments covering the entire sequence of phage 434 Cro protein. *Protein Sci.* 8:1675–1688.

Palczewski K, Kumasaka T, Hori T, Behnke CA, Motoshima H, Fox BA, Le Trong I, Teller DC, Okada T, Stenkamp RE, and others. 2000. Crystal structure of rhodopsin: a G protein-coupled receptor. *Science* 289(5480):739–745.

Papermaster D, Dreyer W. 1974. Rhodopsin content in the outer segment membranes of bovine and frog retinal rods. *Biochemistry* 13:2438–2444.

Pebay-Peyroula E, Rummel G, Rosenbusch JP, Landau EM. 1997. X-ray structure of bacteriorhodopsin at 2.5 angstroms from microcrystals grown in lipidic cubic phases. *Science* 277:1676–1681.

Perozo E, Cortes DM, Cuello LG. 1998. Three-dimensional architecture and gating mechanism of a K channel studied by EPR spectroscopy. *Nature Struct. Biol.* 5:459–469.

Pervushin KV, Orekhov VY, Popov AI, Musina LY, Arseniev AS. 1994. Three-dimensional structure of (1-71)bacterioopsin solubilized in methanol/chloroform and SDS micelles determined by 15N-1H heteronuclear NMR spectroscopy. *Eur. J. Biochem.* 219:571–583.

Pistorius AM, deGrip WJ. 1994. Rhodopsin's secondary structure revisited: assignment of structural elements. *Biochem. Biophys. Res. Commun.* 198:1040–1045.

Poirer MA, Xiao W, Macosko JC, Chan C, Shin Y-K, Bennett MK. 1998. The synaptic SNARE complex is a parallel four-stranded helical bundle. *Nature Struct. Biol.* 5:765–769.

Polozova A, Litman BJ. 2001. Cholesterol dependent recruitment of di22:6-PC by a G protein-coupled receptor into lateral domains. *Biophys. J.* 79:2632–2643.

Popot J-L, Engelman DM. 2000. Helical membrane protein folding, stability, and evolution. *Annu. Rev. Biochem.* 69:881–922.

Ramirez-Alvarado R, Serrano L, Blanco FJ. 1997. Conformational analysis of peptides corresponding to all the secondary structure elements of protein L B1 domain. *Prot. Sci.* 6:162–174.

Reymond MT, Merutka G, Dyson HJ, Wright PE. 1997. Folding propensities of peptide fragments of myoglobin. *Prot. Sci.* 6:706–716.

Rothschild K, DeGrip W, Sanches R. 1980. Fourier transform infrared study of photoreceptor membrane. I. Group assignments based on rhodopsin delipidation and reconstitution. *Biochim. Biophys. Acta* 596:338–351.

Schertler GF, Hargrave PA. 1995. Projection structure of frog rhodopsin in two crystal forms. *Proc. Natl. Acad. Sci U.S.A.* 92:11578–11582.

Schertler GRX, Villa C, Henderson R. 1993. Projection structure of rhodopsin. *Nature* 362:770–772.

Schulte TH, Marchesi VT. 1979. Conformation of the human erythrocyte glycophorin A and its constituent peptides. *Biochemistry* 18:275–280.

Shapiro DA, Kristiansen K, Weiner DM, Kroeze WK, Roth BL. 2002. Evidence for a model of agonist-induced activation of 5-HT2A serotonin receptors which involves the disruption of a strong ionic interactioin between helices 3 and 6. *J. Biol. Chem.* 277:11441–11449.

Sheikh SP, Zvyaga TA, Lichtarge O, Sakmar TP, Bourne HR. 1996. Rhodopsin activation blocked by metal-ion binding sites linking transmembrane helices C and F. *Nature* 383:347–350.

Shichi H. 1971. Circular dichroism of bovine rhodopsin. *Photochem. Photobiol.* 13:499–502.

Shichi H, Shelton E. 1974. Assessment of physiological integrity of sonicated retinal rod membranes. *J. Supramol. Struct.* 2:7–16.

Shnyrov VL, Berman AL. 1988. Calorimetric study of thermal denaturation of vertebrate visual pigments. *Biomed. Biochim. Acta* 47:355–362.

Smith HG, Stubbs GW, Litman BJ. 1975. The isolation and purification of osmotically intact discs from retinal rod outer segments. *Exp. Eye Res.* 20:211–217.

Stenkamp RE, Filipek S, Driessen CA, Teller DC, Palczewski K. 2002. Crystal structure of rhodopsin: a template for cone visual pigments and other G protein-coupled receptors. *Biochim. Biophys. Acta* 1565(2):168–182.

Struthers M, Oprian DD. 2000. Mapping tertiary contacts between amino acid residues within rhodopsin. *Methods Enzymol.* 315:130–143.

Struthers M, Yu H, Kono M, Oprian DD. 1999. Tertiary interactions between the fifth and sixth transmembrane segments of rhodopsin. *Biochemistry* 38(20):6597–6603.

Struthers M, Yu H, Oprian DD. 2000. G protein-coupled receptor activation: analysis of a highly constrained, "straitjacketed" rhodopsin. *Biochemistry* 39(27):7938–7942.

Stubbs GW, Smith HG, Litman BJ. 1976. Alkyl glucosides as effective solubilizing agents for bovine rhodopsin. A comparison with several commonly used detergents. *Biochim. Biophys. Acta* 426:46–56.

Tajima S, Enomoto K, Sato R. 1976. Denaturation of cytochrome b5 by guanidine hydrochloride: evidence for independent folding of the hydrophobic and hydrophilic moieties of the cytochrome molecule. *Arch. Biochem. Biophys.* 172:90–97.

Takemoto DJ, Takemoto LJ, Hansen J, Morrison D. 1985. Regulation of retinal transducin by C-terminal peptides of rhodopsin. *Biochem. J.* 232:669–672.

Unger VM, Hargrave PA, Baldwin JM, Schertler GFX. 1997. Arrangement of rhodopsin transmembrane α-helices. *Nature* 389:203–206.

Voss J, Hubbell WL, Hernandez-Borrell J, Kaback HR. 1997. Site-directed spin-labeling of transmembrane domain VII and the 4B1 antibody epitope in the lactose permease of *Escherichia coli. Biochemistry* 36(49):15055–15061.

White SH, Wimley WC. 1999. Membrane protein folding and stability: physical principles. *Annu. Rev. Biophys. Biomol. Struct.* 28:319–365.

Wilce JA, Salvatore D, Waade JD, Craik DJ. 1999. H-1 NMR structural studies of a cystine-linked peptide containing residues 71-93 of transthyretin and effects of sa ser84 substitution implicated in familial amyloidotic polyneuropathy. *Eur. J. Biochem.* 262:586–594.

Wu J, Voss J, Hubbell WL, Kaback HR. 1996. Site-directed spin labeling and chemical crosslinking demonstrate that helix V is close to helices VII and VIII in the lactose permease of *Escherichia coli. Proc. Natl. Acad. Sci. U.S.A.* 93(19):10123–10127.

Xie HB, Ding FX, Schreiber D, Eng G, Liu SF, Arshava B, Arevalo E, Becker JM, Naider F. 2000. Synthesis and biophysical analysis of transmembrane domains of a *Saccharomyces cerevisiae* G protein-coupled receptor. *Biochemistry* 39:15462–15474.

Yang A-S, Hitz B, Honig B. 1996a. Free energy determinants of secondary structure formation: β turns and their role in protein folding. *J. Mol. Biol.* 259:873–882.

Yang K, Farrens DL, Altenbach C, Farahbakhsh ZT, Hubbell WL, Khorana HG. 1996b. Structure and function in rhodopsin. Cysteines 65 and 316 are in proximity in a rhodopsin mutant as indicated by disulfide formation and interactions between attached spin labels. *Biochemistry* 35:14040–14046.

Yang K, Farrens DL, Hubbell WL, Khorana HG. 1996c. Structure and function in rhodopsin. Single cysteine substitution mutants in the cytoplasmic interhelical E-F loop region show position-specific effects in transducin activation. *Biochemistry* 35:12464–12469.

Yeagle PL, Alderfer JL, Albert AD. 1995. Structure of the carboxyl terminal domain of bovine rhodopsin. *Nature Struct. Biol.* 2:832–834.

Yeagle PL, Alderfer JL, Albert AD. 1996. Structure determination of the fourth cytoplasmic loop and carboxyl terminal domain of bovine rhodopsin. *Molecular Vision* 2: http://www.molvis.org/molvis/v2/p12.

Yeagle PL, Alderfer JL, Albert AD. 1997. The first and second cytoplasmic loops of the G-protein receptor, rhodopsin, independently form β-turns. *Biochemistry* 36:3864–3869.

Yeagle PL, Choi G, Albert AD. 2001. Studies on the structure of the G-protein coupled receptor rhodopsin including the putative G-protein binding site in unactivated and activated forms. *Biochemistry* 40:11932–11937.

Yeagle PL, Danis C, Choi G, Alderfer JL, Albert AD. 2000a. Three-dimensional structure of the seventh transmembrane helical domain of the G-protein receptor, rhodopsin. *Molecular Vision:* www.molvis.org/molvis/v6/a17.

Yeagle PL, Salloum A, Chopra A, Bhawsar N, Ali L, Kuzmanovski G, Alderfer JL, Albert, AD. 2000b. Structures of the intradiskal loops and amino terminus of the G-protein receptor, rhodopsin. *J. Peptide Res.* 55:455–465.

Young JE, Albert AD. 2000. Transducin binding in bovine rod outer segment disk membranes of different age/spatial location. *Exp. Eye Res.* 70:809–812.

Young RW. 1967. The renewal of photoreceptor cell outer segments. *J. Cell Biol.* 33:61–72.

Young RW, Droz B. 1968. *J. Cell Biol.* 39:169–184.

Yu H, Kono M, McKee TD, Oprian DD. 1995. A general method for mapping tertiary contacts between amino acid residues in membrane-embedded proteins. *Biochemistry* 34:14963–14969.

Yu H, Kono M, Oprian DD. 1999. State-dependent disulfide cross-linking in rhodopsin. *Biochemistry* 38:12028–12032.

Yu H, Oprian DD. 1999. Tertiary interactions between transmembrane segments 3 and 5 near the cytoplasmic side of rhodopsin. *Biochemistry* 38(37):12033–12040.

15 Role of Membrane Lipids in Modulating the Activity of Membrane-Bound Enzymes

Richard M. Epand

CONTENTS

15.1 KINETICS

Quantitative analysis of the kinetics of enzyme catalysis in aqueous solution is well developed. Kinetic analysis allows one to formulate a scheme for the reaction, including the formation of intermediates, and, from experimental observation, to assign values of the rate of each step in the process. In addition, from a temperature dependence of the rate, the activation energy of each step can be determined. Kinetics alone cannot determine a reaction path; it can only find the simplest mode that is consistent with the observations. Of course, one of the classic applications of kinetics to enzyme-catalyzed reactions led to the development of the Michaelis–Menten theory, which provided a simple scheme to explain the phenomenon of saturation kinetics and the observation that beyond a certain substrate concentration, the rate of reaction no longer increased. The Michaelis–Menten scheme is

$$\text{Enzyme} + \text{Substrate} \underset{k_{-1}}{\overset{k_1}{\leftrightarrow}} \text{Enzyme} \cdot \text{Substrate Complex} \overset{k_2}{\rightarrow} \text{Enzyme} + \text{Products}$$

There are some enzyme-catalyzed reactions in aqueous solution that cannot be well described by Michaelis–Menten kinetics because of additional complications such as cooperativity, pre–steady-state phenomena, and reversibility of the reaction, among other reasons. Nevertheless, this simple kinetic scheme still proves extremely valuable in the study of enzyme properties and mechanisms. Analysis of the dependence of the reaction rate on substrate concentration allows one

0-8493-1403-8/05/$0.00+$1.50

to calculate two kinetic constants: V_{max} and K_m. Determination of these kinetic constants is useful because these constants have mechanistic implications. V_{max} is the maximal rate of reaction, which is also the rate of conversion of the intermediate enzyme-substrate complex to products. K_m is a kinetic constant equal to $(k_{-1} + k_3)/k_1$. The significance of K_m is that it equals the equilibrium constant for the dissociation of the enzyme-substrate complex when $k_{-1} \gg k_2$. In addition, the ratio of V_{max}/K_m gives the rate at low substrate concentrations and is used as a measure of catalytic efficiency. This ratio is free from complications brought about by the formation of additional intermediates in the path that can affect V_{max} and K_m.

Additional complications arise when enzymes are functioning at the two-dimensional interface of membrane and water. Often, enzymes that function at membranes do not follow a simple Michaelis–Menten scheme, and even when they do, the relationship between the determined values of the constants and the rate of a specific reaction or the substrate binding affinity is not straightforward. This is because there may be additional steps in the process that are not incorporated into the simple, general Michaelis–Menten kinetic scheme. The principal factor that makes the kinetics of reactions catalyzed at membranes different from that of reactions in solution is the nature of the process required to get the enzyme to bind to the substrate. Depending on the specific case, this may include the translocation of the enzyme and/or substrate from the bulk solution to the membrane or diffusion along the plane of the membrane.

The first consideration is the enzyme and its interaction with the membrane. One example is catalysis by an integral membrane protein. For this class of process, the enzyme has no possibility of transferring from one membrane structure to another. Hence, membrane-bound substrates that are not in the same liposome as the enzyme cannot be accessed for reaction. This type of interfacial catalysis has been termed the *scooting mechanism* [1, 2] to indicate that the enzyme must "scoot" along the plane of the membrane in order to access another substrate. This has to be the mode of action of integral membrane proteins, but in addition, it can also correspond to the characteristics of amphitropic proteins that can exist in both membrane and lipid environments, provided that these proteins exchange slowly between membranes. The scooting mechanism is contrasted with the "hopping" mode, in which the enzyme can dissociate from one membrane structure and bind to another. In this fashion, it can access all of the substrate in the reaction vessel, but the kinetics of the process may be modified compared with solution catalysis, depending on the rate of hopping. Under the condition in which the rate of hopping of the enzyme is rapid compared to the rate of catalysis, the system will behave like a soluble enzyme, at least from the point of view of the behavior of the enzyme.

In addition to the enzyme, modifications of the kinetics also result from the partitioning of the substrate between aqueous and membrane environments. It is not the bulk concentration of substrate that will determine the activity of an enzyme; rather, the concentration of substrate within the two-dimensional phase of the membrane will determine the rate of reaction [3]. This can involve the substrate as well as the enzyme, a so-called change of dimensionality, in which the reaction rate is dependent on the concentration of these components per unit area of the membrane, rather than per unit volume in solution. An excellent review of some of the quantitative aspects of the kinetic analysis of interfacial enzymes has recently appeared [4] that also discusses the limitations in the interpretation of the kinetic constants derived from such an analysis.

15.2   INTEGRAL MEMBRANE PROTEINS

Some integral membrane proteins catalyze reactions of water-soluble substrates. One example is acetylcholinesterase, which has been studied extensively because of its role in nerve signal transmission [5, 6]. Another is Na^+K^+-ATPase, in which the importance of membrane properties in the modulation of the enzyme's activity has been demonstrated [7, 8]. In addition, many membrane-bound enzymes act on membrane-bound substrates. The rate of catalysis by these enzymes will be dependent on the rate of collisions between enzyme and substrate, which, in turn, will be determined in part by the rate of lateral diffusion of the enzyme and substrate in the plane of the membrane, as well as their lateral distribution in a membrane containing domains. Both parameters are difficult to determine. Nevertheless, there are some examples in which the properties and kinetics of such an enzyme have been analyzed. For example, the neutral, Mg^{2+}-dependent sphingomyelinases (nSMase) are integral membrane proteins, although there are other forms of SMase that are water soluble [9]. Interest in SMase comes about in part because the product of hydrolysis by these enzymes is ceramide, a lipid that has an important role in signal transduction and more specifically in apoptosis [10]. The kinetics and mechanism of a form of nSMase obtained from *B. cereus* was reported by Fanani and Maggio [11]. Binding to lipids in the form of a monolayer activates the enzyme. The partition coefficient for the transfer of the enzyme to the interface is 7×10^{-4}, indicating a strong affinity of the enzyme for a membrane environment. The enzyme-catalyzed reaction is characterized by a latency period during the adsorption of the enzyme and its activation, a period with pseudo zero-order kinetics and finally a gradual reduction in the rate of product formation as the reaction proceeds. The process was described kinetically by a model that included the formation of dimers [11]. It has been shown that there is a close correlation between the rate of catalysis by SMase and changes in lipid domains in a monolayer, which itself is affected by the production of ceramide as a product of enzyme catalysis [12]. The formation of ceramide-enriched domains as a consequence of the action of SMase on liposomes was demonstrated by microscopy, which also showed vesicular budding [13].

There is considerable evidence to indicate that biological membranes are not homogeneous but contain domains enriched in particular components. One class of domains that has attracted considerable current interest is cholesterol-rich domains. There is evidence that some signal transduction pathways take place in cholesterol-rich domains in membranes, referred to as *rafts* [14]. These pathways often involve a cascade of events in which one protein affects the state of another protein. The sequestering of the proteins involved into the small cross-sectional area of a raft would increase their concentration in the plane of the membrane and hence increase the rate of reaction between them. However, cellular rafts also have a different physical state, referred to as the L_o phase [15], compared with the L_α or liquid–crystalline phase of the bulk of the membrane. The L_o phase is characterized as being less "fluid," as a consequence of both the high cholesterol concentration and the presence of high-melting phospholipids. Thus, there are two features of rafts that may influence the rate of enzyme-catalyzed reactions:

- The fact that they are domains that can concentrate enzymes and substrates
- The difference of their physical state, which would be expected to slow the rate of reaction because of the decreased molecular motion

The enzyme SMase, discussed earlier, is thought to be located in rafts [10] that are enriched in the lipid substrate, sphingomyelin. An example of an increased rate of reaction as a consequence of sequestering into a raft domain is the action of the Tyrosine kinase, Lyn. This protein efficiently catalyzes the phosphorylation of another raft protein, SMase [16]. The tyrosine phosphatase SHP-2 affects different targets, depending on whether or not it is in a raft domain [17]. Although some of the functions of the EGF receptor are facilitated by this protein's being sequestered into rafts, it has also been recently demonstrated that the binding of the EGF ligand and the Tyrosine kinase activity of the EGF receptor are suppressed by being in a raftlike environment [18]. This was based on an observed increase in these activities when cholesterol was depleted by treatment of the membrane with methyl-β-cyclodextrin.

15.3 AMPHITROPIC PROTEINS

There is also a large group of enzymes that are amphitropic, that is, they can exist in both water-soluble and membrane-bound forms. In many cases, these enzymes are activated as a result of binding to a membrane. Thus, in addition to the factors discussed thus far, the activity of these enzymes is also modulated by their translocation from solution to the membrane interface. Because of their water solubility, amphitropic enzymes are generally more amenable to studies of mechanism than enzymes that are integral membrane proteins.

15.3.1 MEMBRANE TRANSLOCATION

Several methods have been used to study the translocation of amphitropic enzymes *in vitro* between aqueous and membrane environments. These methods include those that require a physical separation of the free enzyme and the membrane-bound enzyme, followed by determining the concentration of enzyme in each of the fractions. This is usually done with greatest sensitivity and specificity by following the activity of the enzyme under assay conditions, such as solubilization in detergent, in which the membrane-bound and soluble enzyme fractions could be measured under identical conditions. The physical form of the lipid used for such studies is large unilamellar vesicles (LUVs). Other forms of vesicles are less appropriate, because small unilamellar vesicles (SUVs) are unstable and under strain because of the high curvature, and multilamellar vesicles (MLVs) have only a fraction of the lipid exposed at the outer monolayer. It should also be remembered that only half of the lipid is accessible for binding in LUVs, because the inner monolayer is not in contact with the solution containing the enzyme. Because the partial specific volume of lipids is close to one, LUVs are difficult to separate by centrifugation, either by sedimentation or by flotation. The liposomes, however, are made denser by binding protein, but in order to avoid effects of membrane crowding at the surface of the liposome, the protein concentration is maintained low. Thus, the perturbation of the density of the liposome by the bound protein is small.

Two methods have been introduced to facilitate the separation of free and bound fractions. One is to load the interior of the LUV with a solution of sucrose that will perturb the density [19]. The salt concentration of the extravesicular solution must be increased to maintain an osmotic balance. Nevertheless, sucrose loading does cause an increased difference between the density of the vesicles and the surrounding solution. This results in a facilitation of sedimentation of the liposomes and separation of free and bound enzymes. An alternative to this method is based on increasing the density of the solution with ^2H_2O to cause the liposomes to float [20]. In addition, there has been development of enzyme binding assays to immobilized lipids using surface plasmon resonance [21, 22]. Finally, there are methods based on spectroscopically detectable changes that occur as a result of binding to liposomes. One general method for this, which is often applicable to Trp-containing proteins, is based on resonance energy transfer between the Trp residue(s) of the protein and a dansyl group covalently attached to the amino group of phosphatidylethanolamine. The resonance energy transfer will occur only when the protein is bound to the membrane [23]. It should be remembered, however, that dansyl-phosphatidylethanolamine is an anionic lipid that may itself cause changes in the membrane partitioning of the protein [24].

15.3.2 LIPID BINDING — SPECIFIC INTERACTIONS

The importance of lipid properties in the activity of amphitropic enzymes has been noted [25]. With regard to the lipid requirements for membrane binding, there can be both specific and nonspecific factors. In general, most protein–lipid interactions are nonspecific in nature. However, certain protein domains have binding sites for specific lipid structures [26]. Important examples are protein domains that bind to a phosphorylated form of phosphatidylinositol [27–31]. This specificity is a consequence of the binding with high affinity of specific lipids to a certain protein domain. This phenomenon has important consequences in protein targeting within the cell and in the translocation of certain proteins to the membrane.

15.3.3 MEMBRANE BINDING — NONSPECIFIC INTERACTIONS

Lipids can also have a nonspecific effect on the properties of membrane-bound enzymes as a consequence of their physical properties. This can also be the case for enzymes that are integral membrane proteins, but the effects are generally larger and better understood for amphitropic enzymes, where lipids can play a role both in modulating the catalytic efficiency of the membrane-bound enzyme and in determining the partitioning of the enzyme between aqueous and membrane phases.

15.3.3.1 Electrostatic Charge

One physical feature of the membrane is electrostatic charge. Well-developed theories have been applied to describe the partitioning of charged peptides between the aqueous phase and a charged membrane [32–36]. These methods are based on the application of variations of the Gouy–Chapman–Stern theory of the electrical double layer, in which a charged interface will attract an oppositely charged peptide. The strength of this electrostatic interaction falls off exponentially as the distance from the membrane interface increases; it is also dependent on the salt concentration of the surrounding medium [37]. The difficulty of applying this to proteins is that it requires information about both the structure of the protein and how it associates with membranes. This information is needed to know the distance between the charged groups on the protein and the charged membrane interface.

Nevertheless, one can determine experimentally whether electrostatic interactions are important for the interaction of a particular protein with lipid by changing the charge on the lipid or the ionic strength of the buffer. These parameters will also affect the sequestration of divalent cations to the membrane interface and the decay of the interfacial concentration of these ions with distance from the interface. The distance required for reduction of the concentration of the ion to $1/e$ of its value at the membrane interface is called the *Debye length*. In a case in which the binding of a divalent cation to a site on a protein affects the translocation of the protein to the membrane interface, the location of this binding site on the membrane-bound protein with respect to the interface can be determined. This has been done for the Ca^{2+}-binding site on protein kinase C (PKC), which regulates membrane binding. It was found that, for the membrane-bound PKC, this site is situated at a distance of 0.3 nm from the membrane interface [38]. Thus, in addition to understanding the role of electrostatic interactions in the energy of binding of amphitropic proteins to the membrane interface, such studies can also provide indirectly some structural information.

Other cations besides Ca^{2+} will also be concentrated at the interface of an anionic membrane. This includes protons. As a consequence, the interfacial pH can differ from the bulk pH. An interesting consequence of this is the role of acidic residues in the binding of the enzyme CTP:phosphocholine cytidylyltransferase (CT) to membranes. Not only is this enzyme attracted to anionic membrane interfaces because of interactions with basic amino acid side chains, but the specificity of binding to anionic membranes is maintained by three Glu resides [39]. This is because the acidic residues on the protein will be protonated at anionic membrane interfaces because of the lower interfacial pH. However, with zwitterionic membranes, these groups will remain charged and will inhibit the partitioning of the protein to the membrane, contributing to the specificity of binding to anionic membranes [39]. Another mechanism by which charged membranes can affect the activity of enzymes at membrane interfaces is through the effects of electrostatic fields and dipole potentials [40].

15.3.3.2 Defects

The lipid bilayer stability and lipid packing will also affect the partitioning of amphitropic proteins to the membrane. There are several ways to describe the physical features of a membrane. None of these descriptions is complete, nor are they all independent of each other. The phospholipid bilayer is a complex environment with greater motion near the center of the bilayer [41]. Also, the motion in the bilayer is highly anisotropic. One of the features that would facilitate the binding of amphitropic proteins is "membrane fluidity." This has been a popular concept, in part because it is a parameter that can be measured experimentally using diphenylhexamethyltriene as a fluorescence polarization probe [42]. However, there is no single "fluidity" parameter that can describe the heterogeneity and anisotropy of the molecular motions within a bilayer. Another way to interpret the fluorescence polarization from diphenylhexamethyltriene is to consider that changes in this parameter measure free volume in the membrane [43].

Defects in a membrane provide a site for amphitropic enzymes to bind to membranes. However, many membrane instabilities can be classified as "defects," and they can have several causes and descriptions. One type of defect that is fairly specific in nature and description is a phase boundary defect. This will occur in membranes having two or more coexisting phases that are physically separated into domains. In a liposome composed of a synthetic lipid that can undergo a thermotropic phase transition from an ordered gel phase to a high-temperature liquid–crystalline phase, at the temperature region of the phase transition there will be domains of both phases. Because the packing of each phase is different, the location where the domains meet will have packing imperfections. The presence of such defects has been shown to enhance the activity of phospholipase A_2 at the temperature region of the lipid phase transition [44–48]. The site of action of phospholipase A_2 is at boundary defects in monolayers [49].

There may also be a relationship between the action of SMase and that of phospholipase A_2. The product of SMase action, ceramide, will induce membrane phase separation and the appearance of membrane phase boundary defects; this in turn will lead to an increase in the activity of phospholipase A_2 [50]. There is also product activation of phospholipase A_2, which has been suggested to result from a product-induced phase transition [51]. However, in the case of SMase, the presence of phase boundary defects increases the lag time for reaction [52], possibly as a consequence of slowing the diffusion of the enzyme and substrate from one liquid domain to another, itself a consequence of the presence of the intervening gel phase domains. It has been shown with the activity of pancreatic lipase on monolayers that there is a reduction in rate as a consequence of percolation [53], a conclusion supported by computer simulations [54]. Certain membrane bilayers will also convert to nonlamellar phases. As conditions for the transition to the hexagonal phase are approached, defects will appear in the membrane. In this case, because of the large differences in morphology between the two phases, these defects will appear before phase coexistence occurs. Phospholipase A_2 is also activated by these kinds of defects [55, 56].

15.3.3.3 Curvature Strain and Interfacial Polarity

Changes in the activity of amphitropic enzymes by membranes that are destabilized by a tendency to form nonlamellar phases may be described more specifically in terms of changes in curvature stress and/or changes in interfacial polarity. *Curvature stress* refers to the tendency of one monolayer of a bilayer to bend into a curved shape. A bilayer in a large vesicle or cell membrane will have no physical curvature on a molecular length scale. However, the constituent monolayers of the bilayer may have an intrinsic curvature that is not flat because of the nature of its component lipids and the temperature. The tendency of a monolayer to bend cannot be relieved in a flat bilayer phase; rather, it results in an unstable membrane with defects caused by curvature strain. The extent of this curvature strain can readily be calculated from the intrinsic curvature of the monolayer, obtained from diffraction measurements of the lattice spacing of the H_{II} phase and the elastic bending modulus. A monolayer organized in a structure in which its physical curvature is equal to its intrinsic curvature will not possess any curvature strain. The curvature strain of a monolayer will be equal to the energy required to unbend it from the form in which it has achieved its intrinsic curvature to the flat structure of the bilayer. This energy per unit area of interface is given by

$$\frac{0.5K_c}{R_0^2}$$

where K_c is the elastic bending modulus of the lipid monolayer and R_0 is the lipid monolayer's spontaneous radius of curvature in excess water. The elastic bending modulus is a measure of the stiffness of the monolayer. The easier it is to bend the monolayer, the less curvature strain will be acquired by forming a structure whose curvature is different from the intrinsic curvature. It was found that there was very good correlation between the curvature strain of a membrane and the activation of CT [57, 58].

The concept of curvature stress is useful, but the formulation is relatively simple and does not provide for features that may be specific to certain protein-membrane complexes. An alternative formulation has been proposed by Cantor [59, 60]. This approach focuses on the variation of lateral pressure as a function of the position in the bilayer. In general, lateral pressure profiles and curvature strain are alternative ways of expressing the same phenomenon. If there is a higher lateral pressure at the methyl terminus of the acyl chains than in other regions of the bilayer, this is equivalent to the bilayer's having negative curvature strain. Each of the two formulations has its own advantages. Curvature strain is a simpler idea and is more amenable to direct experimental measurement. However, the concept of lateral pressure profile provides a more detailed molecular description and can be more informative in combination with information about the structure and location of proteins embedded in the membrane [61].

The lateral pressure profile indicates that the curvature properties at a particular location in the bilayer may be different from those describing the bilayer as a whole. In the case of amphitropic proteins, it would be expected that the properties of the membrane interface would be particularly important. Although the activity of both PKC [62, 63] and CT [57] is enhanced by increasing the negative curvature strain of the membrane, in the case of PKC there are two lines of evidence that indicate that this is not a direct consequence of curvature strain [64]:

One indication comes from studies with a series of phosphatidylethanolamines (PE) containing 18:1 acyl chains but differing in the position of unsaturation. The curvature properties of this series of lipids have been determined [65], and the activity of PKC does not correlate with the curvature strain measured with membranes containing this homologous series of lipids [66]. However, the activity of PKC did correlate with properties of an interfacial fluorescent probe, suggesting that the activity of this enzyme was modulated by a property of the membrane interface [67]. The situation with CT is different and indicates that a change in curvature strain is the mechanism by which nonlamellar-forming lipids modulate the activity of this enzyme [58].

The other indication that the activity of PKC is not directly modulated by membrane curvature strain comes from studies of the activity of PKC in the presence of lipids in the cubic phase [68]. Spontaneous conversion of bilayers to a type II cubic phase will result in the relief of negative curvature strain. In order to compare the activity of PKC in the cubic phase with the activity in the lamellar phase, we converted the cubic phase of monoolein to a lamellar phase by the addition of progressively larger amounts of phosphatidylserine. In addition, we compared the activity of PKC using bilayer membranes composed of dielaidoylphosphatidylethanolamine that were converted into a bicontinuous cubic phase with the addition of a small amount of alamethicin. With both systems, the activity of PKC was greater in the cubic phase than in the lamellar phase, and it was shown that this was not due to the small change in membrane composition but rather to the change of phase. Hence, despite the relief of negative curvature strain, the cubic phase is more potent in activating PKC.

Although both PKC and CT are activated by lipids that promote negative curvature strain, the mechanism of this modulation is quite different for the two cases. This is shown directly by comparing the responses of PKC and CT to an oxidized lipid, 1-palmitoyl, 2-(11,15 dihydroxy) eicosatrienoyl PC (diOH-PAPC) [69]. This lipid caused less of an increase in the temperature of the lamellar to H_{II} transition of an unsaturated PE, compared to its parent, PAPC. The activation of CT by diOH-PAPC relative to the parent PAPC is in accord with modulation of the enzyme activity by membrane curvature strain. In contrast, diOH-PAPC inhibits PKC, which is opposite to what one would predict based on curvature properties. Changes in the properties of a polarity-sensitive interfacial fluorescent probe [67] indicated that this oxidized PC increases interfacial packing pressure. Thus, as had been found with the series of homologous PE lipids [66], the modulation of the activity of PKC by this oxidized lipid can be explained by changes in interfacial properties.

The activities of both CT and PKC are known to increase in the presence of nonlamellar-forming lipids. The greater activating effect of diOH-PAPC, compared with PAPC, is consistent with a stimulation of the activity of CT by negative curvature strain. However, this is not the case with PKC, for which we suggest that surface packing pressure is of prime importance. We therefore conclude that although the activity of CT appears to be directly coupled with membrane curvature, nonlamellar-forming lipids modulate the activity of PKC by a less direct mechanism.

15.4 SUMMARY

Interfacial enzyme catalysis plays an important role in many biological processes, including signal transduction and the metabolism of lipids. A quantitative description of the kinetics of these enzymes requires consideration of additional factors, compared with enzyme catalysis in the aqueous phase. Nevertheless, kinetic analysis has been useful to distinguish scooting from hopping modes of action of these enzymes. In addition, the distribution of enzyme and substrate in the plane of the membrane will also affect the catalytic rate. This is thought to be important for increasing the efficiency and rate of signal transduction pathways. There is evidence that many of the enzymes involved in signal transduction pathways are sequestered to cholesterol-rich regions of the membrane, called rafts, and can interact with each other more efficiently by being sequestered into a relatively small area of the membrane.

REFERENCES

1. Gelb, M.H., Min, J.H., and Jain, M.K. Do membrane-bound enzymes access their substrates from the membrane or aqueous phase: interfacial versus non-interfacial enzymes. *Biochim. Biophys. Acta* 1488, 20–27, 2000.
2. Jain, M.K. and Berg, O.G. The kinetics of interfacial catalysis by phospholipase A2 and regulation of interfacial activation: hopping versus scooting. *Biochim. Biophys. Acta* 1002, 127–156, 1989.
3. Dennis, E.A. Phospholipase A2 activity towards phosphatidylcholine in mixed micelles: surface dilution kinetics and the effect of thermotropic phase transitions. *Arch. Biochem. Biophys.* 158, 485–493, 1973.
4. Deems, R.A. Interfacial enzyme kinetics at the phospholipid/water interface: practical considerations. *Anal. Biochem.* 287, 1–16, 2000.
5. Hofer, P. and Fringeli, U.P. Acetylcholinesterase kinetics. *Biophys. Struct. Mech.* 8, 45–59, 1981.
6. Froede, H.C., Wilson, I.B., and Kaufman, H. Acetylcholinesterase: theory of noncompetitive inhibition. *Arch. Biochem. Biophys.* 247, 420–423, 1986.
7. Boldyrev, A.A. Na/K-ATPase as an oligomeric ensemble. *Biochemistry (Mosc.)* 66, 821–831, 2001.
8. Post, R.L. and Klodos, I. Interpretation of extraordinary kinetics of Na(+)-K(+)-ATPase by a phase change. *Am. J. Physiol.* 271, C1415–C1423, 1996.
9. Goñi, F.M. and Alonso, A. Sphingomyelinases: enzymology and membrane activity. *FEBS Lett.* 531, 38–46, 2002.
10. Cremesti, A.E., Goñi, F.M., and Kolesnick, R. Role of sphingomyelinase and ceramide in modulating rafts: do biophysical properties determine biologic outcome? *FEBS Lett.* 531, 47–53, 2002.
11. Fanani, M.L. and Maggio, B. Kinetic steps for the hydrolysis of sphingomyelin by *Bacillus cereus* sphingomyelinase in lipid monolayers. *J. Lipid Res.* 41, 1832–1840, 2000.
12. Fanani, M.L., Hartel, S., Oliveira, R.G., and Maggio, B. Bidirectional control of sphingomyelinase activity and surface topography in lipid monolayers. *Biophys. J.* 83, 3416–3424, 2002.
13. Holopainen, J.M., Angelova, M.I., and Kinnunen, P.K. Vectorial budding of vesicles by asymmetrical enzymatic formation of ceramide in giant liposomes. *Biophys. J.* 78, 830–838, 2000.

14. Bini, L., Pacini, S., Liberatore, S., Valensin, S., Pellegrini, M., Raggiaschi, R., Pallini, V., and Baldari, C.T. Extensive temporally regulated reorganization of the lipid raft proteome following T cell antigen receptor triggering. *Biochem. J.* 369, 301–309, 2003.

15. Brown, D.A. and London, E. Structure and function of sphingolipid- and cholesterol-rich membrane rafts. *J. Biol. Chem.* 275, 17221–17224, 2000.

16. Grazide, S., Maestre, N., Veldman, R.J., Bezombes, C., Maddens, S., Levade, T., Laurent, G., and Jaffrezou, J.P. Ara-C- and daunorubicin-induced recruitment of Lyn in sphingomyelinase-enriched membrane rafts. *FASEB J.* 16, 1685–1687, 2002.

17. Lacalle, R.A., Mira, E., Gomez-Mouton, C., Jimenez-Baranda, S., Martinez, A., and Manes, S. Specific SHP-2 partitioning in raft domains triggers integrin-mediated signaling via Rho activation. *J. Cell Biol.* 157, 277–289, 2002.

18. Pike, L.J. and Casey, L. Cholesterol levels modulate EGF receptor-mediated signaling by altering receptor function and trafficking. *Biochemistry* 41, 10315–10322, 2002.

19. Rebecchi, M., Peterson, A., and McLaughlin, S. Phosphoinositide-specific phospholipase C-delta 1 binds with high affinity to phospholipid vesicles containing phosphatidylinositol 4,5-bisphosphate. *Biochemistry* 31, 12742–12747, 1992.

20. Ostolaza, H. and Goñi, F.M. Interaction of the bacterial protein toxin alpha-haemolysin with model membranes: protein binding does not always lead to lytic activity. *FEBS Lett.* 371, 303–306, 1995.

21. Slater, S.J., Seiz, J.L., Cook, A.C., Buzas, C.J., Malinowski, S.A., Kershner, J.L., Stagliano, B.A., and Stubbs, C.D. Regulation of PKC alpha activity by C1-C2 domain interactions. *J. Biol. Chem.* 277, 15277–15285, 2002.

22. Papo, N. and Shai, Y. 2003. Exploring peptide membrane interaction using surface plasmon resonance: differentiation between pore formation versus membrane disruption by lytic peptides. *Biochemistry.* 42, 458–466, 2003.

23. Hendrickson, H.S., Banovetz, C., Kirsch, M.J., and Hendrickson, E.K. Kinetics of phosphatidylinositol-specific phospholipase C with vesicles of a thiophosphate analogue of phosphatidylinositol. *Chem. Phys. Lipids* 84, 87–92, 1996.

24. Mosior, M., Golini, E.S., and Epand, R.M. Chemical specificity and physical properties of the lipid bilayer in the regulation of protein kinase C by anionic phospholipids: evidence for the lack of a specific binding site for phosphatidylserine. *Proc. Natl. Acad. Sci. U. S. A.* 93, 1907–1912, 1996.

25. Kinnunen, P.K., Koiv, A., Lehtonen, J.Y., Rytomaa, M., and Mustonen, P. Lipid dynamics and peripheral interactions of proteins with membrane surfaces. *Chem. Phys. Lipids* 73, 181–207, 1994.

26. Hurley, J.H. and Meyer, T. Subcellular targeting by membrane lipids. *Curr. Opin. Cell Biol.* 13, 146–152, 2001.

27. Ellson, C.D., Andrews, S., Stephens, L.R., and Hawkins, P.T. The PX domain: a new phosphoinositide-binding module. *J. Cell Sci.* 115, 1099–1105, 2002.

28. Hirata, M., Kanematsu, T., Takeuchi, H., and Yagisawa, H. Pleckstrin homology domain as an inositol compound binding module. *Jpn. J. Pharmacol.* 76, 255–263, 1998.

29. Sato, T.K., Overduin, M., and Emr, S.D. Location, location, location: membrane targeting directed by PX domains. *Science* 294, 1881–1885, 2001.

30. Stenmark, H. and Aasland, R. FYVE-finger proteins — effectors of an inositol lipid. *J. Cell Sci.* 112 (Pt 23), 4175–4183, 1999.

31. Misra, S., Miller, G.J., and Hurley, J.H. Recognizing phosphatidylinositol 3-phosphate. *Cell* 107, 559–562, 2001.

32. Arbuzova, A., Wang, L., Wang, J., Hangyas-Mihalyne, G., Murray, D., Honig, B., and McLaughlin, S. Membrane binding of peptides containing both basic and aromatic residues. Experimental studies with peptides corresponding to the scaffolding region of caveolin and the effector region of MARCKS. *Biochemistry* 39, 10330–10339, 2000.

33. Beschiaschvili, G. and Seelig, J. Melittin binding to mixed phosphatidylglycerol/phosphatidylcholine membranes. *Biochemistry* 29, 52–58, 1990.

34. Seelig, J., Nebel, S., Ganz, P., and Bruns, C. Electrostatic and nonpolar peptide-membrane interactions. Lipid binding and functional properties of somatostatin analogues of charge $z = +1$ to $z = +3$. *Biochemistry* 32, 9714–9721, 1993.

35. Terzi, E., Holzemann, G., and Seelig, J. Alzheimer beta-amyloid peptide 25-35: electrostatic interactions with phospholipid membranes. *Biochemistry* 33, 7434–7441, 1994.

36. Wang, J., Gambhir, A., Hangyas-Mihalyne, G., Murray, D., Golebiewska, U., and McLaughlin, S. Lateral sequestration of phosphatidylinositol 4,5-bisphosphate by the basic effector domain of myristoylated alanine-rich C kinase substrate is due to nonspecific electrostatic interactions. *J. Biol. Chem.* 277, 34401–34412, 2002.

37. McLaughlin, S. and Aderem, A. The myristoyl-electrostatic switch: a modulator of reversible protein-membrane interactions. *Trends Biochem. Sci.* 20, 272–276, 1995.

38. Mosior, M. and Epand, R.M. Characterization of the calcium-binding site that regulates association of protein kinase C with phospholipid bilayers. *J. Biol. Chem.* 269, 13798–13805, 1994.

39. Johnson, J.E., Xie, M., Singh, L.M.R., Edge, R., and Cornell, R.B. Both acidic and basic amino acids in an amphitropic enzyme, CTP:phosphocholine cytidylyltransferase, dictate its selectivity for anionic membranes. *J. Biol. Chem.* 278, 514–522, 2002.

40. Maggio, B. Modulation of phospholipase A2 by electrostatic fields and dipole potential of glycosphingolipids in monolayers. *J. Lipid Res.* 40, 930–939, 1999.

41. Wiener, M.C. and White, S.H. Structure of a fluid dioleoylphosphatidylcholine bilayer determined by joint refinement of x-ray and neutron diffraction data. III. Complete structure. *Biophys. J.* 61, 437–447, 1992.

42. Shinitzky, M. and Barenholz, Y. Fluidity parameters of lipid regions determined by fluorescence polarization. *Biochim. Biophys. Acta* 515, 367–394, 1978.

43. Straume, M. and Litman, B.J. Equilibrium and dynamic bilayer structural properties of unsaturated acyl chain phosphatidylcholine-cholesterol-rhodopsin recombinant vesicles and rod outer segment disk membranes as determined from higher order analysis of fluorescence anisotropy decay. *Biochemistry* 27, 7723–7733, 1988.

44. Wilschut, J.C., Regts, J., Westenberg, H., and Scherphof, G. Action of phospholipases A2 on phosphatidylcholine bilayers. Effects of the phase transition, bilayer curvature and structural defects. *Biochim. Biophys. Acta* 508, 185–196, 1978.

45. Upreti, G.C. and Jain, M.K. Action of phospholipase A2 on unmodified phosphatidylcholine bilayers: organizational defects are preferred sites of action. *J. Membr. Biol.* 55, 113–121, 1980.

46. Lichtenberg, D., Romero, G., Menashe, M., and Biltonen, R.L. Hydrolysis of dipalmitoylphosphatidylcholine large unilamellar vesicles by porcine pancreatic phospholipase A2. *J. Biol. Chem.* 261, 5334–5340, 1986.

47. Menashe, M., Lichtenberg, D., Gutierrez-Merino, C., and Biltonen, R.L. Relationship between the activity of pancreatic phospholipase A2 and the physical state of the phospholipid substrate. *J. Biol. Chem.* 256, 4541–4543, 1981.

48. Romero, G., Thompson, K., and Biltonen, R.L. The activation of porcine pancreatic phospholipase A2 by dipalmitoylphosphatidylcholine large unilamellar vesicles. Analysis of the state of aggregation of the activated enzyme. *J. Biol. Chem.* 262, 13476–13482, 1987.

49. Grainger, D.W., Reichert, A., Ringsdorf, H., and Salesse, C. Hydrolytic action of phospholipase A2 in monolayers in the phase transition region: direct observation of enzyme domain formation using fluorescence microscopy. *Biochim. Biophys. Acta* 1023, 365–379, 1990.

50. Huang, H.W., Goldberg, E.M., and Zidovetzki, R. Ceramide induces structural defects into phosphatidylcholine bilayers and activates phospholipase A2. *Biochem. Biophys. Res. Commun.* 220, 834–838, 1996.

51. Bell, J.D. and Biltonen, R.L. Molecular details of the activation of soluble phospholipase A2 on lipid bilayers. Comparison of computer simulations with experimental results. *J. Biol. Chem.* 267, 11046–11056, 1992.

52. Ruiz-Arguello, M.B., Veiga, M.P., Arrondo, J.L., Goñi, F.M., and Alonso, A. Sphingomyelinase cleavage of sphingomyelin in pure and mixed lipid membranes. Influence of the physical state of the sphingolipid. *Chem. Phys. Lipids* 114, 11–20, 2002.

53. Muderhwa, J.M. and Brockman, H.L. Lateral lipid distribution is a major regulator of lipase activity. Implications for lipid-mediated signal transduction. *J. Biol. Chem.* 267, 24184–24192, 1992.

54. Melo, E.C., Lourtie, I.M., Sankaram, M.B., Thompson, T.E., and Vaz, W.L. Effects of domain connection and disconnection on the yields of in-plane bimolecular reactions in membranes. *Biophys. J.* 63, 1506–1512, 1992.

55. Sen, A., Isac, T.V., and Hui, S.W. Bilayer packing stress and defects in mixed dilinoleoylphosphatidylethanolamine and palmitoyloleoylphosphatidylcholine and their susceptibility to phospholipase A2. *Biochemistry* 30, 4516–4521, 1991.

56. Maggio, B. Control by ganglioside GD1a of phospholipase A2 activity through modulation of the lamellar-hexagonal (HII) phase transition. *Mol. Membr. Biol.* 13, 109–112, 1996.
57. Attard, G.S., Templer, R.H., Smith, W.S., Hunt, A.N., and Jackowski, S. Modulation of CTP:phosphocholine cytidylyltransferase by membrane curvature elastic stress. *Proc. Natl. Acad. Sci. U. S. A.* 97, 9032–9036, 2000.
58. Davies, S.M., Epand, R.M., Kraayenhof, R., and Cornell, R.B. Regulation of CTP: phosphocholine cytidylyltransferase activity by the physical properties of lipid membranes: an important role for stored curvature strain energy. *Biochemistry* 40, 10522–10531, 2001.
59. Cantor, R.S. Lipid composition and the lateral pressure profile in bilayers. *Biophys. J.* 76, 2625–2639, 1999.
60. Cantor, R.S. The influence of membrane lateral pressures on simple geometric models of protein conformational equilibria. *Chem. Phys. Lipids* 101, 45–56, 1999.
61. Cantor, R.S. The lateral pressure profile in membranes: a physical mechanism of general anesthesia. *Biochemistry* 36, 2339–2344, 1997.
62. Epand, R.M. and Lester, D.S. The role of membrane biophysical properties in the regulation of protein kinase C activity. *Trends Pharmacol. Sci.* 11, 317–320, 1990.
63. Epand, R.M. The relationship between the effects of drugs on bilayer stability and protein kinase C activity. *Chem. Biol. Interact.* 63, 239–247, 1987.
64. Mosior, M. and Epand, R.M. Role of the membrane in the modulation of the activity of protein kinase C. *J. Liposome Research* 9, 21–42, 1999.
65. Epand, R.M., Fuller, N., and Rand, R.P. Role of the position of unsaturation on the phase behavior and intrinsic curvature of phosphatidylethanolamines. *Biophys. J.* 71, 1806–1810, 1996.
66. Giorgione, J.R., Kraayenhof, R., and Epand, R.M. Interfacial membrane properties modulate protein kinase C activation: role of the position of acyl chain unsaturation. *Biochemistry* 37, 10956–10960, 1998.
67. Epand, R.F., Kraayenhof, R., Sterk, G.J., Wong Fong Sang, H.W., and Epand, R.M. Fluorescent probes of membrane surface properties. *Biochim. Biophys. Acta* 1284, 191–195, 1996.
68. Giorgione, J.R., Huang, Z., and Epand, R.M. Increased activation of protein kinase C with cubic phase lipid compared with liposomes. *Biochemistry* 37, 2384–2392, 1998.
69. Drobnies, A.E., Davies, S.M., Kraayenhof, R., Epand, R.F., Epand, R.M., and Cornell, R.B. CTP:phosphocholine cytidylyltransferase and protein kinase C recognize different physical features of membranes: differential responses to an oxidized phosphatidylcholine. *Biochim. Biophys. Acta* 1564, 82–90, 2002.

16 Viral Fusion Mechanisms

Aditya Mittal and Joe Bentz

CONTENTS

16.1 INTRODUCTION

Membrane fusion is an essential process in intracellular transport, cellular secretion, and fertilization. All enveloped animal viruses must accomplish membrane fusion with target cells as a precursor to infection. Understanding the general characteristics of protein-mediated membrane fusion is the first step toward developing general strategies to control this ubiquitous biological event. At present, most of the membrane glycoproteins that mediate membrane fusion and entry of the known pathogenic enveloped animal viruses have been identified. Here, we focus primarily on the membrane fusion glycoprotein hemagglutinin (HA) of influenza virus. Its ectodomain, which requires low pH to initiate fusion, was the first membrane fusion protein whose crystal structure was solved (Wilson et al., 1981), and it remains the prototypical fusion protein (Skehel and Wiley, 1998; Sutton et al., 1998; Baker et al., 1999; Bentz, 2000b; Bentz and Mittal, 2003).

The key differences between the "native" structure of HA and the low pH structure appear to be the formation of an extended coiled coil starting from the N-terminus of the native coiled coil and a helix-turn occurring within the C-terminal end of the native coiled coil, near the transmembrane domain (Carr and Kim, 1993; Bullough et al., 1994; Chen et al., 1995). Recently, ectodomain core complexes of other viral membrane fusion proteins have shown remarkably similar equilibrium crystal structures with respect to this coiled coil motif (Fass and Kim, 1995; Chan et al., 1997; Weissenhorn et al., 1997, 1998; Caffrey et al., 1998; Sutton et al., 1998; Skehel and Wiley, 1998; Baker et al., 1999; Singh et al., 1999). This has led to speculation that these proteins share a common molecular mechanism for initiating membrane fusion. It is believed that the energy released by the essential conformational change of the membrane fusion proteins is required to stabilize the initial defects, which begin the transformation from two stable bilayers to the single fusion product.

0-8493-1403-8/05/$0.00+$1.50
© 2005 by CRC Press LLC

As far as we know, biologically relevant membrane fusion involves mixing of the membrane lipids and opening of the fusion pore to join the aqueous volumes initially separated by the membranes. The former is measured as redistribution of membrane probes, and the latter is usually assayed as transfer of aqueous probes that vary in sizes from ions to macromolecules. Hemifusion is defined as mixing only of outer monolayers, without contents mixing or leakage. Many studies on viral fusion used lipid-mixing assays, without knowing whether fusion or hemifusion was observed (see Bentz and Mittal, 2003, for a review). This has muddied the experimental record. Here, we will focus only on studies where it is known whether fusion or hemifusion is observed or, in a few cases, where the ambiguity is not important.

The real problem with membrane fusion is to correlate the "communal" intermediates of the fusion process with the known "individual" low pH conformational changes of HA fragments. There is a wealth of knowledge on the observed changes in fusion/hemifusion phenotypes as a function of site-specific mutations of HA, or added amphipathic molecules or particular IgGs (Chernomordik et al., 1997, 1998; Qiao et al., 1998, 1999; Melikyan et al., 1999; Armstrong et al., 2000; Leikina et al., 2000, 2001; Leikina and Chernomordik, 2000; Markosyan et al., 2001; Gruneke et al., 2002). The conclusions from these studies focus on the implied effect of the individual HA, under the assumption that all HAs behave identically. However, each HA can have one or more of at least four jobs with respect to fusion:

- Binding to target membrane sialates
- Mediating self-aggregation
- Creating the initial bilayer defect
- Allowing closer apposition of the bilayers so that outer monolayer merger can commence

It is essential to know how exogenous agents or mutations affect the aggregate size or the kinetics of the conformational changes. These are the directors of the pathway to be followed.

It has also been speculated that bilayer destabilization is initiated either through a high curvature bending defect (Chernomordik et al., 1998; Kozlov and Chernomordik, 1998; Lentz and Lee, 2000) or a hydrophobic defect Bentz (2000a,b). These two proposals agree on most points, except the direct target of the transduced energy, which is the evolutionary pressure on the conformational change in the first place. So it is actually important to know what is the first step of destabilization. This point has not been resolved and will not be a focus of this review.

16.2 HA CONFORMATIONS

For many viral fusion proteins, the partial ectodomain core complexes have been crystallized, also known as *six-helix bundles,* which correspond to final equilibrium structures (Skehel and Wiley, 1998; Baker et al., 1999). We will begin with a brief review of the known conformations of influenza HA and other viral fusion proteins.

During infection, influenza virus bound to the cell surface is endocytosed and exposed to low pH, which produces at least three new conformations in HA. The native structure of HA is based on the crystal structure of the bromelain-released hemagglutinin ectodomain, BHA (Wilson et al., 1981). HA is a homotrimer, and each monomer is composed of two polypeptide segments, designated HA1 and HA2, connected by a disulfide bond. The HA1 segments contain sialic acid binding sites, which mediate initial HA attachment to the host cell surface. The HA2 segments form the membrane-spanning anchor, the assembly domain of the homotrimer, and its amino-terminal region (Gething et al., 1986).

Upon acidification, the amino terminus of HA2, the fusion peptide, is exposed. This change is rapid compared to fusion and is required to promote fusion between the viral envelope and the target membrane (Skehel et al., 1982; White and Wilson, 1987; Stegmann et al., 1990; Godley et al., 1992; Stegmann and Helenius, 1993; Pak et al., 1994).

The second conformational change leads to the formation of the extended coiled coil of HA2, which was predicted by Carr and Kim (1993), proved for the crystallographic structure of a fragment of BHA (TBHA2, residues 38–175 of HA2 and 1–27 of HA1 held together by the disulfide bond) by Bullough et al. (1994), and morphologically observed on the intact virus by Shangguan et al. (1998), as discussed in Bentz (2000a). In addition, Qiao et al. (1998) and Gruneke et al. (2002) have shown that site-directed point mutations predicted to inhibit the formation of the extended coiled coil did inhibit the fusion of erythrocytes to HA expressing cells.

The crystal structure of TBHA2 (Bullough et al., 1994) also shows that the C-terminal end of HA2, where the native coiled coil flares out to accommodate the fusion peptide in the native state, flips up in helix-turn between residues 106 and 112 of HA2 and forms an antiparallel α-helical annulus from residues 113–129 of HA2, i.e., at the base of the extended coiled coil. Kim et al. (1998) have argued that the helix-turn region, which they term the *kink region* of HA2 (aa 105–113), is important for fusion. Epand et al. (1999) and Leikina et al. (2000) have found that the FHA2 fragment (the equilibrium structure of aa 1–127 of HA2, which runs from the fusion peptide to the end of the annular α-helix, with the extended coiled coil in place and the kink exposed) mediates cell–cell membrane lipid mixing in a pH-dependent fashion, but not contents mixing between cells. FHA2 reversibly aggregates at low pH via the kink region (Yu et al., 1994; Kim et al., 1998; Leikina et al., 2000).

The extended coiled coil and the helix turn together have been termed the six-helix bundle (Skehel and Wiley, 1998), and blocking its formation blocks fusion for sendai virus (Russell et al., 2001) and HA (Gruneke et al., 2002). There is some controversy for HIV, where Melikyan et al. (2000) claim that blocking the six-helix bundle formation blocks fusion and Golding et al. (2002) claim that antibodies binding to six-helix bundles can do so before fusion, i.e., the six-helix bundle formation precedes fusion. Overall, it is clear that both conformational changes appear to be required to complete fusion, as predicted in Bentz (2000b), but not their order or which is more energetic, as discussed later.

16.2.1 THE STAGES OF MEMBRANE FUSION

Most proposed HA-mediated fusion mechanisms contain four distinct "classes" of intermediates, subsequent to close apposition of the membranes and the low pH-induced exposure of the HA2 N-terminus (Bentz, 1992, 2000a,b; Blumenthal et al., 1996; Chernomordik et al., 1998; Mittal et al., 2002b). Currently, these intermediates are

- Aggregates of HA, which are either preformed or form rapidly subsequent to acidification
- The first fusion pore defined by the first conductivity (2–5 nS) across the membranes (Additional flickering pores follow, which lead to the formation of a terminally open pore.)
- The lipidic channel, which is monitored by lipid dye transfer between membranes
- The fusion site, which is monitored by aqueous contents mixing (e.g., fluorophors) and the stable joining of the two membranes and complete aqueous contents mixing

Certainly, subdivisions will be established. For HA, the HA aggregation step has been kinetically proven (Bentz, 2000a; Leikina and Chernomordik, 2000; Mittal and Bentz, 2001). Dutch et al. (1998) found that fusion of SV5 F expressing cells with erythrocyte ghosts was faster when the surface density of SV5 F was increased, indicating that aggregation was important for fusion. Fusion of HIV env fusion proteins (the class of homologues to gp120/41) is faster and/or more extensive with higher surface expression (Kuhmann et al., 2000), likewise suggesting aggregation as an essential step. The other steps are functional steps in membrane destabilization and have been observed, in some fashion, with all fusing systems.

There are many proposed models for the fusion site of HA. Nearly all show apposed bilayers followed by the formation of a lipidic stalk intermediate. Our focus is the steps between these two stable points and the question of the structure of the "transition state." The stalk has been proposed as the first "stable" lipidic connection between the bilayers (Chernomordik et al., 1998; Kuzmin et al., 2001; Kozlovsky and Kozlov, 2002; Kozlovsky et al., 2002; Markin and Albanesi, 2002). Yang and Huang (2002) have recently provided a low-resolution structure of an equilibrium phase of diphytanoyl phosphatidylcholine at low hydration that strongly resembles these theoretically derived structures. That the lipidic stalk is central to one of the intermediates of membrane fusion seems likely. The question is: How does an aggregate of HAs actually cause the lipidic stalk to form?

Current structural techniques cannot resolve the fusion site, because they only assay the average site and, unless one assumes that all HAs refold identically, one cannot even assume that the average site will look like the one that fuses. On virions, there appear to be only a few fusion sites in the area of contact with ganglioside-containing liposomes (Mittal et al., 2002b), but the search for the virions-liposome showing fusion intermediates has been unsuccessful. Micrographs showing structures (Stegmann et al., 1990; Kanaseki et al., 1997) provide no correlation with fusion kinetics, so there is no idea whether the structures form before, during, or after fusion. On HA-expressing cells, only one of hundreds or thousands of HA aggregates — each a potential fusion site — is responsible for the observed lipid mixing (Bentz, 2000a; Frolov et al., 2000; Leikina and Chernomordik, 2000). Before we can propose a structure of the initial site of membrane fusion, we must first have a clear and consistent estimate of how many HAs are at a site leading to fusion or hemifusion, and whether the path to fusion or hemifusion depends on how many HAs undergo the essential conformational change.

16.2.2 THE COMPREHENSIVE MASS ACTION MODEL FOR HA-MEDIATED FUSION

The model for HA-mediated fusion that we have proposed arises from the analysis of fusion kinetics. The advantage of a kinetic analysis of fusion is that only those sites that actually mediate fusion are assayed. Unproductive HA aggregates are silent.

A comprehensive mass action kinetic model of fusion (see Figure 16.1) must contain all of the commonly accepted intermediates and be able to reduce all appropriate data to a consensus set of parameter estimates. For HA-mediated fusion, this means that several parameters and rate constants must be simultaneously fitted (see legend of Figure 16.1 and Mittal and Bentz (2001) for parameter definitions). This requires exhaustive fitting over all parameter values to select the ranges that give best fits to the data. Previous kinetic modeling in the fusion area always assumed some single step to be rate limiting in advance, and the fitting equations were designed for just that step. If no single step is rate limiting, which our analysis shows, then the parameters of the overly simplified models are forced to accommodate the extra information in the data, blurring (at best) their physical meaning.

Exhaustive multiparameter fitting is the key to an unambiguous kinetic analysis. It is important to recognize that our comprehensive kinetic model predicts the ratio of fusogenic aggregates (see legend of Figure 16.1 for definition) between two cell lines without input of relative or absolute HA surface densities. The model only requires the ratio of the average HA surface densities between two cell lines to predict the minimal number of HAs in the aggregate yielding the first fusion pore (or site of a hypothetical intermediate, proposed by Chernomordik et al. [1998], called *restricted hemifusion*). Absolute values of surface density are not required. This means that the extent to which HA is partitioned into lipid rafts (Markovic et al., 2002) is not so crucial, if the two cell lines have similar extents of rafts vs. liquid–crystalline bilayer. The partitioning of HAs into rafts facilitates HA aggregation, and this facilitation is crucial at low HA surface densities. However, forming the HA aggregate required to create a fusion site appears to require an additional, more powerful aggregation mechanism. It is of interest that raft formation in the target membrane appears to have no effect on fusion for influenza (Samsonov et al., 2001) or alpha viruses (Waarts et al., 2002). For HIV, CD4 is initially in rafts, while the cytokine coreceptors are outside of rafts, but

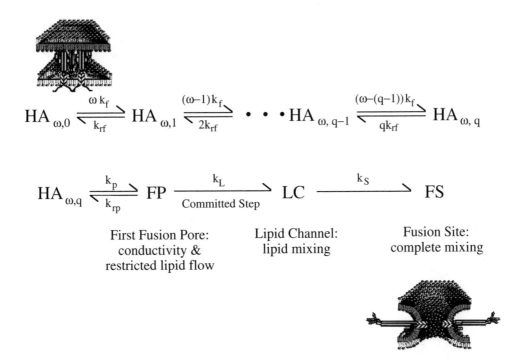

$$HA_{\omega,0} \underset{k_{rf}}{\overset{\omega k_f}{\rightleftharpoons}} HA_{\omega,1} \underset{2k_{rf}}{\overset{(\omega-1)k_f}{\rightleftharpoons}} \cdots \cdot HA_{\omega,q-1} \underset{qk_{rf}}{\overset{(\omega-(q-1))k_f}{\rightleftharpoons}} HA_{\omega,q}$$

$$HA_{\omega,q} \underset{k_{rp}}{\overset{k_p}{\rightleftharpoons}} FP \xrightarrow[\text{Committed Step}]{k_L} LC \xrightarrow{k_S} FS$$

First Fusion Pore: Lipid Channel: Fusion Site:
conductivity &   lipid mixing   complete mixing
restricted lipid flow

FIGURE 16.1 The comprehensive kinetic model for influenza hemagglutinin mediated membrane fusion. Following protonation, the HA aggregate of size ω forms rapidly, denoted as $HA_{\omega,0}$. ω is the minimal size for a fusogenic aggregate (an aggregate of HAs capable of sustaining membrane fusion), in which HAs can undergo the essential conformational change, independently and identically with a rate constant of k_f. Thus, the overall rate constant for the first reaction would be ωk_f. These conformational changes continue for each HA until q of them have occurred, $HA_{\omega,q}$. q is called the minimal fusion unit, as it equals the minimum number of HAs that have undergone the essential conformational change needed to stabilize the first high-energy intermediate for fusion. At this point, the fusogenic aggregate can transform to the first fusion pore, which is observed as the first conductivity across the apposed membranes. The first fusion pore, FP, evolves to the lipid channel, LC, demarked by mixing of lipids, which evolves to the fusion site, FS, demarked by aqueous contents mixing. Pictorial representations are shown of what the fusion site might look like before and after membrane fusion, corresponding to the respective steps in the kinetic model.

binding of the HIV virion to CD4 evidently releases CD4 from the raft into the lipid domain of the coreceptors (Kozac et al., 2002).

Our comprehensive kinetic model is evolving into a refined tool for elucidating the structure of the HA fusion site. The current version is described in detail in Mittal et al. (2002b). It has been built through the postmortem analysis of fusion data from several labs, a meta-analysis that guarantees that the model is not too finicky and is focused on fitting only the most robust parameters. Figure 16.2 shows a compendium of these fits. Mittal and Bentz (2001) showed the collected parameters fitted for HA-expressing cells fusing with either planar bilayers or RBC, which are shown in Table 16.1. The consensus was defined as the subset of values that can fit all data sets simultaneously.

With respect to fusion vs. hemifusion, we note here that the data of Melikyan et al. (1995) followed the conductance of the first fusion pore, whereas Blumenthal et al. (1996) followed transmembrane potential, lipid mixing, and contents mixing. Danieli et al. (1996) followed only lipid mixing; the balance of fusion to hemifusion is not known. However, because the basic architectural parameters fitted by the exhaustive analysis were the same, it appears either that they are not so sensitive to the difference in mechanism between fusion and hemifusion, or that the lipid

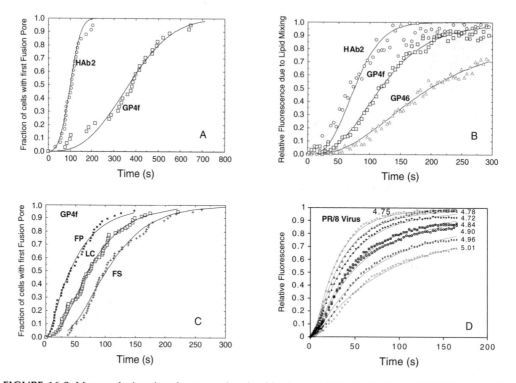

FIGURE 16.2 Meta-analysis using the comprehensive kinetic model shown in Figure 16.1. Representative best fits to the experimental data are shown. (A) The open symbols are the data from Melikyan et al. (1995) showing the cumulant fraction of cells that have achieved their first conductivity channel across the planar bilayer (i.e., the formation of the first fusion pore) as a function of time. The bilayer was composed of dioleoylphosphatidylcholine/bovine brain phosphatidylethanolamine in the ratio 2:1, with 5 mol% gangliosides, equal weights of GD1a and GT1b, added. The cells were prebound to the planar bilayers before acidification. The HAb2 cells (open circles) required about 100 s for half of the cells to achieve their first conductivity channel, whereas the GP4f cells (open squares) required about 360 s. The HAb2 cells express about 1.6 times more HA/μm^2 than GP4f cells. The solid lines show a typical best fit to the data. (B) Lipid channel, LC, formation measured by Danieli et al. (1996) using R18 labeled RBCs fusing with HA-expressing cells. Open symbols show lipid mixing for three cell lines: GP4/6 (Δ), GP4f ($\square$), and HAb2 (O). GP4f cells express 1.6 times more HA/μm^2 than GP4/6 cells (Danieli et al., 1996). Fusion is quantified as relative fluorescence due to lipid mixing (see Equation [1] in "Materials and Methods," Mittal and Bentz, 2001). Solid lines show a typical best fit to the data. (C) Fusion intermediates measured by Blumenthal et al. (1996) for double-labeled RBC ghosts fusing with GP4f cells. Symbols are the waiting times data after acidification, showing the cumulant fraction of cells that have achieved their first conductivity channel FP(x), their first indication for lipid channel formation LC ($\square$) and their first indication for contents mixing FS (+) as functions of time. Solid lines show a typical best fit to the data. (D) Disconnected symbols show pH dependence of influenza A/PR/8/34 virus lipid mixing measured by Shangguan (1995). Concentrated (25x) unlabeled virus and NBD/Rh labeled DOPC/GD1a (90:10) liposomes were mixed and preincubated at 4°C for 30 min to allow binding. A small aliquot of the prebound virus-liposomes was transferred to the preequilibrated cuvette at 37°C to yield 10 μM viral phospholipid and 10 μM liposomal lipid. Lipid mixing was initiated on lowering the pH to the indicated values (next to each curve) by injecting concentrated acetic acid. Solid lines show a typical best fit to the data.

dequenching signal measured by Danieli et al. (1996) was dominated by fusion events rather than by hemifusion (Mittal et al., 2002a).

It was significant that the three independent data sets of HA-expressing cells fusing with target membranes could be explained (Melikyan et al., 1995; Danieli et al., 1996; Blumenthal et al., 1996), i.e., have similar fitted parameters. However, two important questions lingered. First, in

TABLE 16.1

q	Fusion Intermediate (Original Data)	Protein k_f (s^{-1})	Fusion Pore k_p (s^{-1})	Lipid Channel k_l (s^{-1})
2	FP Melikyan et al. (1995)	$(0.3\text{–}2) \times 10^{-4}$	$(0.3\text{–}7) \times 10^{-4}$	n.d.
$\omega = 8$	LC Danieli et al. (1996)	$(4\text{–}4.8) \times 10^{-3}$	$(4.5\text{–}8) \times 10^{-4}$	$(3\text{–}5) \times 10^{-2}$
	LC Blumenthal et al. (1996)	$(4\text{–}4.5) \times 10^{-2}$	$(0.03\text{–}1) \times 10^{-4}$	$(2.5\text{–}2.6) \times 10^{-2}$
	LC Shangguan (1995)	3	5.4×10^{-3}	1.5×10^{-1}

Note: Parameters from exhaustive fitting of different kinetic data on HA mediated fusion using the model in Figure 16.1. For a minimal aggregate size of eight HAs ($\omega = 8$; Bentz 2000a), only a minimal fusion unit of two HAs (q = 2) fit the data of HA-expressing cells (with Japan strain) fusing with (1) ganglioside containing planar bilayers at 37°C (Melikyan et al. 1995), (2) erythrocytes at 28 to 29°C and (3) erythrocyte ghosts at 37°C. The same minimal fusion unit fit the data of PR/8 strain virions fusing with ganglioside containing liposomes (Shangguan, 1995).

terms of these key fusion-site architecture parameters, are the results of HA-expressing cells applicable to the virion fusing with target membranes? Second, although the kinetic analysis assumed a single, homogeneous average surface density for each cell line, because of computational time constraints, the HA-expressing cells probably have an inhomogeneous distribution of HA surface densities. The question, then, was whether the key fusion-site architecture parameters would remain largely unchanged once the distributions were incorporated into the analysis.

Mittal et al. (2002b) showed, for the first time, consensus quantitative agreement on the fusion-site architecture for PR8 influenza virions (Figure 16.2D), whose HA surface density is high and uniform, *and* Japan-influenza HA-expressing cell lines, with lower and heterogeneous surface densities. Evidently, because the minimal HA aggregate size, $\omega = 8$, and the number of these HAs that must undergo the essential conformational change, q = 2, are obtained from ratios of fitted parameters (see Figure 16.2), the effects of the distributions on the HA-expressing cells are not particularly significant. The virus-liposome lipid-mixing data showed complete bilayer mixing, i.e., no hemifusion, because dequenching was quantitative. Evidently, the high PR8 HA surface density does not support hemifusion, although the complete lipid mixing was also lytic for the ganglioside-containing liposomes (Shangguan et al., 1996). Interestingly, it has been reported that alphavirus fusion with liposomes is nonleaky (Smit et al., 2002).

The nucleation model for HA aggregation (simply stating that ω trimers of HA nucleate to form a fusogenic aggregate with a nucleation-equilibrium constant of K_{nuc}) is unlikely to describe accurately the true distribution of HA aggregates over the cell population, but it was used because it yields the minimum estimate for the number of HAs required to form the fusogenic aggregate. In other words, more realistic distributions would require that there are more than eight HAs in a fusogenic aggregate (Bentz, 2000a). The estimate that only two HAs need to undergo the slow essential conformational change to initiate fusion is independent of the HA aggregate size (Bentz, 2000a). We are currently developing extensions needed to fit data from HA-expressing cell populations, taking explicitly into account the distribution of HA surface densities and a more realistic aggregation distribution based on the curvature equations of Kozlov and Chernomordik (1998).

We found that fitting all the PR/8 viral fusion data simultaneously and selecting the best-fit parameter sets yielded only two convergent solutions for parameters (Mittal et al., 2002b), as opposed to ranges for parameters (Bentz, 2000a; Mittal and Bentz, 2001), because more curves were being fitted simultaneously. By providing more data with less experimental noise, the kinetic model could extract very robust estimates for the kinetic parameters. The next section discusses this in detail.

The similarity in rate constants for lipid mixing and contents mixing found in Mittal and Bentz (2001) for HA-mediated fusion and by Lee and Lentz (1998) for PEG-induced fusion of phosphatidylcholine liposomes supports the idea that, subsequent to stable fusion pore formation (e.g., by a lipidic stalk), the evolution of fusion intermediates is determined more by the lipids than by the proteins. However, the rate constants for fusion pore formation were several orders of magnitude faster for the PEG and liposomes, which appears to be due to the different mechanism of formation. PEG forces bilayers together, making the pore easier to form than when the defect formed by HA occurs. The hypothesized formation of the hydrophobic defect within the HA aggregate, where lipid lateral diffusion is blocked by the transmembrane domains and remaining embedded fusion peptides, precedes the formation of stalks and fusion pores (Bentz, 2000b; Bentz and Mittal, 2000, 2003).

16.3 TACTICS OF EXHAUSTIVE FITTING OF KINETIC DATA

Understanding the molecular basis of membrane fusion, be it during viral infection or synaptic transmissions, is difficult because it is a multistep process. Furthermore, it is a highly localized event, wherein a very small fraction of the total proteins and lipids are actually involved in a fusion site. The approach we have taken is to analyze the kinetics of the whole process, which has the virtue of focusing only on the successful fusion sites, rather than the average behavior of all the fusion proteins. We also did not assume that any particular step was rate limiting, thus allowing the kinetic modeling to show which steps are kinetically significant. For HA, this produced a multiparameter model of rate constants, HA aggregate sizes for the fusion sites, and the profoundly kinetic number of how many HAs within the fusion aggregate undergo the essential conformational change before the first fusion site is formed.

The mass-action kinetic model shown in Figure 16.1 has been developed from the commonly accepted steps of the fusion process. Although it includes only essential and/or measurable steps, it is clear that "fitting" the kinetics of a fusion curve involves many parameters and, therefore, requires a tactical approach to find all possible best fits in a reasonable amount of computer time.

Numerical integrations for the model were done using MATLAB (The Math Works) subroutine ODE23s, because the nonstiff routines were not as accurate. Curve fitting was done using the fitting routine *fmins* (or *fminsearch* in the current version) of MATLAB. For any given data set, the goodness-of-fit achieved by a particular set of kinetic parameters was quantitated by the root mean square error (rmse) between the data points and the numerically integrated theoretical curve. A minimum rmse value for each data set was obtained, and best fits were defined as all sets of parameters that were visually indistinguishable from that of the minimal rmse fit. All data-fitting was exhaustive, i.e., the widest possible ranges of initial estimates for the parameters in the kinetic model were tested to ensure that all best fits were found.

We found that all best fits for HA-expressing cells fusing with target membranes were contained within a convex space of the universe of possible parameter values, as shown in the final column of Table 2 of Mittal and Bentz (2001). For the data of Melikyan et al. (1995), Bentz (2000a) found that all of the best-fit parameter sets could be fitted to the equation $k_p k_f{}^q N_\omega (GP4f) = 10^{9.1 \pm 0.5}$ for $q = 2$. In terms of "parameter space," this relation defines a convex volume. However, combinations of parameters that satisfy this relationship would yield good, but not always best, fits. Parts of this parameter space are inaccessible because they would require the number of fusogenic aggregates (i.e., HA ω-mers), denoted by N_ω, to exceed the available number of HAs in the area of apposition for the GP4f cell line. Similar relations were found in Mittal and Bentz (2001) for the best fits to the data of Danieli et al. (1996), where the parameter space is defined by a constant that equals $k_l k_p k_f{}^q N_\omega (GP4f)$, and for the data of Blumenthal et al. (1996), where the parameter space is defined

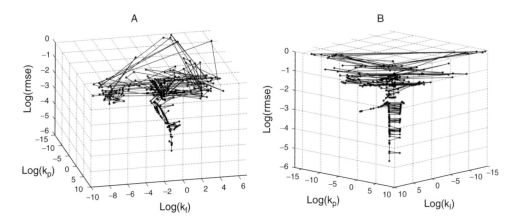

FIGURE 16.3 Exhaustive fitting of kinetic data: searching for the right solution. "Perfect" data were generated by simulating lipid mixing kinetics with rate constants shown for the virus fusing with liposomes in Table 16.1. The simulated kinetic data were then fit with 500 initial conditions for each value of q from 0 to ω and fixing the value of ω = 8 to search for the model parameters in our exhaustive fitting approach. Only q = 2 best fit the data, i.e., gave the least root mean squared error (rmse). Logarithmic scales are shown for convenience. (A) Rate constants k_f and k_p are shown as a function of rmse for all the solutions obtained with q = 2. (B) Rate constants k_f and k_l are shown as a function of rmse for all the solutions obtained with q = 2. Clearly, the search for the "right" solution takes a funnel shape, with a single exact solution obtained at the lowest tip of the funnel corresponding to the lowest rmse. Thus, given clean enough data, the kinetic model extracts the real values of parameters with our multiparameter exhaustive fitting algorithm.

by a constant that equals $k_s k_l k_p k_f{}^q N_\omega(GP4f)$. The parameter space values obtained were very tight, with the standard deviation being less than 1% of the mean values for more than 25 best fits obtained in each case. Note that although the parameter space signifies some sort of relationship between the model parameters, it is probably useful only when multiple solutions are obtained, which is not always the case (as discussed later).

Although the description just given establishes that our approach of exhaustive fitting actually found all possible values for the kinetic parameters, we also found that, for the model in Figure 16.1, our algorithm uniquely fits theoretical data generated from the model. Figure 16.3 shows an example of the results obtained by fitting data generated using the model parameters obtained from Figure 16.2D. Only q = 2 could best fit the data, and the rate constants were extracted exactly at the values that were used for simulating the data in the first place. Further, it is clear from Figure 16.3 that the results obtained from the exhaustive fitting algorithm look analogous to the funnel landscaping theory of protein folding (Dill and Chan, 1997). From all possible values of parameters that can fit the data by minimizing the root mean squared error between individual data points and numerically calculated values, the lowest rmse corresponds to the exact single solution. This also explains parameter ranges obtained for HA-expressing cells fusing with target membranes, in contrast to unique solutions for the virions fusing with liposomes in Table 16.1. With fewer data and with the experimental noise associated with the data (as in the former case), the kinetic analyses could provide ranges, instead of single-value solutions, for the rate constants. Thus, multiple fits are due to experimental noise.

A computational limitation that needed to be dealt with was the time required for exhaustive fitting of all the data in an individual data set *simultaneously*. We found that exhaustive fitting of two fusion curves simultaneously could be achieved in reasonable time. However, each additional fusion curve enormously increased the time required to search for each solution set for the model parameters. To overcome this problem, we compartmentalized the fitting algorithm by utilizing the most robust parameter observed during the initial analysis. The ratio of fusogenic aggregates between cell lines of varying surface density was found to be the most insensitive to noise in the experimental data and does not depend on knowing the real surface densities (Bentz, 2000a; Mittal and Bentz, 2001). Thus, to fit multiple data sets simultaneously, we first fit a pair of two experimental curves exhaustively to obtain the values for rate constants, where the ratio of fusogenic aggregates for the two curves was nearly the same. Then we fit the other experimental curves solely for the number of fusogenic aggregates using the rate constant values obtained from the best fits of the previous steps. Changing the order of fusion curves during this process did not affect the final outcome. This allowed us not only to do the exhaustive kinetic analysis of all the data in a reasonable time, it also provided us with the key ratios of fusogenic aggregates that provide insights into the HA aggregate size at the fusion site.

16.4 FUNCTION OF MEMBRANE FUSION PROTEIN DOMAINS DURING FUSION

It is clear that both viral and intracellular membrane fusion proteins contain a minimal set of domains, which must be deployed at the appropriate time in order to assure a successful fusion event, i.e., transfer of contents between compartments (Bentz and Mittal, 2000). Some of these domains are well known, e.g., the fusion peptide or a binding site for something on the target membrane. Others are recent discoveries and will require greater elucidation. In Table 16.2, we propose seven essential steps for fusion and cite the domains or conformational changes that accomplish these steps for the three best-described viral fusion systems: influenza virus, HIV, and sendai virus. Influenza HA is the only member of this group whose native metastable structure is known (Wilson et al., 1981). Thus, only for HA do we know how it looks before fusion and how a fragment looks after fusion.

TABLE 16.2
Function of Membrane Fusion Protein Domains during Fusion

Function	Fusion Protein		
	Influenza HA	Sendai SV5 F	HIV gp120/41
Binding:			
Virus/vesicle	HA1	HN	gp120
Target Membrane	Sialosides	Sialosides	CD4
Signal for fusion	H+ binding to HA	Could be HN binding to and activating F protein, perhaps due to neuraminidase activity	CD4 binding to gp120 exposes a chemokine receptor binding site which activates gp41
Fusion peptide	N-terminus of HA2	N-terminus of F1	N-terminus of gp41
Aggregation: (possible mechanism)	Tension on membrane from fusion peptide embedded in viral envelope	Could be similar to HA mechanism	Like HA or multiple interactions between gp120, CD4 and CCR5 or CXCR4
High energy conformational change: Extraction of fusion peptide	Formation of extended coiled coil	Expected to be formation of extended coiled coil	Expected to be formation of extended coiled coil
Hydrophobic kink: Stabilizing lipids enroute to hydrophobic defect	Helix turn or "kink" region of low pH structure	Near either the C-terminal end of the N-heptad repeat or the N-terminal end of the C-heptad repeat	Near either the C-terminal end of the N-heptad repeat or the N-terminal end of the C-heptad repeat
Low energy conformational change: Formation of hydrophobic kink or loop	Helix turn of C-terminus of native coiled coil	?	?

16.4.1 BINDING PROTEIN AND TARGET RECEPTOR

HA contains a sialate binding site within the HA1 subunit that can bind to glycosylated proteins and gangliosides (Martin et al., 1998). This provides a wide range of target receptors. Sendai virus also binds to surface sialates via the HN membrane glycoprotein, which has a neuraminidase activity (Dutch et al., 1998). HIV, on the other hand, is quite specific, with gp120 binding first to CD4 on macrophages or T cells and then to a target cell chemokine receptor (Doms and Peiper, 1998).

Influenza HA contains both binding and fusion functions, and there was controversy about whether the same HA can execute both functions (Ellens et al., 1990: Niles and Cohen, 1993; Alford et al., 1994; Stegmann et al., 1995; Millar et al., 1999; Leikina et al., 2000). Leikina et al. (2000) found that soluble sialates inhibit the major conformational change of HA (X31 strain) and that RBC bound to HA(X31)-expressing cells fused faster following a neuraminidase treatment, i.e., with a reduction in HA-sialate contacts. The structural basis for sialate inhibition of the low pH conformational change of HA is unknown at this time.

Although the HAs bound to sialate on glycophorin may be part of the fusogenic aggregate, there must be at least two HAs within the fusogenic aggregate unbound to sialates, so that they can undergo the essential conformational change needed to create the fusogenic defect in a timely fashion (Mittal and Bentz, 2001). This suggested that the relatively weak HA-sialate binding constant could not evolve to a higher affinity, as that would inhibit its ability to mediate fusion. For fusion of the cells with ganglioside planar bilayers, calculations suggested that, on average, fewer than one of the HAs within the fusogenic aggregate are bound. Most likely, the sialates on gangliosides are too close to the bilayer surface to reach effectively the HA1 sialate binding site. This implies that HA binding to sialates is not necessary for fusion.

16.4.2 SIGNAL FOR FUSION

Once bound to the target cell, the influenza virion is endocytosed, wherein acidification is the signal for fusion. Mittal et al. (2002b) has found that the pKa for fusion activation is about 5.7. This initiates the cascade of conformational changes leading to merger of the viral and endosomal membranes. Sendai virus fuses at neutral pH, and most strains require about 1:1 mol ratios of the homotypic HN for maximal fusion or infection (Dutch et al., 1998). It appears that the binding of sialates to HN, which activates its neuraminidase, alters its conformation so that it can activate the F1 protein to start the fusion process. Oddly, SV5 F does not appear to require its homotypic HN for fusion, although the rate is much faster, with about 1:1 mol ratios of its homotypic HN (Dutch et al., 1998). Although the cytoplasmic tail of SV5 F has a role in virion budding, but not fusion (Waning et al., 2002), the related paramyxovirus SER has a longer cytoplasmic tail, which evidently inhibits fusion, relative to SV5 (Tong et al., 2002). Deletion of the "extra" cytoplasmic tail provokes equivalent fusion activity.

HIV is activated at the binding of gp120 to target cell CD4, which exposes an epitope on gp120, which allows it to bind to a chemokine receptor: either CXCR4 or CCR5 (Feng et al., 1996; Doms and Peiper, 1998; Xiao et al., 1999). This ternary structure then signals gp41 to start the fusion process. Hoffman et al. (1999) have isolated a mutant HIV strain that binds directly to either chemokine receptor and mediates fusion, something like SV5 F. For the HIV env mutant, the direct binding to the cytokine receptor is the signal. For SV5 F, the signal is not known in the absence of HN, suggesting that SV5 F must bind directly to an (as yet) unknown target membrane receptor to signal that fusion should start.

16.4.3 FUSION PEPTIDE

After the signal for fusion, at least for HA2, sendai F1, and HIV gp41, the fusion peptide is exposed on the N-terminus of these proteins (Hernandez et al., 1996; Durell et al., 1997; Pecheur et al., 2000). Although there has been much literature proposing that the exposed N-terminal of HA next inserts into the target membrane to start fusion, it is more accurate to say that HAs can have their fusion peptides either suspended between the membranes or embedded in the target bilayer or in their own bilayer (Bentz et al., 1990; Gaudin et al., 1995; Shangguan et al., 1998; Kozlov and Chernomordik, 1998). The fraction in each state probably depends on time, and the real question is: Which state or which sequence of states is on the "fusion pathway?" Kozlov and Chernomordik (1998) and Bentz (2000b) argued that those HAs whose fusion peptides embed initially into the viral/HA-expressing cell envelope are on the fusion pathway. Reaching the target membrane represents a later step in the process.

16.4.4 AGGREGATION

The next step is aggregation of the fusion proteins. Bentz (2000a) showed that rapid aggregation of eight or more HAs followed by a slow "essential" conformational change of two or three of the HAs within the aggregate fitted the kinetics of first fusion pore formation between HA-expressing cells and planar bilayers. This result has been found for a variety of HA fusion systems (Mittal and Bentz, 2001; Mittal et al., 2002b).

The cause of this rapid HA aggregation is not known. It could be adhesion of the fusion peptides within the aqueous space (Ruigrok et al., 1988), although this would render the fusion peptides useless for other functions. Kozlov and Chernomordik (1998) have proposed a mechanism based on membrane curvature minimization, which begins with the fusion peptides embedded in their own (viral or HA-expressing cell) membrane and under tension created by the partial formation of the extended coiled coil. This mechanism for aggregation could function for any of the viral fusion proteins that form an extended coiled coil.

How does a protein create the tension needed to pull at a membrane, causing it to distort, and thereby induce HA aggregation, as proposed by Kozlov and Chernomordik (1998)? The formation of the next heptad repeat of the extended coiled coil should proceed by a somewhat random conformational search, which gets stuck by the formation of heptad repeat. Proteins proceed down a free energy pathway by falling through a sequence of quasi-irreversible steps, like a ratchet. Each step is made essentially irreversible by a large positive activation energy for return. The formation of the extended coiled coil is simply a sequence of heptad repeat binding steps, each of which extends the coiled coil another 10 Å or so and shortens the line to the fusion peptide by as much as 14 Å or so.

This could be accomplished by a bobbing action of the HA in the bilayer, as illustrated by Figure 16.4, where HA1 has been deleted for clarity. When HA bobs down, relative to the bilayer, this would allow enough slack in all three chains of the nascent extended coiled coil domain to form at the same time, as shown in Figure 16.4B. According to Kozlov and Chernomordik (1998), the heptad repeat of a single HA has insufficient binding energy to hold up the bilayer against the curvature energy. However, when several HAs are close, the net curvature to the bilayer needed for all of them to sustain one (or more) heptad repeat binding reaction is sufficiently small to permit the action. Once a cluster of HAs has formed its first heptad repeats of the extended coiled coil, this holds the bilayer curvature in place and the aggregate is formed, as shown in Figure 16.4C. Other HAs can diffuse into the aggregate and form their next heptad repeat, thereby increasing the bilayer curvature. This process would repeat, with HAs in the aggregate forming additional heptad repeats, gradually creating a dimple of bilayer in the center of the HA aggregate.

16.4.5 HIGH-ENERGY CONFORMATIONAL CHANGE

The idea that membrane fusion proteins could share a common mechanism became reasonable when it was discovered that so many systems have ectodomain core complexes composed of a six-helix bundle (Skehel and Wiley, 1998; Baker et al., 1999; Singh et al., 1999). In particular, the energy of the conformational changes could be transduced to create fusion. The initial defect has

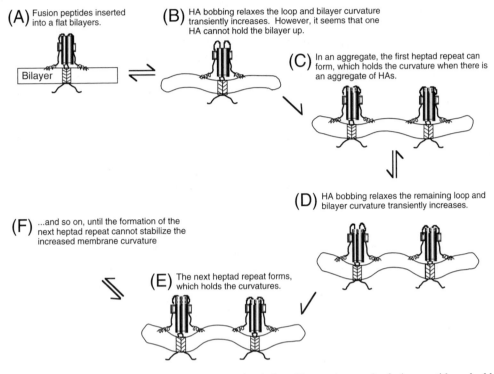

FIGURE 16.4 Proposed mechanism of HA aggregation induced by tension on the fusion peptide embedded in the viral envelope. The globular HA1 headgroups sit on top of the spikelike HA2. Regions of α-helix and coiled coil are shown as cylinders. The bottom aggregate of α-helices in the transmembrane domain are striped and denoted TM. (A) HA with fusion peptide embedded in the viral envelope. (B) When HA bobs down, relative to the bilayer, this would allow slack in all three chains of the nascent extended coiled coil domain to form at the same time. According to Kozlov and Chernomordik (1998), the heptad repeat of a single HA has insufficient binding energy to hold up the bilayer against the curvature energy. (C) However, when several HAs are close, the net curvature to the bilayer needed for all of them to sustain one (or more) heptad repeat binding reactions is sufficiently small to permit the action. (D) Once a cluster of HAs has formed its first heptad repeats of the extended coiled coil, this holds the bilayer curvature in place and the aggregate is formed. (E) Other HAs can diffuse into the aggregate and form their next heptad repeat, thereby increasing the bilayer curvature. (F) This process would repeat, with HAs in the aggregate forming additional heptad repeats, gradually creating a dimple of bilayer in the center of the HA aggregate.

been proposed to be high curvature (Chernomordik et al., 1998; Kozlov and Chernomordik, 1998) or hydrophobic (Bentz, 2000b). The largest bolus of free energy should be used to create the initial defect at the nascent fusion site, so that the elaboration of the fusion site could proceed down the free energy pathway, following the paradigm used by enzymes. This is especially important for the known fusion proteins, wherein the only source of energy to transduce to the formation of bilayer defects is within the conformational change. ATP is used only to lock the fusion proteins into a high energy metastable state.

Because some viral fusion proteins show no evidence of coiled coils, e.g., the E glycoprotein of tick-borne encephalitis virus, TBE-E (Corver et al., 2000), it is worth emphasizing here that coiled coils are not the sole providers of high-energy conformational changes. β-sheets coming together could provide the high-energy conformational change needed for fusion.

16.4.6 Low-Energy Conformational Change

Once the essential defect is formed, it is necessary for the bilayers to approach closely enough for lipids to mix. For HA, it has been argued that the formation of the extended coiled coil would release a greater bolus of free energy than the helix-turn (Bentz, 2000b). It has been argued that the helix-turn releases greater energy (Gruneke et al., 2002). There is no rigorous answer yet. However, the fact that the helix-turn may be necessary for fusion (Gruneke et al., 2002) does not mean it is the essential, i.e., slowest, conformational change. Bentz (2000b) noted that even if the essential defect is formed, unless the bilayers can get close enough to permit membrane merger, the defect will dissipate without fusion. The job of the helix-turn appears to promote close apposition of bilayers and to provide a hydrophobic protein collar between a hydrophobic defect and the target bilayer, without which there is no fusion, as discussed in Bentz (2000b) and in Bentz and Mittal (2000).

The core complexes of SV5 F1 and HIV gp41 show longer annular helices around the N-heptad repeat core than does HA2, suggesting these have more stable binding. Such an extension has been achieved for HA2 using an *E. coli* expression system (Chen et al., 1999). For HA2, the helix-turn could be quite reversible, with much bending up and down. In fact, the formation of the first stable intermediate between the bilayers, presumably the lipidic stalk (Chernomordik et al., 1999; Yang and Huang, 2002), might stabilize the conformational intermediate, and vice versa. Although peptides homologous to C-terminus heptad repeats of SV5 F1 and gp41 will inhibit syncitia formation and membrane destabilization, as discussed in detail in Bentz (2000b), no such reports have been published for HA2. If the helix-turn transition of HA is weaker than those for SV5 F1 and gp41, suggested by the length of the helix contacts to the extended coiled coil, then the C-peptides for HA may not inhibit syncitia formation, due to weak binding. Without the native structures of SV5 F1 and gp41, little can be known about the low-energy conformational change.

16.4.7 Hydrophobic Kink

It has been recently shown that the kink formed by the helix-turn of HA mediates the low pH-dependent aggregation of the HA fragment known as FHA2, which is aa 1–127 of HA2, i.e., the fusion peptide through the C-terminus of the six-helix bundle at equilibrium (Yu et al., 1994; Kim et al., 1998). This peptide aggregation appears to be hydrophobically driven, perhaps following the protonation of aspartates within the kink region. This same fragment induces lipid mixing between liposomes (Epand et al., 1999) and between cells (Leikina et al., 2000). Mutants in the kink region or lacking the fusion peptide do not induce lipid mixing. The function of such a hydrophobic kink is readily apparent in the model of fusion proposed in Bentz (2000b) and by Bentz and Mittal (2000). The formation of one or two helix-turns would bend HA, allowing the close approach of the membranes. Just as importantly, the hydrophobic kink would now form a hydrophobic collar just above the hydrophobic defect. Phospholipids making an excursion from the target membrane would be stabilized by this hydrophobic collar and thereby have a substantially greater chance of reaching the hydrophobic defect, i.e., to begin lipid mixing. As in a proper enzyme, the hydrophobic collar would stabilize the transition state of fusion.

For the viral fusion proteins using coiled coils and folding something like HA (Skehel and Wiley, 1998; Baker et al., 1999), this hydrophobic kink should be near the C-terminus of the N-heptad repeat region or near the N-terminus of the C-heptad repeat region. Interestingly, Peisajovich et al. (2000) have claimed to find a "second fusion peptide" in the sendai virus F1 protein near the C-terminus of the N-heptad repeat domain.

It is worth noting that the various domains discussed here as putative hydrophobic kink formers do not show any obvious sequence homology (data not shown). Tertiary structure could be important. It is also interesting that the fusion protein of the tick-borne encephalitis virus, TBE-E (Rey et al., 1995), not only has two putative amphipathic α-helices just near the N-terminal to its transmembrane domain (Stiasny et al., 1996), which could fuse as a hydrophobic collar, but also a putative internal fusion peptide about 150 aa upstream.

REFERENCES

Alford, D., H. Ellens, and J. Bentz. 1994. Fusion of influenza virus with sialic acid-bearing target membranes. *Biochemistry* 33:1977–1987.

Armstrong, R.T., A.S. Kushnir, and J.M. White. 2000. The transmembrane domain of influenza hemagglutinin exhibits a stringent length requirement to support the hemifusion to fusion transition. *J. Cell. Biol.* 151:425–437.

Baker, K.A., R. Dutch, R.A. Lamb, and T.S. Jardetsky. 1999. Structural basis for paramyxovirus-mediated membrane fusion. *Mol. Cell.* 3:309–319.

Bentz, J. 1992. Intermediates and kinetics of membrane fusion. *Biophys. J.* 63:448–459.

Bentz, J. 2000a. Minimal aggregate size and minimal fusion unit for the first fusion pore of influenza hemagglutinin mediated membrane fusion. *Biophys. J.* 78:227–245.

Bentz, J. 2000b. Membrane fusion mediated by coiled coils: a hypothesis. *Biophys. J.* 78:886–900.

Bentz, J., H. Ellens, and D. Alford. 1990. An architecture for the fusion site of influenza hemagglutinin. *FEBS Lett.* 276:1–5.

Bentz, J. and A. Mittal. 2000. Deployment of membrane fusion protein domains during fusion. *Cell. Biol. Int.* 24(11):819–838.

Bentz, J. and A. Mittal. 2003. Architecture of the influenza hemagglutinin membrane fusion site. *Biochim. Biophys. Acta.* (In press.)

Blumenthal, R., D.P. Sarkar, S. Durell, D.E. Howard, and S.J. Morris. 1996. Dilation of the influenza hemagglutinin fusion pore revealed by the kinetics of individual fusion events. *J. Cell. Biol.* 135:63–71.

Bullough, P.A., F.M. Hughson, J.J. Skehel, and D.C. Wiley. 1994. Structure of influenza haemagglutinin at the pH of membrane fusion. *Nature* 371:37–43.

Caffrey M., M. Cai, J. Kaufman, S.J. Stahl, P.T. Wingfield, D.G. Covell, A.M. Gronenborn, and G.M. Clore. 1998. Three-dimensional solution structure of the 44 kDa ectodomain of SIV gp41. *EMBO J.* 17(16):4572–4584.

Carr, C.M. and P.S. Kim. 1993. A spring-loaded mechanism for the conformational change in influenza hemagglutinin. *Cell* 73:823–832.

Chan, D.C., D. Fass, J.M. Berger, and P.S. Kim. 1997. Core structure of gp41 from the HIV envelope glycoprotein. *Cell* 89:263–273.

Chen, J., S. Wharton, W. Weissenhorn, L. Calder, F. Hughson, J.J. Skehel, and D.C. Wiley. 1995. A soluble domain of the membrane-anchoring chain of influenza virus hemagglutinin (HA2) folds in *Escherichia coli* into the low pH induced conformation. *Proc. Natl. Acad. Sci. U.S.A.* 92:12205–12209.

Chen, J., J.J. Skehel, and D.C. Wiley. 1999. N- and C-terminal residues combine in the fusion-pH influenza hemagglutinin HA(2) subunit to form an N cap that terminates the triple-stranded coiled coil. *Proc. Natl. Acad. Sci. U.S.A.* 96:8967–8972.

Chernomordik, L.V., E. Leikina, V. Frolov, P. Bronk, and J. Zimmerberg. 1997. An early stage of membrane fusion mediated by the low pH conformation of influenza hemagglutinin depends upon membrane lipids. *J. Cell. Biol.* 136(1):81–93.

Chernomordik, L.V., V.A. Frolov, E. Leikina, P. Bronk and J. Zimmerberg. 1998. The pathway of membrane fusion catalyzed by influenza hemagglutinin: restriction of lipids, hemifusion, and lipidic fusion pore formation. *J. Cell. Biol.* 140(6):1369–1382.

Chernomordik, L.V., E. Leikina, M.M. Kozlov, V.A. Frolov, and J. Zimmerberg 1999. Structural intermediates in influenza haemagglutinin-mediated fusion. *Mol. Membr. Biol.* 16(1):33–42.

Corver, J., A. Ortiz, S.L. Allison, J. Schalich, F.X. Heinz, and J. Wilschut. 2000. Membrane fusion activity of tick-borne encephalitis virus and recombinant subviral particles in a liposomal model system. *Virology* 269:37–46.

Danieli, T., S.L. Pelletier, Y.I. Henis, and J.M. White. 1996. Membrane fusion mediated by the influenza virus hemagglutinin requires the concerted action of at least three hemagglutinin trimers. *J. Cell. Biol.* 133:559–569.

Dill, K.A. and H.S. Chan. 1997. From Levinthal to pathways to funnels. *Nat. Struct. Biol.* 4(1):10–19.

Doms, R.W. and S.C. Peiper. 1998. Unwelcome guests with master keys: how HIV uses chemokine receptors for cellular entry. *Virology* 235:179–190.

Durell, S., I. Martin, J-M. Ruysschaert, Y. Shai, and R. Blumenthal. 1997. What studies of fusion peptides tell us about viral envelope glycoprotein-mediated membrane fusion. *Mol. Membr. Biol.* 14:97–112.

Dutch, R.E., S.B. Joshi, and R.A. Lamb. 1998. Membrane fusion promoted by increasing surface densities of the paramyxovirus F and HN proteins: comparison of fusion reactions mediated by simian virus 5 F, human parainfluenza virus type 3 F, and influenza virus HA. *J. Virol.* 72:7745–7753

Ellens, H., J. Bentz, D. Mason, F. Zhang, and J.M. White. 1990. Fusion of influenza hemagglutinin-expressing fibroblasts with glycophorin-bearing liposomes: role of hemagglutinin surface density. *Biochemistry* 29:9697–9707.

Epand, R.F., J.C. Macosko, C.J. Russell, Y.K. Shin, and R.M. Epand. 1999. The ectodomain of HA2 of influenza virus promotes rapid pH dependent membrane fusion. *J. Mol. Biol.* 286:489–503.

Fass, D. and P.S. Kim. 1995. Dissection of a retrovirus envelope protein reveals structural similarity to influenza hemagglutinin. *Curr. Biol.* 5(12):1377–1383.

Feng, Y., C.C. Broder, P.E. Kennedy, and E.A. Berger. 1996. HIV-1 entry cofactor: functional cDNA cloning of a seven-transmembrane, G protein-coupled receptor. *Science* 272(5263):872–877.

Frolov V.A., M.S. Cho, P. Bronk, T.S. Reese, and J. Zimmerberg. 2000. Multiple local contact sites are induced by GPI-linked influenza hemagglutinin during hemifusion and flickering pore formation. *Traffic* 1(8):622–630.

Gaudin, Y., R.W. Ruigrok, and J. Brunner. 1995. Low-pH induced conformational changes in viral fusion proteins: implications for the fusion mechanism. *J. Gen. Virol.* 76(Pt 7):1541–1556.

Gething, M.J., R.W. Doms, D. York, and J. White. 1986. Studies on the mechanism of membrane fusion: site-specific mutagenesis of the hemagglutinin of influenza virus. *J. Cell. Biol.* 102:11–23.

Godley, L., J. Pfeifer, D. Steinhauer, B. Ely, G. Shaw, R. Kaufmann, E. Suchanek, C. Pabo, J.J. Skehel, D.C. Wiley, and S. Wharton. 1992. Introduction of intersubunit disulfide bonds in the membrane-distal region of the influenza hemagglutinin abolishes membrane fusion activity. *Cell* 68(4):635–645.

Gruenke, J.A., R.T. Armstrong, W.W. Newcomb, J.C. Brown, and J.M. White. 2002. New insights into the spring-loaded conformational change of influenza virus hemagglutinin. *J. Virol.* 76:4456–4466.

Hernandez, L.D., L.R. Hoffman, T.G. Wolfsberg, and J.M. White. 1996. Virus-cell and cell-cell fusion. *Annu. Rev. Cell Dev. Biol.* 12:627–661.

Hoffman, T.L., C.C. LaBranche, W. Zhang, G. Canziani, J. Robinson, I. Chaiken, J.A. Hoxie, and R.W. Doms. 1999. Stable exposure of the coreceptor-binding site in a CD4-independent HIV-1 envelope protein. *Proc. Natl. Acad. Sci. U.S.A.* 96:6359–6364.

Kanaseki, T., K. Kawasaki, M. Murata, Y. Ikeuchi, and S. Ohnishi. 1997. Structural features of membrane fusion between influenza virus and liposome as revealed by quick-freezing electron microscopy. *J. Cell. Biol.* 137:1041–1056.

Kim, C., J.C. Macosko, and Y.K. Shin. 1998. The mechanism of low-pH-induced clustering of phospholipid vesicles carrying the HA2 ectodomain of influenza hemagglutinin. *Biochemistry* 37:137–144.

Kozak, S.L., J.M. Heard, and D. Kabat. 2002. Segregation of CD4 and CXCR4 into distinct lipid microdomains in T lymphocytes suggests a mechanism for membrane destabilization by human immunodeficiency virus. *J. Virol.* 76:1802–1815.

Kozlov, M.M. and L.V. Chernomordik. 1998. A mechanism of protein-mediated fusion: coupling between refolding of the influenza hemagglutinin and lipid rearrangements. *Biophys. J.* 75:1384–1396.

Kozlovsky, Y. and M.M. Kozlov. 2002. Stalk model of membrane fusion: solution of the energy crisis. *Biophys. J.*, 82:882–895.

Kozlovsky, Y., L. Chernomordik, and M. Kozlov. 2002. Lipid intermediates in membrane fusion: formation, structure, and decay of hemifusion diaphragm. *Biophys. J.* (In press.)

Kuhmann S.E., E.J. Platt, S.L. Kozak, D. Kabat. 2000. Cooperation of multiple CCR5 coreceptors is required for infections by human immunodeficiency virus type 1. *J. Virol.* 74(15):7005–7015.

Kuzmin, P.I., J. Zimmerberg, Y.A. Chizmadzhev, and F.S. Cohen. 2001. A quantitative model for membrane fusion based on low-energy intermediates. *Proc. Natl. Acad. Sci. U.S.A.* 98:7235–7240.

Lee, J-K. and Lentz, B.R. 1998. Secretory and viral fusion may share mechanistic events with fusion between curved lipid bilayers. *Proc. Natl. Acad. Sci. U.S.A.* 95: 9274–9279.

Leikina, E. and L.V. Chernomordik. 2000. Reversible merger of membranes at the early stage of influenza hemagglutinin-mediated fusion. *Mol. Biol. Cell.* 11:2359–2371.

Leikina, E., I. Markovic, L.V. Chernomordik, and M.M. Kozlov. 2000. Delay of influenza hemagglutinin refolding into a fusion-competent conformation by receptor binding: a hypothesis. *Biophys. J.* 79(3):1415–1427.

Leikina, E., D.L. LeDuc, J.C. Macosko, R. Epand, R. Epand, Y.K. Shin, and L.V. Chernomordik. 2001. The 1-127 HA2 construct of influenza virus hemagglutinin induces cell-cell hemifusion. *Biochemistry* 40:8378–8386.

Lentz, B.R. and J.K. Lee. 2000. Poly(ethylene glycol) (PEG)-mediated fusion between pure lipid bilayers: a mechanism in common with viral fusion and secretory vesicle release? *Mol. Membr. Biol.* 16:279–296.

Markin, V.S. and J. P. Albanesi. 2002. Membrane fusion: stalk model revisited. *Biophys. J.* 82:693–712.

Markosyan, R.M., G.B. Melikyan, and F.S. Cohen. 2001. Evolution of intermediates of influenza virus hemagglutinin-mediated fusion revealed by kinetic measurements of pore formation. *Biophys. J.* 80(2):812–821.

Markovic I., M. Kumar, J. Zimmerberg and L.V. Chernomordik. 2002. Raft association ensures synchronized unfolding of influenza hemagglutinin upon its activation. *Biophys. J.* 82 (1) Part 2:2644a.

Martin, J., S. Wharton, Y.P. Lin, D.K. Takemoto, J.J. Skehel, D.C. Wiley, and D.A. Steinhauer. 1998. Studies of the binding properties of influenza hemagglutinin receptor-site mutants. *Virology* 241:101–111.

Melikyan, G.B., W. Niles, and F.S. Cohen. 1995. The fusion kinetics of influenza hemagglutinin expressing cells to planar bilayer membranes is affected by HA surface density and host cell surface. *J. Gen. Physiol.* 106:783–802.

Melikyan, G.B., S. Lin, M.G. Roth, and F.S. Cohen. 1999. Amino acid sequence requirements of the transmembrane and cytoplasmic domains of influenza virus hemagglutinin for viable membrane fusion. *Mol. Biol. Cell.* 10:1821–1836.

Melikyan, G.B., R.M. Markosyan, H. Hemmati, M.K. Delmedico, D.M. Lambert, and F.S. Cohen. 2000. Evidence that the transition of HIV-1 gp41 into a six-helix bundle, not the bundle configuration, induces membrane fusion. *J. Cell. Biol.* 151:413–423.

Millar, B.M., L.J. Calder, J.J. Skehel, and D.C. Wiley. 1999. Membrane fusion by surrogate receptor-bound influenza haemagglutinin. *Virology* 257:415–423.

Mittal, A. and J. Bentz. 2001. Comprehensive kinetic analysis of influenza hemagglutinin-mediated membrane fusion: role of sialate binding. *Biophys. J.* 81:1521–1535.

Mittal, A., E. Leikina, J. Bentz, and L.V. Chernomordik. 2002a. Kinetics of influenza hemagglutinin-mediated membrane fusion as a function of technique. *Anal. Biochem.* 303:145–152.

Mittal, A., T. Shangguan, and J. Bentz. 2002b. Measuring pKa of activation and pKi of inactivation for influenza hemagglutinin from kinetics of membrane fusion of virions and of HA expressing cells. *Biophys J.* (In press.)

Niles, W.D. and F.S. Cohen. 1993. Single event recording shows that docking onto receptor alters the kinetics of membrane fusion mediated by influenza hemagglutinin. *Biophys. J.* 65(1):171–176.

Pak, C.C., M. Krumbiegel, and R. Blumenthal 1994. Intermediates in influenza virus PR/8 haemagglutinin-induced membrane fusion. *J. Gen. Virol.* 75:395–399.

Pecheur, E.I., I. Martin, A. Bienvenue, J.M. Ruysschaert, and D. Hoekstra. 2000. Protein-induced fusion can be modulated by target membrane lipids through a structural switch at the level of the fusion peptide. *J. Biol. Chem.* 275:3936–3942.

Peisajovich, S.G., O. Samuel, and Y. Shai. 2000. Paramyxovirus F1 protein has two fusion peptides: implications for the mechanism of membrane fusion. *J. Mol. Biol.* 296:1353–1365.

Qiao, H., S. Pelletier, L. Hoffman, J. Hacker, R. Armstrong, and J.M. White. 1998. Specific single or double proline substitutions in the "spring-loaded" coiled coil region of the influenza hemagglutinin impair or abolish membrane fusion activity. *J. Cell. Biol.* 141:1335–1347.

Qiao, H., R.T. Armstrong, G.B. Melikyan, F.S. Cohen, and J.M. White. 1999. A specific point mutant at position 1 of the influenza hemagglutinin fusion peptide displays a hemifusion phenotype. *Mol. Biol. Cell.* 10:2759–2769.

Rey, F.A., F.X. Heinz, C. Mandel, C. Kunz, and S.C. Harrison. 1995. The envelope glycoprotein from tick-borne encephalitis virus at 2 Å resolution. *Nature* 375:291–298.

Ruigrok, R.W.H., A. Aitken, L.J. Calder, S.R. Martin, J.J. Skehel, S.A. Wharton, W. Weis, and D.C. Wiley. 1988. Studies on the structure of the influenza virus hemagglutinin at the pH of membrane fusion. *J. Gen. Virol.* 69:2785–2795.

Russell, C.J., T.S. Jardetzky, and R.A. Lamb. 2001. Membrane fusion machines of paramyxoviruses: capture of intermediates of fusion. *EMBO J.* 20:4024–4034.

Samsonov, A.V., I. Mihalyov, and F.S. Cohen. 2001. Characterization of cholesterol-sphingomyelin domains and their dynamics in bilayer membranes. *Biophys. J.* 81:1486–1500.

Shangguan, T. 1995. Influenza virus fusion mechanisms. Ph. D. dissertation, Drexel University.

Shangguan, T., D. Alford, and J. Bentz. 1996. Influenza virus-liposomes lipid mixing is leaky and largely insensitive to the material properties of the target membrane. *Biochemistry* 35:4956–4965.

Shangguan, T., D. Siegel, J. Lear, P. Axelsen, D. Alford, and J. Bentz 1998. Morphological changes and fusogenic activity of influenza virus hemagglutinin. *Biophys. J.* 74:54–62.

Singh, M., B. Berger, and P.S. Kim. 1999. LearnCoil-VMF: computational evidence for coiled-coil-like motifs in many viral membrane-fusion proteins. *J. Mol. Biol.* 290:1031–1041.

Skehel, J.J., P.M. Bayley, E.B. Brown, S.R. Martin, M.D. Waterfield, J.M. White, I.A. Wilson, and D.C. Wiley. 1982. Changes in the conformation of influenza virus hemagglutinin at the pH optimum of virus-mediated membrane fusion. *Proc. Natl. Acad. Sci. U.S.A.* 79:968–972.

Skehel, J.J. and D.C. Wiley. 1998. Coiled coils in both intracellular vesicle and viral membrane fusion. *Cell* 95(7):871–874.

Smit, J.M., G. Li, P. Schoen, J. Corver, R. Bittman, K.C. Lin, and J. Wilschut. 2002. Fusion of alphaviruses with liposomes is a non-leaky process. *FEBS Lett.* 521:62–66.

Stegmann, T., J.M. White, and A. Helenius. 1990. Intermediates in influenza induced membrane fusion. *EMBO J.* 13:4231–4241.

Stegmann, T. and A. Helenius. 1993. Influenza virus fusion: from models toward a mechanism. In *Viral Fusion Mechanisms*. J. Bentz, editor. CRC Press, Boca Raton, FL, pp. 89–113.

Stegmann, T., I. Bartoldus, and J. Zumbrunn. 1995. Influenza hemagglutinin-mediated membrane fusion: influence of receptor binding on the lag phase preceding fusion. *Biochemistry* 34(6):1825–1832.

Stiasny, K., S.L. Allison, A. Marchler-Bauer, C. Kunz, and F.X. Heinz. 1996. Structural requirements for low-pH-induced rearrangements in the envelope glycoprotein of tick-borne encephalitis virus. *J. Virol.* 70:8142–8147.

Sutton, R.B., D. Fasshauer, R. Jahn, and A.T. Brunger. 1998. Crystal structure of a SNARE complex involved in synaptic exocytosis at 2.4 Å resolution. *Nature* 395:347–353.

Tong, S., M. Li, A. Vincent, R. Compans, F. Fritsch, R. Beier, C. Klenk, M. Ohuchi, and H. Klenk. 2002. Regulation of fusion activity by the cytoplasmic domain of a paramyxovirus f protein. *Virology* 30:322–333.

Waarts, B.L., R. Bittman, and J. Wilschut. 2002. Sphingolipid- and cholesterol-dependence of alphavirus membrane fusion: lack of correlation with lipid raft formation in target liposomes. *J. Biol. Chem.* 277: 38141–38147.

Waning, D.L., A.P Schmitt, G.P. Leser, and R.A. Lamb. 2002 Roles for the cytoplasmic tails of the fusion and hemagglutinin-neuraminidase proteins in budding of the paramyxovirus simian virus 5. *J. Virol.* 76:9284–9297.

Weissenhorn, W., A. Dessen, S.C. Harrison, J.J. Skehel, and D.C. Wiley. 1997. Atomic structure of ectodomain from HIV-1 gp41. *Nature* 371:37–43.

Weissenhorn, W., L.J. Calder, S.A. Wharton, J.J. Skehel, and D.C. Wiley. 1998. The central structural feature of the membrane fusion protein subunit from the ebola virus glycoprotein is a long triple-stranded coiled coil. *Proc. Natl. Acad. Sci. U.S.A.* 95:6032–6036.

White, J. and I.A. Wilson. 1987. Anti-peptide antibodies detect steps in a protein conformational change: low-pH activation of the influenza virus hemagglutinin. *J. Cell. Biol.* 105:2887–2896.

Wilson, I.A., J.J. Skehel, and D.C. Wiley. 1981. Structure of the haemagglutinin membrane glycoprotein of influenza virus at 3 Å resolution. *Nature* 289:366–373.

Xiao, X., L. Wu, T. Stantchev, Y-R. Feng, S. Ugolini, H. Chen, Z. Shen, J.L. Riley, C.C. Broder, Q.J. Sattentau, and D.S. Dimitrov. 1999. Constitutive cell surface association between CD4 and CCR5. *Proc. Natl. Acad. Sci. U.S.A.* 96:7496–7501.

Yang, L. and H.W. Huang. 2002. Observation of a membrane fusion intermediate structure. *Science* 297(5588):1877–1879.

Yu, Y.G., D.S. King, and Y-K. Shin. 1994. Insertion of a coiled coil peptide from influenza virus hemagglutinin into membranes. *Science* 266:274–276.

Index

A

AA, *see* Arachidonic acid
Acetylcholine receptors (AChR), 362, 406, 425, 428
Acholeplasma laidlawii, 99, 100, 194
AChR, *see* Acetylcholine receptors
Activation energies, PC SUV/SUV interactions, 270
Active pump transport, 374
Active transport, 372
Acyl chain
    asymmetry, 123
    inequivalence parameter, 129
    interdigitation, simultaneous headgroup and, 133
    -length asymmetry, 123, 126
Acylglycerols, 9, 12, 30
Adhesion energy, 228
    bilayer, 219, 221
    -induced stress, 296
AE1, *see also* Inorganic anion transporter AE1
    antibodies against, 440
    gene, 436, 438
Alamethicin, 505
Alcohols
    biphasic effect, 136
    DPPC bilayers and, 137
    long-chain, 3, 5
Aldehydes, long-chain, 5
Alkali-cation selectivities
    observed sequences of, 361
    theoretical variation of, 363
Alkylglycerols, 12
Anchorage modulation, membrane protein, 425
Anion exchange
    DIDS-sensitive, 440
    ping-pong catalytic mechanism of, 445
    role of in CO_2 transport, 438
Anion transport, 447
Anisotropy decay
    curves, 402
    effect of melittin on, 404
Ankyrin-binding complex, 443
Antiport, 385
Arachidonic acid (AA), 280
Archaebacteria
    fatty acids in, 107
    glycerol-based lipids in, 16
Arginine residues, 456
Arylsulfonates, 445
A. thaliana, 436, 437
Atherosclerosis, 11
Atomistic molecular dynamics simulation, 294, 295

ATP
    depletion, 447
    hydrolyzing activity, 245
Avogadro's number, 334

B

Bacteria, thermophilic, 143
Bacterial fatty acids, 8
Bacterial membranes, 7, 101
Bacterial photosynthetic reaction centers, 125
Bacteriorhodopsin
    cholesterol and, 404
    hydropathy, 482
BBMs, *see* Brush border membranes
B. cereus, neutral, Mg^{2+}-dependent sphingomyelinase
      obtained from, 501
Benjamin Franklin, observation of, 309
BHA, *see* Bromelain-released hemagglutinin
Bicontinuous phases, 256
Bilayer(s)
    adhesion energies, 219
    –bilayer interaction, 224
    curvature stress, 268, 269
    fluidity parameter, 503
    free volume in, 250
    glucose flux across, 250
    immobilization of, 213
    lamellar stacking of, 178
    measurements of forces between, 203
    permeability, reduction in, 250
    repulsion, amplification of, 223
    surface, lateral tension within, 229
    tension, 209
    thickness, 173
Bile acids, stereochemistry of, 25
Binary system
    eutectic phase diagram for, 129
    induced interdigitation in, 137
Biological membranes, 101
    ethanol effect on, 138
    lipid composition of, 25
Biomembrane(s)
    chain order profiles, 159
    fluid-mosaic model of, 147
    fusion, protein control of, 195
    lipid motions in, 149
    physical parameters, 193
Boltzmann constant, 277, 334
Boltzmann equation, 353